COMPACT WORLD ATLAS

DK
Penguin
Random
House

FOR THE EIGHTH EDITION

DK London

Senior Cartographic Editor Simon Mumford
Jacket Design Development Sophia MTT
Jacket Designer Stephanie Cheng Hui Tan
Senior Production Controller Poppy David
Production Editor Robert Dunn
Publishing Director Jonathan Metcalf
Associate Publishing Director Liz Wheeler
Art Director Karen Self

DK Delhi

Desk Editor Saumya Agarwal
Assistant Editor Agey George
Managing Editor Saloni Singh
Senior Cartographer Subhashree Bharati
Manager Cartography Suresh Kumar
DTP Designer Rakesh Kumar
Senior DTP Designer Pushpak Tyagi
Senior Jackets Coordinator Priyanka Sharma Saddi

Placenames Consultant Juliette Koskinas
Authenticity Reader Bianca Hezekiah

This American Edition, 2023
First American Edition, 2001
Published in the United States by DK Publishing
1745 Broadway, 20th Floor, New York, NY 10019

23 24 25 10 9 8 7 6 5 4 3 2 1
001–334038–Sep/2023

Published in Great Britain by Dorling Kindersley Limited

A catalog record for this book is available from the Library of Congress.
ISBN 978-0-7440-7369-0

Printed and bound in the UAE

www.dk.com

MIX
Paper | Supporting responsible forestry
FSC
www.fsc.org
FSC™ C018179

This book was made with Forest Stewardship Council™ certified paper - one small step in DK's commitment to a sustainable future.
For more information go to www.dk.com/our-green-pledge

Key to map symbols

Physical features
Elevation
19,686ft/6000m
13,124ft/4000m
9843ft/3000m
6562ft/2000m
3281ft/1,000m
1640ft/500m
820ft/250m
0
Below sea level
Mountain
Depression
Volcano
Pass/tunnel
Sandy desert
Drainage features
Major perennial river
Minor perennial river
Seasonal river
Canal
Waterfall
Perennial lake
Seasonal lake
Wetland
Ice features
Permanent ice cap/ice shelf
Winter limit of pack ice
Summer limit of pack ice
Borders
Full international border
Disputed de facto border
Territorial claim border
Cease-fire line
Undefined boundary
Internal administrative boundary
Communications
Major road
Minor road
Railroad
International airport
Settlements
Above 500,000
100,000 to 500,000
50,000 to 100,000
Below 50,000
National capital
Internal administrative capital
Miscellaneous features
Site of interest
Ancient wall
Graticule features
Line of latitude/longitude/ Equator
Tropic/Polar circle
25°
Degrees of latitude/ longitude
Names
Physical features
Andes
Sahara
Ardennes
Landscape features
Land's End
Headland
Mont Blanc 4,807m
Elevation/volcano/pass
Blue Nile
River/canal/waterfall
Ross Ice Shelf
Ice feature
PACIFIC OCEAN
Sulu Sea
Palk Strait
Sea features
Chile Rise
Undersea feature
Regions
FRANCE
Country
BERMUDA (to UK)
Dependent territory
KANSAS
Administrative region
Dordogne
Cultural region
Settlements
PARIS
Capital city
SAN JUAN
Dependent territory capital city
Chicago
Kettering
Burke
Other settlements
Inset map symbols
Urban area
City
Park
Place of interest
Suburb/district

COMPACT WORLD ATLAS

Contents

The Political World 6-7
The Physical World 8-9
Standard Time Zones 10

The World's Regions

North & Central America

North & Central America 12-13
Western Canada & Alaska 14-15
Eastern Canada 16-17
USA: The Northeast 18-19
USA: The Southeast 20-21
Bermuda
USA: Central States 22-23
USA: The West 24-25
Los Angeles & Hawaii
USA: The Southwest 26-27
Mexico 28-29
Central America 30-31
The Caribbean 32-33
Jamaica, St. Lucia & Barbados

South America

South America 34-35
Northern South America 36-37
Western South America 38-39
Galápagos Islands
Brazil 40-41
Southern South America 42-43

The Atlantic Ocean 44-45

Africa

Africa 46-47
Northwest Africa 48-49
Northeast Africa 50-51
West Africa 52-53
Central Africa 54-55
Sao Tome & Principe
Southern Africa 56-57

Europe

Europe 58-59
The North Atlantic 60-61
Scandinavia & Finland 62-63
The Low Countries 64-65
The British Isles 66-67
London
France, Andorra & Monaco 68-69
Paris, Andorra & Monaco
Spain & Portugal 70-71
Azores & Gibraltar

Germany & the Alpine States 72-73
Liechtenstein
Italy 74-75
San Marino & Vatican City
Central Europe 76-77
Southeast Europe 78-79
Bosnia & Herzegovina
The Mediterranean 80-81
Malta & Cyprus
Bulgaria & Greece 82-83
The Baltic States & Belarus 84-85
Ukraine, Moldova & Romania 86-87
European Russia 88-89

North & West Asia

North & West Asia 90-91
Russia & Kazakhstan 92-93
Turkey & the Caucasus 94-95
The Near East 96-97
West Bank
The Middle East 98-99
Central Asia 100-101

South & East Asia

South & East Asia 102-103
Western China & Mongolia 104-105
Eastern China & Korea 106-107
Hong Kong
Japan 108-109
Tokyo & Nansei-Shoto
South India & Sri Lanka 110-111
North India, Pakistan & Bangladesh 112-113
Mainland Southeast Asia 114-115
Maritime Southeast Asia 116-117
Singapore

The Indian Ocean 118-119

Australasia & Oceania

Australasia & Oceania 120-121
The Southwest Pacific 122-123
Western Australia 124-125
Eastern Australia 126-127
Sydney
New Zealand 128-129

The Pacific Ocean 130-131

Antarctica 132

The Arctic Ocean 133

Index – Gazetteer

Countries Factfile 134-150
Overseas Territories & Dependencies 150-151
Geographical Comparisons 152-153
Index 154-192

The Political World

A
B
C
D
1
2
3
4
5
Severnaya Zemlya
Franz Josef Land
SVALBARD (to Norway)
New Siberian Islands
Novaya Zemlya
JAN MAYEN (to Norway)
ICELAND
FAROE ISLANDS (to Denmark)
NORWAY
SWEDEN
FINLAND
R U S S I A
European Russia
Asiatic Russia
EST.
LAT.
LITH.
DENMARK
RUSSIA
UNITED KINGDOM
IRELAND
NETH.
POLAND
BELA.
BELG.
GERMANY
LUX.
CZECHIA
SLVK.
(annexed by Russia, 2014)
FRANCE
LIECH.
UKRAINE
SWITZ.
AUT.
HUNG.
MOLD.
SLVN.
ROM.
MONACO
CRO.
SERBIA
S.M.
B.&H.
KOS. (disputed)
ANDORRA
VAT. CITY
MON.
BULG.
ALB.
GEORGIA
KAZAKHSTAN
MONGOLIA
Azores (to Portugal)
PORT.
SPAIN
ITALY
NORTH MACEDONIA
ARMENIA
AZERB.
TURKEY (TÜRKIYE)
AZ.
UZBEK.
KYRG.
TURKMEN.
TAJIKISTAN
JAPAN
N. KOREA
S. KOREA
GIBRALTAR (to UK)
GREECE
Madeira (to Portugal)
TUNISIA
MALTA
CYPRUS
SYRIA
LEBANON
ISRAEL
IRAQ
IRAN
AFGH.
CHINA
MOROCCO
Canary Islands (to Spain)
ALGERIA
LIBYA
EGYPT
JORDAN
KUWAIT
PAKISTAN
NEPAL
BHUTAN
Ryukyu Islands (to Japan)
WESTERN SAHARA (disputed)
BAHRAIN
QATAR
U.A.E.
BANGLADESH
LAOS
TAIWAN
SAUDI ARABIA
OMAN
INDIA
MYANMAR (BURMA)
CAPE VERDE (CABO VERDE)
MAURITANIA
MALI
NIGER
CHAD
ERITREA
YEMEN
PARACEL ISLANDS (disputed)
NORTHERN MARIANA ISLANDS (to US)
SENEGAL
SUDAN
Socotra (to Yemen)
Laccadive Islands (to India)
Andaman Islands (to India)
THAI.
VIETNAM
THE GAMBIA
BURKINA FASO
DJIBOUTI
CAMB.
PHILIPPINES
GUINEA-BISSAU
GUINEA
NIGERIA
SOUTH SUDAN
ETHIOPIA
SRI LANKA
SPRATLY ISLANDS (disputed)
SIERRA LEONE
BENIN
CAMEROON
C.A.R.
SOMALIA
Nicobar Islands (to India)
LIBERIA
BRUNEI
PALAU
IVORY COAST (CÔTE D'IVOIRE)
GHANA
TOGO
EQ. GUINEA
UGANDA
MALDIVES
MALAYSIA
SINGAPORE
KENYA
GABON
CONGO
RWANDA
BURUNDI
SAO TOME & PRINCIPE
DEM. REP. CONGO
INDONESIA
ASCENSION ISLAND (to St Helena)
Cabinda (to Angola)
TANZANIA
BRITISH INDIAN OCEAN TERRITORY (to UK)
CHRISTMAS ISLAND (to Australia)
SEYCHELLES
EAST TIMOR (TIMOR-LESTE)
Agalega Islands (to Mauritius)
ANGOLA
MALAWI
COMOROS
COCOS (KEELING) ISLANDS (to Australia)
ASHMORE & CARTIER ISLANDS (to Australia)
ST HELENA (to UK)
ZAMBIA
MOZAMBIQUE
MAYOTTE (to France)
MADAGASCAR
ZIMB.
NAMIBIA
MAURITIUS
RÉUNION (to France)
BOTS.
ESWATINI
AUSTRALIA
ATLANTIC OCEAN
INDIAN OCEAN
LESOTHO
SOUTH AFRICA
TRISTAN DA CUNHA (to UK)
Gough Island (to Tristan da Cunha)
Tasmania
Prince Edward Islands (to South Africa)
FRENCH SOUTHERN & ANTARCTIC LANDS (to France)
HEARD ISLAND & McDONALD ISLANDS (to Australia)
Global features
Total number of countries: 196
Largest country: Russia 6,601,668 sq miles (17,098,242 sq km)
Smallest country: Vatican City 0.17 sq miles (0.44 sq km)
Country with most international borders: China 14 / Russia 14
Continental Key
North & Central America
Europe
South America
Asia
Africa
Australasia & Oceania
POLITICAL STATUS:
E.g. MEXICO: independent state
E.g. FAROE ISLANDS (to Denmark): self-governing territory, with parent state indicated
E.g. Andaman Islands (to India): non self-governing territory, with parent stated indicated
ANTARCTICA

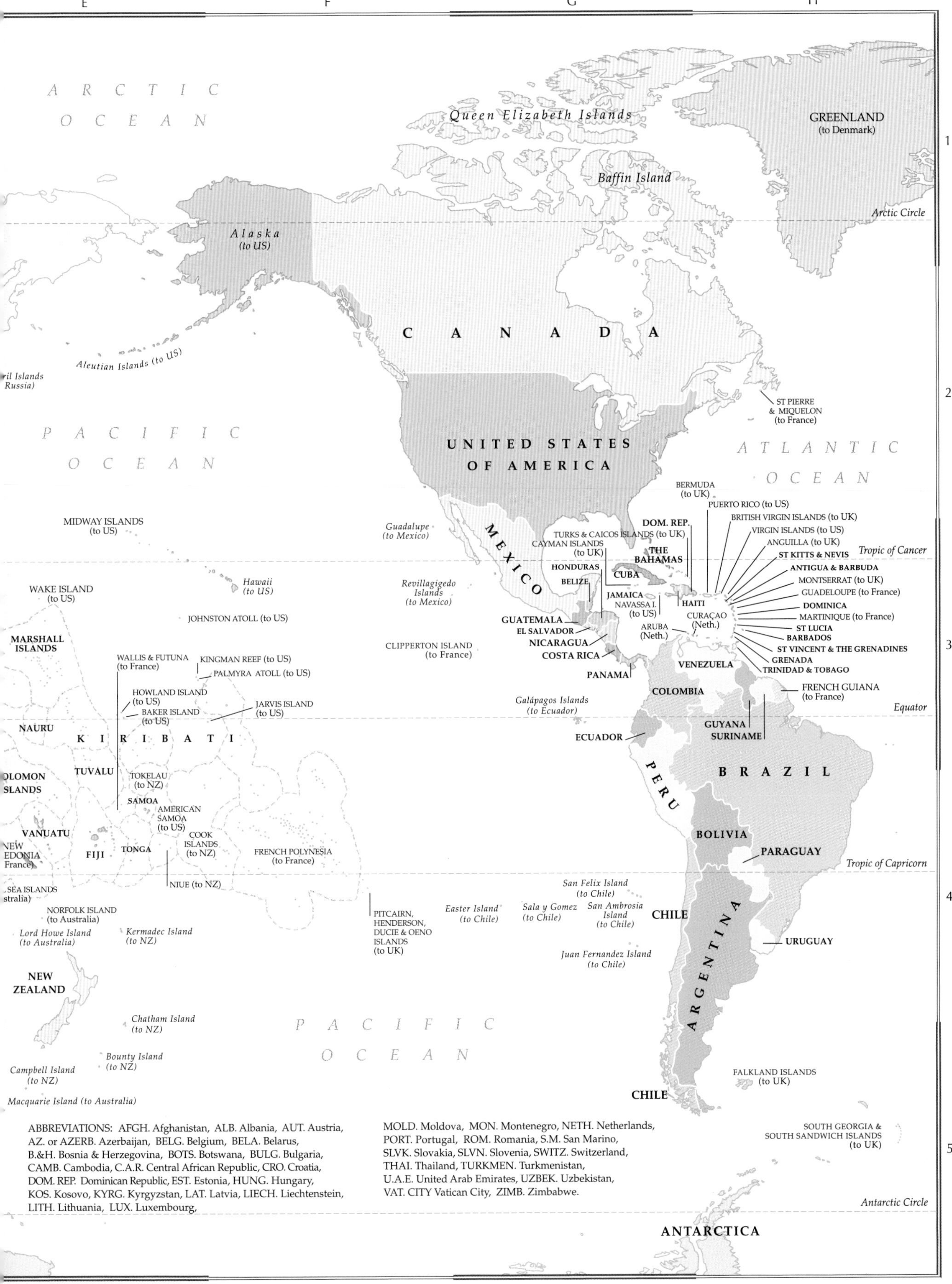
E
F
G
H
1
2
3
4
5
ARCTIC OCEAN
Queen Elizabeth Islands
GREENLAND
(to Denmark)
Baffin Island
Arctic Circle
Alaska
(to US)
CANADA
Aleutian Islands (to US)
ril Islands
Russia)
ST PIERRE
& MIQUELON
(to France)
PACIFIC OCEAN
UNITED STATES OF AMERICA
ATLANTIC OCEAN
BERMUDA
(to UK)
PUERTO RICO (to US)
BRITISH VIRGIN ISLANDS (to UK)
VIRGIN ISLANDS (to US)
ANGUILLA (to UK)
ST KITTS & NEVIS
Tropic of Cancer
ANTIGUA & BARBUDA
MONTSERRAT (to UK)
GUADELOUPE (to France)
DOMINICA
MARTINIQUE (to France)
ST LUCIA
BARBADOS
ST VINCENT & THE GRENADINES
GRENADA
TRINIDAD & TOBAGO
DOM. REP.
TURKS & CAICOS ISLANDS (to UK)
CAYMAN ISLANDS
(to UK)
THE BAHAMAS
HONDURAS
BELIZE
CUBA
JAMAICA
NAVASSA I.
(to US)
HAITI
CURAÇAO
(Neth.)
ARUBA
(Neth.)
MIDWAY ISLANDS
(to US)
Guadalupe
(to Mexico)
MEXICO
Hawaii
(to US)
WAKE ISLAND
(to US)
Revillagigedo
Islands
(to Mexico)
JOHNSTON ATOLL (to US)
GUATEMALA
EL SALVADOR
NICARAGUA
COSTA RICA
PANAMA
MARSHALL ISLANDS
CLIPPERTON ISLAND
(to France)
WALLIS & FUTUNA
(to France)
KINGMAN REEF (to US)
PALMYRA ATOLL (to US)
VENEZUELA
HOWLAND ISLAND
(to US)
BAKER ISLAND
(to US)
JARVIS ISLAND
(to US)
COLOMBIA
FRENCH GUIANA
(to France)
Galápagos Islands
(to Ecuador)
Equator
NAURU
KIRIBATI
ECUADOR
GUYANA
SURINAME
OLOMON
SLANDS
TUVALU
TOKELAU
(to NZ)
PERU
BRAZIL
SAMOA
AMERICAN
SAMOA
(to US)
VANUATU
COOK
ISLANDS
(to NZ)
BOLIVIA
NEW
EDONIA
France)
FIJI
TONGA
FRENCH POLYNESIA
(to France)
PARAGUAY
Tropic of Capricorn
SEA ISLANDS
stralia)
NIUE (to NZ)
San Felix Island
(to Chile)
NORFOLK ISLAND
(to Australia)
PITCAIRN,
HENDERSON,
DUCIE & OENO
ISLANDS
(to UK)
Easter Island
(to Chile)
Sala y Gomez
(to Chile)
San Ambrosia
Island
(to Chile)
CHILE
Lord Howe Island
(to Australia)
Kermadec Island
(to NZ)
ARGENTINA
URUGUAY
Juan Fernandez Island
(to Chile)
NEW
ZEALAND
Chatham Island
(to NZ)
PACIFIC OCEAN
Bounty Island
(to NZ)
Campbell Island
(to NZ)
FALKLAND ISLANDS
(to UK)
CHILE
Macquarie Island (to Australia)
ABBREVIATIONS: AFGH. Afghanistan, ALB. Albania, AUT. Austria, AZ. or AZERB. Azerbaijan, BELG. Belgium, BELA. Belarus, B.&H. Bosnia & Herzegovina, BOTS. Botswana, BULG. Bulgaria, CAMB. Cambodia, C.A.R. Central African Republic, CRO. Croatia, DOM. REP. Dominican Republic, EST. Estonia, HUNG. Hungary, KOS. Kosovo, KYRG. Kyrgyzstan, LAT. Latvia, LIECH. Liechtenstein, LITH. Lithuania, LUX. Luxembourg,
MOLD. Moldova, MON. Montenegro, NETH. Netherlands, PORT. Portugal, ROM. Romania, S.M. San Marino, SLVK. Slovakia, SLVN. Slovenia, SWITZ. Switzerland, THAI. Thailand, TURKMEN. Turkmenistan, U.A.E. United Arab Emirates, UZBEK. Uzbekistan, VAT. CITY Vatican City, ZIMB. Zimbabwe.
SOUTH GEORGIA &
SOUTH SANDWICH ISLANDS
(to UK)
Antarctic Circle
ANTARCTICA
E
F
G
H

The Physical World

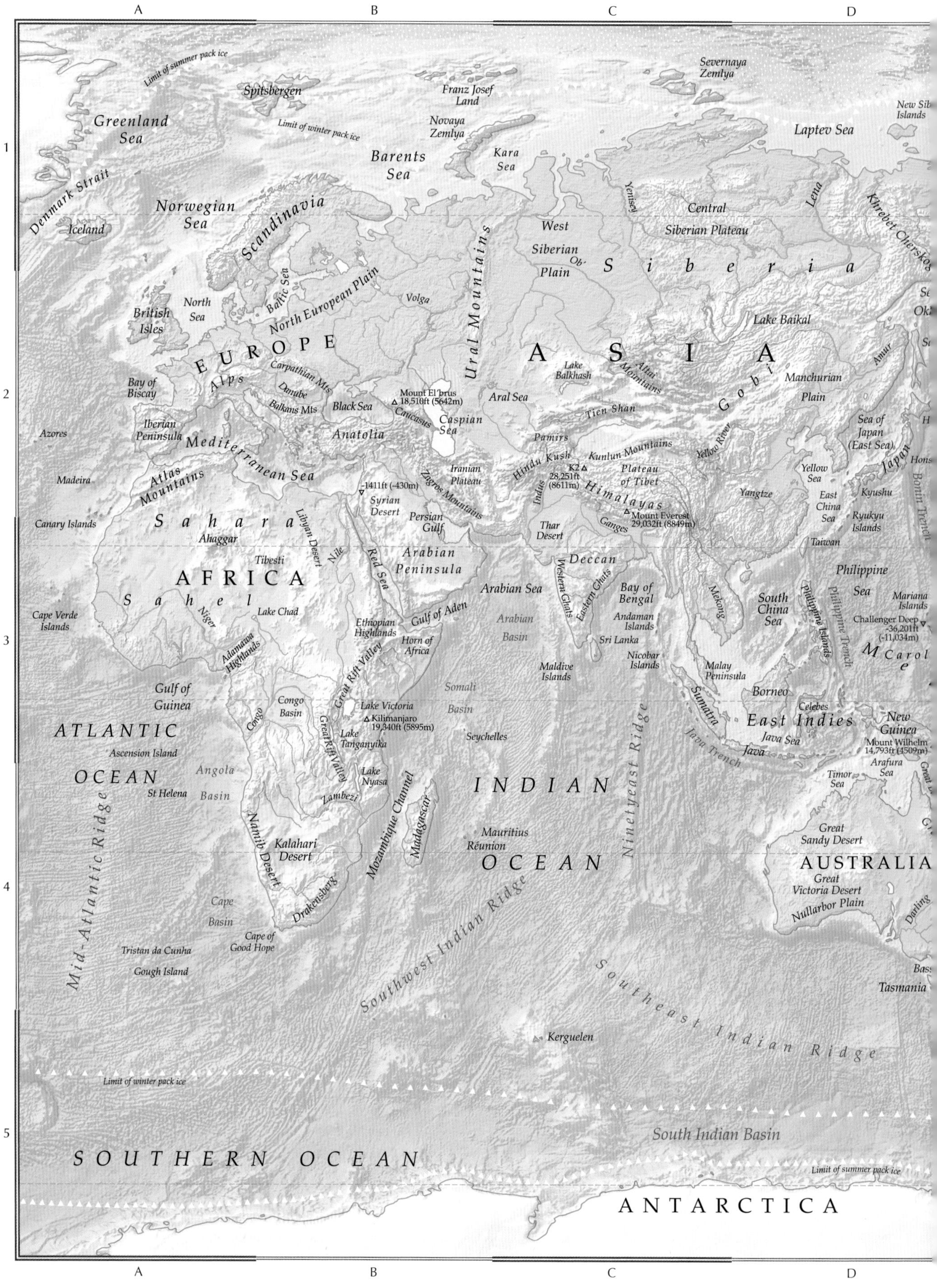

A
B
C
D
1
2
3
4
5
Limit of summer pack ice
Spitsbergen
Franz Josef Land
Severnaya Zemlya
New Sib Islands
Greenland Sea
Limit of winter pack ice
Novaya Zemlya
Laptev Sea
Barents Sea
Kara Sea
Denmark Strait
Norwegian Sea
Scandinavia
Yenisey
Central Siberian Plateau
Lena
Khrebet Chersko
Iceland
West Siberian Plain
Ob'
Siberia
Ural Mountains
Baltic Sea
North European Plain
Volga
British Isles
North Sea
Lake Baikal
EUROPE
ASIA
Lake Balkhash
Altai Mountains
Amur
Carpathian Mts
Alps
Danube
Bay of Biscay
Mount El'brus 18,510ft (5642m)
Aral Sea
Gobi
Manchurian Plain
Balkans Mts
Black Sea
Caucasus
Caspian Sea
Tien Shan
Iberian Peninsula
Anatolia
Sea of Japan (East Sea)
Azores
Mediterranean Sea
Pamirs
Kunlun Mountains
Yellow River
Japan
Hindu Kush
K2 28,251ft (8611m)
Plateau of Tibet
Yellow Sea
Madeira
Atlas Mountains
-1411ft (-430m)
Iranian Plateau
Zagros Mountains
Indus
Himalayas
Yangtze
East China Sea
Kyushu
Syrian Desert
Mount Everest 29,032ft (8849m)
Ryukyu Islands
Canary Islands
Sahara
Persian Gulf
Thar Desert
Ganges
Taiwan
Ahaggar
Libyan Desert
Red Sea
Arabian Peninsula
Tibesti
Nile
Deccan
Philippine Sea
AFRICA
Arabian Sea
Western Ghats
Eastern Ghats
Bay of Bengal
Sahel
Lake Chad
Gulf of Aden
Mekong
South China Sea
Philippine Islands
Philippine Trench
Mariana Islands
Cape Verde Islands
Niger
Ethiopian Highlands
Arabian Basin
Andaman Islands
Challenger Deep -36,201ft (-11,034m)
Horn of Africa
Sri Lanka
Adamawa Highlands
Great Rift Valley
Maldive Islands
Nicobar Islands
Malay Peninsula
Gulf of Guinea
Somali Basin
Borneo
Congo Basin
Lake Victoria
Sumatra
Celebes
East Indies
New Guinea
ATLANTIC OCEAN
Congo
Kilimanjaro 19,340ft (5895m)
Seychelles
Java Sea
Lake Tanganyika
Java
Mount Wilhelm 14,793ft (4509m)
Java Trench
Ascension Island
Angola Basin
Arafura Sea
Lake Nyasa
Timor Sea
St Helena
Zambezi
INDIAN OCEAN
Ninetyeast Ridge
Mozambique Channel
Madagascar
Namib Desert
Kalahari Desert
Mauritius
Réunion
Great Sandy Desert
Mid-Atlantic Ridge
AUSTRALIA
Great Victoria Desert
Drakensberg
Cape Basin
Nullarbor Plain
Southwest Indian Ridge
Cape of Good Hope
Tristan da Cunha
Gough Island
Tasmania
Southeast Indian Ridge
Kerguelen
Limit of winter pack ice
South Indian Basin
SOUTHERN OCEAN
Limit of summer pack ice
ANTARCTICA

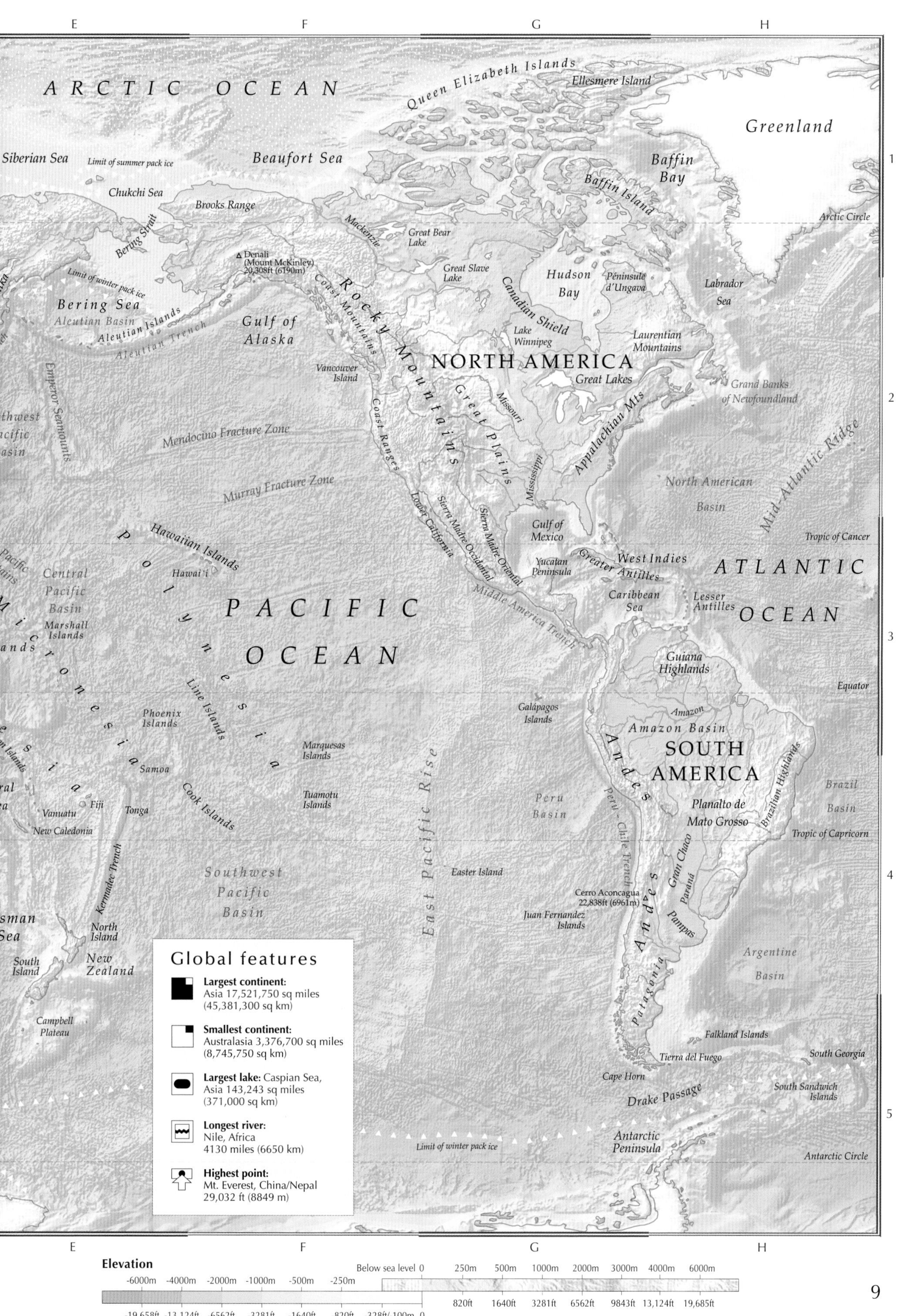
E
F
G
H
ARCTIC OCEAN
Queen Elizabeth Islands
Ellesmere Island
Greenland
Siberian Sea
Limit of summer pack ice
Beaufort Sea
Baffin Bay
Baffin Island
Chukchi Sea
Brooks Range
Arctic Circle
Bering Strait
Mackenzie
Great Bear Lake
Denali (Mount McKinley) 20,308ft (6190m)
Limit of winter pack ice
Great Slave Lake
Hudson Bay
Péninsule d'Ungava
Labrador Sea
Bering Sea
Aleutian Basin
Aleutian Islands
Aleutian Trench
Coast Mountains
Rocky Mountains
Gulf of Alaska
Canadian Shield
Lake Winnipeg
Laurentian Mountains
NORTH AMERICA
Great Lakes
Vancouver Island
Grand Banks of Newfoundland
Emperor Seamounts
Great Plains
Missouri
Appalachian Mts
Coast Ranges
Mendocino Fracture Zone
Mississippi
Mid-Atlantic Ridge
North American Basin
Murray Fracture Zone
Lower California
Sierra Madre Occidental
Sierra Madre Oriental
Gulf of Mexico
Tropic of Cancer
Hawaiian Islands
Polynesia
Yucatan Peninsula
West Indies
Greater Antilles
ATLANTIC OCEAN
Central Pacific Basin
Hawai'i
Caribbean Sea
Lesser Antilles
Micronesia
Marshall Islands
PACIFIC OCEAN
Middle America Trench
Guiana Highlands
Equator
Galápagos Islands
Line Islands
Phoenix Islands
Amazon
Amazon Basin
SOUTH AMERICA
Andes
Marquesas Islands
Samoa
Fiji
Tonga
Vanuatu
New Caledonia
Cook Islands
Tuamotu Islands
East Pacific Rise
Peru Basin
Peru Chile Trench
Planalto de Mato Grosso
Brazilian Highlands
Brazil Basin
Tropic of Capricorn
Kermadec Trench
Southwest Pacific Basin
Easter Island
Cerro Aconcagua 22,838ft (6961m)
Gran Chaco
Paraná
Juan Fernandez Islands
Pampas
North Island
South Island
New Zealand
Argentine Basin
Patagonia
Campbell Plateau
Falkland Islands
Tierra del Fuego
South Georgia
Cape Horn
South Sandwich Islands
Drake Passage
Antarctic Peninsula
Limit of winter pack ice
Antarctic Circle
1
2
3
4
5
Global features
Largest continent: Asia 17,521,750 sq miles (45,381,300 sq km)
Smallest continent: Australasia 3,376,700 sq miles (8,745,750 sq km)
Largest lake: Caspian Sea, Asia 143,243 sq miles (371,000 sq km)
Longest river: Nile, Africa 4130 miles (6650 km)
Highest point: Mt. Everest, China/Nepal 29,032 ft (8849 m)
Elevation
Below sea level 0
-6000m -4000m -2000m -1000m -500m -250m
-19,658ft -13,124ft -6562ft -3281ft -1640ft -820ft -328ft/-100m 0
250m 500m 1000m 2000m 3000m 4000m 6000m
820ft 1640ft 3281ft 6562ft 9843ft 13,124ft 19,685ft

Standard Time Zones

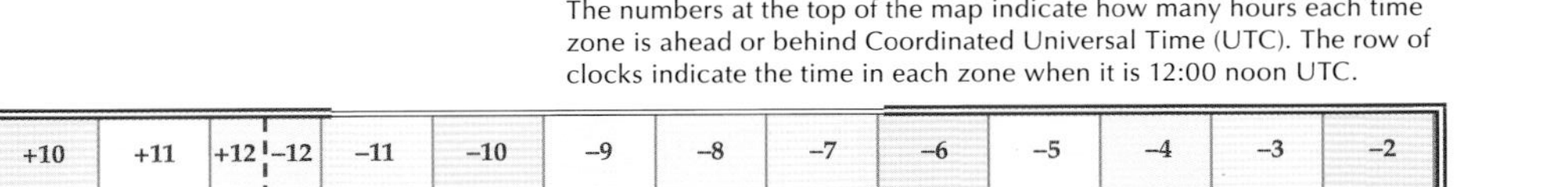

The numbers at the top of the map indicate how many hours each time zone is ahead or behind Coordinated Universal Time (UTC). The row of clocks indicate the time in each zone when it is 12:00 noon UTC.

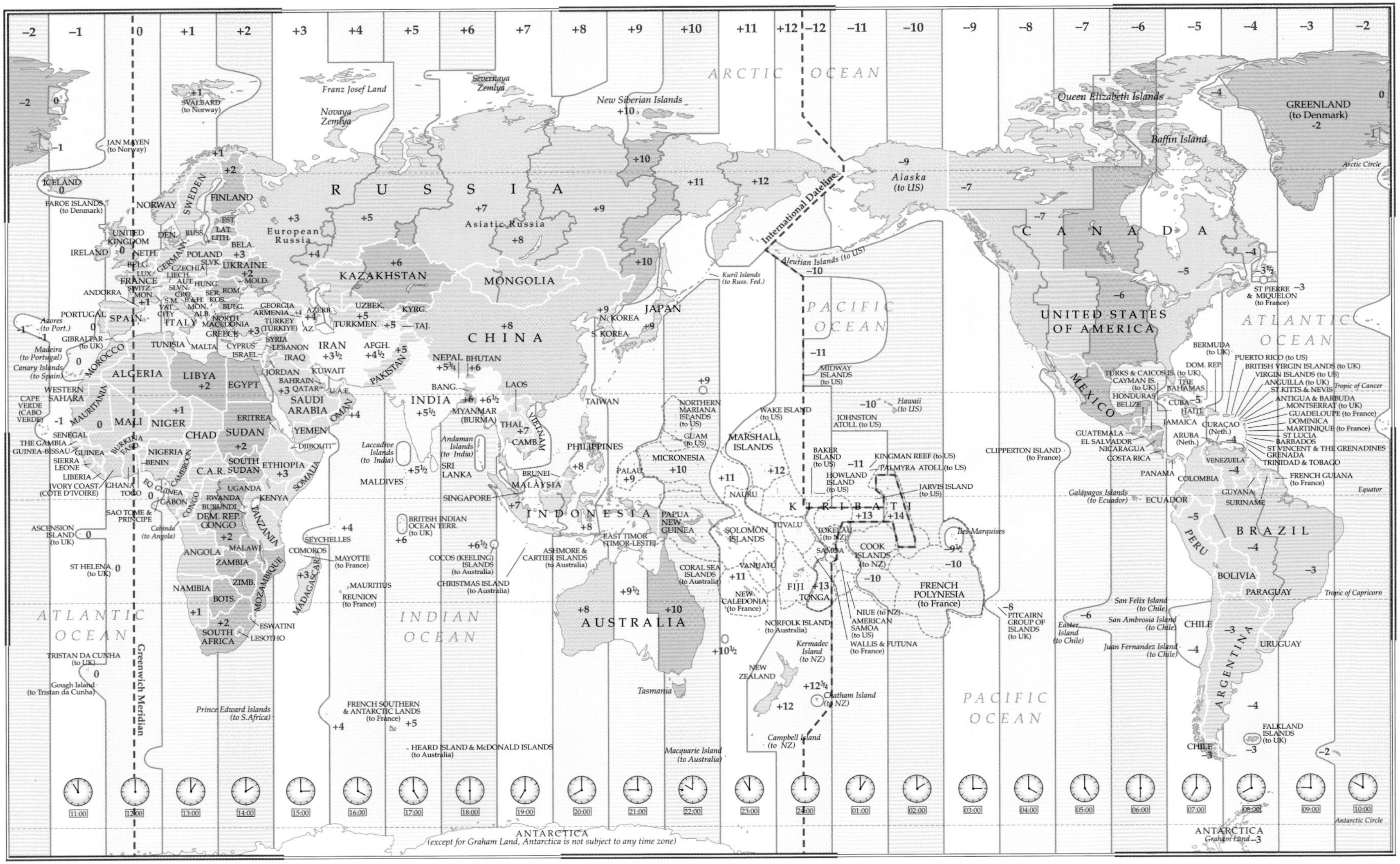

TIME ZONES

Because Earth is a rotating sphere, the Sun shines on only half of its surface at any one time. Thus, it is simultaneously morning, evening, and night time in different parts of the world. Because of these disparities, each country or part of a country adheres to a local time. A region of the Earth's surface within which a single local time is used is called a time zone.

COORDINATED UNIVERSAL TIME (UTC)

Coordinated Universal Time (UTC) is a reference by which the local time in each time zone is set. UTC is a successor to, and closely approximates, Greenwich Mean Time (GMT). However, UTC is based on an atomic clock, whereas GMT is determined by the Sun's position in the sky relative to the 0° longitudinal meridian, which runs through Greenwich, UK.

THE INTERNATIONAL DATELINE

The International Dateline is an imaginary line from pole to pole that roughly corresponds to the 180° longitudinal meridian. It is an arbitrary marker between calendar days. The dateline is needed because of the use of local times around the world rather than a single universal time.

The WORLD ATLAS

THE MAPS IN THIS ATLAS ARE ARRANGED CONTINENT BY CONTINENT, STARTING FROM THE INTERNATIONAL DATE LINE, AND MOVING EASTWARD. THE MAPS PROVIDE A UNIQUE VIEW OF TODAY'S WORLD, COMBINING TRADITIONAL CARTOGRAPHIC TECHNIQUES WITH THE LATEST REMOTE-SENSED AND DIGITAL TECHNOLOGY.

North & Central America

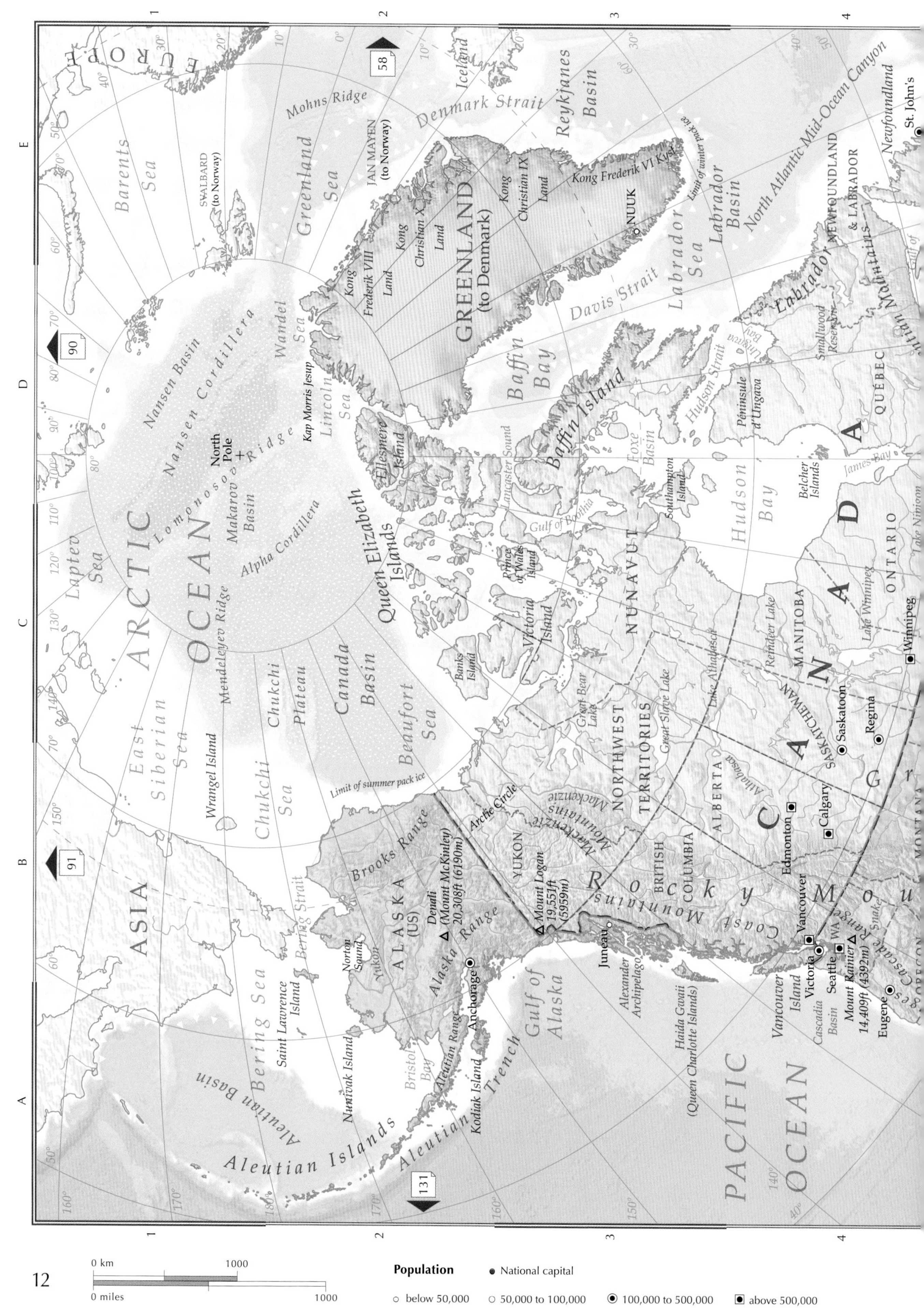

ARCTIC OCEAN
PACIFIC OCEAN
ASIA
EUROPE
GREENLAND (to Denmark)
CANADA
ALASKA (US)
NUUK
Barents Sea
Greenland Sea
Laptev Sea
East Siberian Sea
Chukchi Sea
Beaufort Sea
Bering Sea
Gulf of Alaska
Baffin Bay
Hudson Bay
Labrador Sea
Davis Strait
Denmark Strait
Lincoln Sea
Wandel Sea
Iceland
SVALBARD (to Norway)
JAN MAYEN (to Norway)
Mohns Ridge
Reykjanes Basin
Nansen Basin
Nansen Cordillera
Lomonosov Ridge
Makarov Basin
Alpha Cordillera
Mendeleyev Ridge
Chukchi Plateau
Canada Basin
North Pole
Kap Morris Jesup
Ellesmere Island
Queen Elizabeth Islands
Baffin Island
Victoria Island
Banks Island
Prince of Wales Island
Southampton Island
Belcher Islands
Wrangel Island
Saint Lawrence Island
Nunivak Island
Kodiak Island
Aleutian Islands
Aleutian Basin
Aleutian Trench
Alexander Archipelago
Haida Gwaii (Queen Charlotte Islands)
Vancouver Island
Cascadia Basin
NUNAVUT
NORTHWEST TERRITORIES
YUKON
BRITISH COLUMBIA
ALBERTA
SASKATCHEWAN
MANITOBA
ONTARIO
QUEBEC
NEWFOUNDLAND & LABRADOR
Labrador
Labrador Basin
North Atlantic Mid-Ocean Canyon
Newfoundland
St. John's
Rocky Mountains
Mackenzie Mountains
Brooks Range
Alaska Range
Aleutian Range
Coast Mountains
Cascade Range
Denali (Mount McKinley) 20,308ft (6190m)
Mount Logan 19,551ft (5959m)
Mount Rainier 14,409ft (4392m)
Anchorage
Juneau
Vancouver
Victoria
Seattle
Eugene
Edmonton
Calgary
Saskatoon
Regina
Winnipeg
Great Bear Lake
Great Slave Lake
Lake Athabasca
Reindeer Lake
Lake Winnipeg
James Bay
Foxe Basin
Gulf of Boothia
Lancaster Sound
Hudson Strait
Ungava Bay
Péninsule d'Ungava
Smallwood Reservoir
Norton Sound
Bristol Bay
Bering Strait
Yukon
Mackenzie
Athabasca
Arctic Circle
Limit of summer pack ice
Limit of winter pack ice
Kong Frederik VIII Land
Kong Christian X Land
Kong Christian IX Land
Kong Frederik VI Kyst
58
90
91
131

0 km 1000

0 miles 1000

Population

● National capital

○ below 50,000 ○ 50,000 to 100,000 ◉ 100,000 to 500,000 ▣ above 500,000

UNITED STATES OF AMERICA
MEXICO
ATLANTIC OCEAN
PACIFIC OCEAN
SOUTH AMERICA
Gulf of Mexico
Caribbean Sea
Sargasso Sea
Nares Plain
Hatteras Plain
Bermuda Rise
BERMUDA (to UK)
Blake Plateau
Tropic of Cancer
Equator
Fracture Zone
East Pacific Rise
Middle America Trench
Guatemala Basin
Colón Ridge
Cocos Ridge
Panama Basin
Colombian Basin
Galápagos Islands (to Ecuador)
Andes
Sierra Madre Occidental
Sierra Madre Oriental
Appalachian Mountains
Gulf of California
Lower California
Mount Whitney 14,5050ft (4421m)
Death Valley -282ft (-86m)
Volcán Pico de Orizaba 18,451ft (5700m)
Yucatan Peninsula
Mississippi Delta
Strait of Florida
Greater Antilles
Lesser Antilles
Halifax
Georges Bank
Cape Cod
Boston
New York
Philadelphia
Baltimore
WASHINGTON D.C.
Richmond
Raleigh
Columbia
Jacksonville
Tampa
Miami
Albany
Toronto
Milwaukee
Madison
Chicago
Lansing
Detroit
Cleveland
Columbus
Indianapolis
Springfield
Nashville
Memphis
Atlanta
Montgomery
Jackson
Baton Rouge
New Orleans
Little Rock
Oklahoma City
Dallas
Austin
San Antonio
Houston
El Paso
Phoenix
Denver
Lincoln
Des Moines
Kansas City
San Jose
Los Angeles
San Diego
Monterrey
Guadalajara
MEXICO CITY
Acapulco
THE BAHAMAS
NASSAU
CUBA
HAVANA
Guantanamo Bay (to US)
CAYMAN ISLANDS (to UK)
JAMAICA
KINGSTON
HAITI
PORT-AU-PRINCE
DOMINICAN REPUBLIC
SANTO DOMINGO
PUERTO RICO (to US)
TURKS & CAICOS ISLANDS (to UK)
VIRGIN ISLANDS (to US)
BRITISH VIRGIN ISLANDS (to UK)
ANGUILLA (to UK)
ANTIGUA & BARBUDA
ST KITTS & NEVIS
MONTSERRAT (to UK)
GUADELOUPE (to France)
DOMINICA
MARTINIQUE (to France)
ST LUCIA
ST VINCENT & THE GRENADINES
BARBADOS
GRENADA
PORT OF SPAIN
TRINIDAD & TOBAGO
ARUBA (Neth.)
CURAÇAO (Neth.)
BONAIRE (to Neth.)
BELIZE
BELMOPAN
GUATEMALA
GUATEMALA CITY
HONDURAS
TEGUCIGALPA
EL SALVADOR
SAN SALVADOR
NICARAGUA
MANAGUA
Lake Nicaragua
COSTA RICA
SAN JOSÉ
PANAMA
PANAMA CITY
131
34
44

Political features

Total area:
9,400,000 sq miles
(24,346,000 sq km)

Total number of countries:
23

Total population:
590 million

Largest city with population:
Mexico City, Mexico 24.7 million

Country with highest population density:
Barbados 1730 people per sq mile
(669 people per sq km)

Largest country:
Canada 3,855,171 sq miles
(9,984,670 sq km)

Smallest country:
St. Kitts and Nevis 101 sq miles
(261 sq km)

Physical features

Largest lake:
Lake Superior, Canada/ USA
31,700 sq miles (82,100 sq km)

Longest river:
Mississippi-Missouri, USA
3902 miles (6280 km)

Highest point:
Denali (Mt. McKinley), Alaska, USA
20,310 ft (6190 m)

Lowest point:
Death Valley, California, USA
-282 ft (-86 m) below sea level

Western Canada & Alaska

A
B
C
D
1
2
3
4
5
93
130
131
131
Poluostrov Kamchatka
Arctic Circle
Ostrov Vrangelya
RUSSIA
ARCTIC
Chukchi Sea
Attu Island
Near Islands
Bering Sea
Bering Strait
Wevok
Point Lay
Barrow
Kivalina
Wales
Gambell
Saint Lawrence Island
Colville River
Prudhoe Ba
Rat Islands
Amchitka Island
Deering
Umiat
Norton Sound
Kakt
Brooks Range
Alakanuk
Aleutian Islands
Andreanof Islands
Nunivak Island
Grayling
Yukon River
Kokrines
Fort Yukon
Pribilof Islands
Kwigillingok
ALASKA (to US)
Akla
Atka
Kuskokwim Mts
Fairbanks
Yukon River
McPhers
Platinum
Alaska Range
Denali (Mount McKinley) 6190m
Denali Park
Umnak Island
Dutch Harbor
Bristol Bay
Iliamna Lake
Susitna
Unalaska Island
Anchorage
Unimak Island
Gulkana
Hope
Belkofski
Alaska Peninsula
Valdez
YUKON
Shumagin Islands
Kodiak
Cordova
Chitina
Kodiak Island
Katalla
Mount Logan 5959m
Whitehorse
Yakutat
Gulf of Alaska
Haines
Atlin
Gustavus
Juneau
Kake
Alexander Archipelago
Port Alexander
Ketchikan
PACIFIC OCEAN
Prince Rupert
Kitimat
Haida Gwaii (Queen Charlotte Islands)
Ocean Falls
Queen Charlotte Sound
Mount Waddington 4016
Port Hardy
Campbell River
Vancouver Island
Nanai
Victo
0 km
400
0 miles
400
Population
Internal administrative capital
below 50,000
50,000 to 100,000
100,000 to 500,000
above 500,000

Alert
Ellesmere Island
Nares Strait
Knud Rasmussen Land
GREENLAND
(to Denmark)
OCEAN
Axel Heiberg Island
Queen Elizabeth Islands
Ellef Ringnes Island
Isachsen
Amund Ringnes Island
Prince Patrick Island
Mould Bay
Baffin Bay
Arctic Circle
Bathurst Island
Devon Island
Melville Island
Cornwallis Island
Lancaster Sound
Resolute
(Qausuittuq)
Beaufort Sea
Viscount Melville Sound
Banks Island
Somerset Island
Brodeur Peninsula
Davis Strait
Baffin Island
Prince of Wales Island
M'Clintock Channel
Gulf of Boothia
Boothia Peninsula
Igloolik
Cumberland Sound
Amundsen Gulf
Holman
Victoria Island
Nettilling Lake
Paulatuk
King William Island
Melville Peninsula
Foxe Basin
Iqaluit
(Frobisher Bay)
Cambridge Bay
(Ikaluktutiak)
Gjoa Haven
(Uqsuqtuuq)
Kugaaruk
(Pelly Bay)
Amadjuak Lake
Kugluktuk
(Coppermine)
Naujaat
(Repulse Bay)
Fort Good Hope
(Rádeyilikóé)
Southampton Island
Hudson Strait
Great Bear Lake
Burnside
NUNAVUT
Coral Harbour
(Salliq)
Mackenzie
Echo Bay
Back
Garry Lake
Baker Lake
Péninsule d'Ungava
NORTHWEST TERRITORIES
Rankin Inlet
Coats Island
Mansel Island
Edzo
Yellowknife
Reliance
Whale Cove
(Tikiarjuaq)
QUÉBEC
Fort Simpson
Dubawnt
Great Slave Lake
Lutselk'e
(Snowdrift)
Arviat
Hudson Bay
Fort Providence
Fort Liard
Hay River
Fort Smith
Churchill
Fort Nelson
Lake Athabasca
Belcher Islands
Reindeer Lake
Wollaston Lake
Fort Vermilion
Southern Indian Lake
James Bay
CANADA
Fort St. John
Fort McMurray
Lynn Lake
Nelson
ALBERTA
Buffalo Narrows
Thompson
Grande Prairie
SASKATCHEWAN
Athabasca
Flin Flon
Lake Winnipeg
The Pas
Edmonton
North Saskatchewan
Saskatchewan
MANITOBA
ONTARIO
Mount Robson
3954m
Leduc
Prince Albert
Saskatoon
Red Deer
Kindersley
Yorkton
Lake Manitoba
Kamloops
Calgary
Regina
Qu'Appelle
Winnipeg
Lake of the Woods
Kelowna
Medicine Hat
Brandon
Cranbrook
Weyburn
Lake Superior
Lethbridge
Melita
Lake Huron
Milk River
Estevan
Lake Michigan
UNITED STATES OF AMERICA

133
60
16
23

Elevation

Below sea level	-6000m	-4000m	-2000m	-1000m	-500m	-250m	0	250m	500m	1000m	2000m	3000m	4000m	6000m
	-19,658ft	-13,124ft	-6562ft	-3281ft	-1640ft	-820ft	-328ft/-100m 0	820ft	1640ft	3281ft	6562ft	9843ft	13,124ft	19,685ft

Eastern Canada

A
B
C
D
1
2
3
4
5
15
15
23
18
NORTHWEST TERRITORIES
NUNAVUT
SASKATCHEWAN
MANITOBA
ONTARIO
CANADA
Hudson Bay
James Bay
Coats Island
Mansel Island
Ivujivik
Charles Island
Péninsul d' Ungav
Ottawa Islands
Inukjuak (Port Harrison)
Belcher Islands
Churchill
Southern Indian Lake
Nelson
Hayes
Fort Severn
Severn
Peawanuk
Cedar Lake
Lake Winnipeg
Lake Winnipegosis
Lake Manitoba
Sandy Lake
Winisk
Attawapiskat
Akimiski Island
Fort Albany
Albany
Moosonee
Moose
Harricana
Eastmain
Rivière de Rupert
Chibougamau
Réservoir Gouin
Lac Seul
Armstrong
Kenora
Dryden
Lake of the Woods
Lake Nipigon
Longlac
Hearst
Kapuskasing
Cochrane
Fort Frances
Atikokan
Nipigon
Rainy Lake
Thunder Bay
Marathon
Tip Top Mountain
Δ 640m
Timmins
Amos
Rouyn-Noranda
Val-d'Or
Foleyet
Wawa
Kirkland Lake
Lake Superior
Red River
NORTH DAKOTA
MINNESOTA
SOUTH DAKOTA
UNITED STATES OF AMERICA
WISCONSIN
MICHIGAN
Lake Michigan
IOWA
NEBRASKA
ILLINOIS
INDIANA
OHIO
PENNSYLVANIA
NEW YORK
Mississippi River
Sault Ste.Marie
Sudbury
North Bay
Manitoulin Island
Georgian Bay
Pembroke
Gatineau
Hull
OTTAWA
Lake Huron
Midland
Peterborough
Kingst
Oshawa
Brampton
Toronto
Kitchener
Hamilton
St.Catharines
Sarnia
London
Niagara Falls
Windsor
Leamington
Lake Erie
Lake Onta

0 km 300
0 miles 300

Population
● National capital
● Internal administrative capital
○ below 50,000
○ 50,000 to 100,000
◉ 100,000 to 500,000
▣ above 500,000

Baffin Island
Resolution Island
Button Islands
Akpatok Island
Ungava Bay
Labrador Sea
Nain
Hopedale
Makkovik
Cape Harrison
Cartwright
Labrador
NEWFOUNDLAND & LABRADOR
Schefferville
Smallwood Reservoir
Churchill
Lake Melville
Rivière à la Baleine
Caniapiscau
St.Anthony
Strait of Belle Isle
Gagnon
Réservoir Manicouagan
Laurentian Mountains
Havre-St-Pierre
Sept-Îles
Île d'Anticosti
Corner Brook
Gander
Grand Falls
Newfoundland
St.John's
Cape Race
Channel-Port aux Basques
ST PIERRE & MIQUELON (to France)
Cabot Strait
Gulf of St. Lawrence
Gaspé
Péninsule de Gaspé
Baie-Comeau
St.Lawrence
Matane
Chicoutimi
Rimouski
Bathurst
Îles de la Madeleine
PRINCE EDWARD ISLAND
Charlottetown
Sydney
Glace Bay
Cape Breton Island
Rivière-du-Loup
Edmundston
NEW BRUNSWICK
Moncton
Amherst
New Glasgow
Truro
Oromocto
Fredericton
Saint John
NOVA SCOTIA
Charlesbourg
Québec
St-Georges
Drummondville
Sherbrooke
MAINE
Bay of Fundy
Dartmouth
Halifax
Sable Island
Liverpool
Yarmouth
ATLANTIC OCEAN
NEW HAMPSHIRE
Cape Cod
RHODE ISLAND
60
44
44
44

Elevation

Below sea level: -6000m (-19,658ft), -4000m (-13,124ft), -2000m (-6562ft), -1000m (-3281ft), -500m (-1640ft), -250m (-820ft), -100m (-328ft), 0

Above sea level: 0, 250m (820ft), 500m (1640ft), 1000m (3281ft), 2000m (6562ft), 3000m (9843ft), 4000m (13,124ft), 6000m (19,685ft)

USA: The Northeast

Upper Red Lake
Lower Red Lake
Namakan Lake
16
Isle Royale
Lake Superior
Keweenaw Peninsula
MINNESOTA
ONTARIO
Apostle Islands
Superior
Ashland
Ironwood
Gogebic Range
Houghton
Marquette
Mille Lacs Lake
Saint Croix River
MICHIGAN
Sault Sainte Marie
Woodruff
Rice Lake
Rhinelander
Iron Mountain
Escanaba
Saint Ignace
North Channel
Georgian
Ladysmith
Beaver Island
Cheboygan
WISCONSIN
Door Peninsula
Petoskey
Lake Huron
23
River Falls
Eau Claire
Wausau
Green Bay
Traverse City
Alpena
Mississippi River
Stevens Point
Wisconsin Rapids
Beulah
Roscommon
Appleton
Cadillac
Tomah
Oshkosh
Lake Winnebago
Saginaw Bay
La Crosse
Fond du Lac
Ludington
Lake Michigan
Sheboygan
Midland
West Bend
Mount Pleasant
Bay City
Wisconsin River
Muskegon
Saginaw
Madison
Milwaukee
Grand Rapids
Flint
Waukesha
Wyoming
Port Huron
IOWA
Janesville
Racine
Lansing
Kenosha
Pontiac
Lake Saint Clair
Rockford
Waukegan
Kalamazoo
Livonia
Warren
Detroit
Lake Erie
Elgin
Evanston
Ann Arbor
Sterling
Adrian
Aurora
Chicago
South Bend
Toledo
Rock Island
Joliet
Gary
Elkhart
Cleveland
Euclid
Ottawa
Valparaiso
Bowling Green
Sandusky
Galesburg
Kankakee
Fort Wayne
Findlay
Akron
Youngstown
Peoria
Wabash
Van Wert
Mansfield
Macomb
Bloomington
INDIANA
Marion
Canton
Pekin
Illinois River
Lafayette
Kokomo
OHIO
Quincy
Champaign
Anderson
Muncie
Sidney
Delaware
Wheeling
Carmel
Cambridge
Springfield
Springfield
Indianapolis
23
Jacksonville
Decatur
Dayton
Columbus
Zanesville
Ohio River
ILLINOIS
Terre Haute
Kettering
Wilmington
Effingham
Columbus
Chillicothe
Athens
Alton
Bloomington
Cincinnati
Parkersburg
Lake of the Ozarks
Missouri River
East Saint Louis
Vincennes
Newport
Portsmouth
Belleville
Mount Vernon
Wabash River
Ohio River
Charleston
MISSOURI
New Albany
Louisville
Huntington
Mississippi River
Evansville
Frankfort
Lexington
Saint Albans
Carbondale
Owensboro
Henderson
Elizabethtown
Richmond
Beckley
Ozark Plateau
KENTUCKY
Cairo
Paducah
Green River
Pikeville
Somerset
London
Hopkinsville
Bowling Green
Kentucky Lake
Middlesboro
Bristol
20
ARKANSAS
TENNESSEE
Appalachian

0 km 200
0 miles 200

Population
National capital
Internal administrative capital
below 50,000
50,000 to 100,000
100,000 to 500,000
above 500,000

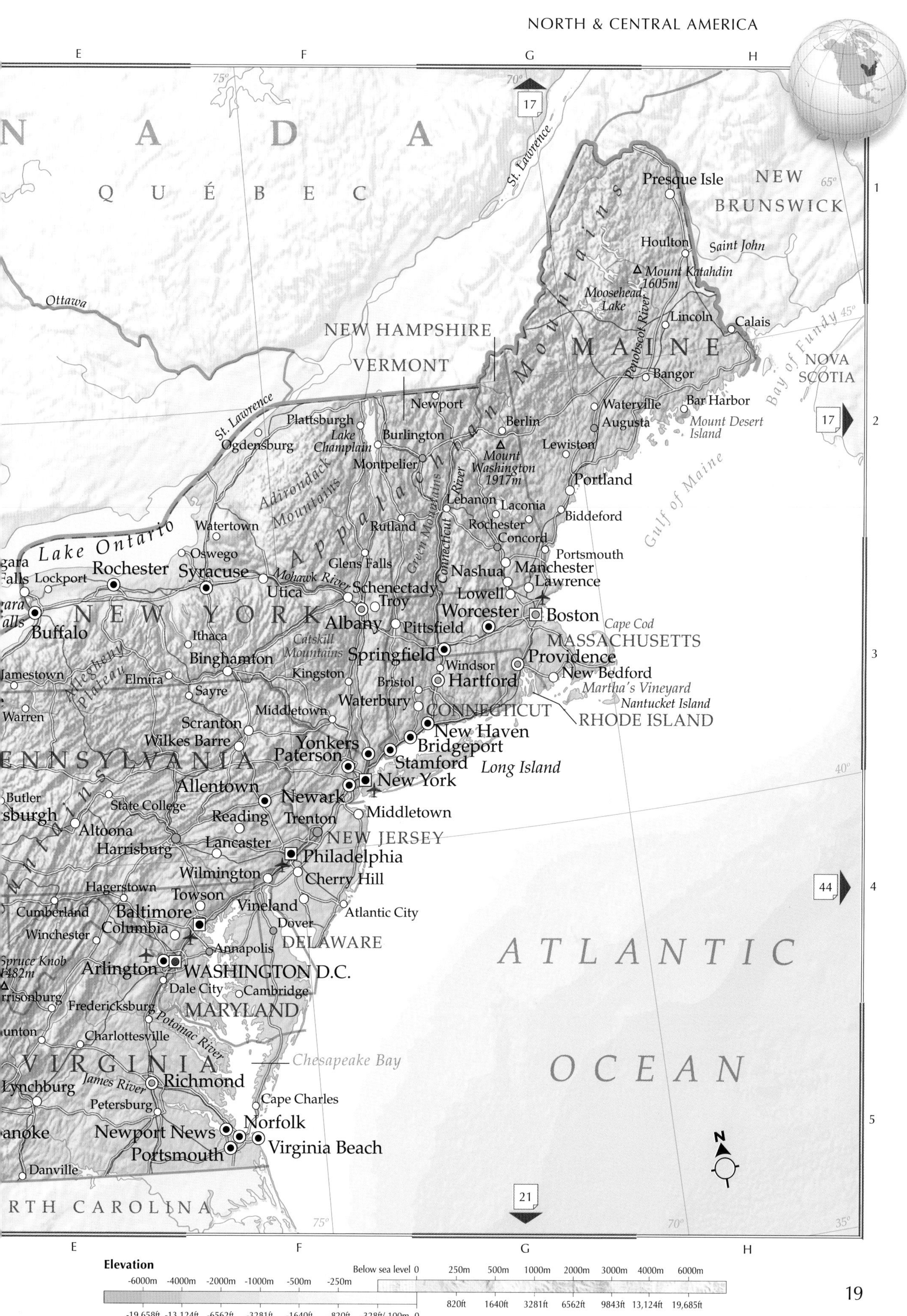
E
F
G
H
1
2
3
4
5
CANADA
QUÉBEC
NEW BRUNSWICK
NOVA SCOTIA
MAINE
NEW HAMPSHIRE
VERMONT
NEW YORK
MASSACHUSETTS
CONNECTICUT
RHODE ISLAND
PENNSYLVANIA
NEW JERSEY
DELAWARE
MARYLAND
VIRGINIA
NORTH CAROLINA
WASHINGTON D.C.
ATLANTIC OCEAN
St. Lawrence
Ottawa
Lake Ontario
Presque Isle
Houlton
Saint John
Mount Katahdin 1605m
Moosehead Lake
Penobscot River
Lincoln
Calais
Bay of Fundy
Bangor
Bar Harbor
Mount Desert Island
Waterville
Augusta
Lewiston
Portland
Gulf of Maine
Biddeford
Portsmouth
Berlin
Mount Washington 1917m
Appalachian Mountains
Newport
Plattsburgh
Lake Champlain
Burlington
Montpelier
Ogdensburg
Adirondack Mountains
Green Mountains
Connecticut River
Lebanon
Laconia
Rochester
Concord
Manchester
Nashua
Lawrence
Lowell
Boston
Cape Cod
Worcester
Rutland
Watertown
Oswego
Syracuse
Rochester
Lockport
Niagara Falls
Buffalo
Mohawk River
Glens Falls
Schenectady
Troy
Utica
Albany
Pittsfield
Springfield
Providence
New Bedford
Martha's Vineyard
Nantucket Island
Windsor
Hartford
Bristol
Waterbury
Ithaca
Catskill Mountains
Binghamton
Kingston
Jamestown
Allegheny Plateau
Elmira
Sayre
Warren
Middletown
Scranton
Wilkes Barre
New Haven
Bridgeport
Stamford
Long Island
Yonkers
Paterson
New York
Newark
Allentown
Butler
State College
Pittsburgh
Altoona
Reading
Trenton
Middletown
Harrisburg
Lancaster
Philadelphia
Wilmington
Cherry Hill
Hagerstown
Towson
Cumberland
Baltimore
Vineland
Atlantic City
Winchester
Columbia
Dover
Annapolis
Spruce Knob 1482m
Arlington
Harrisonburg
Dale City
Cambridge
Fredericksburg
Potomac River
Staunton
Charlottesville
Chesapeake Bay
Lynchburg
James River
Richmond
Petersburg
Cape Charles
Norfolk
Roanoke
Newport News
Virginia Beach
Portsmouth
Danville
17
17
44
21
75°
70°
65°
45°
40°
35°
Elevation
-6000m -4000m -2000m -1000m -500m -250m
Below sea level 0
250m 500m 1000m 2000m 3000m 4000m 6000m
-19,658ft -13,124ft -6562ft -3281ft -1640ft -820ft -328ft/-100m 0
820ft 1640ft 3281ft 6562ft 9843ft 13,124ft 19,685ft

USA: The Southeast

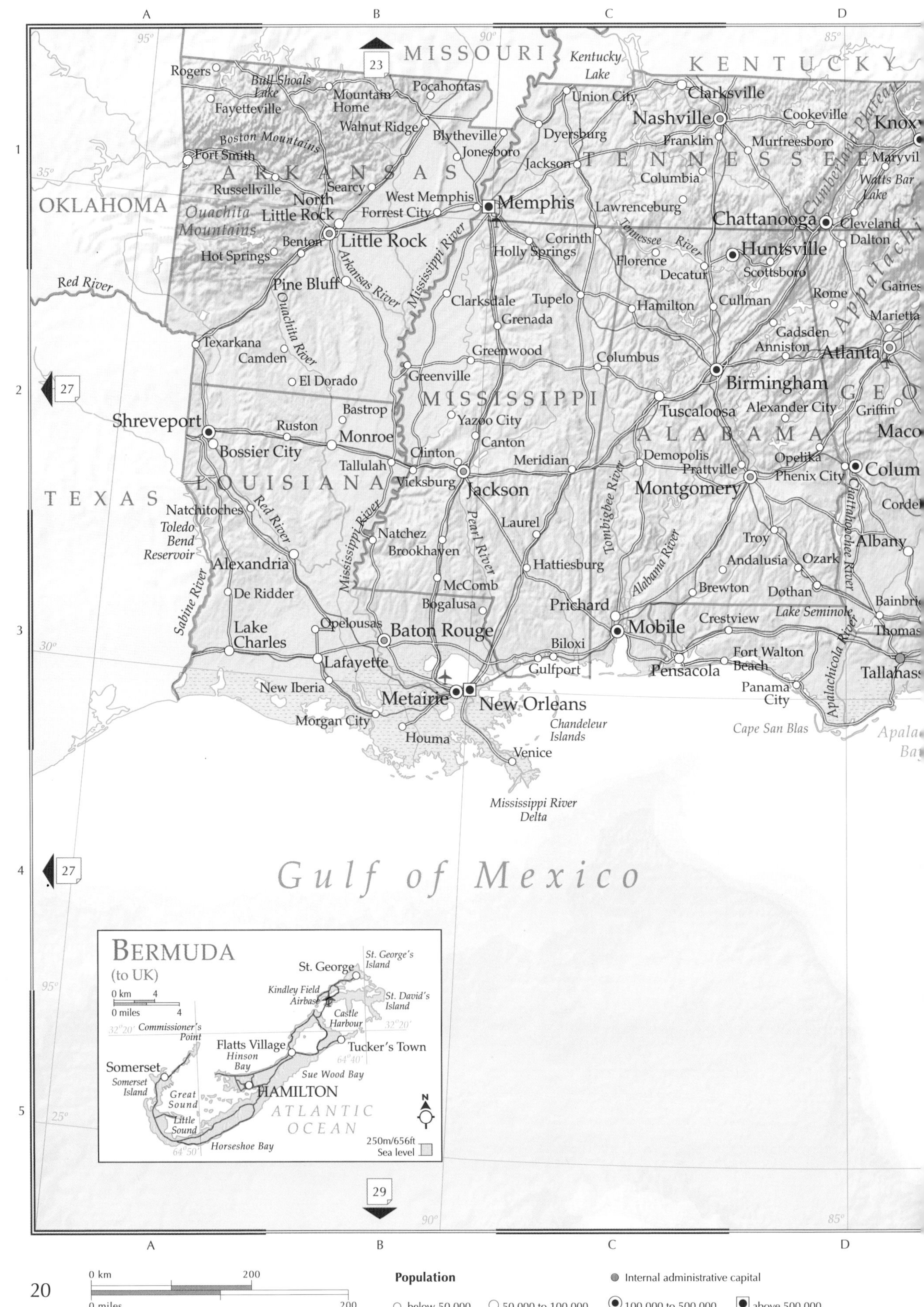

A
B
C
D
1
2
3
4
5
95°
90°
85°
35°
30°
25°
23
27
27
29
MISSOURI
KENTUCKY
TENNESSEE
ARKANSAS
OKLAHOMA
MISSISSIPPI
ALABAMA
LOUISIANA
TEXAS
Kentucky Lake
Rogers
Bull Shoals Lake
Fayetteville
Mountain Home
Pocahontas
Walnut Ridge
Boston Mountains
Fort Smith
Blytheville
Jonesboro
Russellville
Searcy
North Little Rock
Ouachita Mountains
West Memphis
Forrest City
Memphis
Little Rock
Benton
Hot Springs
Red River
Pine Bluff
Arkansas River
Mississippi River
Ouachita River
Texarkana
Camden
El Dorado
Union City
Clarksville
Nashville
Cookeville
Dyersburg
Franklin
Murfreesboro
Jackson
Columbia
Cumberland Plateau
Knox
Maryvil
Watts Bar Lake
Lawrenceburg
Chattanooga
Cleveland
Dalton
Corinth
Holly Springs
Tennessee River
Florence
Decatur
Huntsville
Scottsboro
Appalachian
Clarksdale
Tupelo
Grenada
Hamilton
Cullman
Rome
Gaines
Marietta
Gadsden
Anniston
Atlanta
Greenwood
Columbus
Greenville
Birmingham
Bastrop
Yazoo City
Tuscaloosa
Alexander City
Griffin
Shreveport
Ruston
Monroe
Canton
Bossier City
Tallulah
Clinton
Meridian
Demopolis
Prattville
Opelika
Phenix City
Macon
Colum
Vicksburg
Jackson
Montgomery
Tombigbee River
Natchitoches
Toledo Bend Reservoir
Red River
Mississippi River
Natchez
Brookhaven
Pearl River
Laurel
Alabama River
Troy
Cordele
Albany
Alexandria
Andalusia
Ozark
Hattiesburg
Chattahoochee River
Sabine River
De Ridder
McComb
Brewton
Dothan
Bainbri
Bogalusa
Prichard
Crestview
Lake Seminole
Lake Charles
Opelousas
Baton Rouge
Mobile
Thomas
Lafayette
Biloxi
Gulfport
Pensacola
Fort Walton Beach
Apalachicola River
Tallahass
New Iberia
Metairie
New Orleans
Panama City
Morgan City
Chandeleur Islands
Houma
Cape San Blas
Apala
Bay
Venice
Mississippi River Delta
Gulf of Mexico
BERMUDA
(to UK)
0 km 4
0 miles 4
St. George
St. George's Island
Kindley Field Airbase
St. David's Island
Castle Harbour
32°20'
Commissioner's Point
Flatts Village
Tucker's Town
Hinson Bay
64°40'
Somerset
Somerset Island
Sue Wood Bay
HAMILTON
Great Sound
Little Sound
ATLANTIC OCEAN
N
Horseshoe Bay
64°50'
250m/656ft
Sea level
0 km 200
0 miles 200
Population
Internal administrative capital
below 50,000
50,000 to 100,000
100,000 to 500,000
above 500,000

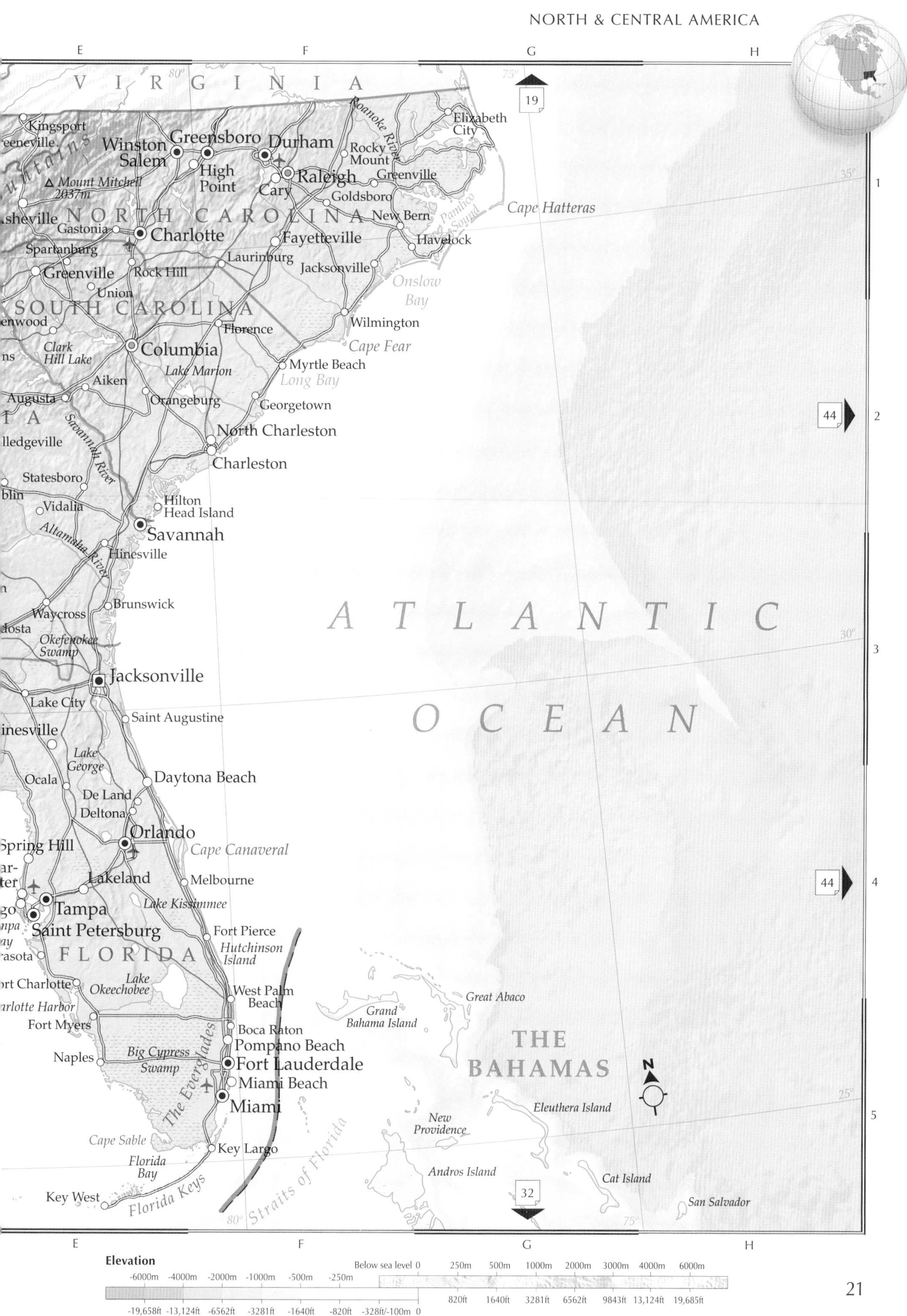
VIRGINIA
Kingsport
Winston Salem
Greensboro
Durham
Rocky Mount
Elizabeth City
High Point
Raleigh
Cary
Greenville
Goldsboro
Mount Mitchell 2037m
NORTH CAROLINA
New Bern
Pamlico Sound
Cape Hatteras
Gastonia
Charlotte
Fayetteville
Havelock
Spartanburg
Laurinburg
Jacksonville
Greenville
Rock Hill
Onslow Bay
Union
SOUTH CAROLINA
Wilmington
Florence
Clark Hill Lake
Columbia
Cape Fear
Myrtle Beach
Lake Marion
Long Bay
Aiken
Augusta
Orangeburg
Georgetown
Savannah River
North Charleston
Charleston
Statesboro
Vidalia
Hilton Head Island
Savannah
Altamaha River
Hinesville
Brunswick
Waycross
ATLANTIC OCEAN
Okefenokee Swamp
Jacksonville
Lake City
Saint Augustine
Lake George
Ocala
Daytona Beach
De Land
Deltona
Orlando
Spring Hill
Cape Canaveral
Lakeland
Melbourne
Tampa
Lake Kissimmee
Saint Petersburg
Fort Pierce
Hutchinson Island
FLORIDA
Lake Okeechobee
West Palm Beach
Great Abaco
Fort Myers
Grand Bahama Island
Boca Raton
Pompano Beach
THE BAHAMAS
Naples
Big Cypress Swamp
Fort Lauderdale
Miami Beach
The Everglades
Miami
Eleuthera Island
New Providence
Cape Sable
Key Largo
Florida Bay
Andros Island
Cat Island
Key West
Florida Keys
Straits of Florida
San Salvador
19
44
32
Elevation
Below sea level 0
250m 500m 1000m 2000m 3000m 4000m 6000m
820ft 1640ft 3281ft 6562ft 9843ft 13,124ft 19,685ft
-6000m -4000m -2000m -1000m -500m -250m
-19,658ft -13,124ft -6562ft -3281ft -1640ft -820ft -328ft/-100m 0

USA: Central States

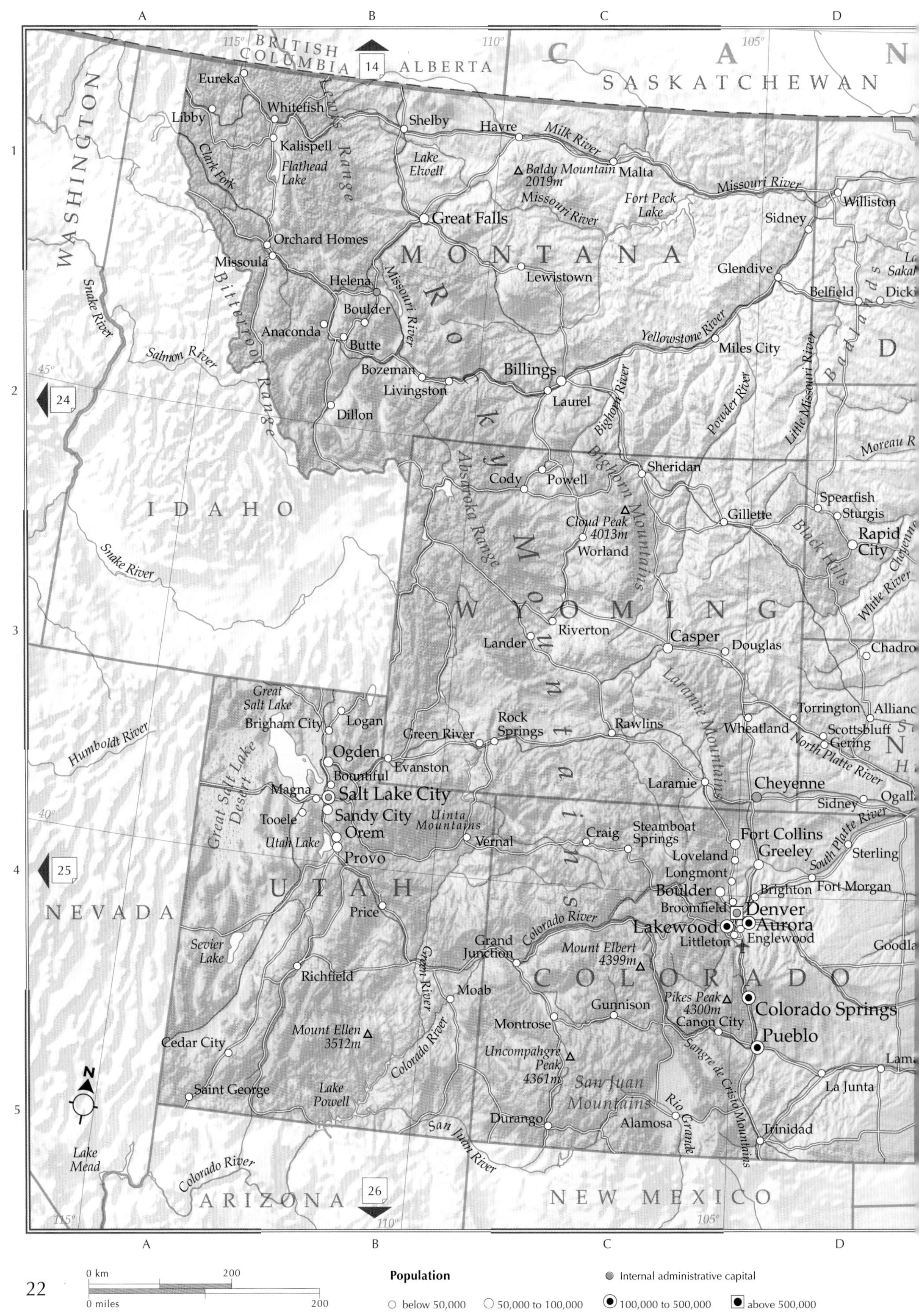

BRITISH COLUMBIA
ALBERTA
SASKATCHEWAN
WASHINGTON
MONTANA
IDAHO
WYOMING
UTAH
NEVADA
COLORADO
ARIZONA
NEW MEXICO
Rocky Mountains
Eureka
Libby
Whitefish
Kalispell
Flathead Lake
Shelby
Havre
Milk River
Malta
Baldy Mountain 2019m
Lake Elwell
Missouri River
Fort Peck Lake
Williston
Sidney
Great Falls
Orchard Homes
Missoula
Lewistown
Glendive
Belfield
Helena
Boulder
Anaconda
Butte
Bozeman
Livingston
Billings
Laurel
Dillon
Yellowstone River
Miles City
Bighorn River
Powder River
Little Missouri River
Badlands
Moreau R
Bitterroot Range
Clark Fork
Lewis Range
Snake River
Salmon River
Sheridan
Cody
Powell
Absaroka Range
Bighorn Mountains
Cloud Peak 4013m
Worland
Gillette
Spearfish
Sturgis
Black Hills
Rapid City
White River
Riverton
Lander
Casper
Douglas
Chadro
Laramie Mountains
Rawlins
Rock Springs
Green River
Evanston
Torrington
Wheatland
Scottsbluff
Gering
North Platte River
Allianc
Great Salt Lake
Brigham City
Logan
Ogden
Bountiful
Magna
Salt Lake City
Tooele
Sandy City
Orem
Utah Lake
Provo
Great Salt Lake Desert
Uinta Mountains
Vernal
Craig
Steamboat Springs
Laramie
Cheyenne
Sidney
Ogall
Fort Collins
Greeley
Loveland
Longmont
Boulder
Broomfield
Denver
Lakewood
Aurora
Littleton
Englewood
Brighton
Fort Morgan
South Platte River
Sterling
Humboldt River
Price
Colorado River
Grand Junction
Mount Elbert 4399m
Sevier Lake
Richfield
Green River
Moab
Gunnison
Pikes Peak 4300m
Colorado Springs
Canon City
Pueblo
Montrose
Mount Ellen 3512m
Cedar City
Uncompahgre Peak 4361m
San Juan Mountains
Sangre de Cristo Mountains
Rio Grande
La Junta
Saint George
Lake Powell
Durango
Alamosa
Trinidad
San Juan River
Lake Mead
Goodla
Lama
14
24
25
26
N

0 km 200
0 miles 200

Population
○ below 50,000 ○ 50,000 to 100,000 ⦿ 100,000 to 500,000 ▣ above 500,000
● Internal administrative capital

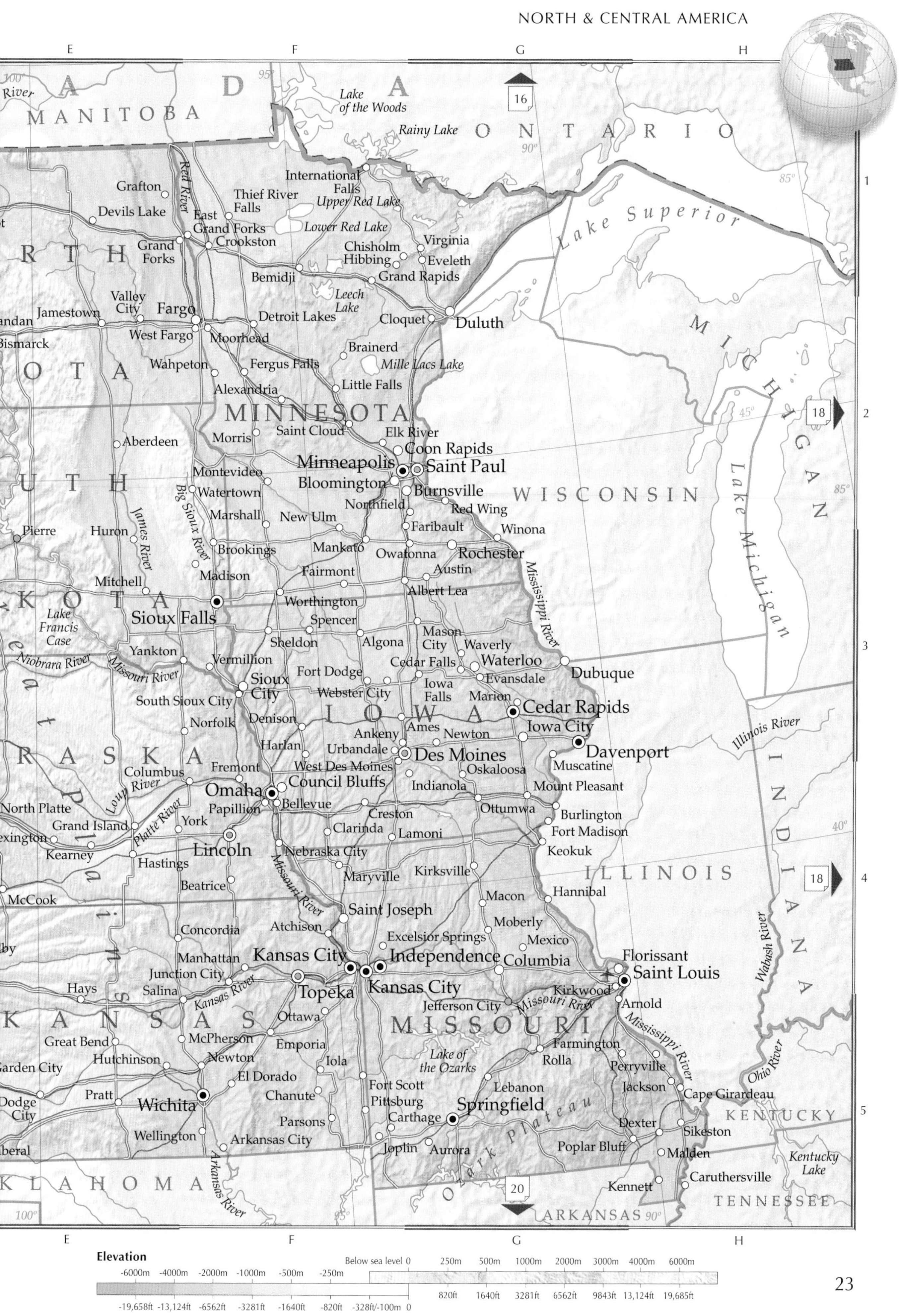
MANITOBA
ONTARIO
Lake of the Woods
Rainy Lake
International Falls
Upper Red Lake
Lower Red Lake
Thief River Falls
Grafton
Devils Lake
East Grand Forks
Grand Forks
Crookston
Red River
Chisholm
Hibbing
Virginia
Eveleth
Grand Rapids
Bemidji
Leech Lake
Lake Superior
Valley City
Jamestown
Fargo
West Fargo
Moorhead
Detroit Lakes
Cloquet
Duluth
Bismarck
Brainerd
Mille Lacs Lake
Wahpeton
Fergus Falls
Little Falls
Alexandria
MINNESOTA
MICHIGAN
Aberdeen
Morris
Saint Cloud
Elk River
Coon Rapids
Minneapolis
Saint Paul
Montevideo
Watertown
Bloomington
Burnsville
Northfield
Red Wing
WISCONSIN
Lake Michigan
Marshall
New Ulm
Faribault
Winona
Pierre
Huron
James River
Big Sioux River
Brookings
Mankato
Owatonna
Rochester
Madison
Fairmont
Austin
Mitchell
Albert Lea
Sioux Falls
Worthington
Lake Francis Case
Spencer
Mississippi River
Sheldon
Algona
Mason City
Waverly
Yankton
Vermillion
Cedar Falls
Waterloo
Niobrara River
Missouri River
Fort Dodge
Evansdale
Dubuque
Sioux City
Webster City
Iowa Falls
Marion
South Sioux City
Cedar Rapids
Norfolk
Denison
IOWA
Ames
Iowa City
Illinois River
Ankeny
Newton
Harlan
Urbandale
Des Moines
Davenport
Muscatine
Fremont
West Des Moines
Columbus
Oskaloosa
Council Bluffs
Loup River
Omaha
Indianola
Mount Pleasant
North Platte
Papillion
Bellevue
Creston
Ottumwa
Burlington
Grand Island
York
Platte River
Clarinda
Lamoni
Fort Madison
Lexington
Kearney
Lincoln
Nebraska City
Keokuk
Hastings
Kirksville
ILLINOIS
INDIANA
Beatrice
Maryville
Missouri River
McCook
Macon
Hannibal
Saint Joseph
Concordia
Atchison
Moberly
Excelsior Springs
Mexico
Wabash River
Manhattan
Kansas City
Independence
Columbia
Florissant
Junction City
Saint Louis
Hays
Salina
Topeka
Kansas City
Kirkwood
Kansas River
Jefferson City
Missouri River
Arnold
KANSAS
Ottawa
MISSOURI
Mississippi River
Great Bend
McPherson
Emporia
Farmington
Hutchinson
Newton
Lake of the Ozarks
Rolla
Perryville
Ohio River
Garden City
Iola
El Dorado
Fort Scott
Lebanon
Jackson
Cape Girardeau
Dodge City
Pratt
Wichita
Chanute
Pittsburg
Springfield
Ozark Plateau
KENTUCKY
Carthage
Dexter
Parsons
Sikeston
Wellington
Arkansas City
Joplin
Aurora
Poplar Bluff
Malden
Liberal
Kentucky Lake
Caruthersville
OKLAHOMA
Arkansas River
Kennett
TENNESSEE
ARKANSAS
16
18
18
20
Elevation
Below sea level 0
250m 500m 1000m 2000m 3000m 4000m 6000m
820ft 1640ft 3281ft 6562ft 9843ft 13,124ft 19,685ft
-6000m -4000m -2000m -1000m -500m -250m
-19,658ft -13,124ft -6562ft -3281ft -1640ft -820ft -328ft/-100m 0

USA: The West

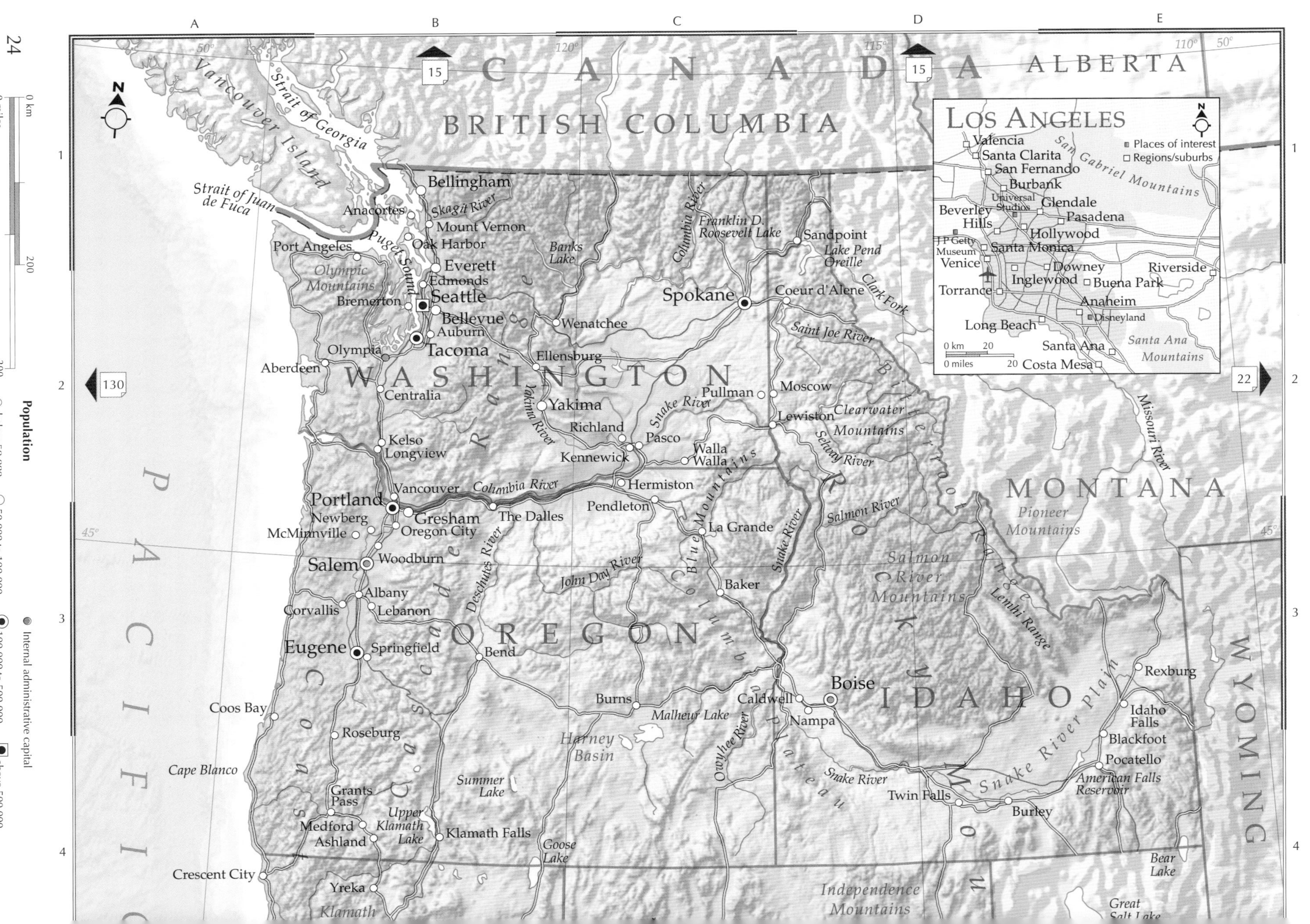

CANADA
ALBERTA
BRITISH COLUMBIA
Vancouver Island
Strait of Georgia
Strait of Juan de Fuca
LOS ANGELES
Places of interest
Regions/suburbs
Valencia
Santa Clarita
San Fernando
San Gabriel Mountains
Burbank
Universal Studios
Glendale
Pasadena
Beverley Hills
Hollywood
J P Getty Museum
Santa Monica
Venice
Downey
Riverside
Inglewood
Buena Park
Torrance
Anaheim
Disneyland
Long Beach
Santa Ana
Santa Ana Mountains
Costa Mesa
0 km 20
0 miles 20
Bellingham
Skagit River
Anacortes
Mount Vernon
Oak Harbor
Port Angeles
Puget Sound
Everett
Olympic Mountains
Edmonds
Seattle
Bremerton
Bellevue
Auburn
Olympia
Tacoma
Aberdeen
WASHINGTON
Centralia
Banks Lake
Wenatchee
Ellensburg
Yakima
Yakima River
Columbia River
Franklin D. Roosevelt Lake
Sandpoint
Lake Pend Oreille
Clark Fork
Spokane
Coeur d'Alene
Saint Joe River
Pullman
Moscow
Snake River
Clearwater Mountains
Lewiston
Selway River
Richland
Pasco
Kennewick
Walla Walla
Kelso
Longview
Vancouver
Columbia River
Hermiston
Portland
Gresham
The Dalles
Pendleton
Newberg
Oregon City
McMinnville
Blue Mountains
La Grande
Salem
Woodburn
Deschutes River
John Day River
Baker
Albany
Corvallis
Lebanon
OREGON
Eugene
Springfield
Bend
Coos Bay
Roseburg
Burns
Malheur Lake
Harney Basin
Owyhee River
Caldwell
Nampa
Boise
IDAHO
Cape Blanco
Summer Lake
Grants Pass
Upper Klamath Lake
Medford
Ashland
Klamath Falls
Goose Lake
Crescent City
Yreka
Klamath
Independence Mountains
Snake River
Twin Falls
Burley
Snake River Plain
Rexburg
Idaho Falls
Blackfoot
Pocatello
American Falls Reservoir
Bear Lake
Great Salt Lake
Salmon River
Salmon River Mountains
Lemhi Range
MONTANA
Pioneer Mountains
Missouri River
WYOMING
PACIFIC
15
22
130

Population

○ below 50,000 ○ 50,000 to 100,000 ◉ 100,000 to 500,000 ■ above 500,000

● Internal administrative capital

0 km 200

0 miles 200

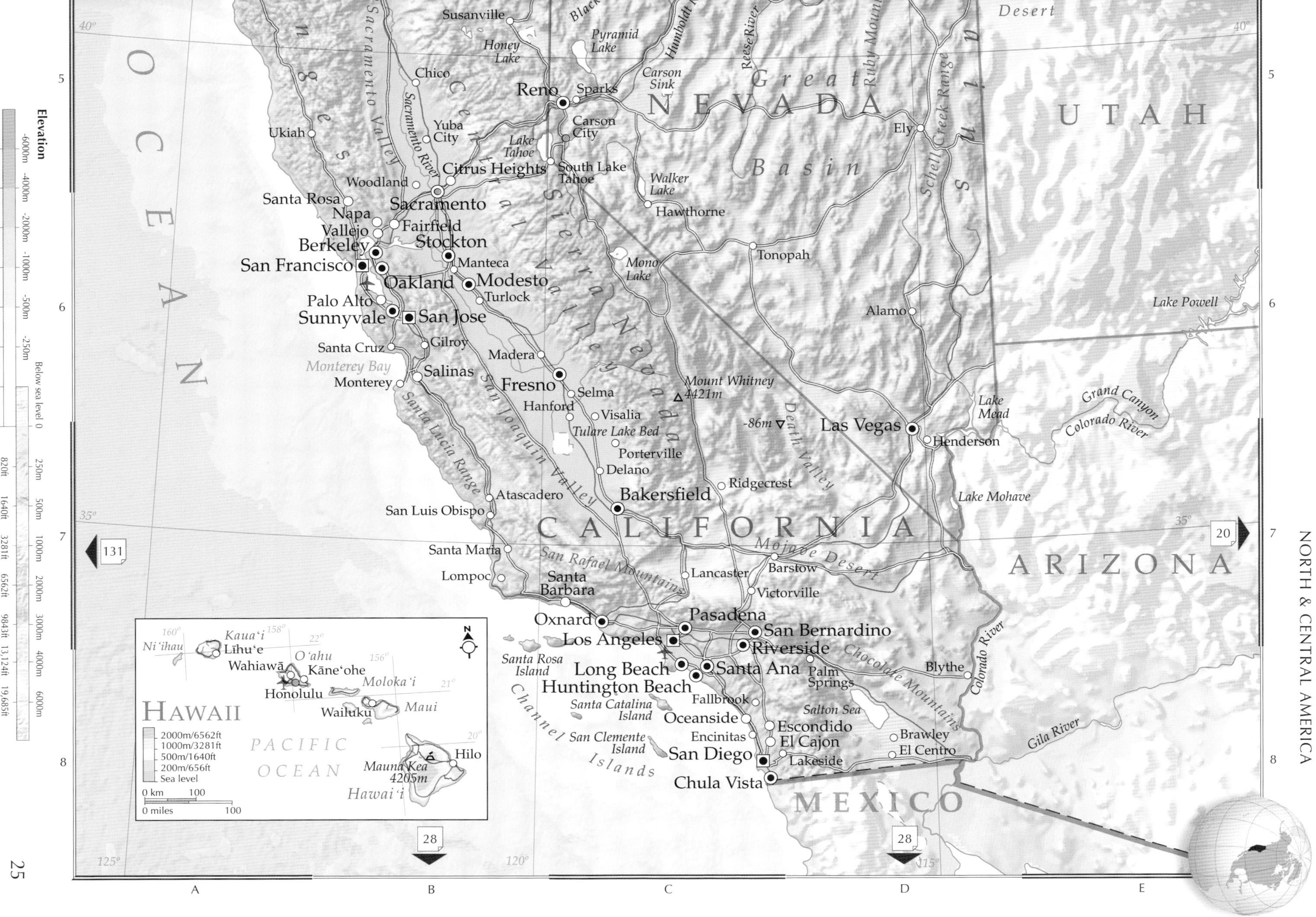
UTAH
NEVADA
ARIZONA
CALIFORNIA
MEXICO
Great Basin
Sierra Nevada
Central Valley
San Joaquin Valley
Sacramento Valley
Death Valley
Mojave Desert
Desert
Lake Powell
Grand Canyon
Colorado River
Lake Mead
Lake Mohave
Gila River
Salton Sea
Chocolate Mountains
Santa Lucia Range
San Rafael Mountains
Schell Creek Range
Channel Islands
Santa Rosa Island
Santa Catalina Island
San Clemente Island
Monterey Bay
Mount Whitney 4421m
-86m
Lake Tahoe
Mono Lake
Walker Lake
Pyramid Lake
Honey Lake
Carson Sink
Tulare Lake Bed
Sacramento River
Reese River
Susanville
Chico
Yuba City
Citrus Heights
Sacramento
Reno
Sparks
Carson City
South Lake Tahoe
Hawthorne
Tonopah
Ely
Alamo
Las Vegas
Henderson
Ukiah
Santa Rosa
Woodland
Napa
Vallejo
Berkeley
San Francisco
Fairfield
Stockton
Oakland
Palo Alto
Sunnyvale
San Jose
Manteca
Modesto
Turlock
Santa Cruz
Gilroy
Salinas
Monterey
Madera
Fresno
Hanford
Selma
Visalia
Porterville
Delano
Bakersfield
Ridgecrest
San Luis Obispo
Atascadero
Santa Maria
Lompoc
Santa Barbara
Oxnard
Los Angeles
Lancaster
Pasadena
Long Beach
Huntington Beach
Santa Ana
Riverside
San Bernardino
Barstow
Victorville
Palm Springs
Blythe
Fallbrook
Oceanside
Encinitas
Escondido
El Cajon
San Diego
Chula Vista
Lakeside
Brawley
El Centro
OCEAN
HAWAII
Ni'ihau
Kaua'i
Līhu'e
Wahiawā
O'ahu
Kāne'ohe
Honolulu
Moloka'i
Maui
Wailuku
Hawai'i
Mauna Kea 4205m
Hilo
PACIFIC OCEAN
2000m/6562ft
1000m/3281ft
500m/1640ft
200m/656ft
Sea level
0 km 100
0 miles 100
20
28
131
Elevation
Below sea level 0
250m 500m 1000m 2000m 3000m 4000m 6000m
820ft 1640ft 3281ft 6562ft 9843ft 13,124ft 19,685ft
-6000m -4000m -2000m -1000m -500m -250m
-19,658ft -13,124ft -6562ft -3281ft -1640ft -820ft -328ft/-100m 0

USA: The Southwest

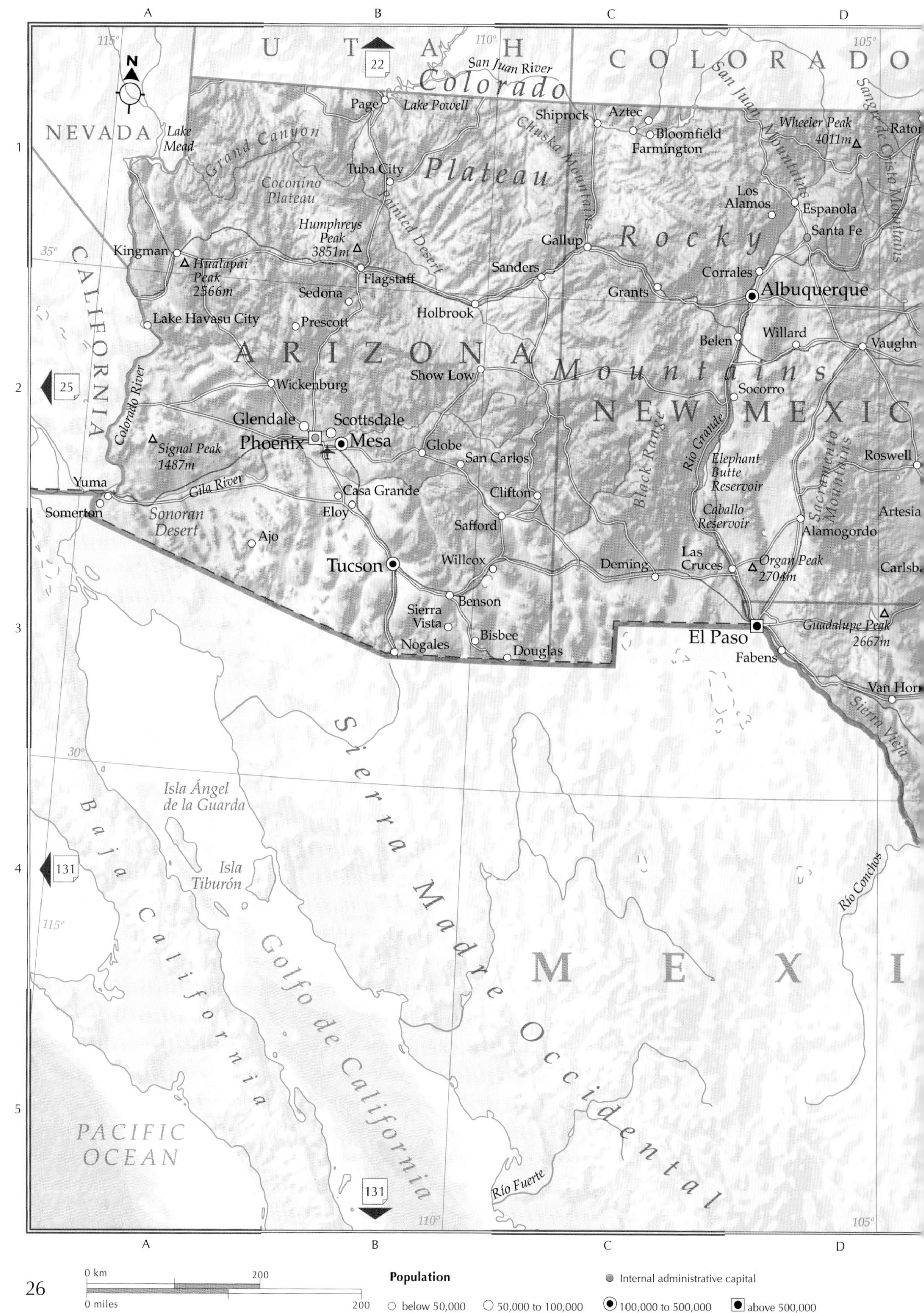

UTAH
COLORADO
NEVADA
CALIFORNIA
ARIZONA
NEW MEXICO
MEXICO
Colorado Plateau
Rocky Mountains
Grand Canyon
Coconino Plateau
Painted Desert
Chuska Mountains
San Juan Mountains
Sangre de Cristo Mountains
Black Range
Sacramento Mountains
Sierra Vieja
Sierra Madre Occidental
Sonoran Desert
Baja California
Golfo de California
PACIFIC OCEAN
Isla Ángel de la Guarda
Isla Tiburón
San Juan River
Lake Powell
Lake Mead
Colorado River
Gila River
Rio Grande
Río Conchos
Río Fuerte
Elephant Butte Reservoir
Caballo Reservoir
Humphreys Peak 3851m
Hualapai Peak 2566m
Signal Peak 1487m
Wheeler Peak 4011m
Organ Peak 2704m
Guadalupe Peak 2667m
Page
Tuba City
Kingman
Flagstaff
Sedona
Prescott
Lake Havasu City
Holbrook
Show Low
Wickenburg
Glendale
Scottsdale
Phoenix
Mesa
Globe
San Carlos
Casa Grande
Eloy
Yuma
Somerton
Ajo
Tucson
Safford
Clifton
Willcox
Benson
Sierra Vista
Nogales
Bisbee
Douglas
Shiprock
Aztec
Bloomfield
Farmington
Gallup
Sanders
Grants
Los Alamos
Espanola
Santa Fe
Corrales
Albuquerque
Belen
Willard
Vaughn
Socorro
Roswell
Artesia
Alamogordo
Deming
Las Cruces
Carlsba
El Paso
Fabens
Van Hor
Rator
22
25
131
131
A
B
C
D
1
2
3
4
5
115°
110°
105°
35°
30°
N

0 km 200
0 miles 200

Population
○ below 50,000
○ 50,000 to 100,000
● 100,000 to 500,000
■ above 500,000
● Internal administrative capital

KANSAS
MISSOURI
OKLAHOMA
ARKANSAS
TEXAS
LOUISIANA
Gulf of Mexico
Sierra Madre Oriental
23
20
32
29
Tulsa
Oklahoma City
Norman
Enid
Lawton
Amarillo
Lubbock
Wichita Falls
Fort Worth
Dallas
Arlington
Plano
Garland
Abilene
Midland
Odessa
San Angelo
Waco
Killeen
Austin
Houston
Pasadena
Beaumont
Port Arthur
San Antonio
Laredo
Corpus Christi
McAllen
Brownsville
Galveston
Victoria
Tyler
Longview
Denton
Bryan
College Station
Edwards Plateau
Llano Estacado
Stockton Plateau
Red River
Brazos River
Colorado River
Rio Grande
Pecos River
Trinity River
Sabine River
Padre Island
Laguna Madre
Emory Peak 2385m
Lake Texoma
Toledo Bend Reservoir
Amistad Reservoir
Lake Tawakoni
Elevation
Below sea level 0
250m 500m 1000m 2000m 3000m 4000m 6000m
820ft 1640ft 3281ft 6562ft 9843ft 13,124ft 19,685ft
-6000m -4000m -2000m -1000m -500m -250m
-19,658ft -13,124ft -6562ft -3281ft -1640ft -820ft -328ft/-100m 0

Mexico

CALIFORNIA
ARIZONA
NEW MEXICO
UNITED STATES
MEXICO
Tijuana
Mexicali
San Luis Río Colorado
Rosarito
Ensenada
Nogales
Agua Prieta
Cananea
Caborca
Magdalena
Cumpas
Samalayuca
Ciudad Juárez
Nuevo Casas Grandes
El Sueco
Ojinaga
El Sáuz
San Pedro de la Cueva
Chihuahua
Cuauhtémoc
Delicias
Ciudad Camargo
Jiménez
San Francisco del Oro
Hidalgo del Parral
Santa Barbara
Hermosillo
Guaymas
Empalme
Esperanza
Ciudad Obregón
Navojoa
Huatabampo
San Blas
Los Mochis
Guasave
Guamúchil
Culiacán
Navolato
El Dorado
Mazatlán
Escuinapa
Acaponeta
Tuxpan
Tepic
Puerto Vallarta
Tlaquepaque
Zamora de Hidal
Ciudad Guzmán
Colima
Manzanillo
Tecomán
Guadalajara
Tequila
Lagos de Mor
Yahualica
Aguascaliente
Jalpa
Villanueva
Zacatecas
Fresnillo
Durango
Miguel Asua
Juan Ald
Gómez Palacio
Torreón
Ciudad Lerdo
Matamoro
San Miguel
Boquillas
Nueva Ro
Monclo
Villa Ac
Guerrero Negro
San Ignacio
Loreto
La Paz
Miraflores
Santa Genoveva 2406m
Colorado River
Desierto de Altar
Sierra San Pedro Mártir
Golfo de California
Baja California
Sierra de la Giganta
Bahía Sebastián Vizcaíno
Isla Cedros
Isla Ángel de la Guarda
Isla Tiburón
Río Bavispe
Río Yaqui
Río Grande
Río Bravo del Norte
Río Conchos
Pecos River
Sierra Madre Occidental
Isla Magdalena
Isla Santa Margarita
Bahía de La Paz
Tropic of Cancer
Isla San Juanito
Isla MaríaMadre
Isla María Magdalena
Isla María Cleofas
Islas Marías
Isla San Benedicto
Isla Roca Partida
Isla Socorro
Isla Clarión
Islas Revillagigedo (to Mexico)
PACIFIC OCEAN
Lázaro Cárd
115°
110°
105°
30°
25°
20°
A
B
C
D
1
2
3
4
5
26
130
130
131
N

0 km 300

0 miles 300

Population

National capital

below 50,000

50,000 to 100,000

100,000 to 500,000

above 500,000

Gulf of Mexico
Bahía de Campeche
Yucatan Channel
Yucatan Peninsula
Tropic of Cancer
LOUISIANA
MISSISSIPPI
ALABAMA
FLORIDA
Mississippi River Delta
Padre Island
Laguna Madre
Laguna de Tamiahua
Sierra Madre Oriental
Sierra Madre del Sur
Istmo de Tehuantepec
Golfo de Tehuantepec
Gulf of Honduras
BELIZE
GUATEMALA
HONDURAS
EL SALVADOR
Nuevo Laredo
Reynosa
Matamoros
Monterrey
Montemorelos
Linares
Ciudad Victoria
Ciudad Madero
Tampico
Ciudad Valles
Tuxpán
Poza Rica
Papantla
Querétaro
Pachuca
CIUDAD DE MÉXICO (MEXICO CITY)
Xalapa
Veracruz
Puebla
Córdoba
Coatzacoalcos
Villahermosa
Minatitlán
Oaxaca
Tuxtla
Chilpancingo
Acapulco
Campeche
Mérida
Cancún
Valladolid
Chetumal
Carmen
Tapachula
Popocatépetl 5452m
20
44
30
131

Elevation

-6000m	-4000m	-2000m	-1000m	-500m	-250m	Below sea level 0	250m	500m	1000m	2000m	3000m	4000m	6000m
-19,658ft	-13,124ft	-6562ft	-3281ft	-1640ft	-820ft	-328ft/-100m 0	820ft	1640ft	3281ft	6562ft	9843ft	13,124ft	19,685ft

Central America

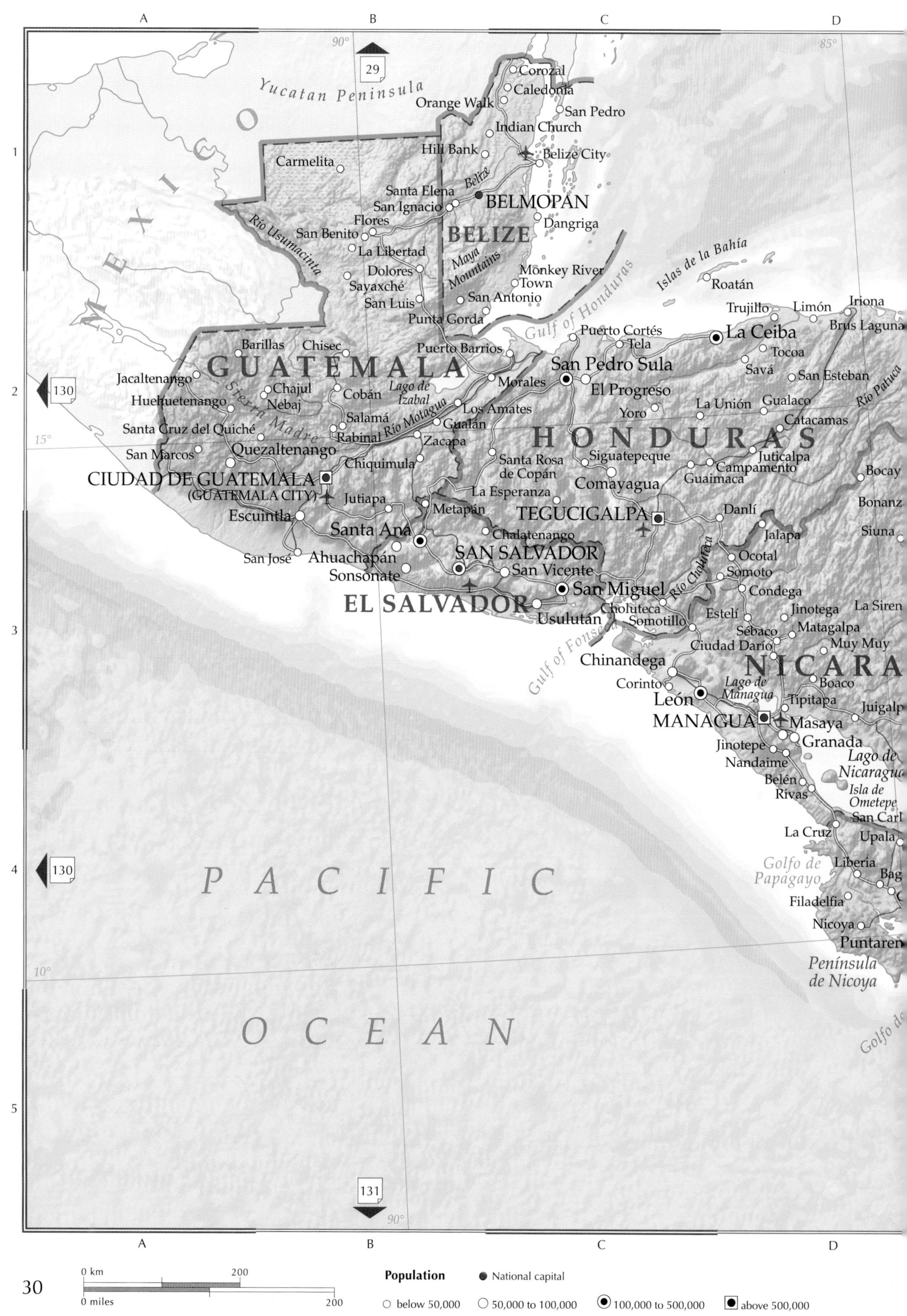
A
B
C
D
1
2
3
4
5
90°
85°
15°
10°
29
130
130
131
Yucatan Peninsula
MEXICO
Río Usumacinta
Carmelita
Corozal
Caledonia
Orange Walk
San Pedro
Indian Church
Hill Bank
Belize City
Santa Elena
San Ignacio
Belize
BELMOPAN
Flores
San Benito
BELIZE
Dangriga
La Libertad
Maya Mountains
Dolores
Sayaxché
San Luis
Monkey River Town
San Antonio
Punta Gorda
Islas de la Bahía
Roatán
Gulf of Honduras
Trujillo
Limón
Iriona
Brus Laguna
Puerto Cortés
Tela
La Ceiba
Tocoa
Savá
San Esteban
Río Patuca
Barillas
Chisec
GUATEMALA
Puerto Barrios
Jacaltenango
Chajul
Nebaj
Cobán
Lago de Izabal
Morales
San Pedro Sula
El Progreso
Huehuetenango
Sierra Madre
Salamá
Río Motagua
Los Amates
Yoro
La Unión
Gualaco
Catacamas
Santa Cruz del Quiché
Rabinal
Gualán
Zacapa
HONDURAS
San Marcos
Quezaltenango
Chiquimula
Santa Rosa de Copán
Siguatepeque
Campamento
Juticalpa
Guaimaca
Bocay
CIUDAD DE GUATEMALA
(GUATEMALA CITY)
Jutiapa
Comayagua
La Esperanza
Bonanz
Escuintla
Metapán
TEGUCIGALPA
Danlí
Santa Ana
Chalatenango
Jalapa
Siuna
San José
Ahuachapán
SAN SALVADOR
Ocotal
Sonsonate
San Vicente
Somoto
Río Choluteca
San Miguel
Condega
EL SALVADOR
Choluteca
Estelí
Jinotega
La Siren
Usulután
Somotillo
Sébaco
Matagalpa
Gulf of Fonseca
Ciudad Darío
Muy Muy
Chinandega
NICARA
Corinto
Lago de Managua
Boaco
León
Tipitapa
Juigalp
MANAGUA
Masaya
Granada
Jinotepe
Nandaime
Lago de Nicaragu
Belén
Rivas
Isla de Ometepe
San Carl
La Cruz
Upala
Golfo de Papagayo
Liberia
Bag
Filadelfia
Nicoya
Puntaren
Península de Nicoya
Golfo de
PACIFIC
OCEAN
0 km
200
0 miles
200
Population
National capital
below 50,000
50,000 to 100,000
100,000 to 500,000
above 500,000

E
F
G
H
32
33
36
131
1
2
3
4
5
Bajo Nuevo
(to Colombia)
Cayo de Serranilla
(to Colombia)
Cayo de Serrana
(to Colombia)
Caribbean
Sea
Isla de Providencia
(to Colombia)
Isla de San Andrés
(to Colombia)
Puerto Lempira
Cayos Miskitos
Tuapi
Puerto Cabezas
Mosquito Coast
Prinzapolka
Barra de Río Grande
Laguna de Perlas
Islas del Maíz
Bluefields
Punta Gorda
San Juan del Norte
COSTA RICA
Siquirres
Heredia
Limón
SAN JOSÉ
Cartago
Guabito
Almirante
Laguna de Chiriquí
Cordillera de Talamanca
Grande 3819m
Buenos Aires
Cortés
Palmar Sur
Volcán Barú 3475m
Boquete
Cordillera Central
La Concepción
David
Golfo de los Mosquitos
Golfo de Chiriquí
Golfo Dulce
Isla de Coiba
Guarumal
Isla Cébaco
Santiago
Ocú
Península de Azuero
Chitré
Las Tablas
Golfo de Panamá
PANAMA
Aguadulce
Penonomé
Capira
Balboa
Lago Gatún
Panama Canal
Cristóbal
Colón
Portobelo
Istmo de Panamá
El Porvenir
Cordillera de San Blas
Aligandí
Lago Bayano
San Miguelito
PANAMÁ
(PANAMA CITY)
Chimán
Archipiélago de las Perlas
Isla del Rey
La Palma
Yaviza
Serranía del Darién
El Real
Garachiné
Jaqué
Puerto Obaldía
Gulf of Darien
COLOMBIA
80°
15°
75°
10°
Elevation
Below sea level 0
-6000m -4000m -2000m -1000m -500m -250m
250m 500m 1000m 2000m 3000m 4000m 6000m
-19,658ft -13,124ft -6562ft -3281ft -1640ft -820ft -328ft/-100m 0
820ft 1640ft 3281ft 6562ft 9843ft 13,124ft 19,685ft

The Caribbean

UNITED STATES OF AMERICA
Gulf of Mexico
The Everglades
Florida Keys
Straits of Florida
Tropic of Cancer
Grand Bahama Island
Freeport
Marsh Harbour
Great Abaco
Bimini Islands
Berry Islands
Northeast Providence Channel
Nicholls Town
NASSAU
New Providence
Eleuthera Island
Rock Sound
Andros Town
Andros Island
Exuma Cays
Exuma Sound
Cat Island
THE BAHAMAS
San Salvador
Rum Cay
George Town
Long Island
Great Exuma Island
Clarence Town
Crooked Island Passage
Crooked Island
Acklins Island
Mayaguana Passage
Caicos Passage
Little Inagua
Lake Rosa
Matthew Town
Great Inag
Cay Sal
Anguilla Cays
LA HABANA (HAVANA)
Guanabacoa
Artemisa
Cárdenas
Matanzas
Pinar del Río
Consolación del Sur
La Fé
Yucatan Channel
Sagua la Grande
Santa Clara
Archipiélago de Camagüey
Cienfuegos
Placetas
Ragged Island Range
Nueva Gerona
Isla de la Juventud
Cayo Largo
Archipiélago de los Canarreos
Bahía de Cochinos
Sancti Spíritus
Morón
Ciego de Ávila
CUBA
Camagüey
Nuevitas
Holguín
Archipiélago de los Jardines de la Reina
Las Tunas
Manzanillo
Bayamo
Palma Soriano
Guantánamo
Santiago de Cuba
Guantánamo Bay (to US)
Windward Passage
Little Cayman
Cayman Brac
GEORGE TOWN
Grand Cayman
CAYMAN ISLANDS (to UK)
NAVASSA ISLAND (to US)
Gonaïves
Île de la Gonâve
Jérémie
PORT-AU-PRINCE
Cayes
Jacm
Haïtie
Montego Bay
Spanish Town
Portmore
KINGSTON
JAMAICA
Pedro Cays
Jamaica Channel
Greater
Caribbean
HONDURAS
NICARAGUA
COSTA RICA
COLOMBI
JAMAICA
Caribbean Sea
Montego Bay
Lucea
Falmouth
Discovery Bay
St Ann's Bay
Ocho Rios
The Cockpit Country
Annotto Bay
Buff Bay
Cambridge
Ewarton
Port Antonio
Savanna-La-Mar
Christiana
Spanish Town
Blue Mountain Peak 2258m
Mandeville
KINGSTON
Black River
May Pen
Old Harbour
Portmore
Morant Bay
Portland Bight
2000m/6562ft
1000m/3281ft
500m/1640ft
200m/656ft
Sea level
0 km 20
0 miles 20
Caribbean Sea
0 km 200
0 miles 200
Population
National capital
below 50,000
50,000 to 100,000
100,000 to 500,000
above 500,000

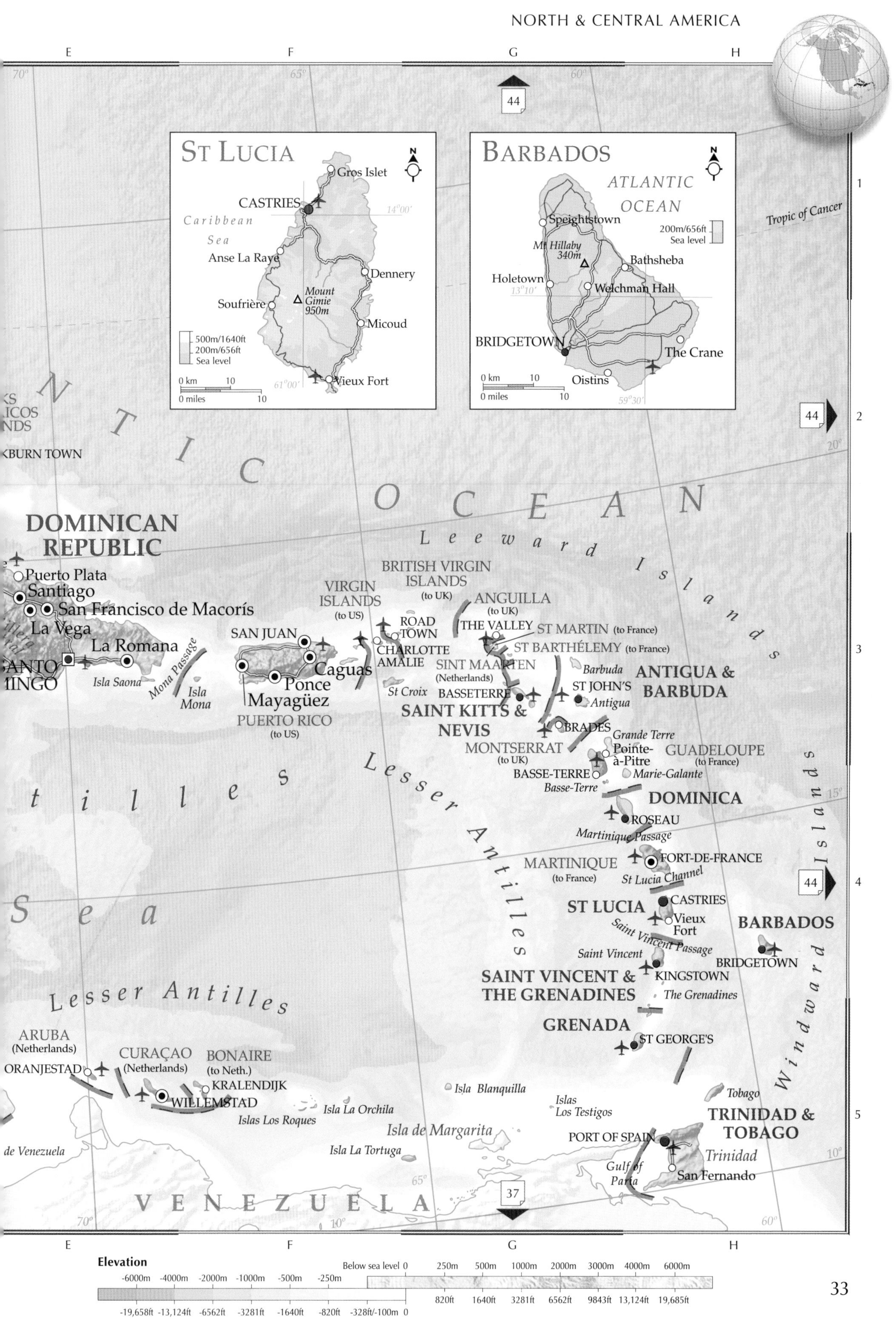
ST LUCIA
Gros Islet
CASTRIES
Caribbean Sea
Anse La Raye
Dennery
Soufrière
Mount Gimie 950m
Micoud
Vieux Fort
500m/1640ft
200m/656ft
Sea level
BARBADOS
ATLANTIC OCEAN
Speightstown
Mt Hillaby 340m
Bathsheba
Holetown
Welchman Hall
BRIDGETOWN
The Crane
Oistins
Tropic of Cancer
DOMINICAN REPUBLIC
Puerto Plata
Santiago
San Francisco de Macorís
La Vega
La Romana
Isla Saona
Mona Passage
Isla Mona
SAN JUAN
Caguas
Ponce
Mayagüez
PUERTO RICO (to US)
VIRGIN ISLANDS (to US)
BRITISH VIRGIN ISLANDS (to UK)
ROAD TOWN
CHARLOTTE AMALIE
St Croix
ANGUILLA (to UK)
THE VALLEY
ST MARTIN (to France)
ST BARTHÉLEMY (to France)
SINT MAARTEN (Netherlands)
BASSETERRE
SAINT KITTS & NEVIS
Barbuda
ANTIGUA & BARBUDA
ST JOHN'S
Antigua
BRADES
MONTSERRAT (to UK)
Grande Terre
Pointe-à-Pitre
GUADELOUPE (to France)
BASSE-TERRE
Basse-Terre
Marie-Galante
DOMINICA
ROSEAU
Martinique Passage
MARTINIQUE (to France)
FORT-DE-FRANCE
St Lucia Channel
ST LUCIA
CASTRIES
Vieux Fort
BARBADOS
BRIDGETOWN
Saint Vincent Passage
Saint Vincent
SAINT VINCENT & THE GRENADINES
KINGSTOWN
The Grenadines
GRENADA
ST GEORGE'S
Tobago
TRINIDAD & TOBAGO
PORT OF SPAIN
Trinidad
San Fernando
Gulf of Paria
ATLANTIC OCEAN
Leeward Islands
Windward Islands
Lesser Antilles
Caribbean Sea
ARUBA (Netherlands)
ORANJESTAD
CURAÇAO (Netherlands)
WILLEMSTAD
BONAIRE (to Neth.)
KRALENDIJK
Islas Los Roques
Isla La Orchila
Isla Blanquilla
Islas Los Testigos
Isla de Margarita
Isla La Tortuga
VENEZUELA
Elevation
-6000m -4000m -2000m -1000m -500m -250m
-19,658ft -13,124ft -6562ft -3281ft -1640ft -820ft -328ft/-100m 0
Below sea level 0
250m 500m 1000m 2000m 3000m 4000m 6000m
820ft 1640ft 3281ft 6562ft 9843ft 13,124ft 19,685ft

South America

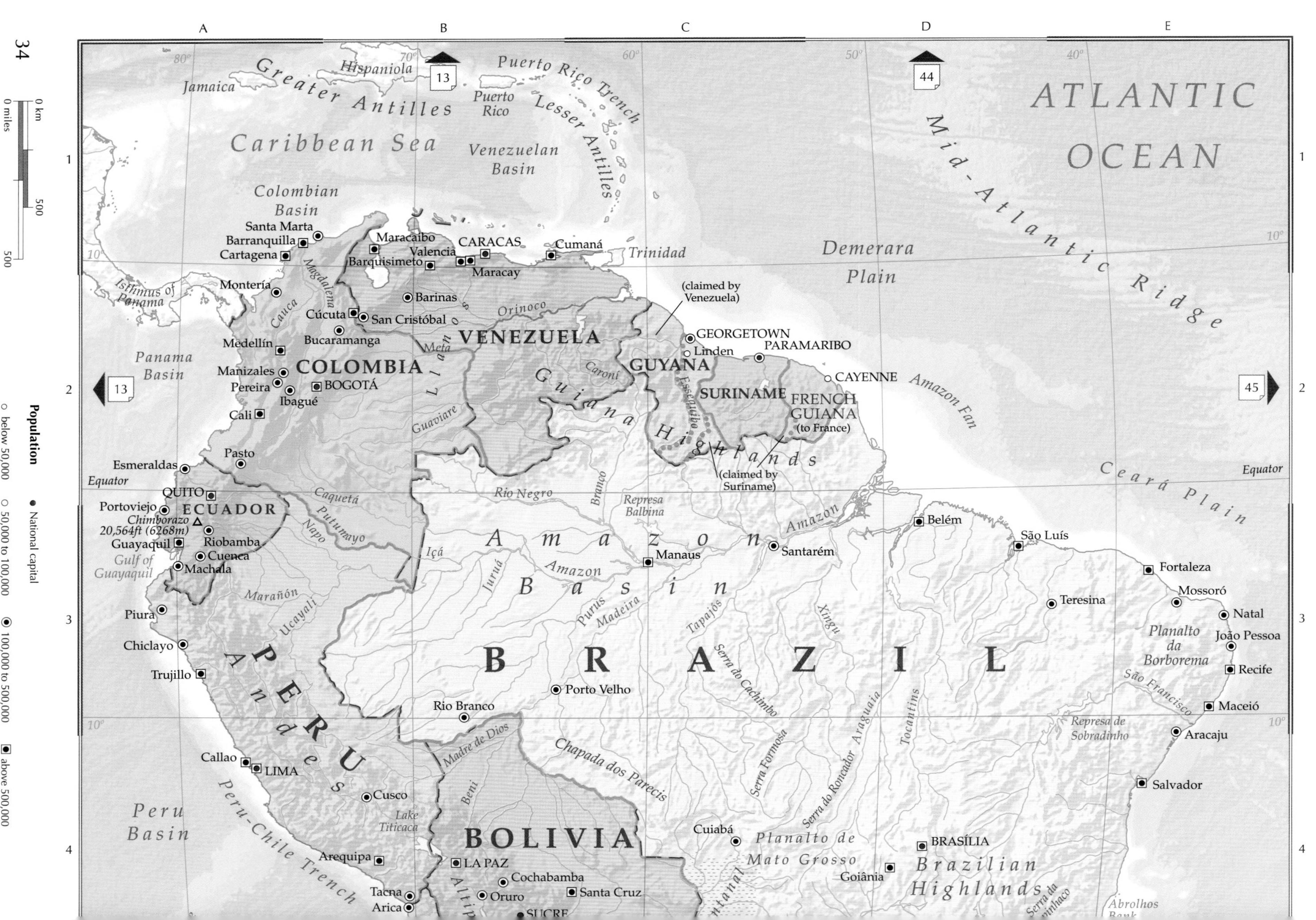

ATLANTIC
OCEAN
Mid-Atlantic Ridge
Ceará Plain
Demerara Plain
Amazon Fan
Equator
Puerto Rico Trench
Lesser Antilles
Greater Antilles
Caribbean Sea
Colombian Basin
Venezuelan Basin
Jamaica
Hispaniola
Puerto Rico
Trinidad
Panama Basin
Isthmus of Panama
Gulf of Guayaquil
Peru Basin
Peru-Chile Trench
13
44
13
45
VENEZUELA
COLOMBIA
GUYANA
SURINAME
FRENCH GUIANA (to France)
ECUADOR
PERU
BOLIVIA
BRAZIL
(claimed by Venezuela)
(claimed by Suriname)
Guiana Highlands
Amazon Basin
Brazilian Highlands
Planalto de Mato Grosso
Planalto da Borborema
Chapada dos Parecis
Serra do Cachimbo
Serra Formosa
Serra do Roncador
Llanos
Andes
Lake Titicaca
Represa Balbina
Represa de Sobradinho
Abrolhos Bank
Orinoco
Caroní
Meta
Guaviare
Magdalena
Cauca
Caquetá
Putumayo
Napo
Marañón
Ucayali
Içá
Juruá
Purus
Madeira
Rio Negro
Branco
Amazon
Tapajós
Xingu
Araguaia
Tocantins
São Francisco
Essequibo
Beni
Madre de Dios
CARACAS
Maracay
Valencia
Barquisimeto
Maracaibo
Barinas
San Cristóbal
Cumaná
Cúcuta
Bucaramanga
BOGOTÁ
Santa Marta
Barranquilla
Cartagena
Montería
Medellín
Manizales
Pereira
Ibagué
Cali
Pasto
QUITO
Esmeraldas
Portoviejo
Chimborazo 20,564ft (6268m)
Guayaquil
Riobamba
Cuenca
Machala
Piura
Chiclayo
Trujillo
Callao
LIMA
Cusco
Arequipa
Tacna
Arica
LA PAZ
Cochabamba
Oruro
Santa Cruz
SUCRE
GEORGETOWN
Linden
PARAMARIBO
CAYENNE
Manaus
Santarém
Porto Velho
Rio Branco
Cuiabá
Goiânia
BRASÍLIA
Belém
São Luís
Teresina
Fortaleza
Mossoró
Natal
João Pessoa
Recife
Maceió
Aracaju
Salvador

0 km 500
0 miles 500

Population

● National capital

○ below 50,000
○ 50,000 to 100,000
◉ 100,000 to 500,000
▣ above 500,000

Political features
Total area:
6,880,000 sq miles
(17,819,000 sq km)
Total number of countries:
12
Total population:
434 million
Largest city with population:
São Paulo, Brazil 22.7 million
Country with highest population density:
Ecuador 157 people per sq mile
(65 people per sq km)
Largest country:
Brazil 3,287,957 sq miles
(8,515,770 sq km)
Smallest country:
Suriname 63,251 sq miles
(163,820 sq km)
Physical features
Largest lake:
Lake Titicaca, Bolivia/Peru
3220 sq miles (8340 sq km)
Longest river:
Amazon, Brazil
3976 miles (6400 km)
Highest point:
Cerro Aconcagua, Argentina
22,838 ft (6961 m)
Lowest point:
Laguna del Carbón, Argentina
-344 ft (-105 m) below sea level
Tropic of Capricorn
PACIFIC OCEAN
ATLANTIC OCEAN
Chile Basin
Chile Rise
Isla San Ambrosio (to Chile)
Isla San Félix (to Chile)
Islas Juan Fernández (to Chile)
Antofagasta
San Salvador de Jujuy
Salta
PARAGUAY
ASUNCIÓN
Formosa
Ciudad del Este
Cerro Ojos del Salado 27,615ft (6893m)
San Miguel de Tucumán
Resistencia
Corrientes
Posadas
Santiago del Estero
La Rioja
La Serena
Coquimbo
CHILE
ARGENTINA
San Juan
Cerro Aconcagua 22,838ft (6961m)
Mendoza
Viña del Mar
Valparaíso
SANTIAGO
Santa Fe
Córdoba
Rosario
Santa Maria
Porto Alegre
Lagoa dos Patos
Mirim Lagoon
URUGUAY
Negro
BUENOS AIRES
MONTEVIDEO
La Plata
Río de la Plata
Mar del Plata
Bahía Blanca
Concepción
Temuco
Valdivia
Neuquén
Colorado
Río Negro
Pampas
Mesopotamia
Paraná
Uruguay
Pilcomayo
Gran Chaco
Atacama Desert
Andes
Serra Geral
Curitiba
Florianópolis
São Paulo
Santos
Rio de Janeiro
Santos Plateau
Puerto Montt
Isla de Chiloé
Golfo San Matías
Rawson
Chubut
Chico
Deseado
Gulf of San Jorge
Patagonia
Laguna del Carbón -344ft (-105m)
Bahía Grande
Argentine Basin
FALKLAND ISLANDS (to UK)
West Falkland
STANLEY
East Falkland
Falkland Plateau
Strait of Magellan
Punta Arenas
Tierra del Fuego
Cape Horn
SOUTH GEORGIA (to UK)
Scotia Sea
South Sandwich Trench
SOUTH SANDWICH ISLANDS (to UK)
Drake Passage
Winter limit of pack ice
South Orkney Islands
South Shetland Islands
ANTARCTICA
Summer limit of pack ice
131
132
45

Northern South America

A
B
C
D
1
2
3
4
5
32
31
38
38
Caribbean Sea
Lesser Antilles
ARUBA (Netherlands)
CURAÇAO (Neth.)
BONAIRE (to Neth.)
Península de la Guajira
Puerto López
Punto Fijo
Islas Los Roques
Ríohacha
Maicao
Golfo de Venezuela
Coro
Puerto Cumarebo
Santa Marta
Barranquilla
Ciénaga
Sabaneta
Dabajuro
Pico Cristóbal Colón 5775m
Soledad
Maracaibo
Puerto Cabello
CARACAS
Cartagena
Sabanalarga
La Concepción
Cabimas
San Felipe
Valledupar
Ciudad Ojeda
Carora
Maracay
Gulf of Darien
El Carmen de Bolívar
Machiques
Barquisimeto
Valencia
San Juan de los Morros
Lago de Maracaibo
Sincelejo
Magangué
San Carlos del Zulia
Valera
Acarigua
PANAMA
Montería
Cereté
Guanare
Calabozo
Valle de la Pascua
El Vigía
Mérida
Planeta Rica
Barinas
Aguachica
Ocaña
Río Guanare
Caucasia
Pico Bolívar 5007m
Golfo de Panamá
Cúcuta
San Cristóbal
San Fernando
Río Apure
Pamplona
Río Cauca
Río Magdalena
Llanos
Yarumal
Río Arauca
Dabeiba
Bucaramanga
VENEZUELA
Barrancabermeja
Arauca
Bello
Puerto Berrío
Medellín
Río Meta
Itagüí
Puerto Carreño
Nuquí
Sogamoso
Quibdó
Tunja
Puerto Ayacucho
Yopal
Manizales
Zipaquirá
PACIFIC OCEAN
Orinoquía
Pereira
Río Meta
Río Orinoco
BOGOTÁ
Armenia
Tuluá
Ibagué
Girardot
Villavicencio
Buenaventura
Buga
Espinal
Río Guaviare
Puerto Inírida
Palmira
COLOMBIA
Cali
Neiva
Cordillera Occidental
Cordillera Central
Cordillera Oriental
Popayán
Garzón
San José del Guaviare
Tumaco
Pitalito
Florencia
Amazonía
Pasto
Río Vaupés
Nevado de Cumbal 4764m
Mocoa
Mitú
Ipiales
Orito
Río Apaporis
Equator
Andes
Río Putumayo
Río Caquetá
Río Japurá
Río Napo
ECUADOR
Río Içá
PERU
Amazon
80°
75°
70°
10°
5°

0 km 200
0 miles 200

Population
● National capital
○ below 50,000
○ 50,000 to 100,000
◉ 100,000 to 500,000
▣ above 500,000

E
F
G
H
SAINT VINCENT & THE GRENADINES
BARBADOS
33
GRENADA
Isla Blanquilla
Isla de Margarita
Islas Los Testigos
La Asunción
Tobago
Carúpano
Güiria
Cariaco
Gulf of Paria
TRINIDAD & TOBAGO
Trinidad
Puerto La Cruz
Barcelona
Maturín
The Serpent's Mouth
San Mateo
Anaco
Cantaura
Tucupita
El Tigre
Río Orinoco
Ciudad Guayana
Ciudad Bolívar
Upata
Embalse de Guri
El Callao
ATLANTIC
OCEAN
45
Matthews Ridge
Charity
Spring Garden
Parika
GEORGETOWN
Aurora
El Dorado
Cuyuni River
New Amsterdam
Peters Mine
Bartica
Rockstone
Linden
Totness
PARAMARIBO
Nieuw Amsterdam
Salto Angel
Kamarang
Nieuw Nickerie
St-Laurent-du-Maroni
Sinnamary
Kourou
Río Paragua
Río Caura
Río Caroní
GUYANA
Mount Roraima 2810m
Orealla
Apoera
Kaaimanston
Maroni River
Montagnes de la Trinité
CAYENNE
W. J. van Blommesteinmeer
Kurupukari
Pakaraima Mountains
SURINAME
Grand-Santi
Montagne Tortue
Ouanary
Juliana Top 1230m
Essequibo River
FRENCH GUIANA (to France)
St-Georges
Lethem
Courantyne River
Camopi
(Venezuela claims all of Guyana west of Essequibo River)
Guiana Highlands
Tumuc-Humac Mountains
Río Orinoco
Acarai Mountains
(claimed by Suriname)
(claimed by Suriname)
40
Equator
Rio Negro
BRAZIL
Amazon
Amazon Basin
Amazon
Amazon
Rio Purus
Rio Tapajós
40
1
2
3
4
5
Elevation
-6000m -4000m -2000m -1000m -500m -250m
-19,658ft -13,124ft -6562ft -3281ft -1640ft -820ft -328ft/-100m 0
Below sea level 0
250m 500m 1000m 2000m 3000m 4000m 6000m
820ft 1640ft 3281ft 6562ft 9843ft 13,124ft 19,685ft

Western South America

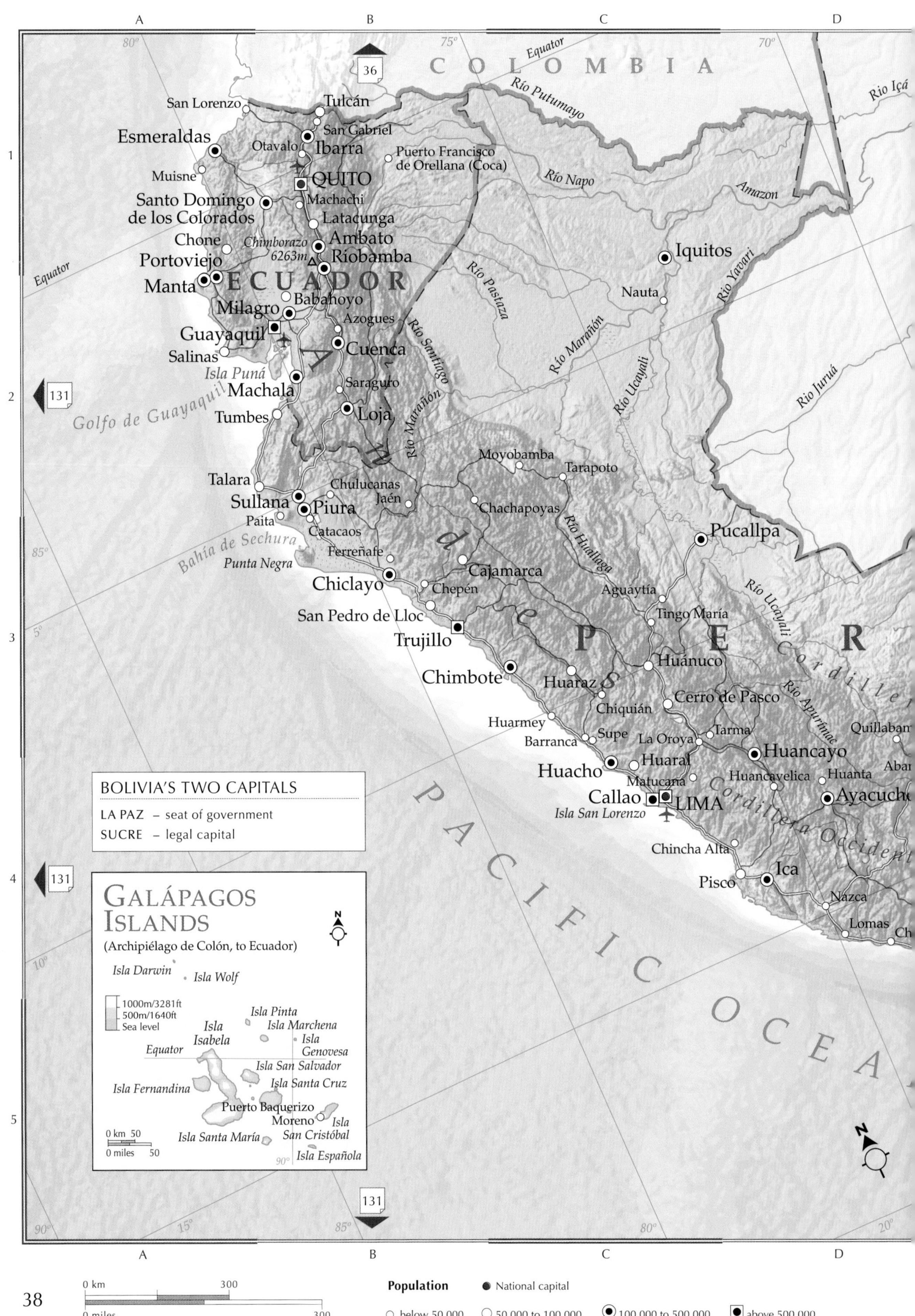

COLOMBIA
ECUADOR
PERU
PACIFIC OCEAN
Andes
QUITO
LIMA
Tulcán
San Lorenzo
Esmeraldas
San Gabriel
Otavalo
Ibarra
Puerto Francisco de Orellana (Coca)
Muisne
Santo Domingo de los Colorados
Machachi
Latacunga
Chone
Chimborazo 6263m
Ambato
Riobamba
Portoviejo
Manta
Babahoyo
Milagro
Guayaquil
Azogues
Cuenca
Salinas
Isla Puná
Machala
Saraguro
Golfo de Guayaquil
Tumbes
Loja
Iquitos
Nauta
Río Putumayo
Río Napo
Amazon
Río Içá
Río Pastaza
Río Santiago
Río Marañón
Río Ucayali
Río Yavari
Río Juruá
Río Huallaga
Río Apurímac
Moyobamba
Tarapoto
Talara
Chulucanas
Jaén
Sullana
Piura
Chachapoyas
Paita
Catacaos
Pucallpa
Bahía de Sechura
Ferreñafe
Punta Negra
Cajamarca
Chiclayo
Chepén
Aguaytía
San Pedro de Lloc
Tingo María
Trujillo
Huánuco
Chimbote
Huaraz
Cerro de Pasco
Chiquián
Huarmey
Supe
La Oroya
Tarma
Quillabamba
Barranca
Huancayo
Huacho
Huaral
Huancavelica
Huanta
Matucana
Callao
Ayacucho
Isla San Lorenzo
Cordillera Occidental
Chincha Alta
Ica
Pisco
Nazca
Lomas
Equator
36
131

BOLIVIA'S TWO CAPITALS

LA PAZ – seat of government

SUCRE – legal capital

GALÁPAGOS ISLANDS

(Archipiélago de Colón, to Ecuador)

Isla Darwin
Isla Wolf
Isla Pinta
Isla Marchena
Isla Genovesa
Isla Isabela
Isla San Salvador
Isla Santa Cruz
Isla Fernandina
Puerto Baquerizo Moreno
Isla San Cristóbal
Isla Santa María
Isla Española
Equator

1000m/3281ft
500m/1640ft
Sea level

0 km 50
0 miles 50

0 km 300
0 miles 300

Population
● National capital
○ below 50,000
○ 50,000 to 100,000
◉ 100,000 to 500,000
▣ above 500,000

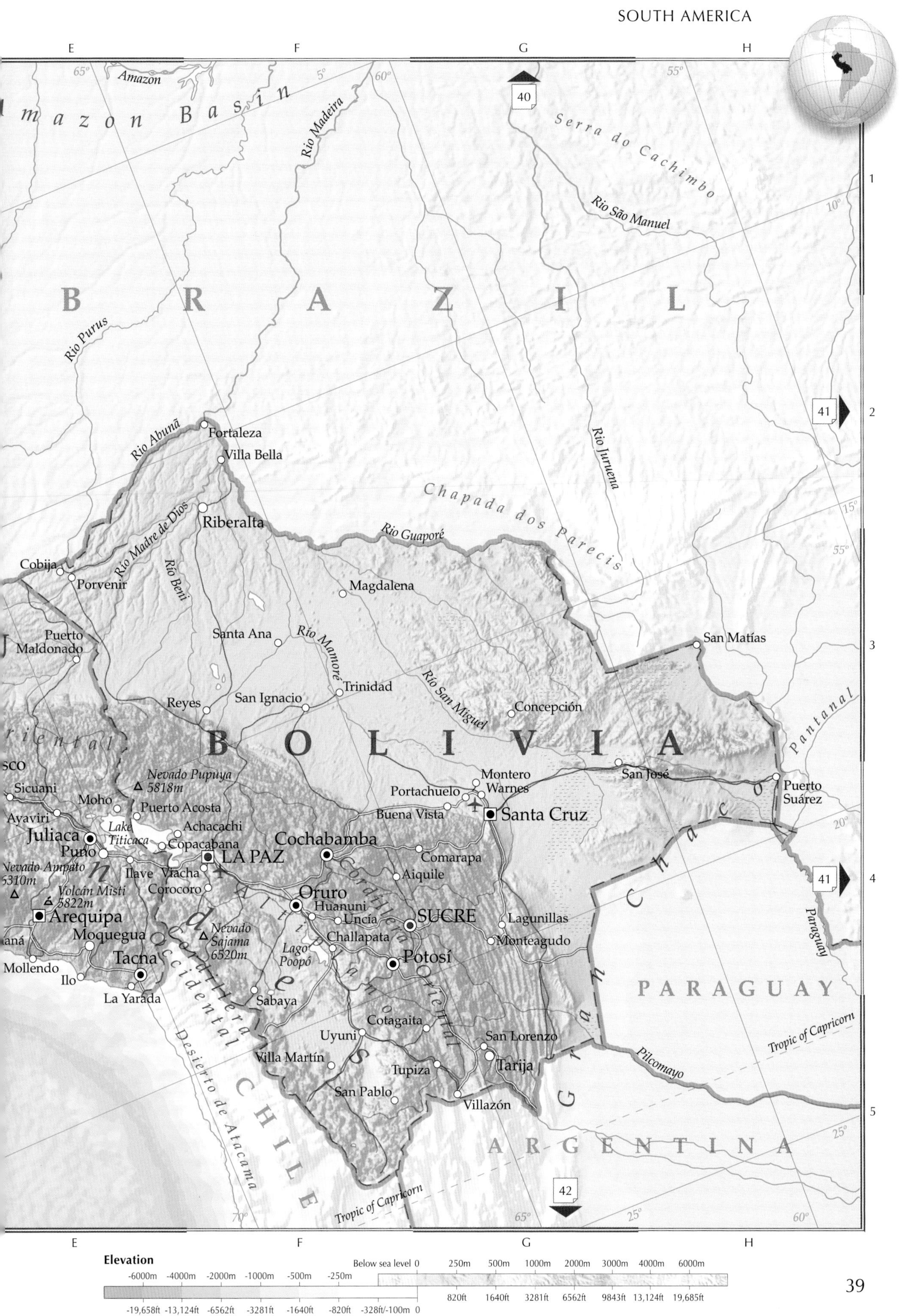
E
F
G
H
Amazon
Amazon Basin
Rio Madeira
Serra do Cachimbo
Rio São Manuel
BRAZIL
Rio Purus
Rio Abunã
Fortaleza
Villa Bella
Rio Juruena
Chapada dos Parecis
Riberalta
Rio Madre de Dios
Rio Guaporé
Cobija
Porvenir
Rio Beni
Magdalena
Puerto Maldonado
Santa Ana
Río Mamoré
San Matías
Trinidad
Río San Miguel
Reyes
San Ignacio
Concepción
BOLIVIA
Pantanal
Nevado Pupuya 5818m
Montero
Warnes
San José
Puerto Suárez
Sicuani
Moho
Portachuelo
Ayaviri
Puerto Acosta
Buena Vista
Santa Cruz
Juliaca
Lake Titicaca
Achacachi
Cochabamba
Puno
Copacabana
LA PAZ
Comarapa
Nevado Ampato 6310m
Ilave
Viacha
Aiquile
Volcán Misti 5822m
Corocoro
Oruro
Huanuni
Uncía
SUCRE
Lagunillas
Arequipa
Nevado Sajama 6520m
Challapata
Moquegua
Monteagudo
Mollendo
Lago Poopó
Potosí
Tacna
Ilo
La Yarada
Sabaya
PARAGUAY
Paraguay
Gran Chaco
Cordillera Occidental
Cordillera Oriental
Altiplano
Andes
Cotagaita
Uyuni
San Lorenzo
Tropic of Capricorn
Pilcomayo
Villa Martín
Tarija
Tupiza
San Pablo
Villazón
Desierto de Atacama
CHILE
ARGENTINA
Tropic of Capricorn
40
41
41
42
1
2
3
4
5
Elevation
Below sea level 0
250m 500m 1000m 2000m 3000m 4000m 6000m
820ft 1640ft 3281ft 6562ft 9843ft 13,124ft 19,685ft
-6000m -4000m -2000m -1000m -500m -250m
-19,658ft -13,124ft -6562ft -3281ft -1640ft -820ft -328ft/-100m 0

Brazil

VENEZUELA
COLOMBIA
ECUADOR
PERU
BOLIVIA
CHILE
ARGENTINA
PACIFIC OCEAN
Andes
Cordillera Occidental
Cordillera Oriental
Galápagos Islands
(Archipiélago de Colón)
(to Ecuador)
Equator
Tropic of Capricorn
Guiana Highlands
Roraima
Uraricoera
Boa Vista
Pico da Neblina
3014m
Rio Negro
Rio Japurá
Río Putumayo
Río Napo
Rio Içá
Manaus
Tefé
Amazon
Coari
Rio Juruá
Rio Purus
Rio Yavari
Río Marañón
Amazon
Humaitá
Japiim
Feijó
Porto Velho
Acre
Rio Abunã
Río Ucayali
Rondônia
Chapada dos Parecis
Rio Guaporé
Río Mamoré
Lake Titicaca
Lago Poopó
Cordillera Occidental
Cordillera Oriental
Desierto de Atacama
Pilcomayo
Río Bermejo
Gran Chaco
Río Salado
A
B
C
D
1
2
3
4
5
36
131

Population
● National capital
○ below 50,000
○ 50,000 to 100,000
⊙ 100,000 to 500,000
▣ above 500,000

0 km 600
0 miles 600

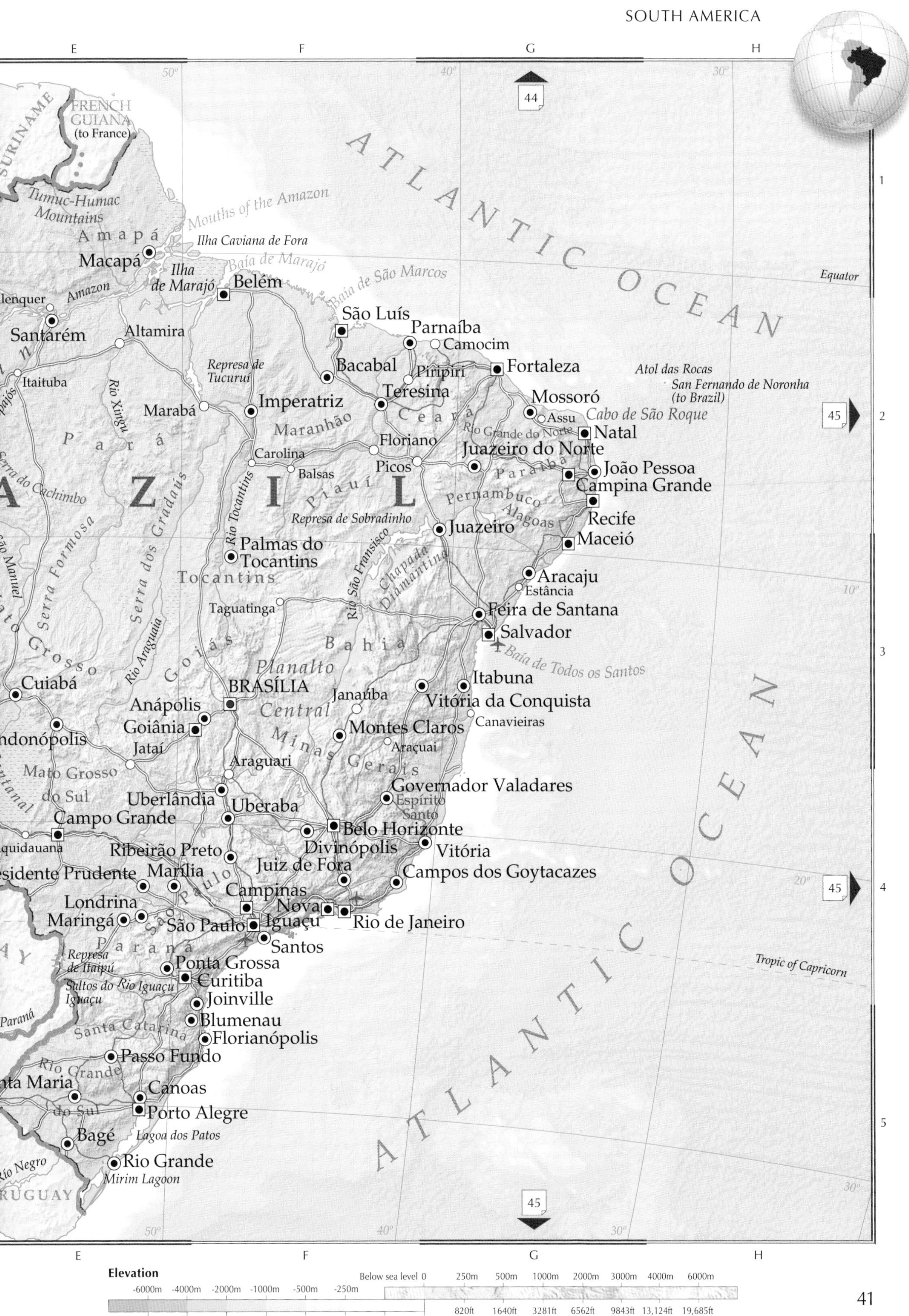
ATLANTIC OCEAN
FRENCH GUIANA (to France)
SURINAME
Tumuc-Humac Mountains
Mouths of the Amazon
Amapá
Macapá
Ilha Caviana de Fora
Ilha de Marajó
Baía de Marajó
Belém
Baía de São Marcos
Amazon
Santarém
Altamira
Itaituba
São Luís
Parnaíba
Camocim
Bacabal
Piripiri
Fortaleza
Teresina
Represa de Tucuruí
Rio Xingu
Marabá
Imperatriz
Maranhão
Ceará
Mossoró
Assu
Cabo de São Roque
Atol das Rocas
San Fernando de Noronha (to Brazil)
Equator
Pará
Rio Grande do Norte
Natal
Carolina
Floriano
Juazeiro do Norte
Balsas
Picos
Paraíba
João Pessoa
Campina Grande
Serra do Cachimbo
BRAZIL
Piauí
Pernambuco
Alagoas
Recife
Rio Tocantins
Represa de Sobradinho
Juazeiro
Maceió
Serra dos Gradaús
Serra Formosa
Palmas do Tocantins
Tocantins
Rio São Fransisco
Chapada Diamantina
Aracaju
Estância
Taguatinga
Feira de Santana
Salvador
Baía de Todos os Santos
Mato Grosso
Bahia
Rio Araguaia
Goiás
Planalto Central
Cuiabá
BRASÍLIA
Janaúba
Itabuna
Vitória da Conquista
Anápolis
Goiânia
Canavieiras
Montes Claros
Araçuai
Jataí
Minas Gerais
Mato Grosso do Sul
Araguari
Governador Valadares
Espírito Santo
Uberlândia
Uberaba
Campo Grande
Belo Horizonte
Divinópolis
Vitória
Ribeirão Preto
Marília
Juiz de Fora
Campos dos Goytacazes
São Paulo
Campinas
Londrina
Nova Iguaçu
Maringá
Rio de Janeiro
Santos
Paraná
Represa de Itaipú
Ponta Grossa
Curitiba
Saltos do Iguaçu
Rio Iguaçu
Joinville
Blumenau
Santa Catarina
Florianópolis
Passo Fundo
Rio Grande do Sul
Canoas
Porto Alegre
Bagé
Lagoa dos Patos
Rio Grande
Mirim Lagoon
Río Negro
URUGUAY
Tropic of Capricorn
E
F
G
H
1
2
3
4
5
50°
40°
30°
10°
20°
44
45
Elevation
Below sea level 0
250m 500m 1000m 2000m 3000m 4000m 6000m
820ft 1640ft 3281ft 6562ft 9843ft 13,124ft 19,685ft
-6000m -4000m -2000m -1000m -500m -250m
-19,658ft -13,124ft -6562ft -3281ft -1640ft -820ft -328ft/-100m 0

Southern South America

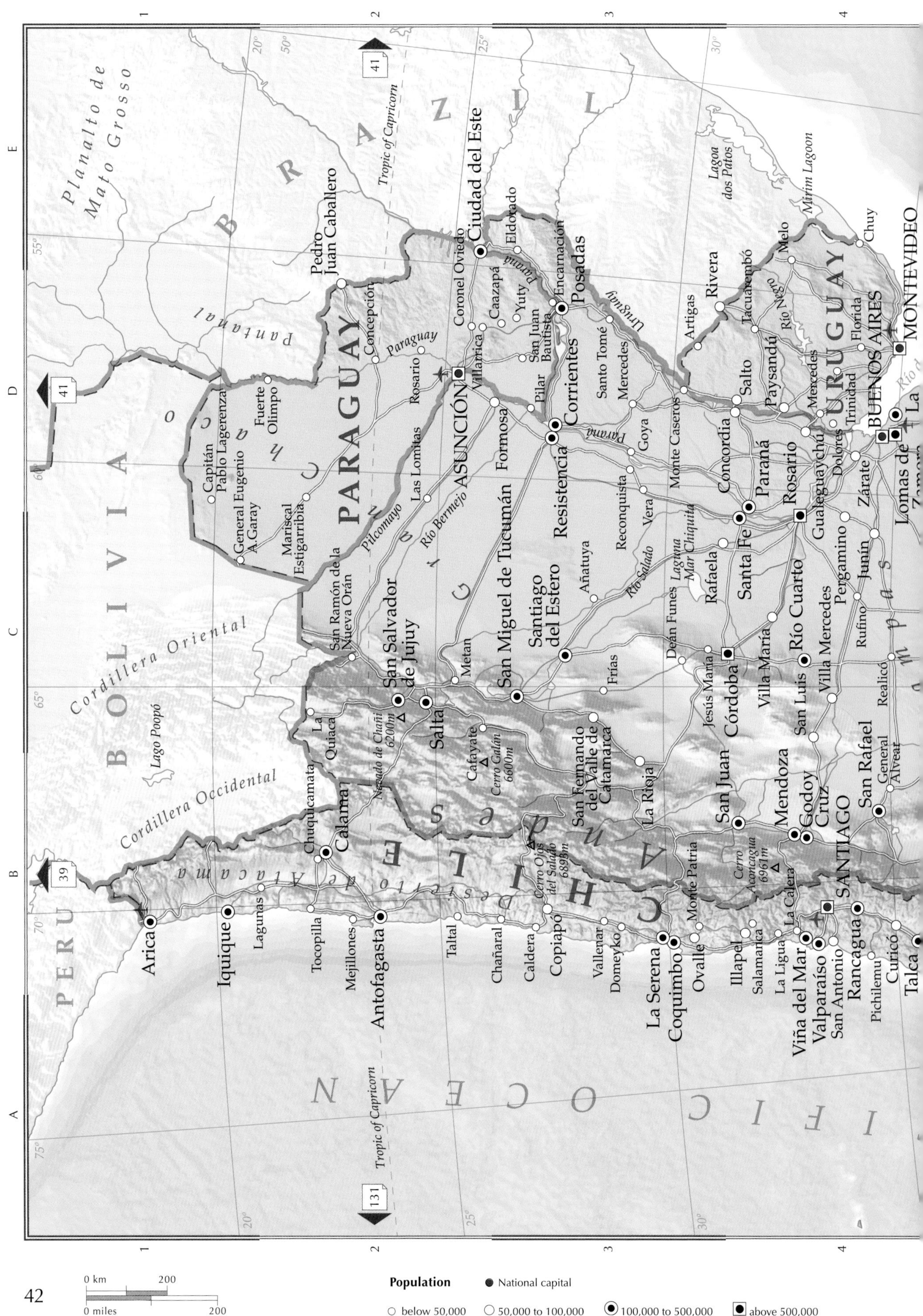

Planalto de Mato Grosso
BRAZIL
Pantanal
Tropic of Capricorn
41
BOLIVIA
Lago Poopó
Cordillera Oriental
Cordillera Occidental
PERU
39
Chaco
PARAGUAY
Pedro Juan Caballero
Concepción
Paraguay
Fuerte Olimpo
Capitán Pablo Lagerenza
General Eugenio A.Garay
Mariscal Estigarribia
Pilcomayo
Rosario
ASUNCIÓN
Villarrica
Coronel Oviedo
Ciudad del Este
Caazapá
Yuty
Eldorado
Paraná
Encarnación
Posadas
San Juan Bautista
Pilar
Las Lomitas
Río Bermejo
Formosa
Resistencia
Corrientes
Santo Tomé
Mercedes
Uruguay
Goya
Reconquista
Vera
Monte Caseros
Artigas
Rivera
Tacuarembó
Melo
Río Negro
Lagoa dos Patos
Mirim Lagoon
Chuy
URUGUAY
Salto
Paysandú
Mercedes
Trinidad
Florida
MONTEVIDEO
BUENOS AIRES
La Plata
Río de la Plata
Concordia
Gualeguaychú
Dolores
Zárate
Lomas de Zamora
Paraná
Santa Fe
Rosario
Rafaela
Laguna Mar Chiquita
Río Salado
Añatuya
Santiago del Estero
San Miguel de Tucumán
Gran Chaco
San Ramón de la Nueva Orán
San Salvador de Jujuy
Metan
Salta
La Quiaca
Nevado de Chañi 6200m
Cafayate
Cerro Galán 6600m
Frías
San Fernando del Valle de Catamarca
La Rioja
Deán Funes
Jesús María
Córdoba
Villa María
Río Cuarto
Villa Mercedes
San Luis
Pergamino
Junín
Rufino
Realicó
Pampas
San Juan
Mendoza
Godoy Cruz
San Rafael
General Alvear
Andes
Chuquicamata
Calama
Desierto de Atacama
Cerro Ojos del Salado 6893m
Cerro Aconcagua 6961m
CHILE
Arica
Iquique
Lagunas
Tocopilla
Mejillones
Antofagasta
Taltal
Chañaral
Caldera
Copiapó
Vallenar
Domeyko
La Serena
Coquimbo
Ovalle
Monte Patria
Illapel
Salamanca
La Ligua
La Calera
Viña del Mar
Valparaíso
San Antonio
SANTIAGO
Rancagua
Pichilemu
Curicó
Talca
PACIFIC OCEAN
131

0 km 200
0 miles 200

Population ● National capital
○ below 50,000 ○ 50,000 to 100,000 ◉ 100,000 to 500,000 ■ above 500,000

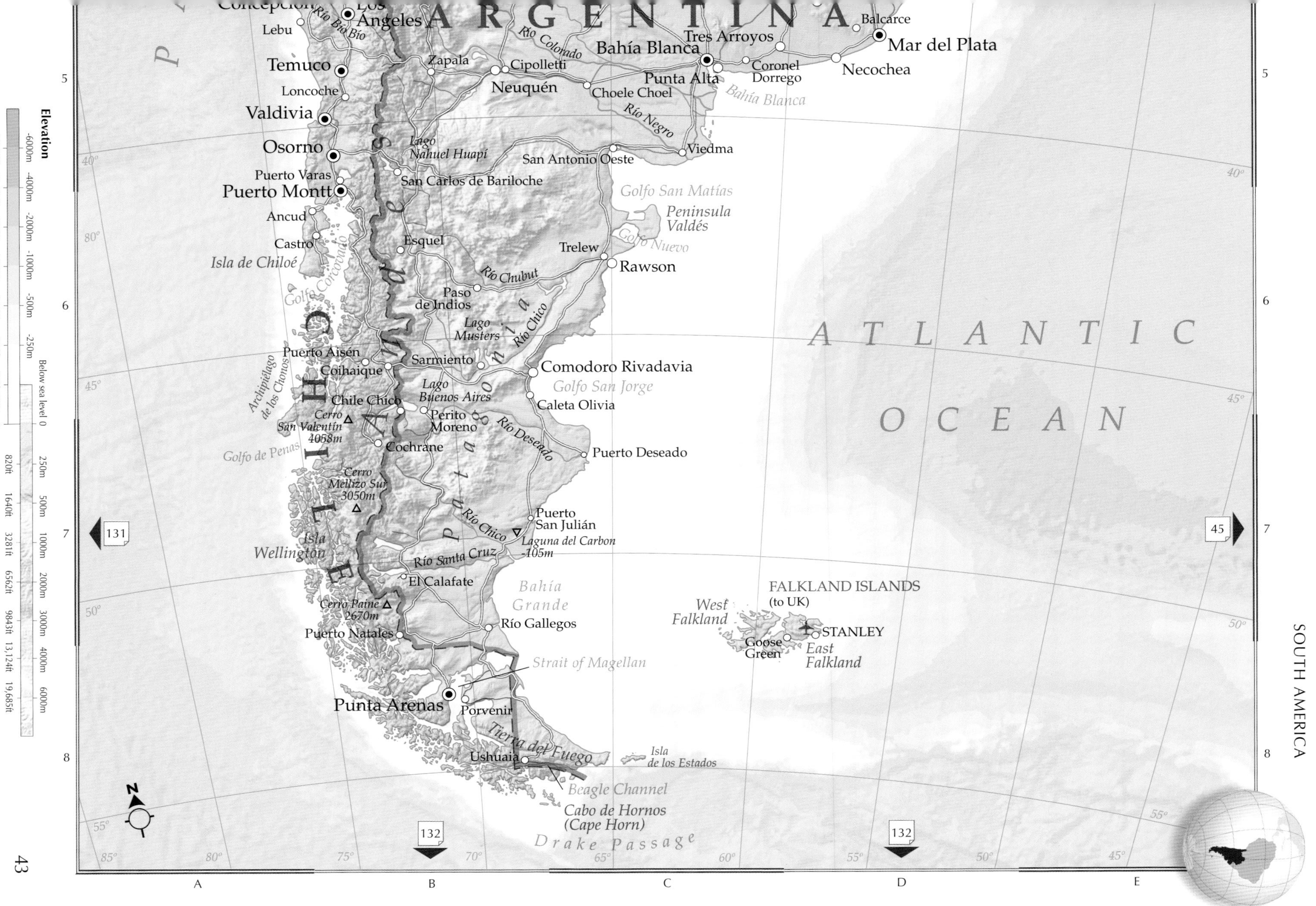
ATLANTIC
OCEAN
ARGENTINA
CHILE
Patagonia
Concepción
Los Ángeles
Lebu
Río Bío Bío
Temuco
Loncoche
Valdivia
Osorno
Puerto Varas
Puerto Montt
Ancud
Castro
Isla de Chiloé
Golfo Corcovado
Puerto Aisén
Coihaique
Archipiélago de los Chonos
Chile Chico
Cerro San Valentín 4058m
Golfo de Penas
Cochrane
Cerro Mellizo Sur 3050m
Isla Wellington
Cerro Paine 2670m
Puerto Natales
Punta Arenas
Porvenir
Tierra del Fuego
Ushuaia
Beagle Channel
Cabo de Hornos (Cape Horn)
Drake Passage
Isla de los Estados
Zapala
Cipolletti
Neuquén
Río Colorado
Lago Nahuel Huapí
San Carlos de Bariloche
Esquel
Paso de Indios
Río Chubut
Lago Musters
Río Chico
Sarmiento
Lago Buenos Aires
Perito Moreno
Río Deseado
Río Chico
Río Santa Cruz
El Calafate
Bahía Blanca
Tres Arroyos
Balcarce
Mar del Plata
Punta Alta
Coronel Dorrego
Necochea
Bahía Blanca
Choele Choel
Río Negro
Viedma
San Antonio Oeste
Golfo San Matías
Peninsula Valdés
Golfo Nuevo
Trelew
Rawson
Comodoro Rivadavia
Golfo San Jorge
Caleta Olivia
Puerto Deseado
Puerto San Julián
Laguna del Carbon -105m
Bahía Grande
Río Gallegos
Strait of Magellan
FALKLAND ISLANDS (to UK)
West Falkland
STANLEY
Goose Green
East Falkland
131
45
132
132
Elevation
Below sea level 0
250m 500m 1000m 2000m 3000m 4000m 6000m
820ft 1640ft 3281ft 6562ft 9843ft 13,124ft 19,685ft
-6000m -4000m -2000m -1000m -500m -250m
-19,658ft -13,124ft -6562ft -3281ft -1640ft -820ft -328ft/-100m 0

The Atlantic Ocean

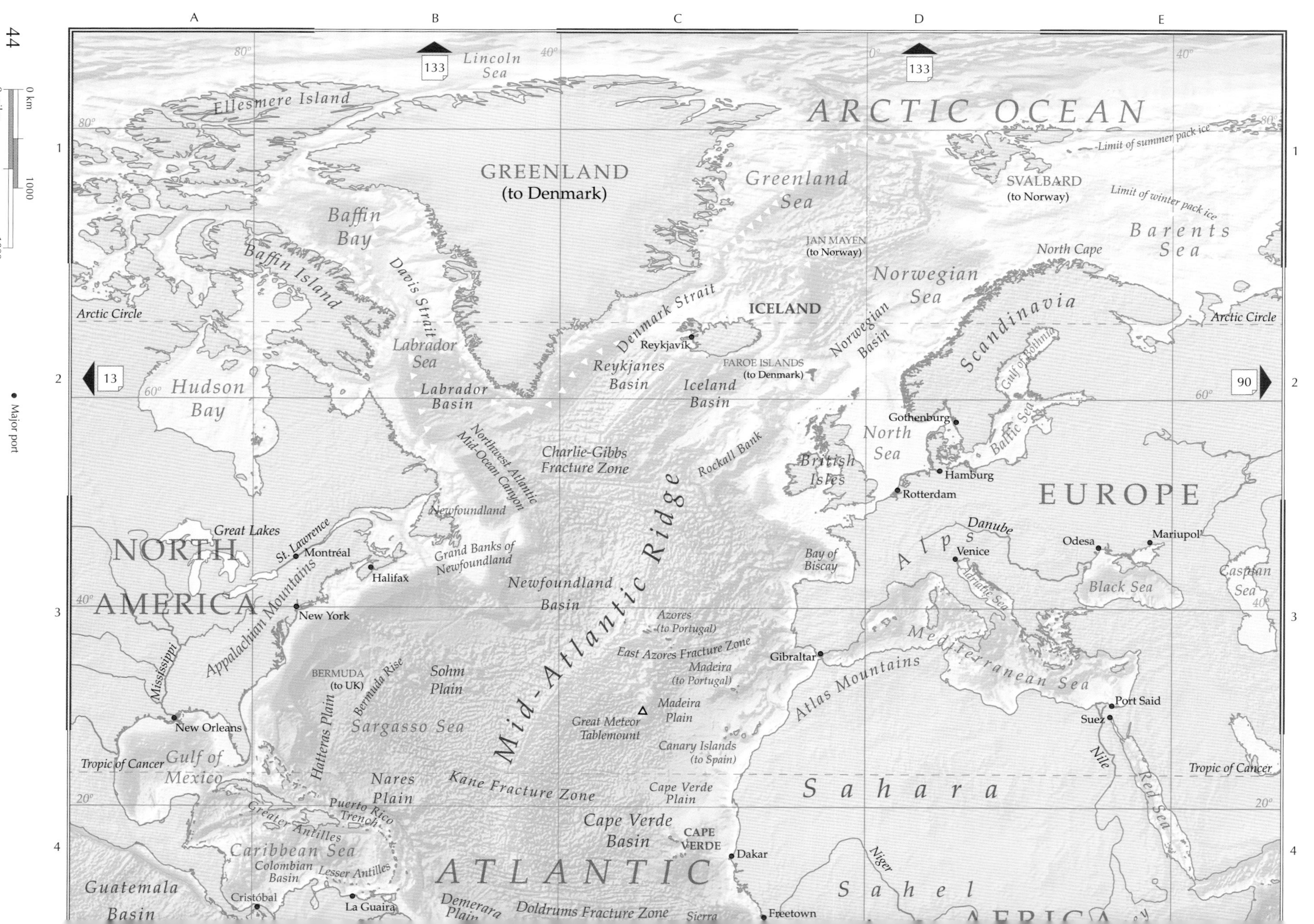

ARCTIC OCEAN
Limit of summer pack ice
Limit of winter pack ice
Barents Sea
SVALBARD (to Norway)
North Cape
Lincoln Sea
Ellesmere Island
GREENLAND (to Denmark)
Greenland Sea
JAN MAYEN (to Norway)
Norwegian Sea
Norwegian Basin
Scandinavia
Gulf of Bothnia
Baltic Sea
Arctic Circle
Baffin Bay
Baffin Island
Davis Strait
Hudson Bay
Labrador Sea
Labrador Basin
Denmark Strait
ICELAND
Reykjavík
Reykjanes Basin
Iceland Basin
FAROE ISLANDS (to Denmark)
Rockall Bank
Gothenburg
Hamburg
Rotterdam
North Sea
British Isles
Bay of Biscay
EUROPE
Mariupol'
Odesa
Black Sea
Caspian Sea
Danube
Alps
Venice
Adriatic Sea
Mediterranean Sea
Port Said
Suez
Nile
Red Sea
Tropic of Cancer
Gibraltar
Atlas Mountains
Sahara
Sahel
Niger
Freetown
Dakar
CAPE VERDE
Northwest Atlantic Mid-Ocean Canyon
Charlie-Gibbs Fracture Zone
Mid-Atlantic Ridge
Newfoundland
Newfoundland Basin
Grand Banks of Newfoundland
Azores (to Portugal)
East Azores Fracture Zone
Madeira (to Portugal)
Madeira Plain
Canary Islands (to Spain)
Great Meteor Tablemount
Cape Verde Plain
Cape Verde Basin
Doldrums Fracture Zone
Kane Fracture Zone
Sohm Plain
Sargasso Sea
Bermuda Rise
BERMUDA (to UK)
Hatteras Plain
Nares Plain
Puerto Rico Trench
Demerara Plain
ATLANTIC
Halifax
Montréal
St. Lawrence
New York
Appalachian Mountains
Great Lakes
NORTH AMERICA
Mississippi
New Orleans
Gulf of Mexico
Greater Antilles
Lesser Antilles
Caribbean Sea
Colombian Basin
Cristóbal
La Guaira
Guatemala Basin

0 km 1000

0 miles 1000

● Major port

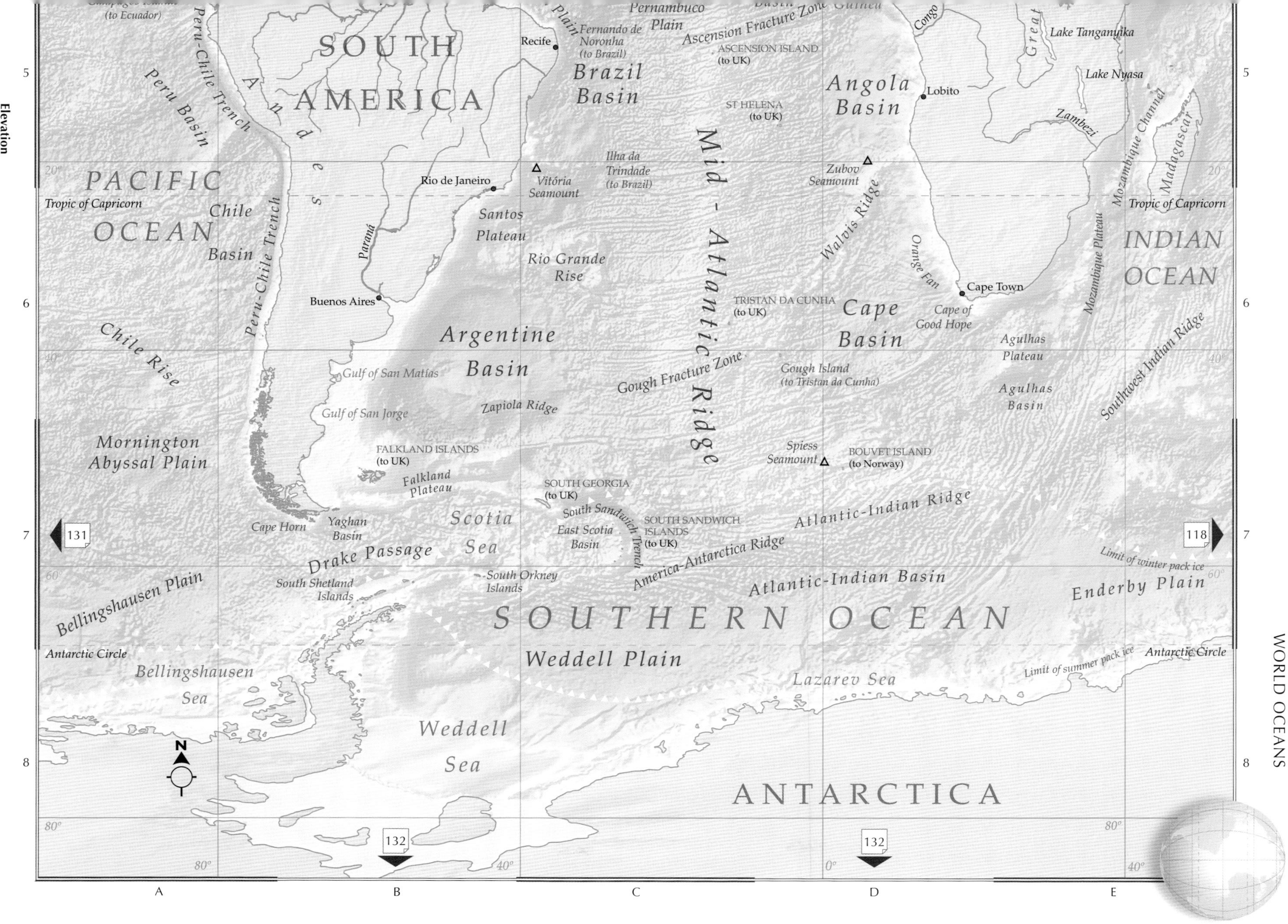
SOUTH AMERICA
Andes
PACIFIC OCEAN
Peru Basin
Peru-Chile Trench
Chile Basin
Chile Rise
Tropic of Capricorn
Mornington Abyssal Plain
Bellingshausen Plain
Antarctic Circle
Bellingshausen Sea
Buenos Aires
Paraná
Rio de Janeiro
Recife
Santos Plateau
Vitória Seamount
Rio Grande Rise
Argentine Basin
Gulf of San Matías
Gulf of San Jorge
FALKLAND ISLANDS (to UK)
Falkland Plateau
Zapiola Ridge
Yaghan Basin
Cape Horn
Drake Passage
Scotia Sea
South Shetland Islands
South Orkney Islands
SOUTH GEORGIA (to UK)
East Scotia Basin
South Sandwich Trench
SOUTH SANDWICH ISLANDS (to UK)
America-Antarctica Ridge
Weddell Plain
Weddell Sea
SOUTHERN OCEAN
Brazil Basin
Fernando de Noronha (to Brazil)
Pernambuco Plain
Ilha da Trindade (to Brazil)
Gough Fracture Zone
Mid - Atlantic Ridge
Ascension Fracture Zone
ASCENSION ISLAND (to UK)
ST HELENA (to UK)
TRISTAN DA CUNHA (to UK)
Gough Island (to Tristan da Cunha)
Spiess Seamount
BOUVET ISLAND (to Norway)
Atlantic-Indian Ridge
Atlantic-Indian Basin
Lazarev Sea
ANTARCTICA
Angola Basin
Zubov Seamount
Walvis Ridge
Cape Basin
Orange Fan
Cape of Good Hope
Cape Town
Lobito
Congo
Agulhas Plateau
Agulhas Basin
Mozambique Plateau
Southwest Indian Ridge
Zambezi
Lake Nyasa
Lake Tanganyika
Mozambique Channel
Madagascar
INDIAN OCEAN
Enderby Plain
Limit of winter pack ice
Limit of summer pack ice
131
132
118

Elevation

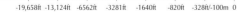

-6000m	-4000m	-2000m	-1000m	-500m	-250m	-100m	0
-19,658ft	-13,124ft	-6562ft	-3281ft	-1640ft	-820ft	-328ft/-100m	0

Africa

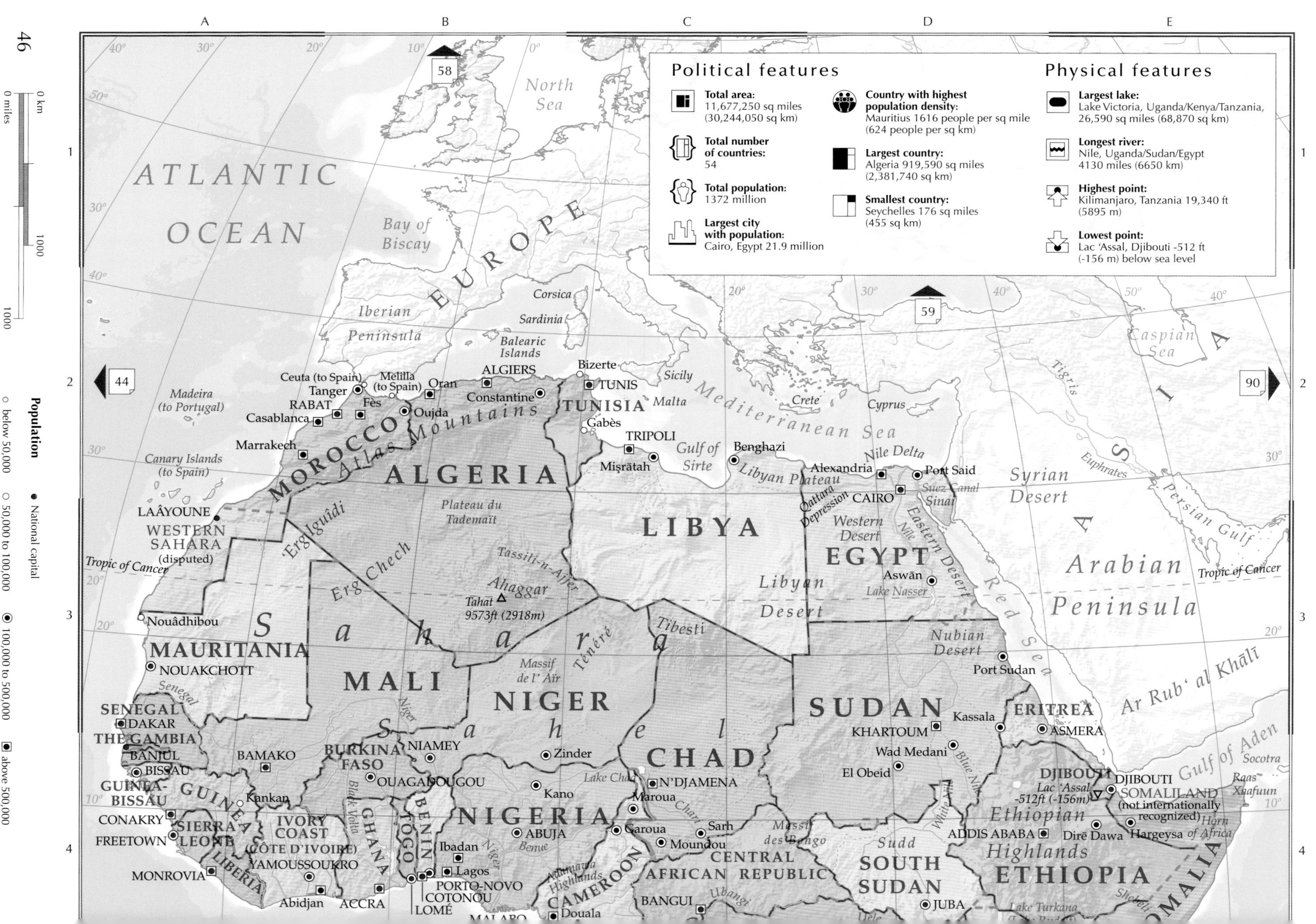

Political features
Total area: 11,677,250 sq miles (30,244,050 sq km)
Total number of countries: 54
Total population: 1372 million
Largest city with population: Cairo, Egypt 21.9 million
Country with highest population density: Mauritius 1616 people per sq mile (624 people per sq km)
Largest country: Algeria 919,590 sq miles (2,381,740 sq km)
Smallest country: Seychelles 176 sq miles (455 sq km)
Physical features
Largest lake: Lake Victoria, Uganda/Kenya/Tanzania, 26,590 sq miles (68,870 sq km)
Longest river: Nile, Uganda/Sudan/Egypt 4130 miles (6650 km)
Highest point: Kilimanjaro, Tanzania 19,340 ft (5895 m)
Lowest point: Lac 'Assal, Djibouti -512 ft (-156 m) below sea level
ATLANTIC OCEAN
EUROPE
ASIA
Mediterranean Sea
Red Sea
Arabian Peninsula
MOROCCO
ALGERIA
TUNISIA
LIBYA
EGYPT
SUDAN
SOUTH SUDAN
ETHIOPIA
ERITREA
DJIBOUTI
SOMALILAND (not internationally recognized)
CHAD
NIGER
MALI
MAURITANIA
WESTERN SAHARA (disputed)
SENEGAL
THE GAMBIA
GUINEA-BISSAU
GUINEA
SIERRA LEONE
LIBERIA
IVORY COAST (CÔTE D'IVOIRE)
BURKINA FASO
GHANA
TOGO
BENIN
NIGERIA
CAMEROON
CENTRAL AFRICAN REPUBLIC
Sahara
Sahel
CAIRO
ALGIERS
TUNIS
TRIPOLI
RABAT
KHARTOUM
ADDIS ABABA
N'DJAMENA
NIAMEY
BAMAKO
NOUAKCHOTT
DAKAR
ABUJA
Tropic of Cancer

Population
• National capital
○ below 50,000
○ 50,000 to 100,000
◉ 100,000 to 500,000
▣ above 500,000

0 km 1000
0 miles 1000

Basin
Mid-Atlantic Ridge
Ascension Fracture Zone
ASCENSION ISLAND (to UK)
GABON
CONGO
BRAZZAVILLE
Cabinda (to Angola)
Matadi
KINSHASA
Congo
Cuango
Kasai
DEM. REP. CONGO
Ilebo
Kananga
Bukavu
Lualaba
Great Rift Valley
Kalemie
Luvua
Lake Mweru
RWANDA
BUJUMBURA
GITEGA
BURUNDI
Lake Tanganyika
DODOMA
TANZANIA
Lake Rukwa
Masai Steppe
NAIROBI
Kilimanjaro 19,340ft (5895m)
Mombasa
Tanga
Pemba
Zanzibar
Dar es Salaam
Somali Basin
Aldabra Group
LUANDA
Cuanza
ANGOLA
Bié Plateau
Môco 8593ft (2619m)
Huambo
Lubango
Namibe
Angola Basin
SAINT HELENA (to UK)
Lubumbashi
Kitwe
Ndola
ZAMBIA
LUSAKA
Zambezi
Cuando
Cubango
Victoria Falls
Lake Kariba
HARARE
MALAWI
Rufiji
Rovuma
Lake Nyasa
LILONGWE
Lúrio
Blantyre
MOZAMBIQUE
Nacala
Nampula
COMOROS
MORONI
MAYOTTE (to France)
Mahajanga
MADAGASCAR
ANTANANARIVO
Fianarantsoa
Toliara
ATLANTIC OCEAN
Cunene
Etosha Pan
NAMIBIA
Okavango Delta
ZIMBABWE
Francistown
Bulawayo
Beira
WINDHOEK
BOTSWANA
Kalahari Desert
Limpopo
Namib Desert
Nossob
GABORONE
PRETORIA
MBABANE
MAPUTO
LOBAMBA
ESWATINI
Johannesburg
Vaal
Tropic of Capricorn
Walvis Ridge
SOUTH AFRICA
LESOTHO
MASERU
Orange River
BLOEMFONTEIN
Durban
Drakensberg
Great Karoo
East London
Orange Fan
CAPE TOWN
Cape of Good Hope
Gqeberha (Port Elizabeth)
Mozambique Channel
Mozambique Plateau
Madagascar Plateau
Madagascar Basin
INDIAN OCEAN
Cape Basin
TRISTAN DA CUNHA (to UK)
Agulhas Plateau
Agulhas Basin
Gough Island (to Tristan da Cunha)
Southwest Indian Ridge
Crozet Plateau
Prince Edward Islands (to South Africa)
Atlantic-Indian Ridge
Winter limit of pack ice
45
119
132
132
A
B
C
D
E
5
6
7
8

Northwest Africa

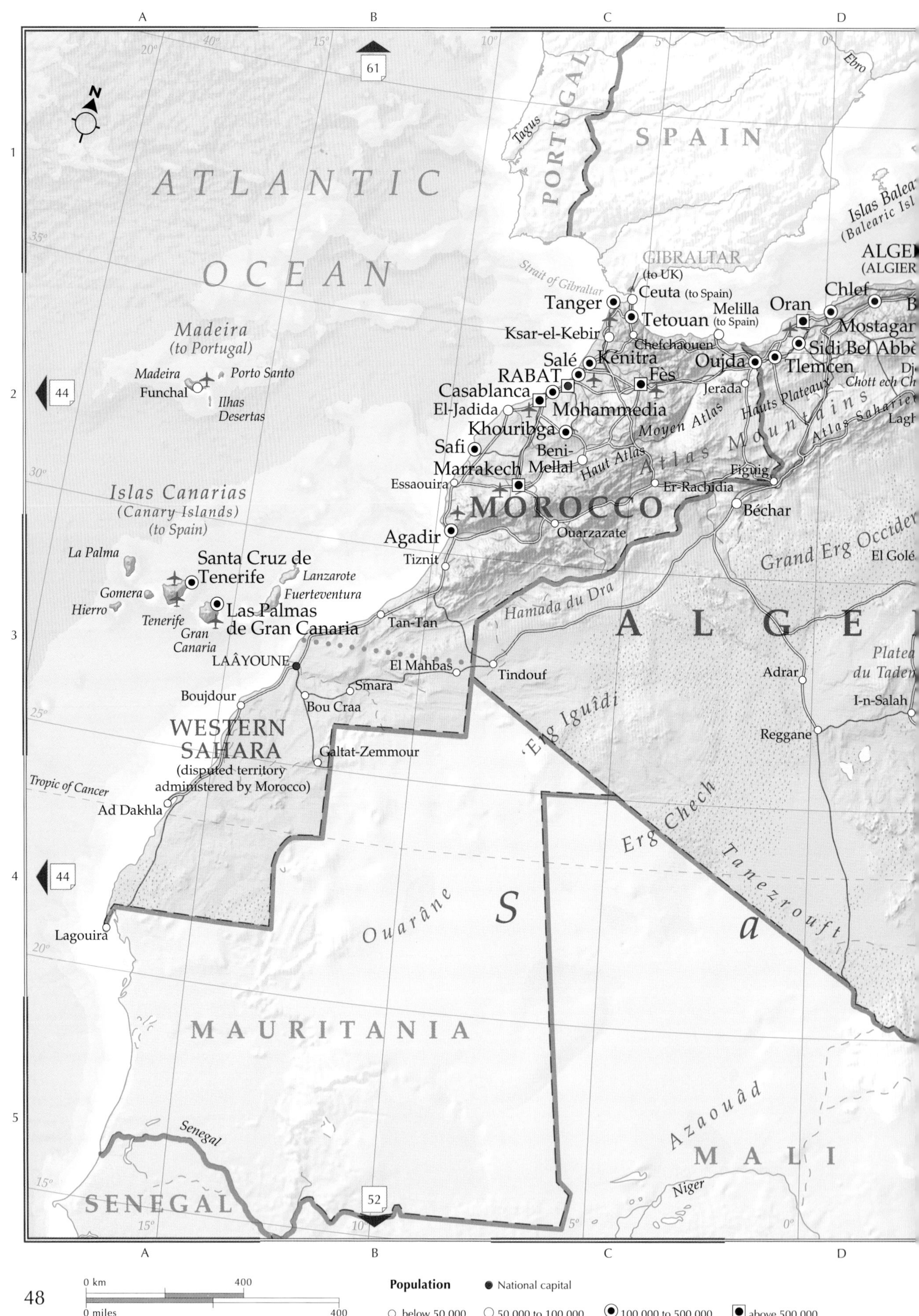
ATLANTIC OCEAN
PORTUGAL
SPAIN
Tagus
Ebro
Islas Baleares (Balearic Islands)
GIBRALTAR (to UK)
Strait of Gibraltar
Ceuta (to Spain)
Melilla (to Spain)
Tanger
Tetouan
Ksar-el-Kebir
Chefchaouen
Salé
Kénitra
RABAT
Fès
Casablanca
Mohammedia
El-Jadida
Khouribga
Safi
Beni-Mellal
Marrakech
Essaouira
MOROCCO
Agadir
Tiznit
Ouarzazate
Er-Rachidia
Figuig
Oujda
Jerada
Oran
Chlef
Mostaganem
Sidi Bel Abbès
Tlemcen
Moyen Atlas
Haut Atlas
Atlas Mountains
Hauts Plateaux
Chott ech Chergui
Atlas Saharien
Béchar
Grand Erg Occidental
El Goléa
Hamada du Dra
ALGERIA
Madeira (to Portugal)
Madeira
Porto Santo
Funchal
Ilhas Desertas
Islas Canarias (Canary Islands) (to Spain)
La Palma
Santa Cruz de Tenerife
Lanzarote
Fuerteventura
Gomera
Hierro
Tenerife
Las Palmas de Gran Canaria
Gran Canaria
Tan-Tan
LAÂYOUNE
El Mahbas
Tindouf
Smara
Boujdour
Bou Craa
WESTERN SAHARA (disputed territory administered by Morocco)
Galtat-Zemmour
Tropic of Cancer
Ad Dakhla
Erg Iguîdi
Erg Chech
Tanezrouft
Adrar
Plateau du Tademaït
I-n-Salah
Reggane
Ouarâne
Sahara
Lagouira
MAURITANIA
Azaouâd
MALI
Niger
Senegal
SENEGAL
61
44
44
52

0 km 400
0 miles 400

Population
● National capital
○ below 50,000
○ 50,000 to 100,000
⊙ 100,000 to 500,000
▣ above 500,000

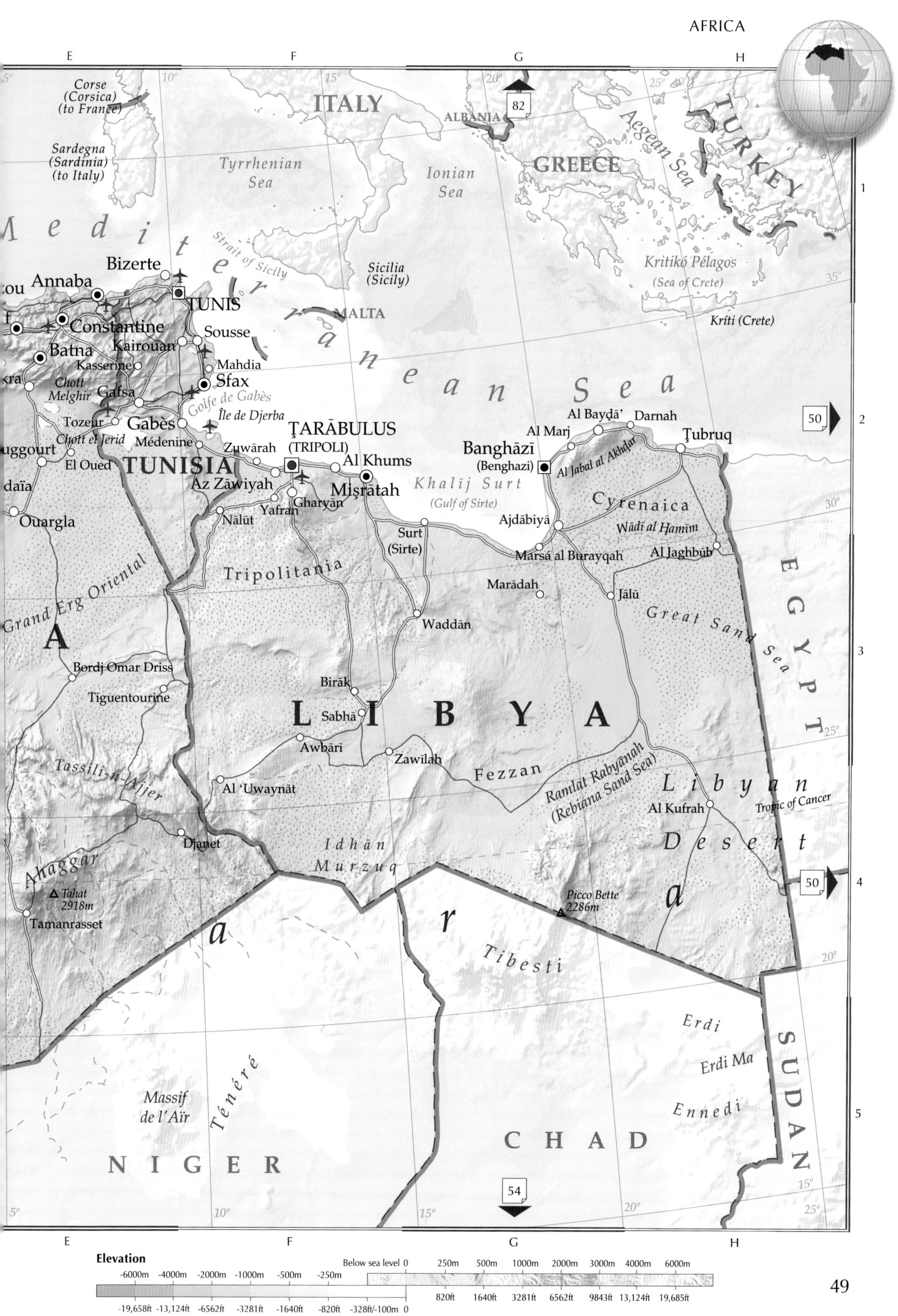
E
F
G
H
Corse
(Corsica)
(to France)
ITALY
ALBANIA
82
TURKEY
Aegean Sea
Sardegna
(Sardinia)
(to Italy)
Tyrrhenian
Sea
Ionian
Sea
GREECE
Mediterranean Sea
Strait of Sicily
Sicilia
(Sicily)
Kritikó Pélagos
(Sea of Crete)
Kríti (Crete)
Bizerte
Annaba
TUNIS
Constantine
MALTA
Sousse
Batna
Kairouan
Kasserine
Mahdia
Sfax
Chott
Melghir
Gafsa
Golfe de Gabès
Tozeur
Gabès
Île de Djerba
Chott el Jerid
Médenine
ȚARĀBULUS
(TRIPOLI)
Zuwārah
Al Khums
El Oued
TUNISIA
Az Zāwiyah
Mişrātah
Banghāzī
(Benghazi)
Al Marj
Al Bayḍā'
Darnah
Ţubruq
50
Al Jabal al Akhḑar
Khalīj Surt
(Gulf of Sirte)
Gharyān
Yafran
Nālūt
Ouargla
Cyrenaica
Ajdābiyā
Wādī al Ḩamīm
Surt
(Sirte)
Marsá al Burayqah
Al Jaghbūb
Tripolitania
Grand Erg Oriental
Marādah
Jālū
EGYPT
Waddān
Great Sand Sea
Bordj Omar Driss
Tiguentourine
Birāk
LIBYA
Sabhā
Awbārī
Zawīlah
Fezzan
Ramlat Rabyānah
(Rebiana Sand Sea)
Libyan Desert
Tassili-n-Ajjer
Al 'Uwaynāt
Al Kufrah
Tropic of Cancer
Djanet
Idhān Murzuq
Ahaggar
Tahat
2918m
Tamanrasset
Picco Bette
2286m
Sahara
Tibesti
Erdi
Erdi Ma
Ennedi
SUDAN
Massif
de l'Aïr
Ténéré
CHAD
NIGER
54
1
2
3
4
5
Elevation
-6000m -4000m -2000m -1000m -500m -250m
-19,658ft -13,124ft -6562ft -3281ft -1640ft -820ft -328ft/-100m 0
Below sea level 0
250m 500m 1000m 2000m 3000m 4000m 6000m
820ft 1640ft 3281ft 6562ft 9843ft 13,124ft 19,685ft

Northeast Africa

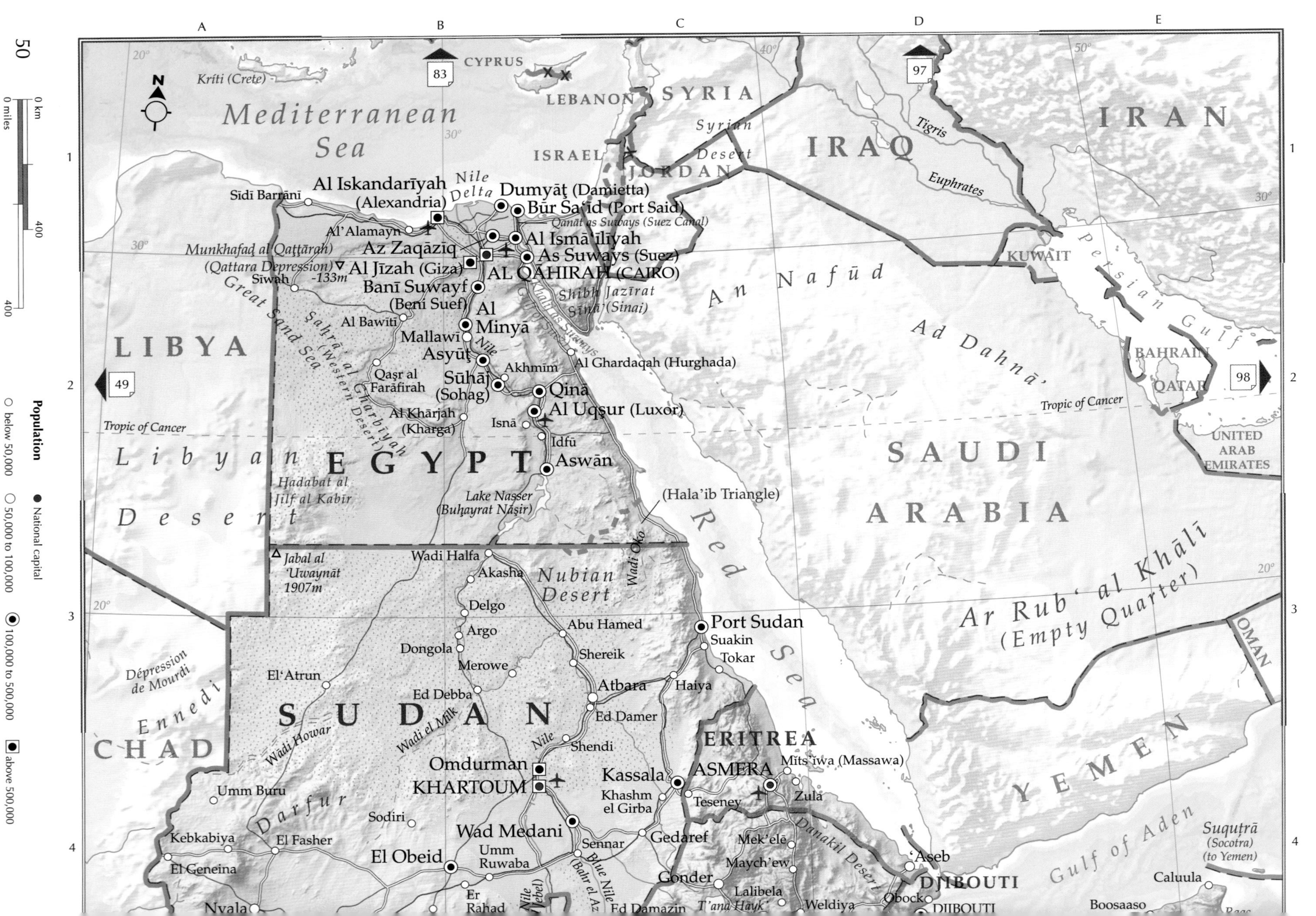

Mediterranean Sea
Kríti (Crete)
CYPRUS
LEBANON
SYRIA
ISRAEL
JORDAN
Syrian Desert
IRAQ
Tigris
Euphrates
IRAN
KUWAIT
Persian Gulf
BAHRAIN
QATAR
UNITED ARAB EMIRATES
OMAN
An Nafūd
Ad Dahnā'
SAUDI ARABIA
Ar Rub' al Khālī (Empty Quarter)
YEMEN
Gulf of Aden
Suquṭrā (Socotra) (to Yemen)
Caluula
Boosaaso
Tropic of Cancer
LIBYA
Libyan Desert
EGYPT
Sīdī Barrānī
Al Iskandarīyah (Alexandria)
Nile Delta
Dumyāṭ (Damietta)
Būr Sa'īd (Port Said)
Qanāt as Suways (Suez Canal)
Al 'Alamayn
Munkhafaḍ al Qaṭṭārah (Qattara Depression)
-133m
Az Zaqāzīq
Al Ismā'īlīyah
As Suways (Suez)
Al Jīzah (Giza)
AL QĀHIRAH (CAIRO)
Sīwah
Great Sand Sea
Banī Suwayf (Beni Suef)
Shibh Jazīrat Sīnā' (Sinai)
Al Bawītī
Al Minyā
Mallawī
Ṣaḥrā' al Gharbīyah (Western Desert)
Asyūṭ
Nile
Qaṣr al Farāfirah
Akhmīm
Al Ghardaqah (Hurghada)
Sūhāj (Sohag)
Qinā
Al Khārjah (Kharga)
Al Uqṣur (Luxor)
Isnā
Idfū
Aswān
Hadabat al Jilf al Kabīr
Lake Nasser (Buḥayrat Nāṣir)
(Hala'ib Triangle)
Red Sea
Wadi Oko
Jabal al 'Uwaynāt 1907m
Wadi Halfa
Akasha
Nubian Desert
Delgo
Argo
Dongola
Merowe
Abu Hamed
Shereik
Port Sudan
Suakin
Tokar
El 'Atrun
Atbara
Haiya
Ed Debba
Ed Damer
SUDAN
Dépression de Mourdi
Ennedi
CHAD
Wadi Howar
Wadi el Milk
Shendi
ERITREA
Omdurman
KHARTOUM
Kassala
ASMERA
Mits'iwa (Massawa)
Khashm el Girba
Teseney
Zula
Umm Buru
Darfur
Sodiri
Wad Medani
Kebkabiya
El Fasher
Sennar
Gedaref
Mek'elē
Danakil Desert
El Geneina
El Obeid
Umm Ruwaba
Blue Nile (Bahr el Azraq)
Gonder
Maych'ew
'Aseb
DJIBOUTI
Nyala
Er Rahad
Nile (Jebel)
Ed Damazin
T'ana Hāyk'
Lalibela
Weldiya
Obock

Population
● National capital
○ below 50,000
○ 50,000 to 100,000
◉ 100,000 to 500,000
▣ above 500,000

0 km 400
0 miles 400

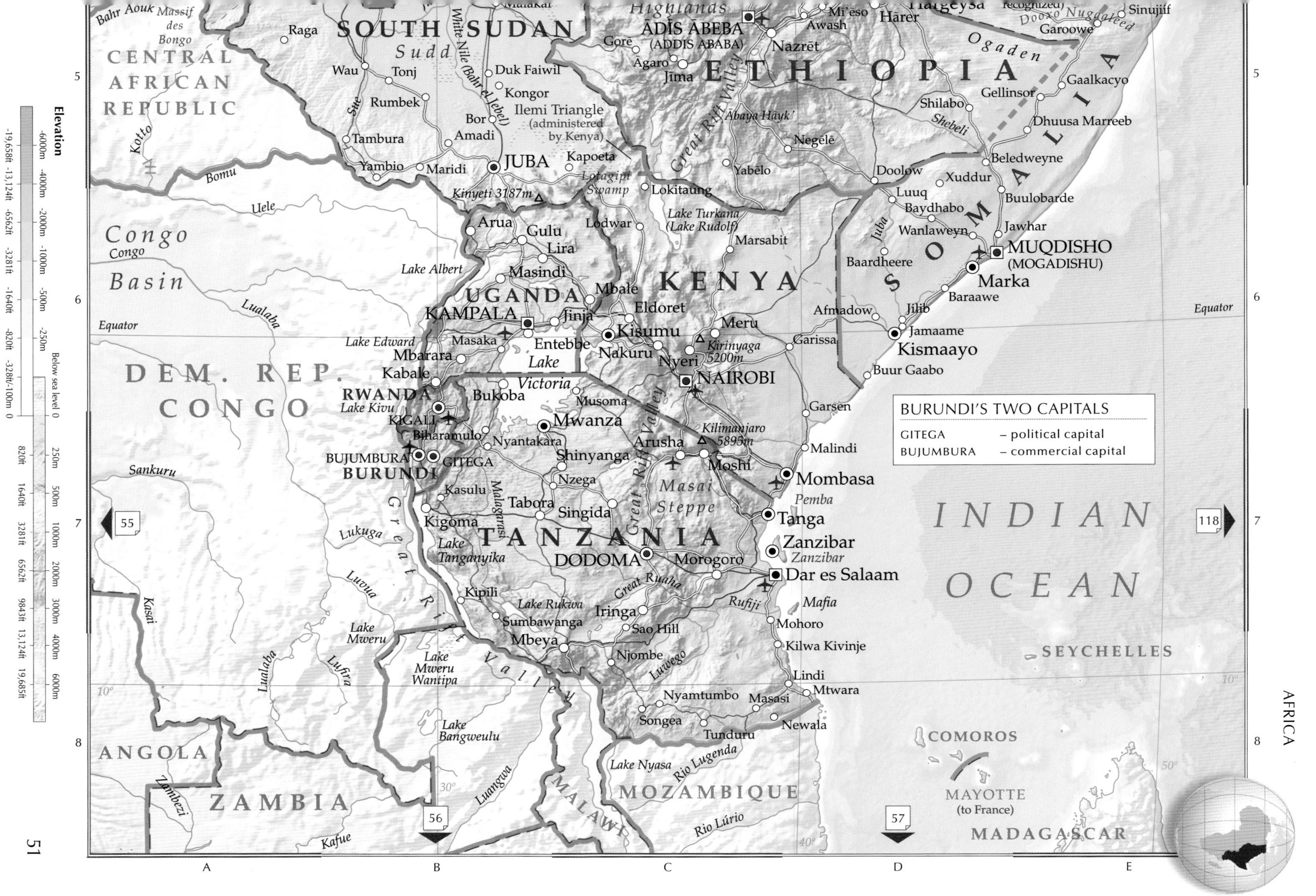
BURUNDI'S TWO CAPITALS
GITEGA – political capital
BUJUMBURA – commercial capital
SOUTH SUDAN
ETHIOPIA
SOMALIA
KENYA
UGANDA
TANZANIA
RWANDA
BURUNDI
DEM. REP. CONGO
CENTRAL AFRICAN REPUBLIC
ANGOLA
ZAMBIA
MALAWI
MOZAMBIQUE
MADAGASCAR
COMOROS
MAYOTTE (to France)
SEYCHELLES
INDIAN OCEAN
ĀDĪS ĀBEBA (ADDIS ABABA)
MUQDISHO (MOGADISHU)
NAIROBI
KAMPALA
JUBA
KIGALI
GITEGA
BUJUMBURA
DODOMA
Dar es Salaam
Mombasa
Kismaayo
Marka
Zanzibar
Tanga
Mwanza
Kisumu
Arusha
Moshi
Kigoma
Tabora
Mbeya
Iringa
Morogoro
Mtwara
Lindi
Malindi
Hārer
Nazrēt
Jīma
Gaalkacyo
Beledweyne
Jawhar
Wau
Tonj
Rumbek
Yambio
Raga
Lake Victoria
Lake Tanganyika
Lake Turkana (Lake Rudolf)
Lake Albert
Lake Edward
Lake Kivu
Lake Rukwa
Lake Nyasa
Lake Mweru
Lake Bangweulu
Great Rift Valley
Ilemi Triangle (administered by Kenya)
Kilimanjaro 5895m
Kirinyaga 5200m
Kinyeti 3187m
Sudd
Ogaden
Congo Basin
Masai Steppe
Equator
Lualaba
Kasai
Zambezi
Luangwa
Rufiji
Shebeli
Juba
White Nile (Bahr el Jebel)
Pemba
Mafia
55
56
57
118
Elevation
-6000m -4000m -2000m -1000m -500m -250m
-19,658ft -13,124ft -6562ft -3281ft -1640ft -820ft -328ft/-100m 0
Below sea level 0
250m 500m 1000m 2000m 3000m 4000m 6000m
820ft 1640ft 3281ft 6562ft 9843ft 13,124ft 19,685ft

West Africa

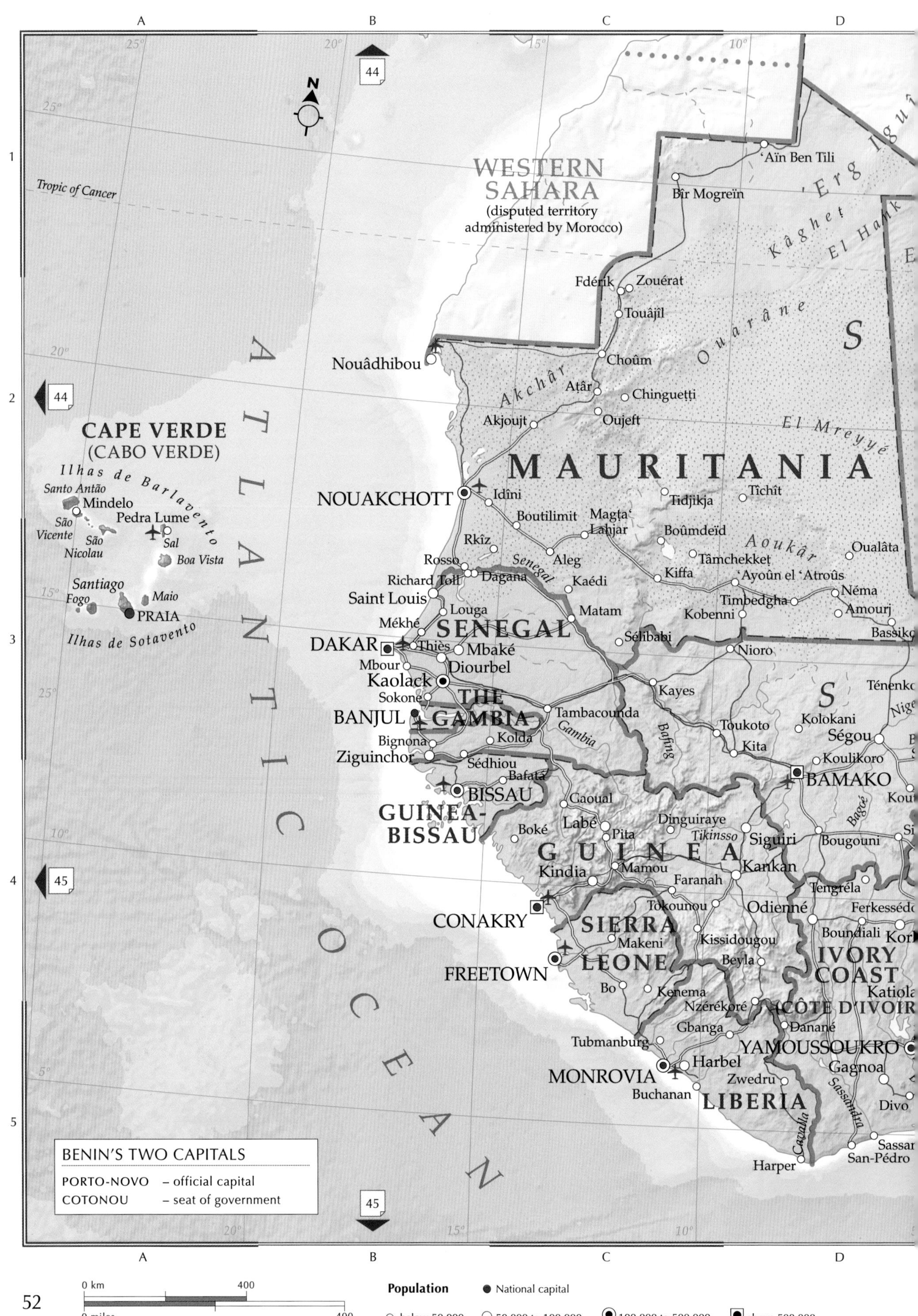

WESTERN SAHARA
(disputed territory administered by Morocco)
MAURITANIA
SENEGAL
THE GAMBIA
GUINEA-BISSAU
GUINEA
SIERRA LEONE
LIBERIA
IVORY COAST
CÔTE D'IVOIR
CAPE VERDE
(CABO VERDE)
ATLANTIC OCEAN
Tropic of Cancer
Ilhas de Barlavento
Ilhas de Sotavento
Santo Antão
Mindelo
São Vicente
São Nicolau
Pedra Lume
Sal
Boa Vista
Santiago
Fogo
Maio
PRAIA
'Aïn Ben Tili
Bîr Mogreïn
Fdérik
Zouérat
Touâjîl
Nouâdhibou
Choûm
Atâr
Chinguetti
Akjoujt
Oujeft
NOUAKCHOTT
Idîni
Tidjikja
Tîchît
Boutilimit
Magta' Lahjar
Boûmdeïd
Rkîz
Rosso
Aleg
Tâmchekket
Oualâta
Richard Toll
Dagana
Kaédi
Kiffa
'Ayoûn el 'Atroûs
Saint Louis
Timbedgha
Néma
Amourj
Louga
Matam
Kobenni
Bassikou
Mékhé
Sélibabi
DAKAR
Thiès
Mbaké
Nioro
Mbour
Diourbel
Kaolack
Kayes
Ténenkou
Sokone
Tambacounda
BANJUL
Toukoto
Kolokani
Ségou
Bignona
Kolda
Kita
Ziguinchor
Sédhiou
Koulikoro
Bafatá
BAMAKO
BISSAU
Gaoual
Boké
Labé
Pita
Dinguiraye
Siguiri
Bougouni
Kindia
Mamou
Kankan
Faranah
Tengréla
CONAKRY
Tokounou
Odienné
Ferkessédougou
Makeni
Kissidougou
Boundiali
FREETOWN
Beyla
Katiola
Bo
Kenema
Nzérékoré
Danané
Gbanga
Tubmanburg
YAMOUSSOUKRO
Harbel
Gagnoa
MONROVIA
Zwedru
Buchanan
Divo
Harper
San-Pédro
Akchâr
Ouarâne
Kâghet
Erg Iguidi
El Hank
El Mreyyé
Aoukâr
Senegal
Gambia
Bafing
Tikinsso
Bagoé
Sassandra
Cavalla
Niger
BENIN'S TWO CAPITALS
PORTO-NOVO – official capital
COTONOU – seat of government
0 km 400
0 miles 400
Population
National capital
below 50,000
50,000 to 100,000
100,000 to 500,000
above 500,000

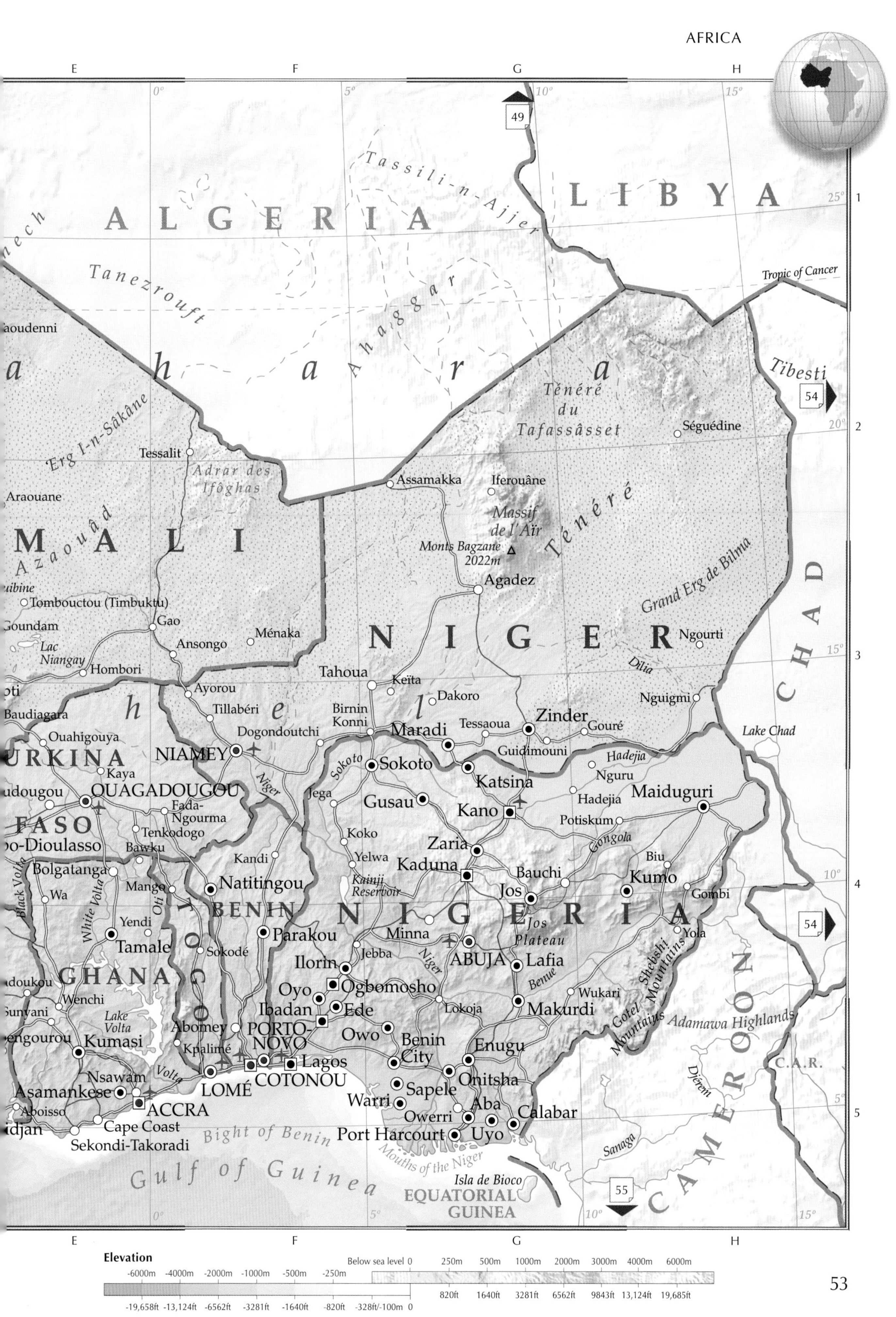

ALGERIA
LIBYA
MALI
NIGER
CHAD
NIGERIA
BENIN
TOGO
GHANA
BURKINA FASO
CAMEROON
EQUATORIAL GUINEA
C.A.R.
Tassili-n-Ajjer
Tanezrouft
Ahaggar
Sahara
Sahel
Ténéré du Tafassâsset
Ténéré
Tibesti
Erg I-n-Sâkâne
Adrar des Ifôghas
Azaouâd
Massif de l'Aïr
Monts Bagzane 2022m
Grand Erg de Bilma
Dilia
Lake Chad
Tropic of Cancer
Séguédine
Assamakka
Iferouâne
Agadez
Ngourti
Nguigmi
Tessalit
Araouane
Tombouctou (Timbuktu)
Goundam
Gao
Ansongo
Ménaka
Lac Niangay
Hombori
Ayorou
Tillabéri
Tahoua
Keïta
Dakoro
Birnin Konni
Dogondoutchi
Maradi
Tessaoua
Zinder
Gouré
Guidimouni
NIAMEY
Baudiagara
Ouahigouya
Kaya
OUAGADOUGOU
Fada-Ngourma
Tenkodogo
Bawku
Bolgatanga
Wa
Mango
Oti
Black Volta
White Volta
Yendi
Tamale
Sokodé
Kandi
Natitingou
Parakou
Wenchi
Lake Volta
Kumasi
Abomey
Kpalimé
PORTO-NOVO
COTONOU
LOMÉ
Lagos
Nsawam
Asamankese
ACCRA
Aboisso
Cape Coast
Sekondi-Takoradi
Bight of Benin
Gulf of Guinea
Mouths of the Niger
Isla de Bioco
Sokoto
Gusau
Jega
Koko
Yelwa
Kainji Reservoir
Katsina
Kano
Hadejia
Nguru
Maiduguri
Potiskum
Gongola
Zaria
Kaduna
Bauchi
Biu
Kumo
Gombi
Jos
Jos Plateau
Minna
ABUJA
Jebba
Ilorin
Ogbomosho
Oyo
Ibadan
Ede
Owo
Lokoja
Lafia
Benue
Makurdi
Wukari
Benin City
Enugu
Onitsha
Sapele
Warri
Owerri
Aba
Port Harcourt
Uyo
Calabar
Yola
Shebshi Mountains
Gotel Mountains
Adamawa Highlands
Djérem
Sanaga
Niger
Volta
49
54
55

Elevation

Below sea level 0 | 250m 500m 1000m 2000m 3000m 4000m 6000m

-6000m -4000m -2000m -1000m -500m -250m

820ft 1640ft 3281ft 6562ft 9843ft 13,124ft 19,685ft

-19,658ft -13,124ft -6562ft -3281ft -1640ft -820ft -328ft/-100m 0

Central Africa

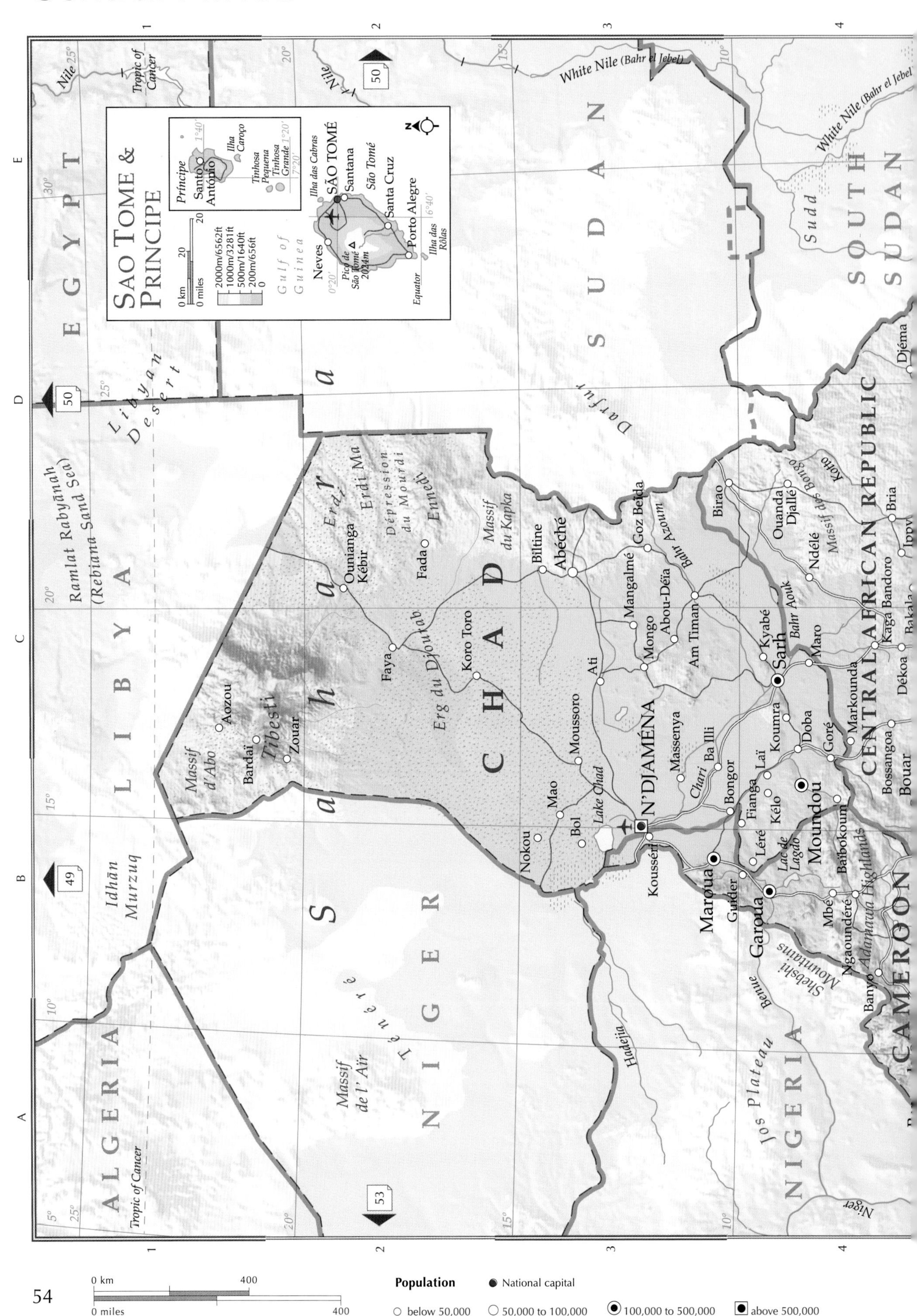

EGYPT
LIBYA
ALGERIA
NIGER
CHAD
SUDAN
SOUTH SUDAN
CENTRAL AFRICAN REPUBLIC
CAMEROON
NIGERIA
Sahara
Libyan Desert
Ramlat Rabyānah (Rebiana Sand Sea)
Idhān Murzuq
Massif d'Abo
Tibesti
Massif de l'Aïr
Ténéré
Erg du Djourab
Erdi
Erdi Ma
Dépression du Mourdi
Ennedi
Massif du Kapka
Darfur
Sudd
Nile
White Nile (Bahr el Jebel)
Tropic of Cancer
Lake Chad
Chari
Bahr Azoum
Bahr Aouk
Kotto
Massif des Bongo
Hadejia
Benue
Niger
Jos Plateau
Shebshi Mountains
Adamawa Highlands
Lac de Lagdo
Aozou
Bardaï
Zouar
Faya
Ouniang Kébir
Fada
Koro Toro
Biltine
Abéché
Goz Beïda
Mangalmé
Abou-Déïa
Am Timan
Mongo
Ati
Moussoro
Mao
Nokou
Bol
N'DJAMÉNA
Kousséri
Massenya
Ba Illi
Bongor
Fianga
Laï
Kélo
Léré
Moundou
Koumra
Doba
Goré
Sarh
Kyabe
Maro
Birao
Ouanda Djallé
Ndélé
Bria
Kaga Bandoro
Markounda
Bossangoa
Bouar
Dékoa
Bakala
Djéma
Maroua
Guider
Garoua
Mbé
Baïbokoum
Ngaoundéré
Banyo
SAO TOME & PRINCIPE
Príncipe
Santo António
Ilha Caroço
Tinhosa Pequena
Tinhosa Grande
Gulf of Guinea
Ilha das Cabras
SÃO TOMÉ
Santana
São Tomé
Santa Cruz
Porto Alegre
Neves
Pico de São Tomé 2024m
Ilha das Rôlas
Equator
2000m/6562ft
1000m/3281ft
500m/1640ft
200m/656ft
0
0 km 20
0 miles 20

0 km 400
0 miles 400

Population ● National capital

○ below 50,000 ○ 50,000 to 100,000 ◉ 100,000 to 500,000 ▣ above 500,000

DEM. REP. CONGO
CONGO
GABON
ANGOLA
ZAMBIA
TANZANIA
UGANDA
RWANDA
BURUNDI
EQUATORIAL GUINEA
SAO TOME & PRINCIPE
ATLANTIC OCEAN
Gulf of Guinea
KINSHASA
BRAZZAVILLE
LIBREVILLE
YAOUNDE
MALABO
SÃO TOMÉ
Kisangani
Mbandaka
Kananga
Mbuji-Mayi
Lubumbashi
Likasi
Kolwezi
Kamina
Kalemie
Bukavu
Goma
Kikwit
Ilebo
Matadi
Boma
Pointe-Noire
Dolisie
Port-Gentil
Cabinda (to Angola)
Congo Basin
Great Rift Valley
Lake Tanganyika
Lake Victoria
Lake Albert
Lake Edward
Lake Kivu
Lake Mweru
Lake Bangweulu
Lake Mweru Wantipa
Mitumba Mountains
Planalto do Bié
Massif du Chaillu
Plateaux Batéké
Elevation
-6000m -4000m -2000m -1000m -500m -250m
-19,658ft -13,124ft -6562ft -3281ft -1640ft -820ft -328ft/-100m 0
Below sea level 0
250m 500m 1000m 2000m 3000m 4000m 6000m
820ft 1640ft 3281ft 6562ft 9843ft 13,124ft 19,685ft

Southern Africa

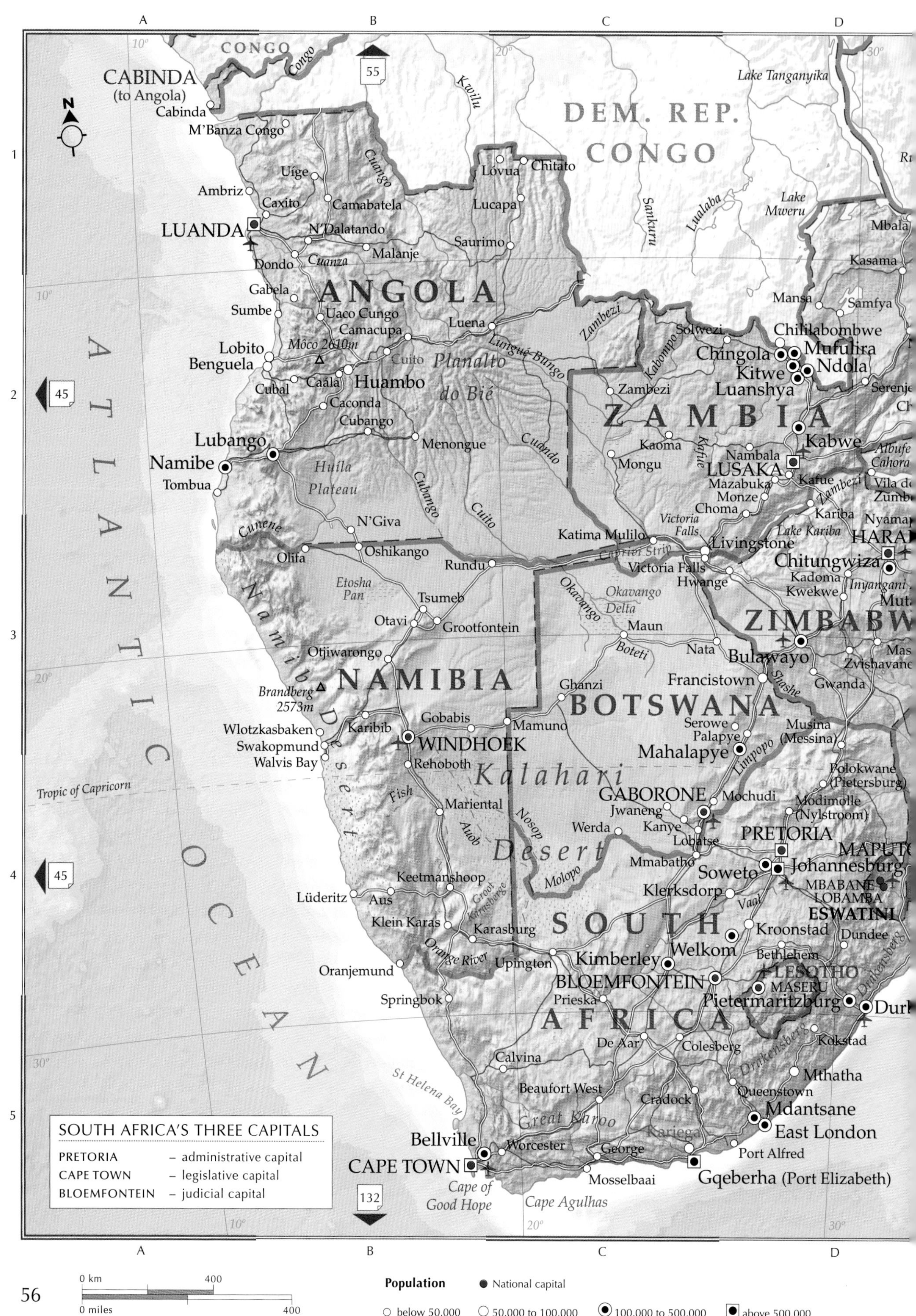

A
B
C
D
1
2
3
4
5
CONGO
Congo
CABINDA
(to Angola)
Cabinda
M'Banza Congo
55
Kwilu
DEM. REP.
CONGO
Lake Tanganyika
Uíge
Cuango
Lóvua
Chitato
Ambriz
Caxito
Camabatela
Lucapa
Sankuru
Lualaba
Lake
Mweru
Mbala
LUANDA
N'Dalatando
Saurimo
Dondo
Cuanza
Malanje
Kasama
Gabela
ANGOLA
Mansa
Samfya
Sumbe
Uaco Cungo
Camacupa
Luena
Zambezi
Solwezi
Chililabombwe
Lobito
Benguela
Môco 2610m
Cuito
Planalto
do Bié
Lungué-Bungo
Kabompo
Chingola
Mufulira
Kitwe
Ndola
Luanshya
Serenje
45
Cubal
Caála
Huambo
Caconda
Cubango
Zambezi
ZAMBIA
Lubango
Menongue
Cuando
Kaoma
Kafue
Kabwe
Namibe
Mongu
Nambala
LUSAKA
Albufeira
Cahora
Tombua
Huíla
Plateau
Cubango
Mazabuka
Monze
Choma
Kafue
Zambezi
Kariba
Cuito
N'Giva
Victoria
Falls
Lake Kariba
Cunene
Katima Mulilo
Livingstone
Olifa
Oshikango
Rundu
Caprivi Strip
Chitungwiza
Victoria Falls
Hwange
Kadoma
Kwekwe
Etosha
Pan
Okavango
Okavango
Delta
Tsumeb
Otavi
Grootfontein
Maun
ZIMBABWE
Nata
Bulawayo
Boteti
Otjiwarongo
Namib Desert
Brandberg
2573m
NAMIBIA
Francistown
Gwanda
Zvishavane
Shashe
Ghanzi
BOTSWANA
Gobabis
Mamuno
Serowe
Palapye
Musina
(Messina)
Wlotzkasbaken
Karibib
WINDHOEK
Swakopmund
Walvis Bay
Rehoboth
Mahalapye
Limpopo
Kalahari
Polokwane
(Pietersburg)
Tropic of Capricorn
Fish
Mariental
GABORONE
Mochudi
Modimolle
(Nylstroom)
Jwaneng
Nosop
Auob
Werda
Kanye
Lobatse
PRETORIA
Desert
Mmabatho
MAPUTO
Johannesburg
Soweto
45
Molopo
MBABANE
LOBAMBA
ESWATINI
Keetmanshoop
Lüderitz
Aus
Groot Karasberge
Klerksdorp
Vaal
Klein Karas
Karasburg
SOUTH
Kroonstad
Dundee
Welkom
Bethlehem
Orange River
Upington
Kimberley
LESOTHO
Oranjemund
BLOEMFONTEIN
MASERU
Drakensberg
Springbok
Prieska
Pietermaritzburg
Durban
AFRICA
De Aar
Colesberg
Kokstad
Drakensberg
Calvina
Mthatha
St Helena Bay
Beaufort West
Cradock
Queenstown
Mdantsane
Great Karoo
East London
Kariega
Bellville
Worcester
George
Port Alfred
CAPE TOWN
Mosselbaai
Gqeberha (Port Elizabeth)
132
Cape of
Good Hope
Cape Agulhas
ATLANTIC OCEAN
10°
20°
30°
SOUTH AFRICA'S THREE CAPITALS
PRETORIA – administrative capital
CAPE TOWN – legislative capital
BLOEMFONTEIN – judicial capital

0 km 400
0 miles 400

Population ● National capital

○ below 50,000 ○ 50,000 to 100,000 ◉ 100,000 to 500,000 ▣ above 500,000

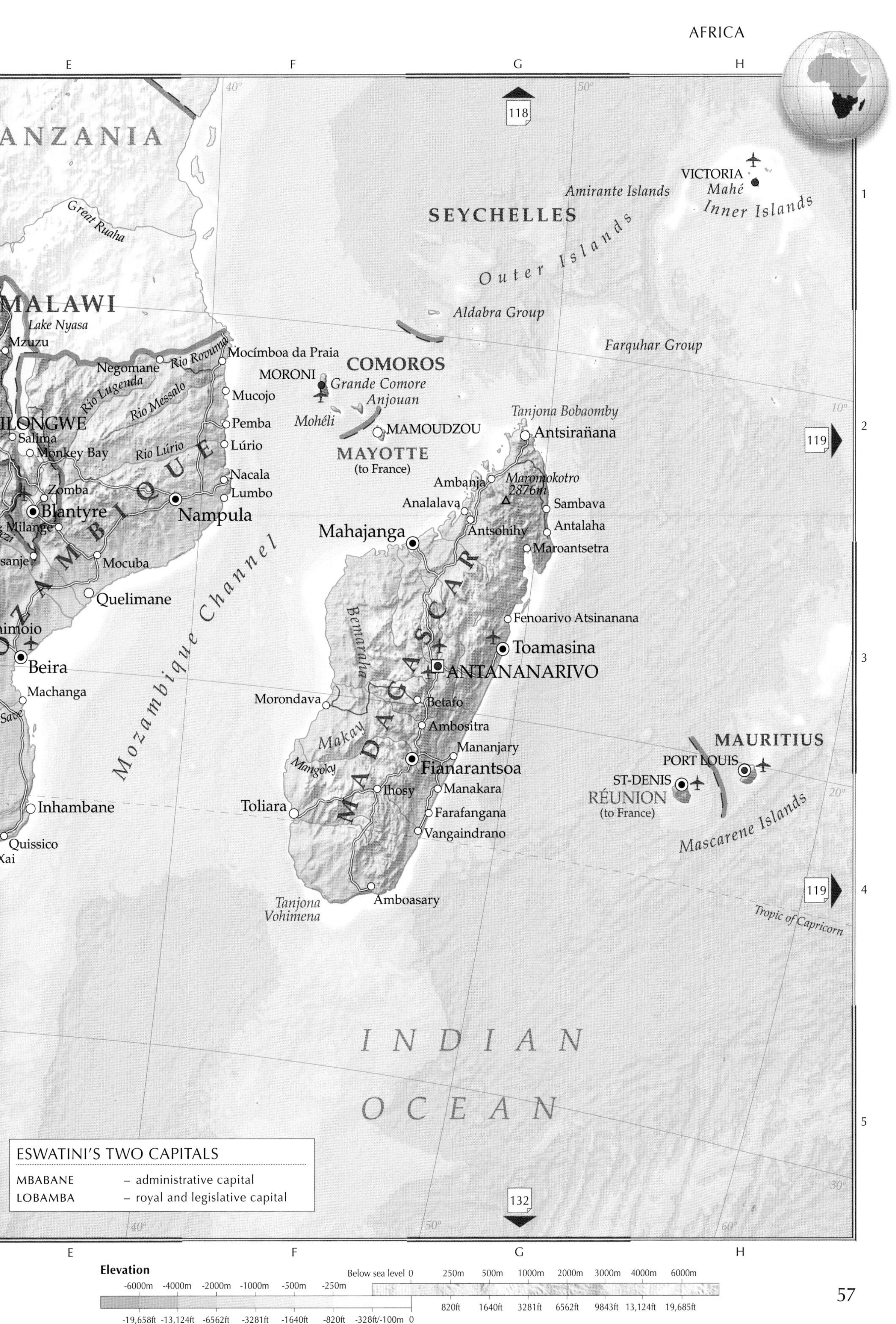

ANZANIA
Great Ruaha
MALAWI
Lake Nyasa
Mzuzu
Negomane
Rio Rovuma
Mocímboa da Praia
Rio Lugenda
Rio Messalo
Mucojo
ILONGWE
Pemba
Salima
Monkey Bay
Rio Lúrio
Lúrio
Nacala
Lumbo
Zomba
Blantyre
Nampula
Milange
Mocuba
Quelimane
Beira
Machanga
Inhambane
Quissico
MOZAMBIQUE
Mozambique Channel
SEYCHELLES
Amirante Islands
VICTORIA
Mahé
Inner Islands
Outer Islands
Aldabra Group
Farquhar Group
COMOROS
MORONI
Grande Comore
Anjouan
Mohéli
MAMOUDZOU
MAYOTTE
(to France)
Tanjona Bobaomby
Antsirañana
Ambanja
Maromokotro 2876m
Analalava
Sambava
Antalaha
Antsohihy
Maroantsetra
Mahajanga
MADAGASCAR
Bemaraha
Fenoarivo Atsinanana
Toamasina
ANTANANARIVO
Morondava
Betafo
Ambositra
Makay
Mananjary
Mangoky
Fianarantsoa
Ihosy
Manakara
Toliara
Farafangana
Vangaindrano
Tanjona Vohimena
Amboasary
MAURITIUS
PORT LOUIS
ST-DENIS
RÉUNION
(to France)
Mascarene Islands
Tropic of Capricorn
INDIAN OCEAN
118
119
119
132

ESWATINI'S TWO CAPITALS

MBABANE	– administrative capital
LOBAMBA	– royal and legislative capital

Elevation

Below sea level	-6000m	-4000m	-2000m	-1000m	-500m	-250m	0
	-19,658ft	-13,124ft	-6562ft	-3281ft	-1640ft	-820ft	-328ft/-100m 0

250m	500m	1000m	2000m	3000m	4000m	6000m
820ft	1640ft	3281ft	6562ft	9843ft	13,124ft	19,685ft

Europe

A
B
C
D
1
2
3
4
5
133
44
44
46
Political features
Total area:
4,809,200 sq miles
(12,456,000 sq km)
Total number of countries:
44
Total population:
723 million
Largest city with population:
Moscow, European Russia 17.4 million
Country with highest population density:
Monaco 50,145 people per sq mile
(19,497 people per sq km)
Largest country:
European Russia 1,527,341 sq miles
(3,955,818 sq km)
Smallest country:
Vatican City, Italy 0.17 sq miles
(0.44 sq km)
Physical features
Largest lake:
Lake Lagoda, European Russia
6800 sq miles (17,700 sq km)
Longest river:
Volga, European Russia
2194 miles (3531 km)
Highest point:
El'brus, Caucasus, European Russia
18,510ft (5642 m)
Lowest point:
Volga Delta, Caspian Sea, European Russia -92ft (-28m) below sea level
Limit of winter pack ice
Arctic Circle
Reykjanes Ridge
REYKJAVÍK
ICELAND
Vatnajökull
Faroe-Iceland Ridge
Norwegian Basin
Norwegian Sea
Iceland Basin
FAROE ISLANDS
(to Denmark)
Hatton Ridge
Faroe-Shetland Trough
Shetland Islands
Trondheim
Rockall Bank
Rockall Trough
Outer Hebrides
Orkney Islands
Bergen
OSLO
Stavanger
British Isles
Glasgow
Edinburgh
Ireland
Belfast
IRELAND
DUBLIN
Isle of Man
UNITED KINGDOM
North Sea
Gothenburg
Aalborg
Jutland
DENMARK
COPENHA
Odense
Liverpool
Manchester
Britain
Porcupine Plain
Celtic Sea
Birmingham
Cardiff
LONDON
NETHERLANDS
THE HAGUE
AMSTERDAM
Rotterdam
Hamburg
Elbe
Hanover
BERLIN
Celtic Shelf
English Channel
Channel Islands
BELGIUM
BRUSSELS
le Havre
Liège
Düsseldorf
Bonn
GERMANY
ATLANTIC
OCEAN
Charcot Seamounts
Biscay Plain
Azores-Biscay Rise
Rennes
Seine
LUXEMBOURG
PARIS
LUXEMBOURG
Rhine
Frankfurt am Main
Stuttgart
Nantes
Loire
Orléans
Strasbourg
CZECHIA
(CZECH REPUB
Munich
Iberian Plain
Galicia Bank
A Coruña
Bay of Biscay
FRANCE
Zurich
BERN
LIECH.
VIENNA
Salzburg
Bordeaux
Lyon
SWITZERLAND
Innsbruck
AUSTRIA
Cordillera Cantábrica
Bilbao
Garonne
Massif Central
Mont Blanc
15,774ft
(4808m)
Porto
Milan
LJUBLJANA
Venice
Trieste
Duero
Pyrenees
Toulouse
Rhône
Turin
PORTUGAL
ANDORRA
Nice
Bologna
Iberian
Zaragoza
Marseille
MONACO
SAN MARINO
Pisa
Tagus Plain
Tagus
MADRID
Ebro
Horseshoe Seamounts
LISBON
SPAIN
Peninsula
Barcelona
Corsica
Apennines
Adriatic Sea
Madeira
(to Portugal)
Guadalquivir
Valencia
VATICAN CITY
ROME
Seville
Strait of Gibraltar
Sardinia
Palma
Balearic Islands
Naples
Bari
Málaga
Algerian Basin
Tyrrhenian Sea
GIBRALTAR
(to UK)
Ceuta
(to Spain)
Cagliari
Mediterranean Sea
Cosenza
Melilla
(to Spain)
Palermo
Mount Etna
10,922ft
(3329m)
Catania
Sicily
Canary Islands
(to Spain)
Atlas Mountains
AFRICA
MALTA
VALLETTA
0 km
500
0 miles
500
Population
National capital
below 50,000
50,000 to 100,000
100,000 to 500,000
above 500,000

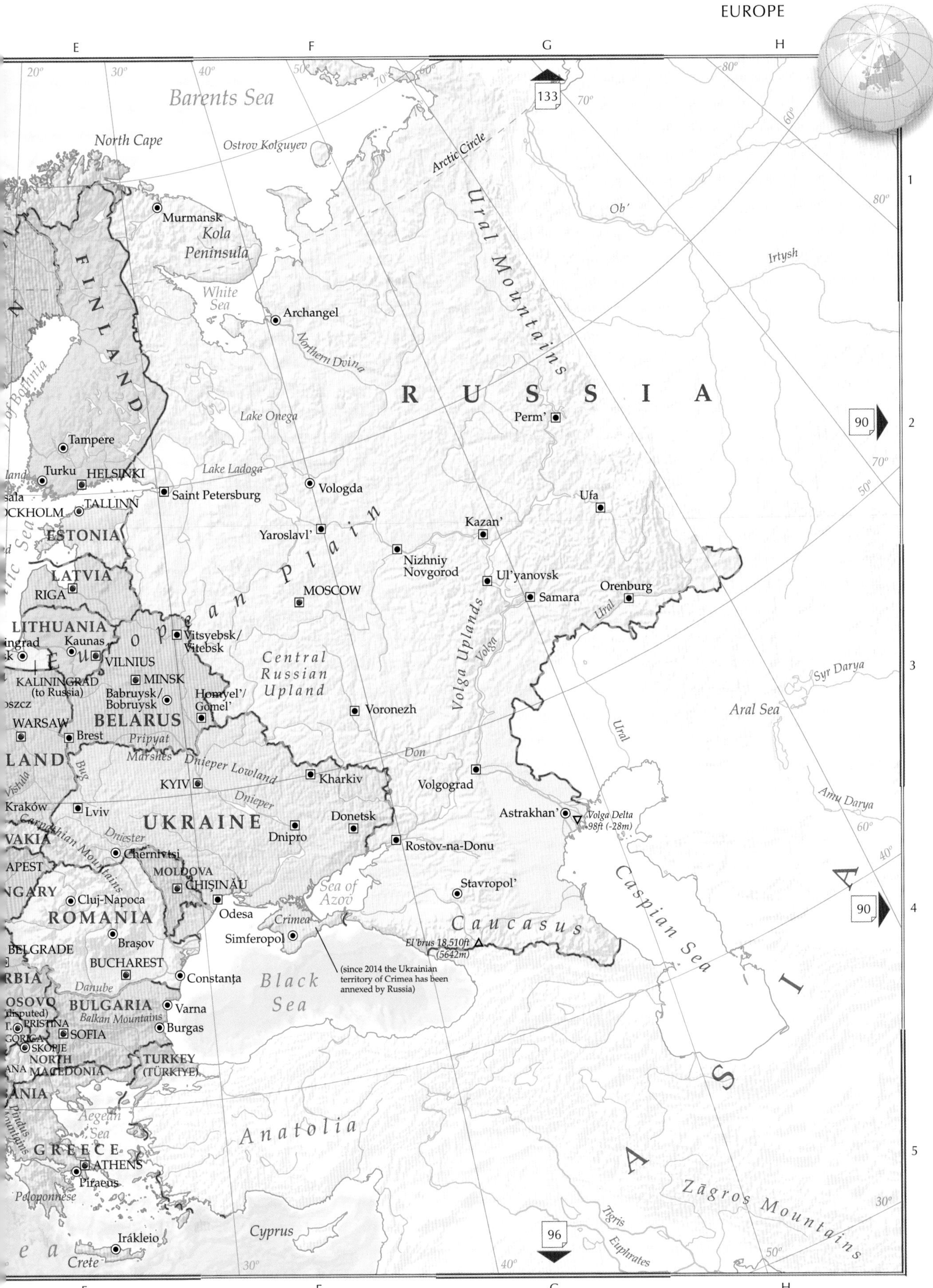
Barents Sea
North Cape
Ostrov Kolguyev
Arctic Circle
Murmansk
Kola Peninsula
White Sea
Archangel
Northern Dvina
Ural Mountains
Ob'
Irtysh
FINLAND
RUSSIA
Lake Onega
Perm'
Tampere
Turku
HELSINKI
Lake Ladoga
Saint Petersburg
Vologda
TALLINN
ESTONIA
Ufa
Yaroslavl'
Kazan'
Nizhniy Novgorod
Ul'yanovsk
LATVIA
RIGA
MOSCOW
Samara
Orenburg
Ural
LITHUANIA
Kaunas
VILNIUS
Vitsyebsk/ Vitebsk
Central Russian Upland
Volga Uplands
Volga
MINSK
KALININGRAD (to Russia)
Babruysk/ Bobruysk
Homyel'/ Gomel'
Voronezh
WARSAW
BELARUS
Brest
Pripyat Marshes
Dnieper Lowland
Don
Syr Darya
Aral Sea
Ural
Kharkiv
KYIV
Volgograd
Dnieper
Kraków
Lviv
UKRAINE
Donetsk
Astrakhan'
Volga Delta 98ft (-28m)
Amu Darya
Dniester
Dnipro
Carpathian Mountains
Chernivtsi
Rostov-na-Donu
MOLDOVA
CHIȘINĂU
Sea of Azov
Stavropol'
Cluj-Napoca
ROMANIA
Odesa
Crimea
Caucasus
Caspian Sea
Brașov
Simferopol
El'brus 18,510ft (5642m)
BELGRADE
BUCHAREST
Constanța
Black Sea
(since 2014 the Ukrainian territory of Crimea has been annexed by Russia)
Danube
BULGARIA
Varna
Balkan Mountains
Burgas
PRISTINA
SOFIA
SKOPJE
NORTH MACEDONIA
TURKEY (TÜRKIYE)
Aegean Sea
Anatolia
GREECE
ATHENS
Piraeus
Peloponnese
ASIA
Zāgros Mountains
Tigris
Euphrates
Cyprus
Irákleio
Crete
133
90
90
96

The North Atlantic

A
B
C
D
1
2
3
4
5
16
17
44
Arctic Circle
Gulf of Boothia
Devon Island
Ellesmere Island
Nares Strait
NUNAVUT
Hudson Bay
Southampton Island
Foxe Basin
CANADA
Baffin Island
Baffin Bay
Qaanaaq
Knud Rasmussen
Innaanganeq
Savissivik
Qimusseriarsuaq
Kullorsuaq
Upernavik
Péninsule d'Ungava
QUÉBEC
Hudson Strait
Limit of summer pack ice
Cumberland Sound
Frobisher Bay
Uummannaq
Qeqertarsuaq
Qeqertarsuaq
Qeqertarsuup Tunua
Davis Strait
Qasigiannguit
Arnaud
Sisimiut
Kong Frederik IX Land
GREENLAND
(to Denmark)
Ungava Bay
Maniitsoq
George
NUUK
Kong Christian IX Land
Gunnbjørn
Mont Forel 3360m
Tasiilaq
Paamiut
Ivittuut
Kong Frederik VI Kyst
NEWFOUNDLAND & LABRADOR
Labrador Sea
Qaqortoq
Nanortalik
Nunap Isua (Kap Farvel)
Limit of winter pack ice
Reykjanes Basin
ATLANTIC OCEAN
0 km
400
0 miles
400
Population
National capital
below 50,000
50,000 to 100,000
100,000 to 500,000
above 500,000

ARCTIC OCEAN
Lincoln Sea
Kap Morris Jesup
Wandel Sea
Independence Fjord
Nord
SVALBARD
(to Norway)
Kvitøya
Nordaustlandet
Kong Karls Land
Barentsøya
Spitsbergen
Edgeøya
LONGYEARBYEN
Barentsburg
Storfjorden
Limit of winter pack ice
Zemlya Frantsa-Iosifa
Novaya Zemlya
Barents Sea
Kong Frederik VIII Land
Greenland Sea
Christian X Land
Limit of summer pack ice
Daneborg
Petermann Bjerg 2940m
Bjørnøya
(to Norway)
Nordkapp
(North Cape)
FINLAND
Mohns Ridge
Kong Oscar Fjord
Ittoqqortoormiit
Kangertittivaq
Kangikajik
JAN MAYEN
(to Norway)
Norwegian Sea
Norwegian Basin
Arctic Circle
Vestfjorden
ICELAND
Siglufjörður
Raufarhöfn
Húsavík
Akureyri
Seyðisfjörður
Stykkishólmur
Neskaupstaður
REYKJAVÍK
Selfoss
Vatnajökull
Djúpivogur
Hvannadalshnúkur 2119m
Vestmannaeyjar
SWEDEN
Gulf of Bothnia
FAROE ISLANDS
(to Denmark)
TÓRSHAVN
NORWAY
Shetland Islands
N
Elevation
-6000m -4000m -2000m -1000m -500m -250m
-19,658ft -13,124ft -6562ft -3281ft -1640ft -820ft -328ft/-100m 0
Below sea level 0
250m 500m 1000m 2000m 3000m 4000m 6000m
820ft 1640ft 3281ft 6562ft 9843ft 13,124ft 19,685ft

133 88 62 63

Scandinavia & Finland

ARCTIC OCEAN
Barents Sea
Norwegian Sea
RUSSIA
FINLAND
Lapland
Nordkapp
(North Cape)
Magerøya
Porsangenfjordan
Varangerhalvøya
Varangerfjorden
Kirkenes
Tana Bru
Deatnu
Lakselv
Alta
Talvik
Sørøya
Finnmarksvidda
Válljohka
Karigasniemi (Gáregasnjárga)
Inarijärvi
Kaamanen
Ivalo
(Avvil, Avveel)
Saariselkä
(Suoločielgi)
Sodankylä
Kitinen
Kemijärvi
Rovaniemi
Kuusamo
Pudasjärvi
Livojoki
Suomussalmi
Kuhmo
Sotkamo
Kajaani (Kajana)
Oulujärvi
Oulujoki
Oulu (Uleåborg)
Kempele
Raahe
Hailuoto
Kokkola
Kemi
Tornio
Haparanda
Kalix
Luleå
Piteå
Skellefteå
Boden
Muonio
Kolari
Tornionjoki
Muonionjoki
Ounasjoki
Kemijoki
Kivalo
Sattanen
Kaaresuvanto
(Gárasavvon)
Visttasjohka
Kiruna
Torneträsk
Gällivare
Malmberget
Luleälven
Jokkmokk
Skalka
Kebnekaise
2117m
Arvidsjaur
Skellefteälven
Umeälven
Lycksele
Storuman
Vilhelmina
Dorotea
Tromsø
Kvaløya
Ringvassøya
Senja
Narvik
Harstad
Andøya
Hinnøya
Vesterålen
Lofoten
Vestfjorden
Fauske
Svartisen
Bodø
Mo i Rana
Mosjøen
Vega
Namsos
Steinkjer
Børgefjell
Kvarnbergsvattnet
Frøya
Arctic Circle

0 km 200
0 miles 200

Population
● National capital
○ below 50,000
○ 50,000 to 100,000
◉ 100,000 to 500,000
▣ above 500,000

NORWAY
SWEDEN
RUSSIA
ESTONIA
LATVIA
LITHUANIA
BELARUS
POLAND
GERMANY
DENMARK
KALININGRAD
(to Russia)
North Sea
Baltic Sea
Skagerrak
Kattegat
Gulf of Finland
Gulf of Riga
Gulf of Gdańsk
Courland Lagoon
Hanöbukten
Ålands Hav
Ladozhskoye Ozero
Lake Peipus
OSLO
STOCKHOLM
HELSINKI/HELSINGFORS
KØBENHAVN
(Copenhagen)
Bergen
Stavanger
Kristiansand
Drammen
Sandvika
Lillestrøm
Göteborg
(Gothenburg)
Malmö
Helsingborg
Uppsala
Västerås
Örebro
Norrköping
Linköping
Jönköping
Borås
Gävle
Sundsvall
Turku
(Åbo)
Tampere (Tammerfors)
Espoo
(Esbo)
Vantaa
(Vanda)
Lahti (Lahtis)
Jyväskylä
Kotka
Lappeenranta
(Villmanstrand)
Pori
(Björneborg)
Aarhus
Aalborg
Odense
Esbjerg
Randers
Kolding
Gotland
Öland
Åland
Bornholm
Sjælland
Fyn
Jylland
Lolland
Falster
Møn
Læsø
Saaremaa
Hiiumaa
Visby
Kalmar
Karlskrona
Kristianstad
Växjö
Halmstad
Varberg
Karlstad
Falun
Borlänge
Mora
Vänern
Vättern
Mälaren
Hjälmaren
Klarälven
Glåma
Mjøsa
Setesdal
Jotunheimen
Hardangervidda
Glittertind 2472m
Yding Skovhøj 173m
Neman
Western Dvina
Wisła
Oder
Elbe
Weser
Ems
Elevation
-6000m -4000m -2000m -1000m -500m -250m
-19,658ft -13,124ft -6562ft -3281ft -1640ft -820ft -328ft/-100m 0
Below sea level 0
250m 500m 1000m 2000m 3000m 4000m 6000m
820ft 1640ft 3281ft 6562ft 9843ft 13,124ft 19,685ft

89
76
72
67

The Low Countries

THE NETHERLAND'S TWO CAPITALS
AMSTERDAM – Capital
THE HAGUE – Seat of Government
NETHERLANDS
North Sea
Waddenzee
Wadden eilanden
IJsselmeer
Texel
Vlieland
Terschelling
Ameland
Schiermonnikoog
AMSTERDAM
'S-GRAVENHAGE (DEN HAAG) (THE HAGUE)
Rotterdam
Utrecht
Groningen
Leeuwarden
Zwolle
Arnhem
Nijmegen
Enschede
Apeldoorn
Amersfoort
Haarlem
Leiden
Dordrecht
Den Helder
Alkmaar
Lelystad
Almere
Deventer
Hengelo
Assen
Emmen
Delft
Gouda
Zoetermeer
Hilversum
's-Hertogenbosch
Population
National capital
below 50,000
50,000 to 100,000
100,000 to 500,000
above 500,000
0 km 50
0 miles 50

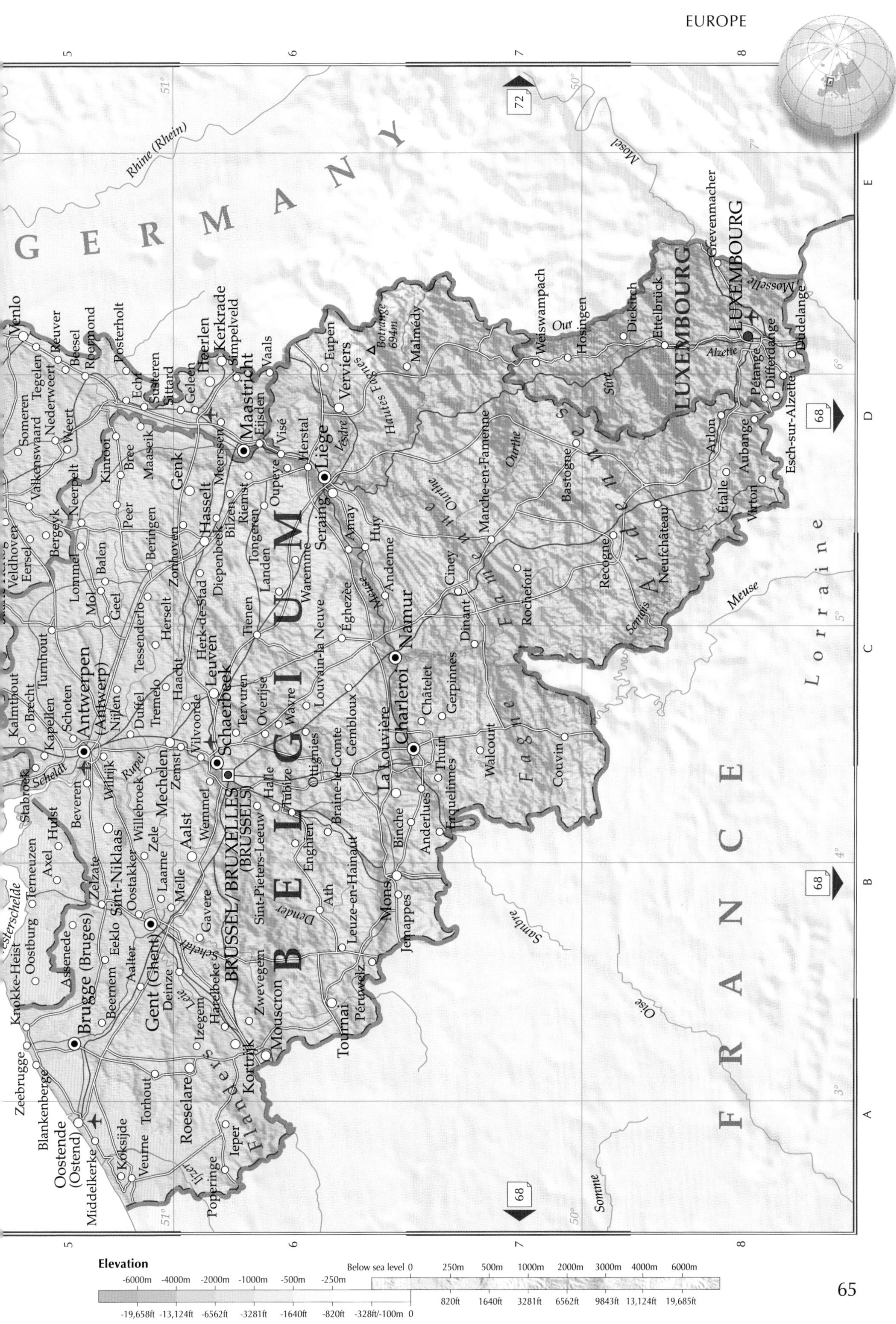
G E R M A N Y
B E L G I U M
LUXEMBOURG
F R A N C E
Lorraine
Flanders
Ardennes
Famenne
Fagne
Hautes Fagnes
Botrange 694m
Rhine (Rhein)
Mosel
Moselle
Meuse
Maas
Sambre
Oise
Somme
Schelde
Scheldt
Westerschelde
Dender
Leie
IJzer
Rupel
Vesdre
Ourthe
Semois
Sûre
Our
Alzette
Venlo
Tegelen
Reuver
Beesel
Roermond
Posterholt
Echt
Susteren
Sittard
Geleen
Heerlen
Kerkrade
Simpelveld
Vaals
Maastricht
Eijsden
Someren
Valkenswaard
Nederweert
Weert
Veldhoven
Eersel
Bergeijk
Neerpelt
Kinrooi
Bree
Maaseik
Genk
Hasselt
Meerssen
Bilzen
Riemst
Tongeren
Diepenbeek
Peer
Beringen
Zonhoven
Lommel
Balen
Mol
Geel
Tessenderlo
Herselt
Herk-de-Stad
Tienen
Landen
Waremme
Oupeye
Visé
Herstal
Liège
Seraing
Eupen
Verviers
Malmédy
Weiswampach
Hosingen
Diekirch
Ettelbrück
Grevenmacher
LUXEMBOURG
Pétange
Differdange
Dudelange
Esch-sur-Alzette
Aubange
Arlon
Étalle
Virton
Bastogne
Neufchâteau
Recogne
Marche-en-Famenne
Rochefort
Ciney
Dinant
Amay
Huy
Andenne
Éghezée
Namur
Louvain-la-Neuve
Wavre
Gembloux
Ottignies
Charleroi
Châtelet
Gerpinnes
Walcourt
Couvin
Thuin
Erquelinnes
Anderlues
Binche
La Louvière
Braine-le-Comte
Tubize
Halle
Enghien
Ath
Leuze-en-Hainaut
Mons
Jemappes
Péruwelz
Tournai
Mouscron
Kortrijk
Zwevegem
Harelbeke
Izegem
Roeselare
Torhout
Ieper
Poperinge
Veurne
Koksijde
Middelkerke
Oostende (Ostend)
Blankenberge
Zeebrugge
Brugge (Bruges)
Knokke-Heist
Oostburg
Assenede
Beernem
Eeklo
Aalter
Gent (Ghent)
Deinze
Zelzate
Terneuzen
Axel
Hulst
Sint-Niklaas
Oostakker
Laarne
Melle
Gavere
Zele
Willebroek
Beveren
Stabroek
Aalst
Wemmel
Sint-Pieters-Leeuw
BRUSSEL/BRUXELLES (BRUSSELS)
Schaerbeek
Vilvoorde
Tervuren
Overijse
Leuven
Haacht
Tremelo
Duffel
Nijlen
Mechelen
Zemst
Wilrijk
Antwerpen (Antwerp)
Kapellen
Kalmthout
Brecht
Schoten
Turnhout
68
72
5
6
7
8
A
B
C
D
E
3°
4°
5°
6°
7°
50°
51°

Elevation

Below sea level 0	250m	500m	1000m	2000m	3000m	4000m	6000m
	820ft	1640ft	3281ft	6562ft	9843ft	13,124ft	19,685ft

-6000m	-4000m	-2000m	-1000m	-500m	-250m	
-19,658ft	-13,124ft	-6562ft	-3281ft	-1640ft	-820ft	-328ft/-100m 0

The British Isles

A
B
C
D
E
1
2
3
4
60
60
44
62
ATLANTIC OCEAN
North Sea
Shetland Islands
Unst
Yell
Fetlar
Mainland
Lerwick
Fair Isle
Orkney Islands
Sanday
Kirkwall
Mainland
Hoy
Thurso
John o'Groats
Ben Hope 927m
Isle of Lewis
Stornoway
The Minch
Ullapool
North West Highlands
Outer Hebrides
St Kilda
Harris
North Uist
South Uist
Barra
The Little Minch
Isle of Skye
Stromeferry
Inverness
Loch Ness
Moray Firth
Elgin
Fraserburgh
Peterhead
Spey
Grampian Mountains
Aberdeen
Dee
Aviemore
Mallaig
SCOTLAND
Rhum
Eigg
Fort William
Ben Nevis 1343 m
Coll
Tiree
Inner Hebrides
Isle of Mull
Oban
Firth of Lorn
Jura
Islay
Kintyre
Isle of Arran
Loch Lomond
Forth
Stirling
Dunfermline
Firth of Forth
Forfar
Tay
Montrose
Arbroath
Dundee
Perth
St Andrews
Greenock
Glasgow
Edinburgh
Paisley
East Kilbride
Hamilton
Clyde
Kilmarnock
Prestwick
Ayr
Galashiels
Hawick
Berwick-upon-Tweed
Newcastle upon Tyne

0 km 100
0 miles 100

Population ● National capital ● Internal administrative capital
○ below 50,000 ○ 50,000 to 100,000 ◉ 100,000 to 500,000 ▣ above 500,000

UNITED KINGDOM
IRELAND
ENGLAND
WALES
FRANCE
LONDON
DUBLIN
DOUGLAS
ISLE OF MAN
(British Crown Dependency)
Irish Sea
Celtic Sea
English Channel
St George's Channel
Bristol Channel
Cardigan Bay
Channel Islands
GUERNSEY
(British Crown Dependency)
JERSEY
(British Crown Dependency)
ST PETER PORT
ST HELIER
Alderney
Sark
Isle of Wight
Isles of Scilly
Land's End
Anglesey
Connaught
Munster
Leinster
Belfast
Cork
Limerick
Galway
Galway Bay
Sligo
Dundalk
Drogheda
Dún Laoghaire
Wexford
Waterford
Kilkenny
Tralee
Killarney
Bantry Bay
Dingle Bay
Liverpool
Manchester
Birmingham
Leeds
Bradford
Sheffield
Nottingham
Leicester
Coventry
Bristol
Cardiff
Swansea
Newport
Plymouth
Southampton
Portsmouth
Brighton
Bournemouth
Norwich
Ipswich
Cambridge
Oxford
Kingston upon Hull
York
Peterborough
Reading
Swindon
Exeter
Torquay
Penzance
The Wash
Channel Tunnel
Seine
Elevation
Below sea level
250m 500m 1000m 2000m 3000m 4000m 6000m
820ft 1640ft 3281ft 6562ft 9843ft 13,124ft 19,685ft
-6000m -4000m -2000m -1000m -500m -250m
-19,658ft -13,124ft -6562ft -3281ft -1640ft -820ft -328ft/-100m
LONDON
Places of interest
Regions/suburbs

France, Andorra & Monaco

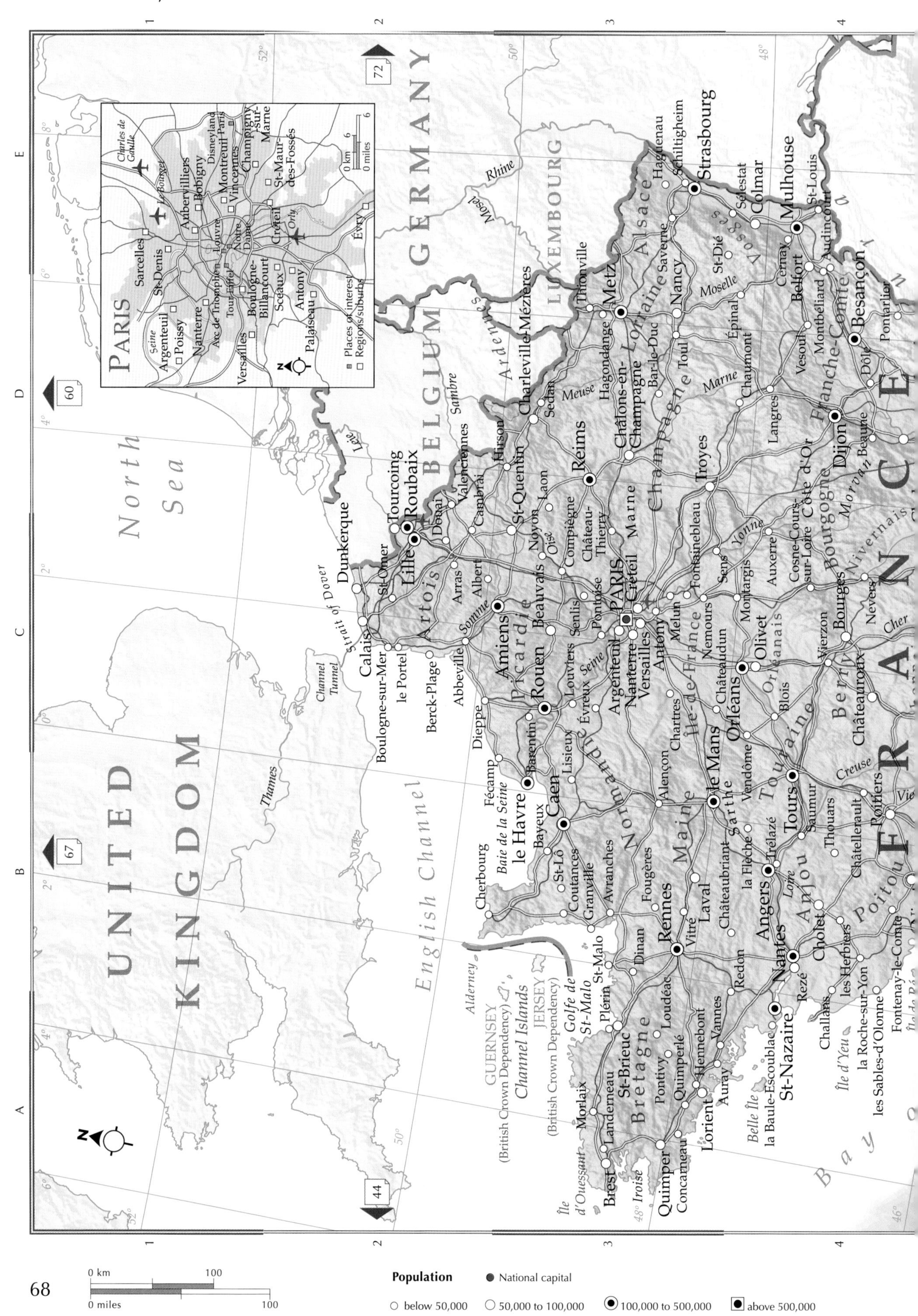

UNITED KINGDOM
GERMANY
BELGIUM
LUXEMBOURG
FRANCE
North Sea
English Channel
Strait of Dover
Channel Tunnel
Channel Islands
GUERNSEY (British Crown Dependency)
JERSEY (British Crown Dependency)
Alderney
Golfe de St-Malo
Baie de la Seine
Île d'Ouessant
Iroise
Belle Île
Île d'Yeu
Thames
Rhine
Mosel
Moselle
Meuse
Sambre
Marne
Seine
Oise
Somme
Yonne
Loire
Cher
Creuse
Ardennes
Vosges
Artois
Picardie
Normandie
Bretagne
Maine
Anjou
Poitou
Touraine
Orléanais
Berry
Île-de-France
Champagne
Lorraine
Alsace
Franche-Comté
Bourgogne
Nivernais
Morvan
Côte d'Or
PARIS
Sarcelles
St-Denis
Argenteuil
Poissy
Nanterre
Arc de Triomphe
Tour Eiffel
Versailles
Boulogne-Billancourt
Sceaux
Antony
Palaiseau
Aubervilliers
Bobigny
Louvre
Notre Dame
Montreuil
Vincennes
Disneyland Paris
Champigny-sur-Marne
St-Maur-des-Fossés
Créteil
Orly
Évry
Charles de Gaulle
Le Bourget
Places of interest
Regions/suburbs
0 km 6
0 miles 6
Dunkerque
Calais
Boulogne-sur-Mer
le Portel
Berck-Plage
Abbeville
St-Omer
Lille
Roubaix
Tourcoing
Douai
Valenciennes
Cambrai
Arras
Albert
Amiens
Hirson
St-Quentin
Laon
Noyon
Compiègne
Beauvais
Senlis
Pontoise
Charleville-Mézières
Sedan
Reims
Château-Thierry
Châlons-en-Champagne
Thionville
Metz
Hagondange
Bar-le-Duc
Toul
Nancy
Saverne
Haguenau
Schiltigheim
Strasbourg
Sélestat
Colmar
St-Dié
Épinal
Mulhouse
St-Louis
Cernay
Belfort
Audincourt
Montbéliard
Besançon
Pontarlier
Vesoul
Dôle
Dijon
Beaune
Langres
Chaumont
Troyes
Sens
Auxerre
Cosne-Cours-sur-Loire
Nevers
Bourges
Vierzon
Châteauroux
Fontainebleau
Melun
Nemours
Montargis
Orléans
Olivet
Blois
Châteaudun
Chartres
Évreux
Louviers
Rouen
Dieppe
Fécamp
Barentin
le Havre
Lisieux
Caen
Bayeux
Cherbourg
St-Lô
Coutances
Granville
Avranches
Alençon
le Mans
Vendôme
Tours
Saumur
Thouars
Châtellerault
Poitiers
la Flèche
Trélazé
Angers
Cholet
Laval
Fougères
Vitré
Rennes
Châteaubriant
Redon
St-Malo
Dinan
Loudéac
Plérin
St-Brieuc
Morlaix
Landerneau
Brest
Quimper
Concarneau
Quimperlé
Pontivy
Hennebont
Lorient
Auray
Vannes
la Baule-Escoublac
St-Nazaire
Nantes
Rezé
Challans
les Herbiers
la Roche-sur-Yon
les Sables-d'Olonne
Fontenay-le-Comte

Population
● National capital
○ below 50,000
○ 50,000 to 100,000
◉ 100,000 to 500,000
▣ above 500,000

0 km 100
0 miles 100

ITALY
MONACO
SPAIN
ANDORRA
ANDORRA LA VELLA
Mediterranean Sea
Ligurian Sea
Golfe du Lion
Côte d'Azur
Corse (Corsica)
Sardinia (to Italy)
Strait of Bonifacio
Bastia
Ajaccio
Monte Cinto 2706m
Monte Incudine 2136m
Sartène
Bonifacio
Massif Central
Auvergne
Limousin
Aquitaine
Gascogne
Landes
Armagnac
Languedoc
Roussillon
Dauphiné
Savoie
Provence
Camargue
Pyrenees
Mont Blanc 4808m
Little St-Bernard Pass 2188m
Col du Mont Cenis 2083m
Col de Montgenèvre 1850m
Lyon
Villeurbanne
St-Étienne
St-Chamond
Clermont-Ferrand
Grenoble
Chambéry
Annecy
Valence
Avignon
Nîmes
Arles
Montpellier
Marseille
Aix-en-Provence
Toulon
Nice
Cannes
Antibes
Perpignan
Toulouse
Bordeaux
Angoulême
Périgueux
Brive-la-Gaillarde
Montauban
Pau
Tarbes
Lourdes
Bayonne
Biarritz
Anglet
Dax
Carcassonne
Béziers
Narbonne
Sète
Îles d'Hyères
Rhône
Garonne
Dordogne
Lot
Tarn
Isère
Durance
Ebro
Po
MONACO
Monte-Carlo
La Condamine
Fontvieille
Larvotto
Monte-Carlo Sporting Club d'Été
Port de Monaco
Grand Prix Circuit
Musée Océanographique
Mediterranean Sea
ANDORRA
El Serrat
Arinsal
Ordino
La Massana
Canillo
Soldeu
Encamp
Escaldes
Sant Julià de Lòria
Port d'Envalira
Pic de Coma Pedrosa 2942m
Valira
FRANCE
SPAIN
70
71
74
80
Elevation
Below sea level 0
250m 500m 1000m 2000m 3000m 4000m 6000m
820ft 1640ft 3281ft 6562ft 9843ft 13,124ft 19,685ft
-6000m -4000m -2000m -1000m -500m -250m
-19,658ft -13,124ft -6562ft -3281ft -1640ft -820ft -328ft/-100m 0

Spain & Portugal

A
B
C
D
1
2
3
4
5
12°
10°
8°
6°
4°
44°
42°
40°
38°
36°
44
48
Bay of Biscay
Costa Verde
ATLANTIC
OCEAN
Baía de Setúbal
Golfo de Cádiz
Costa de la Luz
Strait of Gibraltar
Costa del Sol
PORTUGAL
MOROCCO
Galicia
Asturias
Cantabria
Castilla y León
Extremadura
Andalucía
Algarve
Cordillera Cantábrica
Sistema Central
Sierra de Gredos
Serra da Estrela
Sierra Morena
Serra d'Ossa
Alto da Torre 1993m
Miño
Minho
Douro
Duero
Tagus
Guadiana
Guadalquivir
Embalse de Ricobayo
Embalse de Almendra
Embalse de Alcántara
Embalse de Valdecañas
Barragem do Alqueva
Cabo Fisterra
Cabo de São Vicente
A Coruña
Ferrol
Laracha
Betanzos
Vilalba
Luarca
Avilés
Gijón/Xixón
Villaviciosa
Santander
Llanes
Pravia
Tineo
Oviedo/Uviéu
Mieres del Camín
Torrelavega
La Pola
Cabanaquinta
Reinosa
Santa Cataliña de Armada
Lugo
Outes
Muros
Santiago de Compostela
Chantada
Lalín
Monforte de Lemos
Santa Uxía de Ribeira
Pontevedra
O Carballiño
Ponferrada
León
Astorga
Marín
Vigo
Ourense
Ponteareas
Xinzo de Limia
Burgos
Benavente
Palencia
Ponte da Barca
Bragança
Viana do Castelo
Chaves
Braga
Guimarães
Póvoa de Varzim
Vila do Conde
Vila Real
Valladolid
Zamora
Toro
Matosinhos
Porto (Oporto)
Vila Nova de Gaia
Lamego
Medina del Campo
Ovar
São João da Madeira
Albergaria-a-Velha
Salamanca
Segovia
Aveiro
Ílhavo
Viseu
Guarda
Ciudad-Rodrigo
Ávila
Béjar
Coimbra
Figueira da Foz
Covilhã
Plasencia
Talavera de la Reina
Leiria
Castelo Branco
Coria
Toledo
Tomar
Entroncamento
Abrantes
Cáceres
Peniche
Caldas da Rainha
Trujillo
Santarém
Portalegre
Torres Vedras
Herrera del Duque
Coruche
Sintra
Estremoz
Elvas
Mérida
Villanueva de la Serena
Don Benito
Ciudad Real
Cascais
LISBOA (LISBON)
Badajoz
Almada
Barreiro
Castuera
Puertollano
Setúbal
Évora
Almendralejo
Villafranca de los Barros
Zafra
Alcácer do Sal
Pozoblanco
Jerez de los Caballeros
Azuaga
Sines
Montoro
Beja
Cortegana
Córdoba
Bujalance
Martos
Ourique
Nerva
Palma del Río
Valverde del Camino
La Algaba
Carmona
Écija
Lucena
Sevilla (Seville)
Ayamonte
Lepe
Osuna
Portimão
Faro
Isla Cristina
Huelva
Dos Hermanas
Lagos
Tavira
Olhão
Las Cabezas de San Juan
Antequera
Archidona
Lebrija
Olvera
Álora
Sanlúcar de Barrameda
Ubrique
Ronda
El Puerto de Santa María
Jerez de la Frontera
Coín
Málaga
Cádiz
Fuengirola
San Fernando
Marbella
Vejer de la Frontera
Estepona
Barbate de Franco
Algeciras
GIBRALTAR (to UK)
Ceuta (to Spain)
AZORES (to Portugal)
Corvo
Flores
São Jorge
Graciosa
Terceira
Faial
Pico
São Miguel
Ponta Delgada
Santa Maria
0 km 100
0 miles 100
200m/656ft
Sea level

0 km 100
0 miles 100

Population
- National capital
- below 50,000
- 50,000 to 100,000
- 100,000 to 500,000
- above 500,000

FRANCE
Golfe du Lion
ANDORRA
Pyrenees
Monte Perdido 3348m
Donostia/San Sebastián
Irun
Bermeo
Zarautz
Eibar
Tolosa
Bergara
País Vasco
Pamplona/Iruña
Vitoria-Gasteiz
Miranda de Ebro
Estella
Navarra
Jaca
Logroño
Arnedo
Calahorra
La Rioja
Tudela
Tarazona
Soria
Huesca
Ejea de los Caballeros
Ebro
Barbastro
Monzón
Zaragoza
Lleida (Lérida)
Fraga
La Seu d'Urgell
Berga
Ripoll
Manlleu
Banyoles
Figueres
Girona
Vic
Cataluña
Balaguer
Cervera
Tàrrega
Palafrugell
Palamós
Blanes
Arenys de Mar
Mataró
Sabadell
Terrassa
Barcelona
L'Hospitalet de Llobregat
Costa Brava
Vilafranca del Penedès
Valls
Reus
Sitges
El Vendrell
Tarragona
Calatayud
Aragón
Daroca
Alcañiz
Medinaceli
Sistema Ibérico
Tortosa
Amposta
Sant Carles de la Ràpita
Vinaròs
Guadalajara
Alcalá de Henares
Torrejón de Ardoz
Teruel
Javalambre 2020m
Comunitat Valenciana
Cuenca
Tarancón
Onda
Castellón de la Plana
Borriana
La Vall d'Uixó
Sagunt/Sagunto
Burjassot
Castilla-La Mancha
València
Torrent
Catarroja
Sueca
Cullera
Mota del Cuervo
Campo de Criptana
Socuéllamos
Tomelloso
La Roda
Júcar
Algemesí
Xàtiva
Gandia
Oliva
Dénia
Albacete
Almansa
Ontinyent
Alcoy
Villena
Elda
Benidorm
La Vila Joiosa/Villajoyosa
Sant Joan d'Alacant
Alicante (Alacant)
Elx/Elche
Villanueva de los Infantes
Hellín
Jumilla
Segura
Monòver/Monóvar
Moratalla
Cieza
Villacarrillo
Mula
Cazorla
Callosa de Segura
Orihuela
Murcia
Totana
Huéscar
Lorca
La Unión
Cartagena
Baza
Guadix
Aguilas
Mojácar
Berja
Almería
Adra
Costa del Azahar
Golfo de Valencia
Costa Blanca
Mediterranean Sea
ALGERIA
Ciutadella
Menorca (Minorca)
Maó
Pollença
Sa Pobla
Palma
Llucmajor
Manacor
Felanitx
Mallorca (Majorca)
Illa de Cabrera
Ibiza
Eivissa (Ibiza)
Formentera
Islas Baleares (Balearic Islands)
68
74
75
49
GIBRALTAR (to UK)
SPAIN
Gibraltar Airport
North Mole
Gibraltar Harbour
Catalan Bay
The Rock
Bay of Gibraltar
Summit 426m
Sandy Bay
Rosia
Rosia Bay
Buena Vista
Little Bay
Europa Point
Strait of Gibraltar
200m/656ft
Sea level
0 km 1
0 mile 1
Elevation
Below sea level 0
250m 500m 1000m 2000m 3000m 4000m 6000m
820ft 1640ft 3281ft 6562ft 9843ft 13,124ft 19,685ft
-6000m -4000m -2000m -1000m -500m -250m
-19,658ft -13,124ft -6562ft -3281ft -1640ft -820ft -328ft/-100m 0

Germany & The Alpine States

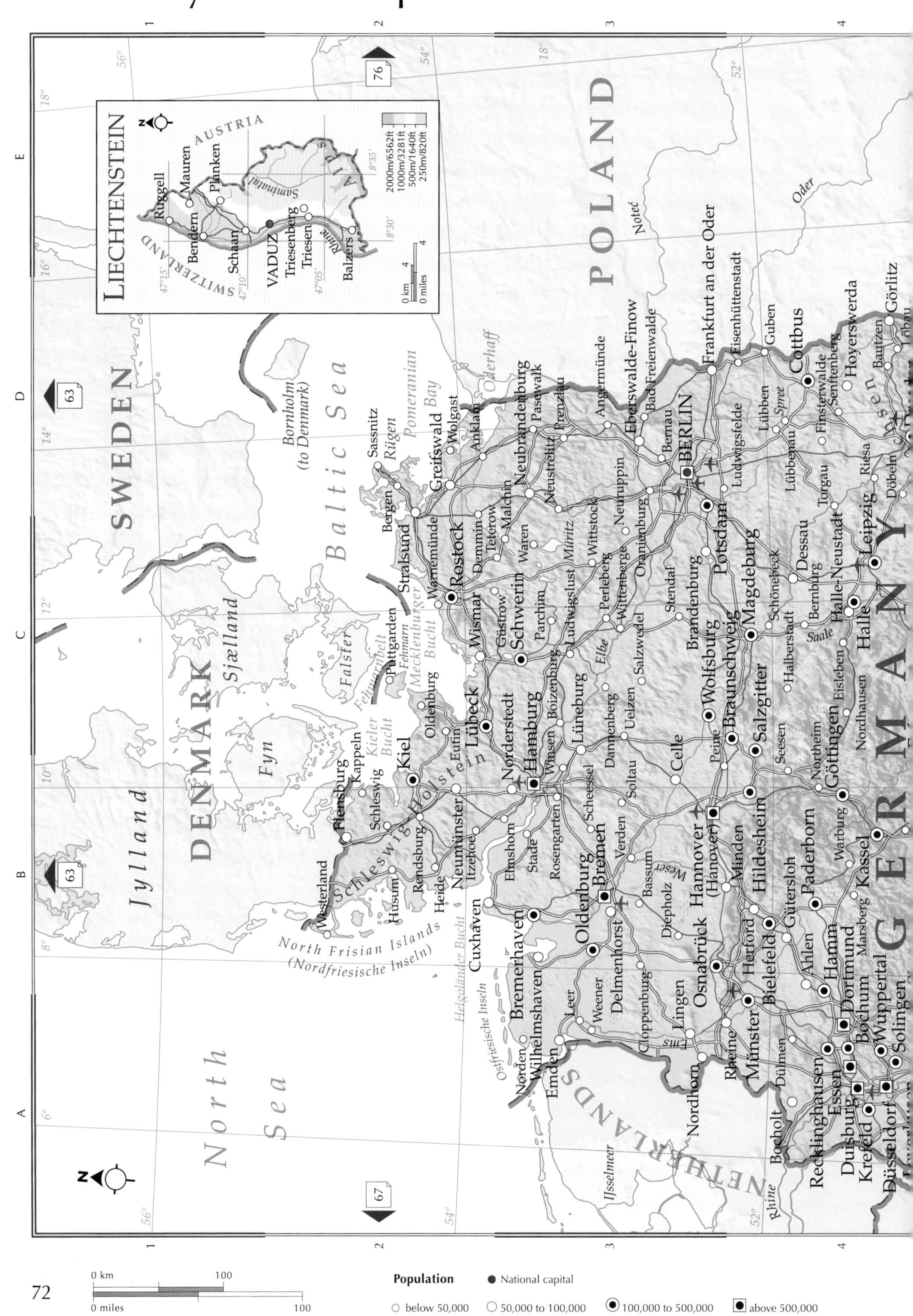

LIECHTENSTEIN
AUSTRIA
SWITZERLAND
Ruggell
Mauren
Planken
Bendern
Schaan
VADUZ
Triesenberg
Triesen
Balzers
Rhine
Samina
Alps
Saminatal
47°15'
47°10'
47°05'
8°30'
8°35'
2000m/6562ft
1000m/3281ft
500m/1640ft
250m/820ft
0 km
0 miles
4
SWEDEN
DENMARK
POLAND
NETHERLANDS
GERMANY
Jylland
Sjælland
Fyn
Falster
Bornholm (to Denmark)
Rügen
North Sea
Baltic Sea
Pomeranian Bay
Mecklenburger Bucht
Kieler Bucht
Fehmarnbelt
Fehmarn
North Frisian Islands (Nordfriesische Inseln)
Ostfriesische Inseln
Helgoländer Bucht
Schleswig-Holstein
Oderhaff
IJsselmeer
Westerland
Husum
Heide
Flensburg
Schleswig
Kappeln
Kiel
Rendsburg
Neumünster
Itzehoe
Eutin
Oldenburg
Puttgarden
Lübeck
Norderstedt
Hamburg
Cuxhaven
Bremerhaven
Wilhelmshaven
Emden
Norden
Leer
Weener
Oldenburg
Delmenhorst
Bremen
Cloppenburg
Lingen
Nordhorn
Rheine
Osnabrück
Münster
Bielefeld
Herford
Gütersloh
Paderborn
Hamm
Ahlen
Dortmund
Bochum
Essen
Duisburg
Krefeld
Düsseldorf
Wuppertal
Solingen
Recklinghausen
Bocholt
Dülmen
Marsberg
Warburg
Kassel
Göttingen
Northeim
Nordhausen
Seesen
Salzgitter
Hildesheim
Minden
Hannover (Hanover)
Diepholz
Bassum
Verden
Rosengarten
Stade
Elmshorn
Scheessel
Soltau
Celle
Peine
Braunschweig
Wolfsburg
Uelzen
Dannenberg
Lüneburg
Winsen
Boizenburg
Ludwigslust
Schwerin
Wismar
Güstrow
Parchim
Rostock
Warnemünde
Stralsund
Bergen
Sassnitz
Greifswald
Wolgast
Anklam
Demmin
Teterow
Malchin
Waren
Müritz
Neubrandenburg
Neustrelitz
Pasewalk
Prenzlau
Angermünde
Eberswalde-Finow
Bad Freienwalde
Bernau
BERLIN
Frankfurt an der Oder
Eisenhüttenstadt
Guben
Cottbus
Lübben
Spree
Lübbenau
Finsterwalde
Senftenberg
Hoyerswerda
Bautzen
Görlitz
Riesa
Döbeln
Torgau
Leipzig
Halle
Halle-Neustadt
Saale
Bernburg
Dessau
Schönebeck
Magdeburg
Halberstadt
Eisleben
Brandenburg
Potsdam
Ludwigsfelde
Oranienburg
Neuruppin
Wittstock
Perleberg
Wittenberge
Stendal
Salzwedel
Elbe
Weser
Ems
Rhine
Oder
Noteć

0 km 100
0 miles 100

Population ● National capital
○ below 50,000 ○ 50,000 to 100,000 ⊙ 100,000 to 500,000 ◘ above 500,000

CZECHIA
(CZECH REPUBLIC)
SLOVAKIA
HUNGARY
AUSTRIA
SLOVENIA
CROATIA
ITALY
SWITZERLAND
LIECHTENSTEIN
FRANCE
BELGIUM
LUX.
Hessen
Bayern
Tirol
Alps
Bohemian Forest
Schwarzwald
Schwäbische Alb
Fränkische Alb
Vosges
Jura
Eifel
Rheinisches
Erzgebirge
Bavarian Alps
Kitzbüheler Alpen
Hohe Tauern
Gurktaler Alpen
Fischbacher Alpen
Berner Alpen
Pennine Alps
Po Valley
Gulf of Venice
Istra
Elbe
Danube (Donau)
Inn
Enns
Mur
Drava
Sava
Main
Neckar
Rhine (Rhein)
Mosel
Aare
Rhône
Po
Lake Constance
Lake Geneva
Lac de Neuchâtel
Thuner See
Zürichsee
Lake Maggiore
Neusiedler See
WIEN (VIENNA)
Mistelbach an der Zaya
Hollabrunn
Tulln
Traiskirchen
Eisenstadt
Perchtoldsdorf
Bad Vöslau
Wiener Neustadt
Sankt Pölten
Zwettl
Linz
Wels
Steyr
Hauzenberg
Vöcklabruck
Ried im Innkreis
Salzburg
Ebensee
Bad Ischl
Liezen
Leoben
Mürzzuschlag
Graz
Judenburg
Wolfsberg
Klagenfurt
Villach
Lienz
Grossglockner 3798m
Plöcken Pass 1357m
Brenner Pass 1374m
Zugspitze 2962m
Loibl Pass 1367m
Maribor
Murska Sobota
Ptuj
Velenje
Celje
Trbovlje
Krško
Novo mesto
Kočevje
LJUBLJANA
Kranj
Jesenice
Tolmin
Nova Gorica
Postojna
Koper
Innsbruck
Schwaz
Passau
Deggendorf
Pocking
Straubing
Regensburg
Regenstauf
Schwandorf
Landshut
München (Munich)
Rosenheim
Ingolstadt
Augsburg
Donauwörth
Marktredwitz
Münchberg
Hof
Plauen
Bayreuth
Kronach
Lichtenfels
Coburg
Forchheim
Bamberg
Schweinfurt
Würzburg
Erlangen
Fürth
Nürnberg (Nuremberg)
Weissenburg in Bayern
Heidenheim an der Brenz
Aalen
Ulm
Neu-Ulm
Mindelheim
Memmingen
Kaufbeuren
Kempten
Füssen
Friedrichshafen
Bregenz
Göppingen
Ludwigsburg
Heilbronn
Sinsheim
Heidelberg
Mannheim
Ludwigshafen
Darmstadt
Pfungstadt
Offenbach
Frankfurt am Main
Bad Homburg vor der Höhe
Wetzlar
Fulda
Wiesbaden
Mainz
Koblenz
Neuwied
Boppard
Worms
Kaiserslautern
Neustadt an der Weinstrasse
Karlsruhe
Pforzheim
Stuttgart
Sindelfingen
Reutlingen
Baden-Baden
Kehl
Offenburg
Lahr
Emmendingen
Freiburg im Breisgau
Bad Krozingen
Müllheim
Lörrach
Rottweil
Villingen-Schwenningen
Stockach
Singen
Konstanz
Trier
Wittlich
Bitburg
Birkenfeld
Merzig
Neunkirchen
Saarbrücken
Blankenheim
Basel
Schaffhausen
Bülach
Winterthur
Zürich
Sankt Gallen
VADUZ
Chur
Klosters
St.Moritz
Zug
Luzern
Schwyz
Biel
BERN
Thun
Brig
Eiger 3970m
Simplon Pass 2005m
Matterhorn 4478m
Great Saint Bernard Pass 2469m
Sion
Monthey
Bellinzona
Locarno
Lugano
La Chaux-de-Fonds
Neuchâtel
Lausanne
Genève (Geneva)
Onex
Vorderrhein
69
74
77
78
A
B
C
D
E
5
6
7
8
6°
8°
10°
12°
14°
16°
46°
48°
50°
Elevation
Below sea level 0
-6000m -4000m -2000m -1000m -500m -250m
-19,658ft -13,124ft -6562ft -3281ft -1640ft -820ft -328ft/-100m 0
250m 500m 1000m 2000m 3000m 4000m 6000m
820ft 1640ft 3281ft 6562ft 9843ft 13,124ft 19,685ft

Italy

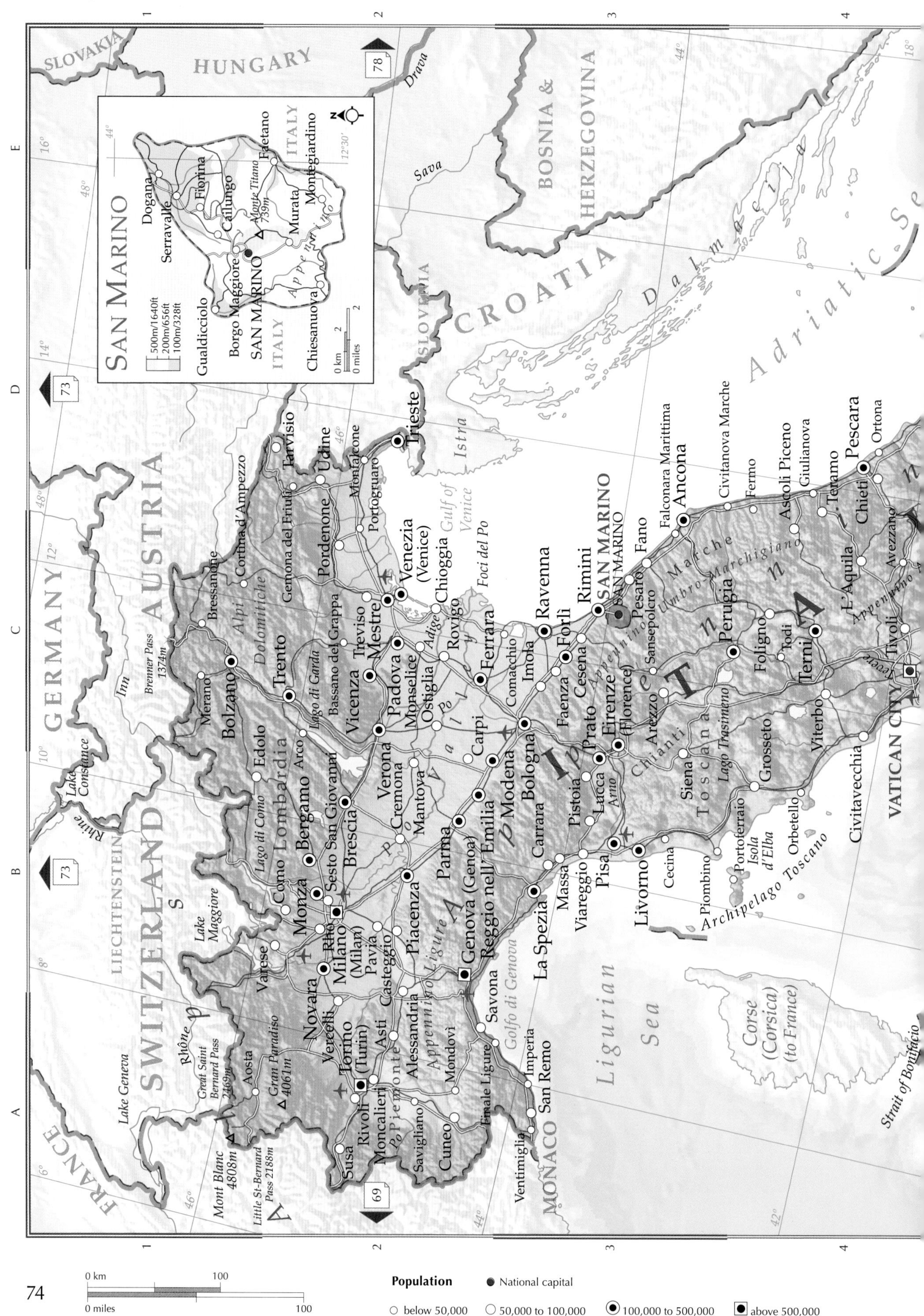

SLOVAKIA
HUNGARY
GERMANY
AUSTRIA
SWITZERLAND
LIECHTENSTEIN
FRANCE
MONACO
SLOVENIA
CROATIA
BOSNIA & HERZEGOVINA
VATICAN CITY
SAN MARINO
Adriatic Sea
Ligurian Sea
Gulf of Venice
Golfo di Genova
Archipelago Toscano
Isola d'Elba
Corse (Corsica) (to France)
Strait of Bonifacio
Lake Geneva
Lake Maggiore
Lake Constance
Lago di Como
Lago di Garda
Lago Trasimeno
Rhine
Rhône
Inn
Drava
Sava
Po
Adige
Arno
Mont Blanc 4808m
Little St-Bernard Pass 2188m
Great Saint Bernard Pass 2469m
Gran Paradiso 4061m
Brenner Pass 1374m
Alpi Dolomitiche
Lombardia
Piemonte
Toscana
Marche
Chianti
Istra
Foci del Po
Dalmacija
Torino (Turin)
Milano (Milan)
Genova (Genoa)
Venezia (Venice)
Firenze (Florence)
Bologna
Verona
Padova
Trieste
Trento
Bolzano
Perugia
Ancona
Pescara
Rimini
Ravenna
Ferrara
Modena
Parma
Piacenza
Brescia
Bergamo
Monza
Como
Varese
Novara
Asti
Alessandria
Cuneo
Savona
La Spezia
Pisa
Livorno
Lucca
Siena
Arezzo
Grosseto
Viterbo
Terni
Tivoli
L'Aquila
Chieti
Teramo
Udine
Pordenone
Treviso
Mestre
Vicenza
Rovigo
Chioggia
Forlì
Cesena
Imola
Faenza
Prato
Pistoia
Massa
Carrara
Viareggio
Reggio nell'Emilia
Cremona
Mantova
Pavia
San Remo
Imperia
Ventimiglia
Aosta
Susa
Rivoli
Moncalieri
Civitavecchia
San Marino
Borgo Maggiore
Serravalle
Dogana
Faetano
Chiesanuova
Fiorentino
Montegiardino
Monte Titano 739m
500m/1640ft
200m/656ft
100m/328ft

Population
● National capital
○ below 50,000 ○ 50,000 to 100,000 ◉ 100,000 to 500,000 ▣ above 500,000

0 km 100
0 miles 100

Sardegna (Sardinia)
Alghero
Ozieri
Siniscola
Nuoro
Macomer
Oristano
Punta La Marmora 1834m
Villacidro
Iglesias
Carbonia
Cagliari
Quartu Sant' Elena
Tyrrhenian Sea
Isole Ponziane
Terracina
Gaeta
Golfo di Gaeta
Caserta
Napoli (Naples)
Vesuvio 1277m
Torre del Greco
Isola di Capri
Campania
Benevento
Avellino
Salerno
Battipaglia
Golfo di Salerno
Agropoli
Sala Consilina
Sapri
Foggia
Cerignola
Andria
Ofanto
Barletta
Molfetta
Bitonto
Bari
Altamura
Potenza
Matera
Puglia
Appennino Lucano
Lauria
Castrovillari
Brindisi
Taranto
Lecce
Manduria
Maglie
Gallipoli
Golfo di Taranto
Strait of Otranto
Rossano
La Sila
Cirò Marina
Crotone
Cosenza
Amantea
Lamezia Terme
Catanzaro
Palmi
Siderno
Reggio di Calabria
Isola Stromboli
Isole Eolie
Isola Lipari
Isola Vulcano
Isola d'Ustica
Stretto di Messina
Messina
Cefalù
Palermo
Trapani
Isole Egadi
Alcamo
Marsala
Castelvetrano
Sicilia (Sicily)
Monte Etna 3329m
Simeto
Catania
Caltanissetta
Agrigento
Gela
Vittoria
Ragusa
Siracusa
Modica
Pozzallo
Mediterranean Sea
Strait of Sicily
Isola di Pantelleria
Ionian Sea
Malta Channel
Gozo
MALTA
VALLETTA
Malta
Isole Pelagie
TUNISIA
VATICAN CITY
0 m 200
0 yds 250
Main Entrance
ROME
Pigna Courtyard
Vatican Museums
Vatican Gardens
Monte Vaticano
Radio Vatican
Sistine Chapel
Raphael Stanza
Papal Apartments
Saint Peter's Basilica
St Peter's Square
Papal Heliport
Vatican Railway Station
ROME
81
49
49
49
Elevation
-6000m -4000m -2000m -1000m -500m -250m
-19,658ft -13,124ft -6562ft -3281ft -1640ft -820ft -328ft/-100m 0
Below sea level 0
250m 500m 1000m 2000m 3000m 4000m 6000m
820ft 1640ft 3281ft 6562ft 9843ft 13,124ft 19,685ft

Central Europe

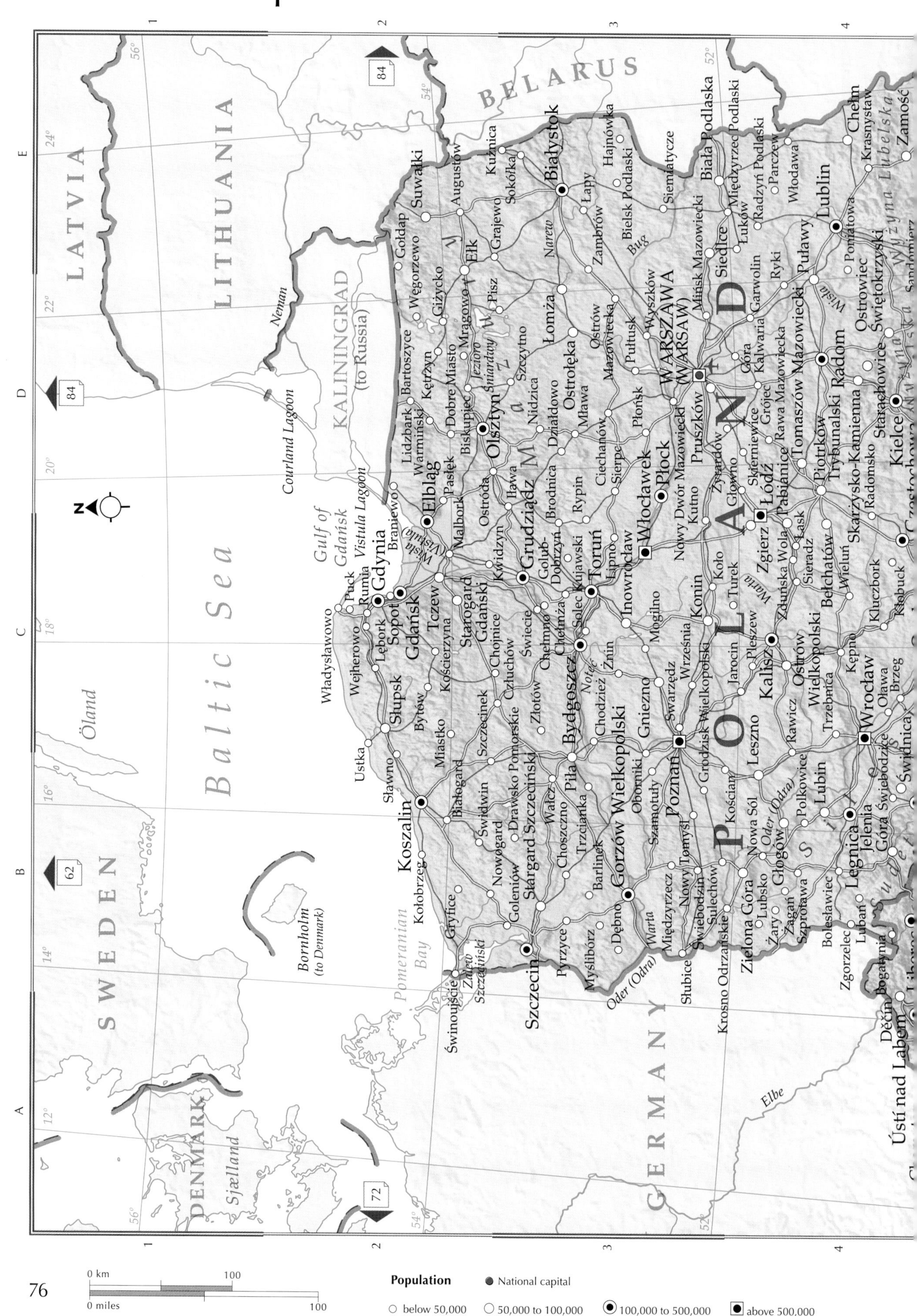
LATVIA
LITHUANIA
BELARUS
KALININGRAD (to Russia)
POLAND
GERMANY
SWEDEN
DENMARK
Baltic Sea
Gulf of Gdańsk
Vistula Lagoon
Courland Lagoon
Pomeranian Bay
Öland
Bornholm (to Denmark)
Sjælland
Neman
Wisła (Vistula)
Oder (Odra)
Warta
Noteć
Narew
Bug
Elbe
84
62
72
WARSZAWA (WARSAW)
Gdańsk
Gdynia
Sopot
Elbląg
Olsztyn
Białystok
Suwałki
Szczecin
Koszalin
Słupsk
Bydgoszcz
Toruń
Grudziądz
Włocławek
Płock
Poznań
Gorzów Wielkopolski
Zielona Góra
Legnica
Wrocław
Łódź
Kalisz
Piotrków Trybunalski
Radom
Kielce
Lublin
Siedlce
Biała Podlaska
Chełm
Zamość
Łomża
Ostrołęka
Ústí nad Labem
Děčín

0 km 100
0 miles 100

Population ● National capital
○ below 50,000 ◎ 50,000 to 100,000 ⊙ 100,000 to 500,000 ▣ above 500,000

UKRAINE
CZECHIA
(CZECH REPUBLIC)
SLOVAKIA
HUNGARY
AUSTRIA
SLOVENIA
CROATIA
BOSNIA & HERZEGOVINA
SERBIA
ROMANIA
ITALY
PRAHA
(PRAGUE)
BRATISLAVA
BUDAPEST
Tachov
Mariánské Lázně
Plzeň
Rokycany
Kolín
Benešov
Čáslav
Pardubice
Zábřeh
Olomouc
Humpolec
Boskovice
Klatovy
Strakonice
Písek
Tábor
Jihlava
Třebíč
Prachatice
České Budějovice
Český Krumlov
Brno
Znojmo
Kyjov
Otrokovice
Hodonín
Prostějov
Přerov
Zlín
Vsetín
Hranice
Ostrava
Opava
Wodzisław Śląski
Żory
Jastrzębie-Zdrój
Tychy
Havířov
Frýdek-Místek
Bielsko-Biała
Wieliczka
Tarnów
Dębica
Rzeszów
Przemyśl
Jasło
Krosno
Sanok
Limanowa
Nowy Sącz
Rabka
Krynica
Ustrzyki Dolne
Čadca
Zakopane
Žilina
Bytča
Považská Bystrica
Martin
Trenčín
Ružomberok
Poprad
Banská Bystrica
Bardejov
Prešov
Snina
Vranov nad Topl'ou
Michalovce
Košice
Trebišov
Senica
Piešťany
Topoľčany
Zvolen
Rožňava
Malacky
Trnava
Nitra
Pezinok
Senec
Galanta
Šurany
Levice
Lučenec
Veľký Krtíš
Encs
Záhony
Sátoraljaújhely
Sajószentpéter
Ózd
Kisvárda
Miskolc
Eger
Fehérgyarmat
Nyíregyháza
Nagykálló
Hajdúhadház
Kolárovo
Mosonmagyaróvár
Neusiedler See
Sopron
Győr
Csorna
Tatabánya
Esztergom
Vác
Gyöngyös
Debrecen
Püspökladány
Berettyóújfalu
Celldömölk
Szombathely
Székesfehérvár
Veszprém
Dunaújváros
Nagykőrös
Szolnok
Mezőtúr
Gyomaendrőd
Körmend
Kecskemét
Tiszakécske
Békéscsaba
Zalaegerszeg
Keszthely
Fonyód
Lenti
Paks
Tolna
Hódmezővásárhely
Nagykanizsa
Kaposvár
Szekszárd
Szeged
Makó
Jánoshalma
Csurgó
Baja
Barcs
Pécs
Siklós
Bohemian Forest
Bohemia
Moravia
Carpathian Mountains
Tatra Mts
Rysy 2499m
Kékes 1014m
Slovenské rudohorie
Little Alföld
Great Hungarian Plain
Bakony
Balaton
Mecsek
Papuk
Vojvodina
Velebit
Alps
Niedere Tauern
Carpații Occidentali
Carpații Meridionali
Danube
Tisza
Drava
Mur
Morava
Váh
Nitra
Ipel'
Ipoly
Rába
Poprad
Laborec
Berettyó
Mureș
Dniester
San
Gulf of Venice
Adriatic Sea
Elevation
-6000m -4000m -2000m -1000m -500m -250m
-19,658ft -13,124ft -6562ft -3281ft -1640ft -820ft -328ft/-100m 0
Below sea level 0
250m 500m 1000m 2000m 3000m 4000m 6000m
820ft 1640ft 3281ft 6562ft 9843ft 13,124ft 19,685ft

Southeast Europe

A
B
C
D
E
1
2
3
4
73
77
74
86
GERMANY
AUSTRIA
SLOVAKIA
UKRAINE
HUNGARY
ITALY
SLOVENIA
CROATIA
ROMANIA
BOSNIA & HERZEGOVINA
REPUBLIKA SRPSKA
FEDERACIJA BOSNE I HERCEGOVINE
Alps
Fischbacher Alpen
Little Alföld
Bakony
Great Hungarian Plain
Mecsek
Transylvania
Carpaţii Meridionali
Papuk
Kozara
Velebit
Vojvodina
Istra
Troglav 1913m
Treskavica
Danube (Donau)
Danube (Dunaj)
Danube (Duna)
Neusiedler See
Balaton
Ipel'
Tisza
Tisa
Mureş
Timiş
Mur
Raab
Drava
Sava
Kolpa
Una
Unac
Sana
Vrbas
Bosna
Drina
Neretva
Velika Morava
Zapadna Morava
Kvarner
Krk
Cres
Lošinj
Pag
Dugi Otok
Brač
Vis
Čakovec
Varaždin
Koprivnica
Križevci
Samobor
Sesvete
ZAGREB
Bjelovar
Virovitica
Poreč
Rovinj
Pula
Opatija
Rijeka
Crikvenica
Ogulin
Karlovac
Petrinja
Sisak
Kutina
Glina
Senj
Slatina
Nova Gradiška
Požega
Bosanska Dubica
Bosanski Novi
Bosanska Gradiška
Slavonski Brod
Cazin
Bihać
Prijedor
Banja Luka
Gospić
Ključ
Zadar
Knin
Jajce
Travnik
Šibenik
Sinj
Trogir
Split
Livno
Makarska
Konjic
Beli Manastir
Osijek
Đakovo
Bosanski Šamac
Derventa
Modriča
Gradačac
Doboj
Brčko
Maglaj
Zavidovići
Zenica
Visoko
SARAJEVO
Rogatica
Goražde
Foča
Tuzla
Bijeljina
Vinkovci
Županja
Vukovar
Borovo
Sombor
Subotica
Kanjiža
Bačka Topola
Senta
Ada
Kikinda
Vrbas
Bečej
Srbobran
Temerin
Bačka Palanka
Futog
Novi Sad
Zrenjanin
Mužlja
Sremska Mitrovica
Ruma
Inđija
Stara Pazova
Batajnica
Šabac
Zemun
BEOGRAD (BELGRADE)
Pančevo
Vršac
Bela Crkva
Loznica
Zvornik
Valjevo
Srebrenica
Smederevo
Požarevac
Mladenovac
Smederevska Palanka
Aranđelovac
Gornji Milanovac
Kragujevac
Užice
Požega
Čačak
Kraljevo
Priboj
Jagodina
Ćuprija
Paraćin
Bor
Zaječar
Negotin

0 km 100
0 miles 100

Population ● National capital ● Internal administrative capital
○ below 50,000 ○ 50,000 to 100,000 ⦿ 100,000 to 500,000 ▣ above 500,000

In February 2008, Kosovo (a UN Protectorate within Serbia since 1999) declared independence. Although recognized by several countries, this decision has proved controversial with other states wary of setting a precedent for separatist groups within their own borders. It is therefore likely to be some time before Kosovo becomes universally recognized.
MONTENEGRO
PODGORICA
Trebinje
Dubrovnik
Mljet
Palagruža
Nikšić
Kotor
Cetinje
Lake Scutari
Bar
Shkodër
Berane
Mitrovicë/Mitrovica
Pejë/Pec
North Albanian Alps
Gjeravicë 2658m
Fushë Kosovë
Bajram Curri
Gjakovë/Đakovica
Rahovec
KOSOVO (disputed)
PRISHTINË/PRIŠTINA (PRISTINA)
Vushtrri/Vucitrn
Podujevë
Kuršumlija
Leskovac
Vlasotince
Pirot
Surdulica
Vranje
Bujanovac
Gjilan/Gnjilane
Ferizaj/Uroševac
Preševo
Prizren
Kukës
Lumi i Drinit
Tetovo
SKOPJE
Kumanovo
BULGARIA
Lezhë
Laç
Peshkopi
Krujë
Burrel
Black Drin
Durrës
TIRANË (TIRANA)
Kavajë
Gostivar
Debar
Kičevo
Veles
Štip
Kočani
Bregalnica
Radoviš
NORTH MACEDONIA
Struga
Prilep
Kavadarci
Strumica
Lumi i Shkubinit
Elbasan
Lushnjë
Lake Ohrid
Ohrid
Bitola
Crna Reka
Gevgelija
Vardar
Kuçovë
Fier
Berat
Pogradec
Lumi i Devollit
Lake Prespa
Strymónas
Vlorë
ALBANIA
Korçë
Lumi i Osumit
Lumi i Vjosës
Tepelenë
Gjirokastër
Sarandë
Konispol
GREECE
Thermaïkós Kólpos
Kérkyra (Corfu)
Píndos (Pindus Mountains)
Pineiós
Aegean Sea
Évvoia (Euboea)
Iónia Nisiá (Ionian Islands)
Kefallonia
Ionian Sea
Strait of Otranto
Sea
ITALY
Appennino Lucano
Golfo di Taranto
BOSNIA & HERZEGOVINA
CROATIA
Sava
Bihač
Banja Luka
Brčko
Bosna
Tuzla
Drina
SERBIA
Split
Sarajevo
Goražde
Mostar
Adriatic Sea
MONTENEGRO
Dubrovnik
0 50 km
0 50 miles
Territorial extent
Republika Srpska
Federacija Bosne i Hercegovine
75
82
81
83
A
B
C
D
E
5
6
7
8

Elevation

-6000m	-4000m	-2000m	-1000m	-500m	-250m	Below sea level 0	250m	500m	1000m	2000m	3000m	4000m	6000m
-19,658ft	-13,124ft	-6562ft	-3281ft	-1640ft	-820ft	-328ft/-100m 0	820ft	1640ft	3281ft	6562ft	9843ft	13,124ft	19,685ft

The Mediterranean

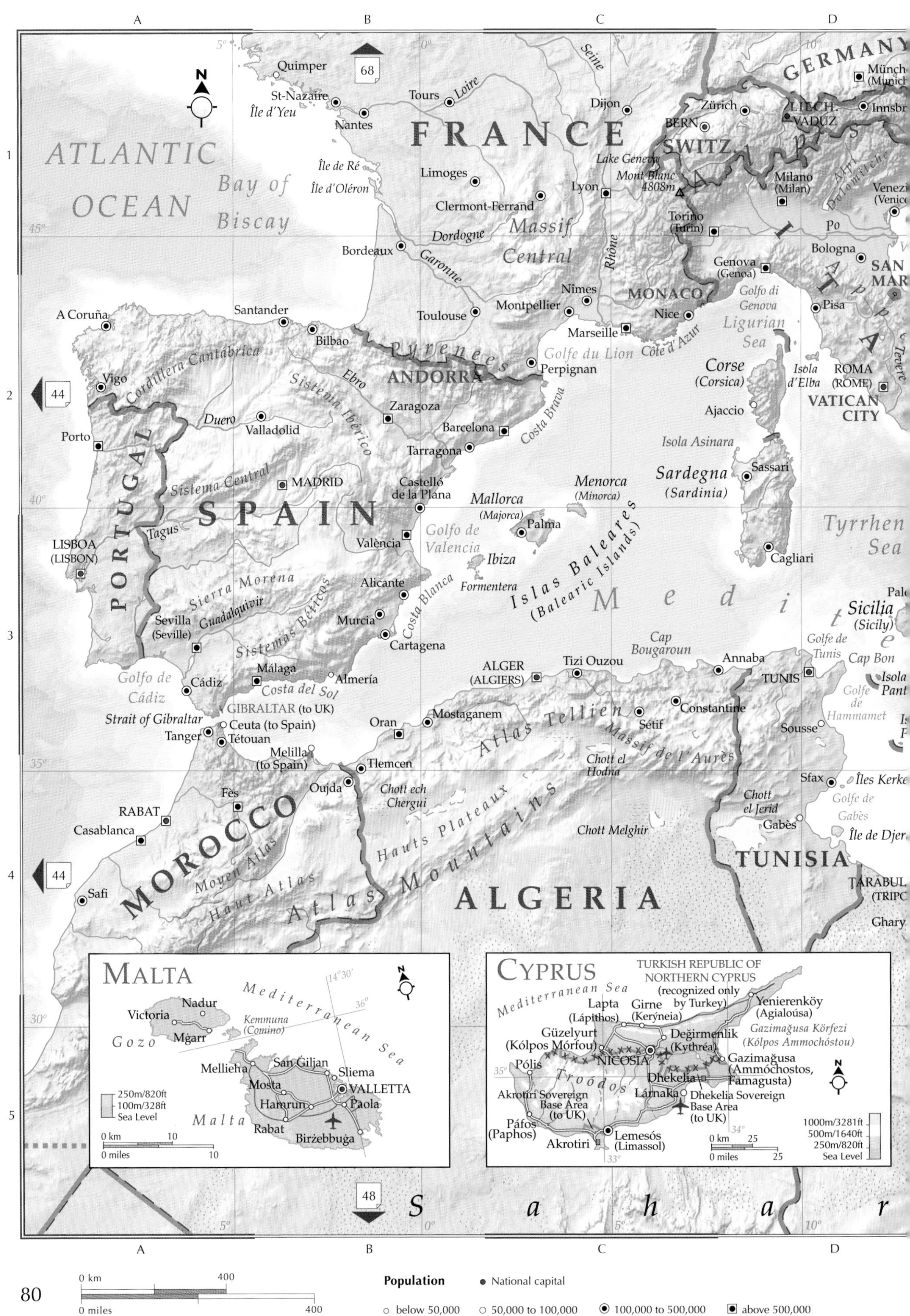
ATLANTIC OCEAN
Bay of Biscay
FRANCE
SPAIN
PORTUGAL
ANDORRA
MOROCCO
ALGERIA
TUNISIA
GERMANY
SWITZ.
LIECH.
VADUZ
BERN
MONACO
VATICAN CITY
Quimper
St-Nazaire
Île d'Yeu
Nantes
Tours
Loire
Seine
Dijon
Zürich
Innsbr
Münch (Munich)
Île de Ré
Île d'Oléron
Limoges
Lyon
Lake Geneva
Mont Blanc 4808m
Milano (Milan)
Venezi (Venice)
Clermont-Ferrand
Massif Central
Dordogne
Torino (Turin)
Po
Bordeaux
Garonne
Rhône
Genova (Genoa)
Bologna
SAN MAR
Nîmes
Montpellier
Nice
Golfo di Genova
Pisa
A Coruña
Santander
Toulouse
Marseille
Ligurian Sea
Bilbao
Pyrenees
Golfe du Lion
Côte d'Azur
Perpignan
Corse (Corsica)
Isola d'Elba
ROMA (ROME)
Vigo
Cordillera Cantábrica
Sistema Ibérico
Ebro
Costa Brava
Ajaccio
Duero
Valladolid
Zaragoza
Porto
Barcelona
Tarragona
Isola Asinara
Sistema Central
MADRID
Castelló de la Plana
Mallorca (Majorca)
Menorca (Minorca)
Sardegna (Sardinia)
Sassari
Tagus
Palma
Tyrrhen Sea
LISBOA (LISBON)
València
Golfo de Valencia
Ibiza
Islas Baleares (Balearic Islands)
Cagliari
Sierra Morena
Alicante
Formentera
Costa Blanca
Mediterranean
Guadalquivir
Sevilla (Seville)
Sistemas Béticos
Murcia
Cartagena
Sicilia (Sicily)
Golfo de Tunis
Cap Bon
Cap Bougaroun
Málaga
Almería
Tizi Ouzou
Annaba
TUNIS
Golfo de Cádiz
Cádiz
Costa del Sol
ALGER (ALGIERS)
Isola Pant
GIBRALTAR (to UK)
Strait of Gibraltar
Ceuta (to Spain)
Tanger
Tétouan
Oran
Mostaganem
Atlas Tellien
Sétif
Constantine
Golfe de Hammamet
Sousse
Melilla (to Spain)
Tlemcen
Massif de l'Aurès
Chott el Hodna
Fès
Oujda
Chott ech Chergui
Sfax
Îles Kerke
Golfe de Gabès
RABAT
Casablanca
Hauts Plateaux
Chott el Jerid
Gabès
Chott Melghir
Île de Djer
Moyen Atlas
Haut Atlas
Atlas Mountains
Safi
TARABUL (TRIPO
Ghary
Sahara
MALTA
Mediterranean Sea
Victoria
Nadur
Gozo
Mġarr
Kemmuna (Comino)
Mellieħa
San Ġiljan
Sliema
Mosta
VALLETTA
Ħamrun
Paola
Malta
Rabat
Birżebbuġa
250m/820ft
100m/328ft
Sea Level
0 km 10
0 miles 10
CYPRUS
TURKISH REPUBLIC OF NORTHERN CYPRUS (recognized only by Turkey)
Mediterranean Sea
Lapta (Lápithos)
Girne (Kerýneia)
Yenierenköy (Agialoúsa)
Güzelyurt (Kólpos Mórfou)
Değirmenlik (Kythréa)
Gazimağusa Körfezi (Kólpos Ammochóstou)
NICOSIA
Gazimağusa (Ammóchostos, Famagusta)
Pólis
Troodos
Dhekelia
Akrotiri Sovereign Base Area (to UK)
Lárnaka
Dhekelia Sovereign Base Area (to UK)
Páfos (Paphos)
Akrotiri
Lemesós (Limassol)
0 km 25
0 miles 25
1000m/3281ft
500m/1640ft
250m/820ft
Sea Level
68
44
48

0 km 400
0 miles 400

Population

- ● National capital
- ○ below 50,000
- ○ 50,000 to 100,000
- ◉ 100,000 to 500,000
- ▣ above 500,000

UKRAINE
MOLD.
CHIŞINĂU
ROMANIA
BUCUREŞTI (BUCHAREST)
HUNGARY
BUDAPEST
SLOVAKIA
CROATIA
ZAGREB
BOSNIA & HERZ.
SARAJEVO
SERBIA
BEOGRAD (BELGRADE)
BULGARIA
SOFIA
MON.
PODGORICA
KOSOVO (disputed)
PRISHTINË/PRIŠTINA (PRISTINA)
NORTH MACEDONIA
SKOPJE
ALBANIA
TIRANË (TIRANA)
GREECE
ATHÍNA (ATHENS)
TURKEY (TÜRKIYE)
ANKARA
RUSS.
SYRIA
LEBANON
BEYROUTH (BEIRUT)
DIMASHQ (DAMASCUS)
ISRAEL
JERUSALEM
'AMMĀN
JORDAN
CYPRUS
NICOSIA
EGYPT
AL QĀHIRAH (CAIRO)
LIBYA
SAUDI ARABIA
Black Sea
Sea of Azov
Aegean Sea
Ionian Sea
Mediterranean Sea
Red Sea
Dead Sea
(since 2014 the Ukrainian territory of Crimea has been annexed by Russia)
86
95
97
50

In 1974 Turkey occupied the northern part of Cyprus while Greek Cypriots remained in control of the south. Cyprus was effectively partitioned and a UN buffer zone currently divides the two areas. In 1983 the north of the island proclaimed itself the Turkish Republic of North Cyprus. It was only recognized by Turkey.

Elevation

Below sea level 0 | 250m 500m 1000m 2000m 3000m 4000m 6000m | 820ft 1640ft 3281ft 6562ft 9843ft 13,124ft 19,685ft

-6000m -4000m -2000m -1000m -500m -250m | -19,658ft -13,124ft -6562ft -3281ft -1640ft -820ft -328ft/-100m 0

Bulgaria & Greece

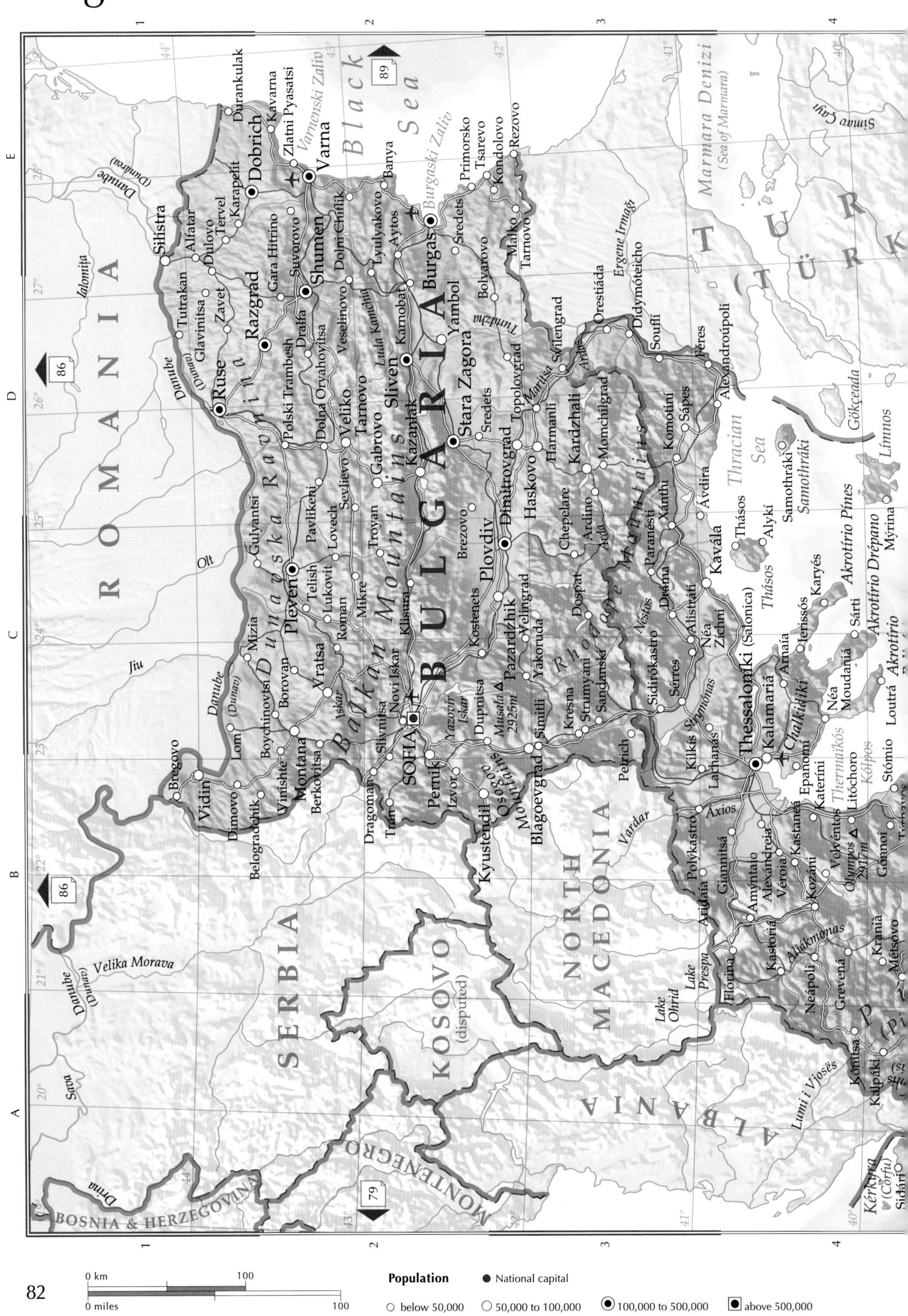

ROMANIA
BULGARIA
SERBIA
KOSOVO
(disputed)
NORTH MACEDONIA
ALBANIA
MONTENEGRO
BOSNIA & HERZEGOVINA
TÜRKIYE
Black Sea
Marmara Denizi
(Sea of Marmara)
Thracian Sea
Balkan Mountains
Rhodope Mountains
Dunavska Ravnina
Danube
(Dunav)
Olt
Jiu
Iskar
Maritsa
Tundzha
Velika Morava
Sava
Drina
Vardar
Strymónas
Néstos
Aliákmonas
Áxios
Ergene Irmaği
Lake Ohrid
Lake Prespa
Thermaïkós Kólpos
Chalkidikí
Thásos
Samothráki
Límnos
Gökçeada
SOFIA
Plovdiv
Varna
Burgas
Ruse
Stara Zagora
Pleven
Sliven
Dobrich
Shumen
Yambol
Pernik
Haskovo
Pazardzhik
Blagoevgrad
Veliko Tarnovo
Vratsa
Gabrovo
Vidin
Montana
Kardzhali
Kyustendil
Kazanlak
Razgrad
Silistra
Lovech
Dimitrovgrad
Thessaloníki (Salonica)
Kavála
Sérres
Dráma
Xánthi
Komotiní
Alexandroúpoli
Kastoriá
Flórina
Kozáni
Véroia
Kateríni
Musala Δ 2925m
Ólympos Δ 2917m
Population
National capital
below 50,000
50,000 to 100,000
100,000 to 500,000
above 500,000
0 km 100
0 miles 100

GREECE
Aegean Sea
Ionian Sea
Mediterranean Sea
Kritikó Pélagos (Sea of Crete)
Myrtóo Pélagos
Kríti (Crete)
Iónioi Nisiá (Ionian Islands)
Pelopónnisos (Peloponnese)
Kykládes (Cyclades)
Dodekánisa (Dodecanese)
ATHÍNA (ATHENS)
Peiraiás (Piraeus)
Pátra
Irákleio
Chaniá
Ródos (Rhodes)
Lésvos (Lesbos)
Chíos
Sámos
Évvoia (Euboea)
Kárpathos
Kýthira
Santorini
Thíra
Náxos
Mýkonos
Gávdos
Elevation
-6000m -4000m -2000m -1000m -500m -250m
-19,658ft -13,124ft -6562ft -3281ft -1640ft -820ft -328ft/-100m 0
Below sea level 0
250m 500m 1000m 2000m 3000m 4000m 6000m
820ft 1640ft 3281ft 6562ft 9843ft 13,124ft 19,685ft

The Baltic States & Belarus

FINLAND
SWEDEN
RUSSIA
ESTONIA
LATVIA
LITHUANIA
KALININGRAD (to Russia)
Gulf of Finland
Gulf of Riga
Baltic Sea
Narva Bay
Lake Peipus
Lake Pskov
Narva Reservoir
Ålands Hav
Skiftet
Gotska Sandön
Gotland
Öland
Hiiumaa
Saaremaa
Vormsi
Muhu
Kihnu
Ruhnu
Väinameri
Suur Väin
Pärnu Laht
Courland Lagoon
TALLINN
RIGA
Tartu
Narva
Sillamäe
Kohtla-Järve
Rakvere
Kunda
Loksa
Maardu
Keila
Paldiski
Haapsalu
Rapla
Paide
Pärnu
Viljandi
Valga
Võru
Põlva
Otepää
Räpina
Valka
Valmiera
Cēsis
Jūrmala
Jelgava
Tukums
Ventspils
Liepāja
Kuldīga
Talsi
Saldus
Bauska
Gulbene
Madona
Rēzekne
Ludza
Balvi
Alūksne
Daugavpils
Jēkabpils
Līvāni
Klaipėda
Šiauliai
Panevėžys
Kaliningrad
Zelenogradsk
Sovetsk
Chernyakhovsk
Gusev
Suur Munamägi 318m
Gaiziņkalns 311m
Western Dvina
Venta
Neman
Žemaičių Aukštumas
63
88
1
2
3
4
A
B
C
D
E

0 km 100
0 miles 100

Population ● National capital

○ below 50,000 ○ 50,000 to 100,000 ◉ 100,000 to 500,000 ▣ above 500,000

BELARUS
RUSSIA
UKRAINE
POLAND
Mazury
Dnieper Lowland
Pripyat Marshes
Byelaruskaya Hrada
Minskaya Wzvyshsha
Wyżyna Lubelska
VILNIUS
MINSK
Alytus
Trakai
Rūdiškės
Veisiejai
Druskininkai
Merkinė
Varėna
Šalčininkai
Hrandzichy
Hrodna/Grodno
Parechcha
Vasilishki
Skidal'
Lida
Voranava
Ashmyany
Shchuchyn
Masty
Orlya
Vawkavysk/Volkoysk
Zel'va
Navahrudak/Novogrudok
Novy Dvor
Slonim
Ruzhany
Pruzhany
Brest
Zhabinka
Ivatsevichy
Kobryn/Kobrin
Haradzyets/Gorodets
Damachava
Makrany
Abrova
Drahichyn/Drogichin
Ivanava
Pinsk
Lyakhavichy
Hantsavichy/Gantsevichi
Lyusina
Salihorsk/Soligorsk
Bastyn'
Luninyets/Luninets
Mikashevichy
Zhytkavichy
Kaptsevichy
Pyetrykaw
Tonyezh
Simanichy
Lyel'chytsy
Milashavichy
Yel'sk
Dabryn'/Dobryn'
Baranavichy/Baranovichi
Stowbtsy
Nyasvizh
Kapyl'
Syemyezhava/Semechev
Slutsk
Starobin
Staryya Darohi
Shyshchytsy
Asipovichy
Mar''ina Horka
Pukhavichy
Tal'ka
Rudzyensk
Chervyen'
Valozhyn
Krasnaye
Maladzyechna/Molodechno
Vilyeyka
Smarhon'/Smorgon'
Myadzyel
Hlybokaye/Glubokoye
Pastavy/Postovy
Plyeshchanitsy
Lyepyel'/Lepel'
Byahoml'/Begoml'
Vyetryna
Navapolatsk/Novopolotsk
Polatsk/Polotsk
Yezyaryshcha
Harany/Goryany
Sarochyna
Obal'
Shumilina
Bacheykava
Chashniki
Bahushewsk
Haradok/Gorodok
Surazh
Vitsyebsk/Vitebsk
Lyozna/Liozno
Barysaw/Borisov
Krupki
Talachyn/Tolochin
Zhodzina/Zhodino
Kruhlaye
Orsha
Shklow
Sava
Horki/Gorki
Byalynichy/Belynichi
Mahilyow/Mogilëv
Yalizava
Chachevichy
Dashkawka
Harbavichy/Gorbovichi
Chavusy/Chausy
Khodasy
Krychaw/Krichëv
Cherykaw
Slawharad
Klimavichy
Kastsyukovichy
Baron'ki
Babruysk/Bobruysk
Abidavichy/Obidovichi
Rahachow/Rogachëv
Brozha
Shchadryn
Aktsyabrski
Zhlobin
Myerkulavichy
Buda-Kashalyova
Svyetlahorsk/Svetlogorsk
Ptsich
Shyichy
Kalinkavichy
Mazyr/Mozyr'
Rechytsa/Rechitsa
Uvaravichy
Bal'shavik
Kastsyukowka
Homyel'/Gomel'
Dobrush
Tsyerakhowka
Narowlya
Khoyniki
Loyew
Byval'ki
Kyivske Vodoskhovyshche
Viliya
Neman
Western Dvina
Dnieper
Dnyapro/Dnepr
Byarezina
Ptsich
Pripyat
Yasyel'da
Horyn'
Styr
Bug
76
86
87
89
A
B
C
D
E
5
6
7
8
22°
24°
26°
28°
30°
32°
50°
52°
54°
56°

Elevation

-6000m	-4000m	-2000m	-1000m	-500m	-250m	Below sea level 0	250m	500m	1000m	2000m	3000m	4000m	6000m
-19,658ft	-13,124ft	-6562ft	-3281ft	-1640ft	-820ft	-328ft/-100m 0	820ft	1640ft	3281ft	6562ft	9843ft	13,124ft	19,685ft

Ukraine, Moldova & Romania

A
B
C
D
1
2
3
4
5
84
77
78
82
N
18°
20°
22°
24°
26°
28°
44°
46°
48°
50°
52°
BELARUS
POLAND
SLOVAKIA
HUNGARY
SERBIA
BULGARIA
UKRAINE
MOLDOVA
ROMANIA
Pripyat
Pripyat Marshes
Małopolska
Wyżyna Lubelska
Wisła
Bug
Styr
Sluch
Carpathian Mountains
Tatra Mountains
Slovenské Rudohorie
Tisza
Great Hungarian Plain
Someș
Mureș
Timiș
Danube
Velika Morava
Danube (Dunărea)
Dunavska Ravnina
Jiu
Olt
Ialomița
Siret
Prut
Dniester
Podilska Vysochyna
Transylvania
Munții Apuseni
Carpații Meridionali
Wallachia
Hora Hoverla 2061m
Vârful Moldoveanu 2544m
Ozero Yalpuh
Lacul Razim
Lacul Sinoie
Kovel
Sarny
Olevsk
Volodymyr-Volynskyi
Novovolynsk
Kivertsi
Lutsk
Rivne
Korosten
Sokal
Dubno
Novohrad-Volynskyi
Malyn
Radomyshl
Chervonohrad
Slavuta
Shepetivka
Zhovkva
Kremenets
Yavoriv
Lviv
Iziaslav
Polonne
Zhytomyr
Horodok
Zolochiv
Sambir
Starokostiantyniv
Berdychiv
Zbarazh
Drohobych
Khodoriv
Berezhany
Boryslav
Ternopil
Khmelnytskyi
Stryi
Zhydachiv
Kalush
Chortkiv
Vinnytsia
Dolyna
Uzhhorod
Ivano-Frankivsk
Zhmerynka
Nadvirna
Kamianets-Podilskyi
Mukachevo
Berehove
Kolomyia
Vynohradiv
Khust
Chernivtsi
Mohyliv-Podilskyi
Negrești-Oaș
Satu Mare
Darabani
Soroca
Carei
Baia Mare
Rădăuți
Dorohoi
Marghita
Baia Sprie
Borșa
Solca
Botoșani
Bălți
Rîbnița
Suceava
Șimleu Silvaniei
Năsăud
Oradea
Zalău
Fălticeni
Aleșd
Dej
Bistrița
Târgu-Neamț
Pașcani
Călărași
Orhei
Salonta
Beiuș
Cluj-Napoca
Toplița
Iași
Ungheni
Strășeni
Bicaz
Reghin
Roman
Curtici
Ineu
Chișinău
Arad
Turda
Luduș
Gheorgheni
Târgu Mureș
Piatra-Neamț
Sânnicolau Mare
Abrud
Bacău
Hîncești
Tiraspol
Lipova
Aiud
Mediaș
Cristuru Secuiesc
Miercurea-Ciuc
Vaslui
Alba Iulia
Târgu Ocna
Jimbolia
Deva
Rupea
Comrat
Timișoara
Bârlad
Lugoj
Hunedoara
Sibiu
Făgăraș
Târgu Secuiesc
Adjud
Oțelu Roșu
Hațeg
Cisnădie
Codlea
Sfântu Gheorghe
Taraclia
Ciadîr-Lunga
Bocșa
Câmpulung
Brașov
Cahul
Artsyz
Râșnov
Focșani
Tecuci
Reșița
Petroșani
Bolhrad
Oravița
Anina
Călimănești
Râmnicu Sărat
Galați
Reni
Târgu Jiu
Curtea de Argeș
Sinaia
Brăila
Moldova Nouă
Râmnicu Vâlcea
Câmpina
Izmayil
Orșova
Moreni
Buzău
Măcin
Isaccea
Tulcea
Motru
Pitești
Mizil
Drobeta-Turnu Severin
Strehaia
Târgoviște
Ploiești
Drăgășani
Babadag
Filiași
Titu
Urziceni
Țăndărei
Hârșova
Buftea
Craiova
Slatina
Balș
Slobozia
București (Bucharest)
Caracal
Fetești
Băilești
Calafat
Medgidia
Roșiori de Vede
Alexandria
Oltenița
Călărași
Constanța
Corabia
Turnu Măgurele
Techirghiol
Giurgiu
Zimnicea
Eforie-Sud
Mangalia
0 km
100
0 miles
100
Population
National capital
below 50,000
50,000 to 100,000
100,000 to 500,000
above 500,000

E
F
G
H
30°
32°
34°
36°
38°
40°
52°
50°
48°
46°
44°
1
2
3
4
5
88
94
Dnieper (Dnyapro)
Desna
Horodnia
Snovsk
Shostka
Hlukhiv
Chernihiv
Krolevets
Konotop
Kyivske Vodoskhovyshche
Oster
Nizhyn
Bakhmach
Nosivka
Romny
Sumy
Dnieper Lowland
Brovary
Pryluky
Psel
Lebedyn
Yahotyn
Pyriatyn
Okhtyrka
Zolochiv
Vasylkiv
Kanivske Vodoskhovyshche
Hrebinka
Lubny
Myrhorod
Derhachi
Liubotyn
Kharkiv
Bila Tserkva
Kaniv
Bohuslav
Merefa
Kupiansk
Oskil
Zolotonosha
Cherkasy
Ilobyne
Poltava
Siverskyi Donets
Horodyshche
Kremenchutske Vodoskhovyshche
Starobilsk
Smila
Shpola
Chyhyryn
Kremenchuk
Izium
Kreminna
Rubizhne
Sievierodonetsk
Talne
Oleksandrivka
Svitlovodsk
Dniprodzerzhynske Vodoskhovyshche
Sloviansk
Lysychansk
Znamianka
Kramatorsk
Zolote
Mala Vyska
Oleksandriia
Kamianske (Dniprodzerzhynsk)
Novomoskovsk
Luhansk
Kostiantynivka
Sorokyne (Krasnodon)
Kropyvnytskyi (Kirovohrad)
Zhovti Vody
Dnipro
Pavlohrad
Horlivka
Kadiivka (Stakhanov)
Piatykhatky
Synelnykove
Yenakiieve
Makiivka
Khrustalnyi (Krasnyi Luch)
Dolynska
Pervomaisk
Bobrynets
Kryvyi Rih
Pokrovske
Chystiakove (Torez)
Kryve Ozero
Arbuzynka
Donetsk
Amvrosiivka
Inhulets
Zaporizhzhia
Novyy Buh
Pokrov (Ordzhonikidze)
Nikopol
Dokuchaievsk
Pivdennyi Buh
Voznesensk
Orikhiv
Volnovakha
Marhanets
Kamianka-Dniprovska
Polohy
Dniprorudne
Don
Tokmak
Kakhovske Vodoskhovyshche
Novoazovsk
Molochansk
Mariupol
Dnieper (Dnipro)
Mykolaiv
Nova Kakhovka
Black Sea Lowland
Gulf of Taganrog
Melitopol
Posad-Pokrovske
Kakhovka
Berdiansk
Yeya
Ochakiv
Kherson
Primorsk
Oleshky (Tsiurupynsk)
Yakymivka
Odesa
Hola Prystan
Chaplynka
Novotroitske
Chornomorsk (Illichivsk)
Kalanchak
Henichesk
Armiansk
Sea of Azov
Bilhorod-Dnistrovskyi
Yany Kapu (Krasnoperekopsk)
Karkinitska Zatoka
Dzhankoi
RUSSIA
Rozdolne
Kurman (Krasnohvardiiske)
Kerch Strait
Nyzhnohirskyi
Chornomorske
Zatoka Syvash
Kerch
Krym (Crimea)
Kuban'
Yevpatoriia
Yedy Kuiu (Lenine)
Saky
Simferopol
Feodosiia
Krymski Hory
Bakhchysarai
Sudak
(since 2014 the Ukrainian territory of Crimea has been annexed by Russia)
Alushta
Sevastopol
Yalta
Alupka
Black Sea
Srednerusskaya Vozvyshennost'
RUSSIA
UKRAINE
Elevation
Below sea level 0
250m
500m
1000m
2000m
3000m
4000m
6000m
-6000m
-4000m
-2000m
-1000m
-500m
-250m
820ft
1640ft
3281ft
6562ft
9843ft
13,124ft
19,685ft
-19,658ft
-13,124ft
-6562ft
-3281ft
-1640ft
-820ft
-328ft/-100m
0

European Russia

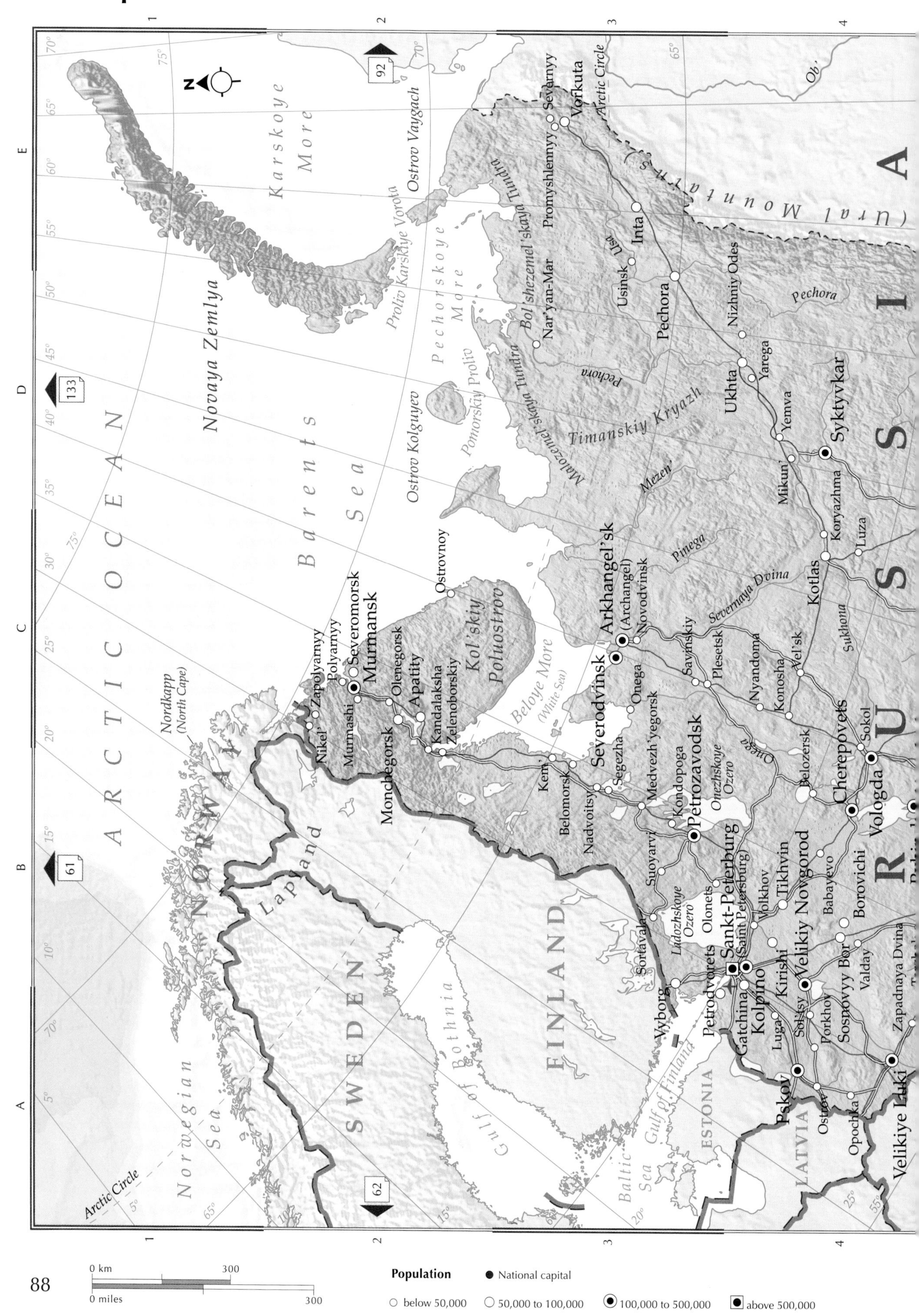

ARCTIC OCEAN
Barents Sea
Karskoye More
Pechorskoye More
Norwegian Sea
Gulf of Bothnia
Gulf of Finland
Baltic Sea
Beloye More (White Sea)
Novaya Zemlya
Ostrov Vaygach
Ostrov Kolguyev
Proliv Karskiye Vorota
Pomorskiy Proliv
Kol'skiy Poluostrov
NORWAY
SWEDEN
FINLAND
ESTONIA
LATVIA
RUSSIA
Lapland
Nordkapp (North Cape)
Arctic Circle
Ural Mountains
Timanskiy Kryazh
Bol'shezemel'skaya Tundra
Malozemel'skaya Tundra
Ladozhskoye Ozero
Onezhskoye Ozero
Pechora
Mezen'
Pinega
Severnaya Dvina
Sukhona
Onega
Usa
Ob'
Zapadnaya Dvina
Nikel'
Zapolyarnyy
Polyarnyy
Severomorsk
Murmansk
Murmashi
Olenegorsk
Apatity
Monchegorsk
Kandalaksha
Zelenoborskiy
Ostrovnoy
Kem'
Belomorsk
Segezha
Nadvoitsy
Medvezh'yegorsk
Kondopoga
Petrozavodsk
Suoyarvi
Sortavala
Olonets
Vyborg
Petrodvorets
Sankt-Peterburg (Saint Petersburg)
Kolpino
Gatchina
Luga
Kirishi
Volkhov
Tikhvin
Velikiy Novgorod
Sol'tsy
Porkhov
Pskov
Ostrov
Opochka
Velikiye Luki
Sosnovyy Bor
Valday
Borovichi
Babayevo
Belozersk
Cherepovets
Sokol
Vologda
Severodvinsk
Arkhangel'sk (Archangel)
Novodvinsk
Onega
Savinskiy
Plesetsk
Nyandoma
Konosha
Vel'sk
Kotlas
Koryazhma
Luza
Mikun'
Syktyvkar
Yemva
Ukhta
Yarega
Nizhniy Odes
Pechora
Usinsk
Inta
Nar'yan-Mar
Promyshlennyy
Vorkuta
Severnyy

0 km 300
0 miles 300

Population
● National capital
○ below 50,000
○ 50,000 to 100,000
◉ 100,000 to 500,000
▣ above 500,000

KAZAKHSTAN
UZBEKISTAN
UKRAINE
GEORGIA
AZERB.
ARM.
TURKMEN.
TURKEY (TÜRKIYE)
Kirghiz Steppe
Kyzylkum Desert
Ustyurt Plateau
Aral Sea
Caspian Sea
Black Sea
Sea of Azov
Caspian Depression
Ural'skiye Gor
El'brus 5642m
Doğu Karadeniz Dağlari
Syr Darya
Amu Darya
Ural
Volga
Don
Dnieper
Desna
Siverskyi Donets
Kuma
Euphrates
Vyatka
Kuybyshevskoye Vodokhranilishche
(since 2014 the Ukrainian territory of Crimea has been annexed by Russia)
(MOSCOW)
Podol'sk
Serpukhov
Kolomna
Vladimir
Dzerzhinsk
Nizhniy Novgorod
Murom
Cheboksary
Novocheboksarsk
Yoshkar-Ola
Kazan'
Naberezhnyye Chelny
Nizhnekamsk
Izhevsk
Glazov
Krasnokamsk
Perm'
Chusovoy
Kungur
Chaykovskiy
Neftekamsk
Ufa
Birsk
Oktyabr'skiy
Al'met'yevsk
Sterlitamak
Salavat
Beloretsk
Kumertau
Sibay
Baymak
Orsk
Novotroitsk
Saraktash
Orenburg
Sol'-Iletsk
Buzuluk
Buguruslan
Samara
Tol'yatti
Syzran'
Chapayevsk
Balakovo
Ul'yanovsk
Dimitrovgrad
Saratov
Vol'sk
Krasnyy Kut
Kamyshin
Volzhskiy
Volgograd
Akhtubinsk
Astrakhan'
Elista
Yaransk
Nolinsk
Kaluga
Tula
Aleksin
Shchëkino
Novomoskovsk
Ryazan'
Sasovo
Sarov
Kanash
Saransk
Penza
Kuznetsk
Michurinsk
Tambov
Orël
Yelets
Lipetsk
Voronezh
Gryazi
Yefremov
Tovarkovskiy
Liski
Borisoglebsk
Balashov
Krasnoarmeysk
Mikhaylovka
Ilovlya
Bryansk
Roslavl'
Pochinok
Klintsy
Zheleznogorsk
Kursk
Belgorod
Gubkin
Shebekino
Staryy Oskol
Rossosh'
Kantemirovka
Millerovo
Kamensk-Shakhtinskiy
Novoshakhtinsk
Taganrog
Rostov-na-Donu
Novocherkassk
Volgodonsk
Zimovniki
Sal'sk
Starominskaya
Tikhoretsk
Kropotkin
Krasnodar
Novorossiysk
Tuapse
Sochi
Maykop
Stavropol'
Svetlograd
Nevinnomyssk
Cherkessk
Kislovodsk
Pyatigorsk
Prokhladnyy
Nal'chik
Vladikavkaz
Groznyy
Buynaksk
Khasavyurt
Makhachkala
Kaspiysk
Derbent
5 6 7 8
A B C D E
35° 40° 45° 50° 55° 60°
87
92
95
100
Elevation
-6000m -4000m -2000m -1000m -500m -250m
Below sea level 0
250m 500m 1000m 2000m 3000m 4000m 6000m
-19,658ft -13,124ft -6562ft -3281ft -1640ft -820ft -328ft/-100m 0
820ft 1640ft 3281ft 6562ft 9843ft 13,124ft 19,685ft

North & West Asia

Franz Josef Land
ARCTIC
Severnaya Zemlya
Ostrov Komsomolets
Ostrov Oktyabr'skoy Revolyutsii
Ostrov Bol'shevik
Summer limit of pack ice
Winter limit of pack ice
Novaya Zemlya
East Novaya Zemlya Trench
Kara Sea
Barents Sea
Norwegian Sea
North Cape
Ostrov Kolguyev
Poluostrov Yamal
Poluostrov Taymyr
North Siberian
Noril'sk
Central Siberian Plateau
Murmansk
Kola Peninsula
Arctic Circle
White Sea
Archangel
Gulf of Bothnia
RUSSIA
West Siberian Plain
Ural Mountains
Lake Onega
Northern Dvina
Lake Ladoga
Saint Petersburg
Vologda
Yaroslavl'
Nizhniy Novgorod
Perm'
Yekaterinburg
Tomsk
Krasnoyarsk
Baltic Sea
Kaliningrad
MOSCOW
Central Russian Upland
Kazan'
Ufa
Chelyabinsk
Omsk
Novosibirsk
Novokuznetsk
KALININGRAD (to Russia)
Ul'yanovsk
Samara
Voronezh
Saratov
Orenburg
ASTANA
Qaraghandy (Karagandy)
Semey
Sayanskiy Khrebet
EUROPE
Oral/Ural'sk
Kazakh Steppe
Kazakh Uplands
Altai Mountains
Volgograd
(since 2014 the Ukrainian territory of Crimea has been annexed by Russia)
Rostov-na-Donu
Aral/Aral'sk
KAZAKHSTAN
Astrakhan'
Stavropol'
Lake Balkhash
Danube
El'brus 18,510ft (5642m)
Aktau
Ustyurt Plateau
Aral Sea
Qyzylorda/Kyzylorda
Black Sea
Caucasus
Caspian Sea
Kyzyl Kum
Taraz
Almaty
GEORGIA
Dasoguz
UZBEKISTAN
BISHKEK
Tien Shan
Jengish Chokusu/Tömür Feng 24,406ft (7439m)
Istanbul
TBILISI
ARMENIA
BAKU
TURKMENISTAN
TASHKENT
KYRGYZSTAN
ANKARA
YEREVAN
AZERB.
DUSHANBE
TURKEY (TÜRKIYE)
TAJIKISTAN
Tabriz
ASHGABAT
Mediterranean Sea
Gaziantep
Adana
Aleppo
Mosul
TEHRAN
Hindu Kush
Kunlun Mountains
CYPRUS
SYRIA
IRAQ
Qom
IRAN
KABUL
Jalalabad
BEIRUT
DAMASCUS
BAGHDAD
Isfahan
Herat
LEBANON
Iranian Plateau
AFGHANISTAN
Khyber Pass
ISRAEL
AMMAN
Syrian Desert
JERUSALEM
Dead Sea -1411ft (-430m)
JORDAN
Basra
Zagros Mountains
Himalayas
KUWAIT
KUWAIT CITY
Shiraz
Zahedan
Thar Desert
SAUDI ARABIA
MANAMA
BAHRAIN
Persian Gulf
Bandar-e 'Abbas
Dubai
Gulf of Oman
RIYADH
DOHA
U.A.E.
ABU DHABI
QATAR
MUSCAT
Sur
Indus Fan
Tropic of Cancer
Jedda
Arabian Peninsula
At Ta'if
Red Sea
Nile
Murray Ridge
Ganges
Ganges Fan
AFRICA
Ar Rub' al Khali
OMAN
YEMEN
SANAA
Ta'izz
Aden
Gulf of Aden
Socotra (to Yemen)
Arabian Sea
Bay of Bengal

133
59
81
47

0 km 800
0 miles 800

Population
• National capital
○ below 50,000
○ 50,000 to 100,000
◉ 100,000 to 500,000
▣ above 500,000

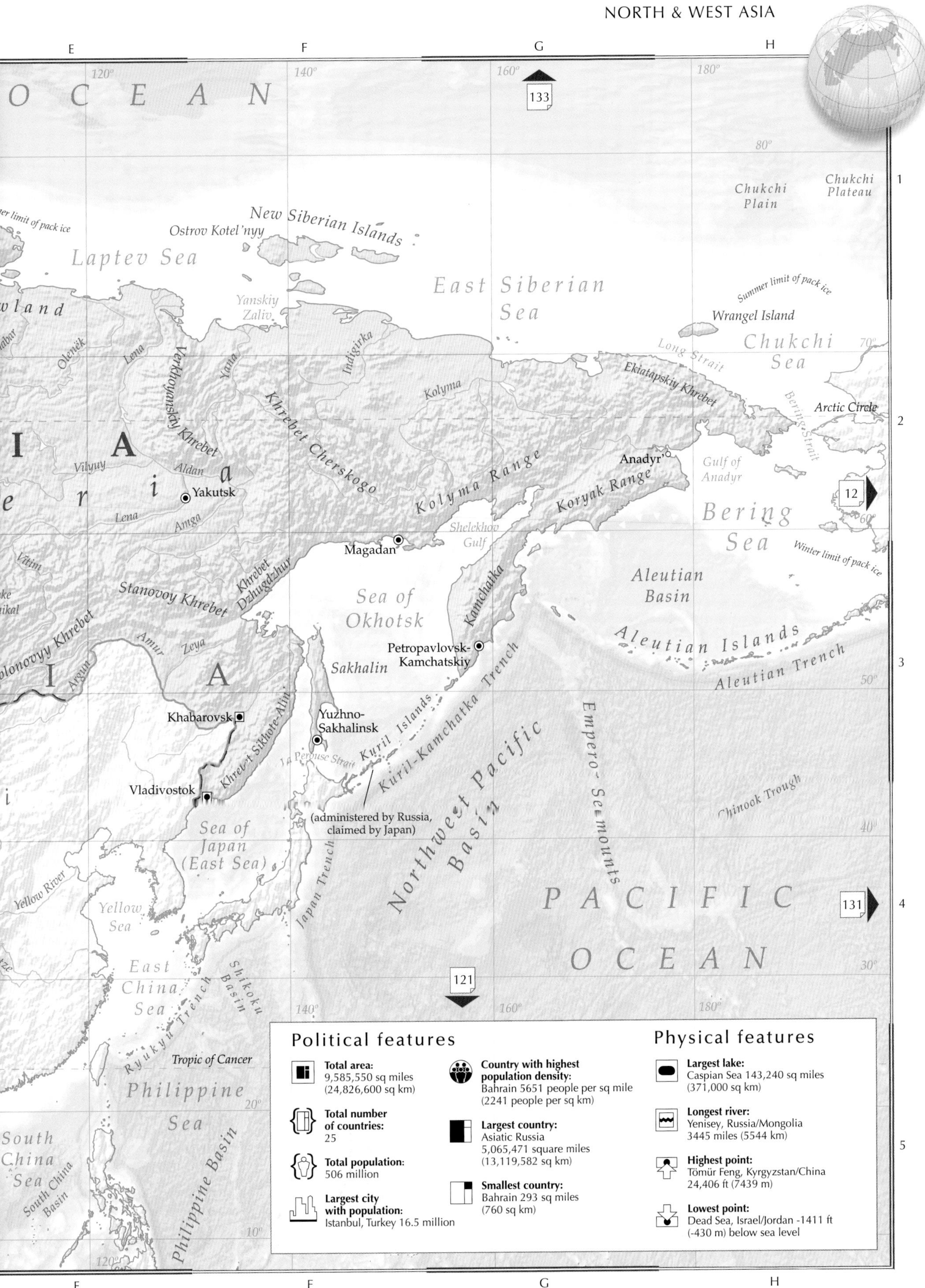
E
F
G
H
120°
140°
160°
180°
133
O C E A N
80°
1
Chukchi Plain
Chukchi Plateau
New Siberian Islands
Ostrov Kotel'nyy
Laptev Sea
Yanskiy Zaliv
East Siberian Sea
Summer limit of pack ice
Wrangel Island
Long Strait
Chukchi Sea
70°
Olenëk
Lena
Verkhoyanskiy Khrebet
Yana
Indigirka
Kolyma
Ekiatapskiy Khrebet
Bering Strait
Arctic Circle
2
Khrebet Cherskogo
Vilyuy
Aldan
Yakutsk
Lena
Amga
Kolyma Range
Anadyr'
Gulf of Anadyr
Koryak Range
12
Bering Sea
60°
Shelekhov Gulf
Magadan
Winter limit of pack ice
Vitim
Khrebet Dzhugdzhur
Stanovoy Khrebet
Sea of Okhotsk
Kamchatka
Aleutian Basin
Aleutian Islands
Amur
Zeya
Argun
Petropavlovsk-Kamchatskiy
Sakhalin
Aleutian Trench
3
50°
Khabarovsk
Yuzhno-Sakhalinsk
Kuril Islands
Kuril-Kamchatka Trench
Khrebet Sikhote-Alin'
La Perouse Strait
Emperor Seamounts
Vladivostok
(administered by Russia, claimed by Japan)
Northwest Pacific Basin
Chinook Trough
Sea of Japan (East Sea)
40°
Yellow River
Japan Trench
Yellow Sea
PACIFIC OCEAN
131
4
East China Sea
Shikoku Basin
30°
121
Ryukyu Trench
Tropic of Cancer
Philippine Sea
20°
South China Sea
South China Basin
Philippine Basin
10°
5
Political features
Total area:
9,585,550 sq miles
(24,826,600 sq km)
Total number of countries:
25
Total population:
506 million
Largest city with population:
Istanbul, Turkey 16.5 million
Country with highest population density:
Bahrain 5651 people per sq mile
(2241 people per sq km)
Largest country:
Asiatic Russia
5,065,471 square miles
(13,119,582 sq km)
Smallest country:
Bahrain 293 sq miles
(760 sq km)
Physical features
Largest lake:
Caspian Sea 143,240 sq miles
(371,000 sq km)
Longest river:
Yenisey, Russia/Mongolia
3445 miles (5544 km)
Highest point:
Tömür Feng, Kyrgyzstan/China
24,406 ft (7439 m)
Lowest point:
Dead Sea, Israel/Jordan -1411 ft
(-430 m) below sea level

Russia & Kazakhstan

NORWAY
SWEDEN
FINLAND
DENMARK
GERMANY
POLAND
LITH.
LAT.
EST.
BELARUS
UKRAINE
MOLDOVA
KALININGRAD (to Russia)
Kaliningrad
SVALBARD (to Norway)
Nordkapp (North Cape)
Barents Sea
Murmansk
Kandalaksha
Sankt-Peterburg
Pskov
Velikiy Novgorod
Petrozavodsk
Severodvinsk
Arkhangel'sk
Smolensk
Cherepovets
MOSKVA (MOSCOW)
Tver'
Vologda
Bryansk
Tula
Yaroslavl'
Kotlas
Ukhta
Vorkuta
Kineshma
Vladimir
Syktyvkar
Salekhard
Belgorod
Ryazan'
Nizhniy Novgorod
Kirov
Voronezh
Tambov
Penza
Kazan'
Glazov
Solikamsk
Perm'
Serov
Ul'yanovsk
Izhevsk
Tol'yatti
Saratov
Naberezhnyye Chelny
Samara
Balakovo
Volgograd
Ufa
Sterlitamak
Yekaterinburg
Tyumen'
Chelyabinsk
Rostov-na-Donu
Krasnodar
Sochi
Stavropol'
Nal'chik
Vladikavkaz
Groznyy
Makhachkala
Astrakhan'
Oral/Ural'sk
Orenburg
Magnitogorsk
Orsk
Rudnyy
Aqtöbe/Aktobe
Atyraū/Atyrau
Aqtaū/Aktau
KAZAKHSTAN
ASTANA
Pavlodar
Qostanay/Kostanay
Petropavlovsk
Omsk
Novosibirsk
Tomsk
Barnaul
Novokuznetsk
Qaraghandy/Karagandy
Temirtau
Zhezqazghan/Zhezkazgan
Qyzylorda/Kyzylorda
Öskemen/Ust'-Kamenogorsk
Semey
Balqash/Balkash
Shymkent
Taraz
Almaty
Taldyqorghan/Taldykorgan
Türkistan/Turkestan
Kentau
UZBEKISTAN
TURKMENISTAN
KYRGYZSTAN
TAJIKISTAN
AFGHANISTAN
IRAN
CHINA
Caspian Sea
Black Sea
Aral Sea
Novaya Zemlya
Karskoye More
Zapadno-Sibirskaya Ravnina
Surgut
Nizhnevartovsk
Noril'sk
Nadym
Nyagan'
Khanty-Mansiysk
(since 2014 the Ukrainian territory of Crimea has been annexed by Russia)
El'brus 5642m
Gora Belukha 4506m
Tien Shan

Population
● National capital
○ below 50,000
○ 50,000 to 100,000
◉ 100,000 to 500,000
▣ above 500,000

0 km 600
0 miles 600

Chukchi Sea
Arctic Circle
Bering Strait
ALASKA (to US)
Ostrov Vrangelya
Proliv Longa
Ekvyvatapskiy Khrebet
Anadyrskiy Zaliv
Pevek
Anadyr'
Anadyr'
Vostochno-Sibirskoye More
OCEAN
Ostrov Komsomolets
Ostrov Oktyabr'skoy Revolyutsii
Severnaya Zemlya
Novosibirskiye Ostrova
Ostrov Novaya Sibir'
Ostrov Kotel'nyy
Ostrov Bol'shoy Lyakhovskiy
Ambarchik
Cherskiy
Bering Sea
Koryakskoye Nagor'ye
More Laptevykh
Alazeya
Indigirka
Kolyma
Ostrov Taymyr
Ozero Taymyr
Ust'-Olenëk
Tiksi
Kazach'ye
Ossora
Ostrov Karaginskiy
Severo-Sibirskaya Nizmennost'
Yana
Khrebet Cherskogo
Adycha
Zaliv Shelikhova
Ust'-Kamchatsk
Vulkan Klyucheyskaya Sopka 4688m
Atlasovo
Anabar
Olenëk
Kotuy
Verkhoyanskiy Khrebet
Susuman
Atka
Magadan
Olenëk
Poluostrov Kamchatka
Mil'kovo
Petropavlovsk-Kamchatskiy
Srednesibirskoye Ploskogor'ye
Lena
Aldan
Okhotsk
Pervyy Kuril'skiy Proliv
Ostrov Paramushir
Vilyuy
Yakutsk
Nyurba
Okhotskoye More
Tunguska
SIBIR'
(SIBERIA)
Mirnyy
Suntar
Lena
Amga
Aldan
Khrebet Dzhugdzhur
Chunya
Olëkminsk
Shantarskiye Ostrova
Kuril'skiye Ostrova (Kuril Islands)
ASIA
Olëkma
Ostrov Sakhalin
Angara
Ust'-Ilimsk
Lena
Bodaybo
Neryungri
Ostrov Urup
Vitim
Tynda
Komsomol'sk-na-Amure
Amur
Ostrov Iturup
Kuril'sk
Ust'-Kut
Skovorodino
Khrebet Sikhote-Alin'
Bratsk
Ozero Baykal
Amur
Svobodnyy
Yuzhno-Sakhalinsk
Tulun
Yablonovyy Khrebet
Khabarovsk
La Perouse Strait
Usol'ye-Sibirskoye
Shilka
Blagoveshchensk
Birobidzhan
Khor
(administered by Russia, claimed by Japan)
Angarsk
Chita
Bikin
Olovyannaya
Irkutsk
Ulan-Ude
Krasnokamensk
Zabaykal'sk
Kyakhta
CHINA
Ussuriysk
Vladivostok
Nakhodka
JAPAN
MONGOLIA
Sea of Japan (East Sea)
NORTH KOREA
Gobi

14
130
130
106

Elevation

Below sea level: -6000m (-19,658ft), -4000m (-13,124ft), -2000m (-6562ft), -1000m (-3281ft), -500m (-1640ft), -250m (-820ft), -100m (-328ft), 0

Above sea level: 0, 250m (820ft), 500m (1640ft), 1000m (3281ft), 2000m (6562ft), 3000m (9843ft), 4000m (13,124ft), 6000m (19,685ft)

Turkey & The Caucasus

ROMANIA
UKRAINE
Krym (Crimea)
(since 2014 the Ukrainian territory of Crimea has been annexed by Russia)
BULGARIA
Black Sea
Danube
Lacul Sinoie
Varnenski Zaliv
Burgaski Zaliv
Maritsa
Edirne
Kırklareli
İstanbul Boğazı (Bosporus)
Ergene Çayi
Çorlu
Tekirdağ
İstanbul
İzmit
Adapazarı
Marmara Denizi (Sea of Marmara)
Yalova
İznik Gölü
Bandırma
Bursa
Bilecik
Çanakkale
Çanakkale Boğazı (Dardanelles)
Balıkesir
Simav Çayi
Bozüyük
Eskişehir
Edremit
Ayvalık
Kütahya
Simav
Akhisar
Gediz
Lésvos
Manisa
Menemen
Gediz Nehri
Uşak
Afyon
Chíos
İzmir
Ödemiş
Alaşehir
Sámos
Aydın
Nazilli
Dinar
Büyükmenderes Nehri
Söke
Denizli
Isparta
Burdur
Burdur Gölü
Tavas
Milas
Muğla
Bodrum
Marmaris
Dalaman
Fethiye
Kaş
Finike
Antalya
Antalya Körfezi
Manavgat
Alanya
Dodekánisa (Dodecánese)
Ródos (Rhodes)
Kárpathos
GREECE
Zonguldak
Bartın
Cide
İnebolu
Sinop
Gerze
Küre Dağları
Kastamonu
Karabük
Devrek
Kargı
Bafra
Samsun
Canik Dağları
Çerkeş
Bolu
Gerede
Çankırı
Kızıl Irmak
Merzifon
Çorum
Kalecik
Alaca
Tokat
ANKARA
Kırıkkale
Sorgun
Yıldızeli
TURKEY (TÜRKIYE)
Polatlı
Hirfanli Baraji
Boğazlıyan
Şarkışla
Kulu
Tuz Gölü
Bünyan
Cihanbeyli
Nevşehir
İncesu
Kayseri
Gürün
Akşehir
Aksaray
Anatolia
Beyşehir Gölü
Konya
Niğde
Göksun
Suğla Gölü
Ereğli
Toros Dağları
Karaman
Kahramanmaraş
Gaziantep
Ceyhan
Tarsus
Adana
Osmaniye
Mersin (İçel)
Mut
İskenderun
Kilis
Kırıkhan
Silifke
Antakya
Anamur
TURKISH REPUBLIC OF NORTHERN CYPRUS (recognized only by Turkey)
CYPRUS
Orontes
Mediterranean Sea
LEBANON
86
82
83
50

Population
● National capital
○ below 50,000 ○ 50,000 to 100,000 ◉ 100,000 to 500,000 ▣ above 500,000

0 km 200
0 miles 200

RUSSIA
Caucasus
Caspian Sea
Abkhazia (Apkhazeti)
Gagra
Gudauta
Sokhumi
Ochamchire
Enguri
Mestia
Kazbek 5047m
South Ossetia
Kutaisi
Samt'redia
GEORGIA
Poti
Gori
Tsalka
TBILISI
Rustavi
Kobuleti
Lesser Caucasus
Batumi
Ajaria (Ach'ara)
Akhaltsikhe
Hopa
Artvin
Pazar
Rize
Of
Trabzon
Giresun
Doğu Karadeniz Dağları
Çoruh Nehri
Gümüşhane
İspir
Sarıkamış
Kars
Gyumri
Art'ik
Vanadzor
Sevan
ARMENIA
YEREVAN
Sevana Lich
Artashat
Aras
Büyük Ağrı Dağı (Mount Ararat) 5137m
Horasan
Pasinler
Aşkale
Erzurum
Ağri
Erzincan
Tercan
Kemah
Doğubayazıt
Patnos
Erciş
Muradiye
Keban Baraji
Bingöl
Muş
Tatvan
Van Gölü
Van
Elazığ
Bitlis
Gevaş
Toroslar
Silvan
Siirt
Diyarbakır
Batman
Silverek
Şırnak
Mardin
Viranşehir
Kurdistan
Nusaybin
Şanlıurfa
Ceylanpınar
Zaqatala
Greater Caucasus
Xaçmaz
Quba
Siyäzän
Şäki
Kura
Mingäçevir
Şamaxi
Sumqayıt
BAKI (BAKU)
Gäncä
Yevlax
AZERBAIJAN
Nagornyy-Karabakh
İmişli
Hacıqabul
Şirvan
Xankändi
Goris
Biläsuvar
Naxçıvan
Länkäran
Daryächeh-ye Orūmīyeh
Reshteh-ye Kuhhā-ye Alborz (Elburz Mountains)
IRAN
Tigris
Al Jazīrah
Euphrates
IRAQ
Jabal Bishrī
Buḩayrat ath Tharthār
Kūhhā-ye Zāgros (Zagros Mountains)
SYRIA
89
100
98
E
F
G
H
1
2
3
4
5
40°
45°
50°
35°
Elevation
Below sea level 0
250m 500m 1000m 2000m 3000m 4000m 6000m
820ft 1640ft 3281ft 6562ft 9843ft 13,124ft 19,685ft
-6000m -4000m -2000m -1000m -500m -250m
-19,658ft -13,124ft -6562ft -3281ft -1640ft -820ft -328ft/-100m 0

The Near East

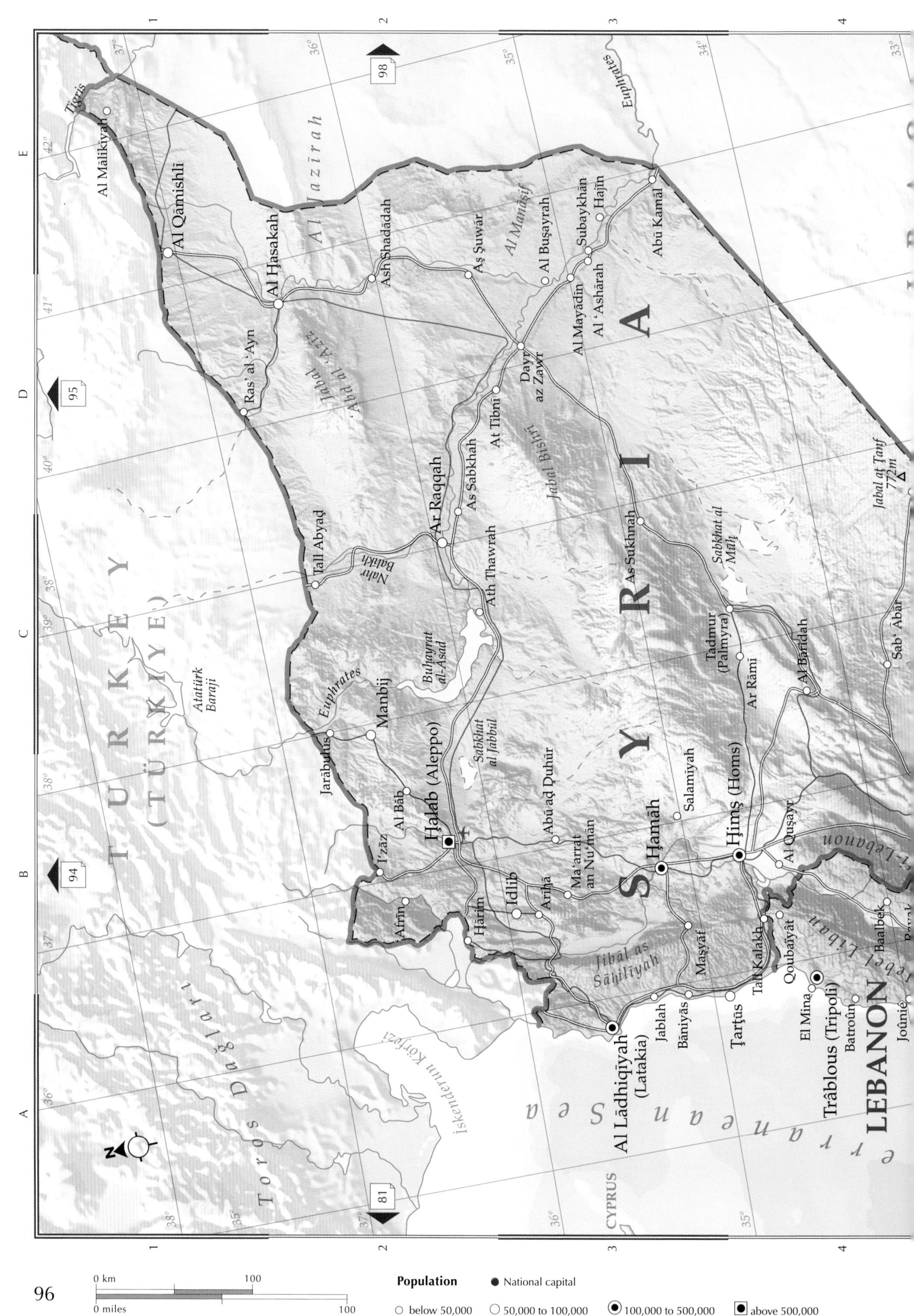
TURKEY
(TÜRKIYE)
SYRIA
LEBANON
CYPRUS
Toros Dağları
İskenderun Körfezi
Atatürk Barajı
Al Jazīrah
Jabal 'Abd al 'Azīz
Jabal Bishrī
Al Manāsif
Buḥayrat al-Asad
Sabkhat al Jabbūl
Sabkhat al Mūḥ
Jibāl as Sāḥilīyah
Jabal at Tanf
772m
Euphrates
Tigris
Nahr Balīkh
Al Mālikīyah
Al Qāmishlī
Al Ḥasakah
Ras' al 'Ayn
Ash Shadādah
Aṣ Ṣuwār
Al Buṣayrah
Subaykhān
Hajīn
Abū Kamāl
Al Mayādīn
Al 'Ashārah
Dayr az Zawr
At Tibnī
As Sabkhah
Ar Raqqah
Tall Abyaḍ
Ath Thawrah
As Sukhnah
Tadmur (Palmyra)
Ar Rāmī
Al Bāridah
Sab' Ābār
Manbij
Jarābulus
Al Bāb
Ḥalab (Aleppo)
I'zāz
'Afrīn
Ḥārim
Idlib
Arīḥā
Abū aḍ Ḍuhūr
Ma'arrat an Nu'mān
Salamīyah
Ḥamāh
Ḥimṣ (Homs)
Al Quṣayr
Maṣyāf
Tall Kalakh
Al Lādhiqīyah (Latakia)
Jablah
Bāniyās
Ṭarṭūs
Qoubaïyât
Baalbek
Trâblous (Tripoli)
El Mina
Batroûn
Joûnié
94
95
98
81

0 km 100
0 miles 100

Population ● National capital
○ below 50,000 ○ 50,000 to 100,000 ◉ 100,000 to 500,000 ▣ above 500,000

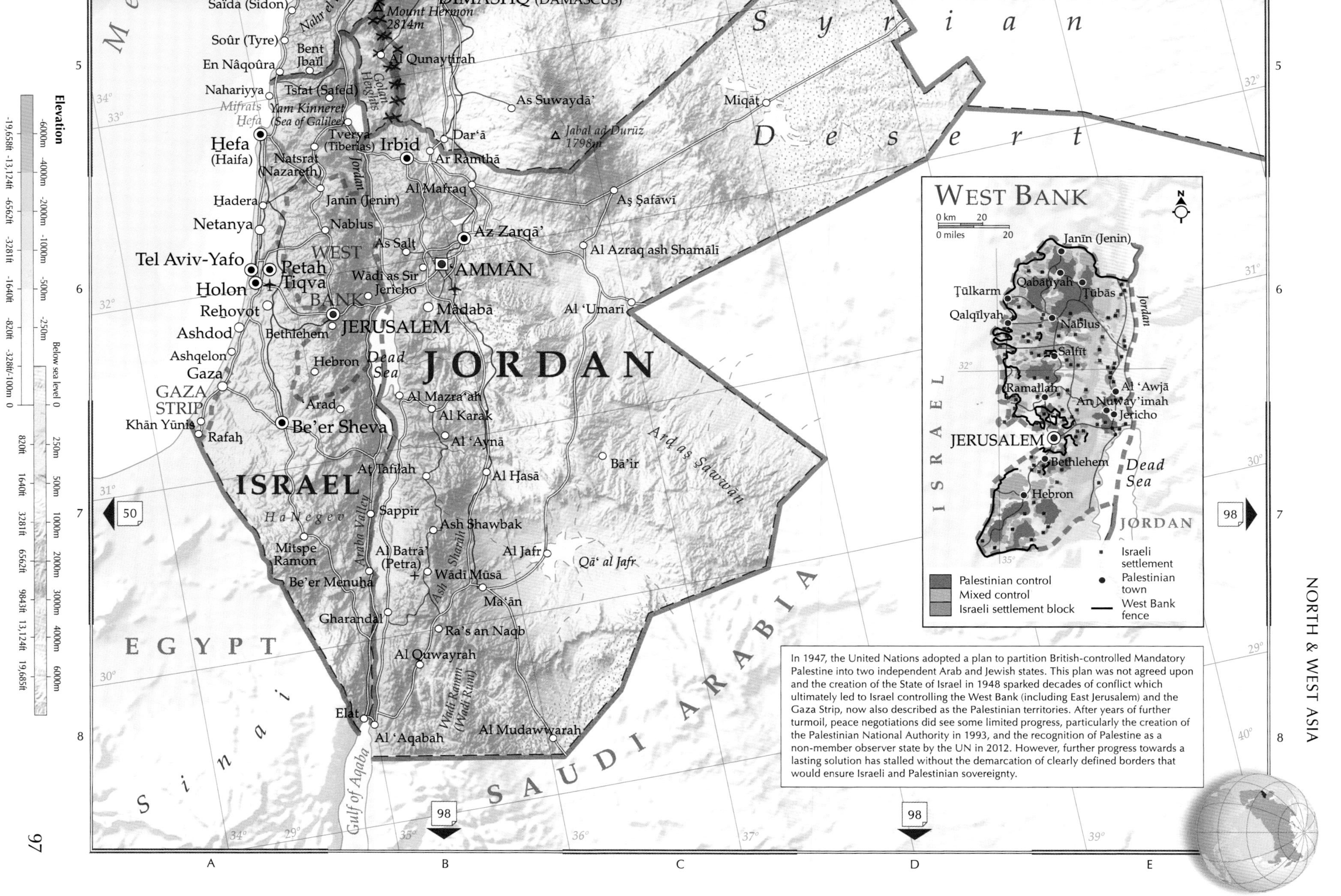
WEST BANK
0 km 20
0 miles 20
Janīn (Jenin)
Qabāţiyah
Ţūbās
Ţūlkarm
Qalqīlyah
Nāblus
Salfīt
Ramallah
Al 'Awjā
An Nuway'imah
Jericho
JERUSALEM
Bethlehem
Hebron
Jordan
Dead Sea
ISRAEL
JORDAN
Palestinian control
Mixed control
Israeli settlement block
Israeli settlement
Palestinian town
West Bank fence
Elevation
-6000m -4000m -2000m -1000m -500m -250m Below sea level 0 250m 500m 1000m 2000m 3000m 4000m 6000m
-19,658ft -13,124ft -6562ft -3281ft -1640ft -820ft -328ft/-100m 0 820ft 1640ft 3281ft 6562ft 9843ft 13,124ft 19,685ft
Syrian Desert
DIMASHQ (DAMASCUS)
Mount Hermon 2814m
Saïda (Sidon)
Soûr (Tyre)
Bent Jbaïl
En Nâqoûra
Al Qunayţirah
Golan Heights
Nahariyya
Tsfat (Safed)
Mifrats Hefa
Yam Kinneret (Sea of Galilee)
As Suwaydā'
Miqāţ
Hefa (Haifa)
Tverya (Tiberias)
Irbid
Dar'ā
Jabal ad Durūz 1798m
Natsrat (Nazareth)
Ar Ramthā
Jordan
Hadera
Janīn (Jenin)
Al Mafraq
Aş Şafāwī
Netanya
Nablus
Az Zarqā'
Al Azraq ash Shamālī
Tel Aviv-Yafo
Petah Tiqva
WEST BANK
As Salt
Wādī as Sīr
AMMĀN
Holon
Jericho
Rehovot
Mādabā
Al 'Umarī
Ashdod
Bethlehem
JERUSALEM
Ashqelon
Gaza
Hebron
Dead Sea
JORDAN
GAZA STRIP
Arad
Al Mazra'ah
Al Karak
Khān Yūnis
Rafah
Be'er Sheva
Al 'Aynā
Ard aş Şawwān
Bā'ir
At Tafilah
ISRAEL
Al Hasā
HaNegev
Sappir
Ash Shawbak
Araba Valley
Mitspe Ramon
Al Batrā' (Petra)
Al Jafr
Qā' al Jafr
Wādī Mūsā
Ash Sharāh
Be'er Menuha
Ma'ān
Gharandal
Ra's an Naqb
EGYPT
Al Quwayrah
Wadi Ramm (Wadi Rum)
Sinai
Elat
Al 'Aqabah
Al Mudawwarah
Gulf of Aqaba
SAUDI ARABIA
50
98

In 1947, the United Nations adopted a plan to partition British-controlled Mandatory Palestine into two independent Arab and Jewish states. This plan was not agreed upon and the creation of the State of Israel in 1948 sparked decades of conflict which ultimately led to Israel controlling the West Bank (including East Jerusalem) and the Gaza Strip, now also described as the Palestinian territories. After years of further turmoil, peace negotiations did see some limited progress, particularly the creation of the Palestinian National Authority in 1993, and the recognition of Palestine as a non-member observer state by the UN in 2012. However, further progress towards a lasting solution has stalled without the demarcation of clearly defined borders that would ensure Israeli and Palestinian sovereignty.

The Middle East

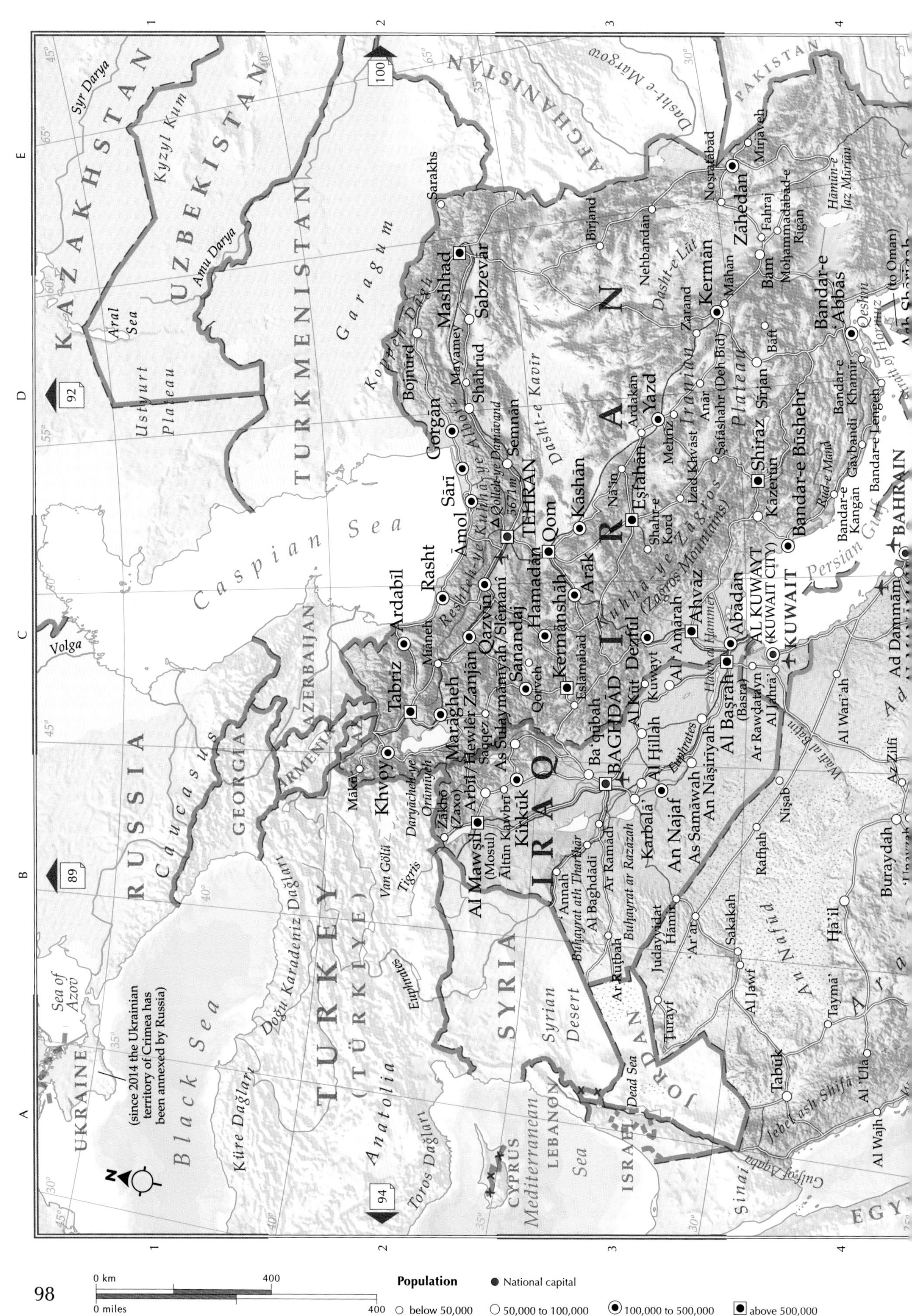

KAZAKHSTAN
UZBEKISTAN
TURKMENISTAN
AFGHANISTAN
PAKISTAN
RUSSIA
UKRAINE
GEORGIA
ARMENIA
AZERBAIJAN
TURKEY
(TÜRKIYE)
SYRIA
IRAQ
IRAN
JORDAN
LEBANON
ISRAEL
CYPRUS
KUWAIT
BAHRAIN
EGY
(since 2014 the Ukrainian territory of Crimea has been annexed by Russia)
Caspian Sea
Black Sea
Sea of Azov
Aral Sea
Mediterranean Sea
Dead Sea
Persian Gulf
Gulf of Aqaba
Strait of Hormuz
Syr Darya
Amu Darya
Volga
Tigris
Euphrates
Kyzyl Kum
Garagum
Ustyurt Plateau
Kopet Dag
Caucasus
Anatolia
Van Gölü
Daryācheh-ye Orūmīyeh
Küre Dağları
Doğu Karadeniz Dağları
Toros Dağları
Syrian Desert
An Nafūd
Jebel ash Shifā
Sinai
Dasht-e Kavīr
Dasht-e Lūt
Dasht-e Mārgow
Iranian Plateau
Kūhhā-ye Alborz
Kūhhā-ye Zāgros (Zagros Mountains)
Qolleh-ye Damāvand 5671m
Hāmūn-e Jaz Mūrīān
TEHRĀN
BAGHDĀD
AL KUWAYT (KUWAIT CITY)
Mashhad
Sarakhs
Sabzevār
Bojnūrd
Mayamey
Shāhrūd
Gorgān
Semnān
Sārī
Āmol
Rasht
Ardabīl
Qazvīn
Tabrīz
Mīāneh
Zanjān
Marāgheh
Khvoy
Mākū
Hamadān
Qom
Kāshān
Arāk
Sanandaj
Kermānshāh
Eslāmābād
Qorveh
Dezfūl
Ahvāz
Ābādān
Eşfahān
Nā'īn
Shahr-e Kord
Yazd
Ardakān
Mehrīz
Anār
Shīrāz
Kāzerūn
Bandar-e Būshehr
Bandar-e Kangān
Bandar-e Lengeh
Bandar-e 'Abbās
Bāft
Sīrjān
Kermān
Zarand
Māhān
Bam
Zāhedān
Nosratābād
Mīrjāveh
Fahraj
Mohammadābād-e Rīgān
Bīrjand
Nehbandān
Al Mawşil (Mosul)
Arbīl / Hewlêr
As Sulaymānīyah / Slêmanî
Kirkūk
Zākhō (Zaxo)
Al Başrah (Başra)
An Najaf
Karbalā'
Al Hillah
Al 'Amārah
Al Kūt
An Nāşirīyah
As Samāwah
Ar Ramādī
Ba'qūbah
Al Jahrā'
Ad Dammām
Buraydah
Ha'il
Tabūk
Al Wajh
Al 'Ulā
Taymā'
Al Jawf
Sakākah
'Ar'ar
Turayf
Rafhāh
Al Wari'ah
Az Zilfī

0 km 400
0 miles 400

Population ● National capital
○ below 50,000 ○ 50,000 to 100,000 ◉ 100,000 to 500,000 ▣ above 500,000

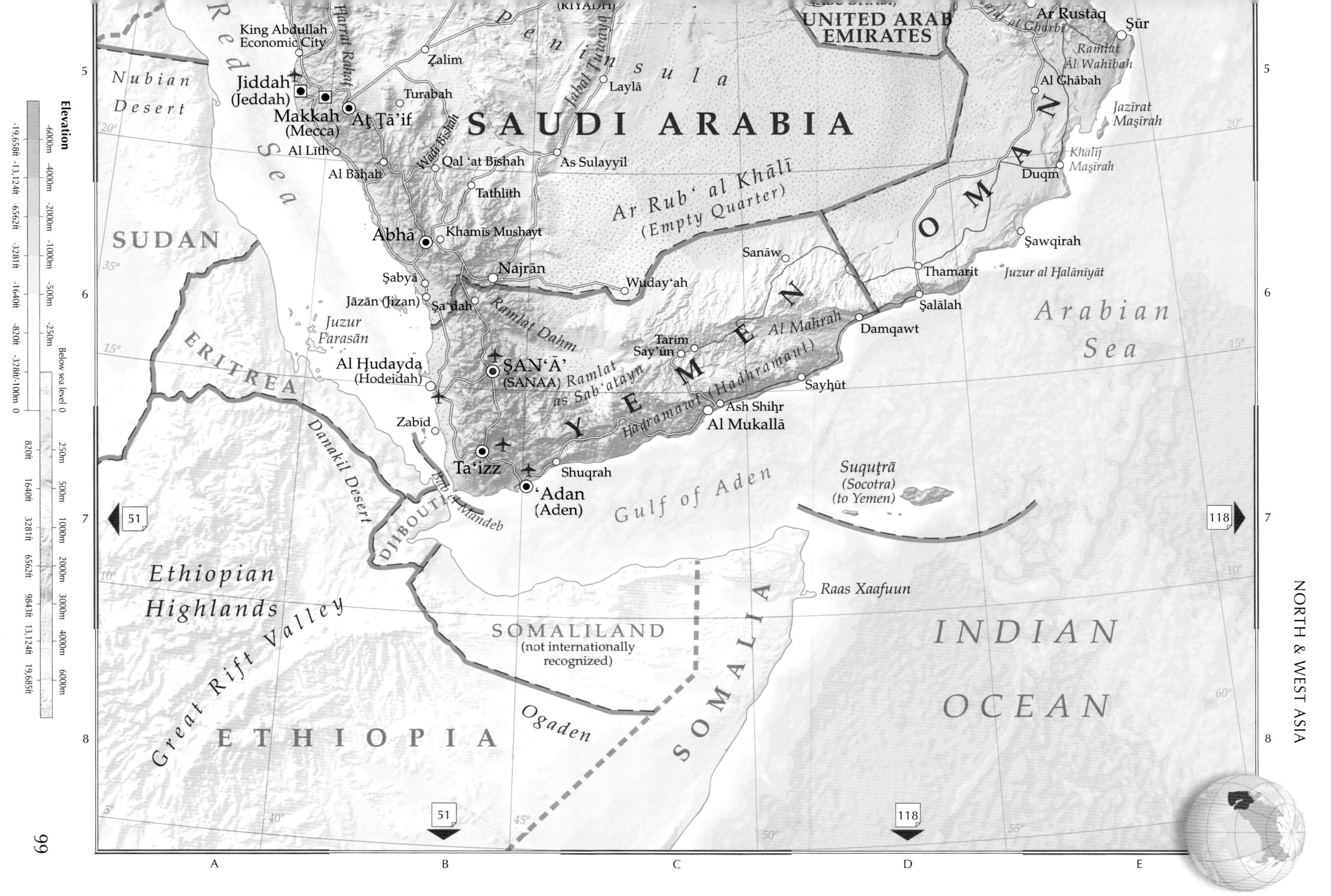
SAUDI ARABIA
YEMEN
OMAN
UNITED ARAB EMIRATES
SUDAN
ERITREA
DJIBOUTI
ETHIOPIA
SOMALIA
SOMALILAND
(not internationally recognized)
Ogaden
INDIAN OCEAN
Arabian Sea
Gulf of Aden
Red Sea
Bab el Mandeb
Peninsula
Ar Rub' al Khālī
(Empty Quarter)
Ramlat Dahm
Ramlat as Sab'atayn
Ḥaḍramawt
(Hadhramaut)
Al Mahrah
Jabal Ṭuwayq
Ḥarrat Rahaṭ
Wādī Bīshah
Nubian Desert
Danakil Desert
Ethiopian Highlands
Great Rift Valley
Juzur Farasān
Juzur al Ḥalānīyāt
Jazīrat Maṣīrah
Khalīj Maṣīrah
Ramlat Al Wahībah
Suquṭrā
(Socotra)
(to Yemen)
Raas Xaafuun
ṢAN'Ā'
(SANAA)
'Adan
(Aden)
Ta'izz
Al Ḥudayda
(Hodeidah)
Zabīd
Ṣa'dah
Najrān
Wuday'ah
Say'ūn
Tarīm
Al Mukallā
Ash Shiḥr
Sayḥūt
Sanāw
Damqawt
Shuqrah
Ṣalālah
Thamarīt
Duqm
Ṣawqirah
Al Ghābah
Ar Rustāq
Ṣūr
Jiddah
(Jeddah)
King Abdullah Economic City
Makkah
(Mecca)
Aṭ Ṭā'if
Al Līth
Al Bāḥah
Abhā
Khamīs Mushayṭ
Ṣabyā
Jāzān (Jīzān)
Turabah
Ẓalim
Qal 'at Bīshah
Tathlīth
As Sulayyil
Laylā
118
51
Elevation
-6000m -4000m -2000m -1000m -500m -250m
Below sea level 0
250m 500m 1000m 2000m 3000m 4000m 6000m
-19,658ft -13,124ft -6562ft -3281ft -1640ft -820ft -328ft/-100m 0
820ft 1640ft 3281ft 6562ft 9843ft 13,124ft 19,685ft

Central Asia

A
B
C
D
1
2
3
4
5
92
98
98
98
RUSSIA
GEORGIA
AZERBAIJAN
Caspian Sea
Ustyurt Plateau
Aral Sea
Mo'ynoq
Chimboy
Taxtako'pir
Turan Lowland
Kyzylk
Köneürgenç
Nukus
Taxiatosh
Gubadag
Gurbansoltan Eje
Uchqudu
Daşoguz
Urganch
UZBEK
Garabogaz Aylagy
Gaplaňgyr Platosy
Uçtagan Gumy
To'rtko'l
Xiva
Gazojak
Lebap
Zarafsh
Üngüz
Angyrsyndaky Garagum
Türkmenbaşy
Türkmenbaşy Aylagy
Balkanabat
Hazar
Derweze
Amu Darya
Bereket
Türkmen Aylagy
TURKMENISTAN
Serdar
Köpetdag Gershi
Magtymguly
Baharly
Garagum
Seydi
Galkynyş
Buxo
Türkmenabat
Esenguly
Abadan
Gökdepe
AŞGABAT (ASHGABAT)
Gora Chapan 2889m
Sayat
Tejen
Mary
Bayramaly
Kaka
Murgap
Garagum
Reshteh-ye Kūhhā-ye Alborz
Tejen
Sarahs
Harīrūd
Garabil
Belentligi
Bala Murghab
Serhetabat
Towraghoudi
Silsilah-ye Safēd Koh
Ghōriyān
Herāt
IRAN
AFGHA
Shīndand
Kūhhā-ye Zāgros
Farah Rod
Farah
Dilārām
Iranian Plateau
Dasht-e Khāsh
Hāmūn-e Şāberī
Lashkar Gāh
Chakhānsūr
Zaranj
Dasht-e Mārgō
Deh-e Shū
Helmand Rod
Chāgai Hills
50°
45°
55°
60°
40°
35°
30°
55°
60°

0 km 200
0 miles 200

Population ● National capital
○ below 50,000 ○ 50,000 to 100,000 ⊙ 100,000 to 500,000 ▣ above 500,000

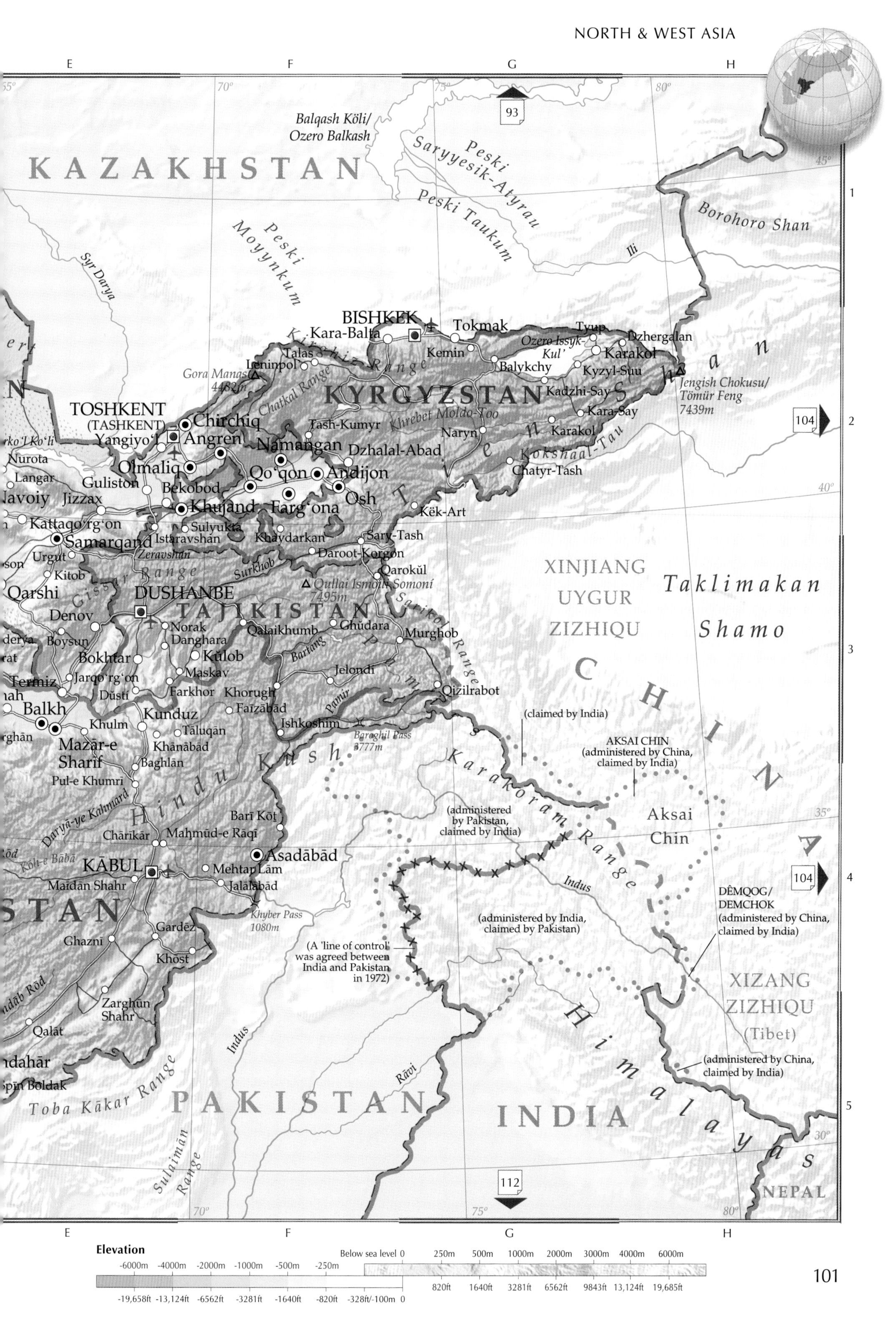

KAZAKHSTAN
Balqash Köli/ Ozero Balkash
Saryyesik-Atyrau Peski
Peski Taukum
Peski Moyynkum
Syr Darya
Borohoro Shan
Ili
BISHKEK
Kara-Balta
Tokmak
Talas
Kemin
Leninpol'
Gora Manas 4482m
Ozero Issyk-Kul'
Tyup
Dzhergalan
Karakol
Balykchy
Kyzyl-Suu
Kadzhi-Say
Kara-Say
Jengish Chokusu/ Tömür Feng 7439m
KYRGYZSTAN
Chatkal Range
Kirghiz Range
Khrebet Moldo-Too
Naryn
Karakol
Kokshaal-Tau
Tien Shan
Chatyr-Tash
TOSHKENT (TASHKENT)
Chirchiq
Yangiyo'l
Angren
Namangan
Dzhalal-Abad
Olmaliq
Qo'qon
Andijon
Nurota
Langar
Guliston
Bekobod
Osh
Navoiy
Jizzax
Khujand
Farg'ona
Këk-Art
Kattaqo'rg'on
Sulyukta
Samarqand
Istaravshan
Khaydarkan
Sary-Tash
Urgut
Zeravshan
Daroot-Korgon
Kitob
Gissar Range
Surkhob
Qarokŭl
Qarshi
DUSHANBE
Qullai Ismoili Somoní 7495m
XINJIANG UYGUR ZIZHIQU
Taklimakan Shamo
Denov
TAJIKISTAN
Norak
Danghara
Qalaikhumb
Ghŭdara
Murghob
Sarikol Range
Boysun
Bokhtar
Kŭlob
Bartang
Maskav
Jelondi
Termiz
Jarqo'rg'on
Dŭstí
Farkhor
Khorugh
Pamir
Qizilrabot
Balkh
Kunduz
Faīzābād
CHINA
Khulm
Tāluqān
Ishkoshim
Baroghil Pass 3777m
Mazār-e Sharīf
Khānābād
(claimed by India)
AKSAI CHIN (administered by China, claimed by India)
Baghlān
Pul-e Khumrī
Hindu Kush
Karakoram Range
Daryā-ye Kahmard
Barī Kōt
(administered by Pakistan, claimed by India)
Aksai Chin
Chārīkār
Maḥmūd-e Rāqī
KĀBUL
Asadābād
Mehtar Lām
Jalālābād
Maīdān Shahr
Indus
DÊMQOG/ DEMCHOK (administered by China, claimed by India)
Khyber Pass 1080m
(administered by India, claimed by Pakistan)
Gardēz
Ghaznī
(A 'line of control' was agreed between India and Pakistan in 1972)
Khōst
XIZANG ZIZHIQU (Tibet)
Zarghūn Shahr
Qalāt
Himalayas
(administered by China, claimed by India)
Indus
Rāvi
Toba Kākar Range
PAKISTAN
INDIA
Sulaimān Range
NEPAL
45°
40°
35°
30°
70°
75°
80°
1
2
3
4
5
E
F
G
H
Elevation
Below sea level 0
250m 500m 1000m 2000m 3000m 4000m 6000m
820ft 1640ft 3281ft 6562ft 9843ft 13,124ft 19,685ft
-6000m -4000m -2000m -1000m -500m -250m
-19,658ft -13,124ft -6562ft -3281ft -1640ft -820ft -328ft/-100m 0

93
104
104
112

South & East Asia

Caspian Sea
Aral Sea
Lake Balkhash
Lake Baikal
Black Sea
Syr Darya
Irtysh
Yenisey
Hovsgol Nuur
Uvs Nuur
Altai Mountains
MONGOLIA
Erdenet
ULAANBAATAR
Plateau of Mongolia
Ürümqi
Tien Shan
Turpan Pendi -505ft (-154m)
Tarim He
Tarim Basin
Gobi
Baotou
ASIA
Iranian Plateau
Hindu Kush
K2 28,251ft (8611m)
Takla Makan Desert
Altun Shan
Qilian Shan
Ordos Desert
Kunlun Mountains
Qaidam Pendi
Xiqing Shan
Lanzhou
Aksai Chin (administered by China, claimed by India)
Demqog/Demchok (administered by China, claimed by India)
Plateau of Tibet
CHINA
Peshawar
ISLAMABAD
Gujranwala
Lahore
Jammu and Kashmir
Quetta
Faisalabad
Multan
Ludhiana
Sutlej
Indus
PAKISTAN
Persian Gulf
Gulf of Oman
Arabian Peninsula
Thar Desert
Yamuna
Himalayas
Brahmaputra
Mekong
Salween
Sichuan Pendi
Chengdu
Chongqing
Delhi
NEW DELHI
Ganges
NEPAL
KATHMANDU
Mount Everest 29,032ft (8849m)
THIMPHU
BHUTAN
Hyderabad
Karachi
Jaipur
Kanpur
Murray Ridge
Mouths of the Indus
Rann of Kachchh
Ahmadabad
Vindhya Range
Patna
Guwahati
Imphal
Chindwinn
Irrawaddy
Kunming
BANGLADESH
DHAKA
Indore
Satpura Range
Khulna
Chattogram (Chittagong)
Nanning
Red River
Narmada
Nagpur
Gulf of Khambhat
Kolkata (Calcutta)
Mouths of the Ganges
Mandalay
VIETNAM
HANOI
Owen Fracture Zone
Arabian Sea
Mumbai (Bombay)
Pune
INDIA
Deccan
Godavari
Solapur
Eastern Ghats
Western Ghats
MYANMAR (BURMA)
Arakan Yoma
LAOS
NAY PYI TAW
Chiang Mai
Louangphabang
Vinh
VIENTIANE
Hyderabad
Bago
Yangon (Rangoon)
Pathein
Vijayawada
Hubballi
Arabian Basin
Bay of Bengal
Mouths of the Irrawaddy
THAILAND
Pakxe
Laccadive Islands (to India)
Bengaluru (Bangalore)
Chennai (Madras)
Mysuru (Mysore)
BANGKOK
Tônlé Sap
CAMBODIA
PHNOM PENH
Carlsberg Ridge
Andaman Islands (to India)
Jaffna
SRI LANKA
Gulf of Mannar
COLOMBO
SRI JAYAWARDENAPURA KOTTE
Nicobar Islands (to India)
Andaman Sea
Gulf of Thailand
Isthmus of Kra
Mouths of the Mekong
MALDIVES
MALE'
Chagos-Laccadive Plateau
Kota Bharu
Ceylon Plain
Equator
Mid-Indian Ridge
Strait of Malacca
Malay Peninsula
Medan
KUALA LUMPUR
PUTRAJAYA
Danau Toba
SINGAPORE
INDIAN OCEAN
Cocos Basin
Ninetyeast Ridge
Sumatra
Pegunungan Barisan
Pekanbaru
Padang
Pontianak
Palembang
BRITISH INDIAN OCEAN TERRITORY (to UK)
Mid-Indian Basin
Mascarene Plateau
Java Trench
Bandung
90
118
119

0 km 1000
0 miles 1000

Population
● National capital
○ below 50,000
○ 50,000 to 100,000
◉ 100,000 to 500,000
▣ above 500,000

Political features
Total area: 7,936,200 sq miles (20,554,700 sq km)
Total number of countries: 24
Total population: 4156 million
Largest city with population: Guangzhou, China 65.1 million
Country with highest population density: Singapore 20,769 people per sq mile (7692 people per sq km)
Largest country: China 3,705,386 sq miles (9,596,960 sq km)
Smallest country: Maldives 120 sq miles (300 sq km)
Physical features
Largest lake: Tônlé Sap, Cambodia 1042 sq miles (2850 sq km)
Longest river: Chang Jiang (Yangtze), China 3917 miles (6300 km)
Highest point: Mount Everest, China/Nepal 29,032ft (8849m)
Lowest point: Turpan Pendi (Turfan Basin), China -505 ft (-154 m) below sea level
130
130
130
120
Sakhalin
Kuril Islands
Kuril-Kamchatka Trench
Qiqihar
Manchuria Plain
Harbin
Lake Khanka
Changchun
Liao He
Shenyang
Great Khingan Range
Sapporo
Hokkaido
Northwest Pacific Basin
Shatskiy Rise
JAPAN
Sea of Japan (East Sea)
Honshu
Japan Trench
Sendai
NORTH KOREA
Dandong
PYONGYANG
Dalian
Tianjin
Bo Hai
SOUTH KOREA
SEOUL
SEJONG
Jinan
Qingdao
Nagoya
Kyoto
TOKYO
Yokohama
Fuji-san 12,389ft (3776m)
Osaka
Hiroshima
Shikoku
Kitakyushu
Kyushu
Korea Strait
Yellow Sea
East China Sea
Shanghai
Hangzhou
Nanchang
Fuzhou
PACIFIC OCEAN
Shikoku Basin
Kyushu Palau Ridge
Ryukyu Islands
Ryukyu Trench
TAIPEI
TAIWAN
Kaohsiung
Taiwan Strait
Hong Kong
Philippine Sea
Luzon Strait
West Mariana Basin
Mariana Trench
Luzon
Baguio
MANILA/MAYNILA
Mindoro
PHILIPPINES
Samar
Panay
Bacolod
Cebu
Negros
Palawan
SPRATLY ISLANDS (disputed)
Philippine Basin
Yap Trench
Micronesia
Equator
Ontong Java Rise
Eauripik Rise
Sulu Sea
Davao
Mindanao
Zamboanga
BANDAR SERI BEGAWAN
Celebes Sea
Manado
Halmahera
Melanesia
Bismarck Archipelago
Solomon Islands
Jayapura
Moluccas
Makassar Strait
Seram
Ambon
Buru
Pegunungan Maoke
New Guinea
Solomon Sea
Borneo
Balikpapan
Banjarmasin
Celebes
Makassar
Banda Sea
Flores Sea
INDONESIA
Lesser Sunda Islands
Bali
Flores
DILI
EAST TIMOR (TIMOR-LESTE)
Timor
Timor Trough
Sumba
Timor Sea
Arafura Sea
AUSTRALIA
Coral Sea

Western China & Mongolia

A
B
C
D
1
2
3
4
5
92
112
113
KAZAKHSTAN
Saryarqa/
Kazakhskiy Melkosopchnik
Balqash Köli
(Ozero Balkash)
Kulunda
Steppe
RUSS
Zapadnyy Sayan
Yenisey
Hövsgöl
Nuur
Uvs Nuur
Ulaangom
Ölgiy
Ozero
Zaysan
Altay
Altai Mountains
Hyargas
Nuur
Har Us Nuur
Har Nuur
Hovd
Hangayn Nuruu
MON
Altay
Bayanhongor
Tsetserleg
Ulungur
Hu
Karamay
Gurbantünggüt
Shamo
Kuytun
Shihezi
Fukang
Jimsar
Yining
Borohoro Shan
Ürümqi
Qitai
Turpan
Hami
Aj Bogd Uul
3802m
Atas Bogd
2695m
G
Ozero Issyk-Kul'
KYRGYZSTAN
Tien Shan
Jengish Chokusu/Tömür Feng
7439m
Turpan
Pendi
Bosten Hu
Korla
Kuruktag
Xingxingxia
Dalain Hob
Kashi
Tarim He
Tarim Basin
Lop Nur
GANSU
TAJIKISTAN
Yengisar
Shache
XINJIANG UYGUR
ZIZHIQU
Yecheng
(claimed
by India)
Pishan
Taklimakan
Shamo
Moyu
Hotan
Qira
Ruoqiang
Altun Shan
Qilian Shan
Danghe Nanshan
Qaidam Pendi
Qinghai
AFGH.
PAKISTAN
Karakoram Range
K2
8611m
Kashmir
Kunlun Shan
AKSAI
CHIN
AKSAI CHIN
(administered by
China, claimed
by India)
Golmud
Dulan
Burhan Budai Shan
Anyêmaqên
JAMMU
AND
KASHMIR
Indus
C
H
QINGHAI
Qingzang Gaoyuan
(Plateau of Tibet)
Tongtian He
Bayan Har Shan
Rutog
DÊMQOG/DEMCHOK
(administered by China,
claimed by India)
Gar Xincun
XIZANG
Zanda
ZIZHIQU
Tanggula Shan
Yushu
Mekong
Gozhê
Siling Co
Amdo
Tangra
Yumco
Gyaring
Co
Nagqu
Qamdo
(Tibet)
Ngangzê
Co
Nam Co
Damxung
Salween
Jinsha Jiang
Brahmaputra
Nyainqêntanglha Shan
Hengduan Shan
Himalayas
NEPAL
Lhazê
Xigazê
Lhasa
Maizhokunggar
Gonggar
ARUNACHAL
PRADESH
(claimed by China)
Mount Everest
8849m
Gyangzê
Yamuna
Ganges
INDIA
BHUTAN
INDIA
MYANMAR
(BURMA)
70°
75°
80°
85°
90°
95°
100°
55°
50°
45°
30°
25°
0 km
400
0 miles
400
Population
National capital
Internal administrative capital
below 50,000
50,000 to 100,000
100,000 to 500,000
above 500,000

E
F
G
H
93
RUSSIA
Ozero Baykal
Shilka
Argun (Ergun He)
Ergun
Jagdaqi
Amur (Heilong Jiang)
Hulun Buir (Hailar)
Manzhouli
Hulun Nur
Da Hinggan Ling
Onon
Selenga
Sühbaatar
Darhan
Onon Gol
Choybalsan
Erdenet
ULAANBAATAR
Dzuunmod
Öndörhaan
Menengiyn Tal
Holin Gol
Baruun-Urt
Kerulen
HEILONGJIANG
Lake Khanka
JILIN
106
Tongliao
Xilinhot
NEI MONGOL ZIZHIQU (Inner Mongolia)
Chifeng (Ulanhad)
Saynshand
Erenhot
Liao He
LIAONING
Sea of Japan (East Sea)
Dalandzadgad
NORTH KOREA
Ulan Qab (Jining)
Korea Bay
Liaodong Wan
Lang Shan
Hohhot
BEIJING
Baotou
Bo Hai
SOUTH KOREA
Huang He (Yellow River)
TIANJIN
Wuhai (Haibowan)
Mu Us Shadi
HEBEI
Tengger Shamo
Great Wall of China
SHANDONG
Yellow Sea
SHANXI
NINGXIA
JAPAN
Huang He (Yellow River)
108
JIANGSU
GANSU
HENAN
East China Sea
SHAANXI
SHANGHAI SHI
Han Shui
ANHUI
HUBEI
ZHEJIANG
Chang Jiang (Yangtze)
SICHUAN
CHONGQING
Nansei-shotō (to Japan)
JIANGXI
HUNAN
FUJIAN
YUNNAN
Tropic of Cancer
107
GUIZHOU
TAIWAN
1
2
3
4
5
Elevation
Below sea level 0
250m 500m 1000m 2000m 3000m 4000m 6000m
820ft 1640ft 3281ft 6562ft 9843ft 13,124ft 19,685ft
-6000m -4000m -2000m -1000m -500m -250m
-19,658ft -13,124ft -6562ft -3281ft -1640ft -820ft -328ft/-100m 0

Eastern China & Korea

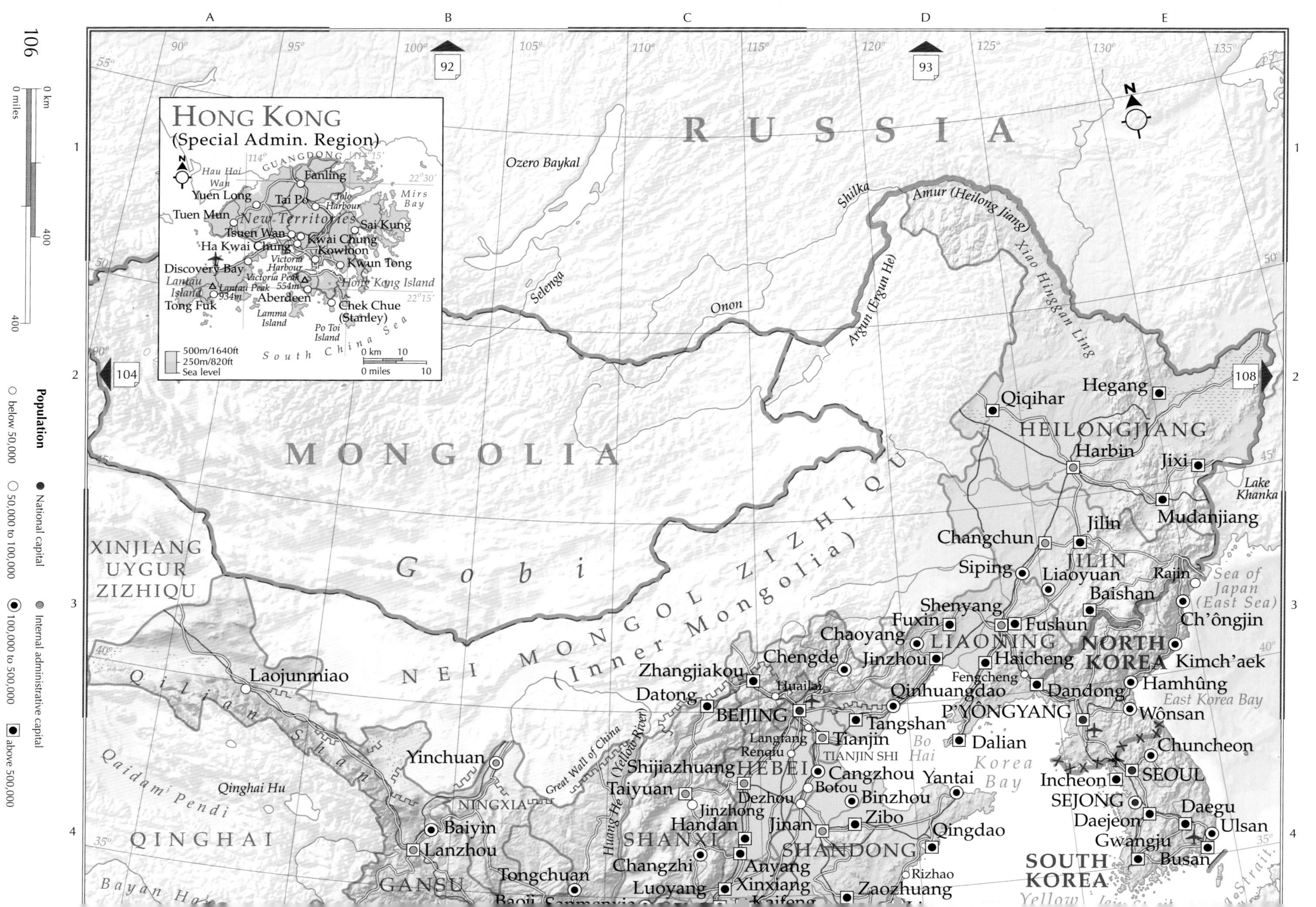

RUSSIA
MONGOLIA
NORTH KOREA
SOUTH KOREA
HEILONGJIANG
JILIN
LIAONING
HEBEI
SHANDONG
SHANXI
NINGXIA
GANSU
QINGHAI
XINJIANG UYGUR ZIZHIQU
NEI MONGOL ZIZHIQU (Inner Mongolia)
TIANJIN SHI
Gobi
Qilian Shan
Qaidam Pendi
Qinghai Hu
Ozero Baykal
Selenga
Onon
Shilka
Amur (Heilong Jiang)
Argun (Ergun He)
Xiao Hinggan Ling
Lake Khanka
Sea of Japan (East Sea)
East Korea Bay
Korea Bay
Bo Hai
Huang He (Yellow River)
Great Wall of China
Qiqihar
Hegang
Harbin
Jixi
Mudanjiang
Jilin
Changchun
Siping
Liaoyuan
Baishan
Rajin
Ch'ôngjin
Shenyang
Fushun
Fuxin
Chaoyang
Jinzhou
Haicheng
Fengcheng
Dandong
Kimch'aek
Hamhûng
Wônsan
P'YÔNGYANG
Chengde
Zhangjiakou
Huailai
Datong
BEIJING
Qinhuangdao
Tangshan
Tianjin
Langfang
Renqiu
Dalian
Shijiazhuang
Cangzhou
Botou
Yantai
Taiyuan
Dezhou
Binzhou
Jinzhong
Handan
Jinan
Zibo
Qingdao
Changzhi
Anyang
Xinxiang
Luoyang
Rizhao
Zaozhuang
Tongchuan
Yinchuan
Baiyin
Lanzhou
Laojunmiao
Chuncheon
SEOUL
Incheon
SEJONG
Daejeon
Daegu
Ulsan
Gwangju
Busan
HONG KONG (Special Admin. Region)
GUANGDONG
Hau Hoi Wan
Fanling
Yuen Long
Tai Po
Tolo Harbour
Mirs Bay
Tuen Mun
New Territories
Sai Kung
Tsuen Wan
Kwai Chung
Ha Kwai Chung
Kowloon
Victoria Harbour
Kwun Tong
Discovery Bay
Lantau Island
Lantau Peak 934m
Victoria Peak 554m
Hong Kong Island
Aberdeen
Tong Fuk
Chek Chue (Stanley)
Lamma Island
Po Toi Island
South China Sea
500m/1640ft
250m/820ft
Sea level
0 km 10
0 miles 10

Population	
●	National capital
◉	Internal administrative capital
○	below 50,000
○	50,000 to 100,000
⊙	100,000 to 500,000
▣	above 500,000

Scale: 0 km – 400; 0 miles – 400

CHINA
SICHUAN
HUBEI
ANHUI
ZHEJIANG
FUJIAN
JIANGXI
HUNAN
GUANGDONG
GUANGXI ZHUANGZU ZIZHIQU
GUIZHOU
CHONGQING
YUNNAN
HAINAN
XIZANG ZIZHIQU (Tibet)
INDIA
MYANMAR (BURMA)
LAOS
THAILAND
CAMBODIA
VIETNAM
PHILIPPINES
TAIWAN
TAIPEI
East China Sea
PACIFIC OCEAN
South China Sea
Gulf of Tonkin
Gulf of Thailand
Luzon Strait
Taiwan Strait
Nansei-shoto (part of Japan)
Okinawa
(China and Taiwan claim all of each other's territory)
Tropic of Cancer
SOUTH KOREA'S TWO CAPITALS
SEOUL – Capital
SEJONG – Administrative capital
PARACEL ISLANDS (disputed by China, Taiwan and Vietnam)
Amphitrite Group
Crescent Group
Triton Island
SPRATLY ISLANDS (disputed by China, Malaysia, Philippines, Taiwan and Vietnam)
Thitu Island
Flat Island
Nanshan Island
Loaita Island
Namyit Island
Len Dao
Spratly Island
Hainan Dao
Hong Kong (S.A.R.)
Macao (Special Administrative Region)
Shanghai
Suzhou
Wuxi
Yangzhou
Nanjing
Hefei
Wuhu
Anqing
Jiaxing
Ningbo
Hangzhou
Jinhua
Wenzhou
Shangrao
Jingdezhen
Huangshi
Wuhan
Xinyang
Yichang
Jiujiang
Nanchang
Fuzhou
Nanping
Yong'an
Quanzhou
Xiamen
Zhangzhou
Longyan
Ganzhou
Shantou
Keelung
Taichung
Chiayi
Tainan
Kaohsiung
Guangzhou
Dongguan
Shaoguan
Jiangmen
Zhaoqing
Maoming
Zhanjiang
Haikou
Xuwen
Danzhou
Dongfang
Suixi
Beihai
Qinzhou
Nanning
Yulin
Liuzhou
Guilin
Yongzhou
Chenzhou
Hengyang
Xiangtan
Changsha
Loudi
Huaihua
Yueyang
Dongting Hu
Lichuan
Wanzhou
Three Gorges Reservoir
Chongqing
Guangyuan
Mianyang
Chengdu
Sichuan Pendi
Neijiang
Zigong
Leshan
Ya'an
Litang
Xichang
Zunyi
Guiyang
Anshun
Kunming
Gejiu
Dali
Baoshan
Jinghong
Chang Jiang
Chang Jiang (Yangtze)
Wuliang Shan
Hengduan Shan
Mekong
Salween
Red River
130
117
114
114
5 6 7 8
A B C D E
Elevation
Below sea level 0
-6000m -4000m -2000m -1000m -500m -250m
-19,658ft -13,124ft -6562ft -3281ft -1640ft -820ft -328ft/-100m 0
250m 500m 1000m 2000m 3000m 4000m 6000m
820ft 1640ft 3281ft 6562ft 9843ft 13,124ft 19,685ft

Japan

RUSSIA
CHINA
Sea of Okhotsk
Sea of Japan
JAPAN
Hokkaidō
Kuril Islands
(administered by Russia, claimed by Japan)
Ostrov Sakhalin
(to Russia)
Ostrov Iturup
Ostrov Kunashir
Ostrov Shikotan
La Pérouse Strait
Tsugaru-kaikyō
Mutsu-wan
Uchiura-wan
Ishikari-wan
Rebun-tō
Rishiri-tō
Okushiri-tō
Asahi-dake 2290m
Horoshiri-dake 2052m
Kuril'sk
Nemuro
Akkeshi
Kushiro
Shari
Kitami
Abashiri
Obihiro
Monbetsu
Nayoro
Shibetsu
Asahikawa
Takikawa
Ebetsu
Chitose
Tomakomai
Noboribetsu
Muroran
Hakodate
Sapporo
Otaru
Iwanai
Wakkanai
Aomori
Hachinohe
Goshogawara
Hirosaki
Odate
Noshiro
Gojome
Akita
Honjō
Sakata
Tsuruoka
Kuji
Iwate
Miyako
Morioka
Yokote
Shinjō
Furukawa
Sendai
Kesennuma
Shizugawa
Ishinomaki
Sendai-wan
TŌKYŌ
Sumitomo Building
Imperial Palace
Tokyo Tower
World Trade Center
Tōkyō University
National Museum
Tōkyō Stock Exchange
Tōkyō Bay
Tama-gawa
Haneda
Chiba
Kawasaki
Yokohama
Yokohama Bay Bridge
Places of interest
Regions/suburbs
NANSEI-SHOTŌ
Nansei-shotō (Ryukyu Islands)
Kyūshū
Ōsumi-shotō
Satsunan-shotō
Amami-guntō
Amami-ō-shima
Naze
Okinawa-shotō
Okinawa
Naha
Sakishima-shotō
Ishigaki-jima
Iriomote-jima
Senkaku-shotō
500m/1640ft
Sea level
Amur

0 km 200
0 miles 200

Population ● National capital

○ below 50,000 ○ 50,000 to 100,000 ◉ 100,000 to 500,000 ▣ above 500,000

Honshū
Iwaki
Hitachi
Utsunomiya
Mito
Chōshi
Oyama
Chiba
Yokohama
Kawagoe
TŌKYŌ
Kawasaki
Maebashi
Kasumiga-ura
Bōsō-hantō
Sagami-nada
Ō-shima
Nii-jima
Miyake-jima
Mikura-jima
Hachijō-jima
Izu-shotō
Kōzu-shima
Izu-hantō
Suruga-wan
Hamamatsu
Fuji
Fuji-san 3776m
Kōfu
Shizuoka
Toyota
Mikuni-sanmyaku
Nagaoka
Shinano-gawa
Jōetsu
Itoigawa
Nagano
Toyama
Matsumoto
Hida-sanmyaku
Nakatsugawa
Okazaki
Ise-wan
Ise
Owase
Shingū
Tanabe
Gobō
Wakayama
Takaoka
Kanazawa
Komatsu
Fukui
Tsuruga
Gifu
Ōgaki
Nagoya
Tsu
Ōtsu
Kyōto
Biwa-ko
Ōsaka
Kōbe
Awaji-shima
Kii-suidō
Toyama-wan
Wakasa-wan
Shikoku
Himeji
Harima-nada
Tokushima
Niihama
Matsuyama
Kōchi
Tosa-wan
Nakamura
Sukumo
Tottori
Yonago
Matsue
Okayama
Kurashiki
Chūgoku-sanchi
Kure
Hiroshima
Iwakuni
Hōfu
Iyo-nada
Gōtsu
Hamada
Masuda
Ube
Yamaguchi
Nagato
Shimonoseki
Kitakyūshū
Fukuoka
Kurume
Ōmuta
Ōita
Kumamoto
Bungo-suidō
Nobeoka
Kyūshū
Miyazaki
Miyakonojō
Yatsushiro
Kagoshima
Kagoshima-wan
Shibushi-wan
Satsuma-Sendai
Amakusa-nada
Sasebo
Nagasaki
Iki
Tsushima
Kō-saki
Gotō-rettō
Koshikijima-rettō
Ōsumi-shotō
Tanega-shima
Yaku-shima
East China Sea
Korea Strait
SOUTH KOREA
Oki-shotō
Dōgo
Dōzen
Liancourt Rocks (under South Korean control)
PACIFIC OCEAN
130
106
Elevation
-6000m -4000m -2000m -1000m -500m -250m
Below sea level 0
250m 500m 1000m 2000m 3000m 4000m 6000m
-19,658ft -13,124ft -6562ft -3281ft -1640ft -820ft -328ft/-100m 0
820ft 1640ft 3281ft 6562ft 9843ft 13,124ft 19,685ft

South India & Sri Lanka

Kalyān
Mumbai (Bombay)
112
Pune
Ahmadnagar
Nānded
Nizāmābād
Karīmnagar
Jagdalpur
Telangana
INDIA
Bārāmati
Solāpur
Hyderābād
Secunderābād
Vizianagaram
Visākhapatnam
Rājahmundry
Kākinada
Sāngli
Kolhāpur
Kalaburagi (Gulbarga)
Deccan
Karnātaka
Rāichūr
Krishna
Vijayawāda
Machilīpatnam
Andhra Pradesh
Arabian Sea
Belagāvi
Gadag
Kurnool
Chīrāla
Panaji
Hubballi
Nandyāl
Ongole
Kāvali
Tungabhadra Reservoir
Tādpatri
Dāvangere
Anantapur
Nellore
Shivamogga
Kadapa (Cuddapah)
Bhadrāvati
Udupi
Tumakūru
Bengalūru (Bangalore)
Chennai (Madras)
Mangalūru (Mangalore)
Mandya
Vellore
Kānchīpuram
Kāsaragod
Krishnagiri
Tiruppattūr
Kannur (Cannanore)
Mysūru (Mysore)
Puducherry (Pondicherry)
Amīndīvi Islands
Erode
Salem
Kozhikode (Calicut)
Neyveli
Coromandel Coast
Coimbatore
Tamil Nādu
Lakshadweep (Laccadive Islands) (to India)
Kavaratti Island
Thrissur (Trichūr)
Tiruchchirāppalli
Ernākulam
Dindigul
Madurai
Kalpeni Island
Kochi (Cochin)
Palk Strait
Jaffna
Nine Degree Channel
Alappuzha (Alleppey)
Rājapālaiyam
SRI LANKA
Kollam (Quilon)
Mannar
Vavuniya
Tuticorin (Thoothukudi)
Minicoy Island
Thiruvananthapuram (Trivandrum)
Trincomalee
Puttalam
Anuradhapura
Eight Degree Channel
Nāgercoil
Gulf of Mannar
Batticaloa
Matale
Negombo
Kandy
COLOMBO
Ihavandhippolhu Atoll
SRI JAYAWARDENAPURA KOTTE
Kalutara
Ratnapura
MALDIVES
Galle
Matara
Faadhippolhu Atoll
Horsburgh Atoll
Male' Atoll
Ari Atoll
MAALE (MALE')
Felidhu Atoll
Mulakatholhu
Kolhumadulu
Hadhdhunmathi Atoll
INDIAN
North Huvadhu Atoll
Equator
South Huvadhu Atoll
Gan
Addu Atoll
118
99
51
0 km 300
0 miles 300
Population
National capital
below 50,000
50,000 to 100,000
100,000 to 500,000
above 500,000

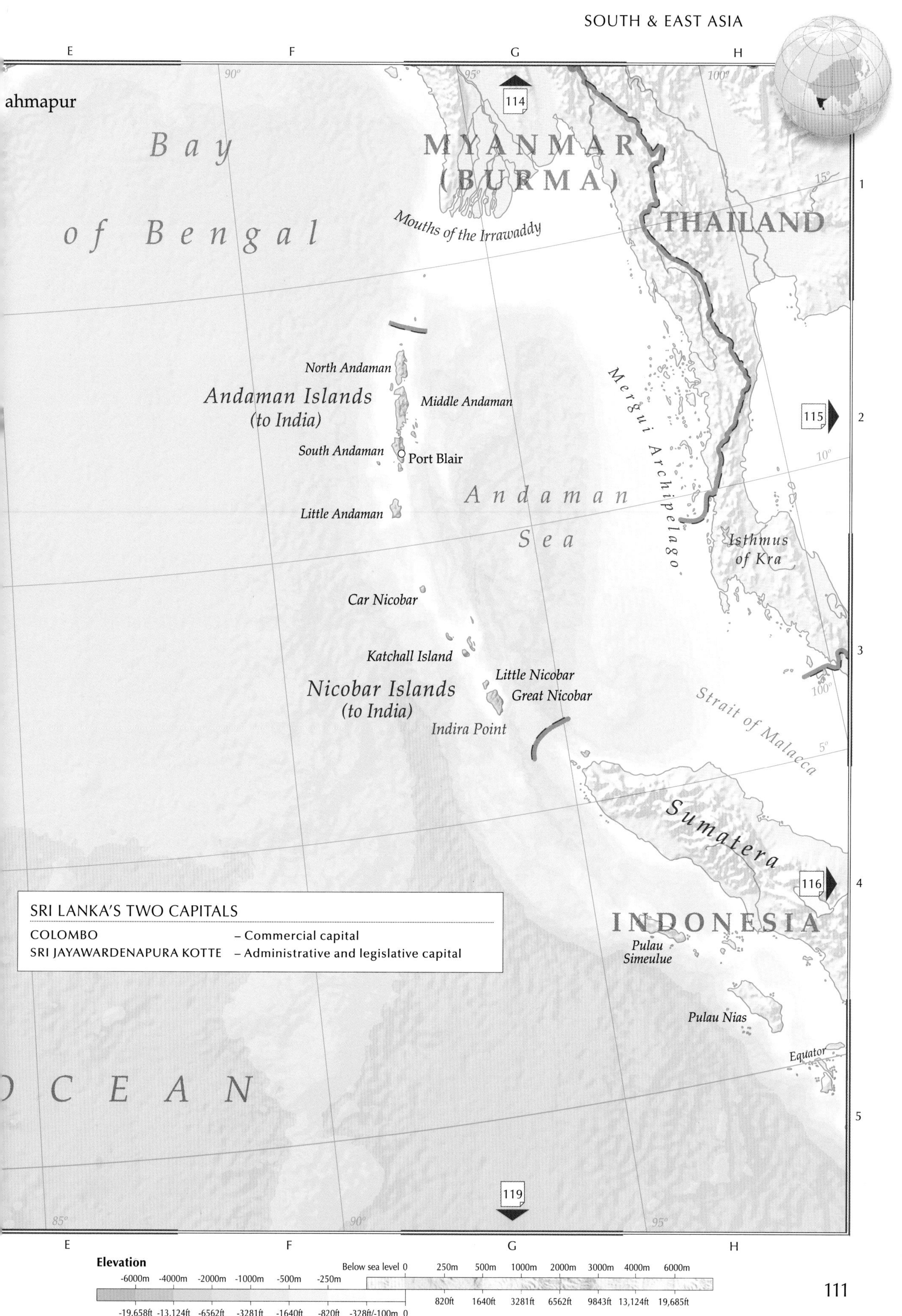
E
F
G
H
ahmapur
Bay of Bengal
MYANMAR (BURMA)
Mouths of the Irrawaddy
THAILAND
114
115
116
119
North Andaman
Andaman Islands (to India)
Middle Andaman
South Andaman
Port Blair
Little Andaman
Andaman Sea
Mergui Archipelago
Isthmus of Kra
Car Nicobar
Katchall Island
Little Nicobar
Nicobar Islands (to India)
Great Nicobar
Indira Point
Strait of Malacca
Sumatera
INDONESIA
Pulau Simeulue
Pulau Nias
Equator
OCEAN
90°
95°
100°
85°
15°
10°
5°
1
2
3
4
5

SRI LANKA'S TWO CAPITALS	
COLOMBO	– Commercial capital
SRI JAYAWARDENAPURA KOTTE	– Administrative and legislative capital

Elevation

-6000m	-4000m	-2000m	-1000m	-500m	-250m	Below sea level 0	250m	500m	1000m	2000m	3000m	4000m	6000m
-19,658ft	-13,124ft	-6562ft	-3281ft	-1640ft	-820ft	-328ft/-100m 0	820ft	1640ft	3281ft	6562ft	9843ft	13,124ft	19,685ft

Northern India, Pakistan & Bangladesh

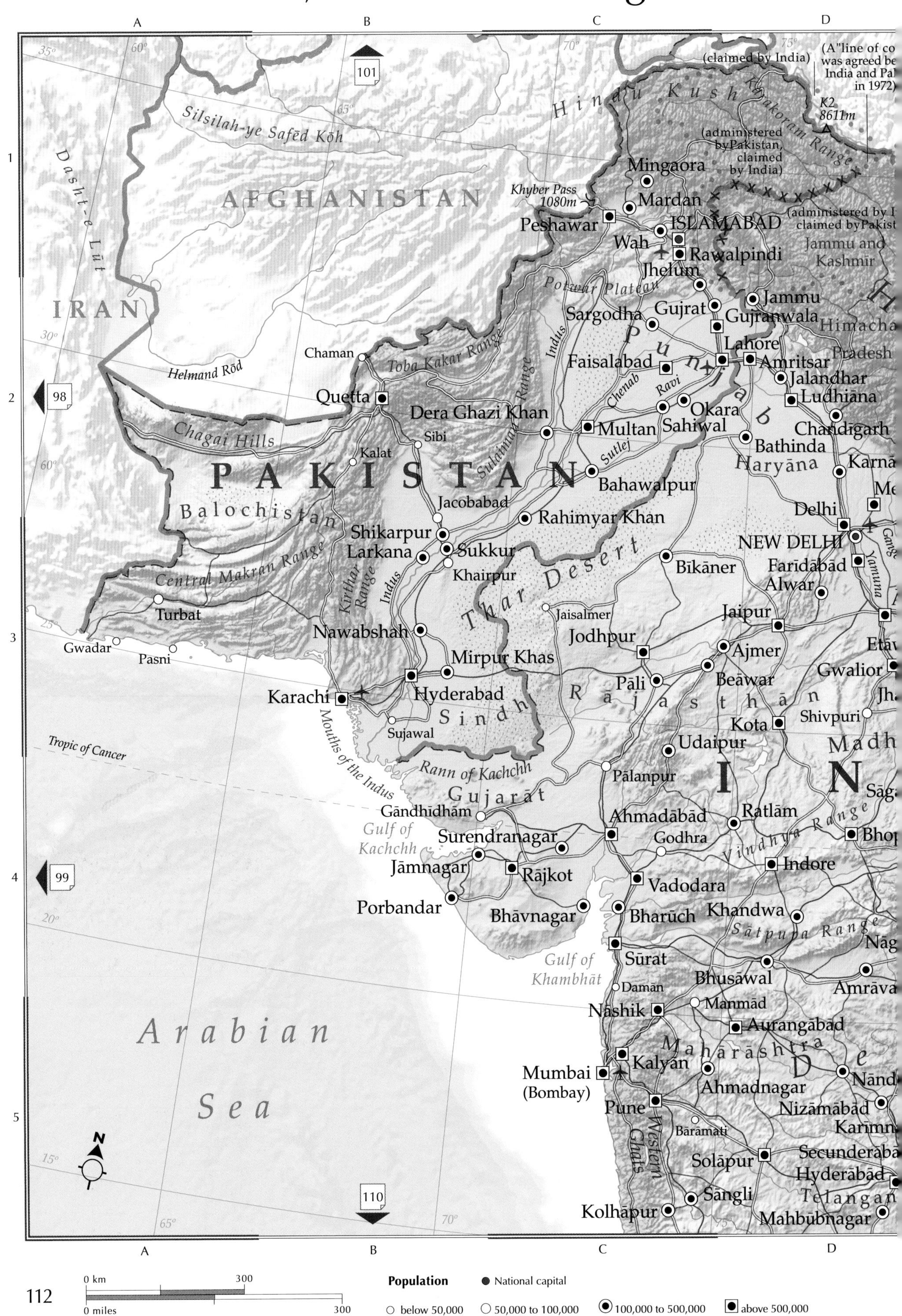

A
B
C
D
1
2
3
4
5
101
98
99
110
AFGHANISTAN
IRAN
PAKISTAN
Dasht-e Lūt
Silsilah-ye Safēd Kōh
Hindu Kush
Karakoram Range
K2 8611m
(claimed by India)
(A"line of co was agreed be India and Pa in 1972)
(administered by Pakistan, claimed by India)
Jammu and Kashmir
Khyber Pass 1080m
Mingaora
Mardan
Peshawar
ISLAMABAD
Wah
Rawalpindi
Jhelum
Potwar Plateau
Jammu
Sargodha
Gujrat
Gujranwala
Himachal Pradesh
Punjab
Lahore
Amritsar
Faisalabad
Jalandhar
Ludhiāna
Chaman
Toba Kakar Range
Helmand Rōd
Quetta
Dera Ghazi Khan
Okara
Sahiwal
Multan
Chandīgarh
Bathinda
Chagai Hills
Sibi
Kalat
Sulaiman Range
Indus
Chenab
Ravi
Sutlej
Haryāna
Karnāl
Bahawalpur
Balochistan
Jacobabad
Rahimyar Khan
Delhi
NEW DELHI
Faridābād
Shikarpur
Larkana
Sukkur
Khairpur
Central Makran Range
Kirthar Range
Thar Desert
Bīkāner
Alwar
Yamuna
Turbat
Jaisalmer
Jaipur
Gwadar
Pasni
Nawabshah
Jodhpur
Ajmer
Mirpur Khas
Beāwar
Gwalior
Pāli
Karachi
Hyderabad
Sindh
Rājasthān
Sujawal
Kota
Shivpuri
Tropic of Cancer
Mouths of the Indus
Udaipur
Rann of Kachchh
Pālanpur
INDIA
Gujarāt
Gāndhīdhām
Ratlām
Ahmadābād
Vindhya Range
Gulf of Kachchh
Surendranagar
Godhra
Bhopal
Jāmnagar
Rājkot
Indore
Vadodara
Porbandar
Bhāvnagar
Bharūch
Khandwa
Sātpura Range
Sūrat
Gulf of Khambhāt
Damān
Bhusāwal
Amrāvati
Nāshik
Manmād
Aurangābād
Arabian Sea
Mahārāshtra
Kalyān
Mumbai (Bombay)
Ahmadnagar
Nānded
Pune
Nizāmābād
Karimnagar
Barāmati
Western Ghats
Solāpur
Secunderābād
Hyderābād
Telangana
Sāngli
Kolhāpur
Mahbūbnagar
N
0 km
300
0 miles
300
Population
National capital
below 50,000
50,000 to 100,000
100,000 to 500,000
above 500,000

XINJIANG UYGUR ZIZHIQU
Kunlun Shan
AKSAI CHIN (administered by China, claimed by India)
CHINA
QINGHAI
SICHUAN
Jinsha Jiang
Mekong (Lancang Jiang)
Qingzang Gaoyuan (Plateau of Tibet)
Tanggula Shan
DÊMQOG/ DEMCHOK (administered by China, claimed by India)
XIZANG ZIZHIQU (Tibet)
Nyainqêntanglha Shan
ARUNACHAL PRADESH (claimed by China)
Brahmaputra
Himalayas
NEPAL
Annapurna 8091m
Mount Everest 8849m
Kula Kangri 7554m
Salyān
Pokhara
KATHMANDU
Bhaktapur
Lalitpur
Bahraich
THIMPHU
BHUTAN
Gangtok
Darjiling
Shiliguri
Biratnager
Dibrugarh
Jorhāt
Assam
Bongaigaon
Koch Bihār
Dispur
Guwāhāti
Shillong
Meghalaya
Kohima
Imphāl
Faizābād
Gorakhpur
Kānpur
Mau
Jaunpur
Chhapra
Bihār
Prayagraj (Allahābād)
Vārānasi
Patna
Bhāgalpur
Bihar Sharīf
Ganges
Gaya
Saidpur
Dinajpur
Rangpur
Jamalpur
Sylhet
Silchar
BANGLADESH
Rajshahi
Pabna
Brahmanbaria
DHAKA
Cumilla
Jharkhand
Dhanbad
Āsansol
Bokāro
Bānkura
Jashore
Khulna
Barishal
Chattogram (Chittagong)
MYANMAR (BURMA)
Tropic of Cancer
Murwāra
Jabalpur
Chota Nāgpur
Rānchi
West Bengal
Jamshedpur
Hāora
Kolkāta (Calcutta)
Chhattisgarh
Rāurkela
Kharagpur
Mouths of the Ganges
Bilāspur
Korba
Gondia
Raipur
Sambalpur
Bāleshwar
Rāj Nāndgaon
Durg
Mahānadi
Odisha (Orissa)
Cuttack
Bhubaneshwar
Puri
Irrawaddy
Jagdalpur
Brahmapur
Bay of Bengal
Godāvari
Andhra Pradesh
Eastern Ghats
Srīkākulam
Vizianagaram
Visākhapatnam
Rajahmundry
Kākināda
Mouths of the Irrawaddy
INDIA
Elevation
Below sea level 0
-6000m -4000m -2000m -1000m -500m -250m
-19,658ft -13,124ft -6562ft -3281ft -1640ft -820ft -328ft/-100m 0
250m 500m 1000m 2000m 3000m 4000m 6000m
820ft 1640ft 3281ft 6562ft 9843ft 13,124ft 19,685ft

104
104
114
111

Mainland Southeast Asia

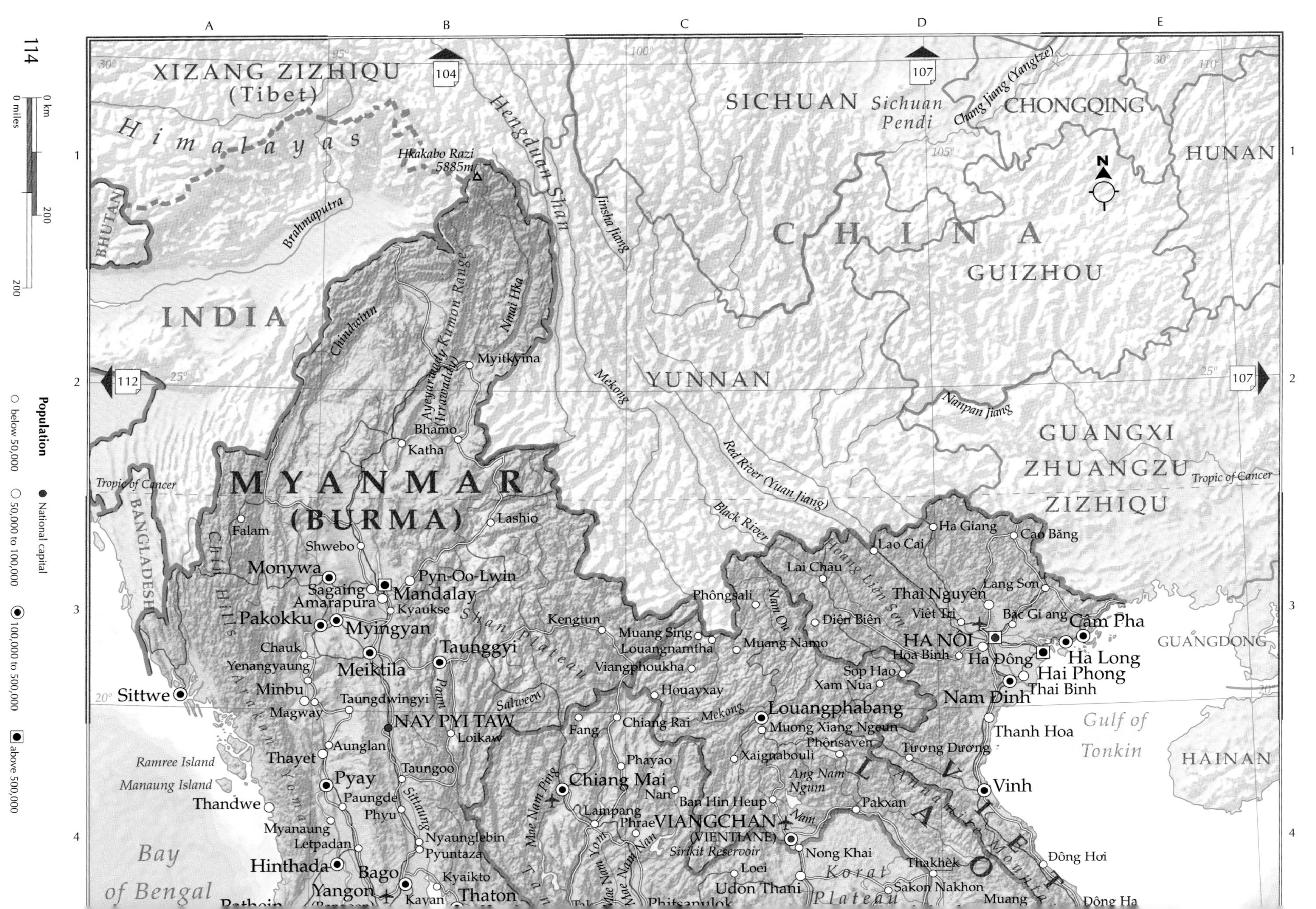

XIZANG ZIZHIQU
(Tibet)
Himalayas
Hkakabo Razi
5885m
BHUTAN
Brahmaputra
INDIA
BANGLADESH
Tropic of Cancer
Sittwe
Ramree Island
Manaung Island
Bay of Bengal
Thandwe
MYANMAR
(BURMA)
Chin Hills
Arakan Yoma
Falam
Monywa
Pakokku
Shwebo
Sagaing
Amarapura
Myingyan
Mandalay
Pyn-Oo-Lwin
Kyaukse
Meiktila
Chauk
Yenangyaung
Minbu
Magway
Taungdwingyi
NAY PYI TAW
Thayet
Aunglan
Pyay
Paungde
Phyu
Taungoo
Sittaung
Hinthada
Letpadan
Myanaung
Yangon
Bago
Kayan
Thaton
Kyaikto
Nyaunglebin
Pyuntaza
Taunggyi
Pawn
Loikaw
Shan Plateau
Salween
Lashio
Katha
Bhamo
Myitkyina
Ayeyarwady (Irrawaddy)
Kumon Range
Nmai Hka
Chindwinn
Hengduan Shan
Jinsha Jiang
Mekong
YUNNAN
SICHUAN
Sichuan Pendi
CHINA
CHONGQING
Chang Jiang (Yangtze)
GUIZHOU
HUNAN
GUANGXI
ZHUANGZU
ZIZHIQU
GUANGDONG
HAINAN
Nanpan Jiang
Red River (Yuan Jiang)
Black River
Gulf of Tonkin
Hoang Liên Sơn
Lao Cai
Ha Giang
Cao Bằng
Lang Sơn
Thai Nguyên
Việt Tri
HA NOI
Hoa Binh
Ha Đông
Bắc Giang
Cẩm Pha
Ha Long
Hai Phong
Thai Binh
Nam Đinh
Thanh Hoa
Vinh
Đông Hơi
Đông Ha
Điên Biên
Lai Châu
Sop Hao
Xam Nua
Phôngsali
Nam Ou
Muang Namo
Louangphabang
Muong Xiang Ngeun
Phônsaven
Xaignabouli
Ang Nam Ngum
Nam
VIANGCHAN
(VIENTIANE)
Nong Khai
Udon Thani
Loei
Sirikit Reservoir
Ban Hin Heup
Pakxan
Thakhèk
Sakon Nakhon
Korat Plateau
Muang
LAOS
VIET
Annamite Mountains
Tương Dương
Muang Sing
Louangnamtha
Viangphoukha
Houayxay
Kengtun
Chiang Rai
Phayao
Nan
Chiang Mai
Lampang
Phrae
Fang
Mae Nam Ping
Mae Nam Yom
Mae Nam Nan
Phitsanulok

114

Population

● National capital

○ below 50,000 ○ 50,000 to 100,000 ◉ 100,000 to 500,000 ◼ above 500,000

0 km 200
0 miles 200

CAMBODIA
MALAYSIA
INDONESIA
INDIAN OCEAN
South China Sea
Andaman Sea
Gulf of Thailand
Strait of Malacca
Malay Peninsula
Myeik Archipelago
Andaman Islands (to India)
North Andaman
Middle Andaman
South Andaman
Little Andaman
Car Nicobar
Katchall Island
Little Nicobar
Nicobar Islands (to India)
Great Nicobar
Pulau Simeulue
Sumatera (Sumatra)
Kepulauan Natuna (to Indonesia)
Mouths of the Mekong
Côn Đao Son
Mekong
KRUNG THEP (BANGKOK)
PHNOM PENH
Hô Chi Minh
Ye
Dawei
Nakhon Pathom
Ratchaburi
Phetchaburi
Mali Kyun
Kadan Kyun
Myeik
Daung Kyun
Tanintharyi
Letsôk-aw Kyun
Lanbi Kyun
Zadetkyi Kyun
Ko Phra Thong
Ranong
Chumphon
Lang Suan
Ko Phangan
Ko Samui
Surat Thani
Sichon
Phang-Nga
Ko Phuket
Phuket
Trang
Ko Lanta
Nakhon Si Thammarat
Pak Phanang
Thung Song
Thale Luang
Phatthalung
Songkhla
Hat Yai
Pattani
Yala
Narathiwat
Pulau Langkawi
Pulau Pinang
Srinagarind Reservoir
Ayutthaya
Nakhon Sawan
Nakhon Ratchasima
Lop Buri
Sara Buri
Samut Prakan
Chon Buri
Pattaya
Rayong
Ao Krung Thep
Ban Hua Hin
Bilauktaung Range
Ko Chang
Chanthaburi
Battambang
Reang Kesei
Ubon Ratchathani
Buriram
Surin
Dangrek Mountains
Samraong
Krâlanh
Tônlé Sap
Moŭng Roessei
Pursat
Chuŏr Phnum Krâvanh
Ôdôngk
Kampong Speu
Sihanoukville
Kampot
Châu Đôc
Rach Gia
Vinh Rach Gia
Ca Mau
Bac Liêu
Soc Trăng
Cân Thơ
Tra Vinh
My Tho
Vung Tau
Long Xuyên
Svay Rieng
Suong
Kampong Cham
Kampong Chhnang
Kampong Thom
Kratie
Kâmpóng Trâbêk
Stung Treng
Muang Không
Champasak
Pakxé
Khôngxédôn
Attapu
Tônle Kông
Virôchey
Tônle Srêpok
Plây Cu
Quy Nhơn
Tuy Hoa
Nha Trang
Đa Lat
Cam Ranh
Di Linh
Phan Rang-Thap Cham
Phan Thiêt
Biên Hoa
Quang Ngai
111
117
116
116
A
B
C
D
E
5
6
7
8

Elevation

-6000m	-4000m	-2000m	-1000m	-500m	-250m	Below sea level 0	250m	500m	1000m	2000m	3000m	4000m	6000m
-19,658ft	-13,124ft	-6562ft	-3281ft	-1640ft	-820ft	-328ft/-100m 0	820ft	1640ft	3281ft	6562ft	9843ft	13,124ft	19,685ft

Maritime Southeast Asia

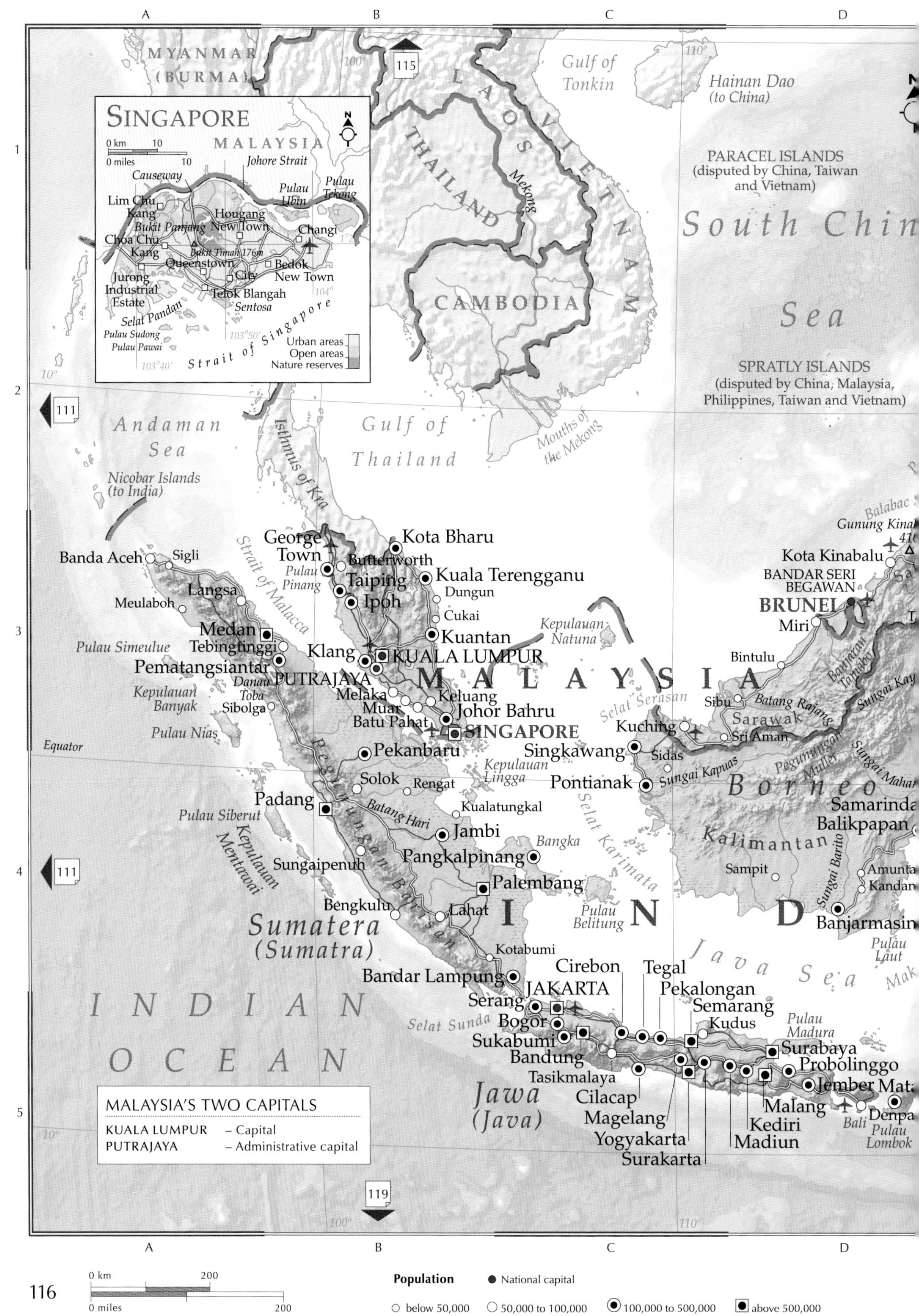

SINGAPORE
MALAYSIA
Johore Strait
Causeway
Pulau Ubin
Pulau Tekong
Lim Chu Kang
Hougang New Town
Bukit Panjang
Changi
Choa Chu Kang
Bukit Timah 176m
Queenstown
Bedok New Town
Jurong Industrial Estate
City
Telok Blangah
Sentosa
Selat Pandan
Pulau Sudong
Pulau Pawai
Strait of Singapore
Urban areas
Open areas
Nature reserves
MYANMAR (BURMA)
LAOS
VIETNAM
THAILAND
CAMBODIA
Mekong
Gulf of Tonkin
Hainan Dao (to China)
PARACEL ISLANDS (disputed by China, Taiwan and Vietnam)
South China Sea
SPRATLY ISLANDS (disputed by China, Malaysia, Philippines, Taiwan and Vietnam)
Andaman Sea
Gulf of Thailand
Mouths of the Mekong
Isthmus of Kra
Nicobar Islands (to India)
Strait of Malacca
George Town
Pulau Pinang
Butterworth
Kota Bharu
Kuala Terengganu
Taiping
Ipoh
Dungun
Cukai
Kuantan
Klang
KUALA LUMPUR
PUTRAJAYA
MALAYSIA
Melaka
Muar
Batu Pahat
Keluang
Johor Bahru
SINGAPORE
Kepulauan Natuna
Banda Aceh
Sigli
Langsa
Meulaboh
Medan
Tebingtinggi
Pematangsiantar
Pulau Simeulue
Danau Toba
Kepulauan Banyak
Sibolga
Pulau Nias
Equator
Pekanbaru
Solok
Rengat
Kepulauan Lingga
Padang
Pulau Siberut
Kepulauan Mentawai
Batang Hari
Kualatungkal
Jambi
Bangka
Pangkalpinang
Sungaipenuh
Palembang
Bengkulu
Lahat
Pegunungan Barisan
Sumatera (Sumatra)
Pulau Belitung
Kotabumi
Bandar Lampung
Selat Sunda
INDIAN OCEAN
Kota Kinabalu
Gunung Kinabalu
Balabac
BANDAR SERI BEGAWAN
BRUNEI
Miri
Bintulu
Sibu
Batang Rajang
Sarawak
Sri Aman
Kuching
Selat Serasan
Singkawang
Sidas
Sungai Kapuas
Pontianak
Selat Karimata
Borneo
Kalimantan
Pegunungan Muller
Samarinda
Balikpapan
Sampit
Sungai Barito
Banjarmasin
Amuntai
Kandangan
Pulau Laut
Java Sea
JAKARTA
Serang
Bogor
Sukabumi
Bandung
Tasikmalaya
Cirebon
Tegal
Pekalongan
Semarang
Kudus
Pulau Madura
Surabaya
Probolinggo
Jember
Jawa (Java)
Cilacap
Magelang
Yogyakarta
Surakarta
Madiun
Kediri
Malang
Bali
Denpasar
Pulau Lombok
INDONESIA
MALAYSIA'S TWO CAPITALS
KUALA LUMPUR – Capital
PUTRAJAYA – Administrative capital

115 111 111 119

0 km 200
0 miles 200

Population ● National capital
○ below 50,000 ○ 50,000 to 100,000 ◉ 100,000 to 500,000 ▣ above 500,000

E
F
G
H
1
2
3
4
5
Luzon Strait
Babuyan Island
Babuyan Channel
Tuguegarao
Ilagan
Luzon
Dagupan
Cabanatuan
MANILA/MAYNILA
Lucena
Naga
PHILIPPINES
Legazpi City
Mindoro
Sibuyan Sea
Calbayog
Samar
Roxas City
Cadiz
Tacloban
Leyte
Panay Island
Iloilo
Bacolod City
Cebu
Palawan
Negros
Bohol Sea
Butuan
Sulu Sea
Iligan
Cagayan de Oro
Bislig
Mindanao
Zamboanga
Moro Gulf
Davao
Basilan
Lebak
Davao Gulf
General Santos
Sulu Archipelago
Philippine Sea
NORTHERN MARIANA ISLANDS (to US)
GUAM (to US)
109
122
Yap
MICRONESIA
PACIFIC OCEAN
Babeldaob
PALAU
Kepulauan Talaud
Kepulauan Sangir
Celebes Sea
Pulau Morotai
Pulau Halmahera
Manado
Bitung
Molucca Sea
Gorontalo
Tomini Teluk
Equator
Pulau Waigeo
Pulau Biak
Manokwari
Sorong
Jazirah Doberai
Pulau Yapen
Laut Halmahera
Selat Dampier
Teluk Cenderawasih
Sungai Mamberamo
Jayapura
Kepulauan Banggai
Sulawesi (Celebes)
Kepulauan Sula
Maluku (Moluccas)
Pulau Misool
Laut Seram
Teluk Berau
Wahai
Danau Towuti
Waflia
Tifu
Puncak Jaya 5030m
Pegunungan Maoke
INDONESIA
Kendari
Pulau Buru
Ambon
Pulau Seram
Papua (Irian Jaya)
PAPUA NEW GUINEA
New Guinea
Kolaka
Pulau Buton
Watampone
Teluk Bone
Makassar
Bulukumba
Kepulauan Kai
Kepulauan Aru
Banda Sea
Sungai Digul
Kepulauan Tanimbar
Pulau Wetar
Pulau Yamdena
Kepulauan Alor
Tenggara
Flores
DILI
Kepulauan Leti
EAST TIMOR (TIMOR-LESTE)
Sumba
Savu Sea
Timor
Nikiniki
Kupang
Arafura Sea
Torres Strait
Timor Sea
126
AUSTRALIA
120°
130°
140°
10°

Elevation

Below sea level: -6000m (-19,658ft), -4000m (-13,124ft), -2000m (-6562ft), -1000m (-3281ft), -500m (-1640ft), -250m (-820ft), -328ft/-100m, 0

Below sea level 0, 250m (820ft), 500m (1640ft), 1000m (3281ft), 2000m (6562ft), 3000m (9843ft), 4000m (13,124ft), 6000m (19,685ft)

The Indian Ocean

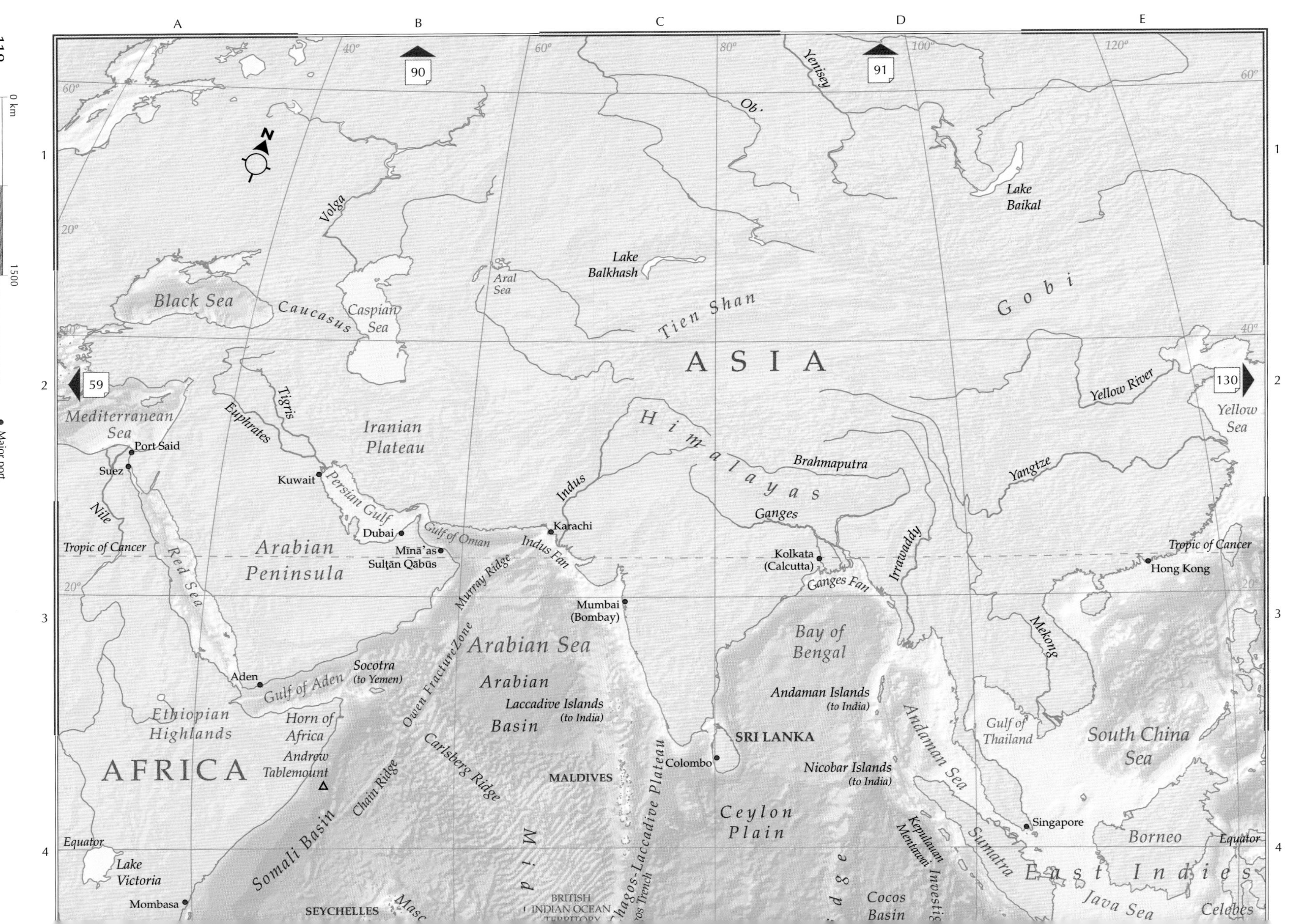

ASIA
AFRICA
Yenisey
Ob'
Lake Baikal
Gobi
Yellow Sea
Yellow River
Yangtze
Tropic of Cancer
Hong Kong
South China Sea
Borneo
Celebes
Java Sea
East Indies
Singapore
Mekong
Gulf of Thailand
Sumatra
Andaman Sea
Kepulauan Mentawai
Cocos Basin
Irrawaddy
Brahmaputra
Himalayas
Ganges
Kolkata (Calcutta)
Ganges Fan
Bay of Bengal
Andaman Islands (to India)
Nicobar Islands (to India)
SRI LANKA
Colombo
Ceylon Plain
Tien Shan
Lake Balkhash
Aral Sea
Indus
Karachi
Indus Fan
Mumbai (Bombay)
Arabian Sea
Arabian Basin
Laccadive Islands (to India)
MALDIVES
Chagos-Laccadive Plateau
BRITISH INDIAN OCEAN TERRITORY
Murray Ridge
Owen Fracture Zone
Carlsberg Ridge
Chain Ridge
Somali Basin
SEYCHELLES
Caspian Sea
Volga
Caucasus
Iranian Plateau
Persian Gulf
Gulf of Oman
Dubai
Mīnā'as Sulṭān Qābūs
Kuwait
Tigris
Euphrates
Arabian Peninsula
Black Sea
Mediterranean Sea
Port Said
Suez
Nile
Red Sea
Aden
Gulf of Aden
Socotra (to Yemen)
Horn of Africa
Andrew Tablemount
Ethiopian Highlands
Lake Victoria
Mombasa
Equator
130
91
90
59

0 km 1500

0 miles 1500

• Major port

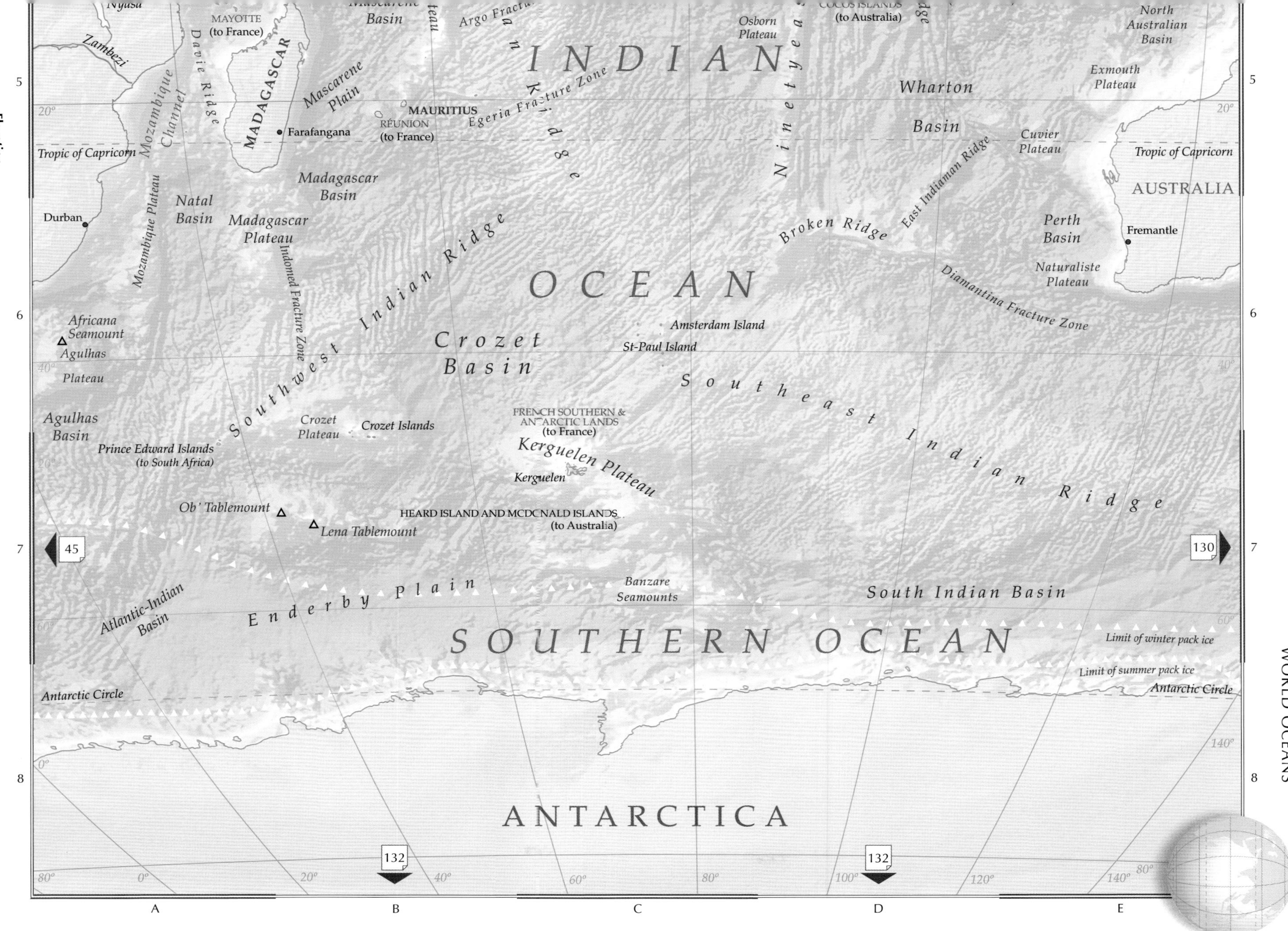
INDIAN OCEAN
SOUTHERN OCEAN
ANTARCTICA
AUSTRALIA
MADAGASCAR
Fremantle
Durban
Farafangana
Zambezi
Tropic of Capricorn
Antarctic Circle
Limit of winter pack ice
Limit of summer pack ice
North Australian Basin
Exmouth Plateau
Cuvier Plateau
Perth Basin
Naturaliste Plateau
Wharton Basin
East Indiaman Ridge
Diamantina Fracture Zone
Broken Ridge
Ninetyeast Ridge
Osborn Plateau
(to Australia)
Amsterdam Island
St-Paul Island
Southeast Indian Ridge
South Indian Basin
Kerguelen Plateau
Kerguelen
FRENCH SOUTHERN & ANTARCTIC LANDS (to France)
HEARD ISLAND AND MCDONALD ISLANDS (to Australia)
Banzare Seamounts
Crozet Basin
Crozet Islands
Crozet Plateau
Indian Ridge
Southwest Indian Ridge
Argo Fracture Zone
Egeria Fracture Zone
MAURITIUS
RÉUNION (to France)
Mascarene Plain
Mascarene Basin
Madagascar Basin
Madagascar Plateau
Indomed Fracture Zone
MAYOTTE (to France)
Davie Ridge
Mozambique Channel
Mozambique Plateau
Natal Basin
Prince Edward Islands (to South Africa)
Ob' Tablemount
Lena Tablemount
Enderby Plain
Atlantic-Indian Basin
Africana Seamount
Agulhas Plateau
Agulhas Basin
130
132
132
45
A B C D E
5 6 7 8

Elevation

-6000m	-4000m	-2000m	-1000m	-500m	-250m	-100m	0
-19,658ft	-13,124ft	-6562ft	-3281ft	-1640ft	-820ft	-328ft/-100m	0

Australasia & Oceania

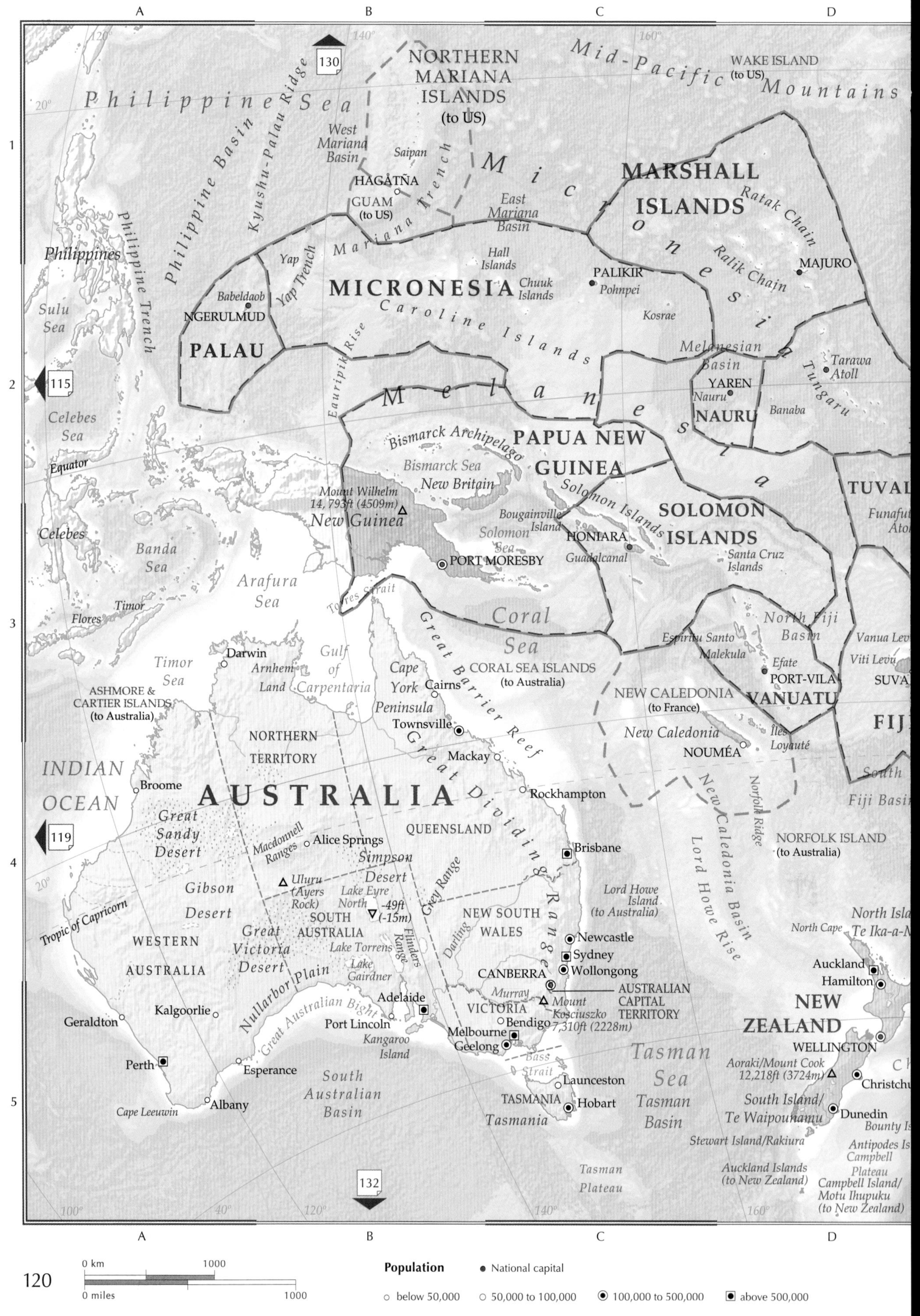

NORTHERN MARIANA ISLANDS (to US)
Philippine Sea
Mid-Pacific Mountains
WAKE ISLAND (to US)
West Mariana Basin
Saipan
HAGÅTÑA
GUAM (to US)
Mariana Trench
East Mariana Basin
MARSHALL ISLANDS
Ratak Chain
Ralik Chain
MAJURO
Micronesia
Philippine Basin
Kyushu-Palau Ridge
Philippine Trench
Philippines
Yap
Yap Trench
Hall Islands
MICRONESIA
Chuuk Islands
PALIKIR
Pohnpei
Kosrae
Sulu Sea
Babeldaob
NGERULMUD
PALAU
Caroline Islands
Eauripik Rise
Melanesian Basin
Tarawa Atoll
YAREN
Nauru
NAURU
Banaba
Tungaru
Celebes Sea
Melanesia
Bismarck Archipelago
PAPUA NEW GUINEA
Bismarck Sea
New Britain
Equator
Mount Wilhelm 14,793ft (4509m)
New Guinea
Bougainville Island
Solomon Islands
SOLOMON ISLANDS
TUVALU
Funafuti Atoll
Celebes
Banda Sea
Solomon Sea
HONIARA
Guadalcanal
Santa Cruz Islands
PORT MORESBY
Arafura Sea
Torres Strait
Timor
Flores
Coral Sea
North Fiji Basin
Espiritu Santo
Malekula
Vanua Levu
Viti Levu
Efate
PORT-VILA
SUVA
Timor Sea
Darwin
Arnhem Land
Gulf of Carpentaria
Cape York Peninsula
Cairns
CORAL SEA ISLANDS (to Australia)
NEW CALEDONIA (to France)
VANUATU
FIJI
ASHMORE & CARTIER ISLANDS (to Australia)
Townsville
Great Barrier Reef
New Caledonia
Îles Loyauté
NOUMÉA
NORTHERN TERRITORY
Mackay
South Fiji Basin
INDIAN OCEAN
Broome
AUSTRALIA
Great Dividing Range
Rockhampton
New Caledonia Basin
Norfolk Ridge
Great Sandy Desert
Macdonnell Ranges
Alice Springs
QUEENSLAND
NORFOLK ISLAND (to Australia)
Simpson Desert
Brisbane
Lord Howe Rise
Uluru (Ayers Rock)
Lake Eyre North -49ft (-15m)
Grey Range
Gibson Desert
SOUTH AUSTRALIA
Lord Howe Island (to Australia)
Tropic of Capricorn
NEW SOUTH WALES
North Island/ Te Ika-a-Māui
North Cape
WESTERN AUSTRALIA
Great Victoria Desert
Lake Torrens
Flinders Range
Darling
Newcastle
Sydney
Wollongong
Nullarbor Plain
Lake Gairdner
CANBERRA
Auckland
Hamilton
Murray
AUSTRALIAN CAPITAL TERRITORY
Adelaide
Mount Kosciuszko 7,310ft (2228m)
NEW ZEALAND
Kalgoorlie
Great Australian Bight
VICTORIA
Geraldton
Port Lincoln
Bendigo
Kangaroo Island
Melbourne
Geelong
WELLINGTON
Tasman Sea
Bass Strait
Aoraki/Mount Cook 12,218ft (3724m)
Christchurch
Perth
Esperance
South Australian Basin
Launceston
TASMANIA
Hobart
Tasmania
Tasman Basin
South Island/ Te Waipounamu
Dunedin
Bounty Is.
Cape Leeuwin
Albany
Stewart Island/Rakiura
Antipodes Is.
Campbell Plateau
Tasman Plateau
Auckland Islands (to New Zealand)
Campbell Island/ Motu Ihupuku (to New Zealand)
A
B
C
D
1
2
3
4
5
130
115
119
132

0 km 1000

0 miles 1000

Population ● National capital

○ below 50,000 ○ 50,000 to 100,000 ◉ 100,000 to 500,000 ▣ above 500,000

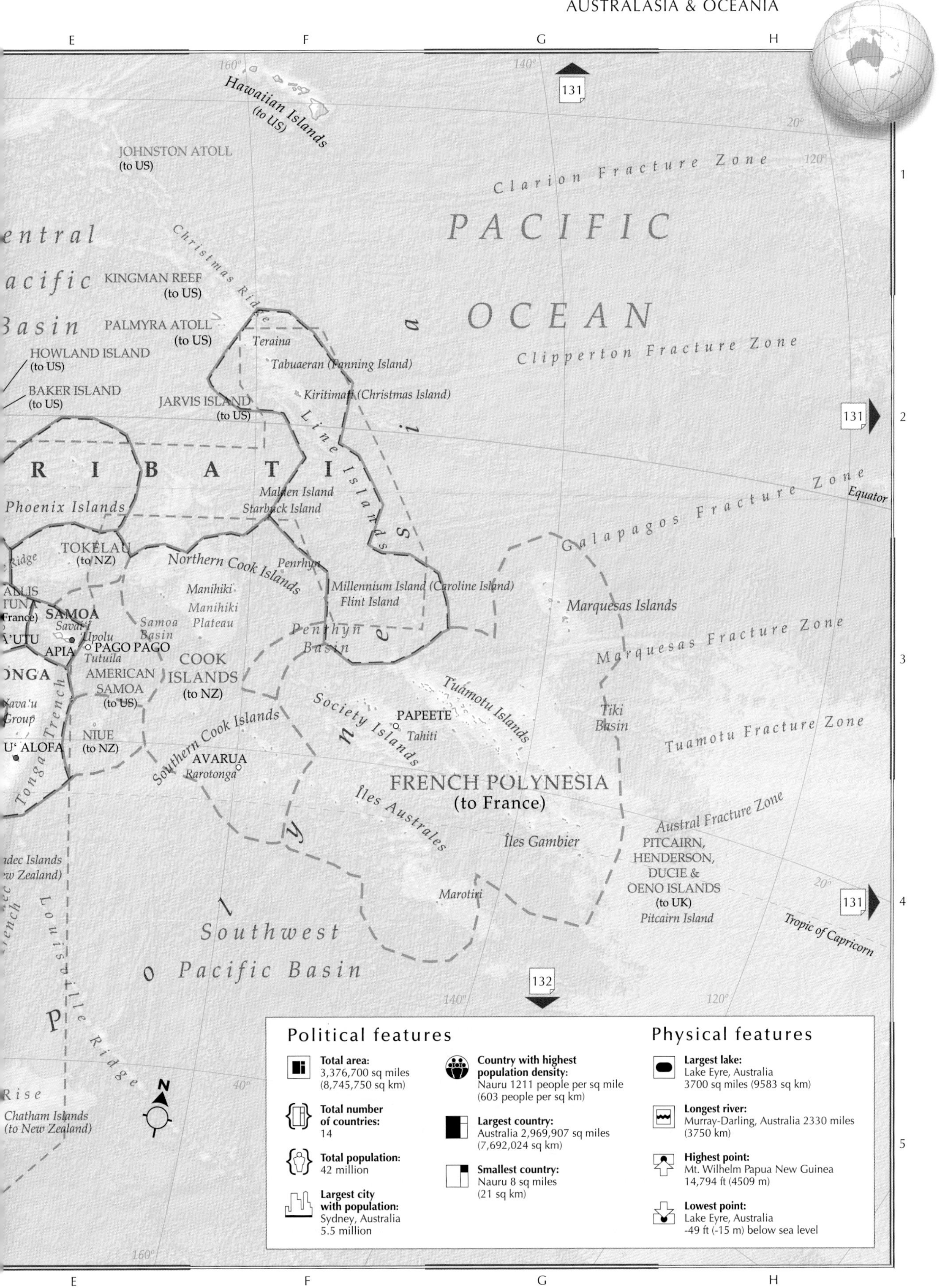

Hawaiian Islands (to US)
JOHNSTON ATOLL (to US)
Clarion Fracture Zone
PACIFIC OCEAN
Central Pacific Basin
Christmas Ridge
KINGMAN REEF (to US)
PALMYRA ATOLL (to US)
HOWLAND ISLAND (to US)
BAKER ISLAND (to US)
Teraina
Tabuaeran (Fanning Island)
Kiritimati (Christmas Island)
JARVIS ISLAND (to US)
Clipperton Fracture Zone
KIRIBATI
Line Islands
Phoenix Islands
Malden Island
Starbuck Island
Galapagos Fracture Zone
Equator
TOKELAU (to NZ)
Northern Cook Islands
Penrhyn
Manihiki
Manihiki Plateau
Millennium Island (Caroline Island)
Flint Island
Marquesas Islands
SAMOA
Savai'i
Upolu
APIA
PAGO PAGO
Tutuila
Samoa Basin
Penrhyn Basin
AMERICAN SAMOA (to US)
COOK ISLANDS (to NZ)
Marquesas Fracture Zone
Tuamotu Islands
Society Islands
PAPEETE
Tahiti
Tiki Basin
Tuamotu Fracture Zone
Southern Cook Islands
NIUE (to NZ)
AVARUA
Rarotonga
Tonga Trench
FRENCH POLYNESIA (to France)
Îles Australes
Îles Gambier
Austral Fracture Zone
PITCAIRN, HENDERSON, DUCIE & OENO ISLANDS (to UK)
Pitcairn Island
Marotiri
Tropic of Capricorn
Southwest Pacific Basin
Louisville Ridge
Chatham Islands (to New Zealand)
N
131
132
Political features
Total area: 3,376,700 sq miles (8,745,750 sq km)
Total number of countries: 14
Total population: 42 million
Largest city with population: Sydney, Australia 5.5 million
Country with highest population density: Nauru 1211 people per sq mile (603 people per sq km)
Largest country: Australia 2,969,907 sq miles (7,692,024 sq km)
Smallest country: Nauru 8 sq miles (21 sq km)
Physical features
Largest lake: Lake Eyre, Australia 3700 sq miles (9583 sq km)
Longest river: Murray-Darling, Australia 2330 miles (3750 km)
Highest point: Mt. Wilhelm Papua New Guinea 14,794 ft (4509 m)
Lowest point: Lake Eyre, Australia -49 ft (-15 m) below sea level

The Southwest Pacific

A
B
C
D
1
2
3
4
5
130
Saipan
Tinian
Rota
NORTHERN MARIANA ISLANDS (to US)
GUAM (to US)
HAGÅTÑA
MARSHALL ISLANDS
Enewetak Atoll
Bikini Atoll
Rongelap Atoll
Ratak Chain
Ralik Chain
Ujelang Atoll
Kwajalein Atoll
Namu Atoll
Ailinglaplap Atoll
Jaluit Atoll
Majuro
Ebon Atoll
Micronesia
MICRONESIA
Yap
Babeldaob
NGERULMUD
PALAU
117
Chuuk Islands
PALIKIR
Pohnpei
Kosrae
Caroline Islands
Equator
YAREN
NAURU
Banaba
Admiralty Islands
St.Matthias Group
Bismarck Archipelago
Bismarck Sea
New Ireland
New Guinea
INDONESIA
PAPUA NEW GUINEA
Madang
Central Range
Mount Wilhelm 4509m
Lae
New Britain
Bougainville Island
Choiseul
Solomon Islands
Solomon Sea
Santa Isabel
SOLOMON ISLANDS
New Georgia Islands
Malaita
HONIARA
Guadalcanal
Melanesia
Arafura Sea
Gulf of Papua
Owen Stanley Range
PORT MORESBY
Torres Strait
D'Entrecasteaux Islands
Louisiade Archipelago
San Cristobal
Rennell
Santa Cruz Islands
Arnhem Land
Groote Eylandt
Gulf of Carpentaria
124
Barkly Tableland
Cape York Peninsula
Great Barrier Reef
Coral Sea
CORAL SEA ISLANDS (to Australia)
Banks Islands
Espiritu Santo
Maéwo
Pentecost
Malekula
Ambrym
Epi
Efate
PORT-VILA
VANUATU
NEW CALEDONIA (to France)
Erromango
Tanna
Aneityum
Ouvéa
Lifou
Maré
Îles Loyauté
New Caledonia
NOUMÉA
Great Dividing Range
NORTHERN TERRITORY
Tropic of Capricorn
Macdonnell Ranges
QUEENSLAND
AUSTRALIA
127
140°
150°
160°
170°
10°
20°

0 km 750
0 miles 750

Population ● National capital
○ below 50,000 ○ 50,000 to 100,000 ◉ 100,000 to 500,000 ▣ above 500,000

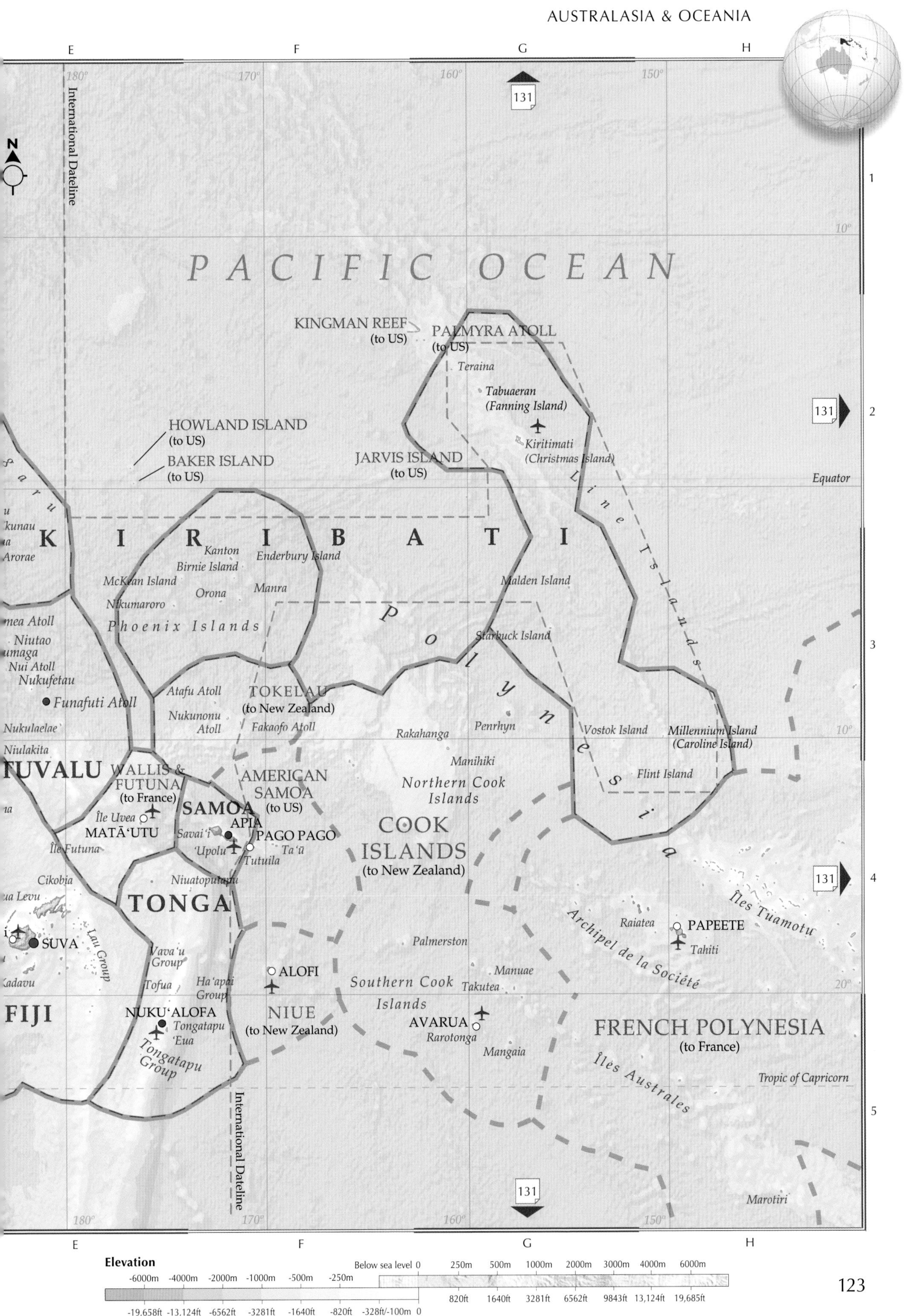
PACIFIC OCEAN
KINGMAN REEF (to US)
PALMYRA ATOLL (to US)
Teraina
Tabuaeran (Fanning Island)
Kiritimati (Christmas Island)
HOWLAND ISLAND (to US)
BAKER ISLAND (to US)
JARVIS ISLAND (to US)
Line Islands
Equator
KIRIBATI
Kanton
Enderbury Island
Birnie Island
McKean Island
Orona
Manra
Nikumaroro
Phoenix Islands
Malden Island
Starbuck Island
Polynesia
Nukufetau
Funafuti Atoll
Nukulaelae
Niulakita
Niutao
Nui Atoll
Arorae
TUVALU
Atafu Atoll
TOKELAU (to New Zealand)
Nukunonu Atoll
Fakaofo Atoll
Rakahanga
Penrhyn
Manihiki
Vostok Island
Millennium Island (Caroline Island)
Flint Island
Northern Cook Islands
WALLIS & FUTUNA (to France)
Île Uvea
MATĀ'UTU
Île Futuna
SAMOA
APIA
Savai'i
'Upolu
AMERICAN SAMOA (to US)
PAGO PAGO
Ta'ū
Tutuila
COOK ISLANDS (to New Zealand)
Cikobia
Niuatoputapu
TONGA
SUVA
Lau Group
Vava'u Group
Tofua
Ha'apai Group
FIJI
NUKU'ALOFA
Tongatapu
'Eua
Tongatapu Group
ALOFI
NIUE (to New Zealand)
Palmerston
Southern Cook Islands
Manuae
Takutea
AVARUA
Rarotonga
Mangaia
Raiatea
PAPEETE
Tahiti
Îles Tuamotu
Archipel de la Société
FRENCH POLYNESIA (to France)
Îles Australes
Tropic of Capricorn
Marotiri
International Dateline
180°
170°
160°
150°
10°
20°
E
F
G
H
1
2
3
4
5
131

Elevation

Below sea level: -6000m (-19,658ft), -4000m (-13,124ft), -2000m (-6562ft), -1000m (-3281ft), -500m (-1640ft), -250m (-820ft), -328ft/-100m, 0

Above sea level: 0, 250m (820ft), 500m (1640ft), 1000m (3281ft), 2000m (6562ft), 3000m (9843ft), 4000m (13,124ft), 6000m (19,685ft)

Western Australia

INDONESIA
Tanimbar Kepulauan
Arafura Sea
EAST TIMOR
Timor
Bali
Flores
Jawa
Pulau Lombok
Pulau Wetar
Pulau Sumba
Timor Sea
Melville Island
Bathurst Island
Croker Island
South Goulburn Island
Van Diemen Gulf
Darwin
Arnhem Land
Pine Creek
Katherine
Daly Waters
Top Springs Roadhouse
Tanami Desert
Tennant Creek
NORTHERN TERRITORY
Cape Londonderry
Joseph Bonaparte Gulf
Wyndham
Kununurra
Victoria River
Bonaparte Archipelago
Bigge Island
Heywood Islands
King Sound
Kimberley Plateau
Fitzroy Crossing
Fitzroy River
Halls Creek
INDIAN OCEAN
Broome
Eighty Mile Beach
Port Hedland
Marble Bar
Great Sandy Desert
Percival Lakes
Lake Mackay
WESTERN
Barrow Island
Dampier
Onslow
Exmouth Gulf
Exmouth
Fortescue River
Hamersley Range
Ashburton River
Newman
Tropic of Capricorn
116
117
119
126

0 km 300
0 miles 300

Population

○ below 50,000 ○ 50,000 to 100,000 ◉ 100,000 to 500,000 ■ above 500,000

● Internal administrative capital

AUSTRALIA
AUSTRALIA
SOUTH AUSTRALIA
INDIAN OCEAN
Great Australian Bight
Nullarbor Plain
Great Victoria Desert
Musgrave Ranges
Uluru (Ayers Rock) 867m
Dorre Island
Carnarvon
Shark Bay
Dirk Hartog Island
Denham
Murchison River
Robinson Range
Meekatharra
Lake Carnegie
Lake Wells
Kalbarri
Mount Magnet
Geraldton
Lake Carey
Lake Barlee
Lake Moore
Lake Rebecca
Coober Pedy
Moora
Kalgoorlie
Zanthus
Reid
Tarcoola
Lake Everard
Gingin
Coolgardie
Southern Cross
Merredin
Perth
Northam
Lake Cowan
Eucla
Penong
Lake Gairdner
Fremantle
Rockingham
Ceduna
Brookton
Norseman
Balladonia
Mandurah
Narrogin
Bunbury
Collie
Wagin
Elliston
Busselton
Katanning
Esperance
Augusta
Manjimup
Port Lincoln
Albany
119
127
132
132
Elevation
-6000m -4000m -2000m -1000m -500m -250m
-19,658ft -13,124ft -6562ft -3281ft -1640ft -820ft -328ft/-100m 0
Below sea level 0
250m 500m 1000m 2000m 3000m 4000m 6000m
820ft 1640ft 3281ft 6562ft 9843ft 13,124ft 19,685ft

Eastern Australia

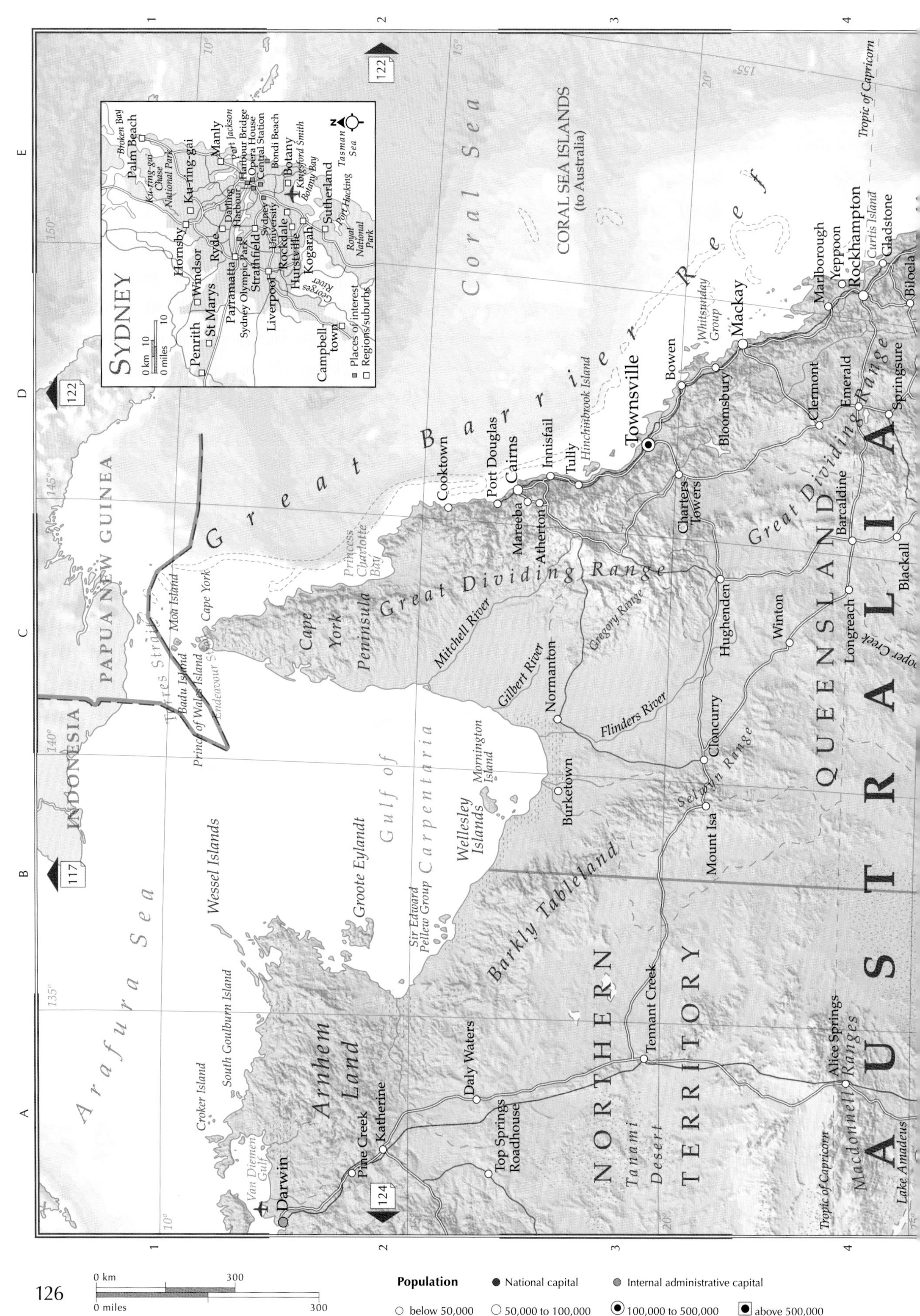

SYDNEY
Palm Beach
Broken Bay
Ku-ring-gai Chase National Park
Ku-ring-gai
Manly
Port Jackson
Harbour Bridge
Opera House
Central Station
Bondi Beach
Botany
Kingsford Smith
Botany Bay
Tasman Sea
Hornsby
Windsor
Ryde
Darling Harbour
Sydney University
Sutherland
Port Hacking
Royal National Park
Penrith
St Marys
Parramatta
Sydney Olympic Park
Strathfield
Liverpool
Rockdale
Hurstville
Kogarah
Georges River
Campbell-town
Places of interest
Regions/suburbs
Arafura Sea
Coral Sea
CORAL SEA ISLANDS (to Australia)
Great Barrier Reef
PAPUA NEW GUINEA
INDONESIA
Torres Strait
Moa Island
Badu Island
Prince of Wales Island
Endeavour Strait
Cape York
Cape York Peninsula
Princess Charlotte Bay
Great Dividing Range
Mitchell River
Gilbert River
Gregory Range
Flinders River
Selwyn Range
Gulf of Carpentaria
Mornington Island
Wellesley Islands
Sir Edward Pellew Group
Groote Eylandt
Wessel Islands
South Goulburn Island
Croker Island
Van Diemen Gulf
Arnhem Land
Barkly Tableland
Tanami Desert
NORTHERN TERRITORY
QUEENSLAND
AUSTRALIA
Macdonnell Ranges
Lake Amadeus
Tropic of Capricorn
Darwin
Pine Creek
Katherine
Daly Waters
Top Springs Roadhouse
Tennant Creek
Alice Springs
Mount Isa
Burketown
Normanton
Cloncurry
Hughenden
Winton
Longreach
Barcaldine
Blackall
Emerald
Springsure
Clermont
Charters Towers
Townsville
Bowen
Bloomsbury
Mackay
Whitsunday Group
Hinchinbrook Island
Tully
Innisfail
Cairns
Port Douglas
Cooktown
Mareeba
Atherton
Marlborough
Yeppoon
Rockhampton
Curtis Island
Gladstone
Biloela

0 km 300
0 miles 300

Population ● National capital ● Internal administrative capital
○ below 50,000 ○ 50,000 to 100,000 ◉ 100,000 to 500,000 ◼ above 500,000

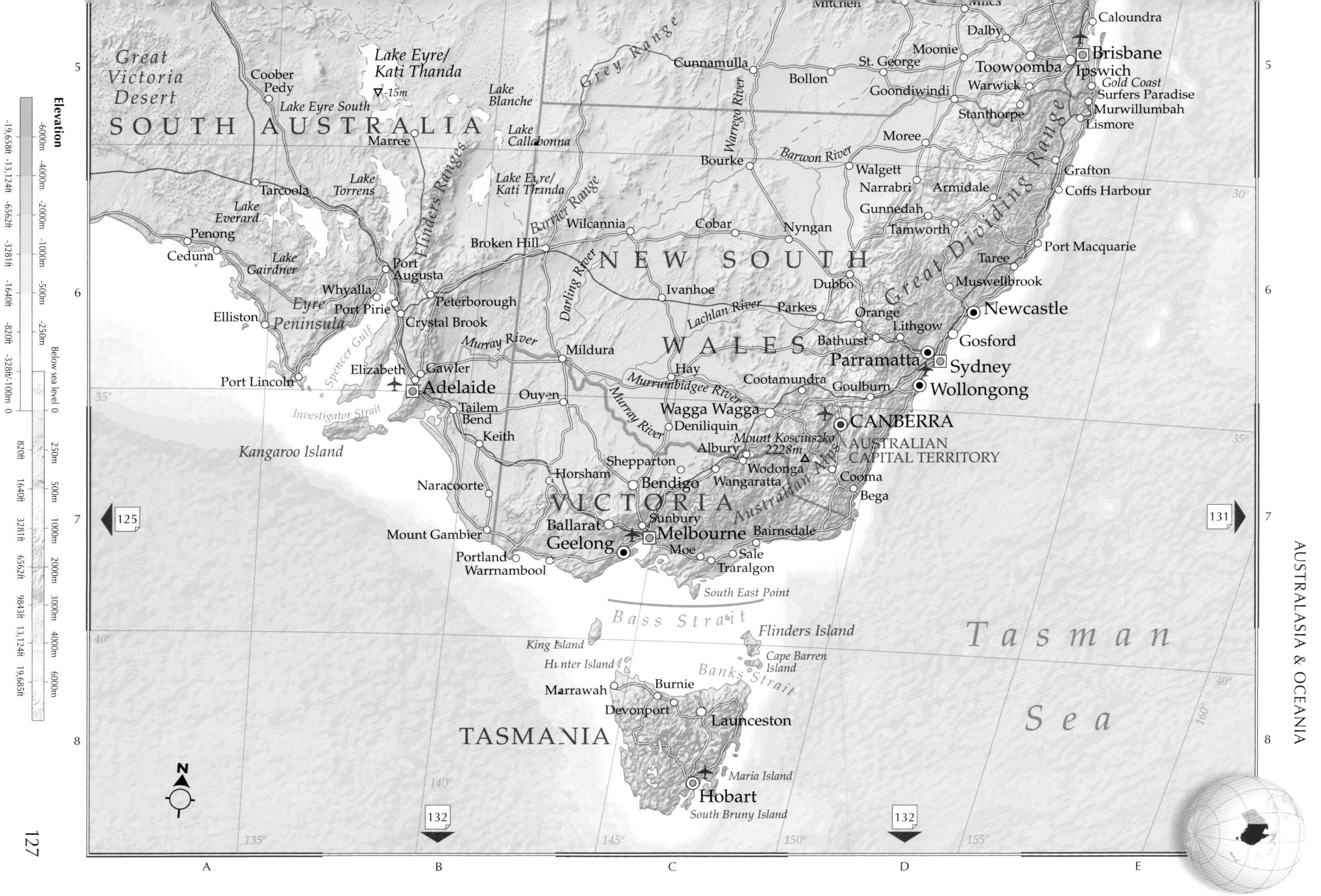
Great Victoria Desert
SOUTH AUSTRALIA
NEW SOUTH WALES
VICTORIA
TASMANIA
AUSTRALIAN CAPITAL TERRITORY
Tasman Sea
Lake Eyre/ Kati Thanda
-15m
Lake Eyre South
Lake Blanche
Lake Callabonna
Lake Eyre/ Kati Thanda
Lake Torrens
Lake Everard
Lake Gairdner
Eyre Peninsula
Spencer Gulf
Investigator Strait
Kangaroo Island
Flinders Ranges
Grey Range
Barrier Range
Great Dividing Range
Australian Alps
Mount Kosciuszko 2228m
Darling River
Warrego River
Barwon River
Lachlan River
Murrumbidgee River
Murray River
Bass Strait
Banks Strait
South East Point
Flinders Island
Cape Barren Island
King Island
Hunter Island
Maria Island
South Bruny Island
Caloundra
Dalby
Moonie
Brisbane
Toowoomba
Ipswich
Gold Coast
Surfers Paradise
Murwillumbah
Lismore
St. George
Bollon
Cunnamulla
Goondiwindi
Warwick
Stanthorpe
Moree
Grafton
Coffs Harbour
Bourke
Walgett
Narrabri
Armidale
Gunnedah
Tamworth
Port Macquarie
Coober Pedy
Marree
Tarcoola
Penong
Ceduna
Wilcannia
Cobar
Nyngan
Broken Hill
Port Augusta
Whyalla
Port Pirie
Peterborough
Crystal Brook
Elliston
Taree
Muswellbrook
Dubbo
Ivanhoe
Parkes
Orange
Lithgow
Newcastle
Gosford
Bathurst
Parramatta
Sydney
Wollongong
Goulburn
Cootamundra
Mildura
Hay
Elizabeth
Gawler
Adelaide
Port Lincoln
Ouyen
Tailem Bend
Wagga Wagga
Deniliquin
CANBERRA
Keith
Albury
Shepparton
Wodonga
Wangaratta
Cooma
Bega
Horsham
Naracoorte
Bendigo
Ballarat
Sunbury
Melbourne
Geelong
Bairnsdale
Mount Gambier
Portland
Warrnambool
Moe
Sale
Traralgon
Marrawah
Burnie
Devonport
Launceston
Hobart
125
131
132
132
5 6 7 8
A B C D E
30° 35° 40° 135° 140° 145° 150° 155° 160°
N
Elevation
-6000m -4000m -2000m -1000m -500m -250m
-19,658ft -13,124ft -6562ft -3281ft -1640ft -820ft -328ft/-100m 0
Below sea level 0
250m 500m 1000m 2000m 3000m 4000m 6000m
820ft 1640ft 3281ft 6562ft 9843ft 13,124ft 19,685ft

New Zealand

Tasman Sea
NEW ZEALAND
North Island/
Te Ika-a-Māui
Manawatāwhi/
Three Kings Islands
Cape Reinga/ Te Rerenga Wairua
North Cape
Great Exhibition Bay
Te Kao
Te-Oneroa-a-Tōhē/
Ninety Mile Beach
Kaitaia
Okaihau
Kaikohe
Kerikeri
Paihia
Hokianga Harbour
Wairoa
Hikurangi
Whangarei
Te Hauturu-o-Toi/
Little Barrier Island
Great Barrier
Island
Colville Channel
Hauraki Gulf/
Tikapa Moana
Coromandel
Peninsula
Coromandel
Whitianga
Ruawai
Wellsford
Warkworth
Kaipara Harbour
Helensville
Takapuna
Auckland
Manurewa
Papakura
Pukekohe
Waiuku
Thames
Paeroa
Mayor
Island
Katikati
Tauranga
Bay of
Plenty
Huntly
Morrinsville
Hamilton
Cambridge
Matamata
Lake
Rotorua
Rotorua
Kawerau
Whakatane
Opotiki
East Cape
Ruatoria
Raukumara Range
Gisborne
Tūranganui-
a-Kiwa/
Poverty Bay
Mahia
Peninsula
Wairoa
Hawke
Bay
Napier
Havelock North
Hastings
Waipawa
Waipukurau
Dannevirke
Lake
Waikaremoana
Murupara
Tokoroa
Otorohanga
Te Kuiti
Lake
Taupo
Taupo
Turangi
Mount Ruapehu
2797m
Waiouru
Taihape
Rangitikei
Ohura
Taumarunui
Raetihi
Marton
Feilding
North Taranaki
Bight
Waitara
New Plymouth
Cape Egmont
Mount Taranaki
(Mount Egmont)
2518m
Stratford
Hawera
Patea
Wanganui
South Taranaki
Bight
Rangitoto
ki te Tonga/
Cape Farewell
A
B
C
D
E
1
2
3
4
121
124
34°
36°
38°
40°
168°
170°
172°
174°
176°
178°

0 km 100
0 miles 100

Population
● National capital
○ below 50,000
○ 50,000 to 100,000
⊙ 100,000 to 500,000
■ above 500,000

PACIFIC OCEAN
South Island/ Te Waipounamu
Lower Hutt
WELLINGTON
Cape Palliser
Cape Campbell
Cook S
Picton
Nelson
Richmond
Richmond Range
Mount Owen 1875m
Blenheim
Wairau
Seddon
Clarence
Clarence
Kaikoura
Kaikoura Peninsula
Hanmer Springs
Hurunui
Waipara
Pegasus Bay
Rangiora
Kaiapoi
Christchurch
Lyttelton
Banks Peninsula
Lake Ellesmere
Canterbury Plains
Canterbury Bight
Ashburton
Hinds
Geraldine
Temuka
Timaru
Studholme
Waimate
Waitaki
Oamaru
Hampden
Otago Peninsula
Dunedin
Mosgiel
Milton
Balclutha
Taieri
Clutha River/ Mata-Au
Alexandra
Cromwell
Queenstown
Wanaka
Lake Hawea
Lake Wanaka
Lake Wakatipu
Lake Pukaki
Fairlie
Mayfield
Aoraki/Mount Cook
Aoraki/Mount Cook 3724m
Rakaia
Darfield
Oxford
Otira
Arthur's Pass 920m
Lake Brunner
Springs Junction
Bight
Seddonville
Westport
Cape Foulwind
Reefton
Runanga
Greymouth
Hokitika
Ross
Abut Head
Whataroa
Fox Glacier
Haast
Jackson Head
Southern Alps
Milford Sound/ Piopiotahi
Milford Sound
Te Houhou/ George Sound
Taitetimu/ Caswell Sound
Lake Te Anau
Te Anau
Livingstone Mts
Eyre Mts
Lumsden
Mataura
Gore
Mataura
Tokanui
Toetoes Bay
Invercargill
Winton
Riverton/Aparima
Lake Manapouri
Waiau
Fiordland
Lake Hauroko
Resolution Island
West Cape
Ta Waewae Bay
Foveaux Strait
Ruapuke Island
Stewart Island/Rakiura
Halfmoon Bay
Codfish Island/ Whenua Hou
Titi/ Muttonbird Islands
South West Cape/ Puhiwaero
131
121
132
132
42°
44°
46°
48°
166°
168°
170°
172°
174°
176°
178°
A
B
C
D
E
5
6
7
8

Elevation

Below sea level	-6000m	-4000m	-2000m	-1000m	-500m	-250m	0
	-19,658ft	-13,124ft	-6562ft	-3281ft	-1640ft	-820ft	-328ft/-100m 0

250m	500m	1000m	2000m	3000m	4000m	6000m
820ft	1640ft	3281ft	6562ft	9843ft	13,124ft	19,685ft

The Pacific Ocean

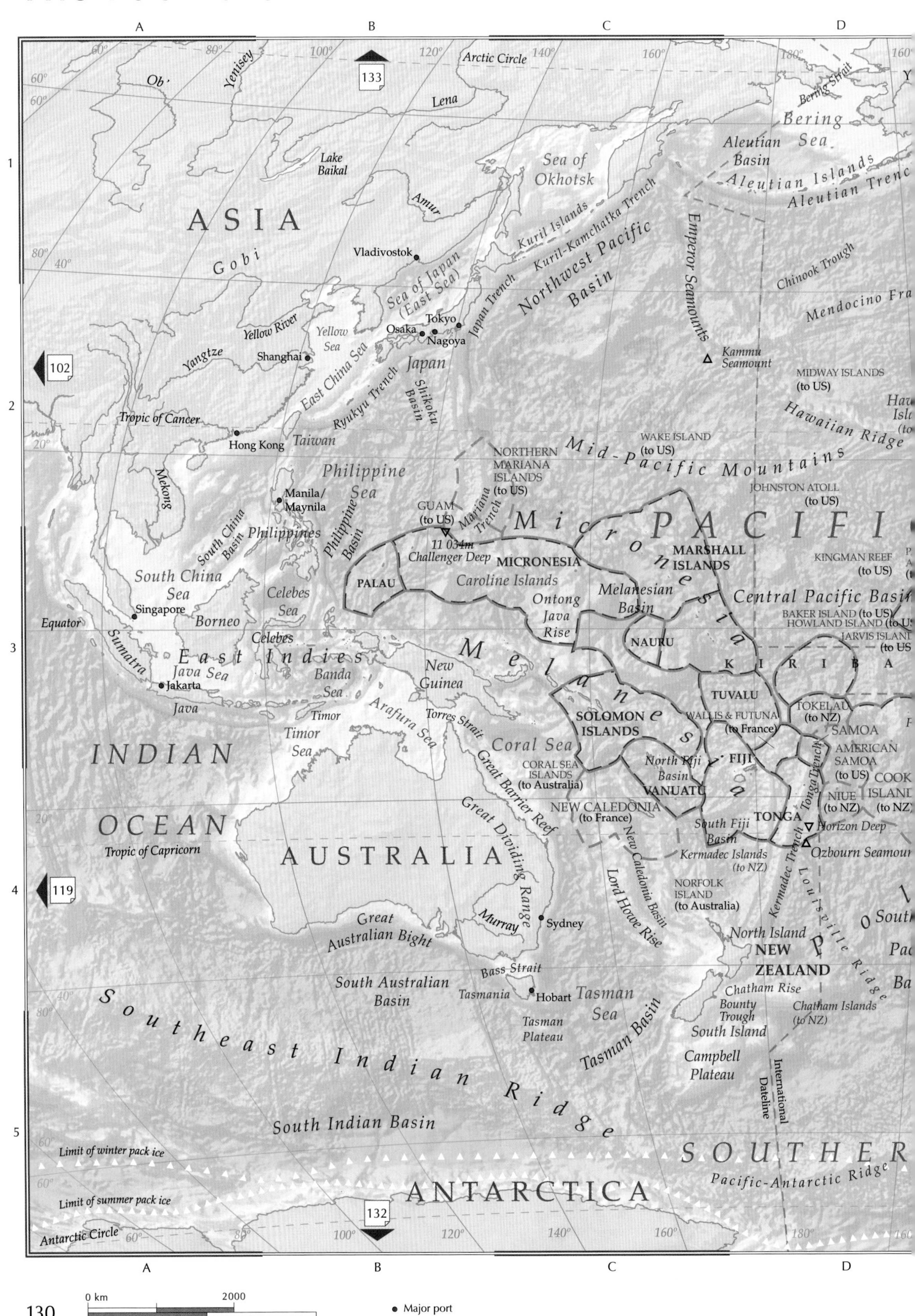
A
B
C
D
1
2
3
4
5
133
102
119
132
Arctic Circle
Ob'
Yenisey
Lena
Lake Baikal
Amur
Sea of Okhotsk
ASIA
Gobi
Vladivostok
Sea of Japan (East Sea)
Kuril Islands
Kuril-Kamchatka Trench
Northwest Pacific Basin
Emperor Seamounts
Bering Strait
Bering Sea
Aleutian Basin
Aleutian Islands
Aleutian Trench
Chinook Trough
Mendocino Fra
Yellow River
Yellow Sea
Tokyo
Osaka
Nagoya
Japan Trench
Japan
Yangtze
Shanghai
East China Sea
Ryukyu Trench
Shikoku Basin
Kammu Seamount
MIDWAY ISLANDS (to US)
Hawaiian Ridge
Tropic of Cancer
Hong Kong
Taiwan
NORTHERN MARIANA ISLANDS (to US)
Mid-Pacific Mountains
WAKE ISLAND (to US)
Mekong
Philippine Sea
Manila/ Maynila
Philippine Basin
GUAM (to US)
Mariana Trench
11 034m Challenger Deep
Micronesia
PACIFI
JOHNSTON ATOLL (to US)
South China Basin
Philippines
MARSHALL ISLANDS
KINGMAN REEF (to US)
South China Sea
PALAU
MICRONESIA
Caroline Islands
Melanesian Basin
Ontong Java Rise
Central Pacific Basin
Singapore
Celebes Sea
Borneo
BAKER ISLAND (to US)
HOWLAND ISLAND
JARVIS ISLAND (to US)
Equator
Sumatra
Celebes
East Indies
NAURU
KIRIBATI
Java Sea
Jakarta
Banda Sea
New Guinea
Melanesia
Polynesia
Java
Timor
Arafura Sea
Torres Strait
SOLOMON ISLANDS
TUVALU
WALLIS & FUTUNA (to France)
TOKELAU (to NZ)
SAMOA
AMERICAN SAMOA (to US)
COOK ISLAND (to NZ)
INDIAN OCEAN
Timor Sea
Coral Sea
CORAL SEA ISLANDS (to Australia)
Great Barrier Reef
North Fiji Basin
FIJI
VANUATU
NIUE (to NZ)
Tonga Trench
TONGA
NEW CALEDONIA (to France)
South Fiji Basin
Horizon Deep
Ozbourn Seamount
Tropic of Capricorn
AUSTRALIA
Great Dividing Range
New Caledonia Basin
Kermadec Islands (to NZ)
Kermadec Trench
Louisville Ridge
Lord Howe Rise
NORFOLK ISLAND (to Australia)
Murray
Sydney
Great Australian Bight
North Island
NEW ZEALAND
Bass Strait
South Australian Basin
Tasmania
Hobart
Tasman Sea
Chatham Rise
Bounty Trough
Chatham Islands (to NZ)
Tasman Plateau
Tasman Basin
South Island
Campbell Plateau
Southeast Indian Ridge
International Dateline
South Indian Basin
Limit of winter pack ice
Limit of summer pack ice
SOUTHER
Pacific-Antarctic Ridge
ANTARCTICA
Antarctic Circle
0 km 2000
0 miles 2000
• Major port

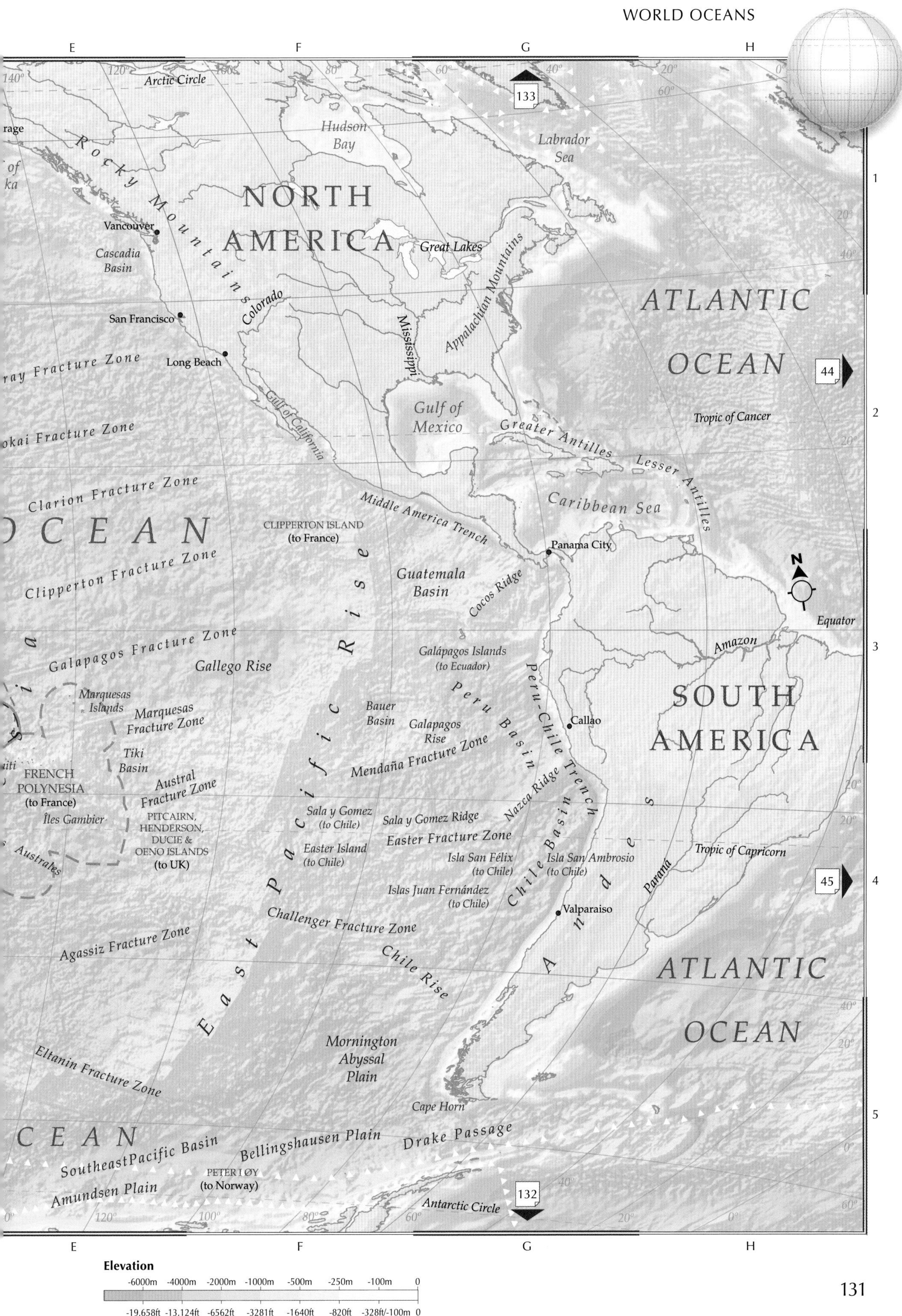

E
F
G
H
Arctic Circle
133
Hudson Bay
Labrador Sea
Rocky Mountains
NORTH AMERICA
Vancouver
Cascadia Basin
Great Lakes
Appalachian Mountains
San Francisco
Colorado
Mississippi
ATLANTIC OCEAN
44
Long Beach
Gulf of California
Gulf of Mexico
Tropic of Cancer
Greater Antilles
Lesser Antilles
Clarion Fracture Zone
Middle America Trench
Caribbean Sea
CLIPPERTON ISLAND (to France)
Panama City
Clipperton Fracture Zone
Guatemala Basin
Cocos Ridge
Equator
Galapagos Fracture Zone
Gallego Rise
Galápagos Islands (to Ecuador)
Amazon
Marquesas Islands
Marquesas Fracture Zone
Peru Basin
Bauer Basin
Galapagos Rise
Peru-Chile Trench
Callao
SOUTH AMERICA
Tiki Basin
Mendaña Fracture Zone
FRENCH POLYNESIA (to France)
Austral Fracture Zone
Nazca Ridge
Îles Gambier
PITCAIRN, HENDERSON, DUCIE & OENO ISLANDS (to UK)
Sala y Gomez (to Chile)
Sala y Gomez Ridge
Easter Fracture Zone
Easter Island (to Chile)
Isla San Félix (to Chile)
Isla San Ambrosio (to Chile)
Tropic of Capricorn
Australes
Chile Basin
Paraná
45
Islas Juan Fernández (to Chile)
Valparaiso
Andes
Challenger Fracture Zone
Agassiz Fracture Zone
East Pacific Rise
Chile Rise
ATLANTIC OCEAN
Mornington Abyssal Plain
Eltanin Fracture Zone
Cape Horn
OCEAN
Bellingshausen Plain
Drake Passage
SoutheastPacific Basin
PETER I ØY (to Norway)
Amundsen Plain
132
Antarctic Circle
1
2
3
4
5
Elevation
-6000m -4000m -2000m -1000m -500m -250m -100m 0
-19,658ft -13,124ft -6562ft -3281ft -1640ft -820ft -328ft/-100m 0

Antarctica

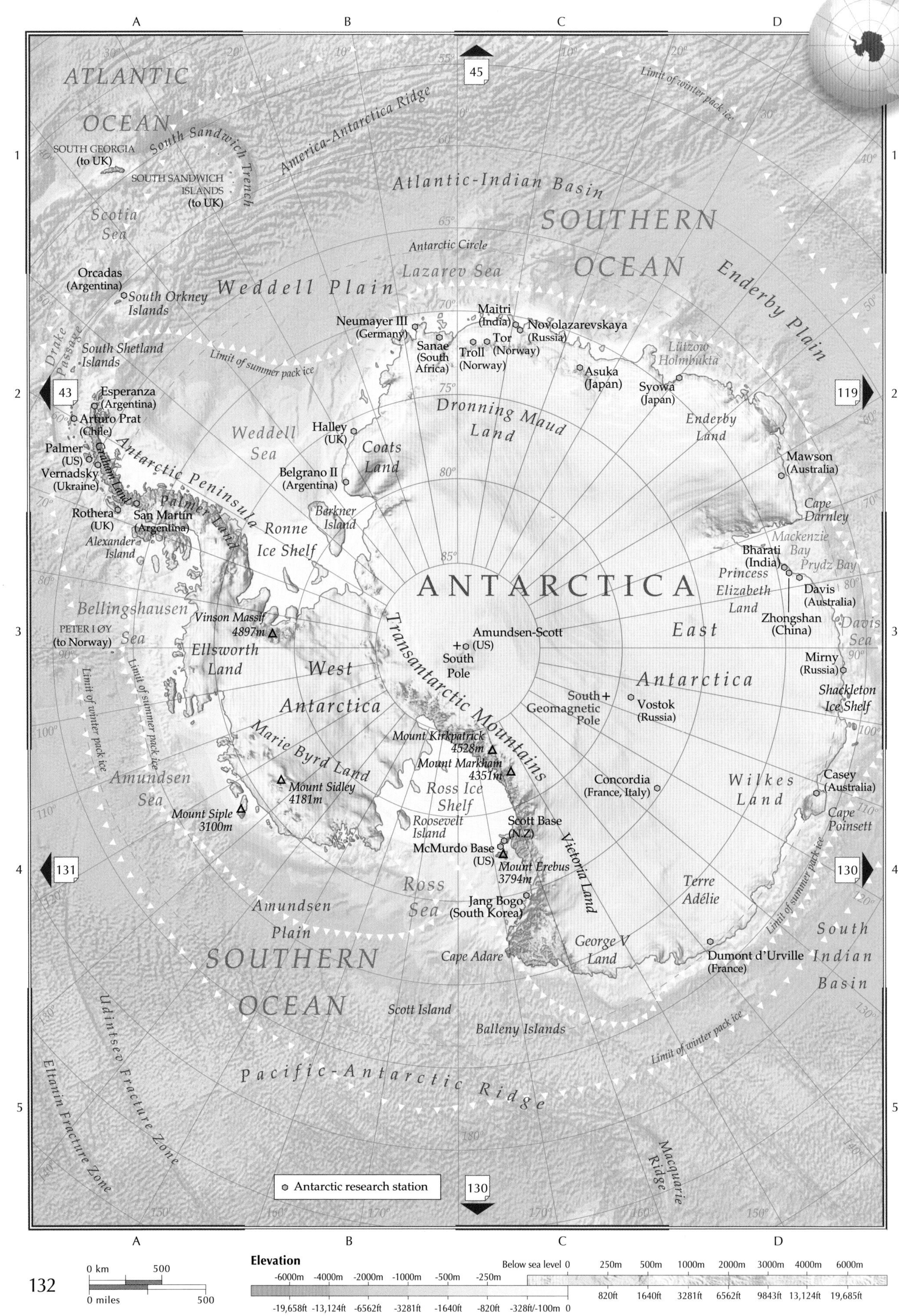

ATLANTIC OCEAN
SOUTH GEORGIA (to UK)
SOUTH SANDWICH ISLANDS (to UK)
South Sandwich Trench
America-Antarctica Ridge
Atlantic-Indian Basin
SOUTHERN OCEAN
Scotia Sea
Antarctic Circle
Lazarev Sea
Weddell Plain
Enderby Plain
Orcadas (Argentina)
South Orkney Islands
Drake Passage
South Shetland Islands
Limit of summer pack ice
Limit of winter pack ice
Neumayer III (Germany)
Sanae (South Africa)
Maitri (India)
Tor (Norway)
Troll (Norway)
Novolazarevskaya (Russia)
Asuka (Japan)
Syowa (Japan)
Lützow Holmbukta
Esperanza (Argentina)
Arturo Prat (Chile)
Palmer (US)
Vernadsky (Ukraine)
Graham Land
Antarctic Peninsula
Palmer Land
Rothera (UK)
San Martín (Argentina)
Alexander Island
Weddell Sea
Halley (UK)
Coats Land
Belgrano II (Argentina)
Berkner Island
Ronne Ice Shelf
Dronning Maud Land
Enderby Land
Mawson (Australia)
Cape Darnley
Mackenzie Bay
Bharati (India)
Prydz Bay
Princess Elizabeth Land
Davis (Australia)
Zhongshan (China)
ANTARCTICA
Bellingshausen Sea
PETER I ØY (to Norway)
Vinson Massif 4897m
Ellsworth Land
West Antarctica
Transantarctic Mountains
Amundsen-Scott (US)
South Pole
East Antarctica
South Geomagnetic Pole
Vostok (Russia)
Davis Sea
Mirny (Russia)
Shackleton Ice Shelf
Marie Byrd Land
Mount Kirkpatrick 4528m
Mount Markham 4351m
Mount Sidley 4181m
Amundsen Sea
Mount Siple 3100m
Ross Ice Shelf
Roosevelt Island
Concordia (France, Italy)
Wilkes Land
Casey (Australia)
Cape Poinsett
Scott Base (N.Z)
McMurdo Base (US)
Mount Erebus 3794m
Victoria Land
Ross Sea
Jang Bogo (South Korea)
Terre Adélie
Amundsen Plain
George V Land
Dumont d'Urville (France)
South Indian Basin
Cape Adare
SOUTHERN OCEAN
Scott Island
Balleny Islands
Pacific-Antarctic Ridge
Udintsev Fracture Zone
Eltanin Fracture Zone
Macquarie Ridge
Antarctic research station
Elevation
Below sea level 0
250m 500m 1000m 2000m 3000m 4000m 6000m
820ft 1640ft 3281ft 6562ft 9843ft 13,124ft 19,685ft
-6000m -4000m -2000m -1000m -500m -250m
-19,658ft -13,124ft -6562ft -3281ft -1640ft -820ft -328ft/-100m 0
0 km 500
0 miles 500

45 43 119 131 130 130

Arctic Ocean

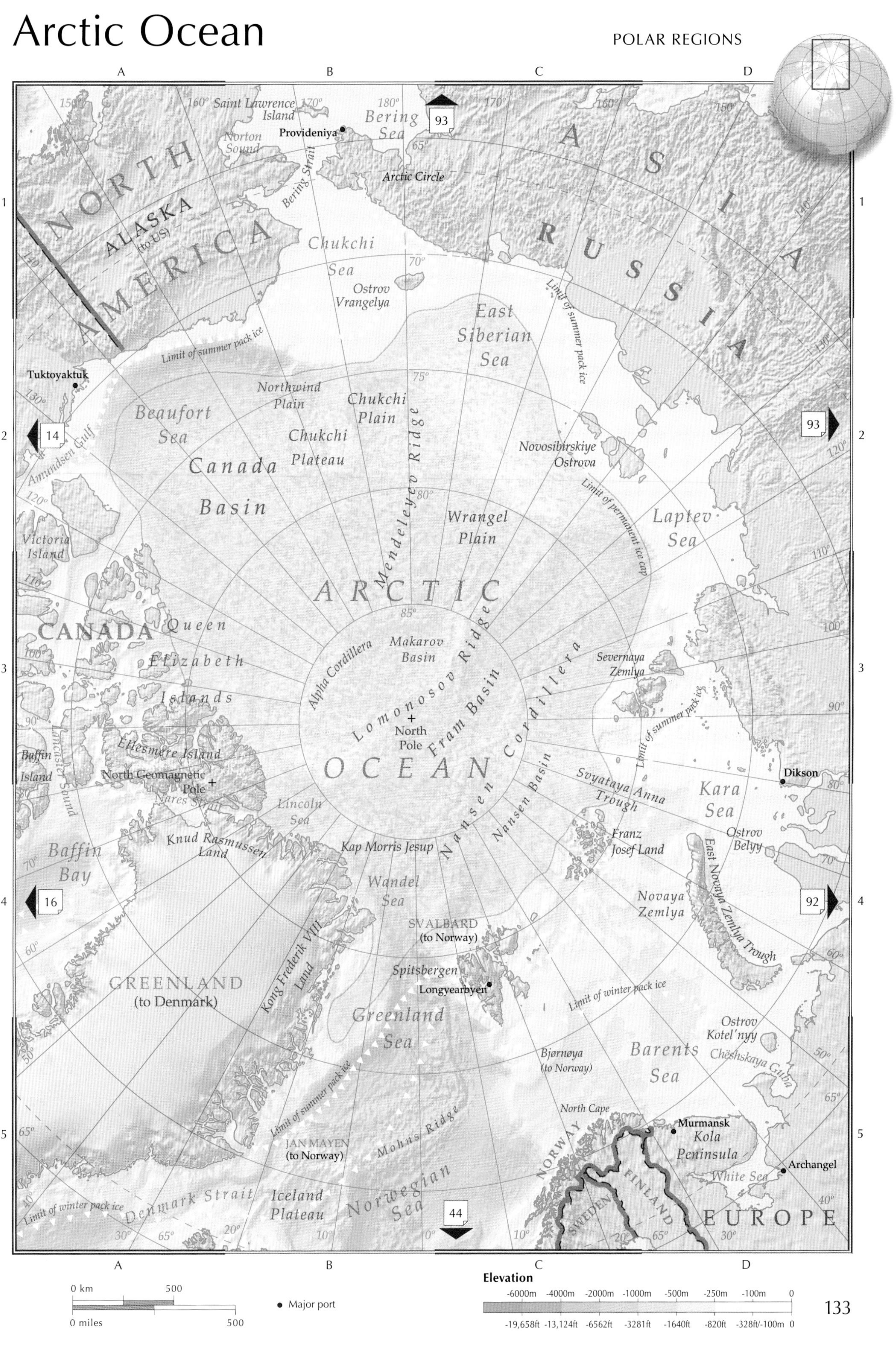

NORTH AMERICA
ALASKA (to US)
Saint Lawrence Island
Norton Sound
Provideniya
Bering Sea
Bering Strait
Arctic Circle
ASIA
RUSSIA
Chukchi Sea
Ostrov Vrangelya
East Siberian Sea
Limit of summer pack ice
Tuktoyaktuk
Beaufort Sea
Northwind Plain
Chukchi Plain
Chukchi Plateau
Mendeleyev Ridge
Canada Basin
Amundsen Gulf
Victoria Island
Novosibirskiye Ostrova
Limit of permanent ice cap
Wrangel Plain
Laptev Sea
ARCTIC OCEAN
CANADA
Queen Elizabeth Islands
Alpha Cordillera
Makarov Basin
Lomonosov Ridge
Fram Basin
North Pole
Nansen Cordillera
Severnaya Zemlya
Baffin Island
Lancaster Sound
Ellesmere Island
North Geomagnetic Pole
Nares Strait
Lincoln Sea
Nansen Basin
Svyataya Anna Trough
Dikson
Kara Sea
Knud Rasmussen Land
Kap Morris Jesup
Franz Josef Land
Ostrov Belyy
East Novaya Zemlya Trough
Baffin Bay
Wandel Sea
Novaya Zemlya
SVALBARD (to Norway)
Kong Frederik VIII Land
Spitsbergen
Longyearbyen
GREENLAND (to Denmark)
Greenland Sea
Limit of winter pack ice
Ostrov Kotel'nyy
Bjørnøya (to Norway)
Barents Sea
Chëshskaya Guba
North Cape
Murmansk
Kola Peninsula
JAN MAYEN (to Norway)
Mohns Ridge
NORWAY
FINLAND
SWEDEN
White Sea
Archangel
Denmark Strait
Iceland Plateau
Norwegian Sea
EUROPE
A B C D
1 2 3 4 5
93 14 16 92 44
0 km 500
0 miles 500
Major port
Elevation
-6000m -4000m -2000m -1000m -500m -250m -100m 0
-19,658ft -13,124ft -6562ft -3281ft -1640ft -820ft -328ft/-100m 0

Country Profiles

This Factfile is intended as a guide to a world that is continually changing as political fashions and personalities come and go. Nevertheless, all the material in these factfiles has been researched from the most up-to-date and authoritative sources to give an incisive portrait of the geographical, political, and social characteristics that make each country so unique.

There are currently 196 independent countries in the world - more than at any previous time - and over 50 dependencies. Antarctica is the only land area on Earth that is not officially part of, and does not belong to, any single country.

Country profile key

Formation Date of formation denotes the date of political origin or independence of a state, i.e. its emergence as a recognizable entity in the modern political world / date current borders were established

Population Total population / population density - based on total *land* area

Languages Main languages, an asterisk (*) denotes the official language(s)

Life expectancy Average number of years a person is expected to live

AFGHANISTAN

Central Asia

Page 100 D4

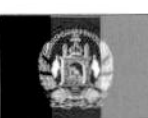

Landlocked in Central Asia, Afghanistan has suffered decades of conflict. After the US-led coalition withdrew troops in 2021, the Taliban movement overthrew the government, leading to global isolation.

Official name Islamic Republic of Afghanistan
Formation 1919 / 1919
Capital Kabul
Population 38.3 million / 152 people per sq mile (59 people per sq km)
Total area 251,827 sq miles (652,230 sq km)
Languages Pashto*, Tajik, Dari*, Farsi, Uzbek, Turkmen, Urdu
Religions Muslim 99% (Sunni 87%, Shi'a 12%), Other 1%
Demographics Pashtun 38%, Tajik 25%, Hazara 19%, Uzbek and Turkmen 15%, Other 3%
Government Nonparty system
Currency Afghani = 100 puls
Literacy rate 37%
Life expectancy 54 years

ALBANIA

Southeast Europe

Page 79 C6

Lying at the southeastern end of the Adriatic Sea, Albania - or the "land of the eagles" - held its first multiparty elections in 1991. It joined NATO in 2009 and became an EU candidate in 2014.

Official name Republic of Albania
Formation 1912 / 1921
Capital Tirana
Population 3 million / 270 people per sq mile (104 people per sq km)
Total area 11,100 sq miles (28,748 sq km)
Languages Albanian*, Greek
Religions Muslim (mainly Sunni) 68%, Roman Catholic 12%, Albanian Orthodox 8%, Nonreligious 6%, Other 6%
Demographics Albanian 83%, Other 17%
Government Parliamentary system
Currency Lek = 100 qindarka (qintars)
Literacy rate 98%
Life expectancy 79 years

ALGERIA

North Africa

Page 48 C3

Set on the Mediterranean coast, Algeria, the largest country in Africa, is rich in natural resources, particularly oil and gas. It has huge potential for solar-power generation from the vast Sahara desert.

Official name People's Democratic Republic of Algeria
Formation 1962 / 1962
Capital Algiers
Population 44.1 million / 48 people per sq mile (19 people per sq km)
Total area 919,590 sq miles (2,381,740 sq km)
Languages Arabic*, Tamazight* (Kabyle, Shawia, Tamashek), French
Religions Muslim (mainly Sunni) 99%, Other 1%
Demographics Arab 75%, Amazigh 24%, European & Jewish 1%
Government Presidential system
Currency Algerian dinar = 100 centimes
Literacy rate 81%
Life expectancy 78 years

ANDORRA

Southwest Europe

Page 69 B6

A tiny landlocked principality, Andorra lies between France and Spain, high in the eastern Pyrenees. Its economy, based on tourism, also features low tax and duty-free shopping.

Official name Principality of Andorra
Formation 1278 / 1278
Capital Andorra la Vella
Population 85,560 / 473 people per sq mile (183 people per sq km)
Total area 181 sq miles (468 sq km)
Languages Spanish, Catalan*, French, Portuguese, Castilian
Religions Christian (mainly Roman Catholic) 90%, Other 10%
Demographics Andorran 48%, Spanish 25%, Portuguese 11%, French 5%, Other 11%
Government Parliamentary system
Currency Euro = 100 cents
Literacy rate 100%
Life expectancy 83 years

ANGOLA

Southern Africa

Page 56 B2

Emerging from decades of devastating civil war in 2002, Angola has enormous natural resources, particularly minerals, oil, and gemstones. It is the third largest producer of diamonds in Africa.

Official name Republic of Angola
Formation 1975 / 1975
Capital Luanda
Population 34.7 million / 72 people per sq mile (28 people per sq km)
Total area 481,354 sq miles (1,246,700 sq km)
Languages Portuguese*, Kimbundu, Umbundu, Chokwe, Kikongo
Religions Roman Catholic 40%, Protestant 38%, Nonreligious 12%, Other (including animist) 10%
Demographics Ovimbundu 37%, Ambundu 25%, Bakongo 13%, Other African 21%, Other 4%
Government Presidential system
Currency Kwanza = 100 centimos
Literacy rate 71%
Life expectancy 62 years

ANTIGUA & BARBUDA

West Indies

Page 33 H3

Lying on the Atlantic edge of the Leeward Islands, tourism is vital to the economy here but it is vulnerable to seasonal storm damage, notably hurricane Irma in 2017 which devastated the island of Barbuda.

Official name Antigua and Barbuda
Formation 1981 / 1981
Capital St. John's
Population 100,335 / 587 people per sq mile (226 people per sq km)
Total area 171 sq miles (443 sq km)
Languages English*, English patois
Religions Other Christian 49%, Anglican 19%, Seventh-day Adventist 13%, Other 19%
Demographics Black 87%, Mixed race 5%, Hispanic 3%, White 2%, Other 3%
Government Parliamentary system
Currency East Caribbean dollar = 100 cents
Literacy rate 98%
Life expectancy 78 years

ARGENTINA

South America

Page 43 B5

This vast country ranges from semiarid lowlands, through fertile grasslands, to the glacial southern tip of South America. This diversity allows for key agricultural exports such as soybeans, corn, and beef.

Official name Argentine Republic
Formation 1816 / 1898
Capital Buenos Aires
Population 46.2 million / 43 people per sq mile (17 people per sq km)
Total area 1,073,518 sq miles (2,780,400 sq km)
Languages Spanish*, Italian, English, German, French, Indigenous (Mapudungun, Quechua)
Religions Roman Catholic 63%, Evangelical 15%, Nonreligious 19%, Other 3%
Demographics European and mixed race 97%, Indigenous 2%, Other 1%
Government Presidential system
Currency Argentine peso = 100 centavos
Literacy rate 99%
Life expectancy 78 years

ARMENIA

Southwest Asia

Page 95 F3

Set at the crossroads between Europe and Asia in the Caucasus Mountains, Armenia is rich in ancient history and culture. It was the first country to adopt Christianity as the state religion, in the 4th century CE.

Official name Republic of Armenia
Formation 1991 / 1991
Capital Yerevan
Population 3 million / 261 people per sq mile (101 people per sq km)
Total area 11,484 sq miles (29,743 sq km)
Languages Armenian*, Azeri, Russian, Kurdish
Religions Orthodox Christian 89%, Nonreligious 2%, Armenian Catholic Church 1%, Other 8%
Demographics Armenian 98%, Yezidi 1%, Other 1%
Government Parliamentary system
Currency Dram = 100 luma
Literacy rate 100%
Life expectancy 76 years

AUSTRALIA
Australasia & Oceania

Page 125 B5

An island continent between the Pacific and Indian oceans, Australia has a highly urbanized population that is concentrated on the coast. The economy is dominated by the service sector underpinned by vast mineral wealth.

Official name Commonwealth of Australia
Formation 1901 / 1901
Capital Canberra
Population 26.1 million / 8 people per sq mile (3 people per sq km)
Total area 2,969,907 sq miles (7,692,024 sq km)
Languages English*, Italian, Cantonese, Greek, Arabic, Vietnamese, Mandarin
Religions Roman Catholic 28%, Nonreligious 24%, Other Christian 20%, Anglican 19%, Other 9%,
Demographics British 33%, Australian 30%, Irish 9%, Scottish 9%, Chinese 5%, Indigenous 3%, Other 11%
Government Parliamentary system
Currency Australian dollar = 100 cents
Literacy rate 99%
Life expectancy 83 years

AUSTRIA
Central Europe

Page 73 D7

Nestled in the eastern Alps of Central Europe, Austria has developed a varied, high-tech, industrialized economy with well over half of its electricity demand now met by renewable resources.

Official name Republic of Austria
Formation 1918 / 1920
Capital Vienna
Population 8.9 million / 275 people per sq mile (106 people per sq km)
Total area 32,383 sq miles (83,871 sq km)
Languages German*, Turkish, Serbian, Croatian, Slovenian, Hungarian (Magyar)
Religions Roman Catholic 55%, None 26%, Other Christian 9%, Muslim 8%, Other 2%
Demographics Austrian 81%, German 3%, Turkish 2%, Other 14%
Government Parliamentary system
Currency Euro = 100 cents
Literacy rate 98%
Life expectancy 82 years

AZERBAIJAN
Southwest Asia

Page 95 G2

Situated on the west coast of the Caspian Sea, Azerbaijan was one of the world's first oil producing nations. The economy is still dominated by oil and gas, accounting for almost 90 percent of total exports.

Official name Republic of Azerbaijan
Formation 1991 / 1991
Capital Baku
Population 10.3 million / 308 people per sq mile (119 people per sq km)
Total area 33,436 sq miles (86,600 sq km)
Languages Azeri*, Russian, Armenian
Religions Muslim (mainly Shi'a) 97%, Christian 3%
Demographics Azeri 92%, Lezgin 2%, Other 6%
Government Presidential system
Currency Manat = 100 gopik
Literacy rate 100%
Life expectancy 74 years

BAHAMAS, THE
West Indies

Page 32 C1

Located off the Florida coast, The Bahamas comprise over 3,000 islands and cays, only 30 of which are inhabited. Tourism and financial services are key sectors in the economy.

Official name Commonwealth of The Bahamas
Formation 1973 / 1973
Capital Nassau
Population 355,608 / 66 people per sq mile (26 people per sq km)
Total area 5359 sq miles (13,880 sq km)
Languages English*, English Creole, French Creole
Religions Baptist 36%, Anglican 14%, Roman Catholic 12%, Pentecostal 9%, Seventh-day Adventist 5%, Methodist 4%, Other 20%
Demographics Black 90%, White 5%, Mixed race 2%, Other 3%
Government Parliamentary system
Currency Bahamian dollar = 100 cents
Literacy rate 96%
Life expectancy 76 years

BAHRAIN
Southwest Asia

Page 98 C4

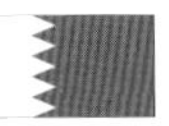

Once famed for its pearl fisheries, this island nation set on the Persian Gulf has relied heavily on petroleum products for its wealth. Steps have been taken to diversify the economy.

Official name Kingdom of Bahrain
Formation 1971 / 2001
Capital Manama
Population 1.5 million / 5651 people per sq mile (2241 people per sq km)
Total area 293 sq miles (760 sq km)
Languages Arabic*, English, Farsi, Urdu
Religions Muslim (mainly Shi'a) 74%, Christian 9%, Other 17%
Demographics Bahraini 46%, Asian 46%, Other Arab 5%, Other 3%
Government Monarchical / parliamentary system
Currency Bahraini dinar = 1000 fils
Literacy rate 100%
Life expectancy 78 years

BANGLADESH
South Asia

Page 113 G3

The geography of this low-lying nation is dominated by the Ganges delta which makes it prone to flooding. However, this also makes it a fertile land, able to produce three rice crops a year in some areas.

Official name People's Republic of Bangladesh
Formation 1971 / 2015
Capital Dhaka
Population 166 million / 3276 people per sq mile (1278 people per sq km)
Total area 57,321 sq miles (148,460 sq km)
Languages Bengali*, Urdu, Chakma, Marma (Magh), Garo, Khasi, Santhali, Tripuri, Mro
Religions Muslim (mainly Sunni) 91%, Hindu 8%, Other 1%
Demographics Bengali 98%, Other indigenous ethnic groups 2%
Government Parliamentary system
Currency Taka = 100 poisha
Literacy rate 78%
Life expectancy 73 years

BARBADOS
West Indies

Page 33 H4

The most easterly of the Windward Islands, Barbados was once a major sugar producer, but this densely populated island now has a diverse economy based on manufacturing, tourism, and financial services.

Official name Barbados
Formation 1966 / 1966
Capital Bridgetown
Population 302,674 / 1730 people per sq mile (669 people per sq km)
Total area 166 sq miles (430 sq km)
Languages Bajan (Barbadian English), English*
Religions Anglican 24%, Nonreligious 21%, Pentecostal 20%, Seventh-day Adventist 6%, Methodist 4%, Roman Catholic 4%, Other 21%
Demographics Black 93%, Mixed race 3%, White 3%, Other 1%
Government Parliamentary system
Currency Barbados dollar = 100 cents
Literacy rate 100%
Life expectancy 81 years

BELARUS
Eastern Europe

Page 85 B6

Landlocked in eastern Europe, Belarus continues to exert heavy state-control over many aspects of media, politics, and the economy. It is closely aligned with Russia, its powerful neighbor and ally.

Official name Republic of Belarus
Formation 1991 / 1991
Capital Minsk
Population 9.4 million / 117 people per sq mile (45 people per sq km)
Total area 80,155 sq miles (207,600 sq km)
Languages Belarussian*, Russian*
Religions Orthodox Christian 73%, Roman Catholic 12%, Nonreligious 3%, Other 12%
Demographics Belarussian 86%, Russian 8%, Polish 3%, Ukrainian 1%, Other 2%
Government Presidential system
Currency Belarussian ruble = 100 kopeks
Literacy rate 100%
Life expectancy 80 years

BELGIUM
Northwest Europe

Page 65 B6

Located in Northwest Europe, Belgium is host to many international organizations and institutions of the European Union (EU). The diamond trade is a key part of a diverse and productive economy.

Official name Kingdom of Belgium
Formation 1830 / 1919
Capital Brussels
Population 11.8 million / 1001 people per sq mile (387 people per sq km)
Total area 11,787 sq miles (30,528 sq km)
Languages Dutch*, French*, German*
Religions Roman Catholic 57%, Nonreligious 30%, Muslim 7%, Other Christian 4%, Other 2%
Demographics Belgian 75%, Italian 4%, Moroccan 4%, French 2%, Turkish 2%, Dutch 2%, Other 11%
Government Parliamentary system
Currency Euro = 100 cents
Literacy rate 99%
Life expectancy 85 years

BELIZE
Central America

Page 30 B1

Lying on the eastern shore of the Yucatan Peninsula, Belize is close to the world's second largest barrier reef, which includes the Great Blue Hole, a giant marine sinkhole popular with recreational divers.

Official name Belize
Formation 1981 / 1981
Capital Belmopan
Population 412,387 / 47 people per sq mile (18 people per sq km)
Total area 8867 sq miles (22,966 sq km)
Languages English Creole, Spanish, English*, Mayan, Garifuna (Carib), German
Religions Roman Catholic 40%, Other Christian 34%, Nonreligious 16%, Other 10%,
Demographics Mixed race 49%, Creole 24%, Maya 10%, Garifuna 6%, Asian Indian 4%, Other 7%
Government Parliamentary system
Currency Belizean dollar = 100 cents
Literacy rate 75%
Life expectancy 76 years

BENIN
West Africa

Page 53 F4

Set on the West African coast, the economy of Benin is dominated by agricultural products. It is the twelfth largest cotton producer and fifth largest cashew nut producer in the world.

Official name Republic of Benin
Formation 1960 / 1960
Capital Porto-Novo; Cotonou
Population 13.7 million / 315 people per sq mile (122 people per sq km)
Total area 43,484 sq miles (112,622 sq km)
Languages Fon, Bariba, Yoruba, Adja, Houeda, Somba, French*
Religions Muslim 28%, Roman Catholic 26%, Other Christian 24%, Vodoun 12%, Other 10%
Demographics Fon 38%, Adja 15%, Yoruba 12%, Bariba 10%, Fulani 9%, Other 16%
Government Presidential system
Currency CFA franc = 100 centimes
Literacy rate 42%
Life expectancy 62 years

BHUTAN
South Asia

Page 113 G3

This landlocked Buddhist kingdom, perched in the eastern Himalayas between India and China, announced the draft of its first constitution in 2005 and held its first parliamentary election in 2008.

Official name Kingdom of Bhutan
Formation 1907 / 2006
Capital Thimphu
Population 867,775 / 59 people per sq mile (23 people per sq km)
Total area 14,824 sq miles (38,394 sq km)
Languages Dzongkha*, Sharchopkha, Lhotshamkha
Religions Mahayana Buddhist 75%, Hindu 22%, Other 3%
Demographics Ngalop 50%, Nepali 35%, Tribal groups 15%
Government Monarchical / parliamentary system
Currency Ngultrum = 100 chetrum
Literacy rate 67%
Life expectancy 72 years

BOLIVIA
South America

Page 39 F3

Landlocked high in central South America, Bolivia is renowned for its vast mineral wealth, which drives the economy, including the world's largest reserves of lithium.

Official name Plurinational State of Bolivia
Formation 1825 / 1938
Capital La Paz (administrative); Sucre (judicial)
Population 12 million / 28 people per sq mile (11 people per sq km)
Total area 424,164 sq miles (1,098,581 sq km)
Languages Aymara*, Quechua*, Spanish*, Guarani
Religions Roman Catholic 70%, Evangelical 15%, Adventist 2%, Church of Jesus Christ 1%, Other 12%
Demographics Mixed race 70%, Indigenous 20%, White 5%, African 1%, Other 4%
Government Presidential system
Currency Boliviano = 100 centavos
Literacy rate 93%
Life expectancy 73 years

BOSNIA & HERZEGOVINA
Southeast Europe

Page 78 B3

Set in the western Balkans, this state has a complex history born out of the differences between its three main ethnic groups. With its spectacular alpine scenery, tourism is a fast growing sector.

Official name Bosnia and Herzegovina
Formation 1992 / 1992
Capital Sarajevo
Population 3.8 million / 192 people per sq mile (74 people per sq km)
Total area 19,767 sq miles (51,197 sq km)
Languages Bosnian*, Serbian*, Croatian*
Religions Muslim (mainly Sunni) 53%, Orthodox Christian 35%, Roman Catholic 8%, Nonreligious 3%, Other 1%
Demographics Bosniak 50%, Serb 31%, Croat 15%, Other 4%
Government Parliamentary system
Currency Marka = 100 pfeninga
Literacy rate 99%
Life expectancy 78 years

BOTSWANA
Southern Africa

Page 56 C3

Botswana lies landlocked in Southern Africa. The Orapa diamond mine is one of the largest in the world and this, along with other mineral resources, helps to provide a relatively prosperous economy.

Official name Republic of Botswana
Formation 1966 / 1966
Capital Gaborone
Population 2.3 million / 13 people per sq mile (5 people per sq km)
Total area 224,607 sq miles (581,730 sq km)
Languages Setswana, English*, Sekalanga, Shona, San, Khoikhoi, isiNdebele
Religions Christian (mainly Protestant) 80%, Nonreligious 15%, Traditional beliefs 4%, Other (including Muslim) 1%
Demographics Tswana (or Setswana) 79%, Kalanga 11%, Basarwa 3%, Other 7%
Government Parliamentary system
Currency Pula = 100 thebe
Literacy rate 89%
Life expectancy 66 years

BRAZIL
South America

Page 40 C2

Brazil covers about half of South America and has immense natural resources. As well as a diverse, high-tech industrial base and a burgeoning service sector, it produces a third of the world's coffee.

Official name Federative Republic of Brazil
Formation 1822 / 1909
Capital Brasília
Population 214 million / 65 people per sq mile (25 people per sq km)
Total area 3,287,957 sq miles (8,515,770 sq km)
Languages Portuguese*, German, Italian, Spanish, Polish, Japanese, Amerindian languages
Religions Roman Catholic 61%, Protestant 26%, Nonreligious 8%, Other 5%
Demographics White 48%, Mixed race 43%, Black 8%, Other 1%
Government Presidential system
Currency Real = 100 centavos
Literacy rate 93%
Life expectancy 76 years

BRUNEI
Southeast Asia

Page 116 D3

Located on the island of Borneo, Brunei has a high standard of living due to oil and gas revenues. The Sultan of Brunei was crowned in 1967, making him one of the world's longest reigning monarchs.

Official name Brunei Darussalam
Formation 1984 / 1984
Capital Bandar Seri Begawan
Population 474,054 / 213 people per sq mile (82 people per sq km)
Total area 2226 sq miles (5765 sq km)
Languages Malay*, English, Chinese dialects
Religions Muslim (mainly Sunni) 79%, Christian 9%, Buddhist 8%, Other 4%
Demographics Malay 66%, Chinese 10%, Indigenous 4%, Other 20%
Government Monarchy
Currency Bruneian dollar = 100 cents
Literacy rate 97%
Life expectancy 78 years

BULGARIA
Southeast Europe

Page 82 C2

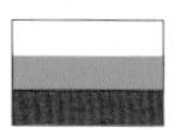

Located on the western shore of the Black Sea, Bulgaria is rich in history and culture. It was here that the Cyrillic script was developed in the 9th century CE. EU membership was achieved in 2007.

Official name Republic of Bulgaria
Formation 1908 / 1947
Capital Sofia
Population 6.8 million / 159 people per sq mile (61 people per sq km)
Total area 42,811 sq miles (110,879 sq km)
Languages Bulgarian*, Turkish, Romani
Religions Orthodox Christian 75%, Muslim 15%, Nonreligious 5%, Protestant 1%, Roman Catholic 1%, Other 3%
Demographics Bulgarian 85%, Turkish 9%, Roma 5%, Other 1%
Government Parliamentary system
Currency Lev = 100 stotinki
Literacy rate 98%
Life expectancy 76 years

BURKINA FASO
West Africa

Page 53 E4

Burkina Faso is landlocked in the semiarid Sahel of West Africa. Often struggling with political instability, it is the fourth largest producer of gold in Africa, which makes up over 75 percent of the country's exports.

Official name Burkina Faso
Formation 1960 / 2016
Capital Ouagadougou
Population 21.9 million / 207 people per sq mile (80 people per sq km)
Total area 105,869 sq miles (274,200 sq km)
Languages Mossi, Fulani, French*, Tuareg, Dyula, Songhai
Religions Muslim 60% Christian 23%, Indigenous beliefs 15%, Other 2%
Demographics Mossi 52%, Fulani 8%, Gurma 7%, Bobo 5%, Gurunsi 5%, Senufo 5%, Bissa 4%, Lobi 2%, Dagara 2%, Other 10%
Government Presidential system
Currency CFA franc = 100 centimes
Literacy rate 39%
Life expectancy 63 years

BURUNDI
Central Africa

Page 51 B7

Landlocked in central Africa, Burundi has a predominantly rural society with an emphasis on agriculture. Ethnic tensions led to brutal conflict but peace efforts have since brought greater stability.

Official name Republic of Burundi
Formation 1962 / 1962
Capital Bujumbura; Gitega
Population 12.6 million / 1173 people per sq mile (453 people per sq km)
Total area 10,745 sq miles (27,830 sq km)
Languages Kirundi*, French*, Kiswahili, English
Religions Roman Catholic 65%, Protestant 23%, Muslim 3%, Seventh-day Adventist 2%, Other 7%
Demographics Hutu 85%, Tutsi 14%, Twa 1%
Government Presidential system
Currency Burundian franc = 100 centimes
Literacy rate 68%
Life expectancy 67 years

CAMBODIA
Southeast Asia

Page 115 D5

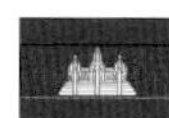

This ancient Southeast Asian kingdom suffered years of brutal totalitarian rule but is now recovering, with recent economic growth helping to dramatically reduce poverty.

Official name Kingdom of Cambodia
Formation 1953 / 1953
Capital Phnom Penh
Population 16.7 million/ 239 people per sq mile (92 people per sq km)
Total area 69,898 sq miles (181,035 sq km)
Languages Khmer*, French, Chinese, Vietnamese, Cham
Religions Buddhist 97%, Muslim 2%, Other (mostly Christian) 1%
Demographics Khmer 95%, Cham 2%, Chinese 2%, Other 1%
Government Parliamentary system
Currency Riel = 100 sen
Literacy rate 81%
Life expectancy 71 years

CAMEROON
Central Africa

Page 54 A4

Set on the central West African coast, Cameroon has a hugely diverse geography, including extensive tropical forests. Small-scale agriculture, forestry, and fishing are key economic activities.

Official name Republic of Cameroon
Formation 1960 / 2006
Capital Yaoundé
Population 29.3 million / 160 people per sq mile (62 people per sq km)
Total area 183,568 sq miles (475,440 sq km)
Languages Bamileke, Fang, Fulani, French*, English*
Religions Roman Catholic 38%, Protestant 26%, Other Christian 7%, Muslim 24%, Other 5%
Demographics Bamileke-Bamu 24%, Beti/Bassa, Mbam 21%, Biu-Mandara 14%, Arab-Choa/Hausa/Kanuri 11%, Adamawa-Ubangi 9%, Other 21%
Government Presidential system
Currency CFA franc = 100 centimes
Literacy rate 77%
Life expectancy 63 years

CANADA
North America

Page 15 E4

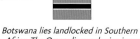

The world's second largest country spans six time zones and is rich in natural resources. Huge oil and gas reserves allow a net export of energy, contibuting to a well-developed and high-tech globalized economy.

Official name Canada
Formation 1867 / 1949
Capital Ottawa
Population 38.2 million / 10 people per sq mile (4 people per sq km)
Total area 3,855,171 sq miles (9,984,670 sq km)
Languages English*, French*, Pubjabi, Cantonese, Spanish, Arabic,Tagalog, Italian, German
Religions Roman Catholic 39%, Other Christian 28%, Nonreligious 24%, Muslim 3%, Other 6%
Demographics European descent 80%, Asian 15%, First Nations, Métis, and Inuit 5%
Government Parliamentary system
Currency Canadian dollar = 100 cents
Literacy rate 99%
Life expectancy 84 years

CAPE VERDE (CABO VERDE)
Atlantic Ocean

Page 52 A2

A group of mainly volcanic islands off the west coast of Africa, Cape Verde has limited natural resources. The economy relies on fisheries and tourists attracted by its warm climate and unspoilt beaches.

Official name Republic of Cabo Verde
Formation 1975 / 1975
Capital Praia
Population 596,707 / 383 people per sq mile (148 people per sq km)
Total area 1557 sq miles (4033 sq km)
Languages Portuguese Creole, Portuguese*
Religions Roman Catholic 77%, Protestant 5%, other Christian 3% , Muslim 2%, Other 13%
Demographics Mixed race 71%, African 28%, European 1%
Government Presidential / parliamentary system
Currency Escudo = 100 centavos
Literacy rate 87%
Life expectancy 74 years

CENTRAL AFRICAN REPUBLIC
Central Africa

Page 54 C4

Set on a landlocked plateau dividing the Chad and Congo river basins, the CAR has suffered from years of political instability and conflict. It is rich in diamonds, gold, oil, and uranium.

Official name Central African Republic
Formation 1960 / 1960
Capital Bangui
Population 5.4 million / 22 people per sq mile (9 people per sq km)
Total area 240,535 sq miles (622,984 sq km)
Languages Sango, Banda, Gbaya, French*
Religions Christian 89%, Muslim 9%, Folk Religion 1%, Unaffiliated 1%
Demographics Baya 29%, Banda 23%, Mandjia 10%, Sara 8%, M'Baka-Bantu 8%, Mbum 6%, Arab-Fulani 6%, Ngbanki 6%, Other 4%
Government Presidential system
Currency CFA franc = 100 centimes
Literacy rate 37%
Life expectancy 56 years

CHAD
Central Africa

Page 54 C3

Landlocked in Central Africa, with desert to the north, Chad has been troubled by periods of civil war. The majority of its people rely on small-scale farming. It became a net oil exporter in 2003.

Official name Republic of Chad
Formation 1960 / 1960
Capital N'Djaména
Population 17.9 million / 36 people per sq mile (14 people per sq km)
Total area 495,755 sq miles (1,284,000 sq km)
Languages French*, Sara, Arabic*, Maba
Religions Muslim 52%, Protestant 24%, Roman Catholic 20%, None 3%, Other 1%
Demographics Sara 31%, Arab 10%, Kanembu 9%, Masalit 7%, Gorane 6%, Other Indigenous groups 34%, Other 3%
Government Presidential system
Currency CFA franc = 100 centimes
Literacy rate 22%
Life expectancy 59 years

CHILE
South America

Page 42 B3

Chile extends in a thin ribbon down the Pacific coast of South America. Rich in mineral resources, it is the world's largest copper producer supplying about a third of the global market.

Official name Republic of Chile
Formation 1810 / 1898
Capital Santiago
Population 18.4 million / 63 people per sq mile (24 people per sq km)
Total area 291,933 sq miles (756,120 sq km)
Languages Spanish*, Amerindian languages
Religions Roman Catholic 60%, Evangelical 18%, Atheist or Agnostic 4%, None 17%
Demographics White and mixed race 89%, Mapuche 9%, Aymara 1%, Other Indigenous 1%
Government Presidential system
Currency Chilean peso = 100 centavos
Literacy rate 96%
Life expectancy 80 years

CHINA
East Asia

Page 104 C4

This vast East Asian country, home to a fifth of the global population, became a communist state in 1949. It has now emerged as one of the world's major political and economic powers.

Official name People's Republic of China
Formation 1949 / 2011
Capital Beijing
Population 1.41 billion / 381 people per sq mile (147 people per sq km)
Total area 3,705,386 sq miles (9,596,960 sq km)
Languages Mandarin*, Wu, Cantonese, Hsiang, Min, Hakka, Kan
Religions Folk religion 22%, Buddhist 18%, Christian 5%, Muslim 2%, Unaffiliated 52%, Hindu, Jewish, and other 1%
Demographics Han Chinese 91%, Ethnic minorities 9%
Government One-party state
Currency Renminbi (or yuan) = 10 jiao = 100 fen
Literacy rate 97%
Life expectancy 78 years

COLOMBIA
South America

Page 36 B3

Lying in northwest South America, Colombia is rich in biodiversity with around 45,000 plant species. Exports of petroleum and agricultural products are driving rapid economic growth.

Official name Republic of Colombia
Formation 1810 / 1903
Capital Bogotá
Population 49.1 million / 112 people per sq mile (43 people per sq km)
Total area 439,736 sq miles (1,138,910 sq km)
Languages Spanish*, Amerindian languages
Religions Christian 92% (mainly Roman Catholic), Unspecified 7%, Other 1%
Demographics Mixed race and White 88%, Indigenous 4%, Afro-Colombian 7%, Unspecified 1%
Government Presidential system
Currency Colombian peso = 100 centavos
Literacy rate 96%
Life expectancy 75 years

COMOROS
Indian Ocean

Page 57 F2

Comoros sits between Mozambique and Madagascar. A volatile political history and limited natural resources have restricted economic growth. Spices, particularly cloves, are the main export.

Official name Union of the Comoros
Formation 1975 / 1975
Capital Moroni
Population 876,437 / 1016 people per sq mile (392 people per sq km)
Total area 863 sq miles (2235 sq km)
Languages Arabic*, Comoran*, French*
Religions Sunni Muslim 98%, Other 2%
Demographics Comoran 97%, Other 3%
Government Presidential system
Currency Comoros franc = 100 centimes
Literacy rate 59%
Life expectancy 67 years

CONGO
Central Africa

Page 55 B5

This country straddles the equator in Central Africa. Decades of political instability have restricted growth. Oil accounts for over 90 percent of export values but there is great untapped mineral wealth.

Official name Republic of the Congo
Formation 1960 / 1960
Capital Brazzaville
Population 5.5 million/ 42 people per sq mile (16 people per sq km)
Total area 132,047 sq miles (342,000 sq km)
Languages Kongo, Teke, Lingala, French*
Religions Roman Catholic 34%, Awakening Churches/Christian Revival 22%, Protestant 20%, None 12%, Other 12%
Demographics Kongo 41%, Teke 17%, Mbochi 13%, Sangha 6%, Mbere/Mbeti/Kele 4%, Punu 4%, Other 15%
Government Presidential system
Currency CFA franc = 100 centimes
Literacy rate 80%
Life expectancy 62 years

CONGO, DEM. REP.
Central Africa

Page 55 C6

Straddling the equator in east Central Africa, this mineral-rich country is the second largest in Africa. Years of political instability and conflict have restricted its potential and led to widespread poverty.

Official name Democratic Republic of the Congo
Formation 1960 / 1960
Capital Kinshasa
Population 108.4 million/ 120 people per sq mile (46 people per sq km)
Total area 905,355 sq miles (2,344,858 sq km)
Languages Kiswahili, Tshiluba, Kikongo, Lingala, French*
Religions Roman Catholic 31%, Protestant 28%, Other Christian 37%, Muslim 1%, Other 3%
Demographics Mongo, Luba, Kongo, and Mangbetu-Azande 45%, Other African ethnic groups 55%
Government Semi-Presidential system
Currency Congolese franc = 100 centimes
Literacy rate 77%
Life expectancy 62 years

COSTA RICA
Central America

Page 31 E4

Set in Central America, Costa Rica abolished its army in 1948 in favour of spending on education and welfare. Its pristine forests and many national parks encourage the principles of ecotourism.

Official name Republic of Costa Rica
Formation 1821 / 1941
Capital San José
Population 5.2 million / 264 people per sq mile (102 people per sq km)
Total area 19,730 sq miles (51,100 sq km)
Languages Spanish*, English Creole, Bribri, Cabecar
Religions Roman Catholic 48%, Evangelical and Pentecostal 20%, Jehovah's Witness 1%, Other Protestant 1%, None 27%, Other 3%
Demographics Mixed race and European 96%, Indigenous 3%, Black 1%
Government Presidential system
Currency Costa Rican colón = 100 céntimos
Literacy rate 98%
Life expectancy 80 years

CROATIA
Southeast Europe

Page 78 B2

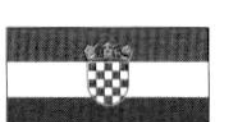

Croatia emerged from bitter conflict in the 1990s to develop into a stable, democratic society, eventually joining the EU in 2013. It is set on the eastern Adriatic coast and tourism is a key part of the economy.

Official name Republic of Croatia
Formation 1991 / 1991
Capital Zagreb
Population 4.1 million / 188 people per sq mile (72 people per sq km)
Total area 21,851 sq miles (56,594 sq km)
Languages Croatian*
Religions Roman Catholic 84%, Nonreligious 7%, Orthodox Christian 4%, Muslim 2%, Other 3%
Demographics Croat 92%, Serb 4%, Bosniak 1%, Other 3%
Government Parliamentary system
Currency Kuna = 100 lipa
Literacy rate 99%
Life expectancy 77 years

CUBA
West Indies

Page 32 C2

The largest of the Caribbean islands, an archipelago of about 1,600 islands, islets, and cays, Cuba is the only communist state in the Americas. A rich cultural history helps to attract millions of tourists.

Official name Republic of Cuba
Formation 1902 / 1902
Capital Havana
Population 11 million / 257 people per sq mile (99 people per sq km)
Total area 42,803 sq miles (110,860 sq km)
Languages Spanish
Religions Christian 58%, Folk religion 16%, Buddhist 1%, Hindu 1%, Jewish 1%, Muslim 1%, None 22%
Demographics White 65%, Mixed race 25%, Black 10%
Government One-party state
Currency Cuban peso = 100 centavos
Literacy rate 100%
Life expectancy 80 years

CYPRUS
Southeast Europe

Page 80 C5

With a strategic location in the eastern Mediterranean, Cyprus has a rich legacy of antiquities and architecture dating back thousands of years. Shipping and tourism are key economic sectors.

Official name Republic of Cyprus
Formation 1960 / 1960
Capital Nicosia
Population 1.2 million / 336 people per sq mile (130 people per sq km)
Total area 3572 sq miles (9251 sq km)
Languages Greek*, Turkish*, English, Romanian, Russian, Bulgarian, Arabic, Filipino
Religions Orthodox Christian 89%, Roman Catholic 3%, Protestant/Anglican 2%, Muslim 2%, Other 4%
Demographics Greek 99%, Other 1%
Government Presidential system
Currency Euro = 100 cents (In TRNC, Turkish lira = 100 kurus)
Literacy rate 99%
Life expectancy 80 years

CZECHIA
Central Europe

Page 77 A5

Landlocked in Central Europe, and formerly part of Czechoslovakia, this country peacefully dissolved its federal union with Slovakia in 1993. It joined NATO in 1999 and the EU in 2004.

Official name Czech Republic
Formation 1993 / 1993
Capital Prague
Population 10.7 million/ 351 people per sq mile (136 people per sq km)
Total area 30,450 sq miles (78,867 sq km)
Languages Czech*, Slovak, Hungarian
Religions Roman Catholic 9%, Other Christian 2%, Nonreligious 57%, Unspecified 32%
Demographics Czech 57%, Moravian 3%, Unspecified 32%, Other 8%
Government Parliamentary system
Currency Czech koruna = 100 haleru
Literacy rate 99%
Life expectancy 80 years

DENMARK
Northern Europe

Page 63 A7

Denmark occupies the Jutland peninsula at the entrance to the Baltic Sea. In the 1930s, it set up one of the first welfare systems supported by a highly diverse and successful industrial economy.

Official name Kingdom of Denmark
Formation 965 / 1944
Capital Copenhagen
Population 5.9 million / 355 people per sq mile (137 people per sq km)
Total area 16,639 sq miles (43,094 sq km)
Languages Danish*,English, Faroese
Religions Evangelical Lutheran 75%, Muslim 5%, Other 20%,
Demographics Danish 86%, Turkish 1%, Other 13%
Government Parliamentary system
Currency Danish krone = 100 øre
Literacy rate 99%
Life expectancy 82 years

DJIBOUTI
East Africa

Page 50 D4

This city state with a desert hinterland lies on the coast of the Horn of Africa. Its economy relies on its Red Sea container port, which handles 95 percent of seaborne trade for neighboring landlocked Ethiopia.

Official name Republic of Djibouti
Formation 1977 / 1977
Capital Djibouti
Population 957,273 / 107 people per sq mile (41 people per sq km)
Total area 8958 sq miles (23,200 sq km)
Languages Somali, Afar, French*, Arabic*
Religions Muslim (mainly Sunni) 94%, Other 6%
Demographics Somali 60%, Afar 35%, Other 5%
Government Presidential system
Currency Djibouti franc = 100 centimes
Literacy rate 68%
Life expectancy 65 years

DOMINICA
West Indies

Page 33 H4

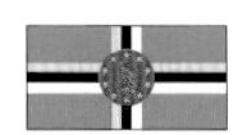

Known for its lush flora and fauna, this Caribbean island is home to a number of endemic species, notably the Sisserou Parrot. Agriculture, particularly bananas and coffee, and tourism are important.

Official name Commonwealth of Dominica
Formation 1978 / 1978
Capital Roseau
Population 72,170 / 249 people per sq mile (96 people per sq km)
Total area 290 sq miles (751 sq km)
Languages French Creole, English*
Religions Roman Catholic 62%, Protestant 30%, Nonreligious 6%, Other 2%
Demographics African 84%, Mixed 9%, Indigenous 4%, Other 3%
Government Parliamentary system
Currency East Caribbean dollar = 100 cents
Literacy rate 94%
Life expectancy 78 years

DOMINICAN REPUBLIC
West Indies

Page 33 E2

Occupying the eastern part of the island of Hispaniola, the Dominican Republic is the Caribbean's top tourist destination and largest economy. The Pueblo Viejo gold mine is the third largest in the world.

Official name Dominican Republic
Formation 1844 / 1936
Capital Santo Domingo
Population 10.6 million / 564 people per sq mile (218 people per sq km)
Total area 18,792 sq miles (48,670 sq km)
Languages Spanish*, French Creole
Religions Roman Catholic 44%, Evangelical 14%, Protestant 8%, None 29%, Unspecified 2%, Other 3%
Demographics Mixed race 70%, Black 16%, White 13%, Other 1%
Government Presidential system
Currency Dominican Republic peso = 100 centavos
Literacy rate 94%
Life expectancy 73 years

EAST TIMOR (TIMOR-LESTE)
Southeast Asia

Page 116 F5

Set on the eastern part of the island of Timor in the East Indies, this nation emerged from a turbulent transition to gain independence from Indonesia in 2002. Exports are dominated by oil and gas.

Official name Democratic Republic of Timor-Leste
Formation 2002 / 2002
Capital Dili
Population 1.4 million / 244 people per sq mile (94 people per sq km)
Total area 5743 sq miles (14,874 sq km)
Languages Tetum* (Portuguese/Austronesian), Bahasa Indonesia, Portuguese*
Religions Roman Catholic 96%, Protestant/ Evangelical 2%, Muslim 1%, Other 1%
Demographics Papuan groups approx. 85%, Indonesian groups approx. 13%, Chinese 2%
Government Semi-presidential system
Currency US dollar = 100 cents
Literacy rate 58%
Life expectancy 70 years

ECUADOR
South America

Page 38 A2

Once part of the Inca heartland on the northwest coast of South America, Ecuador is the world's leading banana exporter. Its territory includes the wildlife-rich Galápagos Islands.

Official name Republic of Ecuador
Formation 1822 / 1942
Capital Quito
Population 17.2 million / 157 people per sq mile (61 people per sq km)
Total area 109,484 sq miles (283,561 sq km)
Languages Spanish*, Quechua, Other Amerindian languages
Religions Roman Catholic 69%, Evangelical 16%, Adventist 1%, Agnostic or Atheist 2%, None 10%, Other 2%
Demographics Mixed race 79%, Black 7%, Indigenous 7%, White 6%, Other 1%
Government Presidential system
Currency US dollar = 100 cents
Literacy rate 94%
Life expectancy 78 years

EGYPT
North Africa

Page 50 B2

Egypt lies in the northeast corner of Africa, where arid deserts are split by the Nile valley. Vast natural gas reserves, agriculture, and tourism underpin the economy. The Suez Canal is a key global transport link.

Official name Arab Republic of Egypt
Formation 1953 / 2017
Capital Cairo
Population 107.7 million/ 279 people per sq mile (108 people per sq km)
Total area 386,662 sq miles (1,001,450 sq km)
Languages Arabic*, French, English, Berber
Religions Muslim (mainly Sunni) 90%, Coptic Christian 9%, Other Christian 1%
Demographics Egyptian 99%, Other 1%
Government Presidential system
Currency Egyptian pound = 100 piastres
Literacy rate 71%
Life expectancy 74 years

EL SALVADOR
Central America

Page 30 B3

El Salvador is Central America's smallest country. Conservation efforts are ongoing to protect the abundant natural biodiversity, including 1,000 species of butterflies and 400 types of orchid.

Official name Republic of El Salvador
Formation 1821 / 1998
Capital San Salvador
Population 6.5 million / 800 people per sq mile (309 people per sq km)
Total area 8124 sq miles (21,041 sq km)
Languages Spanish*, Nawat
Religions Roman Catholic 50%, Protestant 36%, Nonreligious 12%, Other 2%
Demographics Mixed race 86%, White 13%, Indigenous and other 1%
Government Presidential system
Currency Salvadorean colón = 100 centavos; US dollar = 100 cents
Literacy rate 89%
Life expectancy 75 years

EQUATORIAL GUINEA
Central Africa

Page 55 A5

Equatorial Guinea comprises the Rio Muni mainland in west Central Africa and five islands. Large oil and gas reserves dominate exports and the tropical climate helps to support the rural economy.

Official name Republic of Equatorial Guinea
Formation 1968 / 1968
Capital Malabo
Population 1.6 million / 148 people per sq mile (57 people per sq km)
Total area 10,830 sq miles (28,051 sq km)
Languages Spanish*, Fang, Bubi, French*
Religions Roman Catholic 88%, Protestant 5%, Muslim 2%, Other 5%
Demographics Fang 86%, Bubi 6%, Other 8%
Government Presidential system
Currency CFA franc = 100 centimes
Literacy rate 95%
Life expectancy 64 years

ERITREA
East Africa

Page 50 C4

Eritrea lies on the southern shores of the Red Sea. Gold and copper from the Bisha mine currently account for the majority of export values, while 70 percent of the country's workforce is engaged in agriculture.

Official name State of Eritrea
Formation 1993 / 2002
Capital Asmara
Population 6.2 million / 137 people per sq mile (53 people per sq km)
Total area 45,406 sq miles (117,600 sq km)
Languages Tigrinya*, English*, Tigre, Afar, Arabic*, Saho, Bilen, Kunama, Nara, Hadareb
Religions Christian 50%, Muslim 48%, Other 2%
Demographics Tigrinya 50%, Tigre 30%, Saho 4%, Afar 4%, Kunama 4%, Bilen 3%, Other 5%
Government Presidential system
Currency Nakfa = 100 cents
Literacy rate 77%
Life expectancy 70 years

ESTONIA
Northeast Europe

Page 84 D2

Situated toward the eastern end of the Baltic Sea, Estonia's ancient capital city, Tallinn, is rich in medieval architecture. A fast-growing, high-tech economy has seen numerous successful IT start-ups.

Official name Republic of Estonia
Formation 1991 / 1991
Capital Tallinn
Population 1.2 million/ 69 people per sq mile (27 people per sq km)
Total area 17,463 sq miles (45,228 sq km)
Languages Estonian*, Russian, Ukranian
Religions Nonreligious 45%, Orthodox Christian 25%, Lutheran 20%, Other 10%
Demographics Estonian 70%, Russian 25%, Ukrainian 2%, Belarussian 1%, Other 2%
Government Parliamentary system
Currency Euro = 100 cents
Literacy rate 100%
Life expectancy 78 years

ESWATINI
Southern Africa

Page 56 D4

Formerly known as Swaziland, this small, land-locked kingdom is one of the world's last remaining absolute monarchies. The economy is closely linked to South Africa for the majority of its trade.

Official name Kingdom of Eswatini
Formation 1968 / 1968
Capital Mbabane; Lobamba
Population 1.1 million / 164 people per sq mile (63 people per sq km)
Total area 6704 sq miles (17,364 sq km)
Languages English*, siSwati*, isiZulu, Xitsonga
Religions Christian 90%, Muslim 2%, Other 8%
Demographics Swazi 97%, Other 3%
Government Absolute monarchy
Currency Swazi lilangeni = 100 cents; South African rand = 100 cents
Literacy rate 83%
Life expectancy 58 years

ETHIOPIA
East Africa

Page 51 C5

Landlocked Ethiopia is the largest and most populous country in the Horn of Africa. It was a founding member of the UN. Coffee production and extensive hydro-electric schemes bolster the economy.

Official name Federal Democratic Republic of Ethiopia
Formation 1896 / 2002
Capital Addis Ababa
Population 113 million / 265 people per sq mile (102 people per sq km)
Total area 426,373 sq miles (1,104,300 sq km)
Languages Amharic*, Oromo, Tigrinya, Galla, Sidamo, Somali, English, Arabic
Religions Christian 62%, Muslim 34%, Other 4%
Demographics Oromo 36%, Amhara 24%, Somali 7%, Tigray 6%, Sidama 4%, Guragie 3%, Welaita 2%, Afar 2%, Other 16%
Government Parliamentary system
Currency Birr = 100 santim
Literacy rate 57%
Life expectancy 68 years

FIJI
Australasia & Oceania

Page 123 E5

Fiji is a volcanic archipelago of 882 islands in the southern Pacific Ocean. Abundant natural resources, including minerals, forestry, and fisheries, are supplemented by significant levels of tourism.

Official name Republic of Fiji
Formation 1970 / 1970
Capital Suva
Population 943,737 / 134 people per sq mile (52 people per sq km)
Total area 7056 sq miles (18,274 sq km)
Languages Fijian, English*, Hindi, Urdu, Tamil, Telugu
Religions Methodist 35%, Hindu 28%, Other Christian 21%, Roman Catholic 9%, Muslim 6%, Other and nonreligious 1%
Demographics Melanesian 57%, Indian 38%, Other 5%
Government Parliamentary system
Currency Fiji dollar = 100 cents
Literacy rate 99%
Life expectancy 74 years

FINLAND
Northern Europe

Page 62 D4

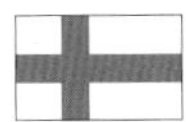

A low-lying country of forests and lakes, Finland was the first to adopt full gender equality, granting men and women the right to both vote and stand for election, in 1906. It joined NATO in 2023.

Official name Republic of Finland
Formation 1917 / 1947
Capital Helsinki / Helsingfors
Population 5.6 million / 43 people per sq mile (17 people per sq km)
Total area 130,127 sq miles (338,145 sq km)
Languages Finnish*, Swedish*, Russian
Religions Lutheran 67%, Greek Orthodox 1%, None 30%, Other 2%
Demographics Finnish 93%, Other (including Sámi) 7%
Government Parliamentary system
Currency Euro = 100 cents
Literacy rate 100%
Life expectancy 82 years

FRANCE
Western Europe

Page 68 B4

Straddling Western Europe from the English Channel to the Mediterranean Sea, France was Europe's first modern republic. It is now one of the world's leading industrial powers.

Official name French Republic
Formation 987 / 1947
Capital Paris
Population 68.3 million / 321 people per sq mile (124 people per sq km)
Total area 212,935 sq miles (551,500 sq km)
Languages French*, Provençal, German, Breton, Catalan, Basque
Religions Roman Catholic 47%, Muslim 4%, None 33%, Protestant 2%, Unspecified 9%, Other 5%
Demographics French 86%, Black 10%, German (Alsace) 2%, Breton 1%, Other 1%
Government Presidential / Parliamentary system
Currency Euro = 100 cents
Literacy rate 99%
Life expectancy 83 years

GABON
Central Africa

Page 55 A5

Gabon straddles the Equator on Africa's west coast. 80 percent of its export revenues come from oil, although planning is now underway to diversify the economy as reserves are declining.

Official name Gabonese Republic
Formation 1960 / 1960
Capital Libreville
Population 2.3 million / 22 people per sq mile (9 people per sq km)
Total area 103,347 sq miles (267,667 sq km)
Languages Fang, French*, Punu, Sira, Nzebi, Mpongwe, Bandjabi
Religions Roman Catholic 42%, Protestant 12%, other Christian 27%, Muslim 10%, Animist 1%, None 7%, Other 1%
Demographics Gabonese 80% , Cameroonian 5%, Malian 2%, Beninese 2%, Togolese 2%, Other 9%
Government Presidential system
Currency CFA franc = 100 centimes
Literacy rate 85%
Life expectancy 70 years

GAMBIA, THE
West Africa

Page 52 B3

A narrow state along the Gambia River on Africa's west coast and surrounded by Senegal, Gambia has a warm, tropical climate which supports a mainly rural economy, with the key activity being nut production.

Official name Republic of The Gambia
Formation 1965 / 1965
Capital Banjul
Population 2.4 million / 550 people per sq mile (212 people per sq km)
Total area 4363 sq miles (11,300 sq km)
Languages Mandinka, Fulani, Wolof, Jola, Soninke, English*
Religions Muslim 96%, Christian 3%, Other 1%
Demographics Mandinka/Jahanka 33%, Fulani/Tukulur/Lorobo 18%, Wolof 13%, Jola/Karoninka 11%, Non-Gambian 10%, Other 15%
Government Presidential system
Currency Dalasi = 100 butut
Literacy rate 51%
Life expectancy 68 years

GEORGIA
Southwest Asia

Page 95 F2

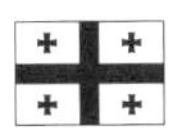

Set in the Caucasus on the Black Sea's eastern shore, Georgia is one of the world's oldest wine producers. Efforts are being made to encourage tourism to numerous spas, ski resorts, and historic sites.

Official name Georgia
Formation 1991 / 1991
Capital Tbilisi
Population 4.9 million / 182 people per sq mile (70 people per sq km)
Total area 26,911 sq miles (69,700 sq km)
Languages Georgian*, Russian, Azeri, Armenian, Mingrelian, Ossetian, Abkhazian (* in Abkhazia)
Religions Orthodox Christian 89%, Muslim 9%, Roman Catholic 1%, Other 1%
Demographics Georgian 87%, Azeri 6%, Armenian 4%, Russian 1%, Other 2%
Government Presidential / Parliamentary system
Currency Lari = 100 tetri
Literacy rate 100%
Life expectancy 78 years

GERMANY
Northern Europe

Page 72 B4

At the heart of Europe, Germany is a global economic power. An efficient, diverse, and highly skilled industrial economy is increasingly using renewable energy sources to reduce dependency on fossil fuels.

Official name Federal Republic of Germany
Formation 1871 / 1990
Capital Berlin
Population 84.3 million / 612 people per sq mile (236 people per sq km)
Total area 137,847 sq miles (357,022 sq km)
Languages German*, Turkish
Religions Roman Catholic 27%, Protestant 24%, Muslim 3%, None 41%, Other 5%
Demographics German 86%, Turkish 2%, Polish 1%, Syrian 1%, Romanian 1%, Other 9%
Government Parliamentary system
Currency Euro = 100 cents
Literacy rate 99%
Life expectancy 82 years

GHANA
West Africa

Page 83 A5

Set on the coast of West Africa, Ghana is a major gold producer. Lake Volta is the largest artificial reservoir in the world and provides enough hydro-electricity to allow exports to neighboring countries.

Official name Republic of Ghana
Formation 1957 / 1957
Capital Accra
Population 33.1 million/ 359 people per sq mile (139 people per sq km)
Total area 92,098 sq miles (238,533 sq km)
Languages Twi, Fanti, Ewe, Ga, Adangbe, Gurma, Dagomba (Dagbani), English*, Asante
Religions Christian 71%, Muslim 20%, Traditionalist 3%, None 1%, Other 5%
Demographics Akan 46%, Mole-Dagbani 18%, Ewe 13%, Ga-Dangme 7%, Other 16%
Government Presidential system
Currency Cedi = 100 pesewas
Literacy rate 79%
Life expectancy 69 years

GREECE
Southeast Europe

Page 83 A5

Situated at the eastern end of the Mediterranean, Greece has more than 2000 islands. It is rich in culture and historic architecture. Shipping and tourism are key economic drivers.

Official name Hellenic Republic
Formation 1830 / 1947
Capital Athens
Population 10.5 million / 206 people per sq mile (80 people per sq km)
Total area 50,949 sq miles (131,957 sq km)
Languages Greek*, Turkish, Macedonian, Albanian
Religions Orthodox Christian 90%, Nonreligious 4%, Muslim 2%, Other 4%
Demographics Greek 92%, Albanian 4%, Other 4%
Government Parliamentary system
Currency Euro = 100 cents
Literacy rate 98%
Life expectancy 80 years

GRENADA
West Indies

Page 33 G5

Also known as the "Island of Spice", Grenada is the most southerly of the Windward Islands. Historically one of the world's largest nutmeg producers, its main economic resource is now tourism.

Official name Grenada
Formation 1974 / 1974
Capital St. George's
Population 113,949 / 857 people per sq mile (331 people per sq km)
Total area 133 sq miles (344 sq km)
Languages English*, English Creole
Religions Protestant 49%, Roman Catholic 36%, Jehovah's Witness 1%, Rastafarian 1%, None 6%, Unspecified 1%, Other 6%
Demographics African 82%, Mixed 14%, East Indian 2%, Unspecified 1%, Other 1%
Government Parliamentary system
Currency East Caribbean dollar = 100 cents
Literacy rate 99%
Life expectancy 76 years

GUATEMALA
Central America

Page 30 A2

Once the heart of the Mayan civilization on the Central American isthmus, Guatemala has emerged from years of conflict into a thriving democracy. Agricultural products and tourism are key economic sectors.

Official name Republic of Guatemala
Formation 1821 / 1838
Capital Guatemala City
Population 17.7 million / 421 people per sq mile (163 people per sq km)
Total area 42,042 sq miles (108,889 sq km)
Languages Quiché, Mam, Cakchiquel, Kekchí, Spanish*, Maya Languages
Religions Roman Catholic 42%, Evangelical 39%, None 14%, Unspecified 2%, Other 3%
Demographics Mixed race 56%, Maya 42%, Xinca 1%, Other 1%
Government Presidential system
Currency Quetzal = 100 centavos
Literacy rate 81%
Life expectancy 73 years

GUINEA
West Africa

Page 52 C4

Set on the west coast of Africa, Guinea is rich in mineral resources, notably bauxite, with over 25 percent of the world's known reserves. Iron-ore, gold, and diamonds are also valuable exports.

Official name Republic of Guinea
Formation 1958 / 1958
Capital Conakry
Population 13.2 million / 139 people per sq mile (54 people per sq km)
Total area 94,926 sq miles (245,857 sq km)
Languages Pulaar, Malinké, Soussou, French*
Religions Muslim 89%, Christian 7%, Nonreligious 2%, Traditional beliefs and other 2%
Demographics Fulani (Peuhl) 33%, Malinke 30%, Susu 21%, Guerze 8%, Kissi 6%, Other 2%
Government Presidential system
Currency Guinea franc = 100 centimes
Literacy rate 40%
Life expectancy 64 years

GUINEA-BISSAU
West Africa

Page 52 B4

Situated on Africa's west coast, Guinea-Bissau has seen considerable political and military upheaval since independence in 1974. Agricultural products, particularly cashew nuts, are the main exports.

Official name Republic of Guinea-Bissau
Formation 1974 / 1974
Capital Bissau
Population 2 million/ 143 people per sq mile (55 people per sq km)
Total area 13,948 sq miles (36,125 sq km)
Languages Portuguese Creole, Balante, Fulani, Malinké, Portuguese*
Religions Muslim 46%, Folk religions 31%, Christian 19%, Unaffiliated or Other 4%
Demographics Balante 30%, Fulani 30%, Papel 7%, Mandyako 14%, Mandinka 13%, Other 6%
Government Semi-presidential system
Currency CFA franc = 100 centimes
Literacy rate 60%
Life expectancy 64 years

GUYANA
South America

Page 37 F3

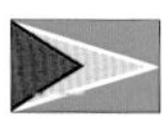

A land of rainforest, mountains, coastal plains, and savanna, Guyana is rich in natural resources that make agriculture a key economic activity. Large oil reserves were discovered off the coast in 2015.

Official name Cooperative Republic of Guyana
Formation 1966 / 1966
Capital Georgetown
Population 789,683 / 10 people per sq mile (4 people per sq km)
Total area 83,000 sq miles (214,969 sq km)
Languages English Creole, Hindi, Tamil, Amerindian languages, English*
Religions Protestant 35%, Hindu 25%, Roman Catholic 7%, Muslim 7%, Other Christian 20%, None 3%, Other 3%
Demographics East Indian 40%, African 29%, Mixed race 20%, Indigenous 11%
Government Parliamentary system
Currency Guyanese dollar = 100 cents
Literacy rate 89%
Life expectancy 72 years

HAITI
West Indies

Page 32 D3

Set on the western side of Hispaniola, Haiti has struggled with both political instability and damage caused by natural disasters. It is a leading producer of vetiver, a plant used in perfume manufacturing.

Official name Republic of Haiti
Formation 1804 / 1936
Capital Port-au-Prince
Population 11.3 million / 1055 people per sq mile (407 people per sq km)
Total area 10,714 sq miles (27,750 sq km)
Languages French Creole*, French*
Religions Catholic 55%, Protestant 28%, Vodou 2%, None 10%, Other 5%
Demographics Black 95%, Mixed race 5%
Government Semi-presidential system
Currency Gourde = 100 centimes
Literacy rate 62%
Life expectancy 66 years

HONDURAS
Central America

Page 30 C2

Straddling the Central American isthmus, Honduras returned to civilian rule in 1984, after a series of military regimes. The country is known for its rich natural resources, including minerals, coffee, and sugarcane.

Official name Republic of Honduras
Formation 1821 / 1998
Capital Tegucigalpa
Population 9.4 million / 217 people per sq mile (84 people per sq km)
Total area 43,278 sq miles (112,090 sq km)
Languages Spanish*, Garífuna (Carib), English Creole
Religions Evangelical/Protestant 48%, Roman Catholic 34%, None 17%, Other 1%
Demographics Mixed race 90%, Black 2%, Indigenous 7%, White 1%
Government Presidential system
Currency Lempira = 100 centavos
Literacy rate 89%
Life expectancy 75 years

HUNGARY
Central Europe

Page 77 C6

This landlocked country is home to the largest lake in Central Europe, Lake Balaton. Hungary returned to democratic rule in 1984, after a period of military government, and joined the EU in 2004.

Official name Hungary
Formation 1918 / 1947
Capital Budapest
Population 9.7 million / 270 people per sq mile (104 people per sq km)
Total area 35,922 sq miles (93,038 sq km)
Languages Hungarian* (Magyar), English, German, Russian
Religions Roman Catholic 56%, Nonreligious 21%, Presbyterian 13%, Other 10%
Demographics Magyar 92%, Roma 3%, German 2%, Other 3%
Government Parliamentary system
Currency Forint = 100 fillér
Literacy rate 99%
Life expectancy 77 years

ICELAND
Northwest Europe

Page 61 E4

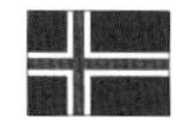

This northerly island outpost of Europe has stunning, sparsely inhabited volcanic terrain. Although its economy crashed heavily in the 2008 global credit crunch, it is now powered by a tourism and construction boom.

Official name Republic of Iceland
Formation 1944 / 1944
Capital Reykjavík
Population 357,603 / 10 people per sq mile (4 people per sq km)
Total area 39,769 sq miles (103,000 sq km)
Languages Icelandic*, English, Nordic languages
Religions The Evangelical Lutheran Church of Iceland 63%, Other Christian 8%, Nonreligious 7%, Roman Catholic 4%, Other 18%
Demographics Icelandic 81%, Polish 6%, Danish 1%, Other 12%
Government Parliamentary system
Currency Icelandic króna = 100 aurar
Literacy rate 99%
Life expectancy 84 years

INDIA
South Asia

Page 112 D4

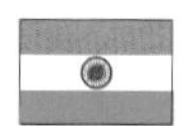

The Indian subcontinent, divided from the rest of Asia by the Himalayas, is home to some of the world's most ancient civilizations. India is the world's largest democracy and second most populous country.

Official name Republic of India
Formation 1947 / 2015
Capital New Delhi
Population 1.39 billion / 1095 people per sq mile (423 people per sq km)
Total area 1,269,219 sq miles (3,287,263 sq km)
Languages Hindi*, English*, Urdu, Bengali, Marathi, Telugu, Tamil, Bihari, Gujarati
Religions Hindu 80%, Muslim 14%, Christian 2%, Sikh 2%, Unspecified or Other 2%
Demographics Indo-Aryan 72%, Dravidian 25%, Other 3%
Government Parliamentary system
Currency Indian rupee = 100 paise
LLiteracy rate 74%
Life expectancy 67 years

INDONESIA
Southeast Asia

Page 116 C4

The world's largest archipelago spans over 3,100 miles (5,000 km), from the Indian to the Pacific Ocean. It has Southeast Asia's largest economy and is home to some of the world's most active volcanoes.

Official name Republic of Indonesia
Formation 1945 / 1999
Capital Jakarta
Population 276 million / 375 people per sq mile (145 people per sq km)
Total area 735,358 sq miles (1,904,569 sq km)
Languages Javanese, Sundanese, Madurese, Bahasa Indonesia*, Dutch, English
Religions Sunni Muslim 87%, Protestant 7%, Roman Catholic 3%, Hindu 2%, Buddhist 1%
Demographics Javanese 40%, Sundanese 16%, Malay 4%, Batak 4%, Madurese 3%, Betawi 3%, Minangkabau 3%, Buginese 3%, Other 24%
Government Presidential system
Currency Rupiah = 100 sen
Literacy rate 96%
Life expectancy 73 years

IRAN
Southwest Asia

Page 98 C3

After the 1979 Islamist revolution deposed the Shah, this ethnically diverse Middle Eastern country became the world's largest theocracy. It has large oil and natural gas reserves.

Official name Islamic Republic of Iran
Formation 1979 / 1990
Capital Tehran
Population 86.7 million / 136 people per sq mile (53 people per sq km)
Total area 636,372 sq miles (1,648,195 sq km)
Languages Farsi*, Azeri, Luri, Gilaki, Mazanderani, Kurdish, Turkmen, Arabic, Baluchi
Religions Shi'a Muslim 90%, Sunni Muslim 9%, Other 1%
Demographics Persian 51%, Azari 24%, Lur and Bakhtiari 8%, Kurdish 7%, Other 10%
Government Islamic theocracy
Currency Iranian rial = 100 dinars
Literacy rate 86%
Life expectancy 75 years

IRAQ
Southwest Asia

Page 98 B3

Political instability in this central Middle Eastern country has hindered its economic growth. Rich with natural resources, Iraq notably has the world's second largest oil reserves.

Official name Republic of Iraq
Formation 1932 / 1991
Capital Baghdad
Population 40.4 million / 239 people per sq mile (92 people per sq km)
Total area 169,235 sq miles (438,317 sq km)
Languages Arabic*, Kurdish*, Turkic languages, Armenian, Assyrian
Religions Muslim 97%, Christian 1%, Other 2%
Demographics Arab 80%, Kurdish 15%, Turkmen 3%, Other 2%
Government Parliamentary system
Currency New Iraqi dinar = 1000 fils
Literacy rate 86%
Life expectancy 73 years

IRELAND
Northwest Europe

Page 67 A6

Ireland transformed from a largely agricultural society into a modern, high-technology economy after joining the European Economic Community in 1973. It has a strong indigenous Celtic culture.

Official name Ireland
Formation 1922 / 1922
Capital Dublin
Population 5.2 million / 192 people per sq mile (74 people per sq km)
Total area 27,133 sq miles (70,273 sq km)
Languages English*, Irish*
Religions Roman Catholic 86%, Other Christian 6%, Nonreligious 6%, Muslim 1%, Other 1%
Demographics Irish 86%, Other White 9%, Asian 2%, Black 1%, Other 2%
Government Parliamentary system
Currency Euro = 100 cents
Literacy rate 99%
Life expectancy 82 years

ISRAEL
Southwest Asia

Page 97 A7

Set on the eastern shore of the Mediterranean Sea, Israel has been in conflict with its Arab neighbors since its inception in 1948. It is home to some of the world's most holy religious sites.

Official name State of Israel
Formation 1948 / 1994
Capital Jerusalem (not internationally recognized)
Population 8.9 million / 1051 people per sq mile (406 people per sq km)
Total area 8470 sq miles (21,937 sq km)
Languages Hebrew*, Arabic*, Yiddish, German, Russian, Polish, Romanian, Persian, English
Religions Jewish 74%, Muslim 18%, Christian 2%, Druze 2%, Other 4%
Demographics Jewish 74%, Arab 21%, Other 5%
Government Parliamentary system
Currency Shekel = 100 agorot
Literacy rate 98%
Life expectancy 83 years

ITALY
Southern Europe

Page 74 B3

Jutting into the central Mediterranean, and once at the heart of the Roman Empire, Italy is a world leader in product design, fashion, and textiles, and is renowned for its art, operas, and architecture.

Official name Italian Republic
Formation 1861 / 1954
Capital Rome
Population 61 million/ 524 people per sq mile (202 people per sq km)
Total area 116,348 sq miles (301,340 sq km)
Languages Italian*, German, French, Rhaeto-Romanic, Sardinian, Slovene
Religions Christian 81%, Muslim 5%, Unaffiliated 13%, Other 1%
Demographics Italian 92%, Other European 5%, African 1%, Other 2%
Government Parliamentary system
Currency Euro = 100 cents
Literacy rate 99%
Life expectancy 83 years

IVORY COAST (CÔTE D'IVOIRE)
West Africa

Page 52 D4

One of the larger countries on the West African coast, Ivory Coast is the world's biggest cocoa producer. Two recent conflicts have damaged its previous reputation for stability.

Official name Republic of Côte d'Ivoire
Formation 1960 / 1960
Capital Yamoussoukro
Population 28.7 million / 231 people per sq mile (89 people per sq km)
Total area 124,504 sq miles (322,463 sq km)
Languages Akan, French*, Krou, Voltaïque, Dioula
Religions Muslim 43%, Nonreligious or traditional beliefs 23%, Roman Catholic 17%, Evangelical 12%, Other Christian 4%, Other 1%
Demographics Akan 29%, Voltaique or Gur 16%, Northern Mande 15%, Kru 8%, Southern Mande 7%, Non-Ivoirian 24%, Other 1%
Government Presidential system
Currency CFA franc = 100 centimes
Literacy rate 90%
Life expectancy 62 years

JAMAICA
West Indies

Page 32 C3

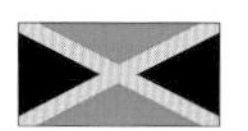

Set in the central Caribbean, Jamaica is the birthplace of Rastafarianism. A strong cultural identity includes reggae and ska music, while tourism, agricultural products, and mining are key economic sectors.

Official name Jamaica
Formation 1962 / 1962
Capital Kingston
Population 2.8 million / 660 people per sq mile (255 people per sq km)
Total area 4244 sq miles (10,991 sq km)
Languages English Creole, English*
Religions Church of God 26%, Nonreligious 22%, Other Christian 21%, Seventh-day Adventist 12%, Pentecostal 11%, Other 8%
Demographics Black 92%, Mixed race 6%, East Indian 1%, Other 1%
Government Parliamentary system
Currency Jamaican dollar = 100 cents
Literacy rate 89%
Life expectancy 76 years

JAPAN
East Asia

Page 108 C4

Japan has four main islands and over 3,000 smaller ones. It has the world's third largest economy, led by high-tech industries and vehicle manufacturing. It retains its emperor as head of state.

Official name Japan
Formation 1890 / 1972
Capital Tokyo
Population 124 million / 850 people per sq mile (328 people per sq km)
Total area 145,914 sq miles (377,915 sq km)
Languages Japanese*, Korean, Chinese
Religions Shintoism 71%, Buddhism 67%, Christianity 2%, Other 6%
Demographics Japanese 92%, Chinese and Korean 1%, Other 1%
Government Parliamentary system
Currency Yen = 100 sen
Literacy rate 99%
Life expectancy 85 years

JORDAN
Southwest Asia

Page 97 B6

This Middle Eastern kingdom provides sanctuary to millions of refugees from neighboring conflicts, which has put great strain on the economy. It is rich in ancient culture and architecture.

Official name Hashemite Kingdom of Jordan
Formation 1946 / 1967
Capital Amman
Population 10.9 million / 316 people per sq mile (122 people per sq km)
Total area 34,495 sq miles (89,342 sq km)
Languages Arabic*, English
Religions Muslim 97% (official; predominantly Sunni), Christian 2% , Other 1%
Demographics Jordanian 69%, Syrian 13%, Palestinian 7%, Egyptian 7%, Iraqi 1%, Other 3%
Government Monarchy
Currency Jordanian dinar = 1000 fils
Literacy rate 98%
Life expectancy 76 years

KAZAKHSTAN
Central Asia

Page 92 B4

Kazakhstan was the last of the former Soviet republics to declare independence in 1991. It has a diverse landscape with vast mineral and oil resources, making it the major economic power in the region.

Official name Republic of Kazakhstan
Formation 1991 / 1991
Capital Astana
Population 19.3 million / 18 people per sq mile (7 people per sq km)
Total area 1,052,089 sq miles (2,724,900 sq km)
Languages Kazakh*, Russian, Ukrainian, German, Uzbek, Tatar, Uighur
Religions Muslim (mainly Sunni) 71%, Christian (mainly Orthodox) 26%, Nonreligious 3%
Demographics Kazakh 68%, Russian 19%, Uzbek 3%, Ukrainian 2%, Uighur 2%, Other 6%
Government Presidential system
Currency Tenge = 100 tiyn
Literacy rate 100%
Life expectancy 73 years

KENYA
East Africa

Page 51 C6

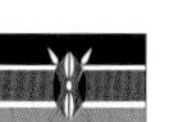

Straddling the equator on Africa's east coast, Kenya is famous for its scenic landscapes and vast wildlife reserves. Tea and coffee production, along with tourism, are important economic activities.

Official name Republic of Kenya
Formation 1963 / 1963
Capital Nairobi
Population 55.8 million / 249 people per sq mile (96 people per sq km)
Total area 224,081 sq miles (580,367 sq km)
Languages Kiswahili*, English*, Kikuyu, Luo, Kalenjin, Kamba
Religions Christian 86%, Muslim 11%, Other 3%
Demographics Kikuyu 17%, Luhya 14%, Kalenjin 13%, Luo 11%, Kamba 10%, Somali 6%, Kisii 6%, Mijikenda 5%, Meru 4%, Maasai 3%, Turkana 2%, Non-Kenyan 1%, Other 8%
Government Presidential system
Currency Kenya shilling = 100 cents
Literacy rate 82%
Life expectancy 70 years

KIRIBATI
Australasia & Oceania

Page 123 F3

Home to the South Pacific's largest marine reserve, the 33 low-lying atolls that make up Kiribati are at risk from rising sea levels. The economy is largely dependent on exports of fish products and coconut oil.

Official name Republic of Kiribati
Formation 1979 / 2012
Capital Tarawa Atoll
Population 114,189 / 365 people per sq mile (141 people per sq km)
Total area 313 sq miles (811 sq km)
Languages English*, Kiribati
Religions Roman Catholic 59%, Kiribati Uniting Church 21%, Kiribati Protestant Church 8%, Church of Jesus Christ 6%, Other 6%
Demographics I-Kiribati 95%, Mixed race 3%, Other 2%
Government Presidential system
Currency Australian dollar = 100 cents
Literacy rate 99%
Life expectancy 68 years

KOSOVO
Southeast Europe

Page 79 D5

NATO intervention in 1999 ended the brutal ethnic violence which ultimately led to a unilateral declaration of independence from Serbia in 2008. Forestry, minerals, and agriculture are key to the economy.

Official name Republic of Kosovo
Formation 2008 / 2008
Capital Prishtinë/Priština (Pristina)
Population 1.9 million / 452 people per sq mile (175 people per sq km)
Total area 4203 sq miles (10,887 sq km)
Languages Albanian*, Serbian*, Bosniak, Gorani, Roma, Turkish
Religions Muslim 96%, Roman Catholic 2%, Orthodox 1%, Other 1%
Demographics Albanians 93%, Bosniaks 1%, Serbs 1%, Turk 1%, Other 4%
Government Parliamentary system
Currency Euro = 100 cents
Literacy rate 92%
Life expectancy 71 years

KUWAIT
Southwest Asia

Page 98 C4

Nestled at the top of the Persian Gulf, Kuwait has large oil and gas reserves which have brought great wealth. An Iraqi invasion in 1990 triggered the Gulf War, but a US-led coalition quickly restored the ruling power.

Official name State of Kuwait
Formation 1961 / 1969
Capital Kuwait City
Population 4.1 million / 596 people per sq mile (230 people per sq km)
Total area 6880 sq miles (17,818 sq km)
Languages Arabic*, English
Religions Muslim (official) 75%, Christian 18%, Unspecified or Other 7%
Demographics Kuwaiti 30%, Other Arab 28%, Asian 40%, African 1%, Other 1%
Government Monarchy
Currency Kuwaiti dinar = 1000 fils
Literacy rate 97%
Life expectancy 79 years

KYRGYZSTAN
Central Asia

Page 101 F2

A mountainous landlocked state in Central Asia, Kyrgyzstan has an economy that revolves around agriculture. Gold and minerals are key exports but it relies on imports for most of its energy needs.

Official name Kyrgyz Republic
Formation 1991 / 1991
Capital Bishkek
Population 6 million / 78 people per sq mile (30 people per sq km)
Total area 77,202 sq miles (199,951 sq km)
Languages Kyrgyz*, Russian*, Uzbek, Tatar, Ukrainian
Religions Muslim 90% (majority Sunni), Christian 7%, Other 3%
Demographics Kyrgyz 74%, Uzbek 15%, Russian 5%, Dungan 1%, Other 5%
Government Presidential / Parliamentary system
Currency Som = 100 tyiyn
Literacy rate 100%
Life expectancy 72 years

LAOS
Southeast Asia

Page 114 D4

Landlocked Laos is one of the world's few remaining communist states. A rural economy supports the majority of its people but a huge capacity for hydro-electric power allows net exports of energy.

Official name Lao People's Democratic Republic
Formation 1953 / 1953
Capital Viangchan (Vientiane)
Population 7.7 million / 84 people per sq mile (33 people per sq km)
Total area 91,429 sq miles (236,800 sq km)
Languages Lao*, Mon-Khmer, Yao, Vietnamese, Chinese, French, English
Religions Buddhist 65%, Christian 2%, None 31%, Other 2%
Demographics Lao 53%, Khmou 11%, Hmong 9%, Phouthay 3%, Tai 3%, Makong 3%, Katong 2%, Lue 2%, Akha 2%, Other 12%
Government One-party state
Currency Kip = 100 att
Literacy rate 85%
Life expectancy 68 years

LATVIA
Northeast Europe

Page 84 C3

Situated on the low-lying eastern shores of the Baltic Sea, Latvia has a diverse economy based on agricultural products and manufacturing. It joined both the EU and NATO in 2004.

Official name Republic of Latvia
Formation 1991 / 1991
Capital Riga
Population 1.8 million/ 72 people per sq mile (28 people per sq km)
Total area 24,938 sq miles (64,589 sq km)
Languages Latvian*, Russian
Religions Lutheran 36%, Roman Catholic 19%, Orthodox 19%, Other Christian 2%, Unspecified 24%
Demographics Latvian 63%, Russian 25%, Belarusian 3%, Ukrainian 2%, Polish 2%, Unspecified 2%, Other 3%
Government Parliamentary system
Currency Euro = 100 cents
Literacy rate 100%
Life expectancy 76 years

LEBANON
Southwest Asia

Page 96 A4

Situated in the eastern Mediterranean, Lebanon has served as a busy commercial and cultural hub for centuries. The country had emerged from decades of conflict but recently, instability has returned.

Official name Lebanese Republic
Formation 1943 / 1941
Capital Beirut
Population 5.2 million/ 1728 people per sq mile (662 people per sq km)
Total area 4015 sq miles (10,400 sq km)
Languages Arabic*, French, Armenian, Assyrian, English
Religions Muslim 64%, Christian 31%, Druze 4%, Other 1%
Demographics Arab 95%, Armenian 4%, Other 1%
Government Parliamentary system
Currency Lebanese pound = 100 piastres
Literacy rate 95%
Life expectancy 79 years

LESOTHO
Southern Africa

Page 56 D5

Lesotho lies within South Africa, on whom it is economically dependent as it has limited agricultural capacity. Hydro-electric power and its diamond industry have helped grow the economy.

Official name Kingdom of Lesotho
Formation 1966 / 1966
Capital Maseru
Population 2.2 million / 188 people per sq mile (72 people per sq km)
Total area 11,720 sq miles (30,355 sq km)
Languages English*, Sesotho*, isiZulu, Xhosa
Religions Protestant 48%, Roman Catholic 40%, Other Christian 9%, Non-Christian 1%, None 2%
Demographics Sotho 99%, European and Asian 1%
Government Parliamentary system
Currency Loti = 100 lisente; South African rand = 100 cents
Literacy rate 79%
Life expectancy 60 years

LIBERIA
West Africa

Page 52 C5

Facing the Atlantic Ocean, Liberia is Africa's oldest republic. Ellen Johnson Sirleaf was the first woman to be elected a head of state in Africa when she was voted president in 2005.

Official name Republic of Liberia
Formation 1847 / 1911
Capital Monrovia
Population 5.3 million / 123 people per sq mile (48 people per sq km)
Total area 43,000 sq miles (111,369 sq km)
Languages Kpelle, Vai, Bassa, Kru, Grebo, Kissi, Gola, Loma, English*
Religions Christian 86%, Muslim 12%, Nonreligious 1%, Traditional beliefs and other 1%
Demographics Indigenous tribes (12 groups) 40%, Kpellé 20%, Bassa 14%, Grebo 10%, Gio 8%, Krou 6%, Other 2%
Government Presidential system
Currency Liberian dollar = 100 cents
Literacy rate 48%
Life expectancy 65 years

LIBYA
North Africa

Page 49 F3

Situated on the Mediterranean coast, most of Libya lies in the Sahara Desert. Emphasis on agricultural and industrial developments aims to reduce the country's dependence on oil revenues.

Official name State of Libya
Formation 1951 / 1951
Capital Tripoli
Population 7.1 million / 11 people per sq mile (4 people per sq km)
Total area 679,362 sq miles (1,759,540 sq km)
Languages Arabic*, Tuareg, English, Italian
Religions Muslim 96%, Christian 2%, Other 2%
Demographics Arab and Berber 97%, Other 3%
Government Transitional regime
Currency Libyan dinar = 1000 dirhams
Literacy rate 91%
Life expectancy 77 years

LIECHTENSTEIN
Central Europe

Page 73 B7

Tucked high in the Alps, Liechtenstein's diverse, industrialized economy is dominated by financial services that benefit from its tax-haven status. Switzerland handles its foreign affairs and defense.

Official name Principality of Liechtenstein
Formation 1719 / 1719
Capital Vaduz
Population 38,250 / 617 people per sq mile (239 people per sq km)
Total area 62 sq miles (160 sq km)
Languages German*, Alemannish dialect, Italian, Turkish, Portuguese
Religions Roman Catholic 78%, Protestant 9%, Muslim 6%, Nonreligious 5%, Orthodox Christian 1%, Other 1%
Demographics Liechtensteiner 65%, German 5% Austrian 6%, Swiss 10%, Italian 3%, Other 11%
Government Monarchy
Currency Swiss franc = 100 rappen/centimes
Literacy rate 100%
Life expectancy 83 years

LITHUANIA
Northeast Europe

Page 84 B4

Mostly low-lying, Lithuania is the largest of the three Baltic states. It was the first former Soviet republic to declare independence in 1990 and went on to join the EU and NATO in 2004.

Official name Republic of Lithuania
Formation 1991 / 2003
Capital Vilnius
Population 2.6 million / 103 people per sq mile (40 people per sq km)
Total area 25,212 sq miles (65,300 sq km)
Languages Lithuanian*, Russian, Polish
Religions Roman Catholic 74%, Russian Orthodox 4%, None 6%, Unspecified 14%, Other 2%
Demographics Lithuanian 85%, Polish 6%, Russian 5%, Belarusian 1%, Other 3%
Government Semi-presidential system
Currency Euro = 100 cents
Literacy rate 100%
Life expectancy 76 years

LUXEMBOURG
Northwest Europe

Page 65 D8

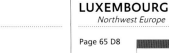

Part of the forested Ardennes plateau in Northwest Europe, Luxembourg is Europe's last independent duchy and one of its richest states. It is a banking center and hosts EU institutions.

Official name Grand Duchy of Luxembourg
Formation 1839 / 1867
Capital Luxembourg
Population 650,364 / 652 people per sq mile (251 people per sq km)
Total area 998 sq miles (2586 sq km)
Languages Luxembourgish*, Portuguese, Italian, German*, French*, English
Religions Christian 71%, Muslim 2%, Unaffiliated 27%
Demographics Luxembourger 53%, Portuguese 15%, French 8%, Italian 4%, Belgian 3%, German 2%, Spanish 1%, Romania 1%, Other 13%
Government Monarchy
Currency Euro = 100 cents
Literacy rate 100%
Life expectancy 83 years

MADAGASCAR
Indian Ocean

Page 79 D6

Off Africa's southeast coast, Madagascar is the world's fourth largest island. An abundance of unique flora and fauna has generated a boom in ecotourism, but economic problems persist.

Official name Republic of Madagascar
Formation 1960 / 1960
Capital Antananarivo
Population 28 million / 124 people per sq mile (48 people per sq km)
Total area 226,658 sq miles (587,042 sq km)
Languages Malagasy*, French*, English*
Religions Traditional beliefs 52%, Christian 41%, Muslim 7%
Demographics Malay 46%, Merina 26%, Betsimisaraka 15%, Betsileo 12%, Other 1%
Government Semi-presidential system
Currency Ariary = 5 iraimbilanja
Literacy rate 77%
Life expectancy 68 years

MALAWI
Southern Africa

Page 57 F4

Landlocked Malawi lies along the Great Rift Valley and Lake Nyasa, Africa's third largest lake. Its rural society depends on the mainly agricultural economy, with tobacco and tea being the principal exports.

Official name Republic of Malawi
Formation 1964 / 1964
Capital Lilongwe
Population 21 million / 459 people per sq mile (177 people per sq km)
Total area 45,746 sq miles (118,484 sq km)
Languages Chewa, Lomwe, Yao, Ngoni, English*
Religions Christian (mainly Protestant) 50%, Other Christian 27%, Muslim 14%, Traditionalist 1%, none 2%, Other 6%
Demographics Chewa 34%, Lomwe 19%, Yao 13%, Ngoni 10%, Tumbuka 9%, Sena 4%, Mang'anja 3%, Tonga 2%, Nyanja 2%, Nkhonde 1%, Other 3%
Government Presidential system
Currency Malawi kwacha = 100 tambala
Literacy rate 62%
Life expectancy 72 years

MALAYSIA
Southeast Asia

Page 57 E1

Spread across two separate regions, Malaysia has built upon its abundant natural resources to create a diverse industrialized economy. Its rich biodiversity attracts huge numbers of tourists.

Official name Malaysia
Formation 1957 / 1965
Capital Kuala Lumpur; Putrajaya (administrative)
Population 33.1 million / 259 people per sq mile (100 people per sq km)
Total area 127,723 sq miles (330,803 sq km)
Languages Bahasa Malaysia*, Malay, Chinese, Tamil, English
Religions Muslim (mainly Sunni) 61%, Buddhist 20%, Christian 9%, Hindu 6%, Other 3%
Demographics Malay 50%, Chinese 22%, Indigenous tribes 12%, Indian 7%, Other 9%
Government Parliamentary system
Currency Ringgit = 100 sen
Literacy rate 95%
Life expectancy 76 years

MALDIVES
Indian Ocean

Page 116 B3

This low-lying group of over 1,000 coral islands in the Indian Ocean is vulnerable to rising sea levels. Idyllic tropical beaches and reefs have boosted growth in luxury tourism.

Official name Republic of Maldives
Formation 1965 / 1965
Capital Maale (Male')
Population 390,164 / 4667 people per sq mile (1812 people per sq km)
Total area 120 sq miles (300 sq km)
Languages Dhivehi* (Maldivian), Sinhala, Tamil, Arabic, English
Religions Sunni Muslim 94%, Hindu 3%, Christian 2%, Buddhist 1%
Demographics Arab-Sinhalese-Malay 100%
Government Presidential system
Currency Rufiyaa = 100 laari
Literacy rate 98%
Life expectancy 77 years

MALI
West Africa

Page 110 A4

Once at the heart of a trans-Saharan trading empire, Mali is geographically dominated by the Sahara Desert. Mining activities yield a third of Africa's gold production while cotton is the main agricultural export.

Official name Republic of Mali
Formation 1960 / 1986
Capital Bamako
Population 20.7 million / 43 people per sq mile (17 people per sq km)
Total area 478,840 sq miles (1,240,192 sq km)
Languages Bambara, Fulani, Senufo, Soninke, French*
Religions Muslim (mainly Sunni) 93%, Traditional beliefs 6%, Christian 4%
Demographics Bambara 33%, Fulani 13%, Sarakole/Soninke/Marka 10%, Senufo/Manianka 10%, Malinke 9%, Dogon 9%, Sonrai 6%, Other 10%
Government Presidential system
Currency CFA franc = 100 centimes
Literacy rate 36%
Life expectancy 62 years

MALTA
Southern Europe

Page 53 E2

A central Mediterranean location has made Malta an important strategic port throughout history. Financial services underpin the economy and the warm climate attracts large numbers of tourists.

Official name Republic of Malta
Formation 1964 / 1964
Capital Valletta
Population 464,186/ 4252 people per sq mile (1615 people per sq km)
Total area 122 sq miles (316 sq km)
Languages Maltese*, English*
Religions Roman Catholic 98%, Nonreligious and other 2%
Demographics Maltese 96%, Other 4%
Government Parliamentary system
Currency Euro = 100 cents
Literacy rate 95%
Life expectancy 83 years

MARSHALL ISLANDS
Australasia & Oceania

Page 80 A5

This group of 34 low-lying coral atolls is spread across a vast area of the Pacific Ocean. The nation was the first to try to make cryptocurrency legal tender by launching the "Sovereign" in 2018.

Official name Republic of the Marshall Islands
Formation 1986 / 1986
Capital Majuro Atoll
Population 59,620 / 852 people per sq mile (329 people per sq km)
Total area 70 sq miles (181 sq km)
Languages Marshallese*, English*, Japanese, German
Religions Protestant 81%, Roman Catholic 8%, Other 11%
Demographics Micronesian 90%, Other 10%
Government Presidential system
Currency US dollar = 100 cents
Literacy rate 98%
Life expectancy 75 years

MAURITANIA
West Africa

Page 122 D1

Mauritania is one of Africa's newest oil producers. Four-fifths of its area lies within the Sahara Desert. Extensive mineral deposits and rich fishing grounds in the Atlantic Ocean are key economic contributors.

Official name Islamic Republic of Mauritania
Formation 1960 / 1960
Capital Nouakchott
Population 4.1 million / 13 people per sq mile (5 people per sq km)
Total area 397,953 sq miles (1,030,700 sq km)
Languages Arabic*, Hassaniyah Arabic, Wolof, Pular, Soninke, French
Religions Sunni Muslim 100%
Demographics Maure 81%, Wolof 7%, Tukolor 5%, Soninka 3%, Other 4%
Government Presidential system
Currency Ouguiya = 5 khoums
Literacy rate 52%
Life expectancy 65 years

MAURITIUS
Indian Ocean

Page 52 C2

This Indian Ocean archipelago has one of Africa's highest per capita incomes. Once dependent on sugar exports, the country now has a diverse economy that includes luxury tourism and financial services.

Official name Republic of Mauritius
Formation 1968 / 1968
Capital Port Louis
Population 1.3 million / 1616 people per sq mile (624 people per sq km)
Total area 787 sq miles (2040 sq km)
Languages French Creole, Hindi, Urdu, Tamil, Chinese, English*, French
Religions Hindu 48%, Roman Catholic 26%, Muslim 17%, Other Christian 7%, Other 2%
Demographics Indo-Mauritian 68%, Creole 27%, Sino-Mauritian 3%, Franco-Mauritian 2%
Government Parliamentary system
Currency Mauritian rupee = 100 cents
Literacy rate 91%
Life expectancy 75 years

MEXICO
North America

Page 57 H3

The third largest country in Latin America, Mexico is a major oil exporter but still suffers from social inequality. The highly diversified industrial economy is the second largest in Latin America.

Official name United Mexican States
Formation 1821 / 1848
Capital Mexico City
Population 130 million / 171 people per sq mile (66 people per sq km)
Total area 758,449 sq miles (1,964,375 sq km)
Languages Spanish*, Nahuatl, Mayan, Zapotec, Mixtec, Otomi, Totonac, Tzotzil, Tzeltal
Religions Roman Catholic 78%, Protestant 11%, Nonreligious 11%, Other 1%
Demographics Mixed race 62%, Indigenous 21%, Other 10%
Government Presidential system
Currency Mexican peso = 100 centavos
Literacy rate 95%
Life expectancy 72 years

MICRONESIA
Australasia & Oceania

Page 28 D3

The Federated States of Micronesia, situated in the western Pacific, comprises 607 islands and atolls. Exports are almost entirely made up of fish products with some tourism, mainly recreational diving.

Official name Federated States of Micronesia
Formation 1986 / 1986
Capital Palikir (Pohnpei Island)
Population 101,009 / 373 people per sq mile (144 people per sq km)
Total area 271 sq miles (702 sq km)
Languages English, Chuukese, Kosraean, Pohnpeian, Yapese, Ulithian, Woleaian, Nukuoro, Kapingamarangi
Religions Roman Catholic 55%, Protestant 41%, Nonreligious 1%, Other 3%
Demographics Chuukese 49%, Pohnpeian 30%, Kosraean 6%, Yapese 5%, Asian 2%, Other 14%
Government Nonparty system
Currency US dollar = 100 cents
Literacy rate 72%
Life expectancy 74 years

MOLDOVA
Southeast Europe

Page 122 B1

The smallest of the ex-Soviet republics, Moldova has strong linguistic and cultural ties with its neighbor, Romania. Although it exports tobacco, wine, and fruit, the service sector drives the economy.

Official name Republic of Moldova
Formation 1991 / 1991
Capital Chisinau
Population 3.2 million / 245 people per sq mile (95 people per sq km)
Total area 13,069 sq miles (33,851 sq km)
Languages Moldovan*, Ukrainian, Russian
Religions Orthodox Christian 90%, Nonreligious 2%, Other 8%
Demographics Moldovan 75%, Ukrainian 7%, Russian 4%, Gagauz 5%, Romanian 7%, Bulgarian 2%, Other 1%
Government Parliamentary system
Currency Moldovan leu = 100 bani
Literacy rate 99%
Life expectancy 72 years

MONACO
Southern Europe

Page 86 D3

This tiny enclave on France's Côte d'Azur has built a thriving financial services and banking sector based on its tax-haven status. It is famous for its expensive property and jet-set lifestyle.

Official name Principality of Monaco
Formation 1419 / 1861
Capital Monaco
Population 39,520 / 50,145 people per sq mile (19,497 people per sq km)
Total area 0.77 sq miles (2 sq km)
Languages French*, Italian, Monégasque, English
Religions Roman Catholic 89%, Protestant 6%, Other 5%
Demographics French 20%, Italian 15%, Monégasque 32%, British 5%, Belgian 2%, Swiss 2%, German 2%, Russian 2%, American 1%, Dutch 1%, Moroccan 1%, Other 17%
Government Monarchical / parliamentary system
Currency Euro = 100 cents
Literacy rate 99%
Life expectancy 90 years

MONGOLIA
East Asia

Page 69 E6

Once the center of the Mongol empire, Mongolia is a sparsely populated, landlocked country on the edge of the Gobi Desert. Huge mineral resources have attracted foreign investment.

Official name Mongolia
Formation 1921 / 1924
Capital Ulaanbaatar
Population 3.3 million / 5 people per sq mile (2 people per sq km)
Total area 603,908 sq miles (1,564,116 sq km)
Languages Khalkha Mongolian*, Kazakh, Chinese, Russian
Religions Tibetan Buddhist 52%, Nonreligious 41%, Muslim 3%, Shamanist 2%, Christian 1%, Other 1%
Demographics Khalkh 84%, Kazakh 4%, Dorvod 3%, Bayad 2%, Other 4%
Government Presidential / Parliamentary system
Currency Tugrik (tögrög) = 100 möngö
Literacy rate 99%
Life expectancy 71 years

MONTENEGRO
Southeast Europe

Page 104 D2

The complex history of this region of SE Europe eventually led to Montenegro declaring independence from Serbia in 2006. The Tara River canyon here is the deepest and longest in Europe.

Official name Montenegro
Formation 2006 / 2006
Capital Podgorica
Population 604,966 / 113 people per sq mile (44 people per sq km)
Total area 5332 sq miles (13,812 sq km)
Languages Montenegrin*, Serbian, Albanian, Bosnian, Croatian
Religions Orthodox Christian 74%, Muslim 20%, Roman Catholic 4%, Nonreligious 1%, Other 1%
Demographics Montenegrin 45%, Serb 29%, Bosniak 9%, Albanian 5%, Other 12%
Government Parliamentary system
Currency Euro = 100 cents
Literacy rate 99%
Life expectancy 78 years

MYANMAR (BURMA)
Southeast Asia

Page 57 E3

Myanmar, set on the Bay of Bengal, underwent gradual liberalization after 2010 before the civilian government was overthrown in a coup in 2021. The rural economy is dominated by rice production.

Official name Republic of the Union of Myanmar
Formation 1948 / 1948
Capital Nay Pyi Taw
Population 57.5 million / 220 people per sq mile (85 people per sq km)
Total area 261,228 sq miles (676,578 sq km)
Languages Burmese* (Myanmar), Shan, Karen, Rakhine, Chin, Yangbye, Kachin, Mon
Religions Buddhist 88%, Christian 6%, Muslim 4%, Animist 1%, Other 1%
Demographics Burman (Bamah) 68%,Chinese 3%, Shan 9%, Karen 7%, Rakhine 4%, Indian 2%, Mon 2%, Other 5%
Government Parliamentary system
Currency Kyat = 100 pyas
Literacy rate 89%
Life expectancy 70 years

NEPAL
South Asia

Page 122 D3

Nestled in the Himalayas, Nepal is home to eight of the world's highest mountains including Mount Everest, known locally as Sagarmatha. Its rural economy is boosted by a rapid growth in tourism.

Official name Nepal
Formation 1768 / 1768
Capital Kathmandu
Population 30 million / 528 people per sq mile (204 people per sq km)
Total area 56,826 sq miles (147,181 sq km)
Languages Nepali*, Maithili, Bhojpuri
Religions Hindu 81%, Buddhist 9%, Muslim 4%, Other 5%
Demographics Chhetri 17%, Hill Brahman 12%, Magar 7%, Tharu 7%, Tamang 6%, Newar 5%, Kami 5%, Muslim 4%, Yadav 4%, Other 33%
Government Parliamentary system
Currency Nepalese rupee = 100 paisa
Literacy rate 68%
Life expectancy 72 years

NICARAGUA
Central America

Page 128 A4

Nicaragua, at the heart of Central America, has traditionally relied on agricultural exports to sustain its economy. Its lush rainforests and spectacular scenery are attracting a large growth in ecotourism.

Official name Republic of Nicaragua
Formation 1821 / 1838
Capital Managua
Population 6.3 million / 125 people per sq mile (48 people per sq km)
Total area 50,336 sq miles (130,370 sq km)
Languages Spanish*, English Creole, Miskito
Religions Roman Catholic 50%, Protestant 33%, Nonreligious 1%, Other 16%
Demographics Mixed race 69%, White 17%, Black 9%, Indigenous 5%
Government Presidential system
Currency Córdoba oro = 100 centavos
Literacy rate 83%
Life expectancy 75 years

MOROCCO
North Africa

Page 79 C5

The mountainous country of Morocco sits on the northwest African coast. Its rich history and culture attract large numbers of tourists. Recent moves have been made to tap the huge solar energy potential.

Official name Kingdom of Morocco
Formation 1956 / 1969
Capital Rabat
Population 36.7 million / 133 people per sq mile (51 people per sq km)
Total area 276,661 sq miles (716,550 sq km)
Languages Arabic*, Tamazight* (Berber), French, Spanish
Religions Muslim (mainly Sunni) 99%, Other 1%
Demographics Arab 70%, Berber 29%, European 1%
Government Monarchical / parliamentary system
Currency Moroccan dirham = 100 centimes
Literacy rate 74%
Life expectancy 74 years

NAMIBIA
Southern Africa

Page 114 A3

On Africa's southwest coast, this large and sparsely populated country is dominated geographically by the Namib Desert. It is rich in minerals, particularly uranium, and is a significant diamond producer.

Official name Republic of Namibia
Formation 1990 / 1994
Capital Windhoek
Population 2.7 million / 8 people per sq mile (3 people per sq km)
Total area 318,260 sq miles (824,292 sq km)
Languages Ovambo, Kavango, English*, Bergdama, German, Afrikaans
Religions Christian 98%, Traditional beliefs 2%
Demographics Ovambo 50%, Other tribes 22%, Kavango 9%, Herero 7%, Damara 7%, Other 5%
Government Presidential system
Currency Namibian dollar = 100 cents; South African rand = 100 cents
Literacy rate 92%
Life expectancy 66 years

NETHERLANDS
Northwest Europe

Page 113 E3

Astride the delta of four major rivers, this low-lying nation (a quarter of its land is below sea-level) is one of the world's foremost maritime trading nations with Rotterdam being Europe's largest port.

Official name Kingdom of the Netherlands
Formation 1648 / 1839
Capital Amsterdam; The Hague (administrative)
Population 17.4 million / 1085 people per sq mile (419 people per sq km)
Total area 16,039 sq miles (41,543 sq km)
Languages Dutch*, Frisian
Religions Roman Catholic 20%, Protestant 15%, Muslim 5%, None 54%, Other 6%
Demographics Dutch 82%, Surinamese 2%, Turkish 2%, Moroccan 2%, Other 12%
Government Parliamentary system
Currency Euro = 100 cents
Literacy rate 99%
Life expectancy 82 years

NIGER
West Africa

Page 30 D3

A vast, arid state on the edge of the Sahara Desert, landlocked Niger is linked to the sea by the Niger River. It is focusing on increased oil exploration and gold mining to help modernize its economy.

Official name Republic of Niger
Formation 1960 / 2016
Capital Niamey
Population 24.4 million / 50 people per sq mile (19 people per sq km)
Total area 489,188 sq miles (1,267,000 sq km)
Languages Hausa, Djerma, Fulani, Tuareg, Teda, French*
Religions Muslim 99%, Other 1%
Demographics Hausa 53%, Djerma and Songhai 21%, Tuareg 11%, Peul 7%, Kanuri 6%, Other 1%
Government Semi-presidential system
Currency CFA franc = 100 centimes
Literacy rate 35%
Life expectancy 60 years

MOZAMBIQUE
Southern Africa

Page 48 C2

Mozambique, on the southeast African coast, is rich in natural resources but remains underdeveloped. The discovery of gas fields off its coast in 2011 has the potential to transform the economy.

Official name Republic of Mozambique
Formation 1975 / 1975
Capital Maputo
Population 31.6 million / 102 people per sq mile (40 people per sq km)
Total area 308,642 sq miles (799,380 sq km)
Languages Makua, Xitsonga, Sena, Nyanja, Chuwabo, Ndau, Tswa, Lomwe, Portuguese*
Religions Roman Catholic 27%, Muslim 19%, Christian 16%, Evangelical/Pentecostal 15%, None 14%, Other 9%
Demographics Makua Lomwe 47%, Tsonga 23%, Malawi 12%, Shona 11%, Yao 4%, Other 3%
Government Presidential system
Currency New metical = 100 centavos
Literacy rate 61%
Life expectancy 57 years

NAURU
Australasia & Oceania

Page 56 B3

This small island nation sits in the Pacific Ocean. It was once dependent on phosphate mining, until viable deposits ran out in the 1980s, causing an economic crisis from which it is only just starting to recover.

Official name Republic of Nauru
Formation 1968 / 1968
Capital None (Yaren *de facto* capital)
Population 10,870 / 1359 people per sq mile (518 people per sq km)
Total area 8 sq miles (21 sq km)
Languages Nauruan*, Kiribati, Chinese, Tuvaluan, English
Religions Nauruan Congregational Church 60%, Roman Catholic 35%, Other 5%
Demographics Nauruan 93%, Chinese 5%, Other Pacific islanders 1%, European 1%
Government Parliamentary system
Currency Australian dollar = 100 cents
Literacy rate 95%
Life expectancy 68 years

NEW ZEALAND
Australasia & Oceania

Page 64 C3

This progressive nation on the southwest Pacific rim was the first in the world to give women the vote, in 1893. The high-tech economy is heavily reliant on international trade, particularly in agricultural products.

Official name New Zealand
Formation 1907 / 1947
Capital Wellington
Population 5 million / 48 people per sq mile (19 people per sq km)
Total area 103,798 sq miles (268,838 sq km)
Languages English*, Maori*
Religions Nonreligious 48%, Christian 15%, Anglican 7%, Roman Catholic 10%, Presbyterian 5%, Other 8%
Demographics European 64%, Maori 17%, Chinese 5%, Samoan 4%, Other 14%
Government Parliamentary system
Currency New Zealand dollar = 100 cents
Literacy rate 99%
Life expectancy 83 years

NIGERIA
West Africa

Page 53 G3

Nigeria has both the largest population and economy in Africa. The Niger Delta holds some of the world's largest proven oil and gas reserves. The cinema industry is one of the largest film producers in the world.

Official name Federal Republic of Nigeria
Formation 1960 / 2006
Capital Abuja
Population 211 million / 592 people per sq mile (228 people per sq km)
Total area 356,667 sq miles (923,768 sq km)
Languages Hausa, English*, Yoruba, Ibo
Religions Muslim 54%, Christian 45%, Traditional beliefs 1%
Demographics Hausa 30%, Yoruba 16%, Ibo 15%, Fulani 6%, Tiv 2%, Kanuri/Beriberi 2%, Ibibio 2%, Ijaw/Izon 2% Other 25%
Government Presidential system
Currency Naira = 100 kobo
Literacy rate 62%
Life expectancy 61 years

NORTH KOREA
East Asia

Page 53 G4

The communist state in Korea's northern half has been largely isolated from the outside world since 1948. The capital and largest city of the country, P'yŏngyang, is a major industrial and transport center.

Official name Democratic People's Republic of Korea
Formation 1945 / 1953
Capital Pyongyang
Population 25.9 million / 557 people per sq mile (215 people per sq km)
Total area 46,539 sq miles (120,538 sq km)
Languages Korean*
Religions Atheist 100%
Demographics Korean 100%
Government One-party state
Currency North Korean won = 100 chon
Literacy rate 100%
Life expectancy 72 years

NORTH MACEDONIA
Southeast Europe

Page 106 E3

Landlocked in the southern Balkans, this state changed its name to North Macedonia in 2019 to settle a long-standing dispute with neighboring Greece and pave the way for EU and NATO candidacy.

Official name Republic of North Macedonia
Formation 1991 / 1991
Capital Skopje
Population 2.1 million / 212 people per sq mile (82 people per sq km)
Total area 9928 sq miles (25,713 sq km)
Languages Macedonian*, Albanian*, Turkish, Romani, Serbian
Religions Orthodox Christian 65%, Muslim 33%, Other 2%
Demographics Macedonian 64%, Albanian 25%, Turkish 4%, Roma 3%, Serb 2%, Other 2%
Government Parliamentary system
Currency Macedonian denar = 100 deni
Literacy rate 98%
Life expectancy 77%

NORWAY
Northern Europe

Page 63 A5

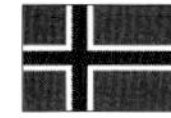

Norway lies on the rugged western coast of Scandinavia, and most people live in southern, coastal areas. Extensive oil and gas reserves give it one of the world's highest standards of living.

Official name Kingdom of Norway
Formation 1905 / 1905
Capital Oslo
Population 5.5 million / 44 people per sq mile (17 people per sq km)
Total area 125,020 sq miles (323,802 sq km)
Languages Norwegian* (Bokmål "book language" and Nynorsk "new Norsk"), Sámi
Religions Evangelical Lutheran 68%, Roman Catholic 3%, Other Christian 4%, Muslim 3%, Unspecified 20%, Other 2%
Demographics Norwegian 81%, Other European 19%
Government Parliamentary system
Currency Norwegian krone = 100 øre
Literacy rate 99%
Life expectancy 83 years

OMAN
Southwest Asia

Page 99 D6

Situated on the eastern corner of the Arabian Peninsula, Oman is the oldest independent state in the Arab world. Oil was discovered here in 1964 and oil and gas products still dominate the economy.

Official name Sultanate of Oman
Formation 1650 / 1955
Capital Muscat
Population 3.7 million / 31 people per sq mile (12 people per sq km)
Total area 119,498 sq miles (309,500 sq km)
Languages Arabic*, Baluchi, Farsi, Hindi, Punjabi
Religions Ibadi Muslim 86%, Other 14%
Demographics Arab 88%, Baluchi 4%, Indian and Pakistani 3%, Persian 3%, African 2%
Government Monarchy
Currency Omani rial = 1000 baisa
Literacy rate 96%
Life expectancy 77 years

PAKISTAN
South Asia

Page 112 B2

Pakistan was created as a Muslim state out of the partition of the Indian subcontinent in 1947. The service sector dominates the economy with burgeoning information technology and internet use.

Official name Islamic Republic of Pakistan
Formation 1947 / 1971
Capital Islamabad
Population 225 million / 732 people per sq mile (283 people per sq km)
Total area 307,373 sq miles (796,095 sq km)
Languages Punjabi, Sindhi, Pashtu, Urdu*, Baluchi, Brahui
Religions Sunni Muslim 77%, Shi'a Muslim 20%, Hindu 2%, Christian 1%
Demographics Punjabi 45%, Pathan (Pashtun) 15%, Sindhi 14%, Mohajir 8%, Baluchi 4%, Other 18%
Government Parliamentary system
Currency Pakistani rupee = 100 paisa
Literacy rate 58%
Life expectancy 70 years

PALAU
Australasia & Oceania

Page 122 A2

This archipelago of over 200 volcanic and coral islands, only ten of which are inhabited, lies in the western Pacific Ocean. The economy relies on fisheries and sustainable tourism.

Official name Republic of Palau
Formation 1994 / 1994
Capital Ngerulmud
Population 18,170 / 103 people per sq mile (40 people per sq km)
Total area 177 sq miles (459 sq km)
Languages Palauan*, English*, Japanese, Angaur, Tobi, Sonsorolese
Religions Roman Catholic 45%, Protestant 35%, Modekngei 6%, Muslim 3%, Other 11%
Demographics Palauan 73%, Filipino 16%, Other Asian 7%, Other Micronesian 3%, Other 2%
Government Nonparty system
Currency US dollar = 100 cents
Literacy rate 97%
Life expectancy 75 years

PANAMA
Central America

Page 31 F5

The southernmost country in Central America, Panama is rich in biodiversity. The Panama Canal, a vital shortcut for shipping between the Atlantic and Pacific oceans, is an important economic contributor.

Official name Republic of Panama
Formation 1903 / 1941
Capital Panama City
Population 4.3 million / 148 people per sq mile (57 people per sq km)
Total area 29,119 sq miles (75,420 sq km)
Languages English Creole, Spanish*, Amerindian languages, Chibchan languages
Religions Roman Catholic 49%, Protestant 30%, Nonreligious 12%, Other 9%
Demographics Mixed race 71%, Black 9%, White 7%, Indigenous 12%, Other 1%
Government Presidential system
Currency Balboa = 100 centésimos; US dollar
Literacy rate 96%
Life expectancy 78 years

PAPUA NEW GUINEA
Australasia & Oceania

Page 122 B3

The world's most linguistically diverse country, mineral-rich PNG occupies the east of the island of New Guinea and several other island groups. Most people make their livelihood from agriculture.

Official name Independent State of Papua New Guinea
Formation 1975 / 1975
Capital Port Moresby
Population 9.5 million / 53 people per sq mile (21 people per sq km)
Total area 178,703 sq miles (462,840 sq km)
Languages Pidgin English, Papuan, English*, Motu, 800 (est.) native languages
Religions Protestant 64%, Roman Catholic 26%, Other 9%
Demographics Melanesian and mixed race 100%
Government Parliamentary system
Currency Kina = 100 toea
Literacy rate 64%
Life expectancy 69 years

PARAGUAY
South America

Page 42 D2

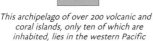

South America's longest dictatorship held power in landlocked Paraguay from 1954 to 1989. Now under democratic government, the country's economy is reliant on agriculture and hydro-electric power.

Official name Republic of Paraguay
Formation 1811 / 1938
Capital Asunción
Population 7.3 million / 46 people per sq mile (18 people per sq km)
Total area 157,047 sq miles (406,752 sq km)
Languages Guaraní*, Spanish*, German
Religions Roman Catholic 90%, Protestant 6%, Other Christian 3%
Demographics Mixed race 95%, Other 5%
Government Presidential system
Currency Guaraní = 100 céntimos
Literacy rate 95%
Life expectancy 78 years

PERU
South America

Page 38 C3

On the Pacific coast of South America, Peru was once the heart of the Inca empire. The country is home to Lake Titicaca, the world's highest navigable lake. Mineral wealth drives the economy.

Official name Republic of Peru
Formation 1821 / 1941
Capital Lima
Population 32.2 million / 65 people per sq mile (25 people per sq km)
Total area 496,224 sq miles (1,285,216 sq km)
Languages Spanish*, Quechua*, Aymara
Religions Roman Catholic 60%, Protestant 15%, Nonreligious 4%, Other 21%
Demographics Indigenous 26%, Mixed race 60%, White 6%, Black 4%, Other 4%
Government Presidential system
Currency New sol = 100 céntimos
Literacy rate 95%
Life expectancy 69 years

PHILIPPINES
Southeast Asia

Page 117 E1

This 7,500-island archipelago between the South China Sea and the Pacific is prone to earthquakes and volcanic activity. Founded in 1611, the University of Santo Tomas in Manila is the oldest in Asia.

Official name Republic of the Philippines
Formation 1946 / 1946
Capital Manila
Population 114 million / 984 people per sq mile (380 people per sq km)
Total area 115,830 sq miles (300,000 sq km)
Languages Filipino*, English*, Tagalog, Cebuano, Ilocano, Hiligaynon, many other local languages
Religions Roman Catholic 80%, Other Christian 6%, Muslim 6%, Other 8%
Demographics Tagalog 24%, Cebuano 10%, Ilocano 9%, Hiligaynon 8%, Bisaya 11%, Bikol 7%, Waray 4%, Other 27%
Government Presidential system
Currency Philippine peso = 100 centavos
Literacy rate 96%
Life expectancy 70 years

POLAND
Northern Europe

Page 76 B3

Extending from the Baltic Sea into the heart of Europe, Poland has a diverse, industrialized economy which supports an extensive health, education, and welfare system. It joined NATO in 1999 and the EU in 2004.

Official name Republic of Poland
Formation 1918 / 1945
Capital Warsaw
Population 38 million / 315 people per sq mile (122 people per sq km)
Total area 120,728 sq miles (312,685 sq km)
Languages Polish*, Silesian
Religions Roman Catholic 85%, Orthodox Christian 1%, Protestant 1%, Other 13%
Demographics Polish 97%, Silesian 1%, Other 2%
Government Parliamentary system
Currency Zloty = 100 groszy
Literacy rate 100%
Life expectancy 79 years

PORTUGAL
Southwest Europe

Page 70 B3 

Portugal, on the Iberian Peninsula, is the westernmost country in mainland Europe. A highly developed, diverse economy is supported by a sizeable tourist industry with over 20 million visitors a year.

Official name Portuguese Republic
Formation 1143 / 1640
Capital Lisbon
Population 10.2 million / 287 people per sq mile (111 people per sq km)
Total area 35,556 sq miles (92,090 sq km)
Languages Portuguese*, Mirandese
Religions Roman Catholic 81%, Nonreligious 7%, Other Christian 3%, Unspecified 8%, Other 1%
Demographics Portuguese 95%, Other 5%
Government Parliamentary system
Currency Euro = 100 cents
Literacy rate 96%
Life expectancy 82 years

QATAR
Southwest Asia

Page 98 C4 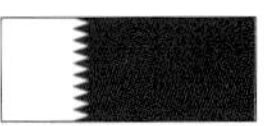

Projecting north from the Arabian Peninsula into the Persian Gulf, Qatar is mostly flat, semiarid desert. Massive reserves of oil and gas have made it one of the world's wealthiest states.

Official name State of Qatar
Formation 1971 / 2001
Capital Doha
Population 2.5 million / 559 people per sq mile (216 people per sq km)
Total area 4473 sq miles (11,586 sq km)
Languages Arabic*
Religions Muslim (mainly Sunni) 65%, Christian 14%, Other 14%
Demographics Qatari 12%, Other Arab 20%, Indian 20%, Nepalese 13%, Filipino 10%, Pakistani 7%, Other 10%
Government Monarchy
Currency Qatar riyal = 100 dirhams
Literacy rate 94%
Life expectancy 80 years

ROMANIA
Southeast Europe

Page 86 B4 

Romania lies on the western shores of the Black Sea. Reforms have allowed a diverse, high-tech economy to develop with an emphasis on electronics and vehicle manufacturing. It joined the EU in 2007.

Official name Romania
Formation 1878 / 2009
Capital Bucharest
Population 18.5 million / 201 people per sq mile (78 people per sq km)
Total area 92,043 sq miles (238,391 sq km)
Languages Romanian*, Hungarian (Magyar), Romani, German
Religions Orthodox Christian 82%, Roman Catholic 4%, Nonreligious 1% , Other 13%
Demographics Romanian 83%, Magyar 6%, Roma 3%, Other 7%
Government Presidential / Parliamentary system
Currency New Romanian leu = 100 bani
Literacy rate 99%
Life expectancy 76 years

RUSSIA
Europe / Asia

Page 92 D4 

Russia is the world's largest country, with vast mineral and energy reserves. It is a major global power with a rich cultural heritage that includes many notable writers, composers, and philosophers.

Official name Russian Federation
Formation 1547 / 2008
Capital Moscow
Population 143 million / 22 people per sq mile (8 people per sq km)
Total area 6,601,668 sq miles (17,098,242 sq km)
Languages Russian*, Tatar, Ukrainian, Chavash, various other national languages
Religions Russian Orthodox 20%, Muslim 15%, Other Christian 2%
Demographics Russian 78%, Tatar 4%, Ukrainian 1%, Bashkir 1%, Chavash 1%, Chechen 1%, Other 10%
Government Presidential / parliamentary system
Currency Russian rouble = 100 kopeks
Literacy rate 100%
Life expectancy 72 years

RWANDA
Central Africa

Page 51 B6 

A small landlocked country, Rwanda lies just south of the Equator in Central Africa. Recovering from ethnic violence that flared into genocide in 1994, the country is now rebuilding both politically and economically.

Official name Republic of Rwanda
Formation 1962 / 1962
Capital Kigali
Population 13.1 million / 1288 people per sq mile (497 people per sq km)
Total area 10,169 sq miles (26,338 sq km)
Languages Kinyarwanda*, French*, Kiswahili, English*
Religions Roman Catholic 38%, Protestant 38%, Seventh-day Adventist 13%, Muslim 2%, Other 2%
Demographics Hutu 85%, Tutsi 14%, Other 1%
Government Presidential system
Currency Rwanda franc = 100 centimes
Literacy rate 73%
Life expectancy 66 years

ST KITTS AND NEVIS
West Indies

Page 33 G3 

Saint Kitts and Nevis are part of the Caribbean Leeward Islands. Once a major sugar exporter, they are now a popular tourist destination while offshore finance is also a key contributor to the economy.

Official name Federation of Saint Christopher and Nevis
Formation 1983 / 1983
Capital Basseterre
Population 53,550 / 530 people per sq mile (205 people per sq km)
Total area 101 sq miles (261 sq km)
Languages English*, English Creole
Religions Protestant 76%, Roman Catholic 6%, Hindu 2%, Jehovah's Witness 1%, Rastafarian 1%, None 9%, Other 5%
Demographics Black 93%, Mixed race 3%, White 2%, Indigenous and other 2%
Government Parliamentary system
Currency East Caribbean dollar = 100 cents
Literacy rate 98%
Life expectancy 77 years

ST LUCIA
West Indies

Page 33 G4 

Part of the Caribbean Windward Islands, Saint Lucia boasts beaches, mountains, exotic plants, and the Qualibou volcano with its boiling sulfur springs. Tourism and fruit production dominate the economy.

Official name Saint Lucia
Formation 1979 / 1979
Capital Castries
Population 167,122 / 705 people per sq mile (271 people per sq km)
Total area 237 sq miles (616 sq km)
Languages English*, French Creole
Religions Roman Catholic 62%, Seventh-day Adventist 10%, Other Christian 3%, Pentecostal 9%, Nonreligious 6%, Rastafarian 2%, Other 1%
Demographics Black 85%, Mixed race 11%, Asian 2%, Other 2%
Government Parliamentary system
Currency East Caribbean dollar = 100 cents
Literacy rate 95%
Life expectancy 79 years

ST VINCENT & THE GRENADINES
West Indies

Page 33 G4 

Home to La Soufriere, an active volcano, this multi-island country forms part of the Caribbean Windward Islands. The economy relies on tourism and banana production, which is susceptible to seasonal storms.

Official name Saint Vincent and the Grenadines
Formation 1979 / 1979
Capital Kingstown
Population 100,969 / 673 people per sq mile (260 people per sq km)
Total area 150 sq miles (389 sq km)
Languages English*, English Creole
Religions Other Christian 48%, Anglican 18%, Pentecostal 18%, Nonreligious 9%, Other 7%
Demographics Black 71%, Mixed race 23%, Carib 4%, Asian 2%, Other 1%
Government Parliamentary system
Currency East Caribbean dollar = 100 cents
Literacy rate %
Life expectancy 77 years

SAMOA
Australasia & Oceania

Page 123 F4 

Samoa is a small island group in the Southern Pacific, and the traditional, communal way of life is still strong here. The economy depends on fishing and agriculture while 60% of its electricity comes from renewables.

Official name Independent State of Samoa
Formation 1962 / 1962
Capital Apia
Population 206,179 / 189 people per sq mile (73 people per sq km)
Total area 1093 sq miles (2,831 sq km)
Languages Samoan*, English*
Religions Other Christian 78%, Roman Catholic 19%, Other 3%
Demographics Samoan 96%, Samoan/New Zealander 2%, Other 2%
Government Parliamentary system
Currency Tala = 100 sene
Literacy rate 99%
Life expectancy 75 years

SAN MARINO
Southern Europe

Page 74 C3 

Perched on the slopes of Monte Titano in the Italian Appennino, San Marino has been a city-state since the 4th century CE. *Beneficial tax arrangements attract businesses, chiefly banking and electronics.*

Official name Republic of San Marino
Formation 301 / 1631
Capital San Marino
Population 34,010 / 1417 people per sq mile (558 people per sq km)
Total area 24 sq miles (61 sq km)
Languages Italian*
Religions Roman Catholic 93%, Nonreligious and other 7%
Demographics Sammarinese 88%, Italian 10%, Other 2%
Government Parliamentary system
Currency Euro = 100 cents
Literacy rate 100%
Life expectancy 84 years

SÃO TOMÉ & PRÍNCIPE
West Africa

Page 55 A5 

São Tomé and Príncipe, once a leading cocoa producer, consists of two main volcanic islands off the west coast of Africa. It has plans to reduce its reliance on imports by capitalizing on offshore oil potential.

Official name Democratic Republic of São Tomé and Príncipe
Formation 1975 / 1975
Capital São Tomé
Population 217,164 / 584 people per sq mile (225 people per sq km)
Total area 372 sq miles (964 sq km)
Languages Portuguese Creole, Portuguese*
Religions Roman Catholic 56%, Nonreligious 21%, Other Christian 16%, Other 7%
Demographics Black 90%, Portuguese and Creole 10%
Government Parliamentary system
Currency Dobra = 100 céntimos
Literacy rate 93%
Life expectancy 67 years

SAUDI ARABIA
Southwest Asia

Page 99 B5 

This vast desert kingdom is one of the world's leading oil and gas producers. It is home to Islam's two holiest cities, Medina and Mecca, the latter of which is the focus of the annual Hajj pilgrimage.

Official name Kingdom of Saudi Arabia
Formation 1932 / 2017
Capital Riyadh
Population 35.3 million / 43 people per sq mile (16 people per sq km)
Total area 829,999 sq miles (2,149,690 sq km)
Languages Arabic*
Religions Sunni Muslim 85%, Shi'a Muslim 15%
Demographics Arab 90%, Afro-Asian 10%
Government Monarchy
Currency Saudi riyal = 100 halalat
Literacy rate 98%
Life expectancy 77 years

SENEGAL
West Africa

Page 52 B3

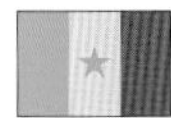

Senegal, and its capital, Dakar, sit on the westernmost coast of Africa. Its mainly rural economy is sensitive to the effects of climate change with fish products, cotton, and groundnuts being the main exports.

Official name Republic of Senegal
Formation 1960 / 1960
Capital Dakar
Population 17.9 million / 236 people per sq mile (91 people per sq km)
Total area 75,954 sq miles (196,722 sq km)
Languages Wolof, Pulaar, Serer, Diola, Mandinka, Malinké, Soninké, French*
Religions Sunni Muslim 97%, Christian 3%
Demographics Wolof 40%, Serer 16%, Peul 14%, Toucouleur 9%, Diola 5%, Other 14%
Government Presidential system
Currency CFA franc = 100 centimes
Literacy rate 52%
Life expectancy 70 years

SERBIA
Southeast Europe

Page 78 D4

Once a key part of the former Yugoslavia, Serbia faced ethnic tensions and conflict in the 1990s, which held back development. Fertile agricultural conditions allow for significant exports of fruit, maize, and wheat.

Official name Republic of Serbia
Formation 2006 / 2008
Capital Belgrade
Population 6.7 million / 224 people per sq mile (86 people per sq km)
Total area 29,912 sq miles (77,474 sq km)
Languages Serbian*, Hungarian (Magyar)
Religions Orthodox Christian 85%, Roman Catholic 5%, Nonreligious 5%, Muslim 3%, Other 1%
Demographics Serb 83%, Magyar 4%, Roma 2%, Bosniak 2%, Croat 1%, Slovak 1%, Other 3%
Government Parliamentary system
Currency Serbian dinar = 100 para
Literacy rate 99%
Life expectancy 74 years

SEYCHELLES
Indian Ocean

Page 57 G1

The tropical location of the Seychelles in the Indian Ocean creates a thriving tourist industry which engenders good health care and education. Its diverse flora includes the world's largest seed, the coco-de-mer.

Official name Republic of Seychelles
Formation 1976 / 1976
Capital Victoria
Population 97,017 / 551 people per sq mile (213 people per sq km)
Total area 176 sq miles (455 sq km)
Languages French Creole*, English*, French*
Religions Roman Catholic 76%, Anglican 6%, Other Christian 7%, Hindu 2%, Muslim 2%, Nonreligious and other 7%
Demographics Creole 89%, Indian 5%, Chinese 2%, Other 4%
Government Presidential system
Currency Seychelles rupee = 100 cents
Literacy rate 96%
Life expectancy 76 years

SIERRA LEONE
West Africa

Page 52 C4

This West African country has undergone significant economic growth recently. With a mainly rural economy, the country is rich in minerals and now carefully managed, ethical diamond mining is on the rise.

Official name Republic of Sierra Leone
Formation 1961 / 1961
Capital Freetown
Population 8.6 million / 310 people per sq mile (120 people per sq km)
Total area 27,698 sq miles (71,740 sq km)
Languages Mende, Temne, Krio, English*
Religions Muslim 77%, Christian 23%,
Demographics Mende 31%, Temne 35%, Limba 9%, Kono 4%, Kurankoh 4%, Fullah 4%, Other 13%
Government Presidential system
Currency Leone = 100 cents
Literacy rate 32%
Life expectancy 59 years

SINGAPORE
Southeast Asia

Page 116 A1

A city state linked to the southern tip of the Malay Peninsula by a causeway, Singapore is one of Asia's major commercial and financial hubs with a high standard of living and generous social care.

Official name Republic of Singapore
Formation 1965 / 1965
Capital Singapore
Population 5.9 million / 20,769 people per sq mile (7692 people per sq km)
Total area 277 sq miles (719 sq km)
Languages Mandarin*, Malay*, Tamil*, English*
Religions Christian 19%, Buddhist 31%, Nonreligious 20%, Muslim 15%, Taoist 9%, Hindu 5%, Other 1%
Demographics Chinese 74%, Malay 14%, Indian 9%, Other 3%
Government Parliamentary system
Currency Singapore dollar = 100 cents
Literacy rate 98%
Life expectancy 86 years

SLOVAKIA
Central Europe

Page 77 C6

Slovakia was part of Czechoslovakia until the "Velvet Divorce" with the Czech Republic in 1993, and went on to join the EU in 2004. Its vibrant economy is one of the world's largest automotive producers.

Official name Slovak Republic
Formation 1993 / 1993
Capital Bratislava
Population 5.4 million / 285 people per sq mile (110 people per sq km)
Total area 18,932 sq miles (49,035 sq km)
Languages Slovak*, Hungarian (Magyar), Czech
Religions Roman Catholic 56%, Nonreligious 24%, Other Christian 7%, Greek Catholic (Uniate) 4%, Other 8%
Demographics Slovak 84%, Magyar 9%, Roma 2%, Other 5%
Government Parliamentary system
Currency Euro = 100 cents
Literacy rate 99%
Life expectancy 78 years

SLOVENIA
Central Europe

Page 73 D8

This small country in Central Europe was the first to break away from the former Yugoslavia in 1991 and was subsequently also the first of the ex-republics to join NATO and the EU in 2004.

Official name Republic of Slovenia
Formation 1991 / 1991
Capital Ljubljana
Population 2.1 million / 268 people per sq mile (104 people per sq km)
Total area 7827 sq miles (20,273 sq km)
Languages Slovenian*
Religions Roman Catholic 58%, Nonreligious 36%, Muslim 2%, Orthodox Christian 2%, Other 1%
Demographics Slovene 83%, Serb 2%, Croat 2%, Bosniak 1%, Other 12%
Government Parliamentary system
Currency Euro = 100 cents
Literacy rate 100%
Life expectancy 82 years

SOLOMON ISLANDS
Australasia & Oceania

Page 122 C3

This archipelago of around 1,000 volcanic islands scattered in the southwest Pacific has mountainous and heavily forested terrain. Most people are involved with the rural economy, particularly fishing.

Official name Solomon Islands
Formation 1978 / 1978
Capital Honiara
Population 702,694 / 63 people per sq mile (24 people per sq km)
Total area 11,156 sq miles (28,896 sq km)
Languages English*, Pidgin English, Melanesian Pidgin, around 120 native languages
Religions Protestant 73%, Roman Catholic 20%, Other Christian 3%, Other 4%
Demographics Melanesian 95%, Polynesian 3%, Micronesian 1%, Other 1%
Government Parliamentary system
Currency Solomon Islands dollar = 100 cents
Literacy rate 77%
Life expectancy 77 years

SOMALIA
East Africa

Page 51 E5

This semiarid state on the Horn of Africa is working toward a democratic system of government after decades of instability. Agriculture dominates the economy, particularly livestock farming and fisheries.

Official name Federal Republic of Somalia
Formation 1960 / 1960
Capital Mogadishu
Population 12.3 million / 50 people per sq mile (19 people per sq km)
Total area 246,199 sq miles (637,657 sq km)
Languages Somali*, Arabic*, English, Italian
Religions Sunni Muslim 99%, Christian 1%
Demographics Somali 85%, Other 15%
Government Parliamentary system
Currency Somali shilin = 100 senti
Literacy rate 38%
Life expectancy 56 years

SOUTH AFRICA
Southern Africa

Page 56 C4

Mineral-rich South Africa has one of the biggest economies on the continent. The country has undergone a dramatic political shift to become a liberal democracy after decades of apartheid rule ended in 1994.

Official name Republic of South Africa
Formation 1910 / 1994
Capital Pretoria; Cape Town; Bloemfontein
Population 57.5 million / 122 people per sq mile (47 people per sq km)
Total area 470,693 sq miles (1,219,090 sq km)
Languages English*, isiZulu*, isiXhosa*, Afrikaans*, Sepedi*, Setswana*, Sesotho*, Xitsonga*, siSwati*, Tshivenda*, isiNdebele*
Religions Christian 86%, Nonreligious 11%, Muslim 2%, Other 1%
Demographics Black 81%, White 8%, Colored 9%, Asian 3%
Government Presidential system
Currency Rand = 100 cents
Literacy rate 95%
Life expectancy 65 years

SOUTH KOREA
East Asia

Page 106 E4

The southern half of the Korean peninsula was separated from the communist North in 1948. The country is one of the world's major economies and a leading exporter of automotive and electronic goods.

Official name Republic of Korea
Formation 1945 / 1953
Capital Seoul; Sejong (administrative)
Population 51 million / 1325 people per sq mile (511 people per sq km)
Total area 38,502 sq miles (99,720 sq km)
Languages Korean*, English
Religions Nonreligious 57%, Mahayana Buddhist 16%, Other Christian 20%, Roman Catholic 8%, Other 1%
Demographics Korean 100%
Government Presidential system
Currency South Korean won = 100 chon
Literacy rate 99%
Life expectancy 83 years

SOUTH SUDAN
East Africa

Page 51 B5

This landlocked country seceded from Sudan in 2011 after years of conflict. Apart from the domestic rural economy, revenues from oil production generate almost the entire government budget.

Official name Republic of South Sudan
Formation 2011 / 2011
Capital Juba
Population 11.5 million / 46 people per sq mile (18 people per sq km)
Total area 248,777 sq miles (644,329 sq km)
Languages Arabic, Dinka, Nuer, Zande, Bari, Shilluk, Lotuko, English*
Religions Christian 60% traditional beliefs 33%, Muslim 6%, Other 1%
Demographics Dinka 40%, Nuer 15%, Shilluk 10%, Azande 10%, Arab 10%, Bari 10%, Other 5%
Government Presidential system
Currency South Sudan Pound = 100 piastres
Literacy rate 35%
Life expectancy 59 years

SPAIN
Southwest Europe

Page 70 D2 

At the gateway to the Mediterranean, Spain has a rich cultural and historical heritage. The high-tech industrialized economy is led by a number of global companies, particularly in telecoms and construction.

Official name Kingdom of Spain
Formation 1492 / 1713
Capital Madrid
Population 47.1 million / 241 people per sq mile (93 people per sq km)
Total area 195,124 sq miles (505,370 sq km)
Languages Spanish*, Catalan*, Galician*, Basque*
Religions Roman Catholic 58%, Nonreligious 16%, Other 3%
Demographics Castilian Spanish 72%, Catalan 17%, Galician 6%, Basque 2%, Roma 1%, Other 2%
Government Parliamentary system
Currency Euro = 100 cents
Literacy rate 99%
Life expectancy 83 years

SRI LANKA
South Asia

Page 110 D3 

Once known as Ceylon, the island republic of Sri Lanka is separated from India by the narrow Palk Strait. A severe economic crisis was declared in 2019, which has led to social unrest.

Official name Democratic Socialist Republic of Sri Lanka
Formation 1948 / 1948
Capital Colombo; Sri Jayawardenapura Kotte
Population 23.1 million / 912 people per sq mile (352 people per sq km)
Total area 25,332 sq miles (65,610 sq km)
Languages Sinhala*, Tamil*, Sinhala-Tamil, English
Religions Buddhist 70%, Hindu 13%, Muslim 10%, Christian (mainly Roman Catholic) 7%
Demographics Sinhalese 75%, Tamil 15%, Moor 9%, Other 1%
Government Presidential system
Currency Sri Lankan rupee = 100 cents
Literacy rate 92%
Life expectancy 78 years

SUDAN
East Africa

Page 50 B4 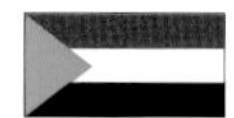

Once the largest country in Africa, Sudan was reduced to two thirds its size following the secession of South Sudan in 2011. Agricultural products and mineral extraction are key economic activities.

Official name Republic of the Sudan
Formation 1956 / 2011
Capital Khartoum
Population 47.9 million / 67 people per sq mile (26 people per sq km)
Total area 718,723 sq miles (1,861,484 sq km)
Languages Arabic*, English, Nubian, Beja, Fur
Religions Muslim (mainly Sunni) 99%, Other 1%
Demographics Arab 70%, Nubian 3%, Beja 6%, Fur 2%, Egyptian 1%, Fulani 1%, Coptic 1%, Other 16%
Government Presidential system
Currency Sudanese pound = 100 piastres
Literacy rate 60%
Life expectancy 67 years

SURINAME
South America

Page 37 G3 

Suriname is one of South America's smallest countries. The tropical rainforest of the Central Suriname Nature Reserve is a World Heritage site. Gold and oil are important contributors to the economy.

Official name Republic of Suriname
Formation 1975 / 1975
Capital Paramaribo
Population 632,000 / 10 people per sq mile (4 people per sq km)
Total area 63,251 sq miles (163,820 sq km)
Languages Sranang Tongo (creole), Dutch*, Caribbean Hindustani , Javanese
Religions Christian 50%, Hindu 23%, Muslim 14%, Other 13%
Demographics East Indian 27%, Creole 16%, Black 37%, Javanese 14%, Mixed race 13%, Other 8%
Government Presidential system
Currency Surinamese dollar = 100 cents
Literacy rate 94%
Life expectancy 72 years

SWEDEN
Northern Europe

Page 62 B4 

Forests cover over two-thirds of this large and densly populated Scandinavian country. A strong economy helps fund the extensive welfare system. It applied to join NATO in 2022.

Official name Kingdom of Sweden
Formation 1523 / 1921
Capital Stockholm
Population 10.4 million / 60 people per sq mile (23 people per sq km)
Total area 173,860 sq miles (450,295 sq km)
Languages Swedish*, Finnish, Sámi
Religions Evangelical Lutheran 58%, Muslim 2%, Other 40%
Demographics Swedish 80%, Syrian 2%, Iraqi 2%, Finnish 1%, Other 15%
Government Parliamentary constitutional monarchy
Currency Swedish krona = 100 öre
Literacy rate 99%
Life expectancy 83 years

SWITZERLAND
Central Europe

Page 73 A7 

One of the world's richest countries with a long tradition of neutrality, this mountainous nation lies at the center of Europe geographically, but outside it politically, having chosen not to join the EU.

Official name Swiss Confederation
Formation 1291 / 1857
Capital Bern
Population 8.5 million / 533 people per sq mile (206 people per sq km)
Total area 15,937 sq miles (41,277 sq km)
Languages German*, Swiss-German, French*, Italian*, Portuguese, Romansch*
Religions Roman Catholic 34%, Protestant 23%, Muslim 5%, Other Christian 6%, Other 32%
Demographics Swiss 69%, German 4%, French 2%, Italian 3%, Portuguese 3%, Kosovo 1%, Turkish 1%, Other 17%
Government Federal republic
Currency Swiss franc = 100 rappen/centimes
Literacy rate 99%
Life expectancy 83 years

SYRIA
Southwest Asia

Page 96 B3 

Situated at the eastern end of the Mediterranean, Syria contains an abundance of important, ancient historical sites. Recent political unrest has led to a brutally oppressed civil war.

Official name Syrian Arab Republic
Formation 1946 / 1967
Capital Damascus
Population 21.5 million / 297 people per sq mile (117 people per sq km)
Total area 72,370 sq miles (184,437 sq km)
Languages Arabic*, French, Kurdish, Armenian, Circassian, Turkic languages, Assyrian, Aramaic
Religions Muslim 87%, Christian 10%, Druze 3%
Demographics Arab 50%, Kurd 10%, Alawite 15%, Levantine 10%, Other 15%
Government Presidential system
Currency Syrian pound = 100 piastres
Literacy rate 86%
Life expectancy 74 years

TAIWAN
East Asia

Page 107 D6 

Following a period of rapid industrial growth in the 1960s, Taiwan is now one of the world's leading economies. Complex relations with China dominate domestic and international politics.

Official name Republic of China (ROC)
Formation 1945 / 1945
Capital Taipei
Population 23.5 million / 1692 people per sq mile (653 people per sq km)
Total area 13,892 sq miles (35,980 sq km)
Languages Amoy Chinese, Mandarin Chinese*, Hakka Chinese
Religions Buddhist 35%, Taoist 33%, Christian 4%, Indigenous 10%, Other 18%
Demographics Han Chinese 95%, Indigenous 2%, Other 3%
Government Semi-presidential system
Currency New Taiwan dollar = 100 cents
Literacy rate 99%
Life expectancy 81 years

TAJIKISTAN
Central Asia

Page 101 F3 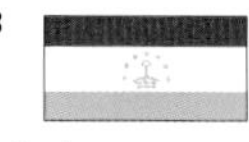

This landlocked ex-Soviet republic lies on the western slopes of the Pamirs in Central Asia. Rich mineral deposits, particularly gold and aluminium, and cotton production are the main economic exports.

Official name Republic of Tajikistan
Formation 1991 / 2011
Capital Dushanbe
Population 9.1 million / 164 people per sq mile (63 people per sq km)
Total area 55,637 sq miles (144,100 sq km)
Languages Tajik*, Uzbek, Russian
Religions Sunni Muslim 95%, Shi'a Muslim 3%, Other 2%
Demographics Tajik 84%, Uzbek 14%, Other 2%
Government Presidential system
Currency Somoni = 100 diram
Literacy rate 100%
Life expectancy 69 years

TANZANIA
East Africa

Page 51 B7 

This East African state was formed in 1964 by the union of Tanganyika and Zanzibar. A third of its area is game reserve or national park, including Africa's highest peak, Mt. Kilimanjaro.

Official name United Republic of Tanzania
Formation 1964 / 1964
Capital Dodoma
Population 63.8 million / 174 people per sq mile (67 people per sq km)
Total area 365,754 sq miles (947,300 sq km)
Languages Kiswahili*, Sukuma, Chagga, Nyamwezi, Hehe, Makonde, Yao, Sandawe, English*
Religions Christian 63%, Muslim 34%, Other 3%
Demographics Indigenous tribes 99%, Other 1%
Government Presidential system
Currency Tanzanian shilling = 100 cents
Literacy rate 78%
Life expectancy 70 years

THAILAND
Southeast Asia

Page 115 C5 

Thailand lies at the heart of the Indochinese Peninsula. Formerly Siam, it has been an independent kingdom for most of its history. The military has frequently intervened in politics.

Official name Kingdom of Thailand
Formation 1238 / 1907
Capital Bangkok
Population 69.6 million / 351 people per sq mile (136 people per sq km)
Total area 198,117 sq miles (513,120 sq km)
Languages Thai*, Chinese, Malay, Khmer, Mon, Karen, Miao
Religions Buddhist 94%, Muslim 4%, Other 2%
Demographics Thai 98%, Burmese 1%, Other 1%
Government Constitutional monarchy
Currency Baht = 100 satang
Literacy rate 94%
Life expectancy 78 years

TOGO
West Africa

Page 53 F4 

Togo lies sandwiched between Ghana and Benin in West Africa. Mining activities, including the world's fourth largest phosphate deposits, and agricultural products underpin the economy.

Official name Togolese Republic
Formation 1960 / 1960
Capital Lomé
Population 8.4 million / 383 people per sq mile (148 people per sq km)
Total area 21,925 sq miles (56,785 sq km)
Languages Ewe and Mina, Kabye, Gurma, French*
Religions Christian 42%, Indigenous 37%, Muslim 14%, Other 7%
Demographics Ewe 42%, Other African 41%, Kabye 12%, Foreigners 5%
Government Presidential system
Currency West African CFA franc = 100 centimes
Literacy rate 67%
Life expectancy 71 years

TONGA
Australasia & Oceania

Page 123 E4 

Northeast of New Zealand, Tonga is a 170-island archipelago, 45 of which are inhabited. In 2010, political reforms allowed more representative elections but the monarchy retains its influence.

Official name Kingdom of Tonga
Formation 1970 / 1970
Capital Nuku'alofa
Population 100,000/ 347 people per sq mile (134 people per sq km)
Total area 288 sq miles (747 sq km)
Languages English*, Tongan*, Other
Religions Free Wesleyan 38%, Roman Catholic 16%, Church of Jesus Christ of Latter-day Saints 17%, Other Christian 16%, Free Church of Tonga 12%, Other 1%
Demographics Tongan 97%, Other 3%
Government Constitutional monarchy
Currency Pa'anga = 100 seniti
Literacy rate 99%
Life expectancy 78 years

TRINIDAD AND TOBAGO
West Indies

Page 33 H5 

The most southerly of the Caribbean islands lie just 9 miles (15km) off the coast of South America. An industrialized economy based on oil and gas generates substantial wealth.

Official name Republic of Trinidad and Tobago
Formation 1962 / 1962
Capital Port of Spain
Population 1.4 million / 707 people per sq mile (273 people per sq km)
Total area 1980 sq miles (5128 sq km)
Languages English Creole, Trinidadian Creole, English*, Caribbean Hindustani, French, Spanish
Religions Protestant 32%, Roman Catholic 22%, Hindu 18%, Muslim 5%, Jehovah's Witness 2%, Other 21%
Demographics East Indian 38%, Black 36%, Mixed race 24%, Other 2%
Government Parliamentary system
Currency Trinidad and Tobago dollar = 100 cents
Literacy rate 99%
Life expectancy 76 years

TUNISIA
North Africa

Page 49 E2 

This northernmost country in Africa has long been an important center of trade and culture. It was the seat of the Arab Spring movement in 2011, and mass civil protests led to successful democratic elections.

Official name Republic of Tunisia
Formation 1956 / 1956
Capital Tunis
Population 11.8 million / 187 people per sq mile (72 people per sq km)
Total area 63,170 sq miles (163,610 sq km)
Languages Arabic*, French, Berber
Religions Muslim (mainly Sunni) 99%, Other 1%
Demographics Arab 98%, European 1%, Jewish and other 1%
Government Parliamentary system
Currency Tunisian dinar = 1000 millimes
Literacy rate 90%
Life expectancy 77 years

TURKEY (TÜRKIYE)
Asia / Europe

Page 94 B3 

With land in Europe and Asia, Turkey controls the entrance to the Black Sea. The balance between secular and religious influences is key to its politics. It has been a NATO member since 1952 with hopes to join the EU.

Official name Republic of Turkey
Formation 1923 / 1939
Capital Ankara
Population 83 million / 274 people per sq mile (106 people per sq km)
Total area 302,535 sq miles (783,562 sq km)
Languages Turkish*, Kurdish, Arabic, Circassian, Armenian, Greek, Georgian, Ladino
Religions Muslim (mainly Sunni) 99%, Other 1%
Demographics Turkish 73%, Kurdish 17%, Other 10%
Government Presidential system
Currency Turkish lira = 100 kuruş
Literacy rate 97%
Life expectancy 76 years

TURKMENISTAN
Central Asia

Page 100 B2 

Stretching from the Caspian Sea to the deserts of central Asia, this vast nation holds the world's sixth largest reserves of natural gas, which was supplied free to its citizens between 1993 and 2017.

Official name Turkmenistan
Formation 1991 / 1991
Capital Ashgabat
Population 5.6 million / 30 people per sq mile (11 people per sq km)
Total area 188,456 sq miles (488,100 sq km)
Languages Turkmen*, Uzbek, Russian, Kazakh, Tatar
Religions Muslim 93%, Christian 6%, Other 2%
Demographics Turkmen 85%, Uzbek 5%, Russian 4%, Other 6%
Government Presidential system / authoritarian
Currency Manat = 100 tenge
Literacy rate 100%
Life expectancy 72 years

TUVALU
Australasia & Oceania

Page 123 E3 

Tuvalu is a chain of nine atolls in the Central Pacific. It has the world's smallest Gross National Income (GNI), but has made substantial earnings leasing its ".tv" internet suffix.

Official name Tuvalu
Formation 1978 / 2012
Capital Funafuti Atoll
Population 11,930 / 1193 people per sq mile (459 people per sq km)
Total area 10 sq miles (26 sq km)
Languages Tuvaluan, Kiribati, Samoan, English*
Religions Church of Tuvalu 91%, Seventh-day Adventist 2%, Baha'i 2%, Other 5%
Demographics Tuvaluan 97%, Other 3%
Government Parliamentary democracy
Currency Australian dollar = 100 cents; Tuvaluan dollar = 100 cents
Literacy rate 95%
Life expectancy 68 years

UGANDA
East Africa

Page 51 B6 

Landlocked and abundantly fertile, Uganda is home to a variety of ecosystems, ranging from volcanic mountains to forested swamps and rainforests. Coffee is the mainstay of its agricultural economy.

Official name Republic of Uganda
Formation 1962 / 1962
Capital Kampala
Population 46.2 million / 496 people per sq mile (192 people per sq km)
Total area 93,065 sq miles (241,038 sq km)
Languages Luganda, Nkole, Chiga, Lango, Acholi, Teso, Lugbara, English*
Religions Roman Catholic 42%, Protestant 42%, Muslim (mainly Sunni) 12%, Other 4%
Demographics Baganda 17%, Banyakole 10%, Basoga 9%, Bakiga 7%, Iteso 7%, Langi 6%, Bagisu 5%, Acholi 4%, Lugbara 3%, Other 32%
Government Presidential system
Currency Ugandan shilling
Literacy rate 77%
Life expectancy 69 years

UKRAINE
Eastern Europe

Page 86 C2 

This vast, fertile nation is one of the world's leading grain producers. In 2022, following years of escalating tension, neighboring Russia launched an unprovoked invasion.

Official name Ukraine
Formation 1991 / 2009
Capital Kyiv
Population 43.5 million / 187 people per sq mile (72 people per sq km)
Total area 233,032 sq miles (603,550 sq km)
Languages Ukrainian*, Russian, Tatar, Other
Religions Orthodox Christian 78%, Roman Catholic 10%, Nonreligious 7%, Other 5%
Demographics Ukrainian 78%, Russian 17%, Belarussian 1%, Other 4%
Government Semi-presidential system
Currency Hryvnia = 100 kopiykas
Literacy rate 100%
Life expectancy 73 years

UNITED ARAB EMIRATES
Southwest Asia

Page 99 D5 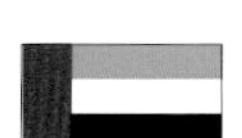

Situated on the Persian Gulf, this federation of seven states is trying to move away from its past reliance on oil and gas exports by encouraging tourism and financial services.

Official name United Arab Emirates
Formation 1971 / 1974
Capital Abu Dhabi
Population 9.9 million / 307 people per sq mile (118 people per sq km)
Total area 32,278 sq miles (83,600 sq km)
Languages Arabic*, Farsi, Indian and Pakistani languages, English
Religions Muslim (mainly Sunni) 76%, Christian 9%, Other 15%
Demographics South Asian 59%, Emirati 12%, Filipino 6%, Egyptian 10%, Other 13%
Government Federation of monarchies
Currency UAE dirham = 100 fils
Literacy rate 98%
Life expectancy 80 years

UNITED KINGDOM
Northwest Europe

Page 67 C5 

The UK comprises England, Scotland, Wales, and Northern Ireland and is one of the world's leading economies. Following a referendum in 2016, the UK left the EU in 2020.

Official name United Kingdom of Great Britain and Northern Ireland
Formation 1707 / 1922
Capital London
Population 67.7 million / 720 people per sq mile (278 people per sq km)
Total area 94,058 sq miles (243,610 sq km)
Languages English*, Welsh (* in Wales), Gaelic, Irish
Religions Christian 64%, Nonreligious 28%, Muslim 5%, Hindu 1%, Other 2%
Demographics White 87%, Indian and Pakistani 4%, Black 3%, Other Asian 2%, Bengali 1%, Other 3%
Government Parliamentary constitutional monarchy
Currency Pound sterling = 100 pence
Literacy rate 99%
Life expectancy 82 years

UNITED STATES
North America

Page 13 B5 

Bestowed with an abundance of natural resources, this vast nation has the world's largest economy, which helps cement its status as a global superpower.

Official name United States of America
Formation 1776 / 1959
Capital Washington D.C.
Population 332 million / 90 people per sq mile (35 people per sq km)
Total area 3,677,649 sq miles (9,525,067 sq km)
Languages English*, Spanish, Chinese, French, Polish, German, Tagalog, Vietnamese, Italian, Korean, Russian
Religions Protestant 47%, Nonreligious 23%, Roman Catholic 21%, Jewish 2%, Muslim 1%, Other 6%
Demographics White 62%, Black 12%, Asian 6%, Indigenous 2%, Mixed race 10%, Other 8%
Government Constitutional federal republic
Currency US dollar = 100 cents
Literacy rate 99%
Life expectancy 81 years

URUGUAY
South America

Page 42 D4 

Due to Uruguay's wealth of fertile pasture and a temperate climate, agriculture is a major part of its economy. After a period of military rule in the 1980s, it transitioned to a thriving democracy.

Official name Oriental Republic of Uruguay
Formation 1828 / 1828
Capital Montevideo
Population 3.4 million / 50 people per sq mile (19 people per sq km)
Total area 68,037 sq miles (176,215 sq km)
Languages Spanish*
Religions Roman Catholic 42%, Nonreligious 37%, Protestant 15%, Other 6%
Demographics White 87%, Black 7%, Mixed race 5%, Other 1%
Government Presidential system
Currency Uruguayan peso = 100 centésimos
Literacy rate 99%
Life expectancy 78 years

UZBEKISTAN
Central Asia

Page 100 D2 

Uzbekistan lies on the ancient Silk Road route between Asia and Europe. Its main exports today are cotton, oil, copper, and gold. It ranks 7th in the world for gold production.

Official name Republic of Uzbekistan
Formation 1991 / 1991
Capital Tashkent
Population 31.1million / 180 people per sq mile (70 people per sq km)
Total area 172,742 sq miles (447,400 sq km)
Languages Uzbek*, Russian, Tajik, Kazakh
Religions Sunni Muslim 88%, Orthodox Christian 9%, Other 3%
Demographics Uzbek 84%, Russian 2%, Tajik 5%, Kazakh 3%, Other 6%
Government Presidential system / authoritarian
Currency So'm = 100 tiyin
Literacy rate 100%
Life expectancy 75 years

VATICAN CITY
Southern Europe

Page 75 A8 

The Vatican City, seat of the Roman Catholic Church, is a walled enclave in Rome. It is the world's smallest country. Its head, the pope, is elected for life by a college of cardinals.

Official name Vatican City State
Formation 1929 / 1929
Capital Vatican City
Population 825 / 4852 people per sq mile (1875 people per sq km)
Total area 0.17 sq miles (0.44 sq km)
Languages Italian*, Latin*, French
Religions Roman Catholic 100%
Demographics Clergy 9%, Diplomats 39%, Swiss Guard 13%, Other 39%
Government Papal state
Currency Euro = 100 cents
Literacy rate 99%
Life expectancy 78 years

VIETNAM
Southeast Asia

Page 114 D4 

After Vietnam emerged from years of bitter conflict in the 1970s, recent economic reforms have resulted in a remarkable growth in GDP, accompanied by a huge surge in tourism.

Official name Socialist Republic of Viet Nam
Formation 1945 / 1976
Capital Hanoi
Population 103.8 million / 812 people per sq mile (313 people per sq km)
Total area 127,881 sq miles (331,210 sq km)
Languages Vietnamese*, English, French, Chinese, Thai, Khmer
Religions Catholic 6%, Buddhist 6%, Protestant 1%, None 86%, Other 1%
Demographics Vietnamese 85%, Tay 2%, Thai 2%, Muong 2%, Other 8%
Government Communist state
Currency Đông
Literacy rate 96%
Life expectancy 76 years

ZAMBIA
Southern Africa

Page 56 C2 

This landlocked nation in the heart of southern Africa is rich in natural resources, including minerals (particularly copper), forestry, and wildlife, all of which have led to economic growth.

Official name Republic of Zambia
Formation 1964 / 1964
Capital Lusaka
Population 19.6 million / 67 people per sq mile (26 people per sq km)
Total area 290,587 sq miles (752,618 sq km)
Languages Bemba, Tonga, Nyanja, Lozi, Lala-Bisa, Nsenga, English*
Religions Protestant 75%, Roman Catholic 20%, Nonreligious 2%, Other 3%
Demographics Bemba 21%, Tonga 14%, Chewa 7%, Lozi 6%, Nsenga 5%, Tumbuka 4%, Ngoni 4%, Lala 3%, Kaonde 3%, Lunda 3%, Other 30%
Government Presidential system
Currency New Zambian kwacha = 100 ngwee
Literacy rate 87%
Life expectancy 66 years

VANUATU
Australasia & Oceania

Page 122 D4 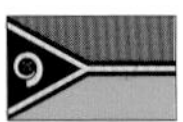

This South Pacific archipelago of 82 islands and islets boasts over 100 indigenous languages, each with an average of only around 2,000 speakers. Its favorable tax status attracts financial services.

Official name Republic of Vanuatu
Formation 1980 / 1980
Capital Port Vila
Population 300,000 / 63 people per sq mile (25 people per sq km)
Total area 4706 sq miles (12,189 sq km)
Languages Bislama*, English*, French*, Other indigenous languages
Religions Protestant 45%, Presbyterian 35%, Roman Catholic 14%. Indigenous 4%, Other 2%
Demographics ni-Vanuatu 99%, Other 1%
Government Parliamentary system
Currency Vanuatu Vatu
Literacy rate 88%
Life expectancy 75 years

VENEZUELA
South America

Page 36 D2 

Located on the Caribbean coast of South America, Venezuela has the continent's most urbanized society. It has the world's largest known oil reserves that have yet to reach their full production potential.

Official name Bolivarian Republic of Venezuela
Formation 1811 / 1811
Capital Caracas
Population 29.7 million / 84 people per sq mile (33 people per sq km)
Total area 352,144 sq miles (912,050 sq km)
Languages Spanish*, Indigenous languages
Religions Roman Catholic 73%, Protestant 17%, Nonreligious 7%, Other 3%
Demographics Mixed race 69%, White 20%, Black 9%, Indigenous 2%
Government Federal presidential republic
Currency Venezuelan bolívar = 100 céntimos
Literacy rate 97%
Life expectancy 73 years

YEMEN
Southwest Asia

Page 99 C7 

Located near the Red Sea, Yemen sits at the crossroads of ancient and modern trade and communications routes. However, it remains one of the world's least developed countries.

Official name Republic of Yemen
Formation 1990 / 2000
Capital Sanaa
Population 30.9 million / 152 people per sq mile (59 people per sq km)
Total area 203,850 sq miles (527,968 sq km)
Languages Arabic*
Religions Muslim 99% (Sunni 65%, Shi'a 35%), Other 1%
Demographics Arab 99%, Other 1%
Government Transitional regime
Currency Yemeni rial = 100 fils
Literacy rate 70%
Life expectancy 68 years

ZIMBABWE
Southern Africa

Page 56 D3 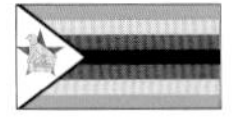

Zimbabwe is rich in natural resources but mismanagement, hyper inflation and drought have caused economic difficulties. Gold and diamonds are key exports while Victoria Falls is a major tourist attraction.

Official name Republic of Zimbabwe
Formation 1980 / 1980
Capital Harare
Population 15.1 million / 100 people per sq mile (39 people per sq km)
Total area 150,872 sq miles (390,757 sq km)
Languages Shona, isiNdebele, English*
Religions Protestant 75%, Roman Catholic 7%, Other Christian 5%, Traditional 2%, None 10%, Other 1%
Demographics Black 99%, Other 1%
Government Presidential system
Currency RTGS dollar = 100 cents; multiple foreign currencies
Literacy rate 87%
Life expectancy 63 years

Overseas Territories and Dependencies

Despite the rapid process of decolonization since the end of the Second World War, around 7.5 million people in more than 50 territories around the world continue to live under the protection of a parent state.

AUSTRALIA

ASHMORE & CARTIER ISLANDS
Indian Ocean
Claimed 1931
Capital not applicable
Area 2 sq miles (5 sq km)
Population None

CHRISTMAS ISLAND
Indian Ocean
Claimed 1958
Capital The Settlement (Flying Fish Cove)
Area 52 sq miles (135 sq km)
Population 1692

COCOS (KEELING) ISLANDS
Indian Ocean
Claimed 1955
Capital West Island
Area 5.5 sq miles (14 sq km)
Population 544

CORAL SEA ISLANDS
Southwest Pacific
Claimed 1969
Capital None
Area Less than 1.2 sq miles (3 sq km)
Population below 10 (scientists)

HEARD ISLAND & McDONALD ISLANDS
Indian Ocean
Claimed 1947
Capital not applicable
Area 159 sq miles (412 sq km)
Population None

NORFOLK ISLAND
Southwest Pacific
Claimed 1914
Capital Kingston
Area 14 sq miles (36 sq km)
Population 2188

DENMARK

FAROE ISLANDS
North Atlantic
Claimed 1380
Capital Tórshavn
Area 538 sq miles (1393 sq km)
Population 52,269

GREENLAND
North Atlantic
Claimed 1380
Capital Nuuk
Area 822,700 sq miles (2,130,783 sq km)
Population 57,792

FRANCE

CLIPPERTON ISLAND
East Pacific
Claimed 1935
Capital not applicable
Area 2 sq miles (6 sq km)
Population None

FRENCH GUIANA
South America
Claimed 1946
Capital Cayenne
Area 32,252 sq miles (83,534 sq km)
Population 304,641

FRENCH POLYNESIA
South Pacific
Claimed 1842
Capital Papeete
Area 1609 sq miles (4167 sq km)
Population 299,356

FRENCH SOUTHERN & ANTARCTIC LANDS
Indian Ocean
Claimed 1924
Capital Port-aux-Français
Area 169,800 sq miles (439,781 sq km)
Population 150

GUADELOUPE
West Indies
Claimed 1635
Capital Basse-Terre
Area 629 sq miles (1628 sq km)
Population 395,752

MARTINIQUE
West Indies
Claimed 1635
Capital Fort-de-France
Area 436 sq miles (1128 sq km)
Population 367,507

MAYOTTE
Indian Ocean
Claimed 1843
Capital Mamoudzou
Area 144 sq miles (374 sq km)
Population 326,206

NEW CALEDONIA
Southwest Pacific
Claimed 1853
Capital Nouméa
Area 7172 sq miles (18,575 sq km)
Population 297,160

RÉUNION
Indian Ocean
Claimed 1638
Capital Saint-Denis
Area 970 sq miles (2511 sq km)
Population 974,157

ST. BARTHÉLEMY
West Indies
Claimed 1878
Capital Gustavia
Area 10 sq miles (25 sq km)
Population 10,000

ST. MARTIN
West Indies
Claimed 1648
Capital Marigot
Area 20 sq miles (53 sq km)
Population 32,500

ST. PIERRE & MIQUELON
North America
Claimed 1604
Capital Saint-Pierre
Area 93 sq miles (242 sq km)
Population 5257

WALLIS & FUTUNA
South Pacific
Claimed 1842
Capital Matā'utu
Area 55 sq miles (142 sq km)
Population 15,891

NETHERLANDS

ARUBA
West Indies
Claimed 1636
Capital Oranjestad
Area 69 sq miles (180 sq km)
Population 122,320

BONAIRE
West Indies
Claimed 1816
Capital Kralendijk
Area 111 sq miles (288 sq km)
Population 22,573

CURAÇAO
West Indies
Claimed 1815
Capital Willemstad
Area 171 sq miles (444 sq km)
Population 152,379

SABA
West Indies
Claimed 1816
Capital The Bottom
Area 5 sq miles (13 sq km)
Population 1911

SINT-EUSTATIUS
West Indies
Claimed 1784
Capital Oranjestad
Area 8 sq miles (21 sq km)
Population 3242

SINT-MAARTEN
West Indies
Claimed 1648
Capital Philipsburg
Area 13 sq miles (34 sq km)
Population 45,126

NEW ZEALAND

COOK ISLANDS
South Pacific
Claimed 1901
Capital Avarua
Area 91 sq miles (236 sq km)
Population 8128

NIUE
South Pacific
Claimed 1901
Capital Alofi
Area 100 sq miles (260 sq km)
Population 2000

TOKELAU
South Pacific
Claimed 1925
Capital not applicable
Area 5 sq miles (12 sq km)
Population 1871

NORWAY

BOUVET ISLAND
South Atlantic
Claimed 1929
Capital not applicable
Area 19 sq miles (49 sq km)
Population None

JAN MAYEN
North Atlantic
Claimed 1929
Capital not applicable
Area 146 sq miles (377 sq km)
Population 35 (military personnel / scientists)

PETER I ISLAND
Antarctica
Claimed 1931
Capital not applicable
Area 60 sq miles (156 sq km)
Population None

SVALBARD
Arctic Ocean
Claimed 1920
Capital Longyearbyen
Area 23,956 sq miles (62,045 sq km)
Population 2926

UNITED KINGDOM

ANGUILLA
West Indies
Claimed 1650
Capital The Valley
Area 35 sq miles (91 sq km)
Population 18,741

ASCENSION ISLAND
South Atlantic
Claimed 1815
Capital Georgetown
Area 34 sq miles (88 sq km)
Population 880

BERMUDA
North Atlantic
Claimed 1612
Capital Hamilton
Area 21 sq miles (54 sq km)
Population 72,337

BRITISH INDIAN OCEAN TERRITORY
Indian Ocean
Claimed 1814
Capital not applicable
Area 23 sq miles (60 sq km)
Population 4000 (UK/US air base)

BRITISH VIRGIN ISLANDS
West Indies
Claimed 1672
Capital Road Town
Area 58 sq miles (151 sq km)
Population 38,632

CAYMAN ISLANDS
West Indies
Claimed 1670
Capital George Town
Area 102 sq miles (264 sq km)
Population 64,309

FALKLAND ISLANDS
South Atlantic
Claimed 1833
Capital Stanley
Area 4699 sq miles (12,173 sq km)
Population 3780

GIBRALTAR
Southwest Europe
Claimed 1713
Capital Gibraltar
Area 3 sq miles (7 sq km)
Population 29,573

GUERNSEY
Northwest Europe
Claimed 1066
Capital St Peter Port
Area 30 sq miles (78 sq km)
Population 67,491

ISLE OF MAN
Northwest Europe
Claimed 1765
Capital Douglas
Area 221 sq miles (572 sq km)
Population 91,382

JERSEY
Northwest Europe
Claimed 1066
Capital St. Helier
Area 45 sq miles (116 sq km)
Population 102,146

MONTSERRAT
West Indies
Claimed 1632
Capital Plymouth (de jure); Brades Estate *(de facto)*
Area 40 sq miles (102 sq km)
Population 5414

PITCAIRN GROUP OF ISLANDS
South Pacific
Claimed 1838
Capital Adamstown
Area 18 sq miles (47 sq km)
Population 50

ST. HELENA
South Atlantic
Claimed 1657
Capital Jamestown
Area 47 sq miles (122 sq km)
Population 4217

SOUTH GEORGIA & THE SOUTH SANDWICH ISLANDS
South Atlantic
Claimed 1775
Capital not applicable
Area 1507 sq miles (3903 sq km)
Population None

TRISTAN DA CUNHA
South Atlantic
Claimed 1816
Capital Edinburgh of the Seven Seas
Area 38 sq miles (98 sq km)
Population 242

TURKS & CAICOS ISLANDS
West Indies
Claimed 1799
Capital Grand Turk (Cockburn Town)
Area 366 sq miles (948 sq km)
Population 58,286

UNITED STATES OF AMERICA

AMERICAN SAMOA
South Pacific
Claimed 1900
Capital Pago Pago
Area 86 sq miles (224 sq km)
Population 45,443

BAKER ISLAND
Central Pacific
Claimed 1856
Capital not applicable
Area 0.8 sq miles (1.2 sq km)
Population None

GUAM
West Pacific
Claimed 1898
Capital Hagåtña
Area 210 sq miles (544 sq km)
Population 169,086

HOWLAND ISLAND
Central Pacific
Claimed 1856
Capital not applicable
Area 1 sq mile (2.6 sq km)
Population None

JARVIS ISLAND
Central Pacific
Claimed 1935
Capital not applicable
Area 2 sq miles (5 sq km)
Population None

JOHNSTON ATOLL
Central Pacific
Claimed 1858
Capital not applicable
Area 1 sq mile (2.6 sq km)
Population None

KINGMAN REEF
Central Pacific
Claimed 1856
Capital not applicable
Area 0.4 sq miles (1 sq km)
Population None

MIDWAY ATOLL
Central Pacific
Claimed 1867
Capital not applicable
Area 2 sq miles (5.2 sq km)
Population 40 (US air base)

NAVASSA ISLAND
Central Pacific
Claimed 1856
Capital not applicable
Area 2 sq miles (5.2 sq km)
Population None

NORTHERN MARIANA ISLANDS
West Pacific
Claimed 1944
Capital Saipan
Area 179 sq miles (464 sq km)
Population 51,475

PALMYRA ATOLL
Central Pacific
Claimed 1898
Capital not applicable
Area 2 sq miles (4 sq km)
Population 20

PUERTO RICO
West Indies
Claimed 1898
Capital San Juan
Area 3515 sq miles (9104 sq km)
Population 3.1 million

VIRGIN ISLANDS
West Indies
Claimed 1917
Capital Charlotte Amalie
Area 134 sq miles (346 sq km)
Population 105,413

WAKE ISLAND
Central Pacific
Claimed 1899
Capital not applicable
Area 3 sq miles (7 sq km)
Population 150 (US air base)

Geographical comparisons

Largest countries

Russia	6,601,668 sq miles	(17,098,242 sq km)
Canada	3,855,171 sq miles	(9,984,670 sq km)
China	3,705,386 sq miles	(9,596,960 sq km)
USA	3,677,649 sq miles	(9,525,067 sq km)
Brazil	3,287,957 sq miles	(8,515,770 sq km)
Australia	2,969,907 sq miles	(7,692,024 sq km)
India	1,269,219 sq miles	(3,287,263 sq km)
Argentina	1,073,518 sq miles	(2,780,400 sq km)
Kazakhstan	1,052,089 sq miles	(2,724,900 sq km)
Algeria	919,590 sq miles	(2,381,740 sq km)

Smallest countries

Vatican City	0.17 sq miles	(0.44 sq km)
Monaco	0.77 sq miles	(2 sq km)
Nauru	8 sq miles	(21 sq km)
Tuvalu	10 sq miles	(26 sq km)
San Marino	24 sq miles	(61 sq km)
Liechtenstein	62 sq miles	(160 sq km)
Marshall Islands	70 sq miles	(181 sq km)
St. Kitts & Nevis	101 sq miles	(261 sq km)
Maldives	120 sq miles	(300 sq km)
Malta	122 sq miles	(316 sq km)

Largest islands

Greenland	822,700 sq miles	(2,130,783 sq km)
New Guinea	303,381 sq miles	(785,753 sq km)
Borneo	288,869 sq miles	(748,167 sq km)
Madagascar	226,658 sq miles	(587,042 sq km)
Baffin Island	195,928 sq miles	(507,451 sq km)
Sumatra	171,068 sq miles	(443,064 sq km)
Honshu	87,200 sq miles	(225,847 sq km)
Victoria Island	83,897 sq miles	(217,292 sq km)
Britain	80,823 sq miles	(209,331 sq km)
Ellesmere Island	75,767 sq miles	(196,236 sq km)

Richest countries

(GNI per capita, in US$)

Liechtenstein	116,540
Switzerland	90,360
Norway	84,490
Luxembourg	81,110
Ireland	74,520
USA	70,430
Denmark	68,110
Iceland	64,410
Singapore	64,010
Sweden	58,890

Poorest countries

(GNI per capita, in US$)

Burundi	240
Somalia	450
Mozambique	480
Madagascar	500
Afghanistan	500
Sierra Leone	510
Central African Republic	530
Congo, Dem. Republic	580
Niger	590
Eritrea	600

Most populous countries

China	1.41 billion
India	1.39 billion
USA	332 million
Indonesia	276 million
Pakistan	225 million
Brazil	214 million

Most populous countries *continued*

Nigeria	211 million
Bangladesh	166 million
Russia	143 million
Mexico	130 million

Least populous countries

Vatican City	825
Nauru	10,870
Tuvalu	11,930
Palau	18,170
San Marino	34,010
Liechtenstein	38,250
Monaco	39,520
St. Kitts & Nevis	53,550
Marshall Islands	59,620
Dominica	72,170

Most densely populated countries

Monaco	50,145 people per sq mile	(19,497 per sq km)
Singapore	20,769 people per sq mile	(7692 per sq km)
Bahrain	5651 people per sq mile	(2241 per sq km)
Vatican City	4852 people per sq mile	(1875 per sq km)
Maldives	4667 people per sq mile	(1812 per sq km)
Malta	4252 people per sq mile	(1615 per sq km)
Bangladesh	3276 people per sq mile	(1278 per sq km)
Barbados	1730 people per sq mile	(669 per sq km)
Lebanon	1728 people per sq mile	(662 per sq km)
Mauritius	1616 people per sq mile	(624 per sq km)

Most sparsely populated countries

Mongolia	5 people per sq mile	(2 per sq km)
Namibia	8 people per sq mile	(3 per sq km)
Australia	8 people per sq mile	(3 per sq km)
Iceland	10 people per sq mile	(4 per sq km)
Guyana	10 people per sq mile	(4 per sq km)
Suriname	10 people per sq mile	(4 per sq km)
Canada	10 people per sq mile	(4 per sq km)
Libya	11 people per sq mile	(4 per sq km)
Botswana	13 people per sq mile	(5 per sq km)
Mauritania	13 people per sq mile	(5 per sq km)

Most widely spoken languages

(Native speakers)

1. Chinese (Mandarin)
2. Spanish
3. English
4. Hindi
5. Arabic
6. Portuguese
7. Bengali
8. Russian
9. Japanese
10. Western Punjabi

Largest conurbations*

Guangzhou (China)	65,100,000
Tokyo (Japan)	40,700,000
Shanghai (China)	39,300,000
Delhi (India)	32,400,000
Jakarta (Indonesia)	28,600,000
Manila (Philippines)	26,400,000
Mumbai (India)	26,100,000
Seoul (South Korea)	24,800,000
Mexico City (Mexico)	24,700,000
New York (USA)	23,000,000
São Paulo (Brazil)	22,700,000
Cairo (Egypt)	21,900,000
Dhaka (Bangladesh)	20,900,000
Lagos (Nigeria)	20,700,000
Beijing (China)	20,500,000
Bangkok (Thailand)	19,900,000
Karachi (Pakistan)	18,600,000
Osaka (Japan)	17,700,000

* Largest conurbations source: Thomas Brinkhoff: City Population, http://www.citypopulation.de

Largest conurbations *continued*

Los Angeles (USA)	17,500,000
Moscow (Russia)	17,400,000
Kolkata (India)	17,200,000
Buenos Aires (Argentina)	16,800,000
Istanbul (Turkey)	16,500,000
Tehran (Iran)	15,800,000
Chengdu (China)	15,200,000

Longest river systems

Nile (Northeast Africa)	4130 miles	(6650 km)
Amazon (South America)	3976 miles	(6400 km)
Yangtze (China)	3917 miles	(6300 km)
Mississippi/Missouri (USA)	3902 miles	(6280 km)
Yenisey (Russia)	3445 miles	(5544 km)
Yellow River (China)	3395 miles	(5464 km)
Ob (Russia)	3364 miles	(5414 km)
Parana (South America)	3030 miles	(4876 km)
Congo (Central Africa)	2922 miles	(4703 km)
Amur (East Asia)	2763 miles	(4447 km)

Highest mountains

(Height above sea level)

Everest	29,032 ft	(8849 m)
K2	28,251 ft	(8611 m)
Kanchenjunga	28,169 ft	(8586 m)
Lhotse	27,940 ft	(8516 m)
Makalu	27,838 ft	(8485 m)
Cho Oyu	26,864 ft	(8188 m)
Dhaulagiri I	26,795 ft	(8167 m)
Manaslu	26,781 ft	(8163 m)
Nanga Parbat	26,660 ft	(8126 m)
Annapurna I	26,545 ft	(8091 m)

Largest bodies of inland water

(Area & depth)

Caspian Sea	143,240 sq miles (371,000 sq km)	3363 ft (1025 m)
Lake Superior	31,700 sq miles (82,100 sq km)	1332 ft (406 m)
Lake Victoria	26,590 sq miles (68,870 sq km)	276 ft (84 m)
Lake Huron	23,000 sq miles (59,600 sq km)	751 ft (229 m)
Lake Michigan	22,400 sq miles (58,000 sq km)	922 ft (281 m)
Lake Tanganyika	12,600 sq miles (32,600 sq km)	4820 ft (1470 m)
Lake Baikal	12,200 sq miles (31,500 sq km)	5371 ft (1637 m)
Great Bear Lake	12,000 sq miles (31,000 sq km)	1463 ft (446 m)
Lake Malawi	11400 sq miles (29,500 sq km)	2316 ft (706 m)
Great Slave Lake	10,000 sq miles (27,000 sq km)	2014 ft (614 m)

Deepest ocean features

Challenger Deep, Mariana Trench (Pacific)	36,197 ft (11,034 m)
Horizon Deep, Tonga Trench (Pacific)	35,702 ft (10,882 m)
Galathea Depth, Philippine Trench (Pacific)	34,580 ft (10,545 m)
Kuril-Kamchatka Trench (Pacific)	34,449 ft (10,542 m)
Kermadec Trench (Pacific)	32,963 ft (10,047 m)
Izu–Ogasawara Trench (Pacific)	32,087 ft (9810 m)
Japan Trench (Pacific)	29,527 ft (9000 m)
Milwaukee Deep, Puerto Rico Trench (Atlantic)	28,232 ft (8605 m)
Yap Trench (Pacific)	27,976 ft (8527 m)
Meteor Deep, South Sandwich Trench (Atlantic)	27,651 ft (8428 m)

Greatest waterfalls

(Mean flow of water)

Boyoma (Congo, Dem. Rep.)	600,000 cu. ft/sec (16,990 cu.m/sec)
Khone (Laos/Cambodia)	410,000 cu. ft/sec (11,600 cu.m/sec)
Pará (Venezuela)	125,000 cu. ft/sec (3540 cu.m/sec)
Paulo Afonso (Brazil)	100,000 cu. ft/sec (2800 cu.m/sec)
Niagara (USA/Canada)	85,000 cu. ft/sec (2407 cu.m/sec)
Vermilion (Canada)	64,000 cu. ft/sec (1800 cu.m/sec)
Iguaçu (Argentina/Brazil)	61,800 cu. ft/sec (1750 cu.m/sec)
Limestone (Canada)	51,600 cu. ft/sec (1460 cu.m/sec)
Pyrite (Canada)	51,600 cu. ft/sec (1460 cu.m/sec)
Victoria (Zimbabwe)	39,000 cu. ft/sec (1100 cu.m/sec)

Greatest waterfalls *continued*

Virginia (Canada)	35,300 cu. ft/sec (1000 cu.m/sec)
Shivanasamudra (India)	33,000 cu. ft/sec (930 cu.m/sec)

Highest waterfalls

Angel (Venezuela)	3212 ft	(979 m)
Tugela (South Africa)	3110 ft	(948 m)
Tres Hermanas (Peru)	2999 ft	(914 m)
Olo'upena (USA)	2953 ft	(900 m)
Yumbilla (Peru)	2940 ft	(896 m)
Skorga (Norway)	2871 ft	(875 m)
Vinnufossen (Norway)	2822 ft	(860 m)
Balåifossen (Norway)	2789 ft	(850 m)
Mattenbachfall (Switzerland)	2756 ft	(840 m)
Pu'uka'oku (USA)	2756 ft	(840 m)
James Bruce (Canada)	2756 ft	(840 m)
Browne (New Zealand)	2743 ft	(836 m)

Largest deserts

Sahara	3,552,140 sq miles (9,200,000 sq km)
Gobi	500,000 sq miles (1,295,000 sq km)
Kalahari	347,492 sq miles (900,000 sq km)
Patagonian	259,847 sq miles (673,000 sq km)
Ar Rub al Khali	250,000 sq miles (650,000 sq km)
Great Basin	190,000 sq miles (492,098 sq km)
Chihuahuan	175,000 sq miles (453,248 sq km)
Karakum	135,136 sq miles (350,000 sq km)
Great Victorian	134,653 sq miles (348,750 sq km)
Sonoran	130,116 sq miles (337,000 sq km)

NB – Most of Antarctica is a polar desert, with only 2 inches (50 mm) of precipitation annually

Hottest inhabited places

(Average annual temperature)

Abéché (Chad)	90.0°F	(32.2°C)
Mecca (Saudi Arabia)	89.8°F	(32.1°C)
Kaédi (Mauritania)	89.2°F	(31.7°C)
Yélimané (Mali)	88.7°F	(31.5°C)
Jizan (Saudi Arabia)	88.0°F	(31.1°C)
Kiffa (Mauritania)	87.9°F	(31.0°C)
Atbara (Sudan)	87.7°F	(30.9°C)
Matam (Senegal)	87.6°F	(30.8°C)
Ayoun al Atrous (Mauritania)	87.2°F	(30.6°C)
Coro (Venezuela)	87.2°F	(30.6°C)

Driest inhabited places

(Average annual rainfall)

Al Jawf (Libya)	<0.10 in	(2.5 mm)
Chimbote (Peru)	<0.10 in	(2.5 mm)
Kaktovik (AK, USA)	<0.10 in	(2.5 mm)
Pisco (Peru)	<0.10 in	(2.5 mm)
Wadi Halfa (Sudan)	<0.10 in	(2.5 mm)
Siwa Oasis (Egypt)	<0.10 in	(2.5 mm)
Kharga Oasis (Egypt)	0.10 in	(2.5 mm)
Aswan (Egypt)	0.10 in	(2.5 mm)
Nok Kundi (Pakistan)	0.10 in	(2.5 mm)
Altos del Mar (Chile)	0.10 in	(2.5 mm)

Wettest inhabited places

(Average annual rainfall)

Mawsynram (India)	467 in	(11871 mm)
Cherrapunji (India)	464 in	(11777 mm)
Tutunendo (Colombia)	463 in	(11770 mm)
San Antonio de Ureca (Equatorial Guinea)	418 in	(10450 mm)
Debundscha (Cameroon)	405 in	(10299 mm)
Quibdó City (Colombia)	289 in	(7328 mm)
Buenaventura (Colombia)	247 in	(6276 mm)
Mawlamyine (Myanmar)	188 in	(4772 mm)
Monrovia (Liberia)	179 in	(4540 mm)
Hilo (Hawaii)	127 in	(3219 mm)

A

Aa *see* Gauja
Aachen 72 A4 Dut. Aken, Fr. Aix-la-Chapelle; anc. Aquae Grani, Aquisgranum. Nordrhein-Westfalen, W Germany
Aaiún *see* Laâyoune
Aalborg 63 B7 var. Ålborg, Ålborg-Nørresundby; anc. Alburgum. Nordjylland, N Denmark
Aalen 73 B6 Baden-Württemberg, S Germany
Aalsmeer 64 C3 Noord-Holland, C Netherlands
Aalst 65 B6 Oost-Vlaanderen, C Belgium
Aalten 64 E4 Gelderland, E Netherlands
Aalter 65 B5 Oost-Vlaanderen, NW Belgium
Aanaarjärvi *see* Inarijärvi
Äänekoski 63 D5 Keski-Suomi, W Finland
Aar *see* Aare
Aare 73 A7 var. Aar. river W Switzerland
Aarhus 63 B7 var. Århus, C Denmark
Aarlen *see* Arlon
Aassi, Nahr El *see* Orontes
Aat *see* Ath
Aba 55 E5 Orientale, NE Dem. Rep. Congo
Aba 53 G5 Abia, S Nigeria
Abā as Su'ūd *see* Najrān
Abaco Island *see* Great Abaco, N The Bahamas
Ābādān 98 C4 Khūzestān, SW Iran
Abadan 100 C3 prev. Bezmein, Büzmeýin, Rus. Byuzmeyin. Ahal Welaýaty, C Turkmenistan
Abai *see* Blue Nile
Abakan 92 D4 Respublika Khakasiya, S Russia
Abancay 38 D4 Apurímac, SE Peru
Abariringa *see* Kanton
Abashiri 108 D2 var. Abasiri. Hokkaidō, NE Japan
Abasiri *see* Abashiri
Ābay Wenz *see* Blue Nile
Abbaia *see* Ābaya Hāyk'
Abbatis Villa *see* Abbeville
Abbazia *see* Opatija
Abbeville 68 C2 anc. Abbatis Villa. Somme, N France
'Abd al 'Azīz, Jabal 96 D2 mountain range NE Syria
Abéché 54 C3 var. Abécher, Abeshr. Ouaddaï, SE Chad
Abécher *see* Abéché
Abela *see* Ávila
Abellinum *see* Avellino
Abemama 122 D2 var. Apamama; prev. Roger Simpson Island. atoll Tungaru, W Kiribati
Abengourou 53 E5 E Ivory Coast
Aberbrothock *see* Arbroath
Abercorn *see* Mbala
Aberdeen 66 D3 anc. Devana. NE Scotland, United Kingdom
Aberdeen 23 E2 South Dakota, N USA
Aberdeen 24 B2 Washington, NW USA
Abergwaun *see* Fishguard
Abertawe *see* Swansea
Aberystwyth 67 C6 W Wales, United Kingdom
Abeshr *see* Abéché
Abhā 99 B6 'Asīr, SW Saudi Arabia
Abidavichy 85 D7 Rus. Obidovichi. Mahilyowskaya Voblasts', E Belarus
Abidjan 53 E5 S Ivory Coast
Abilene 27 F3 Texas, SW USA
Abingdon *see* Pinta, Isla
Abkhazia 95 E1 var. Apkhazeti. autonomous republic NW Georgia
Åbo *see* Turku
Aboisso 53 E5 SE Ivory Coast
Abo, Massif d' 54 B1 mountain range NW Chad
Abomey 53 F5 S Benin
Abou-Déïa 54 C3 Salamat, SE Chad
Aboudouhour *see* Abū aḍ Ḍuhūr
Abou Kémal *see* Abū Kamāl
Abrantes 70 B3 var. Abrántes. Santarém, C Portugal
Abrashlare *see* Brezovo
Abrolhos Bank 34 E4 undersea bank W Atlantic Ocean
Abrova 85 B6 Rus. Obrovo. Brestskaya Voblasts', SW Belarus
Abrud 86 B4 Ger. Gross-Schlatten, Hung. Abrudbánya. Alba, SW Romania
Abrudbánya *see* Abrud
Abruzzese, Appennino 74 C4 mountain range C Italy
Absaroka Range 22 B2 mountain range Montana/Wyoming, NW USA
Abū aḍ Ḍuhūr 96 B3 Fr. Aboudouhour. Idlib, NW Syria
Abu Dhabi *see* Abū Ẓabī
Abu Hamed 50 C3 River Nile, N Sudan
Abū Ḥardān *see* Hajīn
Abuja 53 G4 country capital (Nigeria) Federal Capital District, C Nigeria
Abū Kamāl 96 E3 Fr. Abou Kémal. Dayr az Zawr, E Syria
Abula *see* Ávila
Abunã, Rio 40 C2 var. Río Abuná. river Bolivia/Brazil
Abut Head 129 B6 headland South Island, New Zealand
Abuye Meda 50 D4 mountain C Ethiopia
Abū Ẓabī 99 C5 var. Abū Ẓabī, Eng. Abu Dhabi. country capital (United Arab Emirates) Abū Ẓaby, C United Arab Emirates
Abū Ẓabī *see* Abū Ẓabī
Abyaḍ, Al Baḥr al *see* White Nile
Abyla *see* Ávila
Abyssinia *see* Ethiopia
Acalayong 55 A5 SW Equatorial Guinea
Acaponeta 28 D4 Nayarit, C Mexico
Acapulco 29 E5 var. Acapulco de Juárez. Guerrero, S Mexico
Acapulco de Juárez *see* Acapulco
Acarai Mountains 37 F4 Sp. Serra Acaraí. mountain range Brazil/Guyana
Acaraí, Serra *see* Acarai Mountains
Acarigua 36 D2 Portuguesa, N Venezuela
Accra 53 E5 country capital (Ghana) SE Ghana
Achacachi 39 E4 La Paz, W Bolivia
Ach'ara *see* Ajaria
Acklins Island 32 C2 island SE The Bahamas
Aconcagua, Cerro 42 B4 mountain W Argentina
Açores/Açores, Arquipélago dos/Açores, Ilhas dos *see* Azores
A Coruña 70 B1 Cast. La Coruña, Eng. Corunna; anc. Caronium. Galicia, NW Spain
Acre 40 C2 off. Estado do Acre. state/region W Brazil
Açu *see* Assu
Acunum Acusio *see* Montélimar
Ada 78 D3 Vojvodina, N Serbia
Ada 27 G2 Oklahoma, C USA
Ada Bazar *see* Adapazarı
Adalia *see* Antalya
Adalia, Gulf of *see* Antalya Körfezi
Adama *see* Nazrēt
'Adan 99 B7 Eng. Aden. SW Yemen
Adana 94 D4 var. Seyhan. Adana, S Turkey (Türkiye)
Adâncata *see* Horlivka
Adapazarı 94 B2 prev. Ada Bazar. Sakarya, NW Turkey (Türkiye)
Adare, Cape 132 B4 cape Antarctica
Ad Dahna 98 C4 desert E Saudi Arabia
Ad Dakhla 48 A4 var. Dakhla. SW Western Sahara
Ad Dalanj *see* Dilling
Ad Damar *see* Ed Damer
Ad Damazīn *see* Ed Damazin
Ad Dāmir *see* Ed Damer
Ad Dammām 98 C4 Ash Sharqīyah, NE Saudi Arabia
Ad Dāmūr *see* Damoûr
Ad Dawḥah 98 C4 Eng. Doha. country capital (Qatar) C Qatar
Aḍ Ḍiffah *see* Libyan Plateau
Addis Ababa *see* Ādīs Ābeba
Addoo Atoll *see* Addu Atoll
Addu Atoll 110 A5 var. Addoo Atoll, Seenu Atoll. atoll S Maldives
Adelaide 127 B6 state capital South Australia
Adelsberg *see* Postojna
Aden *see* 'Adan
Aden, Gulf of 99 C7 gulf SW Arabian Sea
Adige 74 C2 Ger. Etsch. river N Italy
Adirondack Mountains 19 F2 mountain range New York, NE USA
Ādīs Ābeba 51 C5 Eng. Addis Ababa. country capital (Ethiopia) Ādīs Ābeba, C Ethiopia
Adıyaman 95 E4 Adıyaman, SE Turkey (Türkiye)
Adjud 86 C4 Vrancea, E Romania
Admiralty Islands 122 B3 island group N Papua New Guinea
Adra 71 E5 Andalucía, S Spain
Adrar 48 D3 C Algeria
Adrian 18 C3 Michigan, N USA
Adrianople/Adrianopolis *see* Edirne
Adriatico, Mare *see* Adriatic Sea
Adriatic Sea 81 E2 Alb. Deti Adriatik, It. Mare Adriatico, Croatian Jadransko More, Slvn. Jadransko Morje. sea N Mediterranean Sea
Adriatik, Deti *see* Adriatic Sea
Adycha 93 F2 river NE Russia
Aegean Sea 83 C5 Gk. Aigaíon Pelagos, Aigaío Pélagos, Turk. Ege Denizi. sea NE Mediterranean Sea
Aegviidu 84 D2 Ger. Charlottenhof. Harjumaa, NW Estonia
Aegyptus *see* Egypt
Aelana *see* Al 'Aqabah
Aelok *see* Ailuk Atoll
Aelōnlaplap *see* Ailinglaplap Atoll
Aemona *see* Ljubljana
Aeolian Islands 75 C6 var. Isole Lipari, Eng. Aeolian Islands, Lipari Islands. island group S Italy
Aeolian Islands *see* Eolie, Isole
Æsernia *see* Isernia
Afar Depression *see* Danakil Desert
Afars et des Issas, Territoire Français des *see* Djibouti
Afghanistan 100 C4 off. Islamic Republic of Afghanistan; var. Dari as Jamhūrī-ye Islāmī-ye Afghānistān, Pash. Dē Afghānistān Islāmī Jumhūriyat, prev. Republic of Afghanistan. country C Asia
Afghānistān, Dari as Jamhūrī-ye Islāmī-ye *see* Afghanistan
Afmadow 51 D6 Jubbada Hoose, S Somalia
Africa 46 continent
Africa, Horn of 46 E4 physical region Ethiopia/Somalia
Africana Seamount 119 A6 seamount SW Indian Ocean
'Afrīn 96 B2 Ḥalab, N Syria
Afyon 94 B3 prev. Afyonkarahisar. Afyon, W Turkey (Türkiye)
Agadès *see* Agadez
Agadez 53 G3 prev. Agadès. Agadez, C Niger
Agadir 48 B3 SW Morocco
Agana/Agaña *see* Hagåtña
Āgaro 51 C5 Oromīya, C Ethiopia
Agassiz Fracture Zone 121 G5 fracture zone S Pacific Ocean
Agatha *see* Agde
Agathónisi 83 D6 island Dodekánisa, Greece, Aegean Sea
Agde 69 C6 anc. Agatha. Hérault, S France
Agedabia *see* Ajdābiyā
Agen 69 B5 anc. Aginnum. Lot-et-Garonne, SW France
Agendicum *see* Sens
Agiá 82 B4 var. Ayiá. Thessalía, C Greece
Agialoúsa *see* Yenierenköy
Agía Marína 83 E6 Léros, Dodekánisa, Greece, Aegean Sea
Aginnum *see* Agen
Ágios Efstrátios 82 D4 var. Áyios Evstrátios, Hagios Evstrátios. island E Greece
Ágios Nikólaos 83 D8 var. Áyios Nikólaos. Kríti, Greece, E Mediterranean Sea
Āgra 112 D3 Uttar Pradesh, N India
Agra and Oudh, United Provinces of *see* Uttar Pradesh
Agram *see* Zagreb
Ağrı 95 F3 var. Karaköse; prev. Karakılısse. Ağrı, NE Turkey (Türkiye)
Ağrı Dağı *see* Büyük Ağrı Dağı
Agrigento 75 C7 Gk. Akragas; prev. Girgenti. Sicilia, Italy, C Mediterranean Sea
Agriovótano 83 C5 Évvoia, C Greece
Agropoli 75 D5 Campania, S Italy
Aguachica 36 B2 Cesar, N Colombia
Aguadulce 31 F5 Coclé, S Panama
Agua Prieta 28 B1 Sonora, NW Mexico
Aguascalientes 28 D4 Aguascalientes, C Mexico
Aguaytía 38 C3 Ucayali, C Peru
Águilas 71 E4 Murcia, SE Spain
Aguililla 28 D4 Michoacán, SW Mexico
Agulhas Basin 47 D8 undersea basin SW Indian Ocean
Agulhas, Cape 56 C5 headland SW South Africa
Agulhas Plateau 45 D6 undersea plateau SW Indian Ocean
Ahaggar 53 F2 high plateau region SE Algeria
Ahlen 72 B4 Nordrhein-Westfalen, W Germany
Ahmadābād 112 C4 var. Ahmedabad. Gujarāt, W India
Ahmadnagar 112 C5 var. Ahmednagar. Mahārāshtra, W India
Ahmedabad *see* Ahmadābād
Ahmednagar *see* Ahmadnagar
Ahuachapán 30 B3 Ahuachapán, W El Salvador
Ahvāz 98 C3 var. Ahwāz; prev. Nāsiri. Khūzestān, SW Iran
Ahvenanmaa *see* Åland
Ahwāz *see* Ahvāz
Aigaíon Pelagos/Aigaío Pélagos *see* Aegean Sea
Aígina 83 C6 var. Aíyina, Egina. Aígina, C Greece
Aígio 83 B5 var. Egio; prev. Aíyion. Dytikí Ellás, S Greece
Aiken 21 E2 South Carolina, SE USA
Ailinglaplap Atoll 122 D2 var. Aelōnlaplap. atoll Ralik Chain, S Marshall Islands
Ailuk Atoll 122 D1 var. Aelok. atoll Ratak Chain, NE Marshall Islands
Ainaži 84 D3 Est. Heinaste, Ger. Hainasch. Limbaži, N Latvia
'Aïn Ben Tili 52 D1 Tiris Zemmour, N Mauritania
Aintab *see* Gaziantep
Aïoun el Atrous/Aïoun el Atroûss *see* 'Ayoûn el 'Atroûs
Aiquile 39 F4 Cochabamba, C Bolivia
Aïr *see* Aïr, Massif de l'
Aïr du Azbine *see* Aïr, Massif de l'
Aïr, Massif de l' 53 G2 var. Aïr, Air du Azbine, Asben. mountain range NC Niger
Aiud 86 B4 Ger. Strassburg, Hung. Nagyenyed; prev. Engeten. Alba, SW Romania
Aix *see* Aix-en-Provence
Aix-en-Provence 69 D6 var. Aix; anc. Aquae Sextiae. Bouches-du-Rhône, SE France
Aix-la-Chapelle *see* Aachen
Aíyina *see* Aígina
Aíyion *see* Aígio
Aizkraukle 84 C4 Aizkraukle, S Latvia
Ajaccio 69 E7 Corse, France, C Mediterranean Sea
Ajaria 95 F2 var. Ach'ara. autonomous republic SW Georgia
Aj Bogd Uul 104 D2 mountain SW Mongolia
Ajdābiyā 49 G2 var. Agedabia, Ajdābiyah. NE Libya
Ajdābiyah *see* Ajdābiyā
Ajjinena *see* El Geneina
Ajmer 112 D3 var. Ajmere. Rājasthān, N India
Ajo 26 A3 Arizona, SW USA
Akaba *see* Al 'Aqabah
Akamagaseki *see* Shimonoseki
Akasha 50 B3 Northern, N Sudan
Akchâr 52 C2 desert W Mauritania
Aken *see* Aachen
Akermanceaster *see* Bath
Akhaltsikhe 95 F2 prev. Akhalts'ikhe. SW Georgia
Akhalts'ikhe *see* Akhaltsikhe
Akhisar 94 A3 Manisa, W Turkey (Türkiye)
Akhmīm 50 B2 var. Akhmîm; anc. Panopolis. C Egypt
Akhtubinsk 89 C7 Astrakhanskaya Oblast', SW Russia
Akhtyrka *see* Okhtyrka
Akimiski Island 16 C3 island Nunavut, C Canada
Akinovka 87 F4 Zaporiz'ka Oblast', S Ukraine
Akita 108 D4 Akita, Honshū, C Japan
Akjoujt 52 C2 prev. Fort-Repoux. Inchiri, W Mauritania
Akkerman *see* Bilhorod-Dnistrovskyi
Akkeshi 108 E2 Hokkaidō, NE Japan
Aklavik 14 D3 Northwest Territories, NW Canada
Akmola *see* Nur-Sultan
Akmolinsk *see* Nur-Sultan
Aknavásár *see* Târgu Ocna
Akpatok Island 17 E1 island Nunavut, E Canada
Akragas *see* Agrigento
Akron 18 D4 Ohio, N USA
Akrotiri 80 C5 var. Akrotírion. UK Sovereign Base Area S Cyprus
Akrotírion *see* Akrotiri
Akrotírion Aksai Chin 102 B2 Chin. Aksayqin. disputed region China/India
Aksaray 94 C4 Aksaray, C Turkey (Türkiye)
Aksayqin *see* Aksai Chin
Akşehir 94 B4 Konya, W Turkey (Türkiye)
Aktash *see* Oqtosh
Aktau *see* Aqtaū
Aktjubinsk/Aktyubinsk *see* Aqtöbe
Aktobe *see* Aqtöbe
Aktsyabrski 85 C7 Rus. Oktyabr'skiy; prev. Karpilovka. Homyel'skaya Voblasts', SE Belarus
Aktyubinsk *see* Aqtöbe
Akula 55 C5 Equateur, NW Dem. Rep. Congo
Akureyri 61 E4 Nordhurland Eystra, N Iceland
Akyab *see* Sittwe
Alabama 20 C2 off. State of Alabama, also known as Camellia State, Heart of Dixie, The Cotton State, Yellowhammer State. state S USA
Alabama River 20 C3 river Alabama, S USA
Alaca 94 C3 Çorum, N Turkey (Türkiye)
Alacant *see* Alicante
Alagoas 41 G2 off. Estado de Alagoas. state/region E Brazil
Alais *see* Alès
Alajuela 31 E4 Alajuela, C Costa Rica
Alakanuk 14 C2 Alaska, USA
Al 'Alamayn 50 B1 var. El 'Alamein. N Egypt
Al 'Amārah 98 C3 var. Amara. Maysān, E Iraq
Alamo 25 D6 Nevada, W USA
Alamogordo 26 D3 New Mexico, SW USA
Alamosa 22 C5 Colorado, C USA
Åland 63 C6 var. Aland Islands, Fin. Ahvenanmaa. island group SW Finland
Aland Islands *see* Åland
Aland Sea *see* Ålands Hav
Ålands Hav 63 C6 var. Aland Sea. strait Baltic Sea/Gulf of Bothnia
Alanya 94 C4 Antalya, S Turkey (Türkiye)
Alappuzha 110 C3 var. Alleppey. Kerala, SW India
Al 'Aqabah 97 B8 var. Akaba, Aqaba, 'Aqaba; anc. Aelana, Elath. Al 'Aqabah, SW Jordan
Al 'Arabīyah as Su'ūdīyah *see* Saudi Arabia
Alasca, Golfo de *see* Alaska, Gulf of
Alaşehir 94 A4 Manisa, W Turkey (Türkiye)
Al 'Ashārah 96 E3 var. Ashara. Dayr az Zawr, E Syria
Alaska 14 C3 off. State of Alaska, also known as Land of the Midnight Sun, The Last Frontier, Seward's Folly; prev. Russian America. state NW USA
Alaska, Gulf of 14 C4 var. Golfo de Alasca. gulf Canada/USA
Alaska Peninsula 14 C3 peninsula Alaska, USA
Alaska Range 12 B2 mountain range Alaska, USA
Al-Asnam *see* Chlef
Alattio *see* Alta
Al Awaynāt *see* Al 'Uwaynāt
Al 'Awjā 97 E7 E West Bank, Middle East
Alaykel'/Alay-Kuu *see* Këk-Art
Al 'Aynā 97 B7 Al Karak, W Jordan
Alazeya 93 G2 river NE Russia
Al Bāb 96 B2 Ḥalab, N Syria
Albacete 71 E3 Castilla-La Mancha, C Spain
Al Baghdādī 98 B3 var. Khān al Baghdādī. Al Anbār, SW Iraq
Al Bāha *see* Al Bāḥah
Al Bāḥah 99 B5 var. Al Bāha. Al BāḤah, SW Saudi Arabia
Al Baḥrayn *see* Bahrain
Alba Iulia 86 B4 Ger. Weissenburg, Hung. Gyulafehérvár; prev. Bălgrad, Karlsburg, Károly-Fehérvár. Alba, W Romania
Albania 79 C7 off. Republic of Albania, Alb. Republika e Shqipqërisë, Shqipëria; prev. People's Socialist Republic of Albania. country SE Europe
Albania *see* Aubagne
Albany 125 B7 Western Australia
Albany 20 D3 Georgia, SE USA
Albany 19 F3 state capital New York, NE USA
Albany 24 B3 Oregon, NW USA
Albany 16 C3 river Ontario, S Canada
Alba Regia *see* Székesfehérvár
Al Bāridah 96 C4 var. Bāridah. Ḥimṣ, C Syria
Al Baṣrah 98 C3 Eng. Basra, hist. Busra, Bussora. Al Baṣrah, SE Iraq
Al Batrūn *see* Batroûn
Al Bawītī 50 B2 var. Bawīṭī C Egypt
Al Bayḍā' 49 G2 var. Beida. NE Libya
Albemarle Island *see* Isabela, Isla
Albemarle Sound 21 G1 inlet W Atlantic Ocean
Albergaria-a-Velha 70 B2 Aveiro, N Portugal
Albert 68 C3 Somme, N France
Alberta 15 E4 province SW Canada
Albert Edward Nyanza *see* Edward, Lake
Albert, Lake 51 B6 var. Albert Nyanza, Lac Mobutu Sese Seko. lake Uganda/Dem. Rep. Congo
Albert Lea 23 F3 Minnesota, N USA
Albert Nyanza *see* Albert, Lake
Albertville *see* Kalemie
Albi 69 C6 anc. Albiga. Tarn, S France
Albiga *see* Albi
Ålborg *see* Aalborg
Ålborg-Nørresundby *see* Aalborg
Alborz, Reshteh-ye Kūhhā-ye 98 C2 Eng. Elburz Mountains. mountain range N Iran
Albuquerque 26 D2 New Mexico, SW USA
Al Burayqah *see* Marsá al Burayqah
Alburgum *see* Aalborg
Albury 127 C7 New South Wales, SE Australia
Al Buṣayrah 96 D3 Dayr az Zawr, E Syria
Alcácer do Sal 70 B4 Setúbal, W Portugal
Alcalá de Henares 71 E3 Ar. Alkal'a; anc. Complutum. Madrid, C Spain
Alcamo 75 C7 Sicilia, Italy, C Mediterranean Sea
Alcañiz 71 F2 Aragón, NE Spain
Alcántara, Embalse de 70 C3 reservoir W Spain
Alcaudete 70 D4 Andalucía, S Spain
Alcázar *see* Ksar-el-Kebir
Alcazarquivir *see* Ksar-el-Kebir
Alcoi *see* Alcoy
Alcoy 71 F4 Cat. Alcoi. Comunitat Valenciana, E Spain
Aldabra Group 57 G2 island group SW Seychelles
Aldan 93 F3 river NE Russia
al Dar al Baida *see* Rabat
Alderney 68 A2 island Channel Islands
Aleg 52 C3 Brakna, SW Mauritania
Aleksandriya *see* Oleksandriia
Aleksandropol' *see* Gyumri
Aleksandrovka *see* Oleksandrivka
Aleksandrovsk *see* Zaporizhzhia
Aleksin 89 B5 Tul'skaya Oblast', W Russia
Aleksinac 78 E4 Serbia, SE Serbia
Alençon 68 B3 Orne, N France
Alenquer 41 E2 Pará, NE Brazil
Alep/Aleppo *see* Ḥalab
Alert 15 F1 Ellesmere Island, Nunavut, N Canada
Alès 69 C6 prev. Alais. Gard, S France
Aleşd 86 B3 Hung. Élesd. Bihor, SW Romania
Alessandria 74 B2 Fr. Alexandrie. Piemonte, N Italy
Ålesund 63 A5 Møre og Romsdal, S Norway
Aleutian Basin 91 G3 undersea basin Bering Sea
Aleutian Islands 14 A3 island group Alaska, USA
Aleutian Range 12 A2 mountain range Alaska, USA
Aleutian Trench 91 H3 trench S Bering Sea
Alexander Archipelago 14 D4 island group Alaska, USA
Alexander City 20 D2 Alabama, S USA
Alexander Island 132 A3 island Antarctica
Alexander Range *see* Kirghiz Range
Alexandra 129 B7 Otago, South Island, New Zealand
Alexándreia 82 B4 var. Alexándria. Kentrikí Makedonía, N Greece
Alexandretta *see* İskenderun
Alexandretta, Gulf of *see* İskenderun Körfezi
Alexandria *see* Al Iskandarīyah
Alexandria 86 C5 Teleorman, S Romania
Alexandria 20 B3 Louisiana, S USA
Alexandria 23 F2 Minnesota, N USA
Alexándria *see* Alexándreia
Alexandrie *see* Alessandria
Alexandroúpoli 82 D3 var. Alexandroúpolis, Turk. Dedeagaç, Dedeagach. Anatolikí Makedonía kai Thráki, NE Greece
Alexandroúpolis *see* Alexandroúpoli
Al Fāshir *see* El Fasher
Alfatar 82 E1 Silistra, NE Bulgaria
Alfeiós 83 B6 prev. Alfiós; anc. Alpheius, Alpheus. river S Greece
Alfiós *see* Alfeiós
Alföld *see* Great Hungarian Plain
Al-Furāt *see* Euphrates
Alga 92 B4 Kaz. Algha. Aktyubinsk, NW Kazakhstan
Algarve 70 B4 cultural region S Portugal
Algeciras 70 C5 Andalucía, SW Spain
Algemesí 71 F3 Comunitat Valenciana, E Spain
Al-Genain *see* El Geneina
Alger 49 E1 var. Algiers, El Djazaïr, Al Jazair. country capital (Algeria) N Algeria
Algeria 48 C3 off. People's Democratic Republic of Algeria. prev. Democratic and Popular Republic of Algeria. country N Africa
Algeria, Democratic and Popular Republic of *see* Algeria
Algeria, People's Democratic Republic of *see* Algeria
Algerian Basin 58 C5 var. Balearic Plain. undersea basin W Mediterranean Sea
Algha *see* Alga
Al Ghābah 99 E5 var. Ghaba. C Oman
Alghero 75 A5 Sardegna, Italy, C Mediterranean Sea
Al Ghardaqah 50 C2 var. Hurghada, Ghurdaqah. E Egypt
Algiers *see* Alger
Al Golea *see* El Goléa
Algona 23 F3 Iowa, C USA
Al Hajar al Gharbī 99 D5 mountain range N Oman
Al Hamad *see* Syrian Desert
Al Ḥasakah 96 D2 var. Al Hasijah, El Haseke, Fr. Hassetché. Al Ḥasakah, NE Syria
Al Hasijah *see* Al Ḥasakah
Al Ḥillah 98 B3 var. Hilla. Bābil, C Iraq
Al Ḥisā 97 B7 Aṭ Ṭafīlah, W Jordan
Al Ḥudayda 99 B6 Eng. Hodeidah. W Yemen
Al Hufūf 98 C4 var. Hofuf. Ash Sharqīyah, NE Saudi Arabia
Aliákmon *see* Aliákmonas
Aliákmonas 82 B4 prev. Aliákmon; anc. Haliacmon. river N Greece
Alíartos 83 C5 Stereá Ellás, C Greece
Äli-Bayramı *see* Şirvan
Alicante 71 F4 Cat. Alacant, Lat. Lucentum. Comunitat Valenciana, SE Spain
Alice 27 G5 Texas, SW USA
Alice Springs 126 A4 Northern Territory, C Australia
Alifu Atoll *see* Ari Atoll
Aligandí 31 G4 Kuna Yala, NE Panama
Aliki *see* Alykí
Alima 55 B6 river C Congo
Al Imārāt al 'Arabīyah al Muttaḥidah *see* United Arab Emirates
Alindao 54 C4 Basse-Kotto, S Central African Republic
Aliquippa 18 D4 Pennsylvania, NE USA
Al Iskandarīyah 50 B1 Eng. Alexandria. N Egypt
Al Ismā'īlīya 50 B1 var. Ismailia, Ismâ'ilîya. N Egypt
Alistráti 82 C3 Kentrikí Makedonía, NE Greece
Alivéri 83 C5 var. Alivérion. Évvoia, C Greece
Alivérion *see* Alivéri
Al Jabal al Akhḍar 49 G2 mountain range NE Libya
Al Jafr 97 B7 Ma'ān, S Jordan
Al Jaghbūb 49 H3 NE Libya
Al Jahrā' 98 C4 var. Al Jahrah, Jahra. C Kuwait
Al Jahrah *see* Al Jahrā'
Al Jawf 98 B4 off. Jauf. Al Jawf, NW Saudi Arabia
Al Jawlān *see* Golan Heights
Al Jazair *see* Alger
Al Jazīrah 96 E2 physical region Iraq/Syria
Al Jīzah 50 B1 var. El Gîza, Gizeh; Eng. Giza. N Egypt
Al Junaynah *see* El Geneina
Alkal'a *see* Alcalá de Henares
Al Karak 97 B7 var. El Kerak, Karak, Kerak; anc. Kir Moab, Kir of Moab. Al Karak, W Jordan
Al-Kasr al-Kebir *see* Ksar-el-Kebir
Al Khalīl *see* Hebron
Al Khārjah 50 B2 var. El Khârga; Eng. Kharga. C Egypt
Al Khums 49 F2 var. Homs, Khoms, Khums. NW Libya
Alkmaar 64 C2 Noord-Holland, NW Netherlands
Al Kufrah 49 H4 SE Libya
Al Kūt 98 C3 var. Kūt al 'Amārah, Kut al Imara. Wāsiṭ, E Iraq
Al-Kuwait *see* Al Kuwayt
Al Kuwayt 98 C4 var. Al-Kuwait, Eng. Kuwait, Kuwait City; prev. Qurein. country capital (Kuwait) E Kuwait
Al Lādhiqīyah 96 A3 Eng. Latakia, Fr. Lattaquié; anc. Laodicea, Laodicea ad Mare. Al Lādhiqīyah, W Syria
Allahābād *see* Prayagraj
Allanmyo *see* Aunglan
Allegheny Plateau 19 E3 mountain range New York/Pennsylvania, NE USA
Allenstein *see* Olsztyn
Allentown 19 F4 Pennsylvania, NE USA
Alleppey *see* Alappuzha
Alliance 22 D3 Nebraska, C USA
Al Līth 99 B5 Makkah, SW Saudi Arabia
Alma-Ata *see* Almaty
Almada 70 B4 Setúbal, W Portugal
Al Madīnah 99 A5 Eng. Medina. Al Madīnah, W Saudi Arabia
Al Mafraq 97 B6 var. Mafraq. Al Mafraq, N Jordan
Al Maghrib *see* Morocco
Al Mahdīyah *see* Mahdia
Al Mahrah 99 C6 mountain range E Yemen
Al Majma'ah 98 B4 Ar Riyāḍ, C Saudi Arabia
Al Mālikīyah 96 E1 var. Malkiye. Al Ḥasakah, N Syria
Almalyk *see* Olmaliq
Al Mamlaka al Urdunīya al Hashemīyah *see* Jordan
Al Manāmah 98 C4 Eng. Manama. country capital (Bahrain) N Bahrain
Al Manāṣif 96 E3 mountain range E Syria
Almansa 71 F4 Castilla-La Mancha, C Spain
Al-Mariyya *see* Almería
Al Marj 49 G2 var. Barka, It. Barce. NE Libya
Almaty 92 C5 var. Alma-Ata. Almaty, SE Kazakhstan
Al Mawṣil 98 B2 Eng. Mosul. Nīnawá, N Iraq
Al Mayādīn 96 D3 var. Mayadin, Fr. Meyadine. Dayr az Zawr, E Syria
Al Mazra' *see* Al Mazra'ah
Al Mazra'ah 97 B6 var. Al Mazra', Mazra'a. Al Karak, W Jordan
Almelo 64 E3 Overijssel, E Netherlands
Almendra, Embalse de 70 C2 reservoir Castilla y León, NW Spain
Almendralejo 70 C4 Extremadura, W Spain

Almere *64 C3 var.* Almere-stad. Flevoland, C Netherlands
Almere-stad *see* Almere
Almería *71 E5 Ar.* Al-Mariyya; *anc.* Unci, *Lat.* Portus Magnus. Andalucía, S Spain
Al'met'yevsk *89 D5* Respublika Tatarstan, W Russia
Al Mīnā' *see* El Mina
Al Minyā *50 B2 var.* El Minya, Minya. C Egypt
Almirante *31 E4* Bocas del Toro, NW Panama
Al Mudawwarah *97 B8* Ma'ān, SW Jordan
Al Mukallā *99 C6 var.* Mukalla. SE Yemen
Al Obayyid *see* El Obeid
Alofi *123 F4 dependent territory capital* (Niue) W Niue
Aloha State *see* Hawaii
Aloja *84 D3* Limbaži, N Latvia
Alónnisos *83 C5 island* Vóreies Sporádes, Greece, Aegean Sea
Álora *70 D5* Andalucía, S Spain
Alor, Kepulauan *117 E5 island group* E Indonesia
Al Oued *see* El Oued
Alpen *see* Alps
Alpena *18 D2* Michigan, N USA
Alpes *see* Alps
Alpha Cordillera *133 B3 var.* Alpha Ridge. *seamount range* Arctic Ocean
Alpha Ridge *see* Alpha Cordillera
Alpheius *see* Alfeiós
Alphen *see* Alphen aan den Rijn
Alphen aan den Rijn *64 C3 var.* Alphen. Zuid-Holland, C Netherlands
Alpheus *see* Alfeiós
Alpi *see* Alps
Alpine *27 E4* Texas, SW USA
Alps *80 C1 Fr.* Alpes, *Ger.* Alpen, *It.* Alpi. *mountain range* C Europe
Al Qaḑārif *see* Gedaref
Al Qāhirah *50 B2 var.* El Qâhira, *Eng.* Cairo. *country capital* (Egypt) N Egypt
Al Qāmishlī *96 E1 var.* Kamishli, Qamishly. Al Ḩasakah, NE Syria
Al Qaşrayn *see* Kasserine
Al Qayrawān *see* Kairouan
Al-Qsar al-Kbir *see* Ksar-el-Kebir
Al Qubayyāt *see* Qoubaïyât
Al Quds/Al Quds ash Sharīf *see* Jerusalem
Alqueva, Barragem do *70 C4 reservoir* Portugal/Spain
Al Qunayţirah *97 B5 var.* El Kuneitra, El Quneitra, Kuneitra, Qunaytra. Al Qunayţirah, SW Syria
Al Quşayr *96 B4 var.* El Quseir, Quşayr, *Fr.* Kousseir. Ḩimş, W Syria
Al Quwayrah *97 B8 var.* El Quweira. Al 'Aqabah, SW Jordan
Alsace *68 E3 Ger.* Elsass; *anc.* Alsatia. *cultural region* NE France
Alsatia *see* Alsace
Alsdorf *72 A4* Nordrhein-Westfalen, W Germany
Alt *see* Olt
Alta *62 D2 Fin.* Alattio. Finnmark, N Norway
Altai *see* Altai Mountains
Altai Mountains *104 C2 var.* Altai, *Chin.* Altay Shan, *Rus.* Altay. *mountain range* Asia/Europe
Altamaha River *21 E3 river* Georgia, SE USA
Altamira *41 E2* Pará, NE Brazil
Altamura *75 E5 anc.* Lupatia. Puglia, SE Italy
Altar, Desierto de *28 A1 var.* Sonoran Desert. *desert* Mexico/USA
Altar, Desierto de *see* Sonoran Desert
Altay *104 C2* Xinjiang Uygur Zizhiqu, NW China
Altay *104 D2 prev.* Yösönbulag. Govĭ-Altay, W Mongolia
Altay Altai Mountains, Asia/Europe
Altay Shan *see* Altai Mountains
Altbetsche *see* Bečej
Altenburg *see* Bucureşti, Romania
Altin Köprü *see* Āltūn Kawbrī
Altiplano *39 F4 physical region* W South America
Altkanischa *see* Kanjiža
Alton *18 B4* Illinois, C USA
Altoona *19 E4* Pennsylvania, NE USA
Alto Paraná *see* Paraná
Altpasua *see* Stara Pazova
Alt-Schwanenburg *see* Gulbene
Altsohl *see* Zvolen
Āltūn Kawbrī *98 B3 var.* Altin Köprü, Altun Kupri. Kirkūk, N Iraq
Altun Kupri *see* Āltūn Kawbrī
Altun Shan *104 C3 var.* Altyn Tagh. *mountain range* NW China
Altus *27 F2* Oklahoma, C USA
Altyn Tagh *see* Altun Shan
Al Ubayyiḑ *see* El Obeid
Alūksne *84 D3 Ger.* Marienburg. Alūksne, NE Latvia
Al 'Ulā *98 A4* Al Madīnah, NW Saudi Arabia
Al 'Umarī *97 C6* 'Ammān, E Jordan
Alupka *87 F5* Respublika Krym, S Ukraine
Al Uqşur *50 B2 Eng.* Luxor. E Egypt
Al Urdunn *see* Jordan
Alushta *87 F5* Respublika Krym, S Ukraine
Al 'Uwaynāt *49 F4 var.* Al Awaynāt. SW Libya
Alva *27 F1* Oklahoma, C USA
Alvarado *29 F4* Veracruz-Llave, E Mexico
Alvin *27 H4* Texas, SW USA
Al Wajh *98 A4* Tabūk, NW Saudi Arabia
Alwar *112 D3* Rājasthān, N India
Al Warī'ah *98 C4* Ash Sharqīyah, N Saudi Arabia
Al Yaman *see* Yemen
Alykí *82 C4 var.* Aliki. Thásos, N Greece
Alytus *85 B5 Pol.* Olita. Alytus, S Lithuania
Alzette *65 D8 river* S Luxembourg
Amadeus, Lake *125 D5 seasonal lake* Northern Territory, C Australia
Amadi *51 B5* W Equatoria, S South Sudan
Amadjuak Lake *15 G3 lake* Baffin Island, Nunavut, N Canada
Amakusa-nada *109 A7 gulf* SW Japan
Åmål *63 B6* Västra Götaland, S Sweden
Amami-gunto *108 A3 island group* SW Japan
Amami-o-shim *108 A3 island* S Japan
Amantea *75 D6* Calabria, SW Italy
Amapá *41 E1 off.* Estado de Amapá; *prev.* Território de Amapá. state/*region* NE Brazil
Amapá, Estado de *see* Amapá
Amapá, Território de *see* Amapá
Amara *see* Al 'Amārah
Amarapura *114 B3* Mandalay, C Myanmar (Burma)
Amarillo *27 E2* Texas, SW USA
Amay *65 C6* Liège, E Belgium
Amazon *41 E1 Sp.* Amazonas. *river* Brazil/Peru
Amazonas *see* Amazon
Amazon Basin *40 D2 basin* N South America
Amazonía *36 B4 region* S Colombia
Amazon, Mouths of the *41 F1 delta* NE Brazil
Ambam *55 B5* Sud, S Cameroon
Ambanja *57 G2* Antsirañana, N Madagascar
Ambarchik *93 G2* Respublika Sakha (Yakutiya), NE Russia
Ambato *38 B1* Tungurahua, C Ecuador
Ambérieu-en-Bugey *69 D5* Ain, E France
Ambianum *see* Amiens
Amboasary *57 F4* Toliara, S Madagascar
Amboina *see* Ambon
Ambon *117 F4 prev.* Amboina, Amboyna. Pulau Ambon, E Indonesia
Ambositra *57 G3* Fianarantsoa, SE Madagascar
Amboyna *see* Ambon
Ambracia *see* Árta
Ambre, Cap d' *see* Bobaomby, Tanjona
Ambrim *see* Ambrym
Ambriz *56 A1* Bengo, NW Angola
Ambrym *122 D4 var.* Ambrim. *island* C Vanuatu
Amchitka Island *14 A2 island* Aleutian Islands, Alaska, USA
Amdo *104 C5* Xizang Zizhiqu, W China
Ameland *64 D1 Fris.* It Amelân. *island* Waddeneilanden, N Netherlands
Amelân, It *see* Ameland
America *see* United States of America
America-Antarctica Ridge *45 C7 undersea ridge* S Atlantic Ocean
America in Miniature *see* Maryland
American Falls Reservoir *24 E4 reservoir* Idaho, NW USA
American Samoa *123 E4 US unincorporated territory* W Polynesia
Amersfoort *64 D3* Utrecht, C Netherlands
Ames *23 F3* Iowa, C USA
Amfilochía *83 A5 var.* Amfilokhía. Dytikí Ellás, C Greece
Amfilokhía *see* Amfilochía
Amga *93 F3 river* NE Russia
Amherst *17 F4* Nova Scotia, SE Canada
Amherst *see* Kyaikkami
Amida *see* Diyarbakır
Amiens *68 C3 anc.* Ambianum, Samarobriva. Somme, N France
Amíndaion/Amindeo *see* Amýntaio
Amindivi Islands *110 A2 island group* Lakshadweep, India, N Indian Ocean
Amirante Islands *57 G1 var.* Amirantes Group. *island group* C Seychelles
Amirantes Group *see* Amirante Islands
Amistad, Presa de la *see* Amistad Reservoir
Amistad Reservoir *27 F4 var.* Presa de la Amistad. *reservoir* Mexico/USA
Amisus *see* Samsun
Ammaia *see* Portalegre
'Ammān *97 B6 anc.* Philadelphia, *Bibl.* Rabbah Ammon, Rabbath Ammon. *country capital* (Jordan) 'Ammān, NW Jordan
Ammassalik *see* Tasiilaq
Ammóchostos *see* Gazimağusa
Ammóchostos, Kólpos *see* Gazimağusa Körfezi
Amnok-kang *see* Yalu
Amoea *see* Portalegre
Amoentai *see* Amuntai
Āmol *98 D2 var.* Amul. Māzandarān, N Iran
Amorgós *83 D6* Amorgós, Kykládes, Greece, Aegean Sea
Amorgós *83 D6 island* Kykládes, Greece, Aegean Sea
Amos *16 D4* Québec, SE Canada
Amourj *52 D3* Hodh ech Chargui, SE Mauritania
Amoy *see* Xiamen
Ampato, Nevado *39 E4 mountain* S Peru
Amposta *71 F2* Cataluña, NE Spain
Amraoti *see* Amrāvati
Amrāvati *112 D4 prev.* Amraoti. Mahārāshtra, C India
Amritsar *112 D2* Punjab, N India
Amstelveen *64 C3* Noord-Holland, C Netherlands
Amsterdam *64 C3 country capital* (Netherlands) Noord-Holland, C Netherlands
Amsterdam Island *119 C6 island* NE French Southern and Antarctic Lands
Am Timan *54 C3* Salamat, SE Chad
Amu Darya *100 D2 Rus.* Amudar'ya, *Taj.* Dar"yoi Amu, *Turkm.* Amyderya, *Uzb.* Amudaryo; *anc.* Oxus. *river* C Asia
Amu-Dar'ya *see* Amyderya
Amudar'ya/Amudaryo/Amu, Dar"yoi *see* Amu Darya
Amul *see* Āmol
Amund Ringnes Island *15 F2 Island* Nunavut, N Canada
Amundsen Basin *see* Fram Basin
Amundsen Plain *132 A4 abyssal plain* S Pacific Ocean
Amundsen-Scott *132 B3 US research station* Antarctica
Amundsen Sea *132 A4 sea* S Pacific Ocean
Amuntai *116 D4 prev.* Amoentai. Borneo, C Indonesia
Amur *93 G4 Chin.* Heilong Jiang. *river* China/Russia
Amvrosiivka *87 H3 prev.* Amvrosiyivka, *Rus.* Amvrosiyevka. Donets'ka Oblast', SE Ukraine
Amvrosiyevka *see* Amvrosiivka
Amvrosiyivka *see* Amvrosiivka
Amyderya *101 E3 Rus.* Amu-Dar'ya. Lebap Welaýaty, NE Turkmenistan
Amyderya *see* Amu Darya
Amýntaio *82 B4 var.* Amindeo; *prev.* Amíndaion. Dytikí Makedonía, N Greece
Anabar *93 E2 river* NE Russia
An Abhainn Mhór *see* Blackwater
Anaco *37 E2* Anzoátegui, NE Venezuela
Anaconda *22 B2* Montana, NW USA
Anacortes *24 B1* Washington, NW USA
Anadolu Dağları *see* Doğu Karadeniz Dağları
Anadyr' *93 H1* Chukotskiy Avtonomnyy Okrug, NE Russia
Anadyr' *93 G1 river* NE Russia
Anadyr, Gulf of *see* Anadyrskiy Zaliv
Anadyrskiy Zaliv *93 H1 Eng.* Gulf of Anadyr. *gulf* NE Russia
Anáfi *83 D7 anc.* Anaphe. *island* Kykládes, Greece, Aegean Sea
'Ānah *see* 'Annah
Anaheim *24 E2* California, W USA
Anaiza *see* 'Unayzah
Analalava *57 G2* Mahajanga, NW Madagascar
Anamur *94 C5* İçel, S Turkey (Türkiye)
Anantapur *110 C2* Andhra Pradesh, S India
Anaphe *see* Anáfi
Anápolis *41 F3* Goiás, C Brazil
Anār *98 D3* Kermān, C Iran
Anatolia *94 C4 plateau* C Turkey (Türkiye)
Anatom *see* Aneityum
Añatuya *42 C3* Santiago del Estero, N Argentina
An Bhearú *see* Barrow
Anchorage *14 C3* Alaska, USA
Ancona *74 C3* Marche, C Italy
Ancud *43 B6 prev.* San Carlos de Ancud. Los Lagos, S Chile
Ancyra *see* Ankara
Åndalsnes *63 A5* Møre og Romsdal, S Norway
Andalucía *70 D4 cultural region* S Spain
Andalusia *20 C3* Alabama, S USA
Andaman Islands *102 B4 island group* India, NE Indian Ocean
Andaman Sea *102 C4 sea* NE Indian Ocean
Andenne *65 C6* Namur, SE Belgium
Anderlues *65 B7* Hainaut, S Belgium
Anderson *18 C4* Indiana, N USA
Andes *42 B3 mountain range* W South America
Andhra Pradesh *113 E5 cultural region* E India
Andijon *101 F2 Rus.* Andizhan. Andijon Viloyati, E Uzbekistan
Andikíthira *see* Antikýthira
Andipaxi *see* Antípaxoi
Andípsara *see* Antípsara
Ándissa *see* Ántissa
Andizhan *see* Andijon
Andkhvoy *100 D3* Fāryāb, N Afghanistan
Andorra *69 A7 off.* Principality of Andorra, *Cat.* Valls d'Andorra, *Fr.* Vallée d'Andorre. *country* SW Europe
Andorra *see* Andorra la Vella
Andorra la Vella *69 A8 var.* Andorra, *Fr.* Andorre la Vielle, *Sp.* Andorra la Vieja. *country capital* (Andorra) C Andorra
Andorra la Vieja *see* Andorra la Vella
Andorra, Principality of *see* Andorra
Andorra, Valls d'/Andorra, Vallée d' *see* Andorra
Andorre la Vielle *see* Andorra la Vella
Andover *67 D7* S England, United Kingdom
Andøya *62 C2 island* C Norway
Andreanof Islands *14 A3 island group* Aleutian Islands, Alaska, USA
Andrews *27 E3* Texas, SW USA
Andrew Tablemount *118 A4 var.* Gora Andryu. *seamount* W Indian Ocean
Andria *75 D5* Puglia, SE Italy
An Droichead Nua *see* Newbridge
Andropov *see* Rybinsk
Ándros *83 D6* Ándros, Kykládes, Greece, Aegean Sea
Ándros *83 C6 island* Kykládes, Greece, Aegean Sea
Andros Island *32 B2 island* NW The Bahamas
Andros Town *32 C1* Andros Island, NW The Bahamas
Andryu, Gora *see* Andrew Tablemount
Aneityum *122 D5 var.* Anatom; *prev.* Kéamu. *island* S Vanuatu
Änewetak *see* Enewetak Atoll
Angara *93 E4 river* C Russia
Angarsk *93 E4* Irkutskaya Oblast', S Russia
Ånge *63 C5* Västernorrland, C Sweden
Ángel de la Guarda, Isla *28 B2 island* NW Mexico
Angeles *117 E1 off.* Angeles City. Luzon, N Philippines
Angeles City *see* Angeles
Angel Falls *37 E3 Eng.* Angel Falls. *waterfall* E Venezuela
Angel Falls *see* Ángel, Salto
Angerburg *see* Węgorzewo
Ångermanälven *62 C4 river* N Sweden
Angermünde *72 D3* Brandenburg, NE Germany
Angers *68 B4 anc.* Juliomagus. Maine-et-Loire, NW France
Anglesey *67 C5 island* NW Wales, United Kingdom
Anglet *69 A6* Pyrénées-Atlantiques, SW France
Angleton *27 H4* Texas, SW USA
Anglia *see* England
Anglo-Egyptian Sudan *see* Sudan
Angmagssalik *see* Tasiilaq
Ang Nam Ngum *114 C4 lake* C Laos
Angola *56 B2 off.* Republic of Angola; *prev.* People's Republic of Angola, Portuguese West Africa. *country* SW Africa
Angola Basin *47 B5 undersea basin* E Atlantic Ocean
Angola, People's Republic of *see* Angola
Angola, Republic of *see* Angola
Angora *see* Ankara
Angostura *see* Ciudad Bolívar
Angostura, Presa de la *29 G5 reservoir* SE Mexico
Angoulême *69 B5 anc.* Iculisma. Charente, W France
Angoumois *69 B5 cultural region* W France
Angra Pequena *see* Lüderitz
Angren *101 F2* Toshkent Viloyati, E Uzbekistan
Anguilla *33 G3 UK Overseas Territory* E West Indies
Anguilla Cays *32 B2 islets* SW The Bahamas
Anhui *106 C5 var.* Anhui Sheng, Anhwei, Wan. *province* E China
AnhuiSheng/Anhwei Wan *see* Anhui
Anicium *see* le Puy
Anina *86 A4 Ger.* Steierdorf, *Hung.* Stájerlakanina; *prev.* Ştaierdorf-Anina, Steierdorf-Anina, Steyerlak-Anina. Caraş-Severin, SW Romania
Anjou *68 B4 cultural region* NW France
Anjouan *57 F2 var.* Ndzouani, Nzwani. *island* SE Comoros
Ankara *94 C3 prev.* Angora; *anc.* Ancyra. *country capital* (Turkey) Ankara, C Turkey (Türkiye)
Ankeny *23 F3* Iowa, C USA
Anklam *72 D2* Mecklenburg-Vorpommern, NE Germany
An Mhuir Cheilteach *see* Celtic Sea
Annaba *49 E1 prev.* Bône. NE Algeria
An Nafud *98 B4 desert* NW Saudi Arabia
'Annah *98 B3 var.* 'Ānah. Al Anbār, NW Iraq
An Najaf *98 B3 var.* Najaf. An Najaf, S Iraq
Annamite Mountains *114 D4 Fr.* Annamitique, Chaîne. *mountain range* C Laos
Annamitique, Chaîne *see* Annamite Mountains
Annapolis *19 F4 state capital* Maryland, NE USA
Annapurna *113 E3 mountain* C Nepal
An Nāqūrah *see* En Nâqoûra
Ann Arbor *18 C3* Michigan, N USA
An Nāşirīyah *98 C3 var.* Nasiriya. Dhī Qār, SE Iraq
Anneciacum *see* Annecy
Annecy *69 D5 anc.* Anneciacum. Haute-Savoie, E France
An Nīl al Abyaḑ *see* White Nile
An Nīl al Azraq *see* Blue Nile
Anniston *20 D2* Alabama, S USA
Annotto Bay *32 B4* C Jamaica
An Nuway'imah *97 E7* E West Bank, Middle East
An Ómaigh *see* Omagh
Anqing *106 D5* Anhui, E China
Anse La Raye *33 F1* NW Saint Lucia
Anshun *106 B6* Guizhou, S China
Ansongo *53 E3* Gao, E Mali
An Srath Bán *see* Strabane
Antakya *94 D4 anc.* Antioch, Antiochia. Hatay, S Turkey (Türkiye)
Antalaha *57 G2* Antsirañana, NE Madagascar
Antalya *94 B4 prev.* Adalia; *anc.* Attaleia, *Bibl.* Attalia. Antalya, SW Turkey (Türkiye)
Antalya, Gulf of *94 B4 var.* Gulf of Adalia, *Eng.* Gulf of Antalya. *gulf* SW Turkey (Türkiye)
Antalya, Gulf of *see* Antalya Körfezi
Antananarivo *57 G3 prev.* Tananarive. *country capital* (Madagascar) Antananarivo, C Madagascar
Antarctica *132 B3 continent*
Antarctic Peninsula *132 A2 peninsula* Antarctica
Antep *see* Gaziantep
Antequera *70 D5 anc.* Anticaria, Antiquaria. Andalucía, S Spain
Antequera *see* Oaxaca
Antibes *69 D6 anc.* Antipolis. Alpes-Maritimes, SE France
Anticaria *see* Antequera
Anticosti, Île d' *17 F3 Eng.* Anticosti Island. *island* Québec, E Canada
Anticosti Island *see* Anticosti, Île d'
Antigua *33 G3 island* S Antigua and Barbuda, Leeward Islands
Antigua and Barbuda *33 G3 country* E West Indies
Antikýthira *83 B7 var.* Andikíthira. *island* S Greece
Anti-Lebanon *96 B4 var.* Jebel esh Sharqi, *Ar.* Al Jabal ash Sharqī, *Fr.* Anti-Liban. *mountain range* Lebanon/Syria
Anti-Liban *see* Anti-Lebanon
Antioch *see* Antakya
Antiochia *see* Antakya
Antípaxoi *83 A5 var.* Andipaxi. *island* Iónia Nisiá, Greece, C Mediterranean Sea
Antipodes Islands *120 D5 island group* S New Zealand
Antipolis *see* Antibes
Antípsara *83 D5 var.* Andípsara. *island* E Greece
Antiquaria *see* Antequera
Ántissa *83 D5 var.* Ándissa. Lésvos, E Greece
An tIúr *see* Newry
Antivari *see* Bar
Antofagasta *42 B2* Antofagasta, N Chile
Antony *68 E2* Hauts-de-Seine, N France
An tSionainn *see* Shannon
Antsirañana *57 G2 province* N Madagascar
Antsohihy *57 G2* Mahajanga, NW Madagascar
An-tung *see* Dandong
Antwerpen *65 C5 Eng.* Antwerp, *Fr.* Anvers. Antwerpen, N Belgium
Anuradhapura *110 D3* North Central Province, C Sri Lanka
Anvers *see* Antwerpen
Anyang *106 C4* Henan, C China
A'nyêmaqên Shan *104 D4 mountain range* C China
Anykščiai *84 C4* Utena, E Lithuania
Anzio *75 C5* Lazio, C Italy
Aomori *108 D3* Aomori, Honshū, C Japan
Aóos *see* Vjosës, Lumi i
Aoraki *129 B6 var.* Mount Cook. South Island, New Zealand
Aoraki *129 B6 prev.* Aorangi, *var.* Mount Cook. *mountain* South Island, New Zealand
Aorangi *see* Aoraki
Aosta *74 A1 anc.* Augusta Praetoria. Valle d'Aosta, NW Italy
Aotearoa *see* New Zealand
Aoukâr *52 D3 var.* Aouker. *plateau* C Mauritania
Aouk, Bahr *54 C4 river* Central African Republic/Chad
Aouker *see* Aoukâr
Aozou *54 C1* Borkou-Ennedi-Tibesti, N Chad
Apalachee Bay *20 D3 bay* Florida, SE USA
Apalachicola River *20 D3 river* Florida, SE USA
Apamama *see* Abemama
Apaporis, Río *36 C4 river* Brazil/Colombia
Aparima *see* Riverton
Apatity *88 C2* Murmanskaya Oblast', NW Russia
Ape *84 D3* Alūksne, NE Latvia
Apeldoorn *64 D3* Gelderland, E Netherlands
Apennines *74 E2 Eng.* Apennines. *mountain range* Italy/San Marino
Apennines *see* Appennino
Āpia *123 F4 country capital* (Samoa) Upolu, SE Samoa
Apkhazeti *see* Abkhazia
Apoera *37 G3* Sipaliwini, NW Suriname
Apostle Islands *18 B1 island group* Wisconsin, N USA
Appalachian Mountains *13 D5 mountain range* E USA
Appingedam *64 E1* Groningen, NE Netherlands
Appleton *18 B2* Wisconsin, N USA
Apulia *see* Puglia
Apure, Río *36 C2 river* W Venezuela
Apurímac, Río *38 D3 river* S Peru
Apuseni, Munţii *86 A4 mountain range* W Romania
Aqaba/'Aqaba *see* Al 'Aqabah
Aqaba, Gulf of *98 A4 var.* Gulf of Elat, *Ar.* Khalīj al 'Aqabah; *anc.* Sinus Aelaniticus. *gulf* NE Red Sea
'Aqabah, Khalīj al *see* Aqaba, Gulf of
Āqchah *101 E3 var.* Āqcheh. Jowzjān, N Afghanistan
Āqcheh *see* Āqchah
Aqmola *see* Nur-Sultan
Aqtaū *92 A4 var.* Aktau; *prev.* Shevchenko. Mangistau, W Kazakhstan
Aqtöbe *92 B4 var.* Aktobe; *prev.* Aktjubinsk, Aktyubinsk. Aktyubinsk, NW Kazakhstan
Aquae Augustae *see* Dax
Aquae Calidae *see* Bath
Aquae Flaviae *see* Chaves
Aquae Grani *see* Aachen
Aquae Sextiae *see* Aix-en-Provence
Aquae Solis *see* Bath
Aquae Tarbelicae *see* Dax
Aquidauana *41 E4* Mato Grosso do Sul, S Brazil
Aquila/Aquila degli Abruzzi *see* L'Aquila
Aquisgranum *see* Aachen
Aquitaine *69 B6 cultural region* SW France
'Arabah, Wadi al *97 B7 Heb.* Ha'Arava. *dry watercourse* Israel/Jordan
Arabian Basin *102 A4 undersea basin* N Arabian Sea
Arabian Desert *see* Sahara el Sharqîya
Arabian Peninsula *99 B5 peninsula* SW Asia
Arabian Sea *102 A3 sea* NW Indian Ocean
Arabicus, Sinus *see* Red Sea
'Arabī, Khalīj al *see* Persian Gulf
'Arabīyah as Su'ūdīyah, Al Mamlakah al *see* Saudi Arabia
'Arabīyah Jumhūrīyat, Mişr al *see* Egypt
Arab Republic of Egypt *see* Egypt
Aracaju *41 G3 state capital* Sergipe, E Brazil
Araçuai *41 F3* Minas Gerais, SE Brazil
Arad *97 B7* Southern, S Israel
Arad *86 A4* Arad, W Romania
Arafura Sea *120 A3 Ind.* Laut Arafuru. *sea* W Pacific Ocean
Arafuru, Laut *see* Arafura Sea
Aragón *71 E2 autonomous community* E Spain
Araguaia, Río *41 E3 var.* Araguaya. *river* C Brazil
Araguari *41 F3* Minas Gerais, SE Brazil
Araguaya *see* Araguaia, Río
Ara Jovis *see* Aranjuez
Arāk *98 C3 prev.* Sultānābād. Markazī, W Iran
Arakan Yoma *114 A3 mountain range* W Myanmar (Burma)
Araks/Arak's *see* Aras
Aral *92 B4 var.* Aral'sk. Kzylorda, SW Kazakhstan
Aral Sea *100 C1 Kaz.* Aral Tengizi, *Rus.* Aral'skoye More, *Uzb.* Orol Dengizi. *inland sea* Kazakhstan/Uzbekistan
Aral'sk *see* Aral
Aral'skoye More/Aral Tengizi *see* Aral Sea
Aranda de Duero *70 D2* Castilla y León, N Spain
Aranđelovac *78 D4 prev.* Arandjelovac. Serbia, C Serbia
Arandjelovac *see* Aranđelovac
Aranjuez *70 D3 anc.* Ara Jovis. Madrid, C Spain
Araouane *53 E2* Tombouctou, N Mali
'Ar'ar *98 B3* Al Ḩudūd ash Shamālīyah, NW Saudi Arabia
Ārārat-e Bozorg, Kūh-e *see* Büyük Ağrı Dağı
Ararat, Mount *see* Büyük Ağrı Dağı
Aras *95 G3 Arm.* Arak's, *Az.* Araz Nehri, *Per.* Rūd-e Aras, *Rus.* Araks; *prev.* Araxes. *river* SW Asia
Aras, Rūd-e *see* Aras
Arauca *36 C2* Arauca, NE Colombia
Arauca, Río *36 C2 river* Colombia/Venezuela
Arausio *see* Orange
Araxes *see* Aras
Araz Nehri *see* Aras
Arbela *see* Arbil
Arbīl *98 B2 var.* Erbil, Irbil, *Kurd.* Hewlêr; *anc.* Arbela. Arbil/Hewlêr, N Iraq
Arbroath *66 D3 anc.* Aberbrothock. E Scotland, United Kingdom
Arbuzinka *see* Arbuzynka
Arbuzynka *87 E3 Rus.* Arbuzinka. Mykolayivs'ka Oblast', S Ukraine
Arcachon *69 B5* Gironde, SW France
Arcae Remorum *see* Châlons-en-Champagne
Arcata *24 A4* California, W USA
Archangel *see* Arkhangel'sk
Archangel Bay *see* Chëshskaya Guba
Archidona *70 D5* Andalucía, S Spain
Arco *74 C2* Trentino-Alto Adige, N Italy
Arctic Mid Oceanic Ridge *see* Nansen Cordillera
Arctic Ocean *133 B3 ocean*
Arda *82 C3 var.* Ardhas, *Gk.* Ardas. *river* Bulgaria/Greece
Ardabīl *98 C2 var.* Ardebil. Ardabīl, NW Iran
Ardakān *98 D3* Yazd, C Iran
Ardas *82 D3 var.* Ardhas, *Bul.* Arda. *river* Bulgaria/Greece
Arḑ aş Şawwān *97 C7 var.* Ardh es Suwwān. *plain* S Jordan
Ardeal *see* Transylvania
Ardebil *see* Ardabīl
Ardèche *69 C5 cultural region* E France
Ardennes *65 C8 physical region* Belgium/France
Ardhas *see* Arda/Ardas
Ardh es Suwwān *see* Arḑ aş Şawwān
Ardino *82 D3* Kardzhali, S Bulgaria
Ard Mhacha *see* Armagh
Ardmore *27 G2* Oklahoma, C USA
Arel *see* Arlon
Arelas/Arelate *see* Arles
Arendal *63 A6* Aust-Agder, S Norway
Arensburg *see* Kuressaare
Arenys de Mar *71 G2* Cataluña, NE Spain
Areópoli *83 B7 prev.* Areópolis. Pelopónnisos, S Greece
Areópolis *see* Areópoli
Arequipa *39 E4* Arequipa, SE Peru
Arezzo *74 C3 anc.* Arretium. Toscana, C Italy
Argalastí *83 C5* Thessalía, C Greece
Argenteuil *68 D1* Val-d'Oise, N France
Argentina *43 B5 off.* Argentine Republic. *country* S South America
Argentina Basin *see* Argentine Basin
Argentine Basin *35 C7 var.* Argentina Basin. *undersea basin* SW Atlantic Ocean
Argentine Republic *see* Argentina
Argentine Rise *see* Falkland Plateau
Argentoratum *see* Strasbourg
Arghandab, Darya-ye *see* Arghandāb Rōd
Arghandāb Rōd *101 E5 var.* Darya-ye Arghandab. *river* SE Afghanistan
Argirocastro *see* Gjirokastër
Argo *50 B3* Northern, N Sudan
Argo Fracture Zone *119 C5 tectonic Feature* C Indian Ocean
Árgos *83 B6* Pelopónnisos, S Greece
Argostóli *83 A5 var.* Argostólion. Kefallonía, Iónia Nísiá, Greece, C Mediterranean Sea
Argostólion *see* Argostóli
Argun *103 E1 Chin.* Ergun He, *Rus.* Argun'. *river* China/Russia
Argyrokastron *see* Gjirokastër
Århus *see* Aarhus
Aria *see* Herāt

Ari Atoll *110 A4 var.* Alifu Atoll. *atoll* C Maldives
Arica *42 B1 hist.* San Marcos de Arica. Tarapacá, N Chile
Aridaía *82 B3 var.* Aridea, Aridhaía. Dytikí Makedonía, N Greece
Aridea *see* Aridaía
Aridhaía *see* Aridaía
Arīḥā *96 B3* Al Karak, W Jordan
Arīḥā *see* Jericho
Ariminum *see* Rimini
Arinsal *69 A7* NW Andorra Europe
Arizona *26 A2 off.* State of Arizona, *also known as* Copper State, Grand Canyon State. *state* SW USA
Arkansas *20 A1 off.* State of Arkansas, *also known as* The Land of Opportunity. *state* S USA
Arkansas City *23 F5* Kansas, C USA
Arkansas River *27 G1 river* C USA
Arkhangel'sk *92 B2 Eng.* Archangel. Arkhangel'skaya Oblast', NW Russia
Arkoí *83 E6 island* Dodekánisa, Greece, Aegean Sea
Arles *69 D6 var.* Arles-sur-Rhône; *anc.* Arelas, Arelate. Bouches-du-Rhône, SE France
Arles-sur-Rhône *see* Arles
Arlington *27 G2* Texas, SW USA
Arlington *19 E4* Virginia, NE USA
Arlon *65 D8 Dut.* Aarlen, *Ger.* Arel, *Lat.* Orolaunum. Luxembourg, SE Belgium
Armagh *67 B5 Ir.* Ard Mhacha. S Northern Ireland, United Kingdom
Armagnac *69 B6 cultural region* S France
Armenia *36 B3* Quindío, W Colombia
Armenia *95 F3 off.* Republic of Armenia, *var.* Hayastan, *Arm.* Hayastani Hanrapetut'yun; *prev.* Armenian Soviet Socialist Republic. *country* SW Asia
Armenian Soviet Socialist Republic *see* Armenia
Armenia, Republic of *see* Armenia
Armiansk *87 F4 prev.* Armyans'k, *Rus.* Armyansk. Respublika Krym, S Ukraine
Armidale *127 D6* New South Wales, SE Australia
Armstrong *16 B3* Ontario, S Canada
Armyansk *see* Armiansk
Armyans'k *see* Armiansk
Arnaía *82 C4 Cont.* Arnea. Kentrikí Makedonía, N Greece
Arnaud *60 A3 river* Québec, E Canada
Arnea *see* Arnaía
Arnedo *71 E2* La Rioja, N Spain
Arnhem *64 D4* Gelderland, SE Netherlands
Arnhem Land *126 A2 physical region* Northern Territory, N Australia
Arno *74 B3 river* C Italy
Arnold *23 G4* Missouri, C USA
Arnswalde *see* Choszczno
Aroe Islands *see* Aru, Kepulauan
Arorae *123 E3 atoll* Tungaru, W Kiribati
Arrabona *see* Győr
Ar Rahad *see* Er Rahad
Ar Ramādī *98 B3 var.* Ramadi, Rumadiya. Al Anbār, SW Iraq
Ar Rāmī *96 C4* Ḥimṣ, C Syria
Ar Ramthā *97 B5 var.* Ramtha. Irbid, N Jordan
Arran, Isle of *66 C4 island* SW Scotland, United Kingdom
Ar Raqqah *96 C2 var.* Rakka; *anc.* Nicephorium. Ar Raqqah, N Syria
Arras *68 C2 anc.* Nemetocenna. Pas-de-Calais, N France
Ar Rawḍatayn *98 C4 var.* Raudhatain. N Kuwait
Arretium *see* Arezzo
Arriaca *see* Guadalajara
Arriaga *29 G5* Chiapas, SE Mexico
Ar Riyāḍ *99 C5 Eng.* Riyadh. *country capital* (Saudi Arabia) Ar Riyāḍ, C Saudi Arabia
Ar Rub 'al Khali *99 C6 Eng.* Empty Quarter, Great Sandy Desert. *desert* SW Asia
Ar Rustāq *99 E5 var.* Rostak, Rustaq. N Oman
Ar Ruṭbah *98 B3 var.* Rutba. Al Anbār, SW Iraq
Árta *83 A5 anc.* Ambracia. Ípeiros, W Greece
Artashat *95 F3* S Armenia
Artemisa *32 B2* La Habana, W Cuba
Artesia *26 D3* New Mexico, SW USA
Arthur's Pass *129 C6 pass* South Island, New Zealand
Artigas *42 D3 prev.* San Eugenio, San Eugenio del Cuareim. Artigas, N Uruguay
Art'ik *95 F2* W Armenia
Artois *68 C2 cultural region* N France
Artsiz *see* Artsyz
Artsyz *86 D4 Rus.* Artsiz. Odes'ka Oblast', SW Ukraine
Arturo Prat *132 A2 Chilean research station* South Shetland Islands, Antarctica
Artvin *95 F2* Artvin, NE Turkey (Türkiye)
Arua *51 B6* NW Uganda
Aruba *36 C1 var.* Oruba. *Dutch self-governing territory* S West Indies
Aru Islands *see* Aru, Kepulauan
Aru, Kepulauan *117 G4 Eng.* Aru Islands; *prev.* Aroe Islands. *island group* E Indonesia
Arunāchal Pradesh *113 G3 prev.* North East Frontier Agency, North East Frontier Agency of Assam. *cultural region* NE India
Arusha *51 C7* Arusha, N Tanzania
Arviat *15 G4 prev.* Eskimo Point. Nunavut, C Canada
Arvidsjaur *62 C4* Norrbotten, N Sweden
Arys' *92 B5 Kaz.* Arys. Türkistan/Turkestan, S Kazakhstan
Arys *see* Arys'
Asadābād *101 F4 var.* Asadābād; *prev.* Chaghasarāy. Konar, E Afghanistan
Asadābād *see* Asadābād
Asad, Buḥayrat al *96 C2 Eng.* Lake Assad. *lake* N Syria
Asahi-dake *108 D2 mountain* Hokkaidō, N Japan
Asahikawa *108 D2* Hokkaidō, N Japan
Asamankese *53 E5* SE Ghana
Āsānsol *113 F4* West Bengal, NE India
Asben *see* Aïr, Massif de l'
Ascension Fracture Zone *47 A5 tectonic Feature* C Atlantic Ocean
Ascension Island *see* St Helena, Ascension and Tristan da Cunha
Ascoli Piceno *74 C4 anc.* Asculum Picenum. Marche, C Italy
Asculum Picenum *see* Ascoli Piceno
'Aseb *50 D4 var.* Assab, *Amh.* Āseb. SE Eritrea
Assen *64 E2* Drenthe, NE Netherlands
Aşgabat *100 C3 prev.* Ashgabat, Ashkhabad, Poltoratsk. *country capital* (Turkmenistan) Ahal Welaýaty, C Turkmenistan
Ashara *see* Al 'Ashārah
Ashburton *129 C6* Canterbury, South Island, New Zealand
Ashburton River *124 A4 river* Western Australia
Ashdod *97 A6 anc.* Azotos, *Lat.* Azotus. Central, W Israel
Asheville *21 E1* North Carolina, SE USA
Ashgabat *see* Aşgabat
Ashkelon *97 A6 prev.* Ashqelon. Southern, C Israel
Ashkhabad *see* Aşgabat
Ashland *24 B4* Oregon, NW USA
Ashland *18 B1* Wisconsin, N USA
Ashmore and Cartier Islands *120 A3 Australian external territory* E Indian Ocean
Ashmyany *85 C5 Rus.* Oshmyany. Hrodzyenskaya Voblasts', W Belarus
Ashqelon *see* Ashkelon
Ash Shadādah *96 D2 var.* Ash Shaddādah, Jisr ash Shadadi, Shaddādī, Shedadi, Tell Shedadi. Al Ḩasakah, NE Syria
Ash Shaddādah *see* Ash Shadādah
Ash Sharah *97 B7 var.* Esh Sharā. *mountain range* W Jordan
Ash Shāriqah *98 D4 Eng.* Sharjah. Ash Shāriqah, NE United Arab Emirates
Ash Shawbak *97 B7* Ma'ān, W Jordan
Ash Shiḥr *99 C6* SE Yemen
Asia *90 continent*
'Āṣī, Nahr Al *see* Orontes
Asinara *74 A4 island* W Italy
Asi Nehri *see* Orontes
Asipovichy *85 D6 Rus.* Osipovichi. Mahilyowskaya Voblasts', C Belarus
Aşkale *95 E3* Erzurum, NE Turkey (Türkiye)
Askersund *63 C6* Örebro, C Sweden
Asmara *see* Asmera
Asmera *50 C4 var.* Asmara. *country capital* (Eritrea) C Eritrea
Aspadana *see* Eşfahān
Asphaltites, Lacus *see* Dead Sea
Aspinwall *see* Colón
Assab *see* 'Aseb
As Sabkhah *96 D2 var.* Sabkha. Ar Raqqah, NE Syria
Assad, Lake *see* Asad, Buḥayrat al
Aş Şafāwī *97 C6* Al Mafraq, N Jordan
Aş Şaḥrā' ash Sharqīyah *see* Sahara el Sharqîya
As Salamīyah *see* Salamīyah
'Assal, Lac *46 E4 lake* C Djibouti
As Salṭ *97 B6 var.* Salt. Al Balqā', NW Jordan
Assamaka *see* Assamakka
Assamakka *53 F2 var.* Assamaka. Agadez, NW Niger
As Samāwah *98 B3 var.* Samawa. Al Muthannā, S Iraq
Assenede *65 B5* Oost-Vlaanderen, NW Belgium
Assiout *see* Asyūṭ
Assiut *see* Asyūṭ
Assling *see* Jesenice
Assouan *see* Aswān
Assu *41 G2 var.* Açu. Rio Grande do Norte, E Brazil
Assuan *see* Aswān
As Sukhnah *96 C3 var.* Sukhne, *Fr.* Soukhné. Ḥimṣ, C Syria
As Sulaymānīyah *98 C3 var.* Sulaimaniya, *Kurd.* Slêmanî. As Sulaymānīyah/Slêmanî, NE Iraq
As Sulayyil *99 B5* Ar Riyāḍ, S Saudi Arabia
Aş Şuwār *96 D2 var.* Şuwār. Dayr az Zawr, E Syria
As Suwaydā' *97 B5 var.* El Suweida, Es Suweida, Suweida, *Fr.* Soueida. As Suwaydā', SW Syria
As Suways *50 B1 var.* Suez; *Ar.* El Suweis. NE Egypt
Asta Colonia *see* Asti
Astacus *see* İzmit
Astana *see* Nur-Sultan
Asta Pompeia *see* Asti
Astarabad *see* Gorgān
Asterābād *see* Gorgān
Asti *74 A2 anc.* Asta Colonia, Asta Pompeia, Hasta Colonia, Hasta Pompeia. Piemonte, NW Italy
Astigi *see* Écija
Astipálaia *see* Astypálaia
Astorga *70 C1 anc.* Asturica Augusta. Castilla y León, N Spain
Astrabad *see* Gorgān
Astrakhan' *89 C7* Astrakhanskaya Oblast', SW Russia
Asturias *70 C1 autonomous community* NW Spain
Asturias *see* Oviedo
Asturica Augusta *see* Astorga
Astypálaia *83 D7 var.* Astipálaia, *It.* Stampalia. *island* Kykládes, Greece, Aegean Sea
Asunción *42 D2 country capital* (Paraguay) Central, S Paraguay
Asuka *132 C2 Japanese research station* Antarctica
Aswān *50 B2 var.* Assouan, Assuan, Aswân; *anc.* Syene. SE Egypt
Aswân *see* Aswān
Asyūṭ *50 B2 var.* Assiout, Assiut, Asyût, Siut; *anc.* Lycopolis. C Egypt
Asyût *see* Asyūṭ
Atacama Desert *42 B2 Eng.* Atacama Desert. *desert* N Chile
Atacama Desert *see* Atacama, Desierto de
Atafu Atoll *123 E3 island* NW Tokelau
Atamyrat *100 D3 prev.* Kerki. Lebap Welaýaty, E Turkmenistan
Aṭâr *52 C2* Adrar, W Mauritania
Atas Bogd *104 D3 mountain* SW Mongolia
Atascadero *25 B7* California, W USA
Atatürk Baraji *95 E4 reservoir* S Turkey (Türkiye)
Atbara *50 C3 var.* 'Aṭbārah. River Nile, NE Sudan
'Aṭbārah/'Aṭbarah, Nahr *see* Atbara
Atbasar *92 C4* Akmola, N Kazakhstan
Atchison *23 F4* Kansas, C USA
Aternum *see* Pescara
Ath *65 B6 var.* Aat. Hainaut, SW Belgium
Athabasca *15 E5* Alberta, SW Canada
Athabasca *15 E5 var.* Athabaska. *river* Alberta, SW Canada
Athabasca, Lake *15 F4 lake* Alberta/Saskatchewan, SW Canada
Athabaska *see* Athabasca
Athenae *see* Athína
Athens *21 E2* Georgia, SE USA
Athens *18 D4* Ohio, N USA
Athens *27 G3* Texas, SW USA
Athens *see* Athína
Atherton *126 D3* Queensland, NE Australia
Athína *83 C6 Eng.* Athens, *prev.* Athínai; *anc.* Athenae. *country capital* (Greece) Attikí, C Greece
Athínai *see* Athína
Athlone *67 B5 Ir.* Baile Átha Luain. C Ireland
Ath Thawrah *96 C2 var.* Madīnat ath Thawrah. Ar Raqqah, N Syria
Ati *54 C3* Batha, C Chad
Atikokan *16 B4* Ontario, S Canada
Atka *93 G3* Magadanskaya Oblast', E Russia
Atka *14 A3* Atka Island, Alaska, USA
Atlanta *20 D2 state capital* Georgia, SE USA
Atlanta *27 H2* Texas, SW USA
Atlantic City *19 F4* New Jersey, NE USA
Atlantic-Indian Basin *45 D7 undersea basin* SW Indian Ocean
Atlantic-Indian Ridge *47 B8 undersea ridge* SW Indian Ocean
Atlantic Ocean *44 B4 ocean*
Atlas Mountains *48 C2 mountain range* NW Africa
Atlasovo *93 H3* Kamchatskaya Oblast', E Russia
Atlas Saharien *48 D2 var.* Saharan Atlas. *mountain range* Algeria/Morocco
Atlas, Tell *see* Atlas Tellien
Atlas Tellien *80 C3 Eng.* Tell Atlas. *mountain range* N Algeria
Atlin *14 D4* British Columbia, W Canada
Aṭ Ṭafīlah *97 B7 var.* Et Tafila, Tafila. Aṭ Ṭafīlah, W Jordan
Aṭ Ṭā'if *99 B5* Makkah, W Saudi Arabia
Attaleia/Attalia *see* Antalya
At Tall al Abyaḍ *see* Tall Abyaḍ
Aṭ Ṭanf *96 D4* Ḥimṣ, S Syria
Attapu *115 E5 var.* Samakhixai, Attopeu. Attapu, S Laos
Attawapiskat *16 C3* Ontario, C Canada
Attawapiskat *16 C3 river* Ontario, S Canada
At Tibnī *96 D2 var.* Tibnī. Dayr az Zawr, NE Syria
Attopeu *see* Attapu
Attu Island *14 A2 island* Aleutian Islands, Alaska, USA
Atyraū *92 B4 var.* Atyrau, *prev.* Gur'yev. Atyrau, W Kazakhstan
Atyrau *see* Atyraū
Aubagne *69 D6 anc.* Albania. Bouches-du-Rhône, SE France
Aubange *65 D8* Luxembourg, SE Belgium
Aubervilliers *68 E1* Seine-St-Denis, Île-de-France, N France Europe
Auburn *24 B2* Washington, NW USA
Auch *69 B6 Lat.* Augusta Auscorum, Elimberrum. Gers, S France
Auckland *128 D2* Auckland, North Island, New Zealand
Auckland Islands *120 C5 island group* S New Zealand
Audern *see* Audru
Audincourt *68 E4* Doubs, E France
Audru *84 D2 Ger.* Audern. Pärnumaa, SW Estonia
Augathella *127 D5* Queensland, E Australia
Augsbourg *see* Augsburg
Augsburg *73 C6 Fr.* Augsbourg; *anc.* Augusta Vindelicorum. Bayern, S Germany
Augusta *125 A7* Western Australia
Augusta *21 E2* Georgia, SE USA
Augusta *19 G2 state capital* Maine, NE USA
Augusta *see* London
Augusta Auscorum *see* Auch
Augusta Emerita *see* Mérida
Augusta Praetoria *see* Aosta
Augusta Trajana *see* Stara Zagora
Augusta Treverorum *see* Trier
Augusta Vangionum *see* Worms
Augusta Vindelicorum *see* Augsburg
Augustobona Tricassium *see* Troyes
Augustodurum *see* Bayeux
Augustoritum Lemovicensium *see* Limoges
Augustów *76 E2 Rus.* Avgustov. Podlaskie, NE Poland
Aulie Ata/Auliye-Ata *see* Taraz
Aunglan *114 B4 var.* Allanmyo, Myaydo. Magway, C Myanmar (Burma)
Auob *56 B4 var.* Oup. *river* Namibia/South Africa
Aurangābād *112 D5* Mahārāshtra, C India
Auray *68 A3* Morbihan, NW France
Aurelia Aquensis *see* Baden-Baden
Aurelianum *see* Orléans
Aurès, Massif de l' *80 C4 mountain range* NE Algeria
Aurillac *69 C5* Cantal, C France
Aurium *see* Ourense
Aurora *37 F2* NW Guyana
Aurora *22 D4* Colorado, C USA
Aurora *18 B3* Illinois, N USA
Aurora *23 G5* Missouri, C USA
Aurora *see* Maéwo, Vanuatu
Aus *56 B4* Karas, SW Namibia
Ausa *see* Vic
Aussig *see* Ústí nad Labem
Austin *23 G3* Minnesota, N USA
Austin *27 G3 state capital* Texas, SW USA
Australes, Archipel des *see* Australes, Îles
Australes et Antarctiques Françaises, Terres *see* French Southern and Antarctic Lands
Australes, Îles *121 F4 var.* Archipel des Australes, Îles Tubuai, Tubuai Islands, *Eng.* Austral Islands. *island group* SW French Polynesia
Austral Fracture Zone *121 H4 tectonic feature* S Pacific Ocean
Australia *120 A4 off.* Commonwealth of Australia. *country*
Australia, Commonwealth of *see* Australia
Australian Alps *127 C7 mountain range* SE Australia
Australian Capital Territory *127 D7 prev.* Federal Capital Territory. *territory* SE Australia
Australie, Bassin Nord de l' *see* North Australian Basin
Austral Islands *see* Australes, Îles
Austrava *see* Ostrov
Austria *73 D7 off.* Republic of Austria, *Ger.* Österreich. *country* C Europe
Austria, Republic of *see* Austria
Autesiodorum *see* Auxerre
Autissiodorum *see* Auxerre
Autricum *see* Chartres
Auvergne *69 C5 cultural region* C France
Auxerre *68 C4 anc.* Autesiodorum, Autissiodorum. Yonne, C France
Avaricum *see* Bourges
Avarua *123 G5 dependent territory capital* (Cook Islands) Rarotonga, S Cook Islands
Avasfelsőfalu *see* Negrești-Oaș
Ávdira *82 C3* Anatolikí Makedonía kai Thráki, NE Greece
Aveiro *70 B2 anc.* Talabriga. Aveiro, W Portugal
Avela *see* Ávila
Avellino *75 D5 anc.* Abellinum. Campania, S Italy
Avenio *see* Avignon
Avesta *63 C6* Dalarna, C Sweden
Aveyron *69 C6 river* S France
Avezzano *74 C4* Abruzzo, C Italy
Avgustov *see* Augustów
Aviemore *66 C3* N Scotland, United Kingdom
Avignon *69 D6 anc.* Avenio. Vaucluse, SE France
Ávila *70 D3 var.* Avila; *anc.* Abela, Abula, Abyla, Avela. Castilla y León, C Spain
Avilés *70 C1* Asturias, NW Spain
Avranches *68 B3* Manche, N France
Avveel *see* Ivalo, Finland
Avvil *see* Ivalo
Awaji-shima *109 C6 island* SW Japan
Āwash *51 D5* Āfar, NE Ethiopia
Awbārī *49 F3* SW Libya
Ax *see* Dax
Axel *65 B5* Zeeland, SW Netherlands
Axel Heiberg Island *15 E1 var.* Axel Heiburg. *island* Nunavut, N Canada
Axel Heiburg *see* Axel Heiberg Island
Áxios *see* Vardar
Ayacucho *38 D4* Ayacucho, S Peru
Ayagoz *92 C5 var.* Ayaguz, *Kaz.* Ayakoz. *river* E Kazakhstan
Ayamonte *70 C4* Andalucía, S Spain
Ayaviri *39 E4* Puno, S Peru
Aydarko'l Ko'li *101 E2 Rus.* Ozero Aydarkul'. *lake* C Uzbekistan
Aydarkul', Ozero *see* Aydarko'l Ko'li
Aydın *94 A4 var.* Aïdin; *anc.* Tralles Aydin. Aydın, SW Turkey (Türkiye)
Ayers Rock *see* Uluru
Ayeyarwady *114 B2 var.* Irrawaddy. *river* W Myanmar (Burma)
Ayiá *see* Agiá
Áyios Evstrátios *see* Ágios Efstrátios
Áyios Nikólaos *see* Ágios Nikólaos
Ayorou *53 E3* Tillabéri, W Niger
'Ayoûn el 'Atroûs *52 D3 var.* Aïoun el Atrous, Aïoun el Atroûss. Hodh el Gharbi, SE Mauritania
Ayr *66 C4* W Scotland, United Kingdom
Ayteke Bi *92 B4 Kaz.* Zhangaqazaly; *prev.* Novokazalinsk. Kzylorda, SW Kazakhstan
Aytos *82 E2* Burgas, E Bulgaria
Ayutthaya *115 C5 var.* Phra Nakhon Si Ayutthaya. Phra Nakhon Si Ayutthaya, C Thailand
Ayvalık *94 A3* Balıkesir, W Turkey (Türkiye)
Azahar, Costa del *71 F3 coastal region* E Spain
Azaouâd *53 E3 desert* C Mali
Azärbaycan/Azärbaycan Respublikasī *see* Azerbaijan
A'zāz *see* I'zāz
Azerbaijan *95 G2 off.* Republic of Azerbaijan, *Az.* Azärbaycan, Azärbaycan Respublikası; *prev.* Azerbaijan SSR. *country* SE Asia
Azerbaijan, Republic of *see* Azerbaijan
Azerbaijan SSR *see* Azerbaijan
Azimabad *see* Patna
Azizie *see* Telish
Azogues *38 B2* Cañar, S Ecuador
Azores *70 A4 var.* Açores, Ilhas dos Açores, *Port.* Arquipélago dos Açores. *island group* Portugal, NE Atlantic Ocean
Azores-Biscay Rise *58 A3 undersea rise* E Atlantic Ocean
Azotos/Azotus *see* Ashdod
Azoum, Bahr *54 C3 seasonal river* SE Chad
Azov, Sea of *81 H1 Rus.* Azovskoye More, *Ukr.* Azovske More. *sea* NE Black Sea
Azovske More/Azovskoye More *see* Azov, Sea of
Azraq, Wāḥat al *97 C6 oasis* N Jordan
Aztec *26 C1* New Mexico, SW USA
Azuaga *70 C4* Extremadura, W Spain
Azuero, Península de *31 F5 peninsula* S Panama
Azul *43 D5* Buenos Aires, E Argentina
Azur, Côte d' *69 E6 coastal region* SE France
'Azza *see* Gaza
Az Zaqāzīq *50 B1 var.* Zagazig. N Egypt
Az Zarqā' *97 B6* NW Jordan
Az Zāwiyah *49 F2 var.* Zawia. NW Libya
Az Zilfī *98 B4* Ar Riyāḍ, N Saudi Arabia

B

Baalbek *96 B4 var.* Ba'labakk; *anc.* Heliopolis. E Lebanon
Baardheere *51 D6 var.* Bardere, *It.* Bardera. Gedo, SW Somalia
Baarle-Hertog *65 C5* Antwerpen, N Belgium
Baarn *64 C3* Utrecht, C Netherlands
Babadag *86 D5* Tulcea, SE Romania
Babahoyo *38 B2 prev.* Bodegas. Los Ríos, C Ecuador
Bābā, Kōh-e *101 E4 mountain range* C Afghanistan
Babayevo *88 B4* Vologodskaya Oblast', NW Russia
Babeldaob *122 A1 var.* Babeldaop, Babelthuap. *island* N Palau
Babeldaop *see* Babeldaob
Bab el Mandeb *99 B7 strait* Gulf of Aden/Red Sea
Babelthuap *see* Babeldaob
Babian Jiang *see* Black River
Babruysk *85 D7 Rus.* Bobruysk. Mahilyowskaya Voblasts', E Belarus
Babuyan Channel *117 E1 channel* N Philippines
Babuyan Islands *117 E1 island group* N Philippines
Bacabal *41 F2* Maranhão, E Brazil
Bacău *86 C4 Hung.* Bákó. Bacău, NE Romania
Bắc Bô, Vịnh *see* Tonkin, Gulf of
Bắc Giang *114 D3* Ha Bắc, N Vietnam
Bacheykava *85 D5 Rus.* Bocheykovo. Vitsyebskaya Voblasts', N Belarus
Back *15 F3 river* Nunavut, N Canada
Bačka Palanka *78 D3 prev.* Palanka. Serbia, NW Serbia
Bačka Topola *78 D3 Hung.* Topolya; *prev. Hung.* Bácstopolya. Vojvodina, N Serbia
Bac Liêu *115 D6 var.* Vinh Loi. Minh Hai, S Vietnam
Bacolod *103 E4 off.* Bacolod City. Negros, C Philippines
Bacolod City *see* Bacolod
Bácsszenttamás *see* Srbobran
Bácstopolya *see* Bačka Topola
Bactra *see* Balkh
Badajoz *70 C4 anc.* Pax Augusta. Extremadura, W Spain
Baden-Baden *73 B6 anc.* Aurelia Aquensis. Baden-Württemberg, SW Germany
Bad Freienwalde *72 D3* Brandenburg, NE Germany
Badger State *see* Wisconsin
Bad Hersfeld *72 B4* Hessen, C Germany
Bad Homburg *see* Bad Homburg vor der Höhe
Bad Homburg vor der Höhe *73 B5 var.* Bad Homburg. Hessen, W Germany
Bad Ischl *73 D7* Oberösterreich, N Austria
Bad Krozingen *73 A6* Baden-Württemberg, SW Germany
Badlands *22 D2 physical region* North Dakota/ South Dakota, N USA
Badu Island *126 C1 island* Queensland, NE Australia
Bad Vöslau *73 E6* Niederösterreich, NE Austria
Baeterrae/Baeterrae Septimanorum *see* Béziers
Baetic Cordillera/Baetic Mountains *see* Béticos, Sistemas
Bafatá *52 C4* C Guinea-Bissau
Baffin Bay *15 G2 bay* Canada/Greenland
Baffin Island *15 G2 island* Nunavut, NE Canada
Bafing *52 C3 river* W Africa
Bafoussam *54 A4* Ouest, W Cameroon
Bafra *94 D2* Samsun, N Turkey (Türkiye)
Bāft *98 D4* Kermān, S Iran
Bagaces *30 D4* Guanacaste, NW Costa Rica
Bagdad *see* Baghdād
Bagé *41 E5* Rio Grande do Sul, S Brazil
Baghdād *98 B3 var.* Bagdad, *country capital* (Iraq) Baghdād, C Iraq
Baghlān *101 E3* Baghlān, NE Afghanistan
Bago *114 B4 var.* Pegu. Bago, SW Myanmar (Burma)
Bagoé *52 D4 river* Ivory Coast/Mali
Bagrationovsk *84 A4 Ger.* Preussisch Eylau. Kaliningradskaya Oblast', W Russia
Bagrax Hu *see* Bosten Hu
Baguio *117 E1 off.* Baguio City. Luzon, N Philippines
Baguio City *see* Baguio
Bagzane, Monts *53 F3 mountain* N Niger
Bahama Islands *see* Bahamas, The
Bahamas, The *32 C2 off.* Commonwealth of The Bahamas. *country* N West Indies
Bahamas *13 D6 var.* Bahama Islands. *island group* N West Indies
Bahamas, Commonwealth of The *see* Bahamas, The
Baharly *100 C3 var.* Bäherden, *Rus.* Bakharden; *prev.* Bakherden. Ahal Welaýaty, C Turkmenistan
Bahawalpur *112 C2* Punjab, E Pakistan
Bäherden *see* Baharly
Bahia *41 F3 off.* Estado da Bahia. *state/region* E Brazil
Bahía Blanca *43 C5* Buenos Aires, E Argentina
Bahia, Estado da *see* Bahia
Bahir Dar *50 C4 var.* Bahr Dar, Bahrdar Giyorgis. Āmara, N Ethiopia
Bahraich *113 E3* Uttar Pradesh, N India
Bahrain *98 C4 off.* Kingdom of Bahrain, Mamlakat al Baḥrayn, *Ar.* Al Baḥrayn, *prev.* Bahrein; *anc.* Tylos, Tyros. *country* SW Asia
Bahrain, Kingdom of *see* Bahrain
Bahrayn, Mamlakat al *see* Bahrain
Bahr Dar/Bahrdar Giyorgis *see* Bahir Dar
Bahrein *see* Bahrain
Bahr el, Azraq *see* Blue Nile
Bahr Tabariya, Sea of *see* Kinneret, Yam
Bahushewsk *85 E6 Rus.* Bogushëvsk. Vitsyebskaya Voblasts', NE Belarus
Baia Mare *86 B3 Ger.* Frauenbach, *Hung.* Nagybánya; *prev.* Neustadt. Maramureș, NW Romania
Baia Sprie *86 B3 Ger.* Mittelstadt, *Hung.* Felsőbánya. Maramureș, NW Romania
Baïbokoum *54 B4* Logone-Oriental, SW Chad
Baidoa *see* Baydhabo
Baie-Comeau *17 E3* Québec, SE Canada
Baikal, Lake *93 E4 Eng.* Lake Baikal. *lake* S Russia
Baikal, Lake *see* Baykal, Ozero
Baile Átha Cliath *see* Dublin
Baile Átha Luain *see* Athlone
Bailén *70 D4* Andalucía, S Spain
Baile na Mainistreach *see* Newtownabbey
Băilești *86 B5* Dolj, SW Romania
Ba Illi *54 B3* Chari-Baguirmi, SW Chad
Bainbridge *20 D3* Georgia, SE USA
Bā'ir *see* Bāyir
Baireuth *see* Bayreuth
Bairnsdale *127 C7* Victoria, SE Australia
Baishan *107 E3 prev.* Hunjiang. Jilin, NE China
Baiyin *106 B4* Gansu, C China
Baja *77 C7* Bács-Kiskun, S Hungary
Baja California *26 A4 Eng.* Lower California. *peninsula* NW Mexico
Baja California Norte *28 B2 state* NW Mexico
Bajo Boquete *see* Boquete
Bajram Curri *79 D5* Kukës, N Albania
Bakala *54 C4* Ouaka, C Central African Republic
Bakan *see* Shimonoseki
Baker *24 C3* Oregon, NW USA
Baker Island *123 E2 US unincorporated territory* W Polynesia
Baker Lake *15 F3* Nunavut, N Canada
Bakersfield *25 C7* California, W USA
Bakharden *see* Baharly
Bakhchisaray *see* Bakhchysarai
Bakhchysarai *87 F5 prev.* Bakhchysaray, *Rus.* Bakhchisaray. Respublika Krym, S Ukraine
Bakhchysaray *see* Bakhchysarai
Bakherden *see* Baharly
Bakhmach *87 F1* Chernihivs'ka Oblast', N Ukraine
Bākhtarān *see* Kermānshāh
Bakı *95 H2 Eng.* Baku. *country capital* (Azerbaijan) E Azerbaijan
Bákó *see* Bacău
Bakony *77 C7 Eng.* Bakony Mountains, *Ger.* Bakonywald. *mountain range* W Hungary
Bakony Mountains/Bakonywald *see* Bakony
Baku *see* Bakı
Bakwanga *see* Mbuji-Mayi
Balabac Island *107 C8 island* W Philippines
Balabac, Selat *see* Balabac Strait
Balabac Strait *116 D2 var.* Selat Balabac. *strait* Malaysia/Philippines
Ba'labakk *see* Baalbek
Balaguer *71 F2* Cataluña, NE Spain
Balakovo *89 C6* Saratovskaya Oblast', W Russia
Bālā Morghāb *see* Bālā Murghāb
Bālā Murghāb *100 D4 prev.* Bālā Morghāb. Laghmān, NW Afghanistan
Balashov *89 B6* Saratovskaya Oblast', W Russia
Balasore *see* Bāleshwar

Balaton, Lake 77 C7 var. Lake Balaton, Ger. Plattensee. lake W Hungary
Balaton, Lake see Balaton
Balbina, Represa 40 D1 reservoir NW Brazil
Balboa 31 G4 Panamá, C Panama
Balcarce 43 D5 Buenos Aires, E Argentina
Balclutha 129 B7 Otago, South Island, New Zealand
Baldy Mountain 22 C1 mountain Montana, NW USA
Bâle see Basel
Balearic Plain see Algerian Basin
Baleares Major see Mallorca
Balearic Islands 71 G3 Eng. Balearic Islands. island group Spain, W Mediterranean Sea
Balearic Islands see Baleares, Islas
Balearis Minor see Menorca
Baleine, Rivière à la 17 E2 river Québec, E Canada
Balen 65 C5 Antwerpen, N Belgium
Bāleshwar 113 F4 prev. Balasore. Odisha, E India
Bălgrad see Alba Iulia
Bali 116 D5 island C Indonesia
Balıkesir 94 A3 Balıkesir, W Turkey (Türkiye)
Balīkh, Nahr 96 C2 river N Syria
Balikpapan 116 D4 Borneo, C Indonesia
Balkanabat 100 B2 Rus. Nebitdag. Balkan Welaýaty, W Turkmenistan
Balkan Mountains 82 C2 Bul./Croatian Stara Planina. mountain range Bulgaria/Serbia
Balkash see Balqash
Balkash, Ozero see Balqash Köli
Balkh 101 E3 anc. Bactra. Balkh, N Afghanistan
Balkhash, Lake see Balqash Köli
Balladonia 125 C6 Western Australia
Ballarat 127 C7 Victoria, SE Australia
Balleny Islands 132 B5 island group Antarctica
Ballinger 27 F3 Texas, SW USA
Balochistan see Baluchistan
Balqash 92 C5 Rus. Balkash. Karagandy, SE Kazakhstan see Balkhash
Balqash Köli 92 C5 Eng. Lake Balkhash, Rus. Ozero Balkash. lake SE Kazakhstan
Balş 86 B5 Olt, S Romania
Balsas 41 F2 Maranhão, E Brazil
Balsas, Río 29 E5 var. Río Mexcala. river S Mexico
Bal'shavik 85 D7 Rus. Bol'shevik. Homyel'skaya Voblasts', SE Belarus
Balta 86 D3 Odes'ka Oblast', SW Ukraine
Bălţi 86 D3 Rus. Bel'tsy. N Moldova
Baltic Port see Paldiski
Baltic Sea 63 C7 Ger. Ostee, Rus. Baltiskoye More. sea N Europe
Baltimore 19 F4 Maryland, NE USA
Baltischport/Baltiski see Paldiski
Baltiskoye More see Baltic Sea
Baltkrievija see Belarus
Baluchistan 112 B3 var. Balochistan, Beluchistan. province SW Pakistan
Balvi 84 D4 Balvi, NE Latvia
Balykchy 101 G2 Kir. Ysyk-Köl; prev. Issyk-Kul', Rybach'ye. Issyk-Kul'skaya Oblast', NE Kyrgyzstan
Balzers 72 E2 S Liechtenstein
Bam 98 E4 Kermān, SE Iran
Bamako 52 D4 country capital (Mali) Capital District, SW Mali
Bambari 54 C4 Ouaka, C Central African Republic
Bamberg 73 C5 Bayern, SE Germany
Bamenda 54 A4 Nord-Ouest, W Cameroon
Banaba 122 D2 var. Ocean Island. island Tungaru, W Kiribati
Banaras see Vārānasi
Bandaaceh 116 A3 var. Banda Atjeh; prev. Koetaradja, Kutaradja, Kutaraja. Sumatera, W Indonesia
Banda Atjeh see Bandaaceh
Banda, Laut see Banda Sea
Bandama 52 D5 var. Bandama Fleuve. river S Ivory Coast
Bandama Fleuve see Bandama
Bandar 'Abbās see Bandar-e 'Abbās
Bandarbeyla 51 E5 var. Bender Beila, Bender Beyla. Bari, NE Somalia
Bandar-e 'Abbās 98 D4 var. Bandar 'Abbās; prev. Gombroon. Hormozgān, S Iran
Bandar-e Būshehr 98 C4 var. Būshehr, Eng. Bushire. Būshehr, S Iran
Bandar-e Kangān 98 D4 var. Kangān. Būshehr, S Iran
Bandar-e Khamīr 98 D4 Hormozgān, S Iran
Bandar-e Langeh see Bandar-e Lengeh
Bandar-e Lengeh 98 D4 var. Bandar-e Langeh, Lingeh. Hormozgān, S Iran
Bandar Kassim see Boosaaso
Bandar Lampung 116 C4 var. Bandarlampung, Tanjungkarang-Telukbetung; prev. Tandjoengkarang, Tanjungkarang, Teloekbetoeng, Telukbetung. Sumatera, W Indonesia
Bandarlampung see Bandar Lampung
Bandar Maharani see Muar
Bandar Masulipatnam see Machilipatnam
Bandar Penggaram see Batu Pahat
Bandar Seri Begawan 116 D3 prev. Brunei Town. country capital (Brunei) N Brunei
Banda Sea 117 F5 var. Laut Banda. sea E Indonesia
Bandiagara 53 E3 Mopti, C Mali
Bandırma 94 A3 var. Penderma. Balıkesir, NW Turkey (Türkiye)
Bandjarmasin see Banjarmasin
Bandoeng see Bandung
Bandundu 55 C6 prev. Banningville. Bandundu, W Dem. Rep. Congo
Bandung 116 C5 prev. Bandoeng. Jawa, C Indonesia
Bangalore see Bengalūru
Bangassou 54 D4 Mbomou, SE Central African Republic
Banggai, Kepulauan 117 E4 island group C Indonesia
Banghāzī 49 G2 Eng. Bengazi, Benghazi, It. Bengasi. NE Libya
Bangka, Pulau 116 C4 island W Indonesia
Bangkok see Krung Thep
Bangkok, Bight of see Krung Thep, Ao
Bangladesh 113 G3 off. People's Republic of Bangladesh; prev. East Pakistan. country S Asia
Bangladesh, People's Republic of see Bangladesh
Bangor 67 C6 NW Wales, United Kingdom
Bangor 67 B5 Ir. Beannchar. E Northern Ireland, United Kingdom
Bangor 19 G2 Maine, NE USA
Bang Pla Soi see Chon Buri
Bangui 55 B5 country capital (Central African Republic) Ombella-Mpoko, SW Central African Republic
Bangweulu, Lake 51 B8 var. Lake Bengweulu. lake N Zambia
Ban Hat Yai see Hat Yai
Ban Hin Heup 114 C4 Viangchan, C Laos
Ban Houayxay/Ban Houei Sai see Houayxay
Ban Hua Hin 115 C6 var. Hua Hin. Prachuap Khiri Khan, SW Thailand
Bani 52 D3 river S Mali
Banias see Bāniyās
Banijska Palanka see Glina
Banī Suwayf 50 B2 var. Beni Suef. N Egypt
Bāniyās 96 B3 var. Banias, Baniyas, Paneas. Tarṭūs, W Syria
Banjak, Kepulauan see Banyak, Kepulauan
Banja Luka 78 B3 Republika Srpska, NW Bosnia and Herzegovina
Banjarmasin 116 D4 prev. Bandjarmasin. Borneo, C Indonesia
Banjul 52 B3 prev. Bathurst. country capital (The Gambia) W The Gambia
Banks, Îles see Banks Islands
Banks Island 15 E2 island Northwest Territories, NW Canada
Banks Islands 122 D4 Fr. Îles Banks. island group N Vanuatu
Banks Lake 24 B1 reservoir Washington, NW USA
Banks Peninsula 129 C6 peninsula South Island, New Zealand
Banks Strait 127 C8 strait SW Tasman Sea
Bānkura 113 F4 West Bengal, NE India
Ban Mak Khaeng see Udon Thani
Banmo see Bhamo
Banningville see Bandundu
Bañolas see Banyoles
Ban Pak Phanang see Pak Phanang
Ban Sichon see Sichon
Banská Bystrica 77 C6 Ger. Neusohl, Hung. Besztercebánya. Banskobystricky Kraj, C Slovakia
Bantry Bay 67 A7 Ir. Bá Bheanntraí. bay SW Ireland
Banya 82 E2 Burgas, E Bulgaria
Banyak, Kepulauan 116 A3 prev. Kepulauan Banjak. island group NW Indonesia
Banyo 54 B4 Adamaoua, NW Cameroon
Banyoles 71 G2 var. Bañolas. Cataluña, NE Spain
Banzare Seamounts 119 C7 seamount range S Indian Ocean
Banzart see Bizerte
Baoji 106 B4 var. Pao-chi, Paoki. Shaanxi, C China
Baoro 54 B4 Nana-Mambéré, W Central African Republic
Baoshan 106 A6 var. Pao-shan. Yunnan, SW China
Baotou 105 F3 var. Pao-t'ou, Paotow. Nei Mongol Zizhiqu, N China
Ba'qūbah 98 B3 var. Qubba. Diyālá, C Iraq
Baquerizo Moreno see Puerto Baquerizo Moreno
Bar 79 C5 It. Antivari. S Montenegro
Baraawe 51 D6 It. Brava. Shabeellaha Hoose, S Somalia
Bārāmati 112 C5 Mahārāshtra, W India
Baranavichy 85 B6 Pol. Baranowicze, Rus. Baranovichi. Brestskaya Voblasts', SW Belarus
Baranovichi/Baranowicze see Baranavichy
Barbados 33 G1 country SE West Indies
Barbastro 71 F2 Aragón, NE Spain
Barbate de Franco 70 C5 Andalucía, S Spain
Barbuda 33 G3 island N Antigua and Barbuda
Barcaldine 126 C4 Queensland, E Australia
Barcarozsnyó see Râşnov
Barcău see Berettyó
Barce see Al Marj
Barcelona 71 G2 anc. Barcino, Barcinona. Cataluña, E Spain
Barcelona 37 E2 Anzoátegui, NE Venezuela
Barcino/Barcinona see Barcelona
Barcoo see Cooper Creek
Barcs 77 C7 Somogy, SW Hungary
Bardaï 54 C1 Borkou-Ennedi-Tibesti, N Chad
Bardejov 77 D5 Ger. Bartfeld, Hung. Bártfa. Presovský Kraj, E Slovakia
Bardera/Bardere see Baardheere
Barduli see Barletta
Bareilly 113 E3 var. Bareli. Uttar Pradesh, N India
Bareli see Bareilly
Barendrecht 64 C4 Zuid-Holland, SW Netherlands
Barentin 68 C3 Seine-Maritime, N France
Barentsburg 61 G2 Spitsbergen, W Svalbard
Barentsevo More/Barents Havet see Barents Sea
Barentsøya 61 G2 island E Svalbard
Barents Sea 88 C2 Nor. Barents Havet, Rus. Barentsevo More. sea Arctic Ocean
Bar Harbor 19 H2 Mount Desert Island, Maine, NE USA
Bari 75 E5 var. Bari delle Puglie; anc. Barium. Puglia, SE Italy
Bāridah see Al Bāridah
Bari delle Puglie see Bari
Barikot see Barī Kōṭ
Barī Kōṭ 101 F4 var. Barikot, Barīkowṭ. Konar, NE Afghanistan
Barīkowṭ see Barī Kōṭ
Barillas 30 A2 var. Santa Cruz Barillas. Huehuetenango, NW Guatemala
Barinas 36 C2 Barinas, W Venezuela
Barisal see Barishal
Barisan, Pegunungan 116 B4 mountain range Sumatera, W Indonesia
Barishal 113 G4 prev. Barisal. Barishal, S Bangladesh
Barito, Sungai 116 D4 river Borneo, C Indonesia
Barium see Bari
Barka see Al Marj
Barkly Tableland 126 B3 plateau Northern Territory/Queensland, N Australia
Bârlad 86 D4 prev. Birlad. Vaslui, E Romania
Barlavento, Ilhas de 52 A2 var. Windward Islands. island group N Cape Verde (Cabo Verde)
Bar-le-Duc 68 D3 var. Bar-sur-Ornain. Meuse, NE France
Barlee, Lake 125 B6 lake Western Australia
Barlee Range 124 A4 mountain range Western Australia
Barletta 75 D5 anc. Barduli. Puglia, SE Italy
Barlinek 76 B3 Ger. Berlinchen. Zachodnio-pomorskie, NW Poland
Barmen-Elberfeld see Wuppertal
Barmouth 67 C6 NW Wales, United Kingdom
Barnaul 92 D4 Altayskiy Kray, C Russia
Barnet 67 A7 United Kingdom
Barnstaple 67 C7 SW England, United Kingdom
Baroda see Vadodara
Baroghil Pass 101 F3 var. Kowtal-e Barowghil. pass Afghanistan/Pakistan
Baron'ki 85 E7 Rus. Boron'ki. Mahilyowskaya Voblasts', E Belarus
Barowghil, Kowtal-e see Baroghil Pass
Barquisimeto 36 C2 Lara, NW Venezuela
Barra 66 B3 island NW Scotland, United Kingdom
Barra de Río Grande 31 E3 Región Autónoma Atlántico Sur, E Nicaragua
Barranca 38 C3 Lima, W Peru
Barrancabermeja 36 B2 Santander, N Colombia
Barranquilla 36 B1 Atlántico, N Colombia
Barreiro 70 B4 Setúbal, W Portugal
Barrier Range 127 C6 hill range New South Wales, SE Australia
Barrow 14 D2 Alaska, USA
Barrow 67 B6 Ir. An Bhearú. river SE Ireland
Barrow-in-Furness 67 C5 NW England, United Kingdom
Barrow Island 124 A4 island Western Australia
Barstow 25 C7 California, W USA
Bar-sur-Ornain see Bar-le-Duc
Bartang 101 F3 river SE Tajikistan
Bartenstein see Bartoszyce
Bártfa/Bartfeld see Bardejov
Bartica 37 F3 N Guyana
Bartın 94 C2 Bartın, NW Turkey (Türkiye)
Bartlesville 27 G1 Oklahoma, C USA
Bartoszyce 76 D2 Ger. Bartenstein. Warmińsko-mazurskie, NE Poland
Baruun-Urt 105 F2 Sühbaatar, E Mongolia
Barú, Volcán 31 E5 var. Volcán de Chiriquí. volcano W Panama
Barwon River 127 D5 river New South Wales, SE Australia
Barysaw 85 D6 Rus. Borisov. Minskaya Voblasts', NE Belarus
Basarabeasca 86 D4 Rus. Bessarabka. SE Moldova
Basel 73 A7 Eng. Basle, Fr. Bâle. Basel-Stadt, NW Switzerland
Basilan 117 E3 island Sulu Archipelago, SW Philippines
Basle see Basel
Basra see Al Baṣrah
Bassano del Grappa 74 C2 Veneto, NE Italy
Bassein see Pathein
Basseterre 33 G4 country capital (Saint Kitts and Nevis) Saint Kitts, Saint Kitts and Nevis
Basse-Terre 33 G3 dependent territory capital (Guadeloupe) Basse Terre, SW Guadeloupe
Basse Terre 33 G4 island W Guadeloupe
Bassikounou 52 D3 Hodh ech Chargui, SE Mauritania
Bass, Îlots de see Marotiri
Bass Strait 127 C7 strait SE Australia
Bassum 72 B3 Niedersachsen, NW Germany
Bastia 69 E7 Corse, France, C Mediterranean Sea
Bastogne 65 D7 Luxembourg, SE Belgium
Bastrop 20 B2 Louisiana, S USA
Bastyn' 85 B7 Rus. Bostyn'. Brestskaya Voblasts', SW Belarus
Basuo see Dongfang
Basutoland see Lesotho
Bata 55 A5 NW Equatorial Guinea
Batae Coritanorum see Leicester
Batajnica 78 D3 Vojvodina, N Serbia
Batangas 117 E2 off. Batangas City. Luzon, N Philippines
Batangas City see Batangas
Batavia see Jakarta
Bătdâmbâng see Battambang
Batéké, Plateaux 55 B6 plateau S Congo
Bath 67 D7 hist. Akermanceaster; anc. Aquae Calidae, Aquae Solis. SW England, United Kingdom
Bathinda 112 D2 Punjab, NW India
Bathsheba 33 G1 E Barbados
Bathurst 127 D6 New South Wales, SE Australia
Bathurst 17 F4 New Brunswick, SE Canada
Bathurst see Banjul
Bathurst Island 124 D2 island Northern Territory, N Australia
Bathurst Island 15 F2 island Parry Islands, Nunavut, N Canada
Batin, Wadi al 98 C4 dry watercourse SW Asia
Batman 95 E4 var. Iluh. Batman, SE Turkey (Türkiye)
Batna 49 E2 NE Algeria
Baton Rouge 20 B3 state capital Louisiana, S USA
Batroûn 96 A4 var. Al Batrūn. N Lebanon
Battambang 115 C5 Khmer. Bătdâmbâng, NW Cambodia
Batticaloa 110 D3 Eastern Province, E Sri Lanka
Battipaglia 75 D5 Campania, S Italy
Battle Born State see Nevada
Batumi 95 F2 prev. Bat'umi. W Georgia
Bat'umi see Batumi
Batu Pahat 116 B3 prev. Bandar Penggaram. Johor, Peninsular Malaysia
Bauchi 53 G4 Bauchi, NE Nigeria
Bauer Basin 131 F3 undersea basin E Pacific Ocean
Bauska 84 C3 Ger. Bauske. Bauska, S Latvia
Bauske see Bauska
Bautzen 72 D4 Lus. Budyšin. Sachsen, E Germany
Bauzanum see Bolzano
Bavaria see Bayern
Bavarian Alps 73 C7 Ger. Bayrische Alpen. mountain range Austria/Germany
Bavière see Bayern
Bavispe, Río 28 C2 river NW Mexico
Bawīṭī see Al Bawīṭī
Bawku 53 E4 N Ghana
Bayamo 32 C3 Granma, E Cuba
Bayan Har Shan 104 D4 var. Bayan Khar. mountain range C China
Bayanhongor 104 D2 Bayanhongor, C Mongolia
Bayan Khar see Bayan Har Shan
Bayano, Lago 31 G4 lake E Panama
Bay City 18 C3 Michigan, N USA
Bay City 27 G4 Texas, SW USA
Baydhabo 51 D6 var. Baydhowa, Isha Baydhabo, It. Baidoa. Bay, SW Somalia
Baydhowa see Baydhabo
Bayern 73 C6 Eng. Bavaria, Fr. Bavière. state SE Germany
Bayeux 68 B3 anc. Augustodurum. Calvados, N France
Bāyir 97 C7 var. Bā'ir. Ma'ān, S Jordan
Bay Islands 30 C1 Eng. Bay Islands. island group N Honduras
Bay Islands see Bahía, Islas de la
Baymak 89 D6 Respublika Bashkortostan, W Russia
Bayonne 69 A6 anc. Lapurdum. Pyrénées-Atlantiques, SW France
Bayou State see Mississippi
Bayram-Ali see Baýramaly
Baýramaly 100 D3 var. Bayramaly; prev. Bayram-Ali. Mary Welaýaty, S Turkmenistan
Bayreuth 73 C5 var. Baireuth. Bayern, SE Germany
Bayrische Alpen see Bavarian Alps
Bayrūt see Beyrouth
Bay State see Massachusetts
Baysun see Boysun
Bayt Laḩm see Bethlehem
Baytown 27 H4 Texas, SW USA
Baza 71 E4 Andalucía, S Spain
Bazargic see Dobrich
Bazin see Pezinok
Beagle Channel 43 C8 channel Argentina/Chile
Béal Feirste see Belfast
Beannchar see Bangor, Northern Ireland, UK
Bear Island see Bjørnøya
Bear Lake 24 E4 lake Idaho/Utah, NW USA
Beas de Segura 71 E4 Andalucía, S Spain
Beata, Isla 33 E3 island SW Dominican Republic
Beatrice 23 F4 Nebraska, C USA
Beaufort Sea 14 D2 sea Arctic Ocean
Beaufort-Wes see Beaufort West
Beaufort West 56 C5 Afr. Beaufort-Wes. Western Cape, SW South Africa
Beaumont 27 H3 Texas, SW USA
Beaune 68 D4 Côte d'Or, C France
Beauvais 68 C3 anc. Bellovacum, Caesaromagus. Oise, N France
Beaver Island 18 C2 island Michigan, N USA
Beaver Lake 27 H1 reservoir Arkansas, C USA
Beaver River 27 F1 river Oklahoma, C USA
Beaver State see Oregon
Beāwar 112 C3 Rājasthān, N India
Bečej 78 D3 Ger. Altbetsche, Hung. Óbecse, Rácz-Becse; prev. Magyar-Becse, Stari Bečej. Vojvodina, N Serbia
Béchar 48 D2 prev. Colomb-Béchar. W Algeria
Beckley 18 D5 West Virginia, NE USA
Bécs see Wien
Bedford 67 D6 E England, United Kingdom
Bedum 64 E1 Groningen, NE Netherlands
Beehive State see Utah
Be'er Menuha 97 B7 prev. Be'ér Menuḩa. Southern, S Israel
Be'ér Menuḩa see Be'er Menuha
Beernem 65 A5 West-Vlaanderen, NW Belgium
Beersheba see Be'er Sheva
Be'er Sheva 97 A7 var. Beersheba, Ar. Bir es Saba; prev. Be'ér Sheva'. Southern, S Israel
Be'ér Sheva' see Be'er Sheva
Beesel 65 D5 Limburg, SE Netherlands
Beeville 27 G4 Texas, SW USA
Bega 127 D7 New South Wales, SE Australia
Begoml' see Byahoml'
Begovat see Bekobod
Behagle see Laï
Behar see Bihār
Beibu Wan see Tonkin, Gulf of
Beida see Al Bayḑā'
Beihai 106 B6 Guangxi Zhuangzu Zizhiqu, S China
Beijing 106 C3 var. Pei-ching, Peking; prev. Pei-p'ing. country capital (China) Beijing Shi, E China
Beilen 64 E2 Drenthe, NE Netherlands
Beira 57 E3 Sofala, C Mozambique
Beirut see Beyrouth
Beit Lekhem see Bethlehem
Beiuş 86 B3 Hung. Belényes. Bihor, NW Romania
Beja 70 B4 anc. Pax Julia. Beja, SE Portugal
Béjar 70 C3 Castilla y León, N Spain
Bejraburi see Phetchaburi
Bekabad see Bekobod
Békás see Bicaz
Bek-Budi see Qarshi
Békéscsaba 77 D7 Rom. Bichiş-Ciaba. Békés, SE Hungary
Bekobod 101 E2 Rus. Bekabad; prev. Begovat. Toshkent Viloyati, E Uzbekistan
Bela Crkva 78 E3 Ger. Weisskirchen, Hung. Fehértemplom. Vojvodina, W Serbia
Belagāvi 110 B1 prev. Belgaum. Karnātaka, W India
Belarus 85 B6 off. Republic of Belarus, var. Belorussia, Latv. Baltkrievija; prev. Belorussian SSR, Rus. Belorusskaya SSR. country E Europe
Belarus, Republic of see Belarus
Belau see Palau
Belaya Tserkov' see Bila Tserkva
Bełchatów 76 C4 var. Belchatow. Łódzski, C Poland
Belchatow see Bełchatów
Belcher, Îles see Belcher Islands
Belcher Islands 16 C2 Fr. Îles Belcher. island group Nunavut, SE Canada
Beledweyne 51 D5 var. Belet Huen, It. Belet Uen. Hiiraan, C Somalia
Belém 41 F1 var. Pará. state capital Pará, N Brazil
Belén 30 D4 Rivas, SW Nicaragua
Belen 26 D2 New Mexico, SW USA
Belényes see Beiuş
Belet Huen/Belet Uen see Beledweyne
Belfast 67 B5 Ir. Béal Feirste. national capital E Northern Ireland, United Kingdom
Belfield 22 D2 North Dakota, N USA
Belfort 68 E4 Territoire-de-Belfort, E France
Belgard see Białogard
Belgaum see Belagāvi
Belgian Congo see Congo (Democratic Republic of)
België/Belgique see Belgium
Belgium 65 B6 off. Kingdom of Belgium, Dut. België, Fr. Belgique. country NW Europe
Belgium, Kingdom of see Belgium
Belgorod 89 A6 Belgorodskaya Oblast', W Russia
Belgorod-Dnestrovskiy see Bilhorod-Dnistrovskyi
Belgrano II 132 A2 Argentinian research station Antarctica
Belice see Belize/Belize City
Beligrad see Berat
Beli Manastir 78 C3 Hung. Pélmonostor; prev. Monostor. Osijek-Baranja, NE Croatia
Bélinga 55 B5 Ogooué-Ivindo, NE Gabon
Belitung, Pulau 116 C4 island W Indonesia
Belize 30 B1 Sp. Belice; prev. British Honduras, Colony of Belize. country Central America
Belize 30 B1 river Belize/Guatemala
Belize see Belize City
Belize City 30 C1 var. Belize, Sp. Belice. Belize, NE Belize
Belize, Colony of see Belize
Beljak see Villach
Belkofski 14 B3 Alaska, USA
Belle Île 68 A4 island NW France
Belle Isle, Strait of 17 G3 strait Newfoundland and Labrador, E Canada
Bellenz see Bellinzona
Belleville 18 B4 Illinois, N USA
Bellevue 23 F4 Iowa, C USA
Bellevue 24 B2 Washington, NW USA
Bellingham 24 B1 Washington, NW USA
Belling Hausen Mulde see Southeast Pacific Basin
Bellingshausen Abyssal Plain see Bellingshausen Plain
Bellingshausen Plain 131 F5 var. Bellingshausen Abyssal Plain. abyssal plain SE Pacific Ocean
Bellingshausen Sea 132 A3 sea Antarctica
Bellinzona 73 B8 Ger. Bellenz. Ticino, S Switzerland
Bello 36 B2 Antioquia, W Colombia
Bello Horizonte see Belo Horizonte
Bellovacum see Beauvais
Bellville 56 B5 Western Cape, SW South Africa
Belmopan 30 C1 country capital (Belize) Cayo, C Belize
Belogradchik 82 B1 Vidin, NW Bulgaria
Belo Horizonte 41 F4 prev. Bello Horizonte. state capital Minas Gerais, SE Brazil
Belomorsk 88 B3 Respublika Kareliya, NW Russia
Beloretsk 89 D6 Respublika Bashkortostan, W Russia
Belorussia/Belorussian SSR see Belarus
Belorusskaya Gryada see Byelaruskaya Hrada
Belorusskaya SSR see Belarus
Beloshchel'ye see Nar'yan-Mar
Belostok see Białystok
Belovár see Bjelovar
Beloye More 88 C3 Eng. White Sea. sea NW Russia
Belozërsk 88 B4 Vologodskaya Oblast', NW Russia
Belton 27 G3 Texas, SW USA
Bel'tsy see Bălţi
Beluchistan see Baluchistan
Belukha, Gora 92 D5 mountain Kazakhstan/Russia
Belynichi see Byalynichy
Belyy, Ostrov 92 D2 island N Russia
Bemaraha 57 F3 var. Plateau du Bemaraha. mountain range W Madagascar
Bemaraha, Plateau du see Bemaraha
Bemidji 23 F1 Minnesota, N USA
Bemmel 64 D4 Gelderland, SE Netherlands
Benaco see Garda, Lago di
Benares see Vārānasi
Benavente 70 D2 Castilla y León, N Spain
Bend 24 B3 Oregon, NW USA
Bender 86 D4 Rus. Bendery; var. Tighina. E Moldova
Bender Beila/Bender Beyla see Bandarbeyla
Bender Cassim/Bender Qaasim see Boosaaso
Bendern 72 E1 NW Liechtenstein Europe
Bendery see Bender
Bendigo 127 C7 Victoria, SE Australia
Beneschau see Benešov
Beneški Zaliv see Venice, Gulf of
Benešov 77 B5 Ger. Beneschau. Středočeský Kraj, W Czechia (Czech Republic)
Benevento 75 D5 anc. Beneventum, Malventum. Campania, S Italy
Beneventum see Benevento
Bengal, Bay of 102 C4 bay N Indian Ocean
Bengalūru 110 C2 prev. Bangalore. state capital Karnātaka, S India
Bengasi see Banghāzī
Bengbu 106 D5 var. Peng-pu. Anhui, E China
Benghazi see Banghāzī
Bengkulu 116 B4 prev. Bengkoeloe, Benkoelen, Benkulen. Sumatera, W Indonesia
Benguela 56 A2 var. Benguella. Benguela, W Angola
Benguella see Benguela
Bengweulu, Lake see Bangweulu, Lake
Ben Hope 66 B2 mountain N Scotland, United Kingdom
Beni 55 E5 Nord-Kivu, NE Dem. Rep. Congo
Benidorm 71 F4 Comunitat Valenciana, SE Spain
Beni-Mellal 48 C2 C Morocco
Benin 53 F4 off. Republic of Benin; prev. Dahomey. country W Africa
Benin, Bight of 53 F5 gulf W Africa
Benin City 53 F5 Edo, SW Nigeria
Benin, Republic of see Benin
Beni, Río 39 E3 river N Bolivia
Beni Suef see Banī Suwayf
Ben Nevis 66 C3 mountain N Scotland, United Kingdom
Bénoué see Benue
Benson 26 B3 Arizona, SW USA
Bent Jbaïl 97 A5 var. Bint Jubayl. S Lebanon
Benton 20 B1 Arkansas, C USA
Benue 54 B4 Fr. Bénoué. river Cameroon/Nigeria
Beograd 78 D3 Eng. Belgrade. Serbia, N Serbia
Berane 79 D5 prev. Ivangrad. E Montenegro
Berat 79 C6 var. Berati, Croatian Beligrad. Berat, C Albania
Berătău see Berettyó
Berati see Berat
Berau, Teluk 117 G4 var. MacCluer Gulf. bay Papua, E Indonesia
Berbera 50 D4 Sahil, NW Somalia
Berbérati 55 B5 Mambéré-Kadéï, SW Central African Republic
Berck-Plage 68 C2 Pas-de-Calais, N France
Berdiansk 87 G4 var. Berdyans'k, Rus. Berdyansk; prev. Osipenko. Zaporiz'ka Oblast', SE Ukraine
Berdichev see Berdychiv
Berdyans'k/Berdyansk see Berdiansk
Berdychiv 86 D2 Rus. Berdichev. Zhytomyrs'ka Oblast', N Ukraine
Beregovo/Beregszász see Berehove
Berehove 86 B3 Cz. Berehovo, Hung. Beregszász, Rus. Beregovo. Zakarpats'ka Oblast', W Ukraine
Berehovo see Berehove
Bereket 100 B2 prev. Rus. Gazandzhyk, Kazandzhik, Turkm. Gazanjyk. Balkan Welaýaty, W Turkmenistan
Beretău see Berettyó
Berettyó 77 D7 Rom. Barcău; prev. Berătău, Beretău. river Hungary/Romania
Berettyóújfalu 77 D6 Hajdú-Bihar, E Hungary

Berezhany 86 C2 Pol. Brzeżany. Ternopil's'ka Oblast', W Ukraine
Berezina see Byerezino
Berezniki 89 D5 Permskaya Oblast', NW Russia
Berga 71 G2 Cataluña, NE Spain
Bergamo 74 B2 anc. Bergomum. Lombardia, N Italy
Bergara 71 E1 País Vasco, N Spain
Bergen 72 D2 Mecklenburg-Vorpommern, NE Germany
Bergen 64 C2 Noord-Holland, NW Netherlands
Bergen 63 A5 Hordaland, S Norway
Bergen see Mons
Bergerac 69 B5 Dordogne, SW France
Bergeyk 65 C5 Noord-Brabant, S Netherlands
Bergomum see Bergamo
Bergse Maas 64 D4 river S Netherlands
Beringen 65 C5 Limburg, NE Belgium
Beringov Proliv see Bering Strait
Bering Sea 14 A2 sea N Pacific Ocean
Bering Strait 14 C2 Rus. Beringov Proliv. strait Bering Sea/Chukchi Sea
Berja 71 E5 Andalucía, S Spain
Berkeley 25 B6 California, W USA
Berkner Island 132 A2 island Antarctica
Berkovitsa 82 C2 Montana, NW Bulgaria
Berlin 72 D3 country capital (Germany) Berlin, NE Germany
Berlin 19 G2 New Hampshire, NE USA
Berlinchen see Barlinek
Bermejo, Río 42 C2 river N Argentina
Bermeo 71 E1 País Vasco, N Spain
Bermuda 13 D6 var. Bermuda Islands, Bermudas; prev. Somers Islands. UK Overseas Territory NW Atlantic Ocean
Bermuda Islands see Bermuda
Bermuda Rise 13 E6 undersea rise C Sargasso Sea
Bermudas see Bermuda
Bern 73 A7 Fr. Berne. country capital (Switzerland) Bern, W Switzerland
Bernau 72 D3 Brandenburg, NE Germany
Bernburg 72 C4 Sachsen-Anhalt, C Germany
Berne see Bern
Berner Alpen 73 A7 var. Berner Oberland, Eng. Bernese Oberland. mountain range SW Switzerland
Berner Oberland/Bernese Oberland see Berner Alpen
Bernier Island 125 A5 island Western Australia
Beroea see Ḩalab
Berry 68 C4 cultural region C France
Berry Islands 32 C1 island group N The Bahamas
Bertoua 55 B5 Est, E Cameroon
Beru 123 E2 var. Peru. atoll Tungaru, W Kiribati
Berwick-upon-Tweed 66 D4 N England, United Kingdom
Berytus see Beyrouth
Besançon 68 D4 anc. Besontium, Vesontio. Doubs, E France
Beskra see Biskra
Besontium see Besançon
Bessarabka see Basarabeasca
Beszterce see Bistriţa
Besztercebánya see Banská Bystrica
Betafo 57 G3 Antananarivo, C Madagascar
Betanzos 70 B1 Galicia, NW Spain
Bethlehem 56 D4 Free State, C South Africa
Bethlehem 97 B6 var. Beit Lekhem, Ar. Bayt Laḩm, Heb. Bet Leḩem. C West Bank, Middle East
Béticos, Sistemas 70 D4 var. Sistema Penibético, Eng. Baetic Cordillera, Baetic Mountains. mountain range S Spain
Bet Leḩem see Bethlehem
Bétou 55 C5 Likouala, N Congo
Bette, Picco 49 G4 var. Bikkū Bittī, It. Picco Bette. mountain S Libya
Bette, Picco see Bette, Picco
Beulah 18 C2 Michigan, N USA
Beuthen see Bytom
Beveren 65 B5 Oost-Vlaanderen, N Belgium
Beverley 67 D5 E England, United Kingdom
Bexley 67 B8 Bexley, SE England, United Kingdom
Beyla 52 D4 SE Guinea
Beyrouth 96 A4 var. Bayrūt, Eng. Beirut; anc. Berytus. country capital (Lebanon) W Lebanon
Beyşehir 94 B4 Konya, SW Turkey (Türkiye)
Beyşehir Gölü 94 B4 lake C Turkey (Türkiye)
Béziers 69 C6 anc. Baeterrae, Baeterrae Septimanorum, Julia Beterrae. Hérault, S France
Bezmein see Abadan
Bezwada see Vijayawāda
Bhadrāvati 110 C2 Karnātaka, SW India
Bhāgalpur 113 F3 Bihār, NE India
Bhaktapur 113 F3 Central, C Nepal
Bhamo 114 B2 var. Banmo. Kachin State, N Myanmar (Burma)
Bhārat see India
Bharati 132 D3 Indian research station. Antarctica
Bharūch 112 C4 Gujarāt, W India
Bhaunagar see Bhāvnagar
Bhāvnagar 112 C4 prev. Bhaunagar. Gujarāt, W India
Bheanntraí, Bá see Bantry Bay
Bhopāl 112 D4 state capital Madhya Pradesh, C India
Bhubaneshwar 113 F5 prev. Bhubaneswar, Bhuvaneshwar. state capital Odisha, E India
Bhubaneswar see Bhubaneshwar
Bhuket see Phuket
Bhusaval see Bhusāwal
Bhusāwal 112 D4 prev. Bhusaval. Mahārāshtra, C India
Bhutan 113 G3 off. Kingdom of Bhutan, var. Druk-yul. country S Asia
Bhutan, Kingdom of see Bhutan
Bhuvaneshwar see Bhubaneshwar
Biak, Pulau 117 G4 island E Indonesia
Biała Podlaska 76 E3 Lubelskie, E Poland
Białogard 76 B2 Ger. Belgard. Zachodnio-pomorskie, NW Poland
Białystok 76 E3 Rus. Belostok, Bielostok. Podlaskie, NE Poland
Bianco, Monte see Blanc, Mont
Biarritz 69 A6 Pyrénées-Atlantiques, SW France
Bicaz 86 C3 Hung. Békás. Neamţ, NE Romania
Bichiş-Ciaba see Békéscsaba
Biddeford 19 G2 Maine, NE USA
Bideford 67 C7 SW England, United Kingdom
Biel 73 A7 Fr. Bienne. Bern, W Switzerland
Bielefeld 72 B4 Nordrhein-Westfalen, NW Germany
Bielitz/Bielitz-Biala see Bielsko-Biała
Bielostok see Białystok
Bielsko-Biała 77 C5 Ger. Bielitz, Bielitz-Biala. Śląskie, S Poland
Bielsk Podlaski 76 E3 Białystok, E Poland
Bien Bien see Điện Biên
Biên Hoa 115 E6 Đông Nai, S Vietnam
Bienne see Biel
Bienville, Lac 16 D2 lake Québec, C Canada
Bié, Planalto do 56 B2 var. Bié Plateau. plateau C Angola
Bié Plateau see Bié, Planalto do
Big Cypress Swamp 21 E5 wetland Florida, SE USA
Bigge Island 124 C2 island Western Australia
Bighorn Mountains 22 C2 mountain range Wyoming, C USA
Bighorn River 22 C2 river Montana/Wyoming, NW USA
Bignona 52 B3 SW Senegal
Bigorra see Tarbes
Bigosovo see Bihosava
Big Sioux River 23 E2 river Iowa/South Dakota, N USA
Big Spring 27 E3 Texas, SW USA
Bihać 78 B3 Federacija Bosna I Hercegovina, NW Bosnia and Herzegovina
Bihār 113 F3 prev. Behar. cultural region N India
Bihār see Bihār Sharīf
Biharamulo 51 B7 Kagera, NW Tanzania
Bihār Sharīf 113 F3 var. Bihār. Bihār, N India
Bihosava 85 D5 Rus. Bigosovo. Vitsyebskaya Voblasts', NW Belarus
Bijeljina 78 C3 Republika Srpska, NE Bosnia and Herzegovina
Bijelo Polje 79 D5 E Montenegro
Bīkāner 112 C3 Rājasthān, NW India
Bikin 93 G4 Khabarovskiy Kray, SE Russia
Bikini Atoll 122 C1 var. Pikinni. atoll Ralik Chain, NW Marshall Islands
Bikkū Bittī see Bette, Picco
Bilāspur 113 E4 Chhattīsgarh, C India
Biläsuvar 95 H3 Rus. Bilyasuvar; prev. Pushkino. SE Azerbaijan
Bila Tserkva 87 E2 Rus. Belaya Tserkov'. Kyivska Oblast, N Ukraine
Bilauktaung Range 115 C6 var. Thanintari Taungdan. mountain range Myanmar (Burma)/ Thailand
Bilbao 71 E1 Basq. Bilbo. País Vasco, N Spain
Bilbo see Bilbao
Bilecik 94 B3 Bilecik, NW Turkey (Türkiye)
Bilhorod-Dnistrovskyi 87 E4 prev. Akkerman, Rus. Belgorod-Dnestrovskiy, Rom. Cetatea Albă, anc. Tyras. SW Ukraine
Billings 22 C2 Montana, NW USA
Bilma, Grand Erg de 53 H3 desert NE Niger
Biloela 126 D4 Queensland, E Australia
Biloxi 20 C3 Mississippi, S USA
Biltine 54 C3 Biltine, E Chad
Bilwi see Puerto Cabezas
Bilyasuvar see Biläsuvar
Bilzen 65 D6 Limburg, NE Belgium
Bimini Islands 32 C1 island group W The Bahamas
Binche 65 B7 Hainaut, S Belgium
Bindloe Island see Marchena, Isla
Bin Ghalfān, Jazā'ir see Ḩalāniyāt, Juzur al
Binghamton 19 F3 New York, NE USA
Bingöl 95 E3 Bingöl, E Turkey (Türkiye)
Bint Jubayl see Bent Jbaïl
Bintulu 116 D3 Sarawak, East Malaysia
Binzhou 106 D4 Shandong, E China
Bío Bío, Río 43 B5 river C Chile
Bioco, Isla de 55 A5 var. Bioko, Eng. Fernando Po, Sp. Fernando Póo; prev. Macías Nguema Biyogo. island NW Equatorial Guinea
Bioko see Bioco, Isla de
Birāk 49 F3 var. Brak. C Libya
Birao 54 D3 Vakaga, NE Central African Republic
Birātnagar 113 F3 Eastern, SE Nepal
Bir es Saba see Be'er Sheva
Birjand 98 E3 Khorāsān-e Janūbī, E Iran
Birkenfeld 73 A5 Rheinland-Pfalz, SW Germany
Birkenhead 67 C5 NW England, United Kingdom
Birlad see Bârlad
Birmingham 67 C6 C England, United Kingdom
Birmingham 20 C2 Alabama, S USA
Bir Moghrein see Bîr Mogreïn
Bîr Mogreïn 52 C1 var. Bir Moghrein; prev. Fort-Trinquet. Tiris Zemmour, N Mauritania
Birnie Island 123 E3 atoll Phoenix Islands, C Kiribati
Birnin Konni 53 F3 var. Birni-Nkonni. Tahoua, SW Niger
Birni-Nkonni see Birnin Konni
Birobidzhan 93 G4 Yevreyskaya Avtonomnaya Oblast', SE Russia
Birsen see Biržai
Birsk 89 D5 Respublika Bashkortostan, W Russia
Biržai 84 C4 Ger. Birsen. Panevėžys, NE Lithuania
Birżebbuġa 80 B5 SE Malta
Bisanthe see Tekirdağ
Bisbee 26 B3 Arizona, SW USA
Biscaia, Baía de see Biscay, Bay of
Biscay, Bay of 58 B4 Sp. Golfo de Vizcaya, Port. Baía de Biscaia. bay France/Spain
Biscay Plain 58 B3 abyssal plain SE Bay of Biscay
Bischofsburg see Biskupiec
Bishah, Wadi 99 B5 dry watercourse C Saudi Arabia
Bishkek 101 G2 var. Pishpek; prev. Frunze. country capital (Kyrgyzstan) Chuyskaya Oblast', N Kyrgyzstan
Bishop's Lynn see King's Lynn
Bishrī, Jabal 96 D3 mountain range E Syria
Biskara see Biskra
Biskra 49 E2 var. Beskra, Biskara. NE Algeria
Biskupiec 76 D2 Ger. Bischofsburg. Warmińsko-Mazurskie, NE Poland
Bislig 117 F2 Mindanao, S Philippines
Bismarck 23 E2 state capital North Dakota, N USA
Bismarck Archipelago 122 B3 island group NE Papua New Guinea
Bismarck Sea 122 B3 sea W Pacific Ocean
Bisnulok see Phitsanulok
Bissau 52 B4 country capital (Guinea-Bissau) W Guinea-Bissau
Bistriţa 86 B3 Ger. Bistritz, Hung. Beszterce; prev. Nösen. Bistriţa-Năsăud, N Romania
Bistritz see Bistriţa
Bitam 55 B5 Woleu-Ntem, N Gabon
Bitburg 73 A5 Rheinland-Pfalz, SW Germany
Bitlis 95 F3 Bitlis, SE Turkey (Türkiye)
Bitoeng see Bitung
Bitola 79 D6 Turk. Monastir; prev. Bitolj. S North Macedonia
Bitolj see Bitola
Bitonto 75 D5 anc. Butuntum. Puglia, SE Italy
Bitterroot Range 24 D2 mountain range Idaho/ Montana, NW USA
Bitung 117 F3 prev. Bitoeng. Sulawesi, C Indonesia
Biu 53 H4 Borno, E Nigeria
Biwa ko 109 C6 lake Honshū, SW Japan
Bizerta see Bizerte
Bizerte 49 E1 Ar. Banzart, Eng. Bizerta. N Tunisia
Bjelovar 78 B2 Hung. Belovár. Bjelovar-Bilogora, N Croatia
Bjeshkët e Namuna see North Albanian Alps
Björneborg see Pori
Bjørnøya 61 F3 Eng. Bear Island. island N Norway
Blackall 126 C4 Queensland, E Australia
Black Drin 79 D6 Alb. Lumi i Drinit të Zi, Croatian Crni Drim. river Albania/ North Macedonia
Blackfoot 24 E4 Idaho, NW USA
Black Forest 73 B6 Eng. Black Forest. mountain range SW Germany
Black Forest see Schwarzwald
Black Hills 22 D3 mountain range South Dakota/ Wyoming, N USA
Blackpool 67 C5 NW England, United Kingdom
Black Range 26 C2 mountain range New Mexico, SW USA
Black River 32 A5 W Jamaica
Black River 114 C3 Chin. Babian Jiang, Lixian Jiang, Fr. Rivière Noire, Vtn. Sông Đa. river China/Vietnam
Black Rock Desert 25 C5 desert Nevada, W USA
Black Sand Desert see Garagum
Black Sea 94 B1 var. Euxine Sea, Bul. Cherno More, Rom. Marea Neagră, Rus. Chernoye More, Turk. Karadeniz, Ukr. Chorne More. sea Asia/Europe
Black Sea Lowland 87 E4 Ukr. Prychornomor'ska Nyzovyna. depression SE Europe
Black Volta 53 E4 var. Borongo, Mouhoun, Moun Hou, Fr. Volta Noire. river W Africa
Blackwater 67 A6 Ir. An Abhainn Mhór. river S Ireland
Blackwater State see Nebraska
Blagoevgrad 82 C3 prev. Gorna Dzhumaya. Blagoevgrad, W Bulgaria
Blagoveshchensk 93 G4 Amurskaya Oblast', SE Russia
Blahovishchenske 87 E3 prev. Ulianovka, Rus. Ulyanovka. Kirovohrads'ka Oblast', C Ukraine
Blake Plateau 13 D6 var. Blake Terrace. undersea plateau W Atlantic Ocean
Blake Terrace see Blake Plateau
Blanca, Bahía 43 C5 bay E Argentina
Blanca, Costa 71 F4 physical region SE Spain
Blanche, Lake 127 B5 lake South Australia
Blanc, Mont 69 D5 It. Monte Bianco. mountain France/Italy
Blanco, Cape 24 A4 headland Oregon, NW USA
Blanes 71 G2 Cataluña, NE Spain
Blankenberge 65 A5 West-Vlaanderen, NW Belgium
Blankenheim 73 A5 Nordrhein-Westfalen, W Germany
Blanquilla, Isla 37 E1 var. La Blanquilla. island N Venezuela
Blanquilla, La see Blanquilla, Isla
Blantyre 57 E2 var. Blantyre-Limbe. Southern, S Malawi
Blantyre-Limbe see Blantyre
Blaricum 64 C3 Noord-Holland, C Netherlands
Blatnitsa see Durankulak
Blenheim 129 C5 Marlborough, South Island, New Zealand
Blesae see Blois
Blida 48 D2 var. El Boulaida, El Boulaïda. N Algeria
Bloemfontein 56 C4 var. Mangaung. country capital (South Africa-judicial capital) Free State, C South Africa
Blois 68 C4 anc. Blesae. Loir-et-Cher, C France
Bloomfield 26 C1 New Mexico, SW USA
Bloomington 18 B4 Illinois, N USA
Bloomington 18 C4 Indiana, N USA
Bloomington 23 F2 Minnesota, N USA
Bloomsbury 126 D3 Queensland, NE Australia
Bluefield 18 D5 West Virginia, NE USA
Bluefields 31 E3 Región Autónoma Atlántico Sur, SE Nicaragua
Bluegrass State see Kentucky
Blue Hen State see Delaware
Blue Law State see Connecticut
Blue Mountain Peak 32 B5 mountain E Jamaica
Blue Mountains 24 C3 mountain range Oregon/ Washington, NW USA
Blue Nile 50 C4 var. Abai, Bahr el, Azraq, Amh. Ābay Wenz, Ar. An Nil al Azraq. river Ethiopia/Sudan
Blumenau 41 E5 Santa Catarina, S Brazil
Blythe 25 D8 California, W USA
Blytheville 20 C1 Arkansas, C USA
Bo 52 C4 S Sierra Leone
Boaco 30 D3 Boaco, S Nicaragua
Boa Vista 40 D1 state capital Roraima, NW Brazil
Boa Vista 52 A3 island Ilhas de Barlavento, E Cape Verde (Cabo Verde)
Bobaomby, Tanjona 57 G2 Fr. Cap d'Ambre. headland N Madagascar
Bobigny 68 E1 Seine-St-Denis, N France
Bobo-Dioulasso 52 D4 SW Burkina
Bobrinets see Bobrynets
Bobruysk see Babruysk
Bobrynets 87 E3 Rus. Bobrinets. Kirovohrads'ka Oblast', C Ukraine
Boca Raton 21 F5 Florida, SE USA
Bocay 30 D2 Jinotega, N Nicaragua
Bocheykovo see Bacheykava
Bocholt 72 A4 Nordrhein-Westfalen, W Germany
Bochum 72 A4 Nordrhein-Westfalen, W Germany
Bocşa 86 A4 Ger. Bokschen, Hung. Boksánbánya. Caraş-Severin, SW Romania
Bodaybo 93 F4 Irkutskaya Oblast', E Russia
Bodegas see Babahoyo
Boden 62 D4 Norrbotten, N Sweden
Bodensee see Constance, Lake, C Europe
Bodmin 67 C7 SW England, United Kingdom
Bodø 62 C3 Nordland, C Norway
Bodrum 94 A4 Muğla, SW Turkey (Türkiye)
Boeloekoemba see Bulukumba
Boende 55 C5 Equateur, C Dem. Rep. Congo
Boeroe see Buru, Pulau
Boetoeng see Buton, Pulau
Bogale 114 B4 Ayeyarwady, SW Myanmar (Burma)
Bogalusa 20 B3 Louisiana, S USA
Bogatynia 76 B4 Ger. Reichenau. Dolnośląskie, SW Poland
Boğazlıyan 94 D3 Yozgat, C Turkey (Türkiye)
Bogendorf see Łuków
Bogor 116 C5 Dut. Buitenzorg. Jawa, C Indonesia
Bogotá 36 B3 prev. Santa Fe, Santa Fe de Bogotá. country capital (Colombia) Cundinamarca, C Colombia
Bogushëvsk see Bahushewsk
Boguslav see Bohuslav
Bo Hai 106 D4 var. Gulf of Chihli. gulf NE China
Bohemia 77 A5 Cz. Čechy, Ger. Böhmen. W Czechia (Czech Republic)
Bohemian Forest 73 C5 Cz. Český Les, Šumava, Ger. Böhmerwald. mountain range C Europe
Böhmen see Bohemia
Böhmerwald see Bohemian Forest
Böhmisch-Krumau see Český Krumlov
Bohol Sea 117 E2 var. Mindanao Sea. sea S Philippines
Bohuslav 87 E2 Rus. Boguslav. Kyivska Oblast, N Ukraine
Boise 24 D3 var. Boise City. state capital Idaho, NW USA
Boise City 27 E1 Oklahoma, C USA
Boise City see Boise
Bois, Lac des see Woods, Lake of the
Bois-le-Duc see 's-Hertogenbosch
Boiarka 87 E2 prev. Boyarka. Kyivska Oblast, N Ukraine
Boizenburg 72 C3 Mecklenburg-Vorpommern, N Germany
Bojador see Boujdour
Bojnūrd 98 D2 var. Bujnurd. Khorāsān-e Shemālī, N Iran
Bokāro 113 F4 Jhārkhand, N India
Boké 52 C4 W Guinea
Bokhara see Buxoro
Bokhtar 101 E3 prev. Qŭrghonteppa, Rus. Kurgan-Tyube. SW Tajikistan
Boksánbánya/Bokschen see Bocşa
Bol 54 B3 Lac, W Chad
Bolgatanga 53 E4 N Ghana
Bolgrad see Bolhrad
Bolhrad 86 D4 Rus. Bolgrad. Odes'ka Oblast', SW Ukraine
Bolívar, Pico 36 C2 mountain W Venezuela
Bolivia 39 F3 off. Plurinational Republic of Bolivia. country W South America
Bolivia, Plurinational Republic of see Bolivia
Bollène 69 D6 Vaucluse, SE France
Bollnäs 63 C5 Gävleborg, C Sweden
Bollon 127 D5 Queensland, C Australia
Bologna 74 C3 Emilia-Romagna, N Italy
Bol'shevik see Bal'shavik
Bol'shevik, Ostrov 93 E2 island Severnaya Zemlya, N Russia
Bol'shezemel'skaya Tundra 88 E3 physical region NW Russia
Bol'shoy Lyakhovskiy, Ostrov 93 F2 island NE Russia
Bolton 67 D5 prev. Bolton-le-Moors. NW England, United Kingdom
Bolton-le-Moors see Bolton
Bolu 94 B3 Bolu, NW Turkey (Türkiye)
Bolungarvík 61 E4 Vestfirdhir, NW Iceland
Bolyarovo 82 D3 prev. Pashkeni. Yambol, E Bulgaria
Bolzano 74 C1 Ger. Bozen; anc. Bauzanum. Trentino-Alto Adige, N Italy
Boma 55 B6 Bas-Congo, W Dem. Rep. Congo
Bombay see Mumbai
Bomu 54 D4 var. Mbomou, Mbomu, M'Bomu. river Central African Republic/Dem. Rep. Congo
Bonaire 33 F5 Dutch special municipality S West Indies
Bonanza 30 D2 Región Autónoma Atlántico Norte, NE Nicaragua
Bonaparte Archipelago 124 C2 island group Western Australia
Bon, Cap 80 D3 headland N Tunisia
Bonda 55 B6 Ogooué-Lolo, C Gabon
Bondoukou 53 E4 E Ivory Coast
Bône see Annaba, Algeria
Bone see Watampone, Indonesia
Bone, Teluk 117 E4 bay Sulawesi, C Indonesia
Bongaigaon 113 G3 Assam, NE India
Bongo, Massif des 54 D4 var. Chaîne des Mongos. mountain range NE Central African Republic
Bongor 54 B3 Mayo-Kébbi, SW Chad
Bonifacio 69 E7 Corse, France, C Mediterranean Sea
Bonifacio, Bocche de/Bonifacio, Bouches de see Bonifacio, Strait of
Bonifacio, Strait of 74 A4 Fr. Bouches de Bonifacio, It. Bocche di Bonifacio. strait C Mediterranean Sea
Bonn 73 A5 Nordrhein-Westfalen, W Germany
Bononia see Vidin, Bulgaria
Bononia see Boulogne-sur-Mer, France
Boosaaso 50 E4 var. Bandar Kassim, Bender Qaasim, Bosaso, It. Bender Cassim. Bari, N Somalia
Boothia Felix see Boothia Peninsula
Boothia, Gulf of 15 F2 gulf Nunavut, NE Canada
Boothia Peninsula 15 F2 prev. Boothia Felix. peninsula Nunavut, NE Canada
Boppard 73 A5 Rheinland-Pfalz, W Germany
Boquete 31 E5 var. Bajo Boquete. Chiriquí, W Panama
Boquillas 28 D2 var. Boquillas del Carmen. Coahuila, NE Mexico
Boquillas del Carmen see Boquillas
Bor 78 E4 Serbia, E Serbia
Bor 51 B5 Jonglei, C South Sudan
Borås 63 B7 Västra Götaland, S Sweden
Borbetomagus see Worms
Borborema, Planalto da 34 E3 plateau NE Brazil
Bordeaux 69 B5 anc. Burdigala. Gironde, SW France
Bordj Omar Driss 49 E3 E Algeria
Borgå see Porvoo
Børgefjelt 62 C4 mountain range C Norway
Borger 64 E2 Drenthe, NE Netherlands
Borger 27 E1 Texas, SW USA
Borgholm 63 C7 Kalmar, S Sweden
Borgo Maggiore 74 E1 NW San Marino
Borislav see Boryslav
Borisoglebsk 89 B6 Voronezhskaya Oblast', W Russia
Borisov see Barysaw
Borlänge 63 C6 Dalarna, C Sweden
Borne 64 E3 Overijssel, E Netherlands
Borneo 116 C4 island Brunei/Indonesia/Malaysia
Bornholm 63 B8 island E Denmark
Borohoro Shan 104 B2 mountain range NW China
Borongo see Black Volta
Boron'ki see Baron'ki
Borosjenő see Ineu
Borovan 82 C2 Vratsa, NW Bulgaria
Borovichi 88 B4 Novgorodskaya Oblast', W Russia
Borovo 78 C3 Vukovar-Srijem, NE Croatia
Borriana 71 F3 var. Burriana. Comunitat Valenciana, E Spain
Borşa 86 C3 Hung. Borsa. Maramureş, N Romania
Boryslav 86 B2 Pol. Borysław, Rus. Borislav. L'vivs'ka Oblast', NW Ukraine
Borysław see Boryslav
Bosanska Dubica 78 B3 var. Kozarska Dubica. Republika Srpska, NW Bosnia and Herzegovina
Bosanska Gradiška 78 B3 var. Gradiška. Republika Srpska, N Bosnia and Herzegovina
Bosanski Novi 78 B3 var. Novi Grad. Republika Srpska, NW Bosnia and Herzegovina
Bosanski Šamac 78 C3 var. Šamac. Republika Srpska, N Bosnia and Herzegovina
Bosaso see Boosaaso
Bösing see Pezinok
Boskovice 77 B5 Ger. Boskowitz. Jihomoravský Kraj, SE Czechia (Czech Republic)
Boskowitz see Boskovice
Bosna 78 C4 river N Bosnia and Herzegovina
Bosnia and Herzegovina 78 B3 country SE Europe
Boso-hanto 109 D6 peninsula Honshū, S Japan
Bosphorus/Bosporus see İstanbul Boğazı
Bosporus Cimmerius see Kerch Strait
Bosporus Thracius see İstanbul Boğazı
Bossangoa 54 C4 Ouham, C Central African Republic
Bossembélé 54 C4 Ombella-Mpoko, C Central African Republic
Bossier City 20 A2 Louisiana, S USA
Bosten Hu 104 C3 var. Bagrax Hu. lake NW China
Boston 67 E6 prev. St.Botolph's Town. E England, United Kingdom
Boston 19 G3 state capital Massachusetts, NE USA
Boston Mountains 20 B1 mountain range Arkansas, C USA
Bostyn' see Bastyn'
Botany 126 E2 New South Wales, E Australia
Botany Bay 126 E2 inlet New South Wales, SE Australia
Boteti 56 C3 var. Botletle. river N Botswana
Bothnia, Gulf of 63 D5 Fin. Pohjanlahti, Swe. Bottniska Viken. gulf N Baltic Sea
Botletle see Boteti
Botoşani 86 C3 Hung. Botosány. Botoşani, NE Romania
Botosány see Botoşani
Botou 106 C4 prev. Bozhen. Hebei, E China
Botrange 65 D6 mountain E Belgium
Botswana 56 C3 off. Republic of Botswana. country S Africa
Botswana, Republic of see Botswana
Bottniska Viken see Bothnia, Gulf of
Bouar 54 B4 Nana-Mambéré, W Central African Republic
Bou Craa 48 B3 var. Bu Craa. NW Western Sahara
Bougainville Island 120 B3 island NE Papua New Guinea
Bougaroun, Cap 80 C3 headland NE Algeria
Bougouni 52 D4 Sikasso, SW Mali
Boujdour 48 A3 var. Bojador. W Western Sahara
Boulder 22 C4 Colorado, C USA
Boulder 22 B2 Montana, NW USA
Boulogne see Boulogne-sur-Mer
Boulogne-Billancourt 68 D1 Île-de-France, N France Europe
Boulogne-sur-Mer 68 C2 var. Boulogne; anc. Bononia, Gesoriacum, Gessoriacum. Pas-de-Calais, N France
Boûmdeïd 52 C3 var. Boumdeït. Assaba, S Mauritania
Boumdeït see Boûmdeïd
Boundiali 52 D4 N Ivory Coast
Bountiful 22 B4 Utah, W USA
Bounty Basin see Bounty Trough
Bounty Islands 120 D5 island group S New Zealand
Bounty Trough 130 C5 var. Bounty Basin. trough S Pacific Ocean
Bourbonnais 68 C4 cultural region C France
Bourbon Vendée see la Roche-sur-Yon
Bourg see Bourg-en-Bresse
Bourgas see Burgas
Bourge-en-Bresse see Bourg-en-Bresse
Bourg-en-Bresse 69 D5 var. Bourg, Bourge-en-Bresse. Ain, E France
Bourges 68 C4 anc. Avaricum. Cher, C France
Bourgogne 68 C4 Eng. Burgundy. cultural region E France
Bourke 127 C5 New South Wales, SE Australia
Bournemouth 67 D7 S England, United Kingdom
Boutilimit 52 C3 Trarza, SW Mauritania
Bouvet Island 45 D7 Norwegian dependency S Atlantic Ocean
Bowen 126 D3 Queensland, NE Australia
Bowling Green 18 B5 Kentucky, S USA
Bowling Green 18 C3 Ohio, N USA
Boxmeer 64 D4 Noord-Brabant, SE Netherlands
Boyarka see Boiarka
Boychinovtsi 82 C2 prev. Lekhchevo. Montana, NW Bulgaria
Boysun 101 E3 Rus. Baysun. Surkhondaryo Viloyati, S Uzbekistan
Bozeman 22 B2 Montana, NW USA
Bozen see Bolzano
Bozhen see Botou
Bozüyük 94 B3 Bilecik, NW Turkey (Türkiye)
Brač 78 B4 var. Brach, It. Brazza; anc. Brattia. island S Croatia
Bracara Augusta see Braga
Brach see Brač
Brades 33 G3 de facto dependent territory capital, de jure capital, Plymouth, destroyed by volcano in 1995 (Montserrat) SW Montserrat
Bradford 67 D5 N England, United Kingdom
Brady 27 F3 Texas, SW USA
Braga 70 B2 anc. Bracara Augusta. Braga, NW Portugal
Bragança 70 C2 Eng. Braganza; anc. Julio Briga. Bragança, NE Portugal
Braganza see Bragança
Brahestad see Raahe
Brahmanbaria 113 G4 Chattogram, E Bangladesh

Brahmapur *113 F5* Odisha, E India
Brahmaputra *113 H3 var.* Padma, Tsangpo, *Ben.* Jamuna, *Chin.* Yarlung Zangbo Jiang, *Ind.* Bramaputra, Dihang, Siang. *river* S Asia
Brăila *86 D4* Brăila, E Romania
Braine-le-Comte *65 B6* Hainaut, SW Belgium
Brainerd *23 F2* Minnesota, N USA
Brak *see* Birāk
Bramaputra *see* Brahmaputra
Brampton *16 D5* Ontario, S Canada
Branco, Rio *34 C3 river* N Brazil
Brandberg *56 A3 mountain* NW Namibia
Brandenburg *72 C3 var.* Brandenburg an der Havel. Brandenburg, NE Germany
Brandenburg an der Havel *see* Brandenburg
Brandon *15 F5* Manitoba, S Canada
Braniewo *76 D2 Ger.* Braunsberg. Warmińsko-mazurskie, N Poland
Brasil *see* Brazil
Brasília *41 F3 country capital* (Brazil) Distrito Federal, C Brazil
Brasil, República Federativa do *see* Brazil
Braşov *86 C4 Ger.* Kronstadt, *Hung.* Brassó; *prev.* Oraşul Stalin. Braşov, C Romania
Brassó *see* Braşov
Bratislava *77 C6 Ger.* Pressburg, *Hung.* Pozsony. *country capital* (Slovakia) Bratislavský Kraj, W Slovakia
Bratsk *93 E4* Irkutskaya Oblast', C Russia
Brattia *see* Brač
Braunsberg *see* Braniewo
Braunschweig *72 C4 Eng./Fr.* Brunswick. Niedersachsen, N Germany
Brava *see* Baraawe
Brava, Costa *71 H2 coastal region* NE Spain
Bravo del Norte, Río/Bravo, Río *see* Grande, Rio
Bravo, Río *28 C1 river* Mexico/USA North America
Brawley *25 D8* California, W USA
Brazil *40 C2 off.* Federative Republic of Brazil, *Port.* República Federativa do Brasil, *Sp.* Brasil; *prev.* United States of Brazil. *country* South America
Brazil Basin *45 C5 var.* Brazilian Basin, Brazil'skaya Kotlovina. *undersea basin* W Atlantic Ocean
Brazil, Federative Republic of *see* Brazil
Brazilian Basin *see* Brazil Basin
Brazilian Highlands *see* Central, Planalto
Brazil'skaya Kotlovina *see* Brazil Basin
Brazil, United States of *see* Brazil
Brazos River *27 G3 river* Texas, SW USA
Brazza *see* Brač
Brazzaville *55 B6 country capital* (Congo) Capital District, S Congo
Brčko *78 C3* Republika Srpska, NE Bosnia and Herzegovina
Brecht *65 C5* Antwerpen, N Belgium
Brecon Beacons *67 C6 mountain range* S Wales, United Kingdom
Breda *64 C4* Noord-Brabant, S Netherlands
Bree *65 D5* Limburg, NE Belgium
Bregalnica *79 E6 river* E North Macedonia
Bregenz *35 B7 anc.* Brigantium. Vorarlberg, W Austria
Bregovo *82 B1* Vidin, NW Bulgaria
Bremen *72 B3 Fr.* Brême. Bremen, NW Germany
Bremerhaven *72 B3* Bremen, NW Germany
Bremerton *24 B2* Washington, NW USA
Brenham *27 G3* Texas, SW USA
Brenner, Col du/Brennero, Passo del *see* Brenner Pass
Brenner Pass *74 C1 var.* Brenner Sattel, *Fr.* Col du Brenner, *Ger.* Brennerpass, *It.* Passo del Brennero. *pass* Austria/Italy
Brennerpass *see* Brenner Pass
Brenner Sattel *see* Brenner Pass
Brescia *74 B2 anc.* Brixia. Lombardia, N Italy
Breslau *see* Wrocław
Bressanone *74 C1 Ger.* Brixen. Trentino-Alto Adige, N Italy
Brest *85 A6 Pol.* Brześć nad Bugiem, *Rus.* Brest-Litovsk; *prev.* Brześć Litewski. Brestskaya Voblasts', SW Belarus
Brest *68 A3* Finistère, NW France
Brest-Litovsk *see* Brest
Bretagne *68 A3 Eng.* Brittany, *Lat.* Britannia Minor. *cultural region* NW France
Brewster, Kap *see* Kangikajik
Brewton *20 C3* Alabama, S USA
Brezhnev *see* Naberezhnyye Chelny
Brezovo *82 D2 prev.* Abrashlare. Plovdiv, C Bulgaria
Bria *54 D4* Haute-Kotto, C Central African Rep.
Briançon *69 D5 anc.* Brigantio. Hautes-Alpes, SE France
Bricgstow *see* Bristol
Bridgeport *19 F3* Connecticut, NE USA
Bridgetown *33 G2 country capital* (Barbados) SW Barbados
Bridlington *67 D5* E England, United Kingdom
Bridport *67 D7* S England, United Kingdom
Brieg *see* Brzeg
Brig *73 A7 Fr.* Brigue, *It.* Briga. Valais, SW Switzerland
Briga *see* Brig
Brigantio *see* Briançon
Brigantium *see* Bregenz
Brigham City *22 B4* Utah, W USA
Brighton *67 E7* SE England, United Kingdom
Brighton *22 D4* Colorado, C USA
Brigue *see* Brig
Brindisi *75 E5 anc.* Brundisium, Brundusium. Puglia, SE Italy
Briovera *see* St-Lô
Brisbane *127 E5 state capital* Queensland, E Australia
Bristol *67 D7 anc.* Bricgstow. SW England, United Kingdom
Bristol *19 F3* Connecticut, NE USA
Bristol *18 D5* Tennessee, S USA
Bristol Bay *14 B3 bay* Alaska, USA
Bristol Channel *67 C7 inlet* England/Wales, United Kingdom
Britain *58 C3 var.* Great Britain. *island* United Kingdom
Britannia Minor *see* Bretagne
British Columbia *14 D4 Fr.* Colombie-Britannique. *province* SW Canada
British Guiana *see* Guyana
British Honduras *see* Belize
British Indian Ocean Territory *119 B5 UK Overseas Territory* C Indian Ocean
British Isles *67 island group* NW Europe
British North Borneo *see* Sabah
British Solomon Islands Protectorate *see* Solomon Islands
British Virgin Islands *33 F3 var.* Virgin Islands. *UK Overseas Territory* E West Indies
Brittany *see* Bretagne
Briva Curretia *see* Brive-la-Gaillarde
Briva Isarae *see* Pontoise
Brive *see* Brive-la-Gaillarde
Brive-la-Gaillarde *69 C5 prev.* Brive; *anc.* Briva Curretia. Corrèze, C France
Brixen *see* Bressanone
Brixia *see* Brescia
Brno *77 B5 Ger.* Brünn. Jihomoravský Kraj, SE Czechia (Czech Republic)
Broceni *84 B3* Saldus, SW Latvia
Brod/Bród *see* Slavonski Brod
Brodeur Peninsula *15 F2 peninsula* Baffin Island, Nunavut, NE Canada
Brod na Savi *see* Slavonski Brod
Brodnica *76 C3 Ger.* Buddenbrock. Kujawski-pomorskie, C Poland
Broek-in-Waterland *64 C3* Noord-Holland, C Netherlands
Broken Arrow *27 G1* Oklahoma, C USA
Broken Bay *126 E1 bay* New South Wales, SE Australia
Broken Hill *127 B6* New South Wales, SE Australia
Broken Ridge *119 D6 undersea plateau* S Indian Ocean
Bromberg *see* Bydgoszcz
Bromley *67 B8* United Kingdom
Brookhaven *20 B3* Mississippi, S USA
Brookings *23 F3* South Dakota, N USA
Brooks Range *14 D2 mountain range* Alaska, USA
Brookton *125 B6* Western Australia
Broome *124 B3* Western Australia
Broomfield *22 D4* Colorado, C USA
Broucsella *see* Brussel/Bruxelles
Brovary *87 E2* Kyivska Oblast, N Ukraine
Brownfield *27 E2* Texas, SW USA
Brownsville *27 G5* Texas, SW USA
Brownwood *27 F3* Texas, SW USA
Brozha *85 D7* Mahilyowskaya Voblasts', E Belarus
Bruges *see* Brugge
Brugge *65 A5 Fr.* Bruges. West-Vlaanderen, NW Belgium
Brummen *64 D3* Gelderland, E Netherlands
Brundisium/Brundusium *see* Brindisi
Brunei *116 D3 off.* Brunei Darussalam, *Mal.* Negara Brunei Darussalam. *country* SE Asia
Brunei Darussalam *see* Brunei
Brunei Town *see* Bandar Seri Begawan
Brünn *see* Brno
Brunner, Lake *129 C5 lake* South Island, New Zealand
Brunswick *21 E3* Georgia, SE USA
Brunswick *see* Braunschweig
Brusa *see* Bursa
Brus Laguna *30 D2* Gracias a Dios, E Honduras
Brussa *see* Bursa
Brussel *65 C6 var.* Brussels, *Fr.* Bruxelles, *Ger.* Brüssel; *anc.* Broucsella. *country capital* (Belgium) Brussels, C Belgium
Brüssel/Brussels *see* Brussel/Bruxelles
Brüx *see* Most
Bruxelles *see* Brussel
Bryan *27 G3* Texas, SW USA
Bryansk *89 A5* Bryanskaya Oblast', W Russia
Brzeg *76 C4 Ger.* Brieg; *anc.* Civitas Altae Ripae. Opolskie, S Poland
Brześć Litewski/Brześć nad Bugiem *see* Brest
Brzeżany *see* Berezhany
Bucaramanga *36 B2* Santander, N Colombia
Buchanan *52 C5 prev.* Grand Bassa. SW Liberia
Buchanan, Lake *27 F3 reservoir* Texas, SW USA
Bucharest *see* Bucureşti
Buckeye State *see* Ohio
Bu Craa *see* Bou Craa
Bucureşti *86 C5 Eng.* Bucharest, *Ger.* Bukarest, *prev.* Altenburg; *anc.* Cetatea Damboviţei. *country capital* (Romania) Bucureşti, S Romania
Buda-Kashalyova *85 D7 Rus.* Buda-Koshelëvo. Homyel'skaya Voblasts', SE Belarus
Buda-Koshelëvo *see* Buda-Kashalyova
Budapest *77 C6 off.* Budapest Főváros, *Croatian* Budimpešta. *country capital* (Hungary) Pest, N Hungary
Budapest Főváros *see* Budapest
Budaun *112 D3* Uttar Pradesh, N India
Buddenbrock *see* Brodnica
Budimpešta *see* Budapest
Budweis *see* České Budějovice
Budyšin *see* Bautzen
Buena Park *24 E2* California, W USA North America
Buenaventura *36 A3* Valle del Cauca, W Colombia
Buena Vista *39 G4* Santa Cruz, C Bolivia
Buena Vista *71 H5* S Gibraltar Europe
Buenos Aires *42 D4 hist.* Santa Maria del Buen Aire. *country capital* (Argentina) Buenos Aires, E Argentina
Buenos Aires *31 E5* Puntarenas, SE Costa Rica
Buenos Aires, Lago *43 B6 var.* Lago General Carrera. *lake* Argentina/Chile
Buffalo *19 E3* New York, NE USA
Buffalo Narrows *15 F4* Saskatchewan, C Canada
Buff Bay *32 B5* E Jamaica
Buftea *86 C5* Ilfov, S Romania
Bug *59 E3 Bel.* Zakhodni Buh, *Eng.* Western Bug, *Rus.* Zapadnyy Bug, *Ukr.* Zakhidnyi Buh. *river* E Europe
Buga *36 B3* Valle del Cauca, W Colombia
Bughotu *see* Santa Isabel
Buguruslan *89 D6* Orenburgskaya Oblast', W Russia
Buitenzorg *see* Bogor
Bujalance *70 D4* Andalucía, S Spain
Bujanovac *79 E5* SE Serbia
Bujnurd *see* Bojnūrd
Bujumbura *51 B7 prev.* Usumbura. *country capital* (Burundi - commercial) W Burundi
Buucureşti
Bukavu *55 E6 prev.* Costermansville. Sud-Kivu, E Dem. Rep. Congo
Bukhara *see* Buxoro
Bukoba *51 B6* Kagera, NW Tanzania
Bülach *73 B7* Zürich, NW Switzerland
Bulawayo *56 D3* Matabeleland North, SW Zimbabwe
Bulgan *105 E2* Bulgan, N Mongolia
Bulgaria *82 C2 off.* Republic of Bulgaria, *Bul.* Republika Bulgaria; *prev.* People's Republic of Bulgaria. *country* SE Europe
Bulgaria, People's Republic of *see* Bulgaria
Bulgaria, Republic of *see* Bulgaria
Bulgaria, Republika *see* Bulgaria
Bullion State *see* Missouri
Bull Shoals Lake *20 B1 reservoir* Arkansas/Missouri, C USA
Bulukumba *117 E4 prev.* Boeloekoemba. Sulawesi, C Indonesia
Bumba *55 D5* Equateur, N Dem. Rep. Congo
Bunbury *125 A7* Western Australia
Bundaberg *126 E4* Queensland, E Australia
Bungo-suido *109 B7 strait* SW Japan
Bunia *55 E5* Orientale, NE Dem. Rep. Congo
Bünyan *94 D3* Kayseri, C Turkey (Türkiye)
Buraida *see* Buraydah
Buraydah *98 B4 var.* Buraida. Al Qaşīm, N Saudi Arabia
Burdigala *see* Bordeaux
Burdur *94 B4 var.* Buldur. Burdur, SW Turkey (Türkiye)
Burdur Gölü *94 B4 salt lake* SW Turkey (Türkiye)
Burē *50 C4* Āmara, N Ethiopia
Burgas *82 E2 var.* Bourgas. Burgas, E Bulgaria
Burgaski Zaliv *82 E2 gulf* E Bulgaria
Burgos *70 D2* Castilla y León, N Spain
Burgundy *see* Bourgogne
Burhan Budai Shan *104 D4 mountain range* C China
Buriram *115 D5 var.* Buri Ram, Puriramya. Buri Ram, E Thailand
Buri Ram *see* Buriram
Burjassot *71 F3* Comunitat Valenciana, E Spain
Burkburnett *27 F2* Texas, SW USA
Burketown *126 B3* Queensland, NE Australia
Burkina *see* Burkina Faso
Burkina Faso *53 E4 off.* Burkina Faso; *var.* Burkina; *prev.* Upper Volta. *country* W Africa
Burley *24 D4* Idaho, NW USA
Burlington *23 G4* Iowa, C USA
Burlington *19 F2* Vermont, NE USA
Burma *see* Myanmar
Burnie *127 C8* Tasmania, SE Australia
Burns *24 C3* Oregon, NW USA
Burnside *15 F3 river* Nunavut, NW Canada
Burnsville *23 F2* Minnesota, N USA
Burrel *79 D6 var.* Burreli. Dibër, C Albania
Burreli *see* Burrel
Burriana *see* Borriana
Bursa *94 B3 var.* Brussa, *prev.* Brusa; *anc.* Prusa. Bursa, NW Turkey (Türkiye)
Būr Saʿīd *50 B1 var.* Port Said. N Egypt
Burtnieks *84 C3 var.* Burtnieks Ezers. *lake* N Latvia
Burtnieks Ezers *see* Burtnieks
Burundi *51 B7 off.* Republic of Burundi; *prev.* Kingdom of Burundi, Urundi. *country* C Africa
Burundi, Kingdom of *see* Burundi
Burundi, Republic of *see* Burundi
Buru, Pulau *117 F4 prev.* Boeroe. *island* E Indonesia
Busan *107 E4 off.* Busan Gwang-yeoksi, *prev.* Pusan, *Jap.* Fusan. SE South Korea
Busan Gwang-yeoksi *see* Busan
Büshehr/Bushire *see* Bandar-e Būshehr
Busra *see* Al Başrah, Iraq
Busselton *125 A7* Western Australia
Bussora *see* Al Başrah
Buta *55 D5* Orientale, N Dem. Rep. Congo
Butembo *55 E5* Nord-Kivu, NE Dem. Rep. Congo
Butler *19 E4* Pennsylvania, NE USA
Buton, Pulau *117 E4 var.* Pulau Butung; *prev.* Boetoeng. *island* C Indonesia
Bütow *see* Bytów
Butte *22 B2* Montana, NW USA
Butterworth *116 B3* Pinang, Peninsular Malaysia
Button Islands *17 E1 island group* Nunavut, NE Canada
Butuan *117 F2 off.* Butuan City. Mindanao, S Philippines
Butuan City *see* Butuan
Butung, Pulau *see* Buton, Pulau
Butuntum *see* Bitonto
Buulobarde *51 D5 var.* Buulo Berde. Hiiraan, C Somalia
Buulo Berde *see* Buulobarde
Buur Gaabo *51 D6* Jubbada Hoose, S Somalia
Buxoro *100 D2 var.* Bokhara, *Rus.* Bukhara. Buxoro Viloyati, C Uzbekistan
Buynaksk *89 B8* Respublika Dagestan, SW Russia
Büyük Ağrı Dağı *95 F3 var.* Ağrı Daği, Kūh-e Ārārat-e Bozorg, Masis, *Eng.* Great Ararat, Mount Ararat. *mountain* E Turkey (Türkiye)
Büyükmenderes Nehri *94 A4 river* SW Turkey (Türkiye)
Buzău *86 C4* Buzău, SE Romania
Büzmeýin *see* Abadan
Buzuluk *89 D6* Orenburgskaya Oblast', W Russia
Byahoml' *85 D5 Rus.* Begoml'. Vitsyebskaya Voblasts', N Belarus
Byalynichy *85 D6 Rus.* Belynichi. Mahilyowskaya Voblasts', E Belarus
Byan Tumen *see* Choybalsan
Byarezina *85 D6 prev.* Byerezino, *Rus.* Berezina. *river* C Belarus
Bydgoszcz *76 C3 Ger.* Bromberg. Kujawski-pomorskie, C Poland
Byelaruskaya Hrada *85 B6 Rus.* Belorusskaya Gryada. *ridge* N Belarus
Byerezino *see* Byarezina
Byron Island *see* Nikunau
Bystrovka *see* Kemin
Bytča *77 C5* Žilinský Kraj, N Slovakia
Bytom *77 C5 Ger.* Beuthen. Śląskie, S Poland
Bytów *76 C2 Ger.* Bütow. Pomorskie, N Poland
Byuzmeyin *see* Abadan
Byval'ki *85 D8* Homyel'skaya Voblasts', SE Belarus
Byzantium *see* İstanbul

C

Caála *56 B2 var.* Kaala, Robert Williams, *Port.* Vila Robert Williams. Huambo, C Angola
Caazapá *42 D3* Caazapá, S Paraguay
Caballo Reservoir *26 C3 reservoir* New Mexico, SW USA
Cabanaquinta *70 D1 var.* Cabañaquinta. Asturias, N Spain
Cabañaquinta *see* Cabanaquinta
Cabanatuan *117 E1 off.* Cabanatuan City. Luzon, N Philippines
Cabanatuan City *see* Cabanatuan
Cabillonum *see* Chalon-sur-Saône
Cabimas *36 C1* Zulia, NW Venezuela
Cabinda *56 A1 var.* Kabinda. Cabinda, NW Angola
Cabinda *56 A1 var.* Kabinda. *province* NW Angola
Cabo Verde, Republic of *see* Cape Verde
Cahora Bassa, Albufeira de *56 D2 var.* Lake Cabora Bassa. *reservoir* NW Mozambique
Cabora Bassa, Lake *see* Cahora Bassa, Albufeira de
Caborca *28 B1* Sonora, NW Mexico
Cabot Strait *17 G4 strait* E Canada
Cabo Verde, Ilhas do *see* Cape Verde
Cabras, Ilha das *54 E2 island* S Sao Tome and Principe, Africa, E Atlantic Ocean
Cabrera, Illa de *71 G3 island* E Spain
Cáceres *70 C3 Ar.* Qazris. Extremadura, W Spain
Cachimbo, Serra do *41 E2 mountain range* C Brazil
Caconda *56 B2* Huíla, C Angola
Čadca *77 C5 Hung.* Csaca. Žilinský Kraj, N Slovakia
Cadillac *18 C2* Michigan, N USA
Cadiz *117 E2 off.* Cadiz City. Negros, C Philippines
Cádiz *70 C5 anc.* Gades, Gadier, Gadir, Gadire. Andalucía, SW Spain
Cadiz City *see* Cadiz
Cádiz, Golfo de *70 B5 Eng.* Gulf of Cadiz. *gulf* Portugal/Spain
Cadiz, Gulf of *see* Cádiz, Golfo de
Cadurcum *see* Cahors
Caen *68 B3* Calvados, N France
Caene/Caenepolis *see* Qinā
Caerdydd *see* Cardiff
Caer Glou *see* Gloucester
Caer Gybi *see* Holyhead
Caerleon *see* Chester
Caer Luel *see* Carlisle
Caesaraugusta *see* Zaragoza
Caesarea Mazaca *see* Kayseri
Caesarobriga *see* Talavera de la Reina
Caesarodunum *see* Tours
Caesaromagus *see* Beauvais
Caesena *see* Cesena
Cafayate *42 C2* Salta, N Argentina
Cagayan de Oro *117 E2 off.* Cagayan de Oro City. Mindanao, S Philippines
Cagayan de Oro City *see* Cagayan de Oro
Cagliari *75 A6 anc.* Caralis. Sardegna, Italy, C Mediterranean Sea
Caguas *33 F3* E Puerto Rico
Cahors *69 C5 anc.* Cadurcum. Lot, S France
Cahul *86 D4 Rus.* Kagul. S Moldova
Caicos Passage *32 D2 strait* The Bahamas/Turks and Caicos Islands
Caiffa *see* Hefa
Cailungo *74 E1* N San Marino
Caiphas *see* Hefa
Cairns *126 D3* Queensland, NE Australia
Cairo *see* Al Qāhirah
Cairo *18 B5* Illinois, N USA
Caisleán an Bharraigh *see* Castlebar
Cajamarca *38 B3 prev.* Caxamarca. Cajamarca, NW Peru
Čakovec *78 B2 Ger.* Csakathurn, *Hung.* Csáktornya; *prev. Ger.* Tschakathurn. Medimurje, N Croatia
Calabar *53 G5* Cross River, S Nigeria
Calabozo *36 D2* Guárico, C Venezuela
Calafat *86 B5* Dolj, SW Romania
Calafate *see* El Calafate
Calahorra *71 E2* La Rioja, N Spain
Calais *68 C2* Pas-de-Calais, N France
Calais *19 H2* Maine, NE USA
Calais, Pas de *see* Dover, Strait of
Calama *42 B2* Antofagasta, N Chile
Calarasi *see* Călăraşi
Călăraşi *86 D3 var.* Călăras, *Rus.* Kalarash. C Moldova
Călăraşi *86 C5* Călăraşi, SE Romania
Calatayud *71 E2* Aragón, NE Spain
Calbayog *117 E2 off.* Calbayog City. Samar, C Philippines
Calbayog City *see* Calbayog
Calcutta *see* Kolkāta
Caldas da Rainha *70 B3* Leiria, W Portugal
Caldera *42 B3* Atacama, N Chile
Caldwell *24 C3* Idaho, NW USA
Caledonia *30 C1* Corozal, N Belize
Caleta Olivia *43 B6* Santa Cruz, SE Argentina
Calgary *15 E5* Alberta, SW Canada
Cali *36 B3* Valle del Cauca, W Colombia
Calicut *see* Kozhikode
California *25 B7 off.* State of California, *also known as* El Dorado, The Golden State. *state* W USA
California, Golfo de *28 B2 Eng.* Gulf of California; *prev.* Sea of Cortez. *gulf* W Mexico
California, Gulf of *see* California, Golfo de
Călimăneşti *86 B4* Vâlcea, SW Romania
Calisia *see* Kalisz
Callabonna, Lake *127 B5 lake* South Australia
Callao *38 C4* Callao, W Peru
Callatis *see* Mangalia
Callosa de Segura *71 F4* Comunitat Valenciana, E Spain
Calmar *see* Kalmar
Caloundra *127 E5* Queensland, E Australia
Caltanissetta *75 C7* Sicilia, Italy, C Mediterranean Sea
Caluula *50 E4* Bari, NE Somalia
Calvinia *56 C5* W South Africa
Camabatela *56 B1* Cuanza Norte, NW Angola
Camacupa *56 B2 var.* General Machado, *Port.* Vila General Machado. Bié, C Angola
Camagüey *32 C2 prev.* Puerto Príncipe. Camagüey, C Cuba
Camagüey, Archipiélago de *32 C2 island group* C Cuba
Camana *39 E4 var.* Camaná. Arequipa, SW Peru
Camargue *69 D6 physical region* SE France
Ca Mau *115 D6 var.* Quan Long. Minh Hai, S Vietnam
Cambay, Gulf of *see* Khambhāt, Gulf of
Camberia *see* Chambéry
Cambodia *115 D5 off.* Kingdom of Cambodia, *var.* Democratic Kampuchea, Roat Kampuchea, *Cam.* Kampuchea; *prev.* People's Democratic Republic of Kampuchea. *country* SE Asia
Cambodia, Kingdom of *see* Cambodia
Cambrai *68 C2 Flem.* Kambryk, *prev.* Cambray; *anc.* Cameracum. Nord, N France
Cambray *see* Cambrai
Cambrian Mountains *67 C6 mountain range* C Wales, United Kingdom
Cambridge *32 A4* W Jamaica
Cambridge *128 D3* Waikato, North Island, New Zealand
Cambridge *67 E6 Lat.* Cantabrigia. E England, United Kingdom
Cambridge *19 F4* Maryland, NE USA
Cambridge *18 D4* Ohio, NE USA
Cambridge Bay *15 F3 var.* Ikaluktutiak. Victoria Island, Nunavut, NW Canada
Camden *20 B2* Arkansas, C USA
Camellia State *see* Alabama
Cameracum *see* Cambrai
Cameroon *54 A4 off.* Republic of Cameroon, *Fr.* Cameroun. *country* W Africa
Cameroon, Republic of *see* Cameroon
Cameroun *see* Cameroon
Camocim *41 F2* Ceará, E Brazil
Camopi *37 H3* E French Guiana
Campamento *30 C2* Olancho, C Honduras
Campania *75 D5 Eng.* Champagne. *region* S Italy
Campbell, Cape *129 D5 headland* South Island, New Zealand
Campbell Island *120 D5 var.* Motu Ihupuku *island* S New Zealand
Campbell Plateau *120 D5 undersea plateau* SW Pacific Ocean
Campbell River *14 D5* Vancouver Island, British Columbia, SW Canada
Campeche *29 G4* Campeche, SE Mexico
Campeche, Bahía de *29 F4 Eng.* Bay of Campeche. *bay* E Mexico
Campeche, Bay of *see* Campeche, Bahía de
Cẩm Phả *114 E3* Quang Ninh, N Vietnam
Câmpina *86 C4 prev.* Cîmpina. Prahova, SE Romania
Campina Grande *41 G2* Paraíba, E Brazil
Campinas *41 F4* São Paulo, S Brazil
Campobasso *75 D5* Molise, C Italy
Campo Criptana *see* Campo de Criptana
Campo de Criptana *71 E3 var.* Campo Criptana. Castilla-La Mancha, C Spain
Campo dos Goytacazes *41 F4 var.* Campos. Rio de Janeiro, SE Brazil
Campo Grande *41 E4 state capital* Mato Grosso do Sul, SW Brazil
Campos *see* Campo dos Goytacazes
Câmpulung *86 B4 prev.* Câmpulung-Muşcel, Cîmpulung. Argeş, S Romania
Câmpulung-Muşcel *see* Câmpulung
Campus Stellae *see* Santiago de Compostela
Cam Ranh *115 E6* Khanh Hoa, S Vietnam
Canada *12 B4 country* N North America
Canada Basin *12 C2 undersea basin* Arctic Ocean
Canadian River *27 E2 river* SW USA
Çanakkale *94 A3 var.* Dardanelli; *prev.* Chanak, Kale Sultanie. Çanakkale, W Turkey (Türkiye)
Cananea *28 B1* Sonora, NW Mexico
Canarreos, Archipiélago de los *32 B2 island group* W Cuba
Canary Islands *48 A2 Eng.* Canary Islands. *island group* Spain, NE Atlantic Ocean
Canary Islands *see* Canarias, Islas
Cañas *30 D4* Guanacaste, NW Costa Rica
Canaveral, Cape *21 E4 headland* Florida, SE USA
Canavieiras *41 G3* Bahia, E Brazil
Canberra *120 C4 country capital* (Australia) Australian Capital Territory, SE Australia
Cancún *29 H3* Quintana Roo, SE Mexico
Candia *see* Irákleio
Canea *see* Chaniá
Cangzhou *106 D4* Hebei, E China
Caniapiscau *17 E2 river* Québec, E Canada
Caniapiscau, Réservoir de *16 D3 reservoir* Québec, C Canada
Canik Dağları *94 D2 mountain range* N Turkey (Türkiye)
Canillo *69 A7* Canillo, C Andorra Europe
Çankırı *94 C3 var.* Chankiri; *anc.* Gangra, Germanicopolis. Çankın, N Turkey (Türkiye)
Cannanore *see* Kannur
Cannes *69 D6* Alpes-Maritimes, SE France
Canoas *41 E5* Rio Grande do Sul, S Brazil
Canon City *22 C5* Colorado, C USA
Cantabria *70 D1 autonomous community* N Spain
Cantábrica, Cordillera *70 C1 mountain range* N Spain
Cantabrigia *see* Cambridge
Cantaura *37 E2* Anzoátegui, NE Venezuela
Canterbury *67 E7 hist.* Cantwaraburh; *anc.* Durovernum, *Lat.* Cantuaria. SE England, United Kingdom
Canterbury Bight *129 C6 bight* South Island, New Zealand
Canterbury Plains *129 C6 plain* South Island, New Zealand
Cần Thơ *115 E6* Cần Thơ, S Vietnam
Canton *20 B2* Mississippi, S USA
Canton *18 D4* Ohio, N USA
Canton *see* Guangzhou
Canton Island *see* Kanton
Cantuaria/Cantwaraburh *see* Canterbury
Canyon *27 E2* Texas, SW USA
Cao Bằng *114 D3 var.* Caobang. Cao Bằng, N Vietnam
Caobang *see* Cao Bằng
Cap-Breton, Île du *see* Cape Breton Island
Cape Barren Island *127 C8 island* Furneaux Group, Tasmania, SE Australia
Cape Basin *47 B7 undersea basin* S Atlantic Ocean
Cape Breton Island *17 G4 Fr.* Île du Cap-Breton. *island* Nova Scotia, SE Canada
Cape Charles *19 F5* Virginia, NE USA
Cape Coast *53 E5 prev.* Cape Coast Castle. S Ghana
Cape Coast Castle *see* Cape Coast
Cape Girardeau *23 H5* Missouri, C USA
Capelle aan den IJssel *64 C4* Zuid-Holland, SW Netherlands
Cape Palmas *see* Harper
Cape Saint Jacques *see* Vung Tau
Cape Town *56 B5 var.* Ekapa, *Afr.* Kaapstad, Kapstad. *country capital* (South Africa-legislative capital) Western Cape, SW South Africa
Cape Verde *52 A2 off.* Republic of Cabo Verde, *Port.* Cabo Verde, Ilhas do Cabo Verde. *country* E Atlantic Ocean
Cape Verde Basin *44 C4 undersea basin* E Atlantic Ocean
Cape Verde Plain *44 C4 abyssal plain* E Atlantic Ocean
Cape York Peninsula *126 C2 peninsula* Queensland, N Australia
Cap-Haïtien *32 D3 var.* Le Cap. N Haiti

Capira 31 G5 Panamá, C Panama
Capitán Pablo Lagerenza 42 D1 var. Mayor Pablo Lagerenza. Chaco, N Paraguay
Capodistria see Koper
Capri 75 C5 island S Italy
Caprivi Concession see Caprivi Strip
Caprivi Strip 56 C3 Ger. Caprivizipfel; prev. Caprivi Concession. cultural region NE Namibia
Caprivizipfel see Caprivi Strip
Cap Saint-Jacques see Vung Tau
Caquetá, Río 36 C5 var. Rio Japurá, Yapurá. river Brazil/Colombia
Caquetá, Río see Japurá, Rio
CAR see Central African Republic
Caracal 86 B5 Olt, S Romania
Caracaraí 40 D1 Rondônia, W Brazil
Caracas 36 D1 country capital (Venezuela) Distrito Federal, N Venezuela
Caralis see Cagliari
Caratasca, Laguna de 31 E2 lagoon NE Honduras
Carballiño see O Carballiño
Carbón, Laguna del 43 B7 physical feature SE Argentina
Carbondale 18 B5 Illinois, N USA
Carbonia 75 A6 var. Carbonia Centro. Sardegna, Italy, C Mediterranean Sea
Carbonia Centro see Carbonia
Carcaso see Carcassonne
Carcassonne 69 C6 anc. Carcaso. Aude, S France
Cardamomes, Chaîne des see Krâvanh, Chuŏr Phnum
Cardamom Mountains see Krâvanh, Chuŏr Phnum
Cárdenas 32 B2 Matanzas, W Cuba
Cardiff 67 C7 Wel. Caerdydd. national capital S Wales, United Kingdom
Cardigan Bay 67 C6 bay W Wales, United Kingdom
Carei 86 B3 Ger. Gross-Karol, Karol, Hung. Nagykároly; prev. Careii-Mari. Satu Mare, NW Romania
Careii-Mari see Carei
Carey, Lake 125 B6 lake Western Australia
Cariaco 37 E1 Sucre, NE Venezuela
Caribbean Sea 32 C4 sea W Atlantic Ocean
Caribrod see Dimitrovgrad
Carlisle 66 C4 anc. Caer Luel, Luguvallium, Luguvallum. NW England, United Kingdom
Carlow 67 B6 Ir. Ceatharlach. SE Ireland
Carlsbad 26 D3 New Mexico, SW USA
Carlsbad see Karlovy Vary
Carlsberg Ridge 118 B4 undersea ridge S Arabian Sea
Carlsruhe see Karlsruhe
Carmana/Carmania see Kermān
Carmarthen 67 C6 SW Wales, United Kingdom
Carmaux 69 C6 Tarn, S France
Carmel 18 C4 Indiana, N USA
Carmelita 30 B1 Petén, N Guatemala
Carmen 29 G4 var. Ciudad del Carmen. Campeche, SE Mexico
Carmona 70 C4 Andalucía, S Spain
Carmona see Uíge
Carnaro see Kvarner
Carnarvon 125 A5 Western Australia
Carnegie, Lake 125 B5 salt lake Western Australia
Car Nicobar 111 F3 island Nicobar Islands, India, NE Indian Ocean
Caroço, Ilha 54 E1 island N Sao Tome and Principe, Africa, E Atlantic Ocean
Carolina 41 F2 Maranhão, E Brazil
Caroline Island see Millennium Island
Caroline Islands 122 B2 island group C Micronesia
Carolopois see Châlons-en-Champagne
Caroní, Río 37 E3 river E Venezuela
Caronium see A Coruña
Carora 36 C1 Lara, N Venezuela
Carpathian Mountains 59 E4 var. Carpathians, Cz./Pol. Karpaty, Ger. Karpaten. mountain range E Europe
Carpathians see Carpathian Mountains
Carpathos/Carpathus see Kárpathos
Carpaţii Meridionali 86 B4 var. Alpi Transilvaniei, Carpaţii Sudici, Eng. South Carpathians, Transylvanian Alps, Ger. Südkarpaten, Transsylvanische Alpen, Hung. Déli-Kárpátok, Erdélyi-Havasok. mountain range C Romania
Carpaţii Sudici see Carpaţii Meridionali
Carpentaria, Gulf of 126 B2 gulf N Australia
Carpi 74 C2 Emilia-Romagna, N Italy
Carrara 74 B3 Toscana, C Italy
Carson City 25 C5 state capital Nevada, W USA
Carson Sink 25 C5 salt flat Nevada, W USA
Carstensz, Puntjak see Jaya, Puncak
Cartagena 36 B1 var. Cartagena de los Indes. Bolívar, NW Colombia
Cartagena 71 F4 anc. Carthago Nova. Murcia, SE Spain
Cartagena de los Indes see Cartagena
Cartago 31 E4 Cartago, C Costa Rica
Carthage 23 F5 Missouri, C USA
Carthago Nova see Cartagena
Cartwright 17 F2 Newfoundland and Labrador, E Canada
Carúpano 37 E1 Sucre, NE Venezuela
Carusbur see Cherbourg
Caruthersville 23 H5 Missouri, C USA
Cary 21 F1 North Carolina, SE USA
Casablanca 48 C2 Ar. Dar-el-Beida. NW Morocco
Casa Grande 26 B2 Arizona, SW USA
Cascade Range 24 B3 mountain range Oregon/Washington, NW USA
Cascadia Basin 12 A4 undersea basin NE Pacific Ocean
Cascais 70 B4 Lisboa, C Portugal
Caserta 75 D5 Campania, S Italy
Casey 132 D4 Australian research station Antarctica
Čáslav 77 B5 Ger. Tschaslau. Střední Čechy, C Czechia (Czech Republic)
Casper 22 C3 Wyoming, C USA
Caspian Depression 89 B7 Kaz. Kaspīy Mangy Oypaty, Rus. Prikaspiyskaya Nizmennost'. depression Kazakhstan/Russia
Caspian Sea 92 A4 Az. Xäzär Dänizi, Kaz. Kaspiy Tengizi, Per. Baḥr-e Khazar, Daryā-ye Khazar, Rus. Kaspiyskoye More. inland sea Asia/Europe
Cassai see Kasai
Cassel see Kassel
Castamoni see Kastamonu
Casteggio 74 B2 Lombardia, N Italy
Castelló de la Plana see Castellón de la Plana
Castellón/Castellón de la Plana see Castelló de la Plana
Castelló de la Plana 71 F3 var. Castellón, Castellón de la Plana. Comunitat Valenciana, E Spain
Castelnaudary 69 C6 Aude, S France
Castelo Branco 70 C3 Castelo Branco, C Portugal
Castelsarrasin 69 B6 Tarn-et-Garonne, S France
Castelvetrano 75 C7 Sicilia, Italy, C Mediterranean Sea
Castilla-La Mancha 71 E3 autonomous community NE Spain
Castilla y León 70 C2 autonomous community NW Spain
Castlebar 67 A5 Ir. Caisleán an Bharraigh. W Ireland
Castleford 67 D5 N England, United Kingdom
Castle Harbour 20 B5 inlet Bermuda, NW Atlantic Ocean
Castra Regina see Regensburg
Castricum 64 C3 Noord-Holland, W Netherlands
Castries 33 F1 country capital (Saint Lucia) N Saint Lucia
Castro 43 B6 Los Lagos, W Chile
Castrovillari 75 D6 Calabria, SW Italy
Castuera 70 D4 Extremadura, W Spain
Caswell Sound see Taitetimu
Catacamas 30 D2 Olancho, C Honduras
Catacaos 38 B3 Piura, NW Peru
Catalan Bay 71 H4 bay E Gibraltar, Mediterranean Sea
Cataluña 71 G2 N Spain
Catamarca see San Fernando del Valle de Catamarca
Catania 75 D7 Sicilia, Italy, C Mediterranean Sea
Catanzaro 75 D6 Calabria, SW Italy
Catarroja 71 F3 Comunitat Valenciana, E Spain
Cat Island 32 C1 island C The Bahamas
Catskill Mountains 19 F3 mountain range New York, NE USA
Cattaro see Kotor
Cauca, Río 36 B2 river N Colombia
Caucasia 36 B2 Antioquia, NW Colombia
Caucasus 59 G4 Rus. Kavkaz. mountain range Georgia/Russia
Caura, Río 37 E3 river C Venezuela
Cavaia see Kavajë
Cavalla 52 D5 var. Cavally, Cavally Fleuve. river Ivory Coast/Liberia
Cavally/Cavally Fleuve see Cavalla
Caviana de Fora, Ilha 41 E1 var. Ilha Caviana. island N Brazil
Caviana, Ilha see Caviana de Fora, Ilha
Cawnpore see Kānpur
Caxamarca see Cajamarca
Caxito 56 B1 Bengo, NW Angola
Cayenne 37 H3 dependent territory/arrondissement capital (French Guiana) NE French Guiana
Cayes 32 D3 var. Les Cayes. SW Haiti
Cayman Brac 32 B3 island E Cayman Islands
Cayman Islands 32 B3 UK Overseas Territory W West Indies
Cayo see San Ignacio
Cay Sal 32 B2 islet SW The Bahamas
Cazin 78 B3 Federacija Bosna I Hercegovina, NW Bosnia and Herzegovina
Cazorla 71 E4 Andalucía, S Spain
Ceadâr-Lunga see Ciadîr-Lunga
Ceará 41 F2 off. Estado do Ceará. state/region C Brazil
Ceará see Fortaleza
Ceara Abyssal Plain see Ceará Plain
Ceará, Estado do see Ceará
Ceará Plain 34 E3 var. Ceara Abyssal Plain. abyssal plain W Atlantic Ocean
Ceatharlach see Carlow
Cébaco, Isla 31 F5 island SW Panama
Cebu 117 E2 off. Cebu City. Cebu, C Philippines
Cebu City see Cebu
Čechy see Bohemia
Cecina 74 B3 Toscana, C Italy
Cedar City 22 A5 Utah, W USA
Cedar Falls 23 G3 Iowa, C USA
Cedar Lake 16 A2 lake Manitoba, C Canada
Cedar Rapids 23 G3 Iowa, C USA
Cedros, Isla 28 A2 island W Mexico
Ceduna 127 A6 South Australia
Cefalù 75 C7 anc. Cephaloedium. Sicilia, Italy, C Mediterranean Sea
Celebes see Sulawesi
Celebes Sea 117 E3 Ind. Laut Sulawesi. sea Indonesia/Philippines
Celje 73 E7 Ger. Cilli. C Slovenia
Celldömölk 77 C6 Vas, W Hungary
Celle 72 B3 var. Zelle. Niedersachsen, N Germany
Celovec see Klagenfurt
Celtic Sea 67 B7 Ir. An Mhuir Cheilteach. sea SW British Isles
Celtic Shelf 58 B3 continental shelf E Atlantic Ocean
Cenderawasih, Teluk 117 G4 var. Teluk Irian, Teluk Sarera. bay W Pacific Ocean
Cenon 69 B5 Gironde, SW France
Centennial State see Colorado
Centrafricaine, République see Central African Republic
Central African Republic 54 C4 var. République Centrafricaine, abbrev. CAR; prev. Ubangi-Shari, Oubangui-Chari, Territoire de l'Oubangui-Chari. country C Africa
Central, Cordillera 36 B3 mountain range W Colombia
Cordillera Central 33 E3 mountain range C Dominican Republic
Cordillera Central 31 F5 mountain range C Panama
Central, Cordillera 117 E1 mountain range Luzon, N Philippines
Central Group see Inner Islands
Centralia 24 B2 Washington, NW USA
Central Indian Ridge see Mid-Indian Ridge
Central Makran Range 112 A3 mountain range W Pakistan
Central Pacific Basin 120 D1 undersea basin C Pacific Ocean
Central, Planalto 41 F3 var. Brazilian Highlands. mountain range E Brazil
Central Provinces and Berar see Madhya Pradesh
Central Range 122 B3 mountain range NW Papua New Guinea
Central Russian Upland see Srednerusskaya Vozvyshennost'
Central Siberian Plateau 92 D3 var. Central Siberian Uplands, Eng. Central Siberian Plateau. mountain range N Russia
Central Siberian Plateau/Central Siberian Uplands see Srednesibirskoye Ploskogor'ye
Central, Sistema 70 D3 mountain range C Spain
Central Valley 25 B6 valley California, W USA
Centum Cellae see Civitavecchia
Ceos see Tziá
Cephaloedium see Cefalù
Ceram see Seram, Pulau
Ceram Sea see Laut Seram
Cerasus see Giresun
Cereté 36 B2 Córdoba, NW Colombia
Cergy-Pontoise see Pontoise
Cerignola 75 D5 Puglia, SE Italy
Çerkeş 94 C2 Çankin, N Turkey (Türkiye)
Cernăuţi see Chernivtsi
Cernay 68 E4 Haut-Rhin, NE France
Cerro de Pasco 38 C3 Pasco, C Peru
Cervera 71 F2 Cataluña, NE Spain
Cervino, Monte see Matterhorn
Cesena 74 C3 anc. Caesena. Emilia-Romagna, N Italy
Cēsis 84 D3 Ger. Wenden. Cēsis, C Latvia
České Budějovice 77 B5 Ger. Budweis. Jihočeský Kraj, S Czechia (Czech Republic)
Česko see Czechia (Czech Republic)
Český Krumlov 77 A5 var. Böhmisch-Krumau, Ger. Krummau. Jihočeský Kraj, S Czech Republic (Czechia)
Český Les see Bohemian Forest
Cetatea Albă see Bilhorod-Dnistrovskyi
Cetatea Damboviţei see Bucureşti
Cetinje 79 C5 It. Cettigne. S Montenegro
Cette see Sète
Cettigne see Cetinje
Ceuta 48 C2 autonomous city of Spain Spain, N Africa
Cévennes 69 C6 mountain range S France
Ceyhan 94 D4 Adana, S Turkey (Türkiye)
Ceylanpınar 95 E4 Şanlıurfa, SE Turkey (Türkiye)
Ceylon see Sri Lanka
Ceylon Plain 102 B4 abyssal plain N Indian Ocean
Ceyre to the Caribs see Marie-Galante
Chachapoyas 38 B2 Amazonas, NW Peru
Chachevichy 85 D6 Rus. Chechevichi. Mahilyowskaya Voblasts', E Belarus
Chaco see Gran Chaco
Chad 54 C3 off. Republic of Chad, Fr. Tchad. country C Africa
Chad, Lake 54 B3 Fr. Lac Tchad. lake C Africa
Chad, Republic of see Chad
Chadron 22 D3 Nebraska, C USA
Chadyr-Lunga see Ciadîr-Lunga
Chagai Hills 112 A2 var. Chah Gay. mountain range Afghanistan/Pakistan
Chaghasarāy see Asadābād
Chagos-Laccadive Plateau 102 B4 undersea plateau N Indian Ocean
Chagos Trench 119 C5 trench N Indian Ocean
Chah Gay see Chagai Hills
Chaillu, Massif du 55 B6 mountain range C Gabon
Chain Ridge 118 B4 undersea ridge W Indian Ocean
Chajul 30 B2 Quiché, W Guatemala
Chakhānsūr 100 D5 Nīmrūz, SW Afghanistan
Chala 38 D4 Arequipa, SW Peru
Chalatenango 30 C3 Chalatenango, N El Salvador
Chalcidice see Chalkidikí
Chalcis see Chalkída
Chalki 83 E7 island Dodekánisa, Greece, Aegean Sea
Chalkída 83 C5 var. Halkida, prev. Khalkís; anc. Chalcis. Evvoia, E Greece
Chalkidikí 82 C4 var. Khalkidhikí; anc. Chalcidice. peninsula NE Greece
Challans 68 B4 Vendée, NW France
Challapata 39 F4 Oruro, SW Bolivia
Challenger Deep 130 B3 trench W Pacific Ocean
Challenger Fracture Zone 131 F4 tectonic feature SE Pacific Ocean
Châlons-en-Champagne 68 D3 prev. Châlons-sur-Marne, hist. Arcae Remorum; anc. Carolopois. Marne, NE France
Châlons-sur-Marne see Châlons-en-Champagne
Chalon-sur-Saône 68 D4 anc. Cabillonum. Saône-et-Loire, C France
Cha Mai see Thung Song
Chaman 112 B2 Baluchistan, SW Pakistan
Chambéry 69 D5 anc. Camberia. Savoie, E France
Champagne 68 D3 cultural region N France
Champagne see Campania
Champaign 18 B4 Illinois, N USA
Champasak 115 D5 Champasak, S Laos
Champlain, Lake 19 F2 lake Canada/USA
Champotón 29 G4 Campeche, SE Mexico
Chanak see Çanakkale
Chañaral 42 B3 Atacama, N Chile
Chan-chiang/Chanchiang see Zhanjiang
Chandeleur Islands 20 C3 island group Louisiana, S USA
Chandigarh 112 D2 state capital Punjab, N India
Chandrapur 113 E5 Mahārāshtra, C India
Changan see Xi'an, Shaanxi, C China
Changane 57 E3 river S Mozambique
Changchun 106 D3 var. Ch'ang-ch'un, Ch'ang-ch'un; prev. Hsinking. province capital Jilin, NE China
Ch'angch'un/Ch'ang-ch'un see Changchun
Chang Jiang 106 A5 Eng. Yangtze; var. Yangtze Kiang. river SW China
Changkiakow see Zhangjiakou
Chang, Ko 115 C6 island S Thailand
Changsha 106 C5 var. Ch'angsha, Ch'ang-sha. province capital Hunan, S China
Ch'angsha/Ch'ang-sha see Changsha
Changzhi 106 C4 Shanxi, C China
Chaniá 83 C7 var. Hania, Khaniá, Eng. Canea; anc. Cydonia. Kríti, Greece, E Mediterranean Sea
Chañi, Nevado de 42 B2 mountain NW Argentina
Chankiri see Çankırı
Channel Islands 67 C8 Fr. Iles Normandes. island group S English Channel
Channel Islands 25 B8 island group California, W USA
Channel-Port aux Basques 17 G4 Newfoundland and Labrador, SE Canada
Channel, The see English Channel
Channel Tunnel 68 C2 tunnel France/United Kingdom
Chantabun/Chantaburi see Chanthaburi
Chantada 70 C1 Galicia, NW Spain
Chanthaburi 115 C6 var. Chantabun, Chantaburi. Chantaburi, S Thailand
Chanute 23 F5 Kansas, C USA
Chaouèn see Chefchaouen
Chaoyang 106 D3 Liaoning, NE China
Chapala, Lago de 28 D4 lake C Mexico
Chapan, Gora 100 B3 mountain C Turkmenistan
Chapayevsk 89 C6 Samarskaya Oblast', W Russia
Chaplynka 87 F4 Khersons'ka Oblast', S Ukraine
Chapra see Chhapra
Charcot Seamounts 58 B3 seamount range E Atlantic Ocean
Chardzhev see Türkmenabat
Chardzhou/Chardzhui see Türkmenabat
Charente 69 B5 cultural region W France
Charente 69 B5 river W France
Chari 54 B3 var. Shari. river Central African Republic/Chad
Chārīkār 101 E4 Parvān, NE Afghanistan
Charity 37 F2 NW Guyana
Chärjew see Türkmenabat
Charkhlik/Charkhliq see Ruoqiang
Charleroi 65 C7 Hainaut, S Belgium
Charlesbourg 17 E4 Québec, SE Canada
Charles de Gaulle 68 E1 (Paris) Seine-et-Marne, N France
Charles Island 16 D1 island Nunavut, NE Canada
Charles Island see Santa María, Isla
Charleston 21 F2 South Carolina, SE USA
Charleston 18 D5 state capital West Virginia, NE USA
Charleville 127 D5 Queensland, E Australia
Charleville-Mézières 68 D3 Ardennes, N France
Charlie-Gibbs Fracture Zone 44 C2 tectonic feature N Atlantic Ocean
Charlotte 21 E1 North Carolina, SE USA
Charlotte Amalie 33 F3 prev. Saint Thomas. dependent territory capital (Virgin Islands (US)) Saint Thomas, N Virgin Islands (US)
Charlotte Harbor 21 E5 inlet Florida, SE USA
Charlottenhof see Aegviidu
Charlottesville 19 E5 Virginia, NE USA
Charlottetown 17 F4 province capital Prince Edward Island, Prince Edward Island, SE Canada
Charlotte Town see Roseau, Dominica
Charsk see Shar
Charters Towers 126 D3 Queensland, NE Australia
Chartres 68 C3 anc. Autricum, Civitas Carnutum. Eure-et-Loir, C France
Chashniki 85 D5 Vitsyebskaya Voblasts', N Belarus
Châteaubriant 68 B4 Loire-Atlantique, NW France
Châteaudun 68 C3 Eure-et-Loir, C France
Châteauroux 68 C4 prev. Indreville. Indre, C France
Château-Thierry 68 C3 Aisne, N France
Châtelet 65 C7 Hainaut, S Belgium
Châtelherault see Châtellerault
Châtellerault 68 B4 var. Châtelherault. Vienne, W France
Chatham Island see San Cristóbal, Isla
Chatham Island Rise see Chatham Rise
Chatham Islands 121 E5 island group New Zealand, SW Pacific Ocean
Chatham Rise 120 D5 var. Chatham Island Rise. undersea rise S Pacific Ocean
Chatkal Range 101 F2 Rus. Chatkal'skiy Khrebet. mountain range Kyrgyzstan/Uzbekistan
Chatkal'skiy Khrebet see Chatkal Range
Chāttagām see Chattogram
Chattahoochee River 20 D3 river SE USA
Chattanooga 20 D1 Tennessee, S USA
Chattogram 113 G4 Ben. Chāttagām, prev. Chittagong. Chattogram, SE Bangladesh
Chatyr-Tash 101 G2 Narynskaya Oblast', C Kyrgyzstan
Châu Đốc 115 D6 var. Chauphu, Chau Phu. An Giang, S Vietnam
Chauk 114 A3 Magway, W Myanmar (Burma)
Chaumont 68 D4 prev. Chaumont-en-Bassigny. Haute-Marne, N France
Chaumont-en-Bassigny see Chaumont
Chau Phu see Châu Đốc
Chausy see Chavusy
Chaves 70 C2 anc. Aquae Flaviae. Vila Real, N Portugal
Chávez, Isla see Santa Cruz, Isla
Chavusy 85 E6 Rus. Chausy. Mahilyowskaya Voblasts', E Belarus
Chaykovskiy 89 D5 Permskaya Oblast', NW Russia
Cheb 77 A5 Ger. Eger. Karlovarský Kraj, W Czechia (Czech Republic)
Cheboksary 89 C5 Chuvashskaya Respublika, W Russia
Cheboygan 18 C2 Michigan, N USA
Chechaouèn see Chefchaouen
Chech, Erg 52 D1 desert Algeria/Mali
Chechevichi see Chachevichy
Che-chiang see Zhejiang
Cheduba Island see Manaung Island)
Chefchaouen 48 C2 var. Chaouèn, Chechaouèn, Sp. Xauen. N Morocco
Chefoo see Yantai
Cheju-do see Jeju-do
Cheju Strait see Jeju Strait
Chekiang see Zhejiang
Cheleken see Hazar
Chelkar see Shalqar
Chełm 76 E4 Rus. Kholm. Lubelskie, SE Poland
Chełmno 76 C3 Ger. Culm, Kulm. Kujawski-pomorskie, C Poland
Chełmża 76 C3 Ger. Culmsee, Kulmsee. Kujawski-pomorskie, C Poland
Cheltenham 67 D6 C England, United Kingdom
Chelyabinsk 92 C3 Chelyabinskaya Oblast', C Russia
Chemnitz 72 D4 prev. Karl-Marx-Stadt. Sachsen, E Germany
Chemulpo see Incheon
Chenab 112 C2 river India/Pakistan
Chengchiatun see Liaoyuan
Ch'eng-chou/Chengchow see Zhengzhou
Chengde 106 D3 var. Jehol. Hebei, E China
Chengdu 106 B5 var. Chengtu, Ch'eng-tu. province capital Sichuan, C China
Chenghsien see Zhengzhou
Chengtu/Ch'eng-tu see Chengdu
Chennai 110 D2 prev. Madras. state capital Tamil Nādu, S India
Chenstokhov see Częstochowa
Chen Xian/Chenxian/Chen Xiang see Chenzhou
Chenzhou 106 C6 var. Chenxian, Chen Xian, Chen Xiang. Hunan, S China
Chepelare 82 C3 Smolyan, S Bulgaria
Chepén 38 B3 La Libertad, C Peru
Cher 68 C4 river C France
Cherbourg 68 B3 anc. Carusbur. Manche, N France
Cherepovets 88 B4 Vologodskaya Oblast', NW Russia
Chergui, Chott ech 48 D2 salt lake NW Algeria
Cherikov see Cherykaw
Cherkassy see Cherkasy
Cherkasy 87 E2 Rus. Cherkassy. Cherkas'ka Oblast', C Ukraine
Cherkessk 89 B7 Karachayevo-Cherkesskaya Respublika, SW Russia
Chernigov see Chernihiv
Chernihiv 87 E1 Rus. Chernigov. Chernihivs'ka Oblast', NE Ukraine
Chernivtsi 86 C3 Ger. Czernowitz, Rom. Cernăuţi, Rus. Chernovtsy. Chernivets'ka Oblast', W Ukraine
Cherno More see Black Sea
Chernomorskoye see Chornomorske
Chernovtsy see Chernivtsi
Chernoye More see Black Sea
Chernyakhovsk 84 A4 Ger. Insterburg. Kaliningradskaya Oblast', W Russia
Cherry Hill 19 F4 New Jersey, NE USA
Cherski Range see Cherskogo, Khrebet
Cherskiy 93 G2 Respublika Sakha (Yakutiya), NE Russia
Cherskogo, Khrebet 93 F2 var. Cherski Range. mountain range NE Russia
Cherso see Cres
Cherven' see Chervyen'
Chervonograd see Chervonohrad
Chervonohrad 86 C2 Rus. Chervonograd. L'vivs'ka Oblast', NW Ukraine
Chervyen' 85 D6 Rus. Cherven'. Minskaya Voblasts', C Belarus
Cherykaw 85 E7 Rus. Cherikov. Mahilyowskaya Voblasts', E Belarus
Chesapeake Bay 19 F5 inlet NE USA
Chesha Bay see Chëshskaya Guba
Chëshskaya Guba 133 D5 var. Archangel Bay, Chesha Bay, Dvina Bay. bay NW Russia
Chester 67 C6 Wel. Caerleon, hist. Legaceaster, Lat. Deva, Devana Castra. C England, United Kingdom
Chetumal 29 H4 var. Payo Obispo. Quintana Roo, SE Mexico
Cheviot Hills 66 D4 hill range England/Scotland, United Kingdom
Cheyenne 22 D4 state capital Wyoming, C USA
Cheyenne River 22 D3 river South Dakota/Wyoming, N USA
Chezdi-Oşorheiu see Târgu Secuiesc
Chhapra 113 F3 prev. Chapra. Bihār, N India
Chhattīsgarh 113 E4 cultural region E India
Chiai see Chiayi
Chia-i see Chiayi
Chiang-hsi see Jiangxi
Chiang Mai 114 B4 var. Chiangmai, Chiengmai, Kiangmai. Chiang Mai, NW Thailand
Chiangmai see Chiang Mai
Chiang Rai 114 C3 var. Chianpai, Chienrai, Muang Chiang Rai. Chiang Rai, NW Thailand
Chiang-su see Jiangsu
Chianning/Chian-ning see Nanjing
Chianpai see Chiang Rai
Chianti 74 C3 cultural region C Italy
Chiapa see Chiapa de Corzo
Chiapa de Corzo 29 G5 var. Chiapa. Chiapas, SE Mexico
Chiayi 106 D6 var. Chiai, Chia-i, Kiayi, Jiayi, Jap. Kagi. C Taiwan
Chiba 108 B1 var. Tiba. Chiba, Honshū, S Japan
Chibougamau 16 D3 Québec, SE Canada
Chicago 18 B3 Illinois, N USA
Ch'i-ch'i-ha-erh see Qiqihar
Chickasha 27 G2 Oklahoma, C USA
Chiclayo 38 B3 Lambayeque, NW Peru
Chico 25 B5 California, W USA
Chico, Río 43 B7 river SE Argentina
Chico, Río 43 B6 river S Argentina
Chicoutimi 17 E4 Québec, SE Canada
Chiengmai see Chiang Mai
Chienrai see Chiang Rai
Chiesanuova 74 D2 SW San Marino
Chieti 74 D4 var. Teate. Abruzzo, C Italy
Chifeng 105 G2 var. Ulanhad. Nei Mongol Zizhiqu, N China
Chigirin see Chyhyryn
Chih-fu see Yantai
Chihli see Hebei
Chihli, Gulf of see Bo Hai
Chihuahua 28 C2 Chihuahua, NW Mexico
Childress 27 F2 Texas, SW USA
Chile 42 B3 off. Republic of Chile. country SW South America
Chile Basin 35 A5 undersea basin E Pacific Ocean
Chile Chico 43 B6 Aisén, W Chile
Chile, Republic of see Chile
Chile Rise 35 A7 undersea rise SE Pacific Ocean
Chilia-Nouă see Kiliya
Chililabombwe 56 D2 Copperbelt, C Zambia
Chi-lin see Jilin
Chillán 43 B5 Bío Bío, C Chile
Chillicothe 18 D4 Ohio, N USA
Chill Mhantáin, Sléibhte see Wicklow Mountains
Chiloé, Isla de 43 A6 var. Isla Grande de Chiloé. island W Chile
Chilpancingo 29 E5 var. Chilpancingo de los Bravos. Guerrero, S Mexico
Chilpancingo de los Bravos see Chilpancingo
Chilung see Keelung
Chimán 31 G5 Panamá, E Panama
Chimbay see Chimboy
Chimborazo 38 A1 volcano C Ecuador
Chimbote 38 C3 Ancash, W Peru
Chimboy 100 D1 Rus. Chimbay. Qoraqalpog'iston Respublikasi, NW Uzbekistan
Chimkent see Shymkent
Chimoio 57 E3 Manica, C Mozambique
China 102 C2 off. People's Republic of China, Chin. Chung-hua Jen-min Kung-ho-kuo, Zhonghua Renmin Gongheguo; prev. Chinese Empire. country E Asia
Chi-nan/Chinan see Jinan
Chinandega 30 C3 Chinandega, NW Nicaragua

China, People's Republic of see China
China, Republic of see Taiwan
Chincha Alta 38 D4 Ica, SW Peru
Chin-chiang see Quanzhou
Chin-chou/Chinchow see Jinzhou
Chindwin see Chindwinn
Chindwinn 114 B2 var. Chindwin. river N Myanmar (Burma)
Chinese Empire see China
Ch'ing Hai see Qinghai Hu, China
Chinghai see Qinghai
Chingola 56 D2 Copperbelt, C Zambia
Ching-Tao/Ch'ing-tao see Qingdao
Chinguețți 52 C2 var. Chinguetti. Adrar, C Mauritania
Chin Hills 114 A3 mountain range W Myanmar (Burma)
Chinhsien see Jinzhou
Chinnereth see Kinneret, Yam
Chinook Trough 91 H4 trough N Pacific Ocean
Chioggia 74 C2 anc. Fossa Claudia. Veneto, NE Italy
Chíos 83 D5 var. Hios, Khíos, It. Scio, Turk. Sakiz-Adasi. Chíos, E Greece
Chíos 83 D5 var. Khíos. island E Greece
Chipata 56 D2 prev. Fort Jameson. Eastern, E Zambia
Chiquián 38 C3 Ancash, W Peru
Chiquimula 30 B2 Chiquimula, SE Guatemala
Chīrāla 110 D1 Andhra Pradesh, E India
Chirchik see Chirchiq
Chirchiq 101 E2 Rus. Chirchik. Toshkent Viloyati, E Uzbekistan
Chiriqui Gulf 31 E5 Eng. Chiriqui Gulf. gulf SW Panama
Chiriqui Gulf see Chiriquí, Golfo de
Chiriquí, Laguna de 31 E5 lagoon NW Panama
Chiriquí, Volcán de see Barú, Volcán
Chirripó, Cerro see Chirripó Grande, Cerro
Chirripó Grande, Cerro 30 D4 var. Cerro Chirripó. mountain SE Costa Rica
Chisec 30 B2 Alta Verapaz, C Guatemala
Chisholm 23 F1 Minnesota, N USA
Chisimaio/Chisimayu see Kismaayo
Chişinău 86 D4 Rus. Kishinev. country capital (Moldova) C Moldova
Chita 93 F4 Chitinskaya Oblast', S Russia
Chitangwiza see Chitungwiza
Chitato 56 C1 Lunda Norte, NE Angola
Chitina 14 D3 Alaska, USA
Chitose 108 D2 var. Titose. Hokkaidō, NE Japan
Chitré 31 F5 Herrera, S Panama
Chittagong see Chattogram
Chitungwiza 56 D3 prev. Chitangwiza. Mashonaland East, NE Zimbabwe
Chkalov see Orenburg
Chlef 48 D2 var. Ech Cheliff, Ech Chleff; prev. Al-Asnam, El Asnam, Orléansville. NW Algeria
Chocolate Mountains 25 D8 mountain range California, W USA
Chodorów see Khodoriv
Chodzież 76 C3 Wielkopolskie, C Poland
Choele Choel 43 C5 Río Negro, C Argentina
Choiseul 122 C3 var. Lauru. island NW Solomon Islands
Chojnice 76 C2 Ger. Konitz. Pomorskie, N Poland
Ch'ok'ē 50 C4 var. Choke Mountains. mountain range NW Ethiopia
Choke Mountains see Ch'ok'ē
Cholet 68 B4 Maine-et-Loire, NW France
Choluteca 30 C3 Choluteca, S Honduras
Choluteca, Río 30 C3 river SW Honduras
Choma 56 D2 Southern, S Zambia
Chomutov 76 A4 Ger. Komotau. Ústecký Kraj, NW Czechia (Czech Republic)
Chona 91 E2 river C Russia
Chon Buri 115 C5 prev. Bang Pla Soi. Chon Buri, S Thailand
Chone 38 A1 Manabí, W Ecuador
Ch'ŏngjin 107 E3 NE North Korea
Chongqing 107 B5 var. Ch'ung-ching, Ch'ung-ch'ing, Chungking, Pahsien, Tchongking, Yuzhou. Chongqing, C China
Chongqing 107 B5 province C China
Chonnacht see Connaught
Chonos, Archipiélago de los 43 A6 island group S Chile
Chóra 83 D7 Kykládes, Greece, Aegean Sea
Chóra Sfakíon 83 C8 var. Sfákia. Kríti, Greece, E Mediterranean Sea
Chorne More see Black Sea
Chornomorsk 87 E4 prev. Illichivsk. Odes'ka Oblast', SW Ukraine
Chornomorske 87 E4 Rus. Chernomorskoye. Respublika Krym, S Ukraine
Chorokh/Chorokhi see Çoruh Nehri
Chortkiv 86 C2 Rus. Chortkov. Ternopil's'ka Oblast', W Ukraine
Chortkov see Chortkiv
Chorzów 77 C5 Ger. Königshütte; prev. Królewska Huta. Śląskie, S Poland
Chośebuz see Cottbus
Chōsen-kaikyō see Korea Strait
Chōshi 109 D5 var. Tyôsi. Chiba, Honshū, S Japan
Chosŏn-minjujuŭi-inmin-kanghwaguk see North Korea
Choszczno 76 B3 Ger. Arnswalde. Zachodnio-pomorskie, NW Poland
Chota Nagpur 113 E4 plateau N India
Choûm 52 C2 Adrar, C Mauritania
Choybalsan 105 F2 prev. Byan Tumen. Dornod, E Mongolia
Christchurch 129 C6 Canterbury, South Island, New Zealand
Christiana 32 B5 C Jamaica
Christiania see Oslo
Christiansand see Kristiansand
Christianshåb see Qasigiannguit
Christiansund see Kristiansund
Christmas Island 119 D5 Australian external territory E Indian Ocean
Christmas Island see Kiritimati
Christmas Ridge 121 E1 undersea ridge C Pacific Ocean
Chuan see Sichuan
Ch'uan-chou see Quanzhou
Chubek see Maskav
Chubut, Río 43 B6 river SE Argentina
Ch'u-chiang see Shaoguan
Chudskoye Ozero see Peipus, Lake
Chugoku-sanchi 109 B6 mountain range Honshū, SW Japan
Chuí see Chuy
Chukai see Cukai
Chukchi Plain 133 B2 abyssal plain Arctic Ocean
Chukchi Plateau 12 C2 undersea plateau Arctic Ocean
Chukchi Sea 12 B2 Rus. Chukotskoye More. sea Arctic Ocean
Chukotskoye More see Chukchi Sea
Chula Vista 25 C8 California, W USA
Chulucanas 38 B2 Piura, NW Peru
Chulym 92 D4 river C Russia
Chumphon 115 C6 var. Jumporn. Chumphon, SW Thailand
Chuncheon 107 E4 prev. Ch'unch'ŏn, Jap. Shunsen. N South Korea
Ch'unch'ŏn see Chuncheon
Ch'ung-ch'ing/Ch'ung-ching see Chongqing
Chung-hua Jen-min Kung-ho-kuo see China
Chungking see Chongqing
Chunya 93 E3 river C Russia
Chuquicamata 42 B2 Antofagasta, N Chile
Chuquisaca see Sucre
Chur 73 B7 Fr. Coire, It. Coira, Rmsch. Cuera, Quera; anc. Curia Rhaetorum. Graubünden, E Switzerland
Churchill 15 G4 Manitoba, C Canada
Churchill 16 B2 river Manitoba/Saskatchewan, C Canada
Churchill 17 F2 river Newfoundland and Labrador, E Canada
Chuska Mountains 26 C1 mountain range Arizona/New Mexico, SW USA
Chusovoy 89 D5 Permskaya Oblast', NW Russia
Chust see Khust
Chuuk Islands 122 B2 var. Hogoley Islands; prev. Truk Islands. island group Caroline Islands, C Micronesia
Chuy 42 E4 var. Chuí. Rocha, E Uruguay
Chyhyryn 87 E2 Rus. Chigirin. Cherkas'ka Oblast', N Ukraine
Chystiakove 87 H3 prev. Chystyakove, Rus. Torez. Donets'ka Oblast', SE Ukraine
Chystyakove see Chystiakove
Ciadîr-Lunga 86 D4 var. Ceadâr-Lunga, Rus. Chadyr-Lunga. S Moldova
Cide 94 C2 Kastamonu, N Turkey (Türkiye)
Ciechanów 76 D3 prev. Zichenau. Mazowieckie, C Poland
Ciego de Ávila 32 C2 Ciego de Ávila, C Cuba
Ciénaga 36 B1 Magdalena, N Colombia
Cienfuegos 32 B2 Cienfuegos, C Cuba
Cieza 71 E4 Murcia, SE Spain
Cihanbeyli 94 C3 Konya, C Turkey (Türkiye)
Cikobia 123 E4 prev. Thikombia. island N Fiji
Cilacap 116 C5 prev. Tjilatjap. Jawa, C Indonesia
Cill Airne see Killarney
Cill Chainnigh see Kilkenny
Cilli see Celje
Cill Mhantáin see Wicklow
Cîmpina see Câmpina
Cîmpulung see Câmpulung
Cincinnati 18 C4 Ohio, N USA
Ciney 65 C7 Namur, SE Belgium
Cinto, Monte 69 E7 mountain Corse, France, C Mediterranean Sea
Cintra see Sintra
Cipolletti 43 B5 Río Negro, C Argentina
Cirebon 116 C4 prev. Tjirebon. Jawa, S Indonesia
Cirkvenica see Crikvenica
Cirò Marina 75 E6 Calabria, S Italy
Cirquenizza see Crikvenica
Cisnădie 86 B4 Ger. Heltau, Hung. Nagydisznód. Sibiu, SW Romania
Citharista see la Ciotat
Citlaltépetl see Orizaba, Volcán Pico de
Citrus Heights 25 B5 California, W USA
Ciudad Acuña see Villa Acuña
Ciudad Bolívar 37 E2 prev. Angostura. Bolívar, E Venezuela
Ciudad Camargo 28 D2 Chihuahua, N Mexico
Ciudad Cortés see Cortés
Ciudad Darío 30 D3 var. Dario. Matagalpa, W Nicaragua
Ciudad de Dolores Hidalgo see Dolores Hidalgo
Ciudad de Guatemala 30 B2 Eng. Guatemala City; prev. Santiago de los Caballeros. country capital (Guatemala) Guatemala, C Guatemala
Ciudad del Carmen see Carmen
Ciudad del Este 42 E2 prev. Ciudad Presidente Stroessner, Presidente Stroessner, Puerto Presidente Stroessner. Alto Paraná, SE Paraguay
Ciudad Delicias see Delicias
Ciudad de México see México
Ciudad de Panama see Panamá
Ciudad Guayana 37 E2 prev. San Tomé de Guayana, Santo Tomé de Guayana. Bolívar, NE Venezuela
Ciudad Guzmán 28 D4 Jalisco, SW Mexico
Ciudad Hidalgo 29 G5 Chiapas, SE Mexico
Ciudad Juárez 28 C1 Chihuahua, N Mexico
Ciudad Lerdo 28 D3 Durango, C Mexico
Ciudad Madero 29 E3 var. Villa Cecilia. Tamaulipas, C Mexico
Ciudad Mante 29 E3 Tamaulipas, C Mexico
Ciudad Miguel Alemán 29 E2 Tamaulipas, C Mexico
Ciudad Obregón 28 B2 Sonora, NW Mexico
Ciudad Ojeda 36 C1 Zulia, NW Venezuela
Ciudad Porfirio Díaz see Piedras Negras
Ciudad Presidente Stroessner see Ciudad del Este
Ciudad Quesada see Quesada
Ciudad Real 70 D3 Castilla-La Mancha, C Spain
Ciudad-Rodrigo 70 C3 Castilla y León, N Spain
Ciudad Trujillo see Santo Domingo
Ciudad Valles 29 E3 San Luis Potosí, C Mexico
Ciudad Victoria 29 E3 Tamaulipas, C Mexico
Ciutadella 71 H3 var. Ciutadella de Menorca. Menorca, Spain, W Mediterranean Sea
Ciutadella de Menorca see Ciutadella
Civitanova Marche 74 D3 Marche, C Italy
Civitas Altae Ripae see Brzeg
Civitas Carnutum see Chartres
Civitas Eburovicum see Évreux
Civitavecchia 74 C4 anc. Centum Cellae, Trajani Portus. Lazio, C Italy
Claremore 27 G1 Oklahoma, C USA
Clarence 129 C5 Canterbury, South Island, New Zealand
Clarence 129 C5 river South Island, New Zealand
Clarence Town 32 D2 Long Island, C The Bahamas
Clarinda 23 F4 Iowa, C USA
Clarion Fracture Zone 131 E2 tectonic feature NE Pacific Ocean
Clarión, Isla 28 A5 island W Mexico
Clark Fork 22 A1 river Idaho/Montana, NW USA
Clark Hill Lake 21 E2 var. J.Storm Thurmond Reservoir. reservoir Georgia/South Carolina, SE USA
Clarksburg 18 D4 West Virginia, NE USA
Clarksdale 20 B2 Mississippi, S USA
Clarksville 20 C1 Tennessee, S USA
Clausentum see Southampton
Clayton 27 E1 New Mexico, SW USA
Clearwater 21 E4 Florida, SE USA
Clearwater Mountains 24 D2 mountain range Idaho, NW USA
Cleburne 27 G3 Texas, SW USA
Clermont 126 D4 Queensland, E Australia
Clermont-Ferrand 69 C5 Puy-de-Dôme, C France
Cleveland 18 D3 Ohio, N USA
Cleveland 20 D1 Tennessee, S USA
Clifton 26 C2 Arizona, SW USA
Clinton 20 B2 Mississippi, S USA
Clinton 27 F1 Oklahoma, C USA
Clipperton Fracture Zone 131 E3 tectonic feature E Pacific Ocean
Clipperton Island 131 F3 administered from France E Pacific Ocean
Cloncurry 126 B3 Queensland, C Australia
Clonmel 67 B6 Ir. Cluain Meala. S Ireland
Cloppenburg 72 B3 Niedersachsen, NW Germany
Cloquet 23 G2 Minnesota, N USA
Cloud Peak 22 C3 mountain Wyoming, C USA
Clovis 27 E2 New Mexico, SW USA
Cluain Meala see Clonmel
Cluj see Cluj-Napoca
Cluj-Napoca 86 B3 Ger. Klausenburg, Hung. Kolozsvár; prev. Cluj. Cluj, NW Romania
Clutha River 129 B7 var. Mata-Au. river South Island, New Zealand
Clyde 66 C4 river W Scotland, United Kingdom
Coari 40 D2 Amazonas, N Brazil
Coast Mountains 14 D4 Fr. Chaîne Côtière. mountain range Canada/USA
Coast Ranges 24 A4 mountain range W USA
Coats Island 15 G3 island Nunavut, NE Canada
Coats Land 132 B2 physical region Antarctica
Coatzacoalcos 29 G4 var. Quetzalcoalco; prev. Puerto México. Veracruz-Llave, E Mexico
Cobán 30 B2 Alta Verapaz, C Guatemala
Cobar 127 C6 New South Wales, SE Australia
Cobija 39 E3 Pando, NW Bolivia
Coblence/Coblenz see Koblenz
Coburg 73 C5 Bayern, SE Germany
Coca see Puerto Francisco de Orellana
Cocanada see Kākināda
Cochabamba 39 F4 hist. Oropeza. Cochabamba, C Bolivia
Cochin see Kochi
Cochinos, Bahía de 32 B2 Eng. Bay of Pigs. bay SE Cuba
Cochrane 16 C4 Ontario, S Canada
Cochrane 43 B7 Aisén, S Chile
Cocibolca see Nicaragua, Lago de
Cockade State see Maryland
Cockburn Town 33 E2 San Salvador, E The Bahamas
Cockpit Country, The 32 A4 physical region W Jamaica
Cocobeach 55 A5 Estuaire, NW Gabon
Coconino Plateau 26 B1 plain Arizona, SW USA
Coco, Río 31 E2 var. Río Wanki, Segoviao Wangkí. river Honduras/Nicaragua
Cocos Basin 102 C5 undersea basin E Indian Ocean
Cocos Island Ridge see Cocos Ridge
Cocos Islands 119 D5 island group E Indian Ocean
Cocos Ridge 13 C8 var. Cocos Island Ridge. undersea ridge E Pacific Ocean
Cod, Cape 19 G3 headland Massachusetts, NE USA
Codfish Island 129 A8 var. Whenua Hou. island SW New Zealand
Codlea 86 C4 Ger. Zeiden, Hung. Feketehalom. Brașov, C Romania
Cody 22 C2 Wyoming, C USA
Coeur d'Alene 24 C2 Idaho, NW USA
Coevorden 64 E2 Drenthe, NE Netherlands
Coffs Harbour 127 E6 New South Wales, SE Australia
Cognac 69 B5 anc. Compniacum. Charente, W France
Cohalm see Rupea
Coiba, Isla de 31 E5 island SW Panama
Coihaique 43 B6 var. Coyhaique. Aisén, S Chile
Coimbatore 110 C3 Tamil Nādu, S India
Coimbra 70 B3 anc. Conimbria, Conimbriga. Coimbra, W Portugal
Coín 70 D5 Andalucía, S Spain
Coira/Coire see Chur
Coirib, Loch see Corrib, Lough
Colby 23 E4 Kansas, C USA
Colchester 67 E6 Connecticut, NE USA
Coleman 27 F3 Texas, SW USA
Coleraine 66 B4 Ir. Cúil Raithin. N Northern Ireland, United Kingdom
Colesberg 56 C5 Northern Cape, C South Africa
Colima 28 D4 Colima, S Mexico
Coll 66 B3 island W Scotland, United Kingdom
College Station 27 G3 Texas, SW USA
Collie 125 A7 Western Australia
Collipo see Leiria
Colmar 68 E4 Ger. Kolmar. Haut-Rhin, NE France
Cöln see Köln
Cologne see Köln
Colomb-Béchar see Béchar
Colombia 36 B3 off. Republic of Colombia. country N South America
Colombian Basin 34 A1 undersea basin SW Caribbean Sea
Colombia, Republic of see Colombia
Colombie-Britannique see British Columbia
Colombo 110 C4 commercial capital (Sri Lanka) Western Province, W Sri Lanka
Colón 31 G4 prev. Aspinwall. Colón, C Panama
Colón, Archipiélago de see Galápagos Islands
Colón Ridge 13 B8 undersea ridge E Pacific Ocean
Colorado 22 C4 off. State of Colorado, also known as Centennial State, Silver State. state C USA
Colorado City 27 F3 Texas, SW USA
Colorado Plateau 26 B1 plateau W USA
Colorado, Río 43 C5 river E Argentina
Colorado, Río see Colorado River
Colorado River 13 B5 var. Río Colorado. river Mexico/USA
Colorado River 27 G4 river Texas, SW USA
Colorado Springs 22 D5 Colorado, C USA
Columbia 19 E4 Maryland, NE USA
Columbia 23 G4 Missouri, C USA
Columbia 21 E2 state capital South Carolina, SE USA
Columbia 20 C1 Tennessee, S USA
Columbia River 24 B3 river Canada/USA
Columbia Plateau 24 C3 plateau Idaho/Oregon, NW USA
Columbus 20 D2 Georgia, SE USA
Columbus 18 C4 Indiana, N USA
Columbus 20 C2 Mississippi, S USA
Columbus 23 F4 Nebraska, C USA
Columbus 18 D4 state capital Ohio, N USA
Colville Channel 128 D2 channel North Island, New Zealand
Colville River 14 D2 river Alaska, USA
Comacchio 74 C3 var. Commachio; anc. Comactium. Emilia-Romagna, N Italy
Comactium see Comacchio
Comalcalco 29 G4 Tabasco, SE Mexico
Coma Pedrosa, Pic de 69 A7 mountain NW Andorra
Comarapa 39 F4 Santa Cruz, C Bolivia
Comayagua 30 C2 Comayagua, W Honduras
Comer See see Como, Lago di
Comilla see Cumilla
Comino see Kemmuna
Comitán 29 G5 var. Comitán de Domínguez. Chiapas, SE Mexico
Comitán de Domínguez see Comitán
Commachio see Comacchio
Commissioner's Point 20 A5 headland W Bermuda
Communism Peak 101 F3 prev. Qullai Kommunizm. mountain E Tajikistan
Como 74 B2 anc. Comum. Lombardia, N Italy
Comodoro Rivadavia 43 B6 Chubut, SE Argentina
Como, Lake 74 B2 var. Lario, Eng. Lake Como, Ger. Comer See. lake N Italy
Como, Lake see Como, Lago di
Comores, Union des see Comoros
Comoros 57 F2 off. Union of the Comoros, Fr. Union des Comores. country W Indian Ocean
Comoros, Union of the see Comoros
Compiègne 68 C3 Oise, N France
Complutum see Alcalá de Henares
Compniacum see Cognac
Compostella see Santiago de Compostela
Comrat 86 D4 Rus. Komrat. S Moldova
Comum see Como
Conakry 52 C4 country capital (Guinea) SW Guinea
Conca see Cuenca
Concarneau 68 A3 Finistère, NW France
Concepción 39 G3 Santa Cruz, E Bolivia
Concepción 43 B5 Bío Bío, C Chile
Concepción 42 D2 var. Villa Concepción. Concepción, C Paraguay
Concepción see La Concepción
Concepción de la Vega see La Vega
Conchos, Río 26 D4 river NW Mexico
Conchos, Río 28 D2 river C Mexico
Concord 19 G3 state capital New Hampshire, NE USA
Concordia 42 D4 Entre Ríos, E Argentina
Concordia 23 E4 Kansas, C USA
Concordia 132 C4 French/Italian research station Antarctica
Côn Đao see Côn Đao Son
Côn Đao Son 115 E7 var. Côn Đao, Con Son. island S Vietnam
Condate see Rennes, Ille-et-Vilaine, France
Condate see St-Claude, Jura, France
Condega 30 D3 Estelí, NW Nicaragua
Condivincum see Nantes
Confluentes see Koblenz
Công Hoa Xa Hôi Chu Nghia Viêt Nam see Vietnam
Congo 55 D5 off. Republic of the Congo, Fr. Moyen-Congo; prev. Middle Congo. country C Africa
Congo 55 C6 off. Democratic Republic of Congo; prev. Zaire, Belgian Congo, Congo (Kinshasa). country C Africa
Congo 55 C6 var. Kongo, Fr. Zaire. river C Africa
Congo Basin 55 C6 drainage basin W Dem. Rep. Congo
Congo/Congo (Kinshasa) see Congo (Democratic Republic of)
Coni see Cuneo
Conimbria/Conimbriga see Coimbra
Conjeeveram see Kānchīpuram
Connacht see Connaught
Connaught 67 A5 var. Connacht, Ir. Chonnacht, Cúige. province W Ireland
Connecticut 19 F3 off. State of Connecticut, also known as Blue Law State, Constitution State, Land of Steady Habits, Nutmeg State. state NE USA
Connecticut 19 G3 river Canada/USA
Conroe 27 G3 Texas, SW USA
Consentia see Cosenza
Consolación del Sur 32 A2 Pinar del Río, W Cuba
Con Son see Côn Đao Son
Constance see Konstanz
Constance, Lake 73 B7 Ger. Bodensee. lake C Europe
Constanța 86 D5 var. Küstendje, Eng. Constanza, Ger. Konstanza, Turk. Küstence. Constanța, SE Romania
Constantia see Coutances
Constantia see Konstanz
Constantine 49 E2 var. Qacentina, Ar. Qoussantîna. NE Algeria
Constantinople see İstanbul
Constantiola see Oltenița
Constanz see Konstanz
Constanza see Constanța
Constitution State see Connecticut
Coo see Kos
Coober Pedy 127 A5 South Australia
Cookeville 20 D1 Tennessee, S USA
Cook Islands 123 F4 self-governing territory in free association with New Zealand S Pacific Ocean
Cook, Mount see Aoraki
Cook Strait 129 D5 var. Raukawa. strait New Zealand
Cooktown 126 D2 Queensland, NE Australia
Coolgardie 125 B6 Western Australia
Cooma 127 D7 New South Wales, SE Australia
Coomassie see Kumasi
Coon Rapids 23 F2 Minnesota, N USA
Cooper Creek 126 C4 var. Barcoo, Cooper's Creek. seasonal river Queensland/South Australia, Australia
Cooper's Creek see Cooper Creek
Coos Bay 24 A3 Oregon, NW USA
Cootamundra 127 D6 New South Wales, SE Australia
Copacabana 39 E4 La Paz, W Bolivia
Copenhagen see København
Copiapó 42 B3 Atacama, N Chile
Copperas Cove 27 G3 Texas, SW USA
Coppermine see Kugluktuk
Copper State see Arizona
Coquilhatville see Mbandaka
Coquimbo 42 B3 Coquimbo, N Chile
Corabia 86 B5 Olt, S Romania
Coral Harbour 15 G3 var. Salliq. Southampton Island, Nunavut, NE Canada
Coral Sea 120 B3 sea SW Pacific Ocean
Coral Sea Islands 122 B4 Australian external territory SW Pacific Ocean
Corantijn Rivier see Courantyne River
Corcovado, Golfo 43 B6 gulf S Chile
Corcyra Nigra see Korčula
Cordele 20 D3 Georgia, SE USA
Córdoba 42 C3 Córdoba, C Argentina
Córdoba 29 F4 Veracruz-Llave, E Mexico
Córdoba 70 D4 var. Cordoba, Eng. Cordova; anc. Corduba. Andalucía, SW Spain
Cordova 14 C3 Alaska, USA
Cordova/Cordoba see Córdoba
Corduba see Córdoba
Corentyne River see Courantyne River
Corfu see Kérkyra
Coria 70 C3 Extremadura, W Spain
Corinth 20 C1 Mississippi, S USA
Corinth see Kórinthos
Corinth, Gulf of/Corinthiacus Sinus see Korinthiakós Kólpos
Corinthus see Kórinthos
Corinto 30 C3 Chinandega, NW Nicaragua
Cork 67 A6 Ir. Corcaigh. S Ireland
Çorlu 94 A2 Tekirdağ, NW Turkey (Türkiye)
Corner Brook 17 G3 Newfoundland, Newfoundland and Labrador, E Canada
Cornhusker State see Nebraska
Corn Islands 31 E3 var. Corn Islands. island group SE Nicaragua
Corn Islands see Maíz, Islas del
Cornwallis Island 15 F2 Island Nunavut, N Canada
Coro 36 C1 prev. Santa Ana de Coro. Falcón, NW Venezuela
Corocoro 39 F4 La Paz, W Bolivia
Coromandel 128 D2 Waikato, North Island, New Zealand
Coromandel Coast 110 D2 coast E India
Coromandel Peninsula 128 D2 peninsula North Island, New Zealand
Coronado, Bahía de 30 D5 bay S Costa Rica
Coronel Dorrego 43 C5 Buenos Aires, E Argentina
Coronel Oviedo 42 D2 Caaguazú, SE Paraguay
Corozal 30 C1 Corozal, N Belize
Corpus Christi 27 G4 Texas, SW USA
Corrales 26 D2 New Mexico, SW USA
Corrib, Lough 67 A5 Ir. Loch Coirib. lake W Ireland
Corrientes 42 D3 Corrientes, NE Argentina
Corriza see Korçë
Corsica 69 E7 Eng. Corsica. island France, C Mediterranean Sea
Corsica see Corse
Corsicana 27 G3 Texas, SW USA
Cortegana 70 C4 Andalucía, S Spain
Cortés 31 E5 var. Ciudad Cortés. Puntarenas, SE Costa Rica
Cortez, Sea of see California, Golfo de
Cortina d'Ampezzo 74 C1 Veneto, NE Italy
Coruche 70 B3 Santarém, C Portugal
Çoruh Nehri 95 E3 Geor. Chorokh, Rus. Chorokhi. river Georgia/Turkey (Türkiye)
Çorum 94 D3 var. Chorum. Çorum, N Turkey (Türkiye)
Corunna see A Coruña
Corvallis 24 B3 Oregon, NW USA
Corvo 70 A5 var. Ilha do Corvo. island Azores, Portugal, NE Atlantic Ocean
Corvo, Ilha do see Corvo
Cos see Kos
Cosenza 75 D6 anc. Consentia. Calabria, SW Italy
Cosne-Cours-sur-Loire 68 C4 Nièvre, Bourgogne, C France Europe
Costa Mesa 24 D2 California, W USA North America
Costa Rica 31 E4 off. Republic of Costa Rica. country Central America
Costa Rica, Republic of see Costa Rica
Costermansville see Bukavu
Cotagaita 39 F5 Potosí, S Bolivia
Côte d'Ivoire see Ivory Coast
Côte d'Ivoire, République de la see Ivory Coast
Côte d'Or 68 D4 cultural region C France
Côte Française des Somalis see Djibouti
Côtière, Chaîne see Coast Mountains
Cotonou 53 F5 var. Kotonu. country capital (Benin - seat of government) S Benin
Cotrone see Crotone
Cotswold Hills 67 D6 var. Cotswolds. hill range S England, United Kingdom
Cotswolds see Cotswold Hills
Cottbus 72 D4 Lus. Chośebuz; prev. Kottbus. Brandenburg, E Germany
Cotton State, The see Alabama
Cotyora see Ordu
Couentrey see Coventry
Council Bluffs 23 F4 Iowa, C USA
Courantyne River 37 G4 var. Corantijn Rivier, Corentyne River. river Guyana/Suriname
Courland Lagoon 84 A4 Ger. Kurisches Haff, Rus. Kurskiy Zaliv. lagoon Lithuania/Russia
Courtrai see Kortrijk
Coutances 68 B3 anc. Constantia. Manche, N France
Couvin 65 C7 Namur, S Belgium
Coventry 67 D6 anc. Couentrey. C England, United Kingdom
Covilhã 70 C3 Castelo Branco, E Portugal
Cowan, Lake 125 B6 lake Western Australia
Coxen Hole see Roatán
Coxin Hole see Roatán
Coyhaique see Coihaique
Coyote State, The see South Dakota
Cozhê 104 C5 Xizang Zizhiqu, W China
Cozumel, Isla 29 H3 island SE Mexico
Cracovia/Cracow see Kraków
Cradock 56 C5 Eastern Cape, S South Africa

Craig 22 C4 Colorado, C USA
Craiova 86 B5 Dolj, SW Romania
Cranbrook 15 E5 British Columbia, SW Canada
Crane see The Crane
Cranz see Zelenogradsk
Crawley 67 E7 SE England, United Kingdom
Cremona 74 B2 Lombardia, N Italy
Creole State see Louisiana
Cres 78 A3 It. Cherso; anc. Crexa. island W Croatia
Crescent City 24 A4 California, W USA
Crescent Group 106 C7 island group C Paracel Islands
Creston 23 F4 Iowa, C USA
Crestview 20 D3 Florida, SE USA
Crete see Kríti
Créteil 68 E2 Val-de-Marne, N France
Crete, Sea of/Creticum, Mare see Kritikó Pélagos
Creuse 68 B4 river C France
Crewe 67 D6 C England, United Kingdom
Crexa see Cres
Crikvenica 78 A3 It. Cirquenizza; prev. Cirkvenica, Crjkvenica. Primorje-Gorski Kotar, NW Croatia
Crimea see Krym
Cristóbal 31 G4 Colón, C Panama
Cristóbal Colón, Pico 36 B1 mountain N Colombia
Cristur/Cristuru Săcuiesc see Cristuru Secuiesc
Cristuru Secuiesc 86 C4 prev. Cristur, Cristuru Săcuiesc, Sitaş Cristuru, Ger. Kreutz, Hung. Székelykeresztúr, Szitás-Keresztúr. Harghita, C Romania
Crjkvenica see Crikvenica
Crna Gora see Montenegro
Crna Reka 79 D6 river S North Macedonia
Crni Drim see Black Drin
Croatia 78 B3 off. Republic of Croatia, Cro. Hrvatska, Republika Hrvatska. country SE Europe
Croatia, Republic of see Croatia
Crocodile see Limpopo
Croia see Krujë
Croker Island 124 E2 island Northern Territory, N Australia
Cromwell 129 B7 Otago, South Island, New Zealand
Crooked Island 32 D2 island SE The Bahamas
Crooked Island Passage 32 D2 channel SE The Bahamas
Crookston 23 F1 Minnesota, N USA
Crossen see Krosno Odrzańskie
Croton/Crotona see Crotone
Crotone 75 E6 var. Cotrone; anc. Croton, Crotona. Calabria, SW Italy
Croydon 67 A8 SE England, United Kingdom
Crozet Basin 119 B6 undersea basin S Indian Ocean
Crozet Islands 119 B7 island group French Southern and Antarctic Lands
Crozet Plateau 119 B7 var. Crozet Plateaus. undersea plateau SW Indian Ocean
Crozet Plateaus see Crozet Plateau
Crystal Brook 127 B6 South Australia
Csaca see Čadca
Csakathurn/Csáktornya see Čakovec
Csíkszereda see Miercurea-Ciuc
Csorna 77 C6 Győr-Moson-Sopron, NW Hungary
Csurgó 77 C7 Somogy, SW Hungary
Cuando 56 C2 var. Kwando. river S Africa
Cuango see Kwango
Cuanza 56 B1 var. Kwanza. river C Angola
Cuauhtémoc 28 C2 Chihuahua, N Mexico
Cuautla 29 E4 Morelos, S Mexico
Cuba 32 B2 off. Republic of Cuba. country W West Indies
Cubal 56 B2 Benguela, W Angola
Cubango 56 B2 var. Kuvango, Port. Vila Artur de Paiva, Vila da Ponte. Huíla, SW Angola
Cubango 56 B2 var. Kavango, Kavengo, Kubango, Okavango, Okavanggo. river S Africa
Cuba, Republic of see Cuba
Cúcuta 36 C2 var. San José de Cúcuta. Norte de Santander, N Colombia
Cuddapah see Kadapa
Cuenca 38 B2 Azuay, S Ecuador
Cuenca 71 E3 anc. Conca. Castilla-La Mancha, C Spain
Cuera see Chur
Cuernavaca 29 E4 Morelos, S Mexico
Cuiabá 41 E3 prev. Cuyabá. state capital Mato Grosso, SW Brazil
Cúige see Connaught
Cúige Laighean see Leinster
Cúige Mumhan see Munster
Cuijck 64 D4 Noord-Brabant, SE Netherlands
Cúil Raithin see Coleraine
Cuito 56 B2 var. Kuito; Port. Silva Porto. Bié, C Angola
Cuito 56 B2 var. Kwito. river SE Angola
Cukai 116 B3 var. Chukai, Kemaman. Terengganu, Peninsular Malaysia
Cularo see Grenoble
Culiacán 28 C3 var. Culiacán Rosales, Culiacán-Rosales. Sinaloa, C Mexico
Culiacán-Rosales/Culiacán Rosales see Culiacán
Cullera 71 F3 Comunitat Valenciana, E Spain
Cullman 20 C2 Alabama, S USA
Culm see Chełmno
Culmsee see Chełmża
Cumaná 37 E1 Sucre, NE Venezuela
Cumbal, Nevado de 36 A4 elevation S Colombia
Cumberland 19 E4 Maryland, NE USA
Cumberland Plateau 20 D1 plateau E USA
Cumberland Sound 15 H3 inlet Baffin Island, Nunavut, NE Canada
Cumilla 113 G4 prev. Comilla, Ben. Kumillā. Chattogram, E Bangladesh
Cumpas 28 B2 Sonora, NW Mexico
Cuneo 74 A2 Fr. Coni. Piemonte, NW Italy
Cunnamulla 127 C5 Queensland, E Australia
Ćuprija 78 E4 Serbia, E Serbia
Curaçao 33 E5 Dutch self-governing territory S West Indies
Curia Rhaetorum see Chur
Curicó 42 B4 Maule, C Chile
Curieta see Krk
Curitiba 41 E4 prev. Curytiba. state capital Paraná, S Brazil
Curtbunar see Tervel
Curtea de Argeş 86 C4 var. Curtea-de-Arges. Argeş, S Romania
Curtea-de-Arges see Curtea de Argeş
Curtici 86 A4 Ger. Kurtitsch, Hung. Kürtös. Arad, W Romania
Curtis Island 126 E4 island Queensland, SE Australia
Curytiba see Curitiba
Curzola see Korčula
Cusco 39 E4 var. Cuzco. Cusco, C Peru
Cusset 69 C5 Allier, C France
Cutch, Gulf of see Kachchh, Gulf of
Cuttack 113 F4 Odisha, E India
Cuvier Plateau 119 E6 undersea plateau E Indian Ocean
Cuxhaven 72 B2 Niedersachsen, NW Germany
Cuyabá see Cuiabá
Cuyuni, Río see Cuyuni River
Cuyuni River 37 F3 var. Río Cuyuni. river Guyana/Venezuela
Cuzco see Cusco
Cyclades see Kykládes
Cydonia see Chaniá
Cymru see Wales
Cyprus 80 C4 off. Republic of Cyprus, Gk. Kypriakí Dimokratía, Kýpros, Turk. Kıbrıs, Kıbrıs Cumhuriyeti. country E Mediterranean Sea
Cyprus, Republic of see Cyprus
Cyrenaica 49 H2 region NE Libya
Cythnos see Kýthnos
Czech Republic see Czechia
Czechia 77 A5 off. Czech Republic, Cz. Česko. country C Europe
Czenstochau see Częstochowa
Czernowitz see Chernivtsi
Częstochowa 76 C4 Ger. Czenstochau, Tschenstochau, Rus. Chenstokhov. Śląskie, S Poland
Człuchów 76 C3 Ger. Schlochau. Pomorskie, NW Poland

D

Dabajuro 36 C1 Falcón, NW Venezuela
Dabeiba 36 B2 Antioquia, NW Colombia
Dąbrowa Tarnowska 77 D5 Małopolskie, S Poland
Dabryn' 85 C8 Rus. Dobryn'. Homyel'skaya Voblasts', SE Belarus
Dacca see Dhaka
Daegu 107 E4 off. Daegu Gwang-yeoksi, prev. Taegu, Jap. Taikyū. SE South Korea
Daegu Gwang-yeoksi see Daegu
Dagana 52 B3 N Senegal
Dagda 84 D4 Krāslava, SE Latvia
Dagden see Hiiumaa
Dagenham 67 B8 United Kingdom
Dağlıq Qarabağ see Nagornyy-Karabakh
Dagö see Hiiumaa
Dagupan 117 E1 off. Dagupan City. Luzon, N Philippines
Dagupan City see Dagupan
Da Hinggan Ling 105 G1 Eng. Great Khingan Range. mountain range NE China
Dahm, Ramlat 99 B6 desert NW Yemen
Dahomey see Benin
Daihoku see Taibei
Daimiel 70 D3 Castilla-La Mancha, C Spain
Daimoniá 83 B7 Pelopónnisos, S Greece
Dainan see Tainan
Daingin, Bá an see Dingle Bay
Dairen see Dalian
Daejeon 107 E4 off. Daejeon Gwang-yeoksi, prev. Taejŏn, Jap. Taiden. C South Korea
Daejeon Gwang-yeoksi see Daejeon
Dakar 52 B3 country capital (Senegal) W Senegal
Dakhla see Ad Dakhla
Dakoro 53 G3 Maradi, S Niger
Đakovica see Gjakovë
Đakovo 78 C3 var. Djakovo, Hung. Diakovár. Osijek-Baranja, E Croatia
Dakshin see Deccan
Dalain Hob 104 D3 var. Ejin Qi. Nei Mongol Zizhiqu, N China
Dalai Nor see Hulun Nur
Dalaman 94 A4 Muğla, SW Turkey (Türkiye)
Dalandzadgad 105 E3 Ömnögovĭ, S Mongolia
Đa Lat 115 E6 Lâm Đông, S Vietnam
Dalby 127 D5 Queensland, E Australia
Dale City 19 E4 Virginia, NE USA
Dalhart 27 E1 Texas, SW USA
Dali 106 A6 var. Xiaguan. Yunnan, SW China
Dalian 106 D4 var. Dairen, Dalien, Jay Dairen, Lüda, Ta-lien, Rus. Dalny. Liaoning, NE China
Dalien see Dalian
Dallas 27 G2 Texas, SW USA
Dalmacija 78 B4 Eng. Dalmatia, Ger. Dalmatien, It. Dalmazia. cultural region S Croatia
Dalmatia/Dalmatien/Dalmazia see Dalmacija
Dalny see Dalian
Dalton 20 D1 Georgia, SE USA
Dálvvadis see Jokkmokk
Daly Waters 126 A2 Northern Territory, N Australia
Damachava 85 A6 var. Damachova, Pol. Domaczewo, Rus. Domachëvo. Brestskaya Voblasts', SW Belarus
Damachova see Damachava
Damān 112 C4 Damān and Diu, W India
Damara 54 C4 Ombella-Mpoko, S Central African Republic
Damas see Dimashq
Damasco see Dimashq
Damascus see Dimashq
Damavand, Qolleh-ye 98 D3 mountain N Iran
Damietta see Dumyāṭ
Dammām see Ad Dammān
Damoûr 97 A5 var. Ad Dāmūr. W Lebanon
Dampier 124 A4 Western Australia
Dampier, Selat 117 F4 strait Papua, E Indonesia
Damqawt 99 D6 var. Damqut. E Yemen
Damqut see Damqawt
Damxung 104 C5 var. Gongtang. Xizang Zizhiqu, W China
Danakil Desert 50 D4 var. Afar Depression, Danakil Plain. desert E Africa
Danakil Plain see Danakil Desert
Danané 52 D5 W Ivory Coast
Đă Năng 115 E5 prev. Tourane. Quang Nam-Đa Năng, C Vietnam
Danborg see Daneborg
Dandong 106 D3 var. Tan-tung; prev. An-tung. Liaoning, NE China
Daneborg 61 E3 var. Danborg. NE Greenland
Dänew see Galkynyş
Dangara see Danghara
Dangerous Archipelago see Tuamotu, Îles
Danghara 101 E3 Rus. Dangara. SW Tajikistan
Danghe Nanshan 104 D3 mountain range W China
Dangrek, Chaîne des see Dangrek Mountains
Dângrêk, Chuŏr Phnum see Dangrek Mountains
Dangrek Mountains 115 D5 Cam. Chuŏr Phnum Dângrêk, Thai. Thiu Khao Phanom Dong Rak, Fr. Chaîne des Dangrek. mountain range Cambodia/Thailand
Dangriga 30 C1 prev. Stann Creek. Stann Creek, E Belize
Danish West Indies see Virgin Islands (US)
Danlí 30 D2 El Paraíso, S Honduras
Danmark see Denmark
Danmarksstraedet see Denmark Strait
Dannenberg 72 C3 Niedersachsen, N Germany
Dannevirke 128 D4 Manawatu-Wanganui, North Island, New Zealand
Dantzig see Gdańsk
Danube 59 E4 Bul., Croatian, Serb. Dunav, Cz., Slov. Dunaj, Ger. Donau, Hung. Duna, Rom. Dunărea, Ukr. Dunai. river C Europe
Danubian Plain see Dunavska Ravnina
Danum see Doncaster
Danville 19 E5 Virginia, NE USA
Danxian/Dan Xian see Danzhou
Danzhou 106 C7 prev. Danxian, Dan Xian, Nada. Hainan, S China
Danzig see Gdańsk
Danziger Bucht see Gdańsk, Gulf of
Danzig, Gulf of see Gdańsk, Gulf of
Daqm see Duqm
Dar'ā 97 B5 var. Der'a, Fr. Déraa. Dar'ā, SW Syria
Darabani 86 C3 Botoşani, NW Romania
Daraut-Kurgan see Daroot-Korgon
Dardanelles 94 A2 Eng. Dardanelles. strait NW Turkey (Türkiye)
Dardanelles see Çanakkale Boğazı
Dardanelli see Çanakkale
Dar-el-Beida see Casablanca
Dar es Salaam 51 C7 Dar es Salaam, E Tanzania
Darfield 129 C6 Canterbury, South Island, New Zealand
Darfur 50 A4 var. Darfur Massif. cultural region W Sudan
Darfur Massif see Darfur
Darhan 105 E2 Darhan Uul, N Mongolia
Darién, Golfo del see Darien, Gulf of
Darien, Gulf of 36 A2 Sp. Golfo del Darién. gulf S Caribbean Sea
Darien, Isthmus of see Panama, Istmo de
Darién, Serranía del 31 H5 mountain range Colombia/Panama
Dario see Ciudad Darío
Dariorigum see Vannes
Darjeeling see Dārjiling
Dārjiling 113 F3 prev. Darjeeling. West Bengal, NE India
Darling River 127 C6 river New South Wales, SE Australia
Darlington 67 D5 N England, United Kingdom
Darmstadt 73 B5 Hessen, SW Germany
Darnah 49 G2 var. Dérna. NE Libya
Darnley, Cape 132 D2 cape Antarctica
Daroca 71 E2 Aragón, NE Spain
Daroot-Korgon 101 F3 var. Daraut-Kurgan. Oshskaya Oblast', SW Kyrgyzstan
Dartford 67 B8 SE England, United Kingdom
Dartmoor 67 C7 moorland SW England, United Kingdom
Dartmouth 17 F4 Nova Scotia, SE Canada
Darvaza see Derweze, Turkmenistan
Darwin 124 D2 prev. Palmerston, Port Darwin. territory capital Northern Territory, N Australia
Darwin, Isla 38 A4 island Galápagos Islands, W Ecuador
Dashhowuz see Daşoguz
Dashkawka 85 D6 Rus. Dashkovka. Mahilyowskaya Voblasts', E Belarus
Dashkovka see Dashkawka
Daşoguz 100 C2 Rus. Dashkhovuz, Turkm. Dashhowuz; prev. Tashauz. Daşoguz Welaýaty, N Turkmenistan
Đa, Sông see Black River
Datong 106 C3 var. Tatung, Ta-t'ung. Shanxi, C China
Daugava see Western Dvina
Daugavpils 84 D4 Ger. Dünaburg; prev. Rus. Dvinsk. Daugvapils, SE Latvia
Daung Kyun 115 B6 island S Myanmar (Burma)
Dauphiné 69 D5 cultural region E France
Dāvangere 110 C2 Karnātaka, W India
Davao 117 F3 off. Davao City. Mindanao, S Philippines
Davao City see Davao
Davao Gulf 117 F3 gulf Mindanao, S Philippines
Davenport 23 G3 Iowa, C USA
David 31 E5 Chiriquí, W Panama
Davie Ridge 119 A5 undersea ridge W Indian Ocean
Davis 132 D3 Australian research station Antarctica
Davis Sea 132 D3 sea Antarctica
Davis Strait 60 B3 strait Baffin Bay/Labrador Sea
Dawei 115 B5 var. Tavoy, Htawei. Tanintharyi, S Myanmar (Burma)
Dawlat Qaṭar see Qatar
Dax 69 B6 var. Ax; anc. Aquae Augustae, Aquae Tarbelicae. Landes, SW France
Dayr az Zawr 96 D3 var. Deir ez Zor. Dayr az Zawr, E Syria
Dayton 18 C4 Ohio, N USA
Daytona Beach 21 E4 Florida, SE USA
De Aar 56 C5 Northern Cape, C South Africa
Dead Sea 97 B6 var. Bahret Lut, Lacus Asphaltites, Ar. Al Baḥr al Mayyit, Baḥrat Lūt, Heb. Yam HaMelaḥ. salt lake Israel/Jordan
Dê Afghānistān Islāmī Jumhūriyat see Afghanistan
Deán Funes 42 C3 Córdoba, C Argentina
Death Valley 25 C7 valley California, W USA
Deatnu 62 D2 Fin. Tenojoki, Nor. Tana. river Finland/Norway
Debar 79 D6 Ger. Dibra, Turk. Debre. W North Macedonia
De Behagle see Laï
Dębica 77 D5 Podkarpackie, SE Poland
De Bildt see De Bilt
De Bilt 64 C3 var. De Bildt. Utrecht, C Netherlands
Dębno 76 B3 Zachodnio-pomorskie, NW Poland
Debre see Debar
Debrecen 77 D6 Ger. Debreczin, Rom. Debreţin; prev. Debreczen. Hajdú-Bihar, E Hungary
Debreczen/Debreczin see Debrecen
Debreţin see Debrecen
Decatur 20 C1 Alabama, S USA
Decatur 18 B4 Illinois, N USA
Deccan 112 D5 Hind. Dakshin. plateau C India
Děčín 76 B4 Ger. Tetschen. Ústecký Kraj, NW Czechia (Czech Republic)
Dedeagaç/Dedeagach see Alexandroúpoli
Dedemsvaart 64 E3 Overijssel, E Netherlands
Dee 66 C3 river NE Scotland, United Kingdom
Deering 14 C2 Alaska, USA
Deés see Dej
Deggendorf 73 D6 Bayern, SE Germany
Değirmenlik 80 C5 Gk. Kythréa. N Cyprus
Deh Bīd see Şafāshahr
Dehli see Delhi
Deh-e Shū 100 D5 prev. Deh Shū, var. Deshu, Dishū. Helmand, S Afghanistan
Deh Shū see Deh-e Shū
Deinze 65 B5 Oost-Vlaanderen, NW Belgium
Deir ez Zor see Dayr az Zawr
Deirgeirt, Loch see Derg, Lough
Dej 86 B3 Hung. Dés; prev. Deés. Cluj, NW Romania
De Jouwer see Joure
Dekéleia see Dhekelia
Dékoa 54 C4 Kémo, C Central African Republic
De Land 21 E4 Florida, SE USA
Delano 25 C7 California, W USA
Delārām see Dilārām
Delaware 18 D4 Ohio, N USA
Delaware 19 F4 off. State of Delaware, also known as Blue Hen State, Diamond State, First State. state NE USA
Delft 64 B4 Zuid-Holland, W Netherlands
Delfzijl 64 E1 Groningen, NE Netherlands
Delgo 50 B3 Northern, N Sudan
Delhi 112 D3 var. Dehli, Hind. Dilli, hist. Shahjahanabad. union territory capital Delhi, N India
Delicias 28 D2 var. Ciudad Delicias. Chihuahua, N Mexico
Déli-Kárpátok see Carpaţii Meridionali
Delmenhorst 72 B3 Niedersachsen, NW Germany
Del Rio 27 F4 Texas, SW USA
Deltona 21 E4 Florida, SE USA
Demba 55 D6 Kasai-Occidental, C Dem. Rep. Congo
Dembia 54 D4 Mbomou, SE Central African Rep.
Demchok see Dêmqog
Demerara Plain 34 C2 abyssal plain W Atlantic Ocean
Deming 26 C3 New Mexico, SW USA
Demmin 72 C2 Mecklenburg-Vorpommern, NE Germany
Demopolis 20 C2 Alabama, S USA
Dêmqog 104 A5 var. Demchok. disputed region China/India
Denali 14 C3 var. Mount McKinley. mountain Alaska, USA
Denali Park 14 C3 prev. McKinley Park, Alaska, USA
Denau see Denov
Dender 65 B6 Fr. Dendre. river W Belgium
Dendre see Dender
Denekamp 64 E3 Overijssel, E Netherlands
Den Haag see 's-Gravenhage
Denham 125 A5 Western Australia
Den Ham 64 E3 Overijssel, E Netherlands
Den Helder 64 C2 Noord-Holland, NW Netherlands
Dénia 71 F4 Comunitat Valenciana, E Spain
Deniliquin 127 C7 New South Wales, SE Australia
Denison 23 F3 Iowa, C USA
Denison 27 G2 Texas, SW USA
Denizli 94 B4 Denizli, SW Turkey (Türkiye)
Denmark 63 A7 off. Kingdom of Denmark, Dan. Danmark; anc. Hafnia. country N Europe
Denmark, Kingdom of see Denmark
Denmark Strait 60 D4 var. Danmarksstraedet. strait Greenland/Iceland
Dennery 33 F1 E Saint Lucia
Denov 101 E3 Rus. Denau. Surkhondaryo Viloyati, S Uzbekistan
Denpasar 116 D5 prev. Paloe. Bali, C Indonesia
Denton 27 G2 Texas, SW USA
D'Entrecasteaux Islands 122 B3 island group SE Papua New Guinea
Denver 22 D4 state capital Colorado, C USA
Der'a/Derá/Déraa see Dar'ā
Dera Ghazi Khan 112 C2 var. Dera Ghazikhan. Punjab, C Pakistan
Dera Ghazikhan see Dera Ghazi Khan
Đeravica 79 D5 mountain S Serbia
Derbent 89 B8 Respublika Dagestan, SW Russia
Derby 67 D6 C England, United Kingdom
Dereli see Gónnoi
Dergachi see Derhachi
Derg, Lough 67 A6 Ir. Loch Deirgeirt. lake W Ireland
Derhachi 87 G2 Rus. Dergachi. Kharkivs'ka Oblast', E Ukraine
De Ridder 20 A3 Louisiana, S USA
Dérna see Darnah
Derry see Londonderry
Dertosa see Tortosa
Derventa 78 B3 Republika Srpska, N Bosnia and Herzegovina
Derweze 100 C2 Rus. Darvaza. Ahal Welaýaty, C Turkmenistan
Dés see Dej
Deschutes River 24 B3 river Oregon, NW USA
Desē 50 C4 var. Desse, It. Dessie. Āmara, N Ethiopia
Deseado, Río 43 B7 river S Argentina
Desertas, Ilhas 48 A2 island group Madeira, Portugal, NE Atlantic Ocean
Deshu see Deh-e Shū
Des Moines 23 F3 state capital Iowa, C USA
Desna 87 E2 river Russia/Ukraine
Dessau 72 C4 Sachsen-Anhalt, E Germany
Desse see Desē
Dessie see Desē
Destêrro see Florianópolis
Detroit 18 D3 Michigan, N USA
Detroit Lakes 23 F2 Minnesota, N USA
Deurne 65 D5 Noord-Brabant, SE Netherlands
Deutschendorf see Poprad
Deutsch-Eylau see Iława
Deutsch Krone see Wałcz
Deutschland/Deutschland, Bundesrepublik see Germany
Deutsch-Südwestafrika see Namibia
Deva 86 B4 Ger. Diemrich, Hung. Déva. Hunedoara, W Romania
Déva see Deva
Deva see Chester
Devana see Aberdeen
Devana Castra see Chester
Đevđelija see Gevgelija
Deventer 64 D3 Overijssel, E Netherlands
Devils Lake 23 E1 North Dakota, N USA
Devoll see Devollit, Lumi i
Devollit, Lumi i 79 D6 var. Devoll. river SE Albania
Devon Island 15 F2 prev. North Devon Island. island Parry Islands, Nunavut, NE Canada
Devonport 127 C8 Tasmania, SE Australia
Devrek 94 C2 Zonguldak, N Turkey (Türkiye)
Dexter 23 H5 Missouri, C USA
Deynau see Galkynyş
Dezfūl 98 C3 var. Dizful. Khūzestān, SW Iran
Dezhou 106 D4 Shandong, E China
Dhaka 113 G4 prev. Dacca. country capital (Bangladesh) Dhaka, C Bangladesh
Dhanbād 113 F4 Jhārkhand, NE India
Dhekelia 80 C5 Gk. Dekéleia. UK Sovereign Base Area SE Cyprus
Dhidhimótikhon see Didymóteicho
Dhíkti Ori see Díkti
Dhodhekánisos see Dodekánisa
Dhomokós see Domokós
Dhráma see Dráma
Dhrepanon, Akrotírio see Drépano, Akrotírio
Dhún na nGall, Bá see Donegal Bay
Dhuusa Marreeb 51 E5 var. Dusa Marreb, It. Dusa Mareb. Galguduud, C Somalia
Diakovár see Đakovo
Diamantina, Chapada 41 F3 mountain range E Brazil
Diamantina Fracture Zone 119 E6 tectonic feature E Indian Ocean
Diamond State see Delaware
Diaoyutai Lieyu/ Diaoyutai Qundao see Senkaku-shoto
Diarbekr see Diyarbakır
Dibio see Dijon
Dibra see Debar
Dibrugarh 113 H3 Assam, NE India
Dickinson 22 D2 North Dakota, N USA
Dicle see Tigris
Didimotiho see Didymóteicho
Didymóteicho 82 D3 var. Dhidhimótikhon, Didimotiho. Anatolikí Makedonía kai Thráki, NE Greece
Diedenhofen see Thionville
Diekirch 65 D7 Diekirch, C Luxembourg
Diemrich see Deva
Điện Biên 114 D3 var. Bien Bien, Dien Bien Phu. Lai Châu, N Vietnam
Dien Bien Phu see Điện Biên
Diepenbeek 65 D6 Limburg, NE Belgium
Diepholz 72 B3 Niedersachsen, NW Germany
Dieppe 68 C2 Seine-Maritime, N France
Dieren 64 D4 Gelderland, E Netherlands
Differdange 65 D8 Luxembourg, SW Luxembourg
Digne 69 D6 var. Digne-les-Bains. Alpes-de-Haute-Provence, SE France
Digne-les-Bains see Digne
Digoel see Digul, Sungai
Digoin 68 C4 Saône-et-Loire, C France
Digul, Sungai 117 H5 prev. Digoel. river Papua, E Indonesia
Dihang see Brahmaputra
Dijlah see Tigris
Dijon 68 D4 anc. Dibio. Côte d'Or, C France
Dikhil 50 D4 SW Djibouti
Dikson 92 D2 Taymyrskiy (Dolgano-Nenetskiy) Avtonomnyy Okrug, N Russia
Díkti 83 D8 var. Dhíkti Ori. mountain range Kríti, Greece, E Mediterranean Sea
Dilārām 100 D5 prev. Delārām. Nīmrūz, SW Afghanistan
Dili 117 F5 var. Dilli, Dilly. country capital (East Timor) N East Timor (Timor-Leste)
Dilia 53 G3 var. Dillia. river SE Niger
Di Linh 115 E6 Lâm Đông, S Vietnam
Dilli see Dili, East Timor (Timor-Leste)
Dilli see Delhi, India
Dillia see Dilia
Dilling 50 B4 var. Ad Dalanj. Southern Kordofan, C Sudan
Dillon 22 B2 Montana, NW USA
Dilly see Dili
Dilolo 55 D7 Katanga, S Dem. Rep. Congo
Dimashq 97 B5 var. Ash Shām, Esh Sham, Eng. Damascus, Fr. Damas, It. Damasco. country capital (Syria) Dimashq, SW Syria
Dimitrovgrad 82 D3 Haskovo, S Bulgaria
Dimitrovgrad 89 C6 prev. Caribrod. Serbia, SE Serbia
Dimitrovo see Pernik
Dimovo 82 B1 Vidin, NW Bulgaria
Dinajpur 113 F3 Rajshahi, NW Bangladesh
Dinan 68 B3 Côtes d'Armor, NW France
Dinant 65 C7 Namur, S Belgium
Dinar 94 B4 Afyon, SW Turkey (Türkiye)
Dinara see Dinaric Alps
Dinaric Alps 78 C4 var. Dinara. mountain range Bosnia and Herzegovina/Croatia
Dindigul 110 C3 Tamil Nādu, SE India
Dingle Bay 67 A6 Ir. Bá an Daingin. bay SW Ireland
Dinguiraye 52 C4 N Guinea
Diourbel 52 B3 W Senegal
Dirē Dawa 51 D5 Dirē Dawa, E Ethiopia
Dirk Hartog Island 125 A5 island Western Australia
Dirschau see Tczew
Disappointment, Lake 124 C4 salt lake Western Australia
Discovery Bay 32 B4 Middlesex, Jamaica, Greater Antilles, C Jamaica Caribbean Sea
Dishū see Deh-e Shū
Disko Bugt see Qeqertarsuup Tunua
Dispur 113 G3 state capital Assam, NE India
Divinópolis 41 F4 Minas Gerais, SE Brazil
Divo 52 D5 S Ivory Coast
Divodurum Mediomatricum see Metz
Diyarbakır 95 E4 var. Diarbekr; anc. Amida. Diyarbakır, SE Turkey (Türkiye)

Dizful see Dezfūl
Djailolo see Halmahera, Pulau
Djajapura see Jayapura
Djakarta see Jakarta
Djakovo see Đakovo
Djambala 55 B6 Plateaux, C Congo
Djambi see Jambi
Djambi see Hari, Batang
Djanet 49 E4 prev. Fort Charlet. SE Algeria
Djéblé see Jablah
Djelfa 48 D2 var. El Djelfa. N Algeria
Djéma 54 D4 Haut-Mbomou, E Central African Republic
Djember see Jember
Djérablous see Jarābulus
Djerba, Île de 49 F2 var. Djerba, Jazīrat Jarbah. island E Tunisia
Djerba see Djerba, Île de
Djérem 54 B4 river C Cameroon
Djevdjelija see Gevgelija
Djibouti 50 D4 var. Jibuti. country capital (Djibouti) E Djibouti
Djibouti 50 D4 off. Republic of Djibouti, var. Jibuti; prev. French Somaliland, French Territory of the Afars and Issas, Fr. Côte Française des Somalis, Territoire Français des Afars et des Issas. country E Africa
Djibouti, Republic of see Djibouti
Djokjakarta see Yogyakarta
Djourab, Erg du 54 C2 desert N Chad
Djúpivogur 61 E5 Austurland, SE Iceland
Dmitriyevsk see Makiivka
Dnepr see Dnieper
Dneprodzerzhinskoye Vodokhranilishche see Dniprodzerzhynske Vodoskhovyshche
Dneprorudnoye see Dniprorudne
Dnestr see Dniester
Dnieper 59 F4 Bel. Dnyapro, Rus. Dnepr, Ukr. Dnipro. river E Europe
Dnieper Lowland 87 E2 Bel. Prydnyaprowskaya Nizina, Ukr. Prydniprovska Nyzovyna. lowlands Belarus/Ukraine
Dniester 59 E4 Rom. Nistru, Rus. Dnestr, Ukr. Dnister; anc. Tyras. river Moldova/Ukraine
Dnipro see Dnieper
Dnipro 87 F3 prev. Dnipropetrovsk, Yekaterinoslav. Dnipropetrovska Oblast, E Ukraine
Dniprodzerzhyns'k see Kamianske
Dniprodzerzhynske Vodoskhovyshche 87 F3 Rus. Dneprodzerzhinskoye Vodokhranilishche. reservoir C Ukraine
Dnipropetrovsk see Dnipro
Dniprorudne 87 F3 Rus. Dneprorudnoye. Zaporiz'ka Oblast', SE Ukraine
Dnister see Dniester
Dnyapro see Dnieper
Doba 54 C4 Logone-Oriental, S Chad
Döbeln 72 D4 Sachsen, E Germany
Doberai Peninsula 117 G4 Dut. Vogelkop. peninsula Papua, E Indonesia
Doboj 78 C3 Republiks Srpska, N Bosnia and Herzegovina
Dobre Miasto 76 D2 Ger. Guttstadt. Warmińsko-mazurskie, NE Poland
Dobrich 82 E1 Rom. Bazargic; prev. Tolbukhin. Dobrich, NE Bulgaria
Dobrush 85 D7 Homyel'skaya Voblasts', SE Belarus
Dobryn' see Dabryn'
Dodecanese see Dodekánisa
Dodekánisa 83 D6 var. Nóties Sporádes, Eng. Dodecanese; prev. Dhodhekánisos, Dodekanisos. island group SE Greece
Dodekanisos see Dodekánisa
Dodge City 23 E5 Kansas, C USA
Dodoma 47 D5 country capital (Tanzania) Dodoma, C Tanzania
Dogana 74 E1 NE San Marino Europe
Dogo 109 B6 island Oki-shotō, SW Japan
Dogondoutchi 53 F3 Dosso, SW Niger
Dogrular see Pravda
Doğubayazıt 95 F3 Ağrı, E Turkey (Türkiye)
Doğu Karadeniz Dağları 95 E3 var. Anadolu Dağları. mountain range NE Turkey (Türkiye)
Doha see Ad Dawḩah
Doire see Londonderry
Dokdo see Liancourt Rocks
Dokkum 64 D1 Fryslân, N Netherlands
Dokuchaievsk 87 G3 prev. Dokuchayevs'k, var. Dokuchayevsk. Donets'ka Oblast', SE Ukraine
Dokuchayevsk/Dokuchayevs'k see Dokuchaievsk
Doldrums Fracture Zone 44 C4 fracture zone W Atlantic Ocean
Dôle 68 D4 Jura, E France
Dolina see Dolyna
Dolinskaya see Dolynska
Dolisie 55 B6 prev. Loubomo. Niari, S Congo
Dolna Oryahovitsa 82 D2 prev. Polikrayshte. Veliko Tarnovo, N Bulgaria
Dolni Chiflik 82 E2 prev. Rudnik. Varna, E Bulgaria
Dolomites 74 C1 var. Dolomiti, Eng. Dolomites. mountain range NE Italy
Dolomites/Dolomiti see Dolomitiche, Alpi
Dolores 42 D4 Buenos Aires, E Argentina
Dolores 30 B1 Petén, N Guatemala
Dolores 42 D4 Soriano, SW Uruguay
Dolores Hidalgo 29 E4 var. Ciudad de Dolores Hidalgo. Guanajuato, C Mexico
Dolyna 86 B2 Rus. Dolina. Ivano-Frankivs'ka Oblast', W Ukraine
Dolynska 87 F3 Rus. Dolinskaya. Kirovohrads'ka Oblast', S Ukraine
Domachëvo/Domaczewo see Damachava
Dombås 63 B5 Oppland, S Norway
Domel Island see Letsôk-aw Kyun
Domesnes, Cape see Kolkasrags
Domeyko 42 B3 Atacama, N Chile
Dominica 33 H4 off. Commonwealth of Dominica. country E West Indies
Dominica Channel see Martinique Passage
Dominica, Commonwealth of see Dominica
Dominican Republic 33 E2 country C West Indies
Domokós 83 B5 var. Dhomokós. Stereá Ellás, C Greece
Don 89 B6 var. Duna, Tanais. river SW Russia
Donau see Danube
Donauwörth 73 C6 Bayern, S Germany
Don Benito 70 C3 Extremadura, W Spain
Doncaster 67 D5 anc. Danum. N England, United Kingdom
Dondo 56 B1 Cuanza Norte, NW Angola
Donegal 67 B5 Ir. Dún na nGall. Donegal, NW Ireland
Donegal Bay 67 A5 Ir. Bá Dhún na nGall. bay NW Ireland
Donets see Siverskyi Donets
Donetsk 87 G3 prev. Stalino. Donets'ka Oblast', E Ukraine
Dongfang 106 B7 var. Basuo. Hainan, S China
Dongguan 106 C6 Guangdong, S China
Đông Ha 114 E4 Quang Tri, C Vietnam
Dong Hai see East China Sea
Đông Hơi 114 D4 Quang Binh, C Vietnam
Dongliao see Liaoyuan
Dongola 50 B3 var. Donqola, Dunqulah. Northern, N Sudan
Dongou 55 C5 Likouala, NE Congo
Dong Rak, Thiu Khao Phanom see Dangrak Mountains
Dongting Hu 106 C5 var. Tung-t'ing Hu. lake S China
Donostia 71 E1 País Vasco, N Spain see also San Sebastián
Donqola see Dongola
Doolow 51 D5 Sumalē, E Ethiopia
Doornik see Tournai
Door Peninsula 18 C2 peninsula Wisconsin, N USA
Dooxo Nugaaleed 51 E5 var. Nogal Valley. valley E Somalia
Dordogne 69 B5 cultural region SW France
Dordogne 69 B5 river W France
Dordrecht 64 C4 var. Dordt, Dort. Zuid-Holland, SW Netherlands
Dordt see Dordrecht
Dorohoi 86 C3 Botoșani, NE Romania
Dorotea 62 C4 Västerbotten, N Sweden
Dorpat see Tartu
Dorre Island 125 A5 island Western Australia
Dort see Dordrecht
Dortmund 72 A4 Nordrhein-Westfalen, W Germany
Dos Hermanas 70 C4 Andalucía, S Spain
Dospad Dagh see Rhodope Mountains
Dospat 82 C3 Smolyan, S Bulgaria
Dothan 20 D3 Alabama, S USA
Dotnuva 84 B4 Kaunas, C Lithuania
Douai 68 C2 prev. Douay; anc. Duacum. Nord, N France
Douala 55 A5 var. Duala. Littoral, W Cameroon
Douay see Douai
Douglas 67 C5 dependent territory capital (Isle of Man) E Isle of Man
Douglas 26 C3 Arizona, SW USA
Douglas 22 D3 Wyoming, C USA
Douma see Dūmā
Douro see Duero
Douvres see Dover
Dover 67 E7 Fr. Douvres, Lat. Dubris Portus. SE England, United Kingdom
Dover 19 F4 state capital Delaware, NE USA
Dover, Strait of 68 C2 var. Straits of Dover, Fr. Pas de Calais. strait England, United Kingdom/France
Dover, Straits of see Dover, Strait of
Dovrefjell 63 B5 plateau S Norway
Downpatrick 67 B5 Ir. Dún Pádraig. SE Northern Ireland, United Kingdom
Dozen 109 B6 island Oki-shotō, SW Japan
Drâa, Hammada du see Dra, Hamada du
Drač/Draç see Durrës
Drachten 64 D2 Fryslân, N Netherlands
Drăgășani 86 B5 Vâlcea, SW Romania
Dragoman 82 B2 Sofiya, W Bulgaria
Dra, Hamada du 48 C3 var. Hammada du Drâa, Haut Plateau du Dra. plateau W Algeria
Dra, Haut Plateau du see Dra, Hamada du
Drahichyn 85 B6 Pol. Drohiczyn Poleski, Rus. Drogichin. Brestskaya Voblasts', SW Belarus
Drakensberg 56 D5 mountain range Lesotho/South Africa
Drake Passage 35 B8 passage Atlantic Ocean/Pacific Ocean
Dralfa 82 D2 Tŭrgovishte, N Bulgaria
Dráma 82 C3 var. Dhráma. Anatolikí Makedonía kai Thráki, NE Greece
Dramburg see Drawsko Pomorskie
Drammen 63 B6 Buskerud, S Norway
Drau see Drava
Drava 78 C3 var. Drau, Eng. Drave, Hung. Dráva. river C Europe
Dráva/Drave see Drau/Drava
Drawsko Pomorskie 76 B3 Ger. Dramburg. Zachodnio-pomorskie, NW Poland
Drépano, Akrotírio 82 C4 var. Akrotírio Dhrepanon. headland N Greece
Drepanum see Trapani
Dresden 72 D4 Sachsen, E Germany
Drin see Drinit, Lumi i
Drina 78 C3 river Bosnia and Herzegovina/Serbia
Drinit, Lumi i 79 D5 var. Drin. river NW Albania
Drinit të Zi, Lumi i see Black Drin
Drissa see Drysa
Drobeta-Turnu Severin 86 B5 prev. Turnu Severin. Mehedinți, SW Romania
Drogheda 67 B5 Ir. Droichead Átha. NE Ireland
Drogichin see Drahichyn
Drogobych see Drohobych
Drohiczyn Poleski see Drahichyn
Drohobych 86 B2 Pol. Drohobycz, Rus. Drogobych. L'vivs'ka Oblast', NW Ukraine
Drohobycz see Drohobych
Droichead Átha see Drogheda
Drôme 69 D5 cultural region E France
Dronning Maud Land 132 B2 physical region Antarctica
Drontheim see Trondheim
Drug see Durg
Druk-yul see Bhutan
Drummondville 17 E4 Québec, SE Canada
Druskienniki see Druskininkai
Druskininkai 85 B5 Pol. Druskienniki. Alytus, S Lithuania
Dryden 16 B3 Ontario, C Canada
Drysa 85 D5 Rus. Drissa. river N Belarus
Duacum see Douai
Duala see Douala
Dubai see Dubayy
Dubăsari 86 D3 Rus. Dubossary. NE Moldova
Dubawnt 15 F4 river Nunavut, NW Canada
Dubayy 98 D4 Eng. Dubai. Dubayy, NE United Arab Emirates
Dubbo 127 D6 New South Wales, SE Australia
Dublin 67 B5 Ir. Baile Átha Cliath; anc. Eblana. country capital (Ireland) Dublin, E Ireland
Dublin 21 E2 Georgia, SE USA
Dubno 86 C2 Rivnens'ka Oblast', NW Ukraine
Dubossary see Dubăsari
Dubris Portus see Dover
Dubrovnik 79 B5 It. Ragusa. Dubrovnik-Neretva, SE Croatia
Dubuque 23 G3 Iowa, C USA
Dudelange 65 D8 var. Forge du Sud, Ger. Dudelingen. Luxembourg, S Luxembourg
Dudelingen see Dudelange
Duero 70 D2 Port. Douro. river Portugal/Spain
Duesseldorf see Düsseldorf
Duffel 65 C5 Antwerpen, C Belgium
Dugi Otok 78 A4 var. Isola Grossa, It. Isola Lunga. island W Croatia
Duinekerke see Dunkerque
Duisburg 72 A4 prev. Duisburg-Hamborn. Nordrhein-Westfalen, W Germany
Duisburg-Hamborn see Duisburg
Duiven 64 D4 Gelderland, E Netherlands
Duk Faiwil 51 B5 Jonglei, C South Sudan
Dulan 104 D4 var. Qagan Us. Qinghai, C China
Dulce, Golfo 31 E5 gulf S Costa Rica
Dulce, Golfo see Izabal, Lago de
Dülmen 72 A4 Nordrhein-Westfalen, W Germany
Dulovo 82 E1 Silistra, NE Bulgaria
Duluth 23 G2 Minnesota, N USA
Dūmā 97 B5 Fr. Douma. Dimashq, SW Syria
Dumas 27 E1 Texas, SW USA
Dumfries 66 C4 S Scotland, United Kingdom
Dumont d'Urville 132 C4 French research station Antarctica
Dumyāṭ 50 B1 Eng. Damietta. N Egypt
Duna see Danube, C Europe
Düna see Western Dvina
Duna see Don
Dünaburg see Daugavpils
Dunai see Danube
Dunaj see Wien, Austria
Dunaj see Danube, C Europe
Dunapentele see Dunaújváros
Dunărea see Danube
Dunaújváros 77 C7 prev. Dunapentele, Sztálinváros. Fejér, C Hungary
Dunav see Danube
Dunavska Ravnina 82 C2 Eng. Danubian Plain. lowlands N Bulgaria
Duncan 27 G2 Oklahoma, C USA
Dundalk 67 B5 Ir. Dún Dealgan. Louth, NE Ireland
Dún Dealgan see Dundalk
Dundee 56 D4 KwaZulu/Natal, E South Africa
Dundee 66 C4 E Scotland, United Kingdom
Dunedin 129 B7 Otago, South Island, New Zealand
Dunfermline 66 C4 C Scotland, United Kingdom
Dungu 55 E5 Orientale, NE Dem. Rep. Congo
Dungun 116 B3 var. Kuala Dungun. Terengganu, Peninsular Malaysia
Dunholme see Durham
Dunkerque 68 C2 Eng. Dunkirk, Flem. Duinekerke; prev. Dunquerque. Nord, N France
Dunkirk see Dunkerque
Dún Laoghaire 67 B6 Eng. Dunleary; prev. Kingstown. E Ireland
Dunleary see Dún Laoghaire
Dún Pádraig see Downpatrick
Dunquerque see Dunkerque
Dunqulah see Dongola
Dupnitsa 82 C2 prev. Marek, Stanke Dimitrov. Kyustendil, W Bulgaria
Duqm 99 E5 var. Daqm. E Oman
Durance 69 D6 river SE France
Durango 28 D3 var. Victoria de Durango. Durango, W Mexico
Durango 22 C5 Colorado, C USA
Durankulak 82 E1 Rom. Răcari; prev. Blatnitsa, Duranulac. Dobrich, NE Bulgaria
Durant 27 G2 Oklahoma, C USA
Duranulac see Durankulak
Durazzo see Durrës
Durban 56 D4 var. Port Natal. KwaZulu/Natal, E South Africa
Durbe 84 B3 Ger. Durben. Liepāja, W Latvia
Durben see Durbe
Durg 113 E4 prev. Drug. Chhattisgarh, C India
Durham 67 D5 hist. Dunholme. N England, United Kingdom
Durham 21 F1 North Carolina, SE USA
Durocortorum see Reims
Durostorum see Silistra
Durovernum see Canterbury
Durrës 79 C6 var. Durrësi, Dursi, It. Durazzo, Croatian Drač, Turk. Draç. Durrës, W Albania
Durrësi see Durrës
Dursi see Durrës
Durūz, Jabal ad 97 C5 mountain SW Syria
D'Urville Island see Rangitoto ki te Tonga
Dusa Mareb/Dusa Marreb see Dhuusa Marreeb
Dushanbe 101 E3 var. Dyushambe; prev. Stalinabad, Taj. Stalinobod. country capital (Tajikistan) W Tajikistan
Düsseldorf 72 A4 var. Duesseldorf. Nordrhein-Westfalen, W Germany
Dŭsti 101 E3 Rus. Dusti. SW Tajikistan
Dutch East Indies see Indonesia
Dutch Guiana see Suriname
Dutch Harbor 14 B3 Unalaska Island, Alaska, USA
Dutch New Guinea see Papua
Duzdab see Zāhedān
Dvina Bay see Chëshskaya Guba
Dvinsk see Daugavpils
Dyanev see Galkynyș
Dyersburg 20 C1 Tennessee, S USA
Dyushambe see Dushanbe
Dza Chu see Mekong
Dzaudzhikau see Vladikavkaz
Dzerzhinsk 89 C5 Nizhegorodskaya Oblast', W Russia
Dzerzhinskiy see Nar'yan-Mar
Dzhalal-Abad 101 F2 Kir. Jalal-Abad. Dzhalal-Abadskaya Oblast', W Kyrgyzstan
Dzhambul see Taraz
Dzhankoi 87 F4 prev. Dzhankoy, Respublika Krym, S Ukraine
Dzhankoy see Dzhankoi
Dzharkurgan see Jarqo'rg'on
Dzhelandy see Jelondi
Dzhergalan 101 G2 Kir. Jyrgalan. Issyk-Kul'skaya Oblast', NE Kyrgyzstan
Dzhezkazgan see Zhezqazghan
Dzhizak see Jizzax
Dzhugdzhur, Khrebet 93 G3 mountain range E Russia
Dzhusaly see Zhosaly
Działdowo 76 D3 Warmińsko-Mazurskie, C Poland
Dzuunmod 105 E2 Töv, C Mongolia
Dzüün Soyonï Nuruu see Vostochnyy Sayan
Dzvina see Western Dvina

E

Eagle Pass 27 F4 Texas, SW USA
East Açores Fracture Zone see East Azores Fracture Zone
East Antarctica 132 C3 var. Greater Antarctica. physical region Antarctica
East Australian Basin see Tasman Basin
East Azores Fracture Zone 44 C3 var. East Açores Fracture Zone. tectonic feature E Atlantic Ocean
Eastbourne 67 E7 SE England, United Kingdom
East Cape 128 E3 headland North Island, New Zealand
East China Sea 103 E2 Chin. Dong Hai. sea W Pacific Ocean
Easter Fracture Zone 131 G4 tectonic feature E Pacific Ocean
Easter Island see Pascua, Isla de
Eastern Desert see Ṣaḥrā' ash Sharqīyah
Eastern Ghats 102 B3 mountain range SE India
Eastern Sayans see Vostochnyy Sayan
Eastern Sierra Madre see Madre Oriental, Sierra
East Falkland 43 D8 var. Isla Soledad. island E Falkland Islands
East Frisian Islands 72 A3 Eng. East Frisian Islands. island group NW Germany
East Frisian Islands see Ostfriesische Inseln
East Grand Forks 23 E1 Minnesota, N USA
East Indiaman Ridge 119 D6 undersea ridge E Indian Ocean
East Indies 130 A3 island group SE Asia
East Kilbride 66 C4 S Scotland, United Kingdom
East Korea Bay 107 E3 bay E North Korea
Eastleigh 67 D7 S England, United Kingdom
East London 56 D5 Afr. Oos-Londen; prev. Emonti, Port Rex. Eastern Cape, S South Africa
Eastmain 16 D3 river Québec, C Canada
East Mariana Basin 120 B1 undersea basin W Pacific Ocean
East Novaya Zemlya Trough 90 C1 var. Novaya Zemlya Trough. trough W Kara Sea
East Pacific Rise 131 F4 undersea rise E Pacific Ocean
East Pakistan see Bangladesh
East Saint Louis 18 B4 Illinois, N USA
East Scotia Basin 45 C7 undersea basin SE Scotia Sea
East Sea 108 A4 var. Sea of Japan, Rus. Yapanskoye More. Sea NW Pacific Ocean
East Siberian Sea see Vostochno-Sibirskoye More
East Timor 117 F5 var. Timor Lorosa'e, Port. Timor-Leste, prev. Portuguese Timor, Timor Timur. country S Indonesia
Eau Claire 18 A2 Wisconsin, N USA
Eau Claire, Lac à L' see St. Clair, Lake
Eauripik Rise 120 B2 undersea rise W Pacific Ocean
Ebensee 73 D6 Oberösterreich, N Austria
Eberswalde-Finow 72 D3 Brandenburg, E Germany
Ebetsu 108 D2 var. Ebetu. Hokkaidō, NE Japan
Ebetu see Ebetsu
Eblana see Dublin
Ebolowa 55 A5 Sud, S Cameroon
Ebon Atoll 122 D2 var. Epoon. atoll Ralik Chain, S Marshall Islands
Ebora see Évora
Eboracum see York
Ebro 71 E2 river NE Spain
Eburacum see York
Ebusus see Eivissa
Ecbatana see Hamadān
Ech Cheliff/Ech Chleff see Chlef
Echo Bay 15 E3 Northwest Territories, NW Canada
Echt 65 D5 Limburg, SE Netherlands
Écija 70 D4 anc. Astigi. Andalucía, SW Spain
Eckengraf see Viesīte
Ecuador 38 B1 off. Republic of Ecuador. country NW South America
Ecuador, Republic of see Ecuador
Ed Da'ein 50 A4 Southern Darfur, W Sudan
Ed Damazin 50 C4 var. Ad Damazīn. Blue Nile, E Sudan
Ed Damer 50 C3 var. Ad Dāmir, Ad Damar. River Nile, NE Sudan
Ed Debba 50 B3 Northern, N Sudan
Ede 64 D4 Gelderland, C Netherlands
Ede 53 F5 Osun, SW Nigeria
Edéa 55 A5 Littoral, SW Cameroon
Edessa see Şanlıurfa
Edfu see Idfū
Edgeøya 61 G2 island S Svalbard
Edgware 67 A7 Harrow, SE England, United Kingdom
Edinburg 27 G5 Texas, SW USA
Edinburgh 66 C4 national capital S Scotland, United Kingdom
Edingen see Enghien
Edirne 94 A2 Eng. Adrianople; anc. Adrianopolis, Hadrianopolis. Edirne, NW Turkey (Türkiye)
Edmonds 24 B2 Washington, NW USA
Edmonton 15 E5 province capital Alberta, SW Canada
Edmundston 17 E4 New Brunswick, SE Canada
Edna 27 G4 Texas, SW USA
Edolo 74 B1 Lombardia, N Italy
Edremit 94 A3 Balıkesir, NW Turkey (Türkiye)
Edward, Lake 55 E5 var. Albert Edward Nyanza, Edward Nyanza, Lac Idi Amin, Lake Rutanzige. lake Uganda/Dem. Rep. Congo
Edward Nyanza see Edward, Lake
Edwards Plateau 27 F3 plain Texas, SW USA
Edzo 31 E4 prev. Rae-Edzo. Northwest Territories, NW Canada
Eeklo 65 B5 var. Eekloo. Oost-Vlaanderen, NW Belgium
Eekloo see Eeklo
Eems see Ems
Eersel 65 C5 Noord-Brabant, S Netherlands
Eesti Vabariik see Estonia
Efate 122 D4 var. Efate, Fr. Vaté; prev. Sandwich Island. island C Vanuatu
Efate see Efate
Effingham 18 B4 Illinois, N USA
Eforie-Sud 86 D5 Constanța, E Romania
Egadi Is. 75 B7 island group S Italy
Ege Denizi see Aegean Sea
Eger 77 D6 Ger. Erlau. Heves, NE Hungary
Eger see Cheb, Czechia (Czech Republic)
Egeria Fracture Zone 119 C5 tectonic feature W Indian Ocean
Éghezèe 65 C6 Namur, C Belgium
Egina see Aígina
Egio see Aígio
Egmont see Taranaki, Mount
Egmont, Cape 128 C4 headland North Island, New Zealand
Egoli see Johannesburg
Egypt 50 B2 off. Arab Republic of Egypt, Ar. Jumhūrīyat Miṣr al 'Arabīyah, Miṣr, prev. United Arab Republic; anc. Aegyptus. country NE Africa
Eibar 71 E1 País Vasco, N Spain
Eibergen 64 E3 Gelderland, E Netherlands
Eidfjord 63 A5 Hordaland, S Norway
Eier-Berg see Suur Munamägi
Eifel 73 A5 plateau W Germany
Eiger 73 B7 mountain C Switzerland
Eigg 66 B3 island W Scotland, United Kingdom
Eight Degree Channel 110 B3 channel India/Maldives
Eighty Mile Beach 124 B4 beach Western Australia
Eijsden 65 D6 Limburg, SE Netherlands
Eilat see Elat
Eindhoven 65 D5 Noord-Brabant, S Netherlands
Eipel see Ipel'
Éire see Ireland
Éireann, Muir see Irish Sea
Eisenhüttenstadt 72 D4 Brandenburg, E Germany
Eisenmarkt see Hunedoara
Eisenstadt 73 E6 Burgenland, E Austria
Eisleben 72 C4 Sachsen-Anhalt, C Germany
Eivissa 71 G3 var. Iviza, Cast. Ibiza; anc. Ebusus. Ibiza, Spain, W Mediterranean Sea
Ejea de los Caballeros 71 E2 Aragón, NE Spain
Ejin Qi see Dalain Hob
Ekapa see Cape Town
Ekaterinodar see Krasnodar
Ekvyvatapskiy Khrebet 93 G1 mountain range NE Russia
El 'Alamein see Al 'Alamayn
El Asnam see Chlef
Elat 97 B8 var. Eilat, Elath. Southern, S Israel
Elat, Gulf of see Aqaba, Gulf of
Elath see Elat, Israel
Elath see Al 'Aqabah, Jordan
El'Atrun 50 B3 Northern Darfur, NW Sudan
Elâzığ 95 E3 var. Elazig, Elâziz. Elâzığ, E Turkey (Türkiye)
Elba 74 B4 island Archipelago Toscano, C Italy
Elbasan 79 D6 var. Elbasani. Elbasan, C Albania
Elbasani see Elbasan
Elbe 58 D3 Cz. Labe. river Czechia (Czech Republic) /Germany
Elbert, Mount 22 C4 mountain Colorado, C USA
Elbing see Elbląg
Elbląg 76 C2 var. Elbląg, Ger. Elbing. Warmińsko-Mazurskie, NE Poland
El Boulaida/El Boulaïda see Blida
El'brus 89 A8 var. Gora El'brus. mountain SW Russia
El'brus, Gora see El'brus
El Burgo de Osma 71 E2 Castilla y León, C Spain
Elburz Mountains see Alborz, Reshteh-ye Kūhhā-ye
El Cajon 25 C8 California, W USA
El Calafate 43 B7 var. Calafate. Santa Cruz, S Argentina
El Callao 37 E2 Bolívar, E Venezuela
El Campo 27 G4 Texas, SW USA
El Carmen de Bolívar 36 B2 Bolívar, NW Colombia
El Cayo see San Ignacio
El Centro 25 D8 California, W USA
Elche see Elx
Elda 71 F4 Comunitat Valenciana, E Spain
El Djazaïr see Alger
El Djelfa see Djelfa
Eldorado 42 E3 Misiones, NE Argentina
El Dorado 28 C3 Sinaloa, C Mexico
El Dorado 20 B2 Arkansas, C USA
El Dorado 23 F5 Kansas, C USA
El Dorado 37 F2 Bolívar, E Venezuela
El Dorado see California
Eldoret 51 C6 Rift Valley, W Kenya
Elektrostal' 89 B5 Moskovskaya Oblast', W Russia
Elephant Butte Reservoir 26 C2 reservoir New Mexico, SW USA
Élesd see Aleșd
Eleuthera Island 32 C1 island N The Bahamas
El Fasher 50 A4 var. Al Fāshir. Northern Darfur, W Sudan
El Ferrol/El Ferrol del Caudillo see Ferrol
El Gedaref see Gedaref
El Geneina 50 A4 var. Ajjinena, Al-Genain, Al Junaynah. Western Darfur, W Sudan
Elgin 66 C3 NE Scotland, United Kingdom
Elgin 18 B3 Illinois, N USA
El Gîza see Al Jīzah
El Goléa 48 D3 var. Al Golea. C Algeria
El Ḩank 52 D1 cliff N Mauritania
El Haseke see Al Ḩasakah
Elimberrum see Auch
Eliocroca see Lorca
Élisabethville see Lubumbashi
Elista 89 B7 Respublika Kalmykiya, SW Russia
Elizabeth 127 B6 South Australia
Elizabeth City 21 G1 North Carolina, SE USA
Elizabethtown 18 C5 Kentucky, S USA
El-Jadida 48 C2 prev. Mazagan. W Morocco
Ełk 76 E2 Ger. Lyck. Warmińsko-mazurskie, NE Poland
Elk City 27 F1 Oklahoma, C USA
El Khalil see Hebron
El Khârga see Al Khārjah
Elkhart 18 C3 Indiana, N USA
El Khartûm see Khartoum
Elk River 23 F2 Minnesota, N USA
El Kuneitra see Al Qunayṭirah
Ellás see Greece
Elláda see Greece
Ellef Ringnes Island 15 E1 island Nunavut, N Canada
Ellen, Mount 22 B5 mountain Utah, W USA
Ellensburg 24 B2 Washington, NW USA
Ellesmere Island 15 F1 island Queen Elizabeth Islands, Nunavut, N Canada
Ellesmere, Lake 129 C6 lake South Island, New Zealand
Ellice Islands see Tuvalu

Ellinikí Dimokratía see Greece
Elliston 127 A6 South Australia
Ellsworth Land 132 A3 physical region Antarctica
El Mahbas 48 B3 var. Mahbés. SW Western Sahara
El Mina 96 B4 var. Al Mīnā'. N Lebanon
El Minya see Al Minyā
Elmira 19 E3 New York, NE USA
El Mreyyé 52 D2 desert E Mauritania
Elmshorn 72 B3 Schleswig-Holstein, N Germany
El Muglad 50 B4 Western Kordofan, C Sudan
El Obeid 50 B4 var. Al Obayyid, Al Ubayyiḑ. Northern Kordofan, C Sudan
El Ouâdi see El Oued
El Oued 49 E2 var. Al Oued, El Ouâdi, El Wad. NE Algeria
Eloy 26 B2 Arizona, SW USA
El Paso 26 D3 Texas, SW USA
El Porvenir 31 G4 Kuna Yala, N Panama
El Progreso 30 C2 Yoro, NW Honduras
El Puerto de Santa María 70 C5 Andalucía, S Spain
El Qâhira see Al Qāhirah
El Quneitra see Al Qunayṭirah
El Quseir see Al Quṣayr
El Quweira see Al Quwayrah
El Rama 31 E3 Región Autónoma Atlántico Sur, SE Nicaragua
El Real 31 H5 var. El Real de Santa María. Darién, SE Panama
El Real de Santa María see El Real
El Reno 27 F1 Oklahoma, C USA
El Salvador 30 B3 off. Republic of El Salvador, Sp. República de El Salvador. country Central America
El Salvador, Republic of see El Salvador
El Salvador, República de see El Salvador
Elsass see Alsace
El Sáuz 28 C2 Chihuahua, N Mexico
El Serrat 69 A7 N Andorra Europe
Elst 64 D4 Gelderland, E Netherlands
El Sueco 28 C2 Chihuahua, N Mexico
El Suweida see As Suwaydā'
El Suweis see As Suways
Eltanin Fracture Zone 131 E5 tectonic feature SE Pacific Ocean
El Tigre 37 E2 Anzoátegui, NE Venezuela
Elvas 70 C4 Portalegre, C Portugal
El Vendrell 71 G2 Cataluña, NE Spain
El Vigía 36 C2 Mérida, NW Venezuela
El Wad see El Oued
Elwell, Lake 22 B1 reservoir Montana, NW USA
Elx 71 F4 var. Elche; anc. Ilici, Lat. Illicis. Comunitat Valenciana, E Spain
Ely 25 D5 Nevada, W USA
El Yopal see Yopal
Emajõgi 84 D3 Ger. Embach. river SE Estonia
Emāmrūd see Shāhrūd
Emāmshahr see Shāhrūd
Emba 92 B4 Kaz. Embi. Aktyubinsk, W Kazakhstan
Embach see Emajõgi
Embi see Emba
Emden 72 A3 Niedersachsen, NW Germany
Emerald 126 D4 Queensland, E Australia
Emerald Isle see Montserrat
Emesa see Ḩimş
Emmaste 84 C2 Hiiumaa, W Estonia
Emmeloord 64 D2 Flevoland, N Netherlands
Emmen 64 E2 Drenthe, NE Netherlands
Emmendingen 73 A6 Baden-Württemberg, SW Germany
Emona see Ljubljana
Emonti see East London
Emory Peak 27 E4 mountain Texas, SW USA
Empalme 28 B2 Sonora, NW Mexico
Emperor Seamounts 91 G3 seamount range NW Pacific Ocean
Empire State of the South see Georgia
Emporia 23 F5 Kansas, C USA
Empty Quarter see Ar Rub 'al Khālī
Ems 72 A3 Dut. Eems. river NW Germany
Enareträsk see Inarijärvi
Encamp 69 A8 Encamp, C Andorra Europe
Encarnación 42 D3 Itapúa, S Paraguay
Encinitas 25 C8 California, W USA
Encs 77 D6 Borsod-Abaúj-Zemplén, NE Hungary
Endeavour Strait 126 C1 strait Queensland, NE Australia
Enderbury Island 123 F3 atoll Phoenix Islands, C Kiribati
Enderby Land 132 C2 physical region Antarctica
Enderby Plain 132 D2 abyssal plain S Indian Ocean
Endersdorf see Jędrzejów
Enewetak Atoll 122 C1 var. Ānewetak, Eniwetok. atoll Ralik Chain, W Marshall Islands
Enfield 67 A7 United Kingdom
Engeten see Aiud
Enghien 65 B6 Dut. Edingen. Hainaut, SW Belgium
England 67 D5 Lat. Anglia. cultural region England, United Kingdom
Englewood 22 D4 Colorado, C USA
English Channel 67 D8 var. The Channel, Fr. la Manche. channel NW Europe
Engure 84 C3 Tukums, W Latvia
Engures Ezers 84 B3 lake NW Latvia
Enguri 95 F1 Rus. Inguri. river NW Georgia
Enid 27 F1 Oklahoma, C USA
Enikale Strait see Kerch Strait
Eniwetok see Enewetak Atoll
En Nâqoûra 97 A5 var. An Nāqūrah. SW Lebanon
En Nazira see Natzrat
Ennedi 54 D2 plateau E Chad
Ennis 67 A6 Ir. Inis. Clare, W Ireland
Ennis 27 G3 Texas, SW USA
Enniskillen 67 B5 var. Inniskilling, Ir. Inis Ceithleann. SW Northern Ireland, United Kingdom
Enns 73 D6 river C Austria
Enschede 64 E3 Overijssel, E Netherlands
Ensenada 28 A1 Baja California Norte, NW Mexico
Entebbe 51 B6 S Uganda
Entroncamento 70 B3 Santarém, C Portugal
Enugu 53 G5 Enugu, S Nigeria
Epanomí 82 B4 Kentrikí Makedonía, N Greece
Epéna 55 B5 Likouala, NE Congo
Eperies/Eperjes see Prešov
Epi 122 D4 var. Épi. island C Vanuatu
Épi see Epi
Épinal 68 D4 Vosges, NE France
Epiphania see Ḩamāh
Epoon see Ebon Atoll
Epsom 67 A8 United Kingdom
Equality State see Wyoming
Equatorial Guinea 55 A5 off. Equatorial Guinea, Republic of. country C Africa
Equatorial Guinea, Republic of see Equatorial Guinea
Erautini see Johannesburg
Erbil see Arbīl
Erciş 95 F3 Van, E Turkey (Türkiye)
Erdély see Transylvania
Erdélyi-Havasok see Carpaţii Meridionali
Erdenet 105 E2 Orhon, N Mongolia
Erdi 54 C2 plateau NE Chad
Erdi Ma 54 D2 desert NE Chad
Erebus, Mount 132 B4 volcano Ross Island, Antarctica
Ereğli 94 C4 Konya, S Turkey (Türkiye)
Erenhot 105 F2 var. Erlian. Nei Mongol Zizhiqu, NE China
Erfurt 72 C4 Thüringen, C Germany
Ergene Çayı see Ergene Irmağī
Ergene Irmağī 94 A2 var. Ergene Çayı. river NW Turkey (Türkiye)
Ergun 105 F1 var. Labudalin; prev. Ergun Youqi. Nei Mongol Zizhiqu, N China
Ergun He see Argun
Ergun Youqi see Ergun
Erie 18 D3 Pennsylvania, NE USA
Érié, Lac see Erie, Lake
Erie, Lake 18 D3 Fr. Lac Érié. lake Canada/USA
Eritrea 50 C4 off. State of Eritrea, Iertra. country E Africa
Eritrea, State of see Eritrea
Erivan see Yerevan
Erlangen 73 C5 Bayern, S Germany
Erlau see Eger
Erlian see Erenhot
Ermelo 64 D3 Gelderland, C Netherlands
Ermióni 83 C6 Pelopónnisos, S Greece
Ermoúpoli 83 D6 var. Hermoupolis; prev. Ermoúpolis. Sýyros, Kykládes, Greece, Aegean Sea
Ermoúpolis see Ermoúpoli
Ernākulam 110 C3 Kerala, SW India
Erode 110 C2 Tamil Nādu, SE India
Erquelinnes 65 B7 Hainaut, S Belgium
Er-Rachidia 48 C2 var. Ksar al Soule. E Morocco
Er Rahad 50 B4 var. Ar Rahad. Northern Kordofan, C Sudan
Erromango 122 D4 island S Vanuatu
Ertis see Irtysh, C Asia
Erzerum see Erzurum
Erzgebirge 73 C5 Cz. Krušné Hory, Eng. Ore Mountains. mountain range Czechia (Czech Republic) /Germany
Erzincan 95 E3 var. Erzinjan. Erzincan, E Turkey (Türkiye)
Erzinjan see Erzincan
Erzurum 95 E3 prev. Erzerum. Erzurum, NE Turkey (Türkiye)
Esbjerg 63 A7 Ribe, W Denmark
Esbo see Espoo
Escaldes 69 A8 Escaldes Engordany, C Andorra Europe
Escanaba 18 C2 Michigan, N USA
Escaut see Scheldt
Esch-sur-Alzette 65 D8 Luxembourg, S Luxembourg
Esclaves, Grand Lac des see Great Slave Lake
Escondido 25 C8 California, W USA
Escuinapa 28 D3 var. Escuinapa de Hidalgo. Sinaloa, C Mexico
Escuinapa de Hidalgo see Escuinapa
Escuintla 30 B2 Escuintla, S Guatemala
Escuintla 29 G5 Chiapas, SE Mexico
Esenguly 100 B3 Rus. Gasan-Kuli. Balkan Welaýaty, W Turkmenistan
Eşfahān 98 C3 Eng. Isfahan; anc. Aspadana. Eşfahān, C Iran
Esh Sharā see Ash Sharāh
Esil see Ishim, Kazakhstan/Russia
Eskimo Point see Arviat
Eskişehir 94 B3 var. Eskishehr. Eskişehir, W Turkey (Türkiye)
Eskishehr see Eskişehir
Eslāmābād 98 C3 var. Eslāmābād-e Gharb; prev. Harunabad, Shāhābād. Kermānshāhān, W Iran
Eslāmābād-e Gharb see Eslāmābād
Esmeraldas 38 A1 Esmeraldas, N Ecuador
Esna see Isnā
España see Spain
Espanola 26 D1 New Mexico, SW USA
Esperance 125 B7 Western Australia
Esperanza 28 B2 Sonora, NW Mexico
Esperanza 132 A2 Argentinian research station Antarctica
Espinal 36 B3 Tolima, C Colombia
Espinhaço, Serra do 34 D4 mountain range SE Brazil
Espírito Santo 41 F4 off. Estado do Espírito Santo. region E Brazil
Espírito Santo 41 F4 off. Estado do Espírito Santo. state E Brazil
Espírito Santo, Estado do see Espírito Santo
Espiritu Santo 122 C4 var. Santo. island W Vanuatu
Espoo 63 D6 Swe. Esbo. Uusimaa, S Finland
Esquel 43 B6 Chubut, SW Argentina
Essaouira 48 B2 prev. Mogador. W Morocco
Esseg see Osijek
Es Semara see Smara
Essen 65 C5 Antwerpen, N Belgium
Essen 72 A4 var. Essen an der Ruhr. Nordrhein-Westfalen, W Germany
Essen an der Ruhr see Essen
Essequibo River 37 F3 river C Guyana
Es Suweida see As Suwaydā'
Estacado, Llano 27 E2 plain New Mexico/Texas, SW USA
Estados, Isla de los 43 C8 prev. Eng. Staten Island. island S Argentina
Estância 41 G3 Sergipe, E Brazil
Estelí 30 D3 Estelí, NW Nicaragua
Estella 71 E1 Bas. Lizarra. Navarra, N Spain
Estepona 70 D5 Andalucía, S Spain
Estevan 15 F5 Saskatchewan, S Canada
Estland see Estonia
Estonia 84 D2 off. Republic of Estonia, Est. Eesti Vabariik, Ger. Estland, Latv. Igaunija; prev. Estonian SSR, Rus. Estonskaya SSR. country NE Europe
Estonian SSR see Estonia
Estonia, Republic of see Estonia
Estonskaya SSR see Estonia
Estrela, Serra da 70 C3 mountain range C Portugal
Estremadura see Extremadura
Estremoz 70 C4 Évora, S Portugal
Eswatini 56 D4 off. Kingdom of Eswatini; prev. Swaziland. country S Africa
Eswatini, Kingdom of see Eswatini
Eszék see Osijek
Esztergom 77 C6 Ger. Gran; anc. Strigonium. Komárom-Esztergom, N Hungary
Ettelle 65 D8 Luxembourg, SE Belgium
Etāwah 112 D3 Uttar Pradesh, N India
Ethiopia 51 C5 off. Federal Democratic Republic of Ethiopia; prev. Abyssinia, People's Democratic Republic of Ethiopia. country E Africa
Ethiopia, Federal Democratic Republic of see Ethiopia
Ethiopian Highlands 51 C5 var. Ethiopian Plateau. plateau N Ethiopia
Ethiopian Plateau see Ethiopian Highlands
Ethiopia, People's Democratic Republic of see Ethiopia
Etna, Monte 75 C7 Eng. Mount Etna. volcano Sicilia, Italy, C Mediterranean Sea
Etna, Mount see Etna, Monte
Etosha Pan 56 B3 salt lake N Namibia
Etoumbi 55 B5 Cuvette Ouest, NW Congo
Etsch see Adige
Et Tafila see Aţ Ţafilah
Ettelbrück 65 D8 Diekirch, C Luxembourg
'Eua 123 E5 prev. Middleburg Island. island Tongatapu Group, SE Tonga
Euboea see Évvoia
Euboea see Évvoia
Eucla 125 D6 Western Australia
Euclid 18 D3 Ohio, N USA
Eufaula Lake 27 G1 var. Eufaula Reservoir. reservoir Oklahoma, C USA
Eufaula Reservoir see Eufaula Lake
Eugene 24 B3 Oregon, NW USA
Eumolpias see Plovdiv
Eupen 65 D6 Liège, E Belgium
Euphrates 90 B4 Ar. Al-Furāt, Turk. Firat Nehri. river SW Asia
Eureka 25 A5 California, W USA
Eureka 22 A1 Montana, NW USA
Europa Point 71 H5 headland S Gibraltar
Europe 58 continent
Eutin 72 C2 Schleswig-Holstein, N Germany
Euxine Sea see Black Sea
Evansdale 23 G3 Iowa, C USA
Evanston 18 B3 Illinois, N USA
Evanston 22 B4 Wyoming, C USA
Evansville 18 B5 Indiana, N USA
Eveleth 23 G1 Minnesota, N USA
Everard, Lake 127 A6 salt lake South Australia
Everest, Mount 104 B5 Chin. Qomolangma Feng, Nep. Sagarmāthā. mountain China/Nepal
Everett 24 B2 Washington, NW USA
Everglades, The 21 F5 wetland Florida, SE USA
Evje 63 A6 Aust-Agder, S Norway
Evmolpia see Plovdiv
Évora 70 B4 anc. Ebora, Lat. Liberalitas Julia. Évora, C Portugal
Évreux 68 C3 anc. Civitas Eburovicum. Eure, N France
Évros see Maritsa
Évry 68 E2 Essonne, N France
Évvoia 83 C5 var. Euboea, Lat. Euboea. island C Greece
Ewarton 32 B5 C Jamaica
Excelsior Springs 23 F4 Missouri, C USA
Exe 67 C7 river SW England, United Kingdom
Exeter 67 C7 anc. Isca Damnoniorum. SW England, United Kingdom
Exmoor 67 C7 moorland SW England, United Kingdom
Exmouth 124 A4 Western Australia
Exmouth 67 C7 SW England, United Kingdom
Exmouth Gulf 124 A4 gulf Western Australia
Exmouth Plateau 119 E5 undersea plateau E Indian Ocean
Extremadura 70 C3 var. Estremadura. autonomous community W Spain
Exuma Cays 32 C1 islets C The Bahamas
Exuma Sound 32 C1 sound C The Bahamas
Eyre Mountains 129 A7 mountain range South Island, New Zealand
Eyre, Lake 127 A5 var. Kati Thanda. salt lake South Australia
Eyre Peninsula 127 A6 peninsula South Australia
Eyre South, Lake 127 A5 salt lake South Australia
Ezo see Hokkaidō

F

Faadhippolhu Atoll 110 B4 var. Fadiffolu, Lhaviyani Atoll. atoll N Maldives
Fabens 26 D3 Texas, SW USA
Fada 54 C2 Borkou-Ennedi-Tibesti, E Chad
Fada-Ngourma 53 E4 E Burkina
Fadiffolu see Faadhippolhu Atoll
Faenza 74 C3 anc. Faventia. Emilia-Romagna, N Italy
Faeroe Islands see Faero Islands
Færøerne see Faroe Islands
Faetano 74 E2 E San Marino
Făgăraş 86 C4 Ger. Fogarasch, Hung. Fogaras. Braşov, C Romania
Fagibina, Lake see Faguibine, Lac
Fagne 65 C7 hill range S Belgium
Faguibine, Lac 53 E3 var. Lake Fagibina. lake NW Mali
Fahlun see Falun
Fahraj 98 E4 Kermān, SE Iran
Faial 70 A5 var. Ilha do Faial. island Azores, Portugal, NE Atlantic Ocean
Faial, Ilha do see Faial
Faifo see Hôi An
Fairbanks 14 D3 Alaska, USA
Fairfield 25 B6 California, W USA
Fair Isle 66 D2 island NE Scotland, United Kingdom
Fairlie 129 B6 Canterbury, South Island, New Zealand
Fairmont 23 F3 Minnesota, N USA
Faisalabad 112 C2 prev. Lyallpur. Punjab, NE Pakistan
Faizābād 113 E3 Uttar Pradesh, N India
Faiẓābād 101 F3 var. Feyẓābād, Fyzabad. Badakhshān, NE Afghanistan
Fakaofo Atoll 123 F3 island SE Tokelau
Falam 114 A3 Chin State, W Myanmar (Burma)
Falconara Marittima 74 C3 Marche, C Italy
Falkenau an der Eger see Sokolov
Falkland Islands 43 D7 var. Falklands, Islas Malvinas. UK Overseas Territory SW Atlantic Ocean
Falkland Plateau 35 D7 var. Argentine Rise. undersea feature SW Atlantic Ocean
Falklands see Falkland Islands
Falknov nad Ohří see Sokolov
Fallbrook 25 C8 California, W USA
Falmouth 32 A4 W Jamaica
Falmouth 67 C7 SW England, United Kingdom
Falster 63 B8 island SE Denmark
Fălticeni 86 C3 Hung. Falticsén. Suceava, NE Romania
Falticsén see Fălticeni
Falun 63 C6 var. Fahlun. Kopparberg, C Sweden
Famagusta see Gazimağusa
Famagusta Bay see Gazimağusa Körfezi
Famenne 65 C7 physical region SE Belgium
Fang 114 C3 Chiang Mai, NW Thailand
Fanning Island see Tabuaeran
Fanø 74 C3 island W Denmark
Farafangana 57 G4 Fianarantsoa, SE Madagascar
Farāh 100 D4 var. Farah, Fararud. Farāh, W Afghanistan
Farah Rōd 100 D4 river W Afghanistan
Faranah 52 C4 Haute-Guinée, S Guinea
Fararud see Farāh
Farasan Islands see Farasān, Jazur
Farasān, Jazur 99 A6 var. Farasan Islands. island group SW Saudi Arabia
Farewell, Cape 128 C4 headland South Island, New Zealand
Farewell, Cape see Nunap Isua
Fargo 23 F2 North Dakota, N USA
Farg'ona 101 F2 Rus. Fergana; prev. Novyy Margilan. Farg'ona Viloyati, E Uzbekistan
Faribault 23 F2 Minnesota, N USA
Farīdābād 112 D3 Haryāna, N India
Farkhor 101 E3 Rus. Parkhar. SW Tajikistan
Farmington 23 G5 Missouri, C USA
Farmington 26 C1 New Mexico, SW USA
Faro 70 B5 Faro, S Portugal
Faroe-Iceland Ridge 58 C1 undersea ridge NW Norwegian Sea
Faroe Islands 61 E5 var. Faero Islands, Dan. Færøerne, Far. Føroyar. Self-governing territory of Denmark N Atlantic Ocean
Faroe-Shetland Trough 58 C2 trough NE Atlantic Ocean
Farquhar Group 57 G2 island group S Seychelles
Fars, Khalij-e see Persian Gulf
Farvel, Kap see Nunap Isua
Fastiv 87 E2 Rus. Fastov. Kyivska Oblast, NW Ukraine
Fastov see Fastiv
Fauske 62 C3 Nordland, C Norway
Faventia see Faenza
Faxa Bay see Faxaflói
Faxaflói 60 D5 Eng. Faxa Bay. bay W Iceland
Faya 54 C2 prev. Faya-Largeau, Largeau. Borkou-Ennedi-Tibesti, N Chad
Faya-Largeau see Faya
Fayetteville 20 A1 Arkansas, C USA
Fayetteville 21 F1 North Carolina, SE USA
Fdérick see Fdérik
Fdérik 52 C2 var. Fdérick, Fr. Fort Gouraud. Tiris Zemmour, NW Mauritania
Fear, Cape 21 F2 headland Bald Head Island, North Carolina, SE USA
Fécamp 68 B3 Seine-Maritime, N France
Fédala see Mohammedia
Federal Capital Territory see Australian Capital Territory
Fehérgyarmat 77 E6 Szabolcs-Szatmár-Bereg, E Hungary
Fehértemplom see Bela Crkva
Fehmarn 72 C2 island N Germany
Fehmarn Belt 72 C2 Dan. Femern Bælt, Ger. Fehmarnbelt. strait Denmark /Germany
Fehmarnbelt Fehmarn Belt/Femer Bælt
Feijó 40 C2 Acre, W Brazil
Feilding 128 D4 Manawatu-Wanganui, North Island, New Zealand
Feira see Feira de Santana
Feira de Santana 41 G3 var. Feira. Bahia, E Brazil
Feketehalom see Codlea
Felanitx 71 G3 Mallorca, Spain, W Mediterranean Sea
Felicitas Julia see Lisboa
Felidhu Atoll 110 B4 atoll C Maldives
Felipe Carrillo Puerto 29 H4 Quintana Roo, SE Mexico
Felixstowe 67 E6 E England, United Kingdom
Fellin see Viljandi
Felsőbánya see Baia Sprie
Felsőmuzslya see Mužlja
Femunden 63 B5 lake S Norway
Fénérive see Fenoarivo Atsinanana
Fengcheng 106 D3 var. Feng-cheng, Fenghwangcheng. Liaoning, NE China
Feng-cheng see Fengcheng
Fenghwangcheng see Fengcheng
Fengtien see Shenyang, China
Fengtien see Liaoning, China
Fenoarivo Atsinanana 57 G3 Fr. Fénérive. Toamasina, E Madagascar
Fens, The 67 E6 wetland E England, United Kingdom
Feodosiia 87 F5 prev. Feodosiya, var. Kefe, It. Kaffa; anc. Theodosia. Respublika Krym, S Ukraine
Feodosiya see Feodosiia
Ferdinand see Montana, Bulgaria
Ferdinandsberg see Oţelu Roşu
Féres 82 D3 Anatolikí Makedonía kai Thráki, NE Greece
Fergana see Farg'ona
Fergus Falls 23 F2 Minnesota, N USA
Ferizaj 79 D5 Serb. Uroševac. C Kosovo
Ferkessédougou 52 D4 N Ivory Coast
Fermo 74 C4 anc. Firmum Picenum. Marche, C Italy
Fernandina, Isla 38 A5 var. Narborough Island. island Galápagos Islands, Ecuador, E Pacific Ocean
Fernando de Noronha 41 H2 island E Brazil
Fernando Po/Fernando Póo see Bioco, Isla de
Ferrara 74 C2 anc. Forum Alieni. Emilia-Romagna, N Italy
Ferreñafe 38 B3 Lambayeque, W Peru
Ferro see Hierro
Ferrol 70 B1 var. El Ferrol; prev. El Ferrol del Caudillo. Galicia, NW Spain
Fertő see Neusiedler See
Ferwerd see Ferwert
Ferwert 64 D1 Dutch. Ferwerd. Fryslân, N Netherlands
Fès 48 C2 Eng. Fez. N Morocco
Feteşti 86 D5 Ialomiţa, SE Romania
Fethiye 94 B4 Muğla, SW Turkey (Türkiye)
Fetlar 66 D1 island NE Scotland, United Kingdom
Feuilles, Rivière aux 16 D2 river Québec, E Canada
Feyẓābād see Faiẓābād
Fez see Fès
Fezzan 49 G4 region S Libya
Fianarantsoa 57 F3 Fianarantsoa, C Madagascar
Fianga 54 B4 Mayo-Kébbi, SW Chad
Fier 79 C6 var. Fieri. Fier, SW Albania
Fieri see Fier
Figeac 69 C5 Lot, S France
Figig see Figuig
Figueira da Foz 70 B3 Coimbra, W Portugal
Figueres 71 G2 Cataluña, E Spain
Figuig 48 D2 var. Figig. E Morocco
Fiji 123 E5 off. Republic of Fiji, Fij. Viti. country SW Pacific Ocean
Fiji, Republic of see Fiji
Filadelfia 30 D4 Guanacaste, W Costa Rica
Filiaşi 86 B5 Dolj, SW Romania
Filipstad 63 B6 Värmland, C Sweden
Finale Ligure 74 A3 Liguria, NW Italy
Finchley 67 A7 United Kingdom
Findlay 18 C4 Ohio, N USA
Finike 94 B4 Antalya, SW Turkey (Türkiye)
Finland 62 D4 off. Republic of Finland, Fin. Suomen Tasavalta, Suomi. country N Europe
Finland, Gulf of 63 D6 Est. Soome Laht, Fin. Suomenlahti, Ger. Finnischer Meerbusen, Rus. Finskiy Zaliv, Swe. Finska Viken. gulf E Baltic Sea
Finland, Republic of see Finland
Finnischer Meerbusen see Finland, Gulf of
Finnmarksvidda 62 D2 physical region N Norway
Finska Viken/Finskiy Zaliv see Finland, Gulf of
Finsterwalde 72 D4 Brandenburg, E Germany
Fiordland 129 A7 physical region South Island, New Zealand
Fiorina 74 E1 NE San Marino
Firat Nehri see Euphrates
Firenze 74 C3 Eng. Florence; anc. Florentia. Toscana, C Italy
Firmum Picenum see Fermo
First State see Delaware
Fischbacher Alpen 73 E7 mountain range E Austria
Fischhausen see Primorsk
Fish 56 B4 var. Vis. river S Namibia
Fishguard 67 C6 Wel. Abergwaun. SW Wales, United Kingdom
Fisterra, Cabo 70 B1 headland NW Spain
Fitzroy Crossing 124 C3 Western Australia
Fitzroy River 124 C3 river Western Australia
Fiume see Rijeka
Flagstaff 26 B2 Arizona, SW USA
Flanders 65 A6 Dut. Vlaanderen, Fr. Flandre. cultural region Belgium/France
Flandre see Flanders
Flathead Lake 22 B1 lake Montana, NW USA
Flat Island 106 C8 island NE Spratly Islands
Flatts Village 20 B5 var. The Flatts Village. C Bermuda
Flensburg 72 B2 Schleswig-Holstein, N Germany
Flessingue see Vlissingen
Flickertail State see North Dakota
Flinders Island 127 C8 island Furneaux Group, Tasmania, SE Australia
Flinders Ranges 127 B6 mountain range South Australia
Flinders River 126 C3 river Queensland, NE Australia
Flin Flon 15 F5 Manitoba, C Canada
Flint 18 C3 Michigan, N USA
Flint Island 123 G4 island Line Islands, E Kiribati
Floreana, Isla see Santa María, Isla
Florence 20 C1 Alabama, S USA
Florence 21 F2 South Carolina, SE USA
Florence see Firenze
Florencia 36 B4 Caquetá, S Colombia
Florentia see Firenze
Flores 30 B1 Petén, N Guatemala
Flores 117 E5 island Nusa Tenggara, C Indonesia
Flores 70 A5 island Azores, Portugal, NE Atlantic Ocean
Flores, Laut see Flores Sea
Flores Sea 116 D5 Ind. Laut Flores. sea C Indonesia
Floriano 41 F2 Piauí, E Brazil
Florianópolis 41 F5 prev. Destêrro. state capital Santa Catarina, S Brazil
Florida 42 D4 Florida, S Uruguay
Florida 21 E4 off. State of Florida, also known as Peninsular State, Sunshine State. state SE USA
Florida Bay 21 E5 bay Florida, SE USA
Florida Keys 21 E5 island group Florida, SE USA
Florida, Straits of 32 B1 strait Atlantic Ocean/ Gulf of Mexico
Flórina 82 B4 var. Phlórina. Dytikí Makedonía, N Greece
Florissant 23 G4 Missouri, C USA
Floúda, Akrotírio 83 D7 headland Astypálaia, Kykládes, Greece, Aegean Sea
Flushing see Vlissingen
Flylân see Vlieland
Foča 78 C4 var. Srbinje. SE Bosnia and Herzegovina
Focşani 86 C4 Vrancea, E Romania
Fogaras/Fogarasch see Făgăraş
Foggia 75 D5 Puglia, SE Italy
Fogo 52 A3 island Ilhas de Sotavento, SW Cape Verde (Cabo Verde)
Foix 69 B6 Ariège, S France
Folégandros 83 C7 island Kykládes, Greece, Aegean Sea
Foleyet 16 C4 Ontario, S Canada
Foligno 74 C4 Umbria, C Italy
Folkestone 67 E7 SE England, United Kingdom
Fond du Lac 18 B2 Wisconsin, N USA
Fonseca, Golfo de see Fonseca, Gulf of
Fonseca, Gulf of 30 C3 Sp. Golfo de Fonseca. gulf C Central America
Fontainebleau 68 C3 Seine-et-Marne, N France
Fontenay-le-Comte 68 B4 Vendée, NW France
Fontvieille 69 B8 SW Monaco Europe
Fonyód 77 C7 Somogy, W Hungary

Foochow *see* Fuzhou
Forchheim *73 C5* Bayern, SE Germany
Forel, Mont *60 D4 mountain* SE Greenland
Forfar *66 C3* E Scotland, United Kingdom
Forge du Sud *see* Dudelange
Forlì *74 C3 anc.* Forum Livii. Emilia-Romagna, N Italy
Formentera *71 G4 anc.* Ophiusa, *Lat.* Frumentum. *island* Islas Baleares, Spain, W Mediterranean Sea
Formosa *42 D2* Formosa, NE Argentina
Formosa/Formo'sa *see* Taiwan
Formosa, Serra *41 E3 mountain range* C Brazil
Formosa Strait *see* Taiwan Strait
Føroyar *see* Faroe Islands
Forrest City *20 B1* Arkansas, C USA
Fort Albany *16 C3* Ontario, C Canada
Fortaleza *39 F2* Pando, N Bolivia
Fortaleza *41 G2 prev.* Ceará. *state capital* Ceará, NE Brazil
Fort-Archambault *see* Sarh
Fort-Bayard *see* Zhanjiang
Fort-Cappolani *see* Tidjikja
Fort Charlet *see* Djanet
Fort-Chimo *see* Kuujjuaq
Fort Collins *22 D4* Colorado, C USA
Fort-Crampel *see* Kaga Bandoro
Fort Davis *27 E3* Texas, SW USA
Fort-de-France *33 H4 prev.* Fort-Royal. *dependent territory capital* (Martinique) W Martinique
Fort Dodge *23 F3* Iowa, C USA
Fortescue River *124 A4 river* Western Australia
Fort-Foureau *see* Kousséri
Fort Frances *16 B4* Ontario, S Canada
Fort Good Hope *15 E3 var.* Rádeyilikóé. Northwest Territories, NW Canada
Fort Gouraud *see* Fdérik
Forth *66 C4 river* C Scotland, United Kingdom
Forth, Firth of *66 C4 estuary* E Scotland, United Kingdom
Fortín General Eugenio Garay *see* General Eugenio A. Garay
Fort Jameson *see* Chipata
Fort-Lamy *see* Ndjamena
Fort Lauderdale *21 F5* Florida, SE USA
Fort Liard *15 E4 var.* Liard. Northwest Territories, W Canada
Fort Madison *23 G4* Iowa, C USA
Fort McMurray *15 E4* Alberta, C Canada
Fort McPherson *14 D3 var.* McPherson. Northwest Territories, NW Canada
Fort Morgan *22 D4* Colorado, C USA
Fort Myers *21 E5* Florida, SE USA
Fort Nelson *15 E4* British Columbia, W Canada
Fort Peck Lake *22 C1 reservoir* Montana, NW USA
Fort Pierce *21 F4* Florida, SE USA
Fort Providence *15 E4 var.* Providence. Northwest Territories, W Canada
Fort-Repoux *see* Akjoujt
Fort Rosebery *see* Mansa
Fort Rousset *see* Owando
Fort-Royal *see* Fort-de-France
Fort St. John *15 E4* British Columbia, W Canada
Fort Scott *23 F5* Kansas, C USA
Fort Severn *16 C2* Ontario, C Canada
Fort-Shevchenko *92 A4* Mangistau, W Kazakhstan
Fort-Sibut *see* Sibut
Fort Simpson *15 E4 var.* Simpson. Northwest Territories, W Canada
Fort Smith *15 E4* Northwest Territories, W Canada
Fort Smith *20 B1* Arkansas, C USA
Fort Stockton *27 E3* Texas, SW USA
Fort-Trinquet *see* Bîr Mogreïn
Fort Vermilion *15 E4* Alberta, W Canada
Fort Victoria *see* Masvingo
Fort Walton Beach *20 C3* Florida, SE USA
Fort Wayne *18 C4* Indiana, N USA
Fort William *66 C3* N Scotland, United Kingdom
Fort Worth *27 G2* Texas, SW USA
Fort Yukon *14 D3* Alaska, USA
Forum Alieni *see* Ferrara
Forum Livii *see* Forlì
Fossa Claudia *see* Chioggia
Fougamou *55 A6* Ngounié, C Gabon
Fougères *68 B3* Ille-et-Vilaine, NW France
Fou-hsin *see* Fuxin
Foulwind, Cape *129 B5 headland* South Island, New Zealand
Foumban *54 A4* Ouest, NW Cameroon
Fou-shan *see* Fushun
Foveaux Strait *129 A8 strait* S New Zealand
Foxe Basin *15 G3 sea* Nunavut, N Canada
Fox Glacier *129 B6* West Coast, South Island, New Zealand
Fraga *71 F2* Aragón, NE Spain
Fram Basin *133 C3 var.* Amundsen Basin. *undersea basin* Arctic Ocean
France *68 B4 off.* French Republic, *It./Sp.* Francia; *prev.* Gaul, Gaule, *Lat.* Gallia. *country* W Europe
Franceville *55 B6 var.* Massoukou, Masuku. Haut-Ogooué, E Gabon
Francfort *see* Frankfurt am Main
Franche-Comté *68 D4 cultural region* E France
Francia *see* France
Francis Case, Lake *23 E3 reservoir* South Dakota, N USA
Francisco Escárcega *29 G4* Campeche, SE Mexico
Francistown *56 D3* North East, NE Botswana
Franconian Jura *see* Fränkische Alb
Frankenalb *see* Fränkische Alb
Frankenstein/Frankenstein in Schlesien *see* Ząbkowice Śląskie
Frankfort *18 C5 state capital* Kentucky, S USA
Frankfort on the Main *see* Frankfurt am Main
Frankfurt *see* Frankfurt am Main, Germany
Frankfurt *see* Słubice, Poland
Frankfurt am Main *73 B5 var.* Frankfurt, *Fr.* Francfort; *prev. Eng.* Frankfort on the Main. Hessen, SW Germany
Frankfurt an der Oder *72 D3* Brandenburg, E Germany
Fränkische Alb *73 C6 var.* Frankenalb, *Eng.* Franconian Jura. *mountain range* S Germany
Franklin *20 C1* Tennessee, S USA
Franklin D. Roosevelt Lake *24 C1 reservoir* Washington, NW USA
Franz Josef Land *92 D1 Eng.* Franz Josef Land. *island group* N Russia
Franz Josef Land *see* Frantsa-Iosifa, Zemlya
Fraserburgh *66 D3* NE Scotland, United Kingdom
Fraser Island *126 E4 var.* Great Sandy Island. *island* Queensland, E Australia
Frauenbach *see* Baia Mare
Frauenburg *see* Saldus, Latvia
Fredericksburg *19 E5* Virginia, NE USA
Fredericton *17 F4 province capital* New Brunswick, SE Canada
Frederikshåb *see* Paamiut
Fredrikshald *see* Halden
Fredrikstad *63 B6* Østfold, S Norway
Freeport *32 C1* Grand Bahama Island, N The Bahamas
Freeport *27 H4* Texas, SW USA
Free State *see* Maryland
Freetown *52 C4 country capital* (Sierra Leone) W Sierra Leone
Freiburg *see* Freiburg im Breisgau, Germany
Freiburg im Breisgau *73 A6 var.* Freiburg, *Fr.* Fribourg-en-Brisgau. Baden-Württemberg, SW Germany
Freiburg in Schlesien *see* Świebodzice
Fremantle *125 A6* Western Australia
Fremont *23 F4* Nebraska, C USA
French Guiana *37 H3 var.* Guiana, Guyane. *French overseas department* N South America
French Guinea *see* Guinea
French Polynesia *121 F4 Overseas Country of France* S Pacific Ocean
French Republic *see* France
French Somaliland *see* Djibouti
French Southern and Antarctic Lands *119 B7 Fr.* Terres Australes et Antarctiques Françaises. *French overseas territory* S Indian Ocean
French Sudan *see* Mali
French Territory of the Afars and Issas *see* Djibouti
French Togoland *see* Togo
Fresnillo *28 D3 var.* Fresnillo de González Echeverría. Zacatecas, C Mexico
Fresnillo de González Echeverría *see* Fresnillo
Fresno *25 C6* California, W USA
Frías *42 C3* Catamarca, N Argentina
Fribourg-en-Brisgau *see* Freiburg im Breisgau
Friedek-Mistek *see* Frýdek-Místek
Friedrichshafen *73 B7* Baden-Württemberg, S Germany
Friendly Islands *see* Tonga
Frisches Haff *see* Vistula Lagoon
Frobisher Bay *60 B3 inlet* Baffin Island, Nunavut, NE Canada
Frobisher Bay *see* Iqaluit
Frohavet *62 B4 sound* C Norway
Frome, Lake *127 B6 salt lake* South Australia
Frontera *29 G4* Tabasco, SE Mexico
Frontignan *69 C6* Hérault, S France
Frostviken *see* Kvarnbergsvattnet
Frøya *62 A4 island* W Norway
Frumentum *see* Formentera
Frunze *see* Bishkek
Frýdek-Místek *77 C5 Ger.* Friedek-Mistek. Moravskoslezský Kraj, E Czechia (Czech Republic)
Fu-chien *see* Fujian
Fu-chou *see* Fuzhou
Fuengirola *70 D5* Andalucía, S Spain
Fuerte Olimpo *42 D2 var.* Olimpo. Alto Paraguay, NE Paraguay
Fuerte, Río *26 C5 river* C Mexico
Fuerteventura *48 B3 island* Islas Canarias, Spain, NE Atlantic Ocean
Fuhkien *see* Fujian
Fu-hsin *see* Fuxin
Fuji *109 D6 var.* Huzi. Shizuoka, Honshū, S Japan
Fujian *106 D6 var.* Fu-chien, Fuhkien, Fukien, Min, Fujian Sheng. *province* SE China
Fujian Sheng *see* Fujian
Fuji, Mount/Fujiyama *see* Fuji-san
Fuji-san *109 C6 var.* Fujiyama, *Eng.* Mount Fuji. *mountain* Honshū, SE Japan
Fukang *104 C2* Xinjiang Uygur Zizhiqu, W China
Fukien *see* Fujian
Fukui *109 C6 var.* Hukui. Fukui, Honshū, SW Japan
Fukuoka *109 A7 var.* Hukuoka, *hist.* Najima. Fukuoka, Kyūshū, SW Japan
Fukushima *108 D4 var.* Hukusima. Fukushima, Honshū, C Japan
Fulda *73 B5* Hessen, C Germany
Funafuti Atoll *123 E3 atoll and capital* (Tuvalu) C Tuvalu
Funchal *48 A2* Madeira, Portugal, NE Atlantic Ocean
Fundy, Bay of *17 F5 bay* Canada/USA
Fünen *see* Fyn
Fünfkirchen *see* Pécs
Furnes *see* Veurne
Fürth *73 C5* Bayern, S Germany
Furukawa *108 D4 var.* Hurukawa, Ōsaki. Miyagi, Honshū, C Japan
Fusan *see* Busan
Fushë Kosovë *79 D5 Serb.* Kosovo Polje. C Kosovo
Fushun *106 D3 var.* Fou-shan, Fu-shun. Liaoning, NE China
Fu-shun *see* Fushun
Fusin *see* Fuxin
Füssen *73 C7* Bayern, S Germany
Futog *78 D3* Vojvodina, NW Serbia
Futuna, Île *123 E4 island* S Wallis and Futuna
Fuxin *106 D3 var.* Fou-hsin, Fu-hsin, Fusin. Liaoning, NE China
Fuzhou *106 D6 var.* Foochow, Fu-chou. *province capital* Fujian, SE China
Fyn *63 B8 Ger.* Fünen. *island* C Denmark
Fyzabad *see* Faiẓābād

G

Gaafu Alifu Atoll *see* North Huvadhu Atoll
Gaalkacyo *51 E5 var.* Galka'yo, *It.* Galcaio. Mudug, C Somalia
Gabela *56 B2* Cuanza Sul, W Angola
Gaberones *see* Gaborone
Gabès *49 E2 var.* Qābis. E Tunisia
Gabès, Golfe de *49 F2 Ar.* Khalīj Qābis. *gulf* E Tunisia
Gabon *55 B6 off.* Gabonese Republic. *country* C Africa
Gabonese Republic *see* Gabon
Gaborone *56 C4 prev.* Gaberones. *country capital* (Botswana) South East, SE Botswana
Gabrovo *82 D2* Gabrovo, N Bulgaria
Gadag *110 C1* Karnātaka, W India
Gades/Gadier/Gadir/Gadire *see* Cádiz
Gadsden *20 D2* Alabama, S USA
Gaeta *75 C5* Lazio, C Italy
Gaeta, Golfo di *75 C5 var.* Gulf of Gaeta. *gulf* C Italy
Gaeta, Gulf of *see* Gaeta, Golfo di
Gäfle *see* Gävle
Gafsa *49 E2 var.* Qafṣah. W Tunisia
Gagnoa *52 D5* C Ivory Coast
Gagra *95 E1* NW Georgia
Gaillac *69 C6 var.* Gaillac-sur-Tarn. Tarn, S France
Gaillac-sur-Tarn *see* Gaillac
Gaillimh *see* Galway
Gaillimhe, Cuan na *see* Galway Bay
Gainesville *21 E3* Florida, SE USA
Gainesville *20 D2* Georgia, SE USA
Gainesville *27 G2* Texas, SW USA
Lake Gairdner *127 A6 salt lake* South Australia
Gaizina Kalns *see* Gaiziņkalns
Gaiziņkalns *84 C3 var.* Gaizina Kalns. *mountain* E Latvia
Galán, Cerro *42 B3 mountain* NW Argentina
Galanta *77 C6 Hung.* Galánta. Trnavský Kraj, W Slovakia
Galapagos Fracture Zone *131 E3 tectonic feature* E Pacific Ocean
Galápagos Islands *131 F3 var.* Islas de los Galápagos, *Sp.* Archipiélago de Colón, *Eng.* Galapagos Islands, Tortoise Islands. *island group* Ecuador, E Pacific Ocean
Galapagos Islands *see* Galápagos Islands
Galápagos, Islas de los *see* Galápagos Islands
Galapagos Rise *131 F3 undersea rise* E Pacific Ocean
Galashiels *66 C4* SE Scotland, United Kingdom
Galați *86 D4 Ger.* Galatz. Galați, E Romania
Galatz *see* Galați
Galcaio *see* Gaalkacyo
Galesburg *18 B3* Illinois, N USA
Galicia *70 B1 anc.* Gallaecia. *autonomous community* NW Spain
Galicia Bank *58 B4 undersea bank* E Atlantic Ocean
Galilee, Sea of *see* Kinneret, Yam
Galka'yo *see* Gaalkacyo
Galkynyş *100 D3 prev. Rus.* Deynau, Dyanev, *Turkm.* Dänew. Lebap Welaýaty, NE Turkmenistan
Gallaecia *see* Galicia
Galle *110 D4 prev.* Point de Galle. Southern Province, SW Sri Lanka
Gallego Rise *131 F3 undersea rise* E Pacific Ocean
Gallegos *see* Río Gallegos
Gallia *see* France
Gallipoli *75 E6* Puglia, SE Italy
Gällivare *62 C3 Lapp.* Váhtjer. Norrbotten, N Sweden
Gallup *26 C1* New Mexico, SW USA
Galtat-Zemmour *48 B3* C Western Sahara
Galveston *27 H4* Texas, SW USA
Galway *67 A5 Ir.* Gaillimh. W Ireland
Galway Bay *67 A6 Ir.* Cuan na Gaillimhe. *bay* W Ireland
Gâmas *see* Kaamanen
Gambell *14 C2* Saint Lawrence Island, Alaska, USA
Gambia *see* Gambia, The
Gambia *52 C3 Fr.* Gambie. *river* W Africa
Gambia, Republic of The *see* Gambia, The
Gambia, The *52 B3 off.* Republic of The Gambia *var.* Gambia. *country* W Africa
Gambie *see* Gambia, The
Gambier, Îles *121 G4 island group* E French Polynesia
Gamboma *55 B6* Plateaux, E Congo
Gamlakarleby *see* Kokkola
Gan *110 B5* Addu Atoll, C Maldives
Gan *see* Gansu, China
Gan *see* Jiangxi, China
Ganaane *see* Juba
Gäncä *95 G2 Rus.* Gyandzha; *prev.* Kirovabad, Yelisavetpol. W Azerbaijan
Gand *see* Gent
Gandajika *55 D7* Kasai-Oriental, S Dem. Rep. Congo
Gander *17 G3* Newfoundland and Labrador, SE Canada
Gāndhīdhām *112 C4* Gujarāt, W India
Gandia *71 F3 prev.* Gandía. Comunitat Valenciana, E Spain
Gandía *see* Gandia
Ganges *113 F3 Ben.* Padma. *river* Bangladesh/India
Ganges Cone *see* Ganges Fan
Ganges Fan *118 D3 var.* Ganges Cone. *undersea fan* N Bay of Bengal
Ganges, Mouths of the *113 G4 delta* Bangladesh/India
Gangra *see* Çankırı
Gangtok *113 F3 state capital* Sikkim, N India
Gansos, Lago dos *see* Goose Lake
Gansu *106 B4 var.* Gan, Gansu Sheng, Kansu. *province* N China
Gansu Sheng *see* Gansu
Gantsevichi *see* Hantsavichy
Ganzhou *106 D6* Jiangxi, S China
Gao *53 E3* Gao, E Mali
Gaocheng *see* Litang
Gaoual *52 C4* N Guinea
Gaoxiong *see* Kaohsiung
Gap *69 D5 anc.* Vapincum. Hautes-Alpes, SE France
Gaplaňgyr Platosy *100 C2 Rus.* Plato Kaplangky. *ridge* Turkmenistan/Uzbekistan
Gar *see* Gar Xincun
Garabil Belentligi *100 D3 Rus.* Vozvyshennost' Karabil'. *mountain range* S Turkmenistan
Garabogaz Aylagy *100 B2 Rus.* Zaliv Kara-Bogaz-Gol. *bay* NW Turkmenistan
Garachiné *31 G5* Darién, SE Panama
Garagum *100 C3 var.* Garagumy, Qara Qum, *Eng.* Black Sand Desert, Kara Kum; *prev.* Peski Karakumy. *desert* C Turkmenistan
Garagum Canal *100 D3 var.* Kara Kum Canal, *Rus.* Karagumskiy Kanal, Karakumskiy Kanal. *canal* C Turkmenistan
Garagumy *see* Garagum
Gara Hitrino *82 D2 var.* Gara Khitrino. Shumen, NE Bulgaria
Gara Khitrino *see* Gara Hitrino
Gárasavon *see* Kaaresuvanto
Garda, Lago di *74 C2 var.* Benaco, *Eng.* Lake Garda, *Ger.* Gardasee. *lake* NE Italy
Garda, Lake *see* Garda, Lago di
Gardasee *see* Garda, Lago di
Garden City *23 E5* Kansas, C USA
Garden State, The *see* New Jersey
Gardēz *101 E4 prev.* Gardīz. E Afghanistan
Gardīz *see* Gardēz
Gardner Island *see* Nikumaroro
Garegegasnjárga *see* Karigasniemi
Gargždai *84 B3* Klaipėda, W Lithuania
Garissa *51 D6* Coast, E Kenya
Garland *27 G2* Texas, SW USA
Garoe *see* Garoowe
Garonne *69 B5 anc.* Garumna. *river* S France
Garoowe *51 E5 var.* Garoe. Nugaal, N Somalia
Garoua *54 B4 var.* Garua. Nord, N Cameroon
Garrygala *see* Magtymguly
Garry Lake *15 F3 lake* Nunavut, N Canada
Garsen *51 D6* Coast, S Kenya
Garua *see* Garoua
Garumna *see* Garonne
Garwolin *76 D4* Mazowieckie, E Poland
Gar Xincun *104 A4 prev.* Gar. Xizang Zizhiqu, W China
Gary *18 B3* Indiana, N USA
Garzón *36 B4* Huila, S Colombia
Gasan-Kuli *see* Esenguly
Gascogne *69 B6 Eng.* Gascony. *cultural region* S France
Gascony *see* Gascogne
Gascoyne River *125 A5 river* Western Australia
Gaspé *17 F3* Québec, SE Canada
Gaspé, Péninsule de *17 E4 var.* Péninsule de la Gaspésie. *peninsula* Québec, SE Canada
Gaspésie, Péninsule de la *see* Gaspé, Péninsule de
Gastonia *21 E1* North Carolina, SE USA
Gastoúni *83 B6* Dytikí Elláda, S Greece
Gatchina *88 B4* Leningradskaya Oblast', NW Russia
Gatineau *16 D4* Québec, SE Canada
Gatooma *see* Kadoma
Gatún, Lake *31 F4 reservoir* C Panama
Gauhāti *see* Guwāhāti
Gauja *84 D3 Ger.* Aa. *river* Estonia/Latvia
Gaul/Gaule *see* France
Gauteng *see* Johannesburg, South Africa
Gāvbandī *98 D4* Hormozgān, S Iran
Gávdos *83 C8 island* SE Greece
Gavere *65 B6* Oost-Vlaanderen, NW Belgium
Gävle *63 C6 var.* Gäfle; *prev.* Gefle. Gävleborg, C Sweden
Gawler *127 B6* South Australia
Gaya *113 F3* Bihār, N India
Gaya *see* Kyjov
Gayndah *127 E5* Queensland, E Australia
Gaysin *see* Haisyn
Gaza *97 A6 Ar.* Ghazzah, *Heb.* 'Azza. NE Gaza Strip
Gaz-Achak *see* Gazojak
Gazandzhyk/Gazanjyk *see* Bereket
Gaza Strip *97 A7 Ar.* Qita Ghazzah. *disputed region* SW Asia
Gaziantep *94 D4 var.* Gazi Antep; *prev.* Aintab, Antep. Gaziantep, S Turkey (Türkiye)
Gazi Antep *see* Gaziantep
Gazimağusa *80 D5 var.* Famagusta, *Gk.* Ammóchostos. E Cyprus
Gazimağusa Körfezi *80 C5 var.* Famagusta Bay, *Gk.* Kólpos Ammóchostos. *bay* E Cyprus
Gazli *100 D2* Buxoro Viloyati, C Uzbekistan
Gazojak *100 D2 Rus.* Gaz-Achak. Lebap Welaýaty, NE Turkmenistan
Gbanga *52 D5 var.* Gbarnga. N Liberia
Gbarnga *see* Gbanga
Gdańsk *76 C2 Fr.* Dantzig, *Ger.* Danzig. Pomorskie, N Poland
Gdańsk, Gulf of *76 C2 var.* Gulf of Danzig, *Ger.* Danziger Bucht, *Pol.* Zakota Gdańska, *Rus.* Gdan'skaya Bukhta. *gulf* N Poland
Gdan'skaya Bukhta *see* Gdańsk, Gulf of
Gdańska, Zakota *see* Gdańsk, Gulf of
Gdingen *see* Gdynia
Gdynia *76 C2 Ger.* Gdingen. Pomorskie, N Poland
Gedaref *50 C4 var.* Al Qaḍārif, El Gedaref. Gedaref, E Sudan
Gediz *94 B3* Kütahya, W Turkey (Türkiye)
Gediz Nehri *94 A3 river* W Turkey (Türkiye)
Geel *65 C5 var.* Gheel. Antwerpen, N Belgium
Geelong *127 C7* Victoria, SE Australia
Ge'e'mu *see* Golmud
Gefle *see* Gävle
Geilo *63 A5* Buskerud, S Norway
Gejiu *106 B6 var.* Kochiu. Yunnan, S China
Gëkdepe *see* Gökdepe
Gela *75 C7 prev.* Terranova di Sicilia. Sicilia, Italy, C Mediterranean Sea
Geldermalsen *64 C4* Gelderland, C Netherlands
Geleen *65 D6* Limburg, SE Netherlands
Gelib *see* Jilib
Gellinsor *51 E5* Mudug, C Somalia
Gembloux *65 C6* Namur, Belgium
Gemena *55 C5* Equateur, NW Dem. Rep. Congo
Gem of the Mountains *see* Idaho
Gemona del Friuli *74 D2* Friuli-Venezia Giulia, NE Italy
Gem State *see* Idaho
Genalē Wenz *see* Juba
Genck *see* Genk
General Alvear *42 B4* Mendoza, W Argentina
General Carrera, Lago *see* Buenos Aires, Lago
General Eugenio A. Garay *42 C1 var.* Fortín General Eugenio Garay; *prev.* Yrendagüé. Nueva Asunción, NW Paraguay
General José F.Uriburu *see* Zárate
General Machado *see* Camacupa
General Santos *117 F3 off.* General Santos City. Mindanao, S Philippines
General Santos City *see* General Santos
Gênes *see* Genova
Geneva *see* Genève
Geneva, Lake *73 A7 Fr.* Lac de Genève, Lac Léman, le Léman, *Ger.* Genfer See. *lake* France/Switzerland
Genève *73 A7 Eng.* Geneva, *Ger.* Genf, *It.* Ginevra. Genève, SW Switzerland
Genève, Lac de *see* Geneva, Lake
Genf *see* Genève
Genfer See *see* Geneva, Lake
Genichesk *see* Henichesk
Genk *65 D6 var.* Genck. Limburg, NE Belgium
Gennep *64 D4* Limburg, SE Netherlands
Genoa *see* Genova
Genoa, Gulf of *74 A3 Eng.* Gulf of Genoa. *gulf* NW Italy
Genoa, Gulf of *see* Genova, Golfo di
Genova *80 D1 Eng.* Genoa; *anc.* Genua, *Fr.* Gênes. Liguria, NW Italy
Genovesa, Isla *38 B5 var.* Tower Island. *island* Galápagos Islands, Ecuador, E Pacific Ocean
Gent *65 B5 Eng.* Ghent, *Fr.* Gand. Oost-Vlaanderen, NW Belgium
Genua *see* Genova
Geok-Tepe *see* Gökdepe
George *56 C5* Western Cape, S South Africa
George *60 A4 river* Newfoundland and Labrador/Québec, E Canada
George, Lake *21 E3 lake* Florida, SE USA
Georgenburg *see* Jurbarkas
Georges Bank *13 D5 undersea bank* W Atlantic Ocean
George Sound *see* Te Houhou
Georges River *126 D2 river* New South Wales, E Australia
Georgetown *37 F2 country capital* (Guyana) N Guyana
George Town *32 C2* Great Exuma Island, C The Bahamas
George Town *32 B3 var.* Georgetown. *dependent territory capital* (Cayman Islands) Grand Cayman, SW Cayman Islands
George Town *116 B3 var.* Penang, Pinang. Pinang, Peninsular Malaysia
Georgetown *21 F2* South Carolina, SE USA
Georgetown *see* George Town
George V Land *132 C4 physical region* Antarctica
Georgia *95 F2 off.* Georgia, *Geor.* Sakartvelo, *Rus.* Gruzinskaya SSR, Gruziya. *country* SW Asia
Georgia *20 D2 off.* State of Georgia, *also known as* Empire State of the South, Peach State. *state* SE USA
Georgian Bay *18 D2 lake bay* Ontario, S Canada
Georgia, Strait of *24 A1 strait* British Columbia, W Canada
Georgi Dimitrov *see* Kostenets
Georgiu-Dezh *see* Liski
Gera *72 C4* Thüringen, E Germany
Geráki *83 B6* Pelopónnisos, S Greece
Geraldine *129 B6* Canterbury, South Island, New Zealand
Geraldton *125 A6* Western Australia
Geral, Serra *35 D5 mountain range* S Brazil
Gerede *94 C2* Bolu, N Turkey (Türkiye)
Gereshk *see* Girishk
Gering *22 D3* Nebraska, C USA
German East Africa *see* Tanzania
Germanicopolis *see* Çankırı
Germanicum, Mare/German Ocean *see* North Sea
German Southwest Africa *see* Namibia
Germany *72 B4 off.* Federal Republic of Germany, Bundesrepublik Deutschland, *Ger.* Deutschland. *country* N Europe
Germany, Federal Republic of *see* Germany
Geroliménas *83 B7* Pelopónnisos, S Greece
Gerona *see* Girona
Gerpinnes *65 C7* Hainaut, S Belgium
Gerunda *see* Girona
Gerze *94 D2* Sinop, N Turkey (Türkiye)
Gesoriacum *see* Boulogne-sur-Mer
Gessoriacum *see* Boulogne-sur-Mer
Getafe *70 D3* Madrid, C Spain
Gevaş *95 F3* Van, SE Turkey (Türkiye)
Gevgeli *see* Gevgelija
Gevgelija *79 E6 var.* Đevđelija, Djevdjelija, *Turk.* Gevgeli. SE North Macedonia
Ghaba *see* Al Ghābah
Ghana *53 E5 off.* Republic of Ghana. *country* W Africa
Ghanzi *56 C3 var.* Khanzi. Ghanzi, W Botswana
Gharandal *97 B7* Al'Aqabah, SW Jordan
Gharbt, Jabal al *see* Lîban, Jebel
Ghardaïa *48 D2* N Algeria
Gharvān *see* Gharyān
Gharyān *49 F2 var.* Gharvān. NW Libya
Ghawdex *see* Gozo
Ghaznī *101 E4 var.* Ghazni. Ghaznī, E Afghanistan
Ghazzah *see* Gaza
Gheel *see* Geel
Ghent *see* Gent
Gheorgheni *86 C4 prev.* Gheorghieni, Sîn-Miclăuş, *Ger.* Niklasmarkt, *Hung.* Gyergyószentmiklós. Harghita, C Romania
Gheorghieni *see* Gheorgheni
Ghōriān *see* Ghōriyān
Ghōriyān *100 D4 var.* Ghōriān, Ghūrīān. Herāt, W Afghanistan
Ghŭdara *101 F3 var.* Gudara, *Rus.* Kudara. SE Tajikistan
Ghurdaqah *see* Al Ghurdaqah
Ghūrīān *see* Ghōriyān
Giamame *see* Jamaame
Giannitsá *82 B4 var.* Yiannitsá. Kentrikí Makedonía, N Greece
Gibraltar *71 G4 UK Overseas Territory* SW Europe
Gibraltar, Bay of *71 G5 bay* Gibraltar/Spain Europe Mediterranean Sea Atlantic Ocean
Gibraltar, Détroit de/Gibraltar, Estrecho de *see* Gibraltar, Strait of
Gibraltar, Strait of *70 C5 Fr.* Détroit de Gibraltar, *Sp.* Estrecho de Gibraltar. *strait* Atlantic Ocean/Mediterranean Sea
Gibson Desert *125 B5 desert* Western Australia
Giedraičiai *85 C5* Utena, E Lithuania
Giessen *73 B5* Hessen, W Germany
Gifu *109 C6 var.* Gihu. Gifu, Honshū, SW Japan
Giganta, Sierra de la *28 B3 mountain range* NW Mexico
Gihu *see* Gifu
G'ijduvon *100 D2 Rus.* Gizhduvon. Buxoro Viloyati, C Uzbekistan
Gijón *70 D1 var.* Xixón. Asturias, NW Spain
Gila River *26 A2 river* Arizona, SW USA
Gilbert Islands *see* Tungaru
Gilbert River *126 C3 river* Queensland, NE Australia
Gilf Kebir Plateau *see* Hadabat al Jilf al Kabīr
Gillette *22 D3* Wyoming, C USA
Gilolo *see* Halmahera, Pulau
Gilroy *25 B6* California, W USA
Gimie, Mount *33 F1 mountain* C Saint Lucia
Gimma *see* Jīma
Ginevra *see* Genève
Gingin *125 A6* Western Australia
Giohar *see* Jawhar
Gipeswic *see* Ipswich
Girardot *36 B3* Cundinamarca, C Colombia
Giresun *95 E2 var.* Kerasunt; *anc.* Cerasus, Pharnacia. Giresun, NE Turkey (Türkiye)

Girgenti see Agrigento
Girin see Jilin
Girishk 100 D5 var. Gereshk. Helmand, SW Afghanistan
Girne 80 C5 Gk. Kerýneia, Kyrenia. N Cyprus
Giron see Kiruna
Girona 71 G2 var. Gerona; anc. Gerunda. Cataluña, NE Spain
Gisborne 128 E3 Gisborne, North Island, New Zealand
Gissar Range 101 E3 Rus. Gissarskiy Khrebet. mountain range Tajikistan/Uzbekistan
Gissarskiy Khrebet see Gissar Range
Gitega 51 B7 prev. Kitega. country capital (Burundi - political) W Burundi
Githio see Gýtheio
Giulianova 74 D4 Abruzzi, C Italy
Giumri see Gyumri
Giurgiu 86 C5 Giurgiu, S Romania
Giza see Al Jīzah
Gizeh see Al Jīzah
Gizhduvon see G'ijduvon
Giżycko 76 D2 Ger. Lötzen. Warmińsko-Mazurskie, NE Poland
Gjakovë 79 D5 Serb. Đakovica. W Kosovo
Gjilan 79 D5 Serb. Gnjilane. E Kosovo
Gjinokastër see Gjirokastër
Gjirokastër 79 C7 var. Gjirokastra; prev. Gjinokastër, Gk. Argyrokastron, It. Argirocastro. Gjirokastër, S Albania
Gjirokastra see Gjirokastër
Gjoa Haven 15 F3 var. Uqsuqtuuq. King William Island, Nunavut, NW Canada
Gjøvik 63 B5 Oppland, S Norway
Glace Bay 17 G4 Cape Breton Island, Nova Scotia, SE Canada
Gladstone 126 E4 Queensland, E Australia
Glåma 63 B5 var. Glommen. river S Norway
Glasgow 66 C4 S Scotland, United Kingdom
Glavinitsa 82 D1 prev. Pravda, Dogrular. Silistra, NE Bulgaria
Glavn'a Morava see Velika Morava
Glazov 89 D5 Udmurtskaya Respublika, NW Russia
Gleiwitz see Gliwice
Glendale 26 B2 Arizona, SW USA
Glendive 22 D2 Montana, NW USA
Glens Falls 19 F3 New York, NE USA
Glevum see Gloucester
Glina 78 B3 var. Banijska Palanka. Sisak-Moslavina, NE Croatia
Glittertind 63 A5 mountain S Norway
Gliwice 77 C5 Ger. Gleiwitz. Śląskie, S Poland
Globe 26 B2 Arizona, SW USA
Globino see Hlobyne
Glogau see Głogów
Głogów 76 B4 Ger. Glogau, Glogow. Dolnośląskie, SW Poland
Glogow see Głogów
Glomma see Glåma
Glommen see Glåma
Gloucester 67 D6 hist. Caer Glou, Lat. Glevum. C England, United Kingdom
Głowno 76 D4 Łódź, C Poland
Glubokoye see Hlybokaye
Glukhov see Hlukhiv
Gnesen see Gniezno
Gniezno 76 C3 Ger. Gnesen. Weilkopolskie, C Poland
Gnjilane see Gjilan
Gobabis 56 B3 Omaheke, E Namibia
Gobi 104 D3 desert China/Mongolia
Gobō 109 C6 Wakayama, Honshū, SW Japan
Godāvari 102 B3 var. Godavari. river C India
Godavari see Godāvari
Godhavn see Qeqertarsuaq
Godhra 112 C4 Gujarāt, W India
Göding see Hodonín
Godoy Cruz 42 B4 Mendoza, W Argentina
Godthaab/Godthåb see Nuuk
Godwin Austen, Mount see K2
Goede Hoop, Kaap de see Good Hope, Cape of
Goeie Hoop, Kaap die see Good Hope, Cape of
Goeree 64 B4 island SW Netherlands
Goes 65 B5 Zeeland, SW Netherlands
Goettingen see Göttingen
Gogebic Range 18 B1 hill range Michigan/Wisconsin, N USA
Goiânia 41 E3 prev. Goyania. state capital Goiás, C Brazil
Goiás 41 E3 off. Estado de Goiás; prev. Goiaz, Goyaz. state/region C Brazil
Goiás, Estado de see Goiás
Goiaz see Goiás
Goidhoo Atoll see Horsburgh Atoll
Gojōme 108 D4 Akita, Honshū, NW Japan
Gökçeada 82 D4 var. Imroz Adasī, Gk. Imbros. island NW Turkey (Türkiye)
Gökdepe 100 C3 Rus. Gëkdepe, Geok-Tepe. Ahal Welaýaty, C Turkmenistan
Göksun 94 D4 Kahramanmaraş, C Turkey (Türkiye)
Gol 63 A5 Buskerud, S Norway
Golan Heights 97 B5 Ar. Al Jawlān, Heb. HaGolan. mountain range SW Syria
Golaya Pristan see Hola Prystan
Gołdap 76 E2 Ger. Goldap. Warmińsko-Mazurskie, NE Poland
Gold Coast 127 E5 cultural region Queensland, E Australia
Golden Bay 128 C4 var. Mohua. bay South Island, New Zealand
Golden State, The see California
Goldingen see Kuldīga
Goldsboro 21 F1 North Carolina, SE USA
Goleniów 76 B3 Ger. Gollnow. Zachodnio-pomorskie, NW Poland
Gollnow see Goleniów
Golmo see Golmud
Golmud 104 D4 var. Ge'e'mu, Golmo, Chin. Ko-erh-mu. Qinghai, C China
Golovanevsk see Holovanivsk
Golub-Dobrzyń 76 C3 Kujawski-pomorskie, C Poland
Goma 55 E6 Nord-Kivu, NE Dem. Rep. Congo
Gombi 53 H4 Adamawa, E Nigeria
Gombroon see Bandar-e 'Abbās
Gomel' see Homyel'
Gomera 48 A3 island Islas Canarias, Spain, NE Atlantic Ocean
Gómez Palacio 28 D3 Durango, C Mexico
Gonaïves 32 D3 var. Les Gonaïves. N Haiti
Gonâve, Île de la 32 D3 island C Haiti
Gondar see Gonder
Gonder 50 C4 var. Gondar. Āmara, NW Ethiopia
Gondia 113 E4 Mahārāshtra, C India
Gonggar 104 C5 var. Gyixong. Xizang Zizhiqu, W China
Gongola 53 G4 river E Nigeria
Gongtang see Damxung
Gonni/Gónnos see Gónnoi
Gónnoi 82 B4 var. Gonni, Gónnos; prev. Dereli. Thessalía, C Greece
Good Hope, Cape of 56 B5 Afr. Kaap de Goede Hoop, Kaap die Goeie Hoop. headland SW South Africa
Goodland 22 D4 Kansas, C USA
Goondiwindi 127 D5 Queensland, E Australia
Goor 64 E3 Overijssel, E Netherlands
Goose Green 43 D7 var. Prado del Ganso. East Falkland, Falkland Islands
Goose Lake 24 B4 var. Lago dos Gansos. lake California/Oregon, W USA
Gopher State see Minnesota
Göppingen 73 B6 Baden-Württemberg, SW Germany
Góra Kalwaria 92 D4 Mazowieckie, C Poland
Gorakhpur 113 E3 Uttar Pradesh, N India
Gorany see Harany
Goražde 78 C4 Federacija Bosna I Hercegovina, SE Bosnia and Herzegovina
Gorbovichi see Harbavichy
Goré 54 C4 Logone-Oriental, S Chad
Gorē 51 C5 Oromīya, C Ethiopia
Gore 129 B7 Southland, South Island, New Zealand
Gorgān 98 D2 var. Astarabad, Astrabad, Gurgan, prev. Asterābād; anc. Hyrcania. Golestān, N Iran
Gori 95 F2 C Georgia
Gorinchem 64 C4 var. Gorkum. Zuid-Holland, C Netherlands
Goris 95 G3 SE Armenia
Gorki see Horki
Gor'kiy see Nizhniy Novgorod
Gorkum see Gorinchem
Görlitz 72 D4 Sachsen, E Germany
Görlitz see Zgorzelec
Gorlovka see Horlivka
Gorna Dzhumaya see Blagoevgrad
Gornja Mužlja see Mužlja
Gornji Milanovac 78 C4 Serbia, C Serbia
Gorodets see Haradzyets
Gorodishche see Horodyshche
Gorodnya see Horodnia
Gorodok see Haradok
Gorodok/Gorodok Yagellonski see Horodok
Gorontalo 117 E4 Sulawesi, C Indonesia
Gorontalo, Teluk see Tomini, Gulf of
Gorssel 64 D3 Gelderland, E Netherlands
Goryn see Horyn'
Gorzów Wielkopolski 76 B3 Ger. Landsberg, Landsberg an der Warthe. Lubuskie, W Poland
Gosford 127 D6 New South Wales, SE Australia
Goshogawara 108 D3 var. Gosyogawara. Aomori, Honshū, C Japan
Gospić 78 A3 Lika-Senj, C Croatia
Gostivar 79 D6 W Macedonia
Gosyogawara see Goshogawara
Göteborg 63 B7 Eng. Gothenburg. Västra Götaland, S Sweden
Gotel Mountains 53 G5 mountain range E Nigeria
Gotha 72 C4 Thüringen, C Germany
Gothenburg see Göteborg
Gotland 63 C7 island SE Sweden
Goto-retto 109 A7 island group SW Japan
Gotska Sandön 84 B1 island SE Sweden
Gōtsu 109 B6 var. Gôtu. Shimane, Honshū, SW Japan
Göttingen 72 B4 var. Goettingen. Niedersachsen, C Germany
Gottschee see Kočevje
Gottwaldov see Zlín
Gôtu see Gōtsu
Gouda 64 C4 Zuid-Holland, C Netherlands
Gough Fracture Zone 45 C6 tectonic feature S Atlantic Ocean
Gough Island 47 B8 island Tristan da Cunha, S Atlantic Ocean
Gouin, Réservoir 16 D4 reservoir Québec, SE Canada
Goulburn 127 D6 New South Wales, SE Australia
Goundam 53 E3 Tombouctou, NW Mali
Gouré 53 G3 Zinder, SE Niger
Goverla, Gora see Hoverla, Hora
Governador Valadares 41 F4 Minas Gerais, SE Brazil
Govi Altayn Nuruu 105 E3 mountain range S Mongolia
Goya 42 D3 Corrientes, NE Argentina
Goyania see Goiânia
Goyaz see Goiás
Goz Beïda 54 C3 Ouaddaï, SE Chad
Gozo 75 C8 var. Ghawdex. island N Malta
Gqeberha 56 C5 prev. Port Elizabeth. Eastern Cape, S South Africa
Graciosa 70 A5 var. Ilha Graciosa. island Azores, Portugal, NE Atlantic Ocean
Graciosa, Ilha see Graciosa
Gradačac 78 C3 Federacija Bosna I Hercegovina, N Bosnia and Herzegovina
Gradaús, Serra dos 41 E3 mountain range C Brazil
Gradiška see Bosanska Gradiška
Grafton 127 E5 New South Wales, SE Australia
Grafton 23 E1 North Dakota, N USA
Graham Land 132 A2 physical region Antarctica
Grajewo 76 E3 Podlaskie, NE Poland
Grampian Mountains 66 C3 mountain range C Scotland, United Kingdom
Gran see Esztergom, Hungary
Granada 30 D3 Granada, SW Nicaragua
Granada 70 D5 Andalucía, S Spain
Gran Canaria 48 A3 var. Grand Canary. island Islas Canarias, Spain, NE Atlantic Ocean
Gran Chaco 42 D2 var. Chaco. lowland plain South America
Grand Bahama Island 32 B1 island N The Bahamas
Grand Banks of Newfoundland 12 E4 undersea basin NW Atlantic Ocean
Grand Bassa see Buchanan
Grand Canary see Gran Canaria
Grand Canyon 26 A1 canyon Arizona, SW USA
Grand Canyon State see Arizona
Grand Cayman 32 B3 island SW Cayman Islands
Grand Duchy of Luxembourg see Luxembourg
Grande, Bahía 43 B7 bay S Argentina
Grande-Comor see Ngazidja
Grande de Chiloé, Isla see Chiloé, Isla de
Grande Prairie 15 E4 Alberta, W Canada
Grand Erg Occidental 48 D3 desert W Algeria
Grand Erg Oriental 49 E3 desert Algeria/Tunisia
Rio Grande 29 E2 var. Río Bravo, Sp. Río Bravo del Norte, Bravo del Norte. river Mexico/USA
Grande Terre 33 G3 island E West Indies
Grand Falls 17 G3 Newfoundland, Newfoundland and Labrador, SE Canada
Grand Forks 23 E1 North Dakota, N USA
Grandichi see Hrandzichy
Grand Island 23 E4 Nebraska, C USA
Grand Junction 22 C4 Colorado, C USA
Grand Paradis see Gran Paradiso
Grand Rapids 18 C3 Michigan, N USA
Grand Rapids 23 F1 Minnesota, N USA
Grand-Saint-Bernard, Col du see Great Saint Bernard Pass
Grand-Santi 37 G3 W French Guiana
Granite State see New Hampshire
Gran Lago see Nicaragua, Lago de
Gran Malvina see West Falkland
Gran Paradiso 74 A2 Fr. Grand Paradis. mountain NW Italy
Gran San Bernardo, Passo di see Great Saint Bernard Pass
Gran Santiago see Santiago
Grants 26 C2 New Mexico, SW USA
Grants Pass 24 B4 Oregon, NW USA
Granville 68 B3 Manche, N France
Gratianopolis see Grenoble
Gratz see Graz
Graudenz see Grudziądz
Graulhet 69 C6 Tarn, S France
Grave 64 D4 Noord-Brabant, SE Netherlands
Grayling 14 C2 Alaska, USA
Graz 73 E7 prev. Gratz. Steiermark, SE Austria
Great Abaco 32 C1 var. Abaco Island. island N The Bahamas
Great Alfold see Great Hungarian Plain
Great Ararat see Büyük Ağrı Dağı
Great Australian Bight 125 D7 bight S Australia
Great Barrier Island 128 D2 island N New Zealand
Great Barrier Reef 126 D2 reef Queensland, NE Australia
Great Basin 25 C5 basin W USA
Great Bear Lake 15 E3 Fr. Grand Lac de l'Ours. lake Northwest Territories, NW Canada
Great Belt 63 B8 var. Store Bælt, Eng. Great Belt, Storebelt. channel Baltic Sea/Kattegat
Great Belt see Storebælt
Great Bend 23 E5 Kansas, C USA
Great Britain see Britain
Great Dividing Range 126 D4 mountain range NE Australia
Greater Antilles 32 D3 island group West Indies
Greater Caucasus 95 G2 mountain range Azerbaijan/Georgia/Russia Asia/Europe
Greater Sunda Islands 102 D5 var. Sunda Islands. island group Indonesia
Great Exhibition Bay 128 C1 inlet North Island, New Zealand
Great Exuma Island 32 C2 island C The Bahamas
Great Falls 22 B1 Montana, NW USA
Great Grimsby see Grimsby
Great Hungarian Plain 77 C7 var. Great Alfold, Plain of Hungary, Hung. Alföld. plain SE Europe
Great Inagua 32 D2 var. Inagua Islands. island S The Bahamas
Great Indian Desert see Thar Desert
Great Khingan Range see Da Hinggan Ling
Great Lake see Tônlé Sap
Great Lakes 13 C5 lakes Ontario, Canada/USA
Great Lakes State see Michigan
Great Meteor Seamount see Great Meteor Tablemount
Great Meteor Tablemount 44 B3 var. Great Meteor Seamount. seamount E Atlantic Ocean
Great Nicobar 111 G3 island Nicobar Islands, India, NE Indian Ocean
Great Plain of China 103 E2 plain E China
Great Plains 23 E3 var. High Plains. plains Canada/USA
Great Rift Valley 51 C5 var. Rift Valley. depression Asia/Africa
Great Ruaha 51 C7 river S Tanzania
Great Saint Bernard Pass 74 A1 Fr. Col du Grand-Saint-Bernard, It. Passo del Gran San Bernardo. pass Italy/Switzerland
Great Salt Desert see Kavīr, Dasht-e
Great Salt Lake 22 A3 salt lake Utah, W USA
Great Salt Lake Desert 22 A4 plain Utah, W USA
Great Sand Sea 49 H3 desert Egypt/Libya
Great Sandy Desert 124 C4 desert Western Australia
Great Sandy Desert see Ar Rub 'al Khālī
Great Sandy Island see Fraser Island
Great Slave Lake 15 E4 Fr. Grand Lac des Esclaves. lake Northwest Territories, NW Canada
Great Socialist People's Libyan Arab Jamahiriya see Libya
Great Sound 20 A5 sound Bermuda, NW Atlantic Ocean
Great Victoria Desert 125 C5 desert South Australia/Western Australia
Great Wall of China 106 C4 ancient monument N China Asia
Great Yarmouth 67 E6 var. Yarmouth. E England, United Kingdom
Grebenka see Hrebinka
Gredos, Sierra de 70 D3 mountain range W Spain
Greece 83 A5 off. Hellenic Republic, Gk. Ellinikí Dimokratía, Elláda, Ellás; anc. Hellas. country SE Europe
Greeley 22 D4 Colorado, C USA
Green Bay 18 B2 Wisconsin, N USA
Green Bay 18 B2 lake bay Michigan/Wisconsin, N USA
Greeneville 21 E1 Tennessee, S USA
Greenland 60 D3 Dan. Grønland, Inuit Kalaallit Nunaat. Danish self-governing territory NE North America
Greenland Sea 61 F2 sea Arctic Ocean
Green Mountains 19 G2 mountain range Vermont, NE USA
Green Mountain State see Vermont
Greenock 66 C4 W Scotland, United Kingdom
Green River 22 B3 Wyoming, C USA
Green River 18 C5 river Kentucky, C USA
Green River 22 B4 river Utah, W USA
Greensboro 21 F1 North Carolina, SE USA
Greenville 20 B2 Mississippi, S USA
Greenville 21 F1 North Carolina, SE USA
Greenville 21 E1 South Carolina, SE USA
Greenville 27 G2 Texas, SW USA
Greenwich 67 B8 United Kingdom
Greenwood 20 B2 Mississippi, S USA
Greenwood 21 E2 South Carolina, SE USA
Gregory Range 126 C3 mountain range Queensland, E Australia
Greifenberg/Greifenberg in Pommern see Gryfice
Greifswald 72 D2 Mecklenburg-Vorpommern, NE Germany
Grenada 20 C2 Mississippi, S USA
Grenada 33 G5 country SE West Indies
Grenadines, The 33 H4 island group Grenada/St Vincent and the Grenadines
Grenoble 69 D5 anc. Cularo, Gratianopolis. Isère, E France
Gresham 24 B3 Oregon, NW USA
Grevená 82 B4 Dytikí Makedonía, N Greece
Grevenmacher 65 E8 Grevenmacher, E Luxembourg
Greymouth 129 B5 West Coast, South Island, New Zealand
Grey Range 127 C5 mountain range New South Wales/Queensland, E Australia
Greytown see San Juan del Norte
Griffin 20 D2 Georgia, SE USA
Grimari 54 C4 Ouaka, C Central African Republic
Grimsby 67 E5 prev. Great Grimsby. E England, United Kingdom
Grobin see Grobiņa
Grobiņa 84 B3 Ger. Grobin. Liepāja, W Latvia
Gródek Jagielloński see Horodok
Grodno see Hrodna
Grodzisk Wielkopolski 76 B3 Wielkopolskie, C Poland
Groesbeek 64 D4 Gelderland, SE Netherlands
Grójec 76 D4 Mazowieckie, C Poland
Groningen 64 E1 Groningen, NE Netherlands
Grønland see Greenland
Groote Eylandt 126 B2 island Northern Territory, N Australia
Grootfontein 56 B3 Otjozondjupa, N Namibia
Groot Karasberge 56 B4 mountain range S Namibia
Gros Islet 33 F1 N Saint Lucia
Grossa, Isola see Dugi Otok
Grossbetschkerek see Zrenjanin
Grosse Morava see Velika Morava
Grosser Sund see Suur Väin
Grosseto 74 B4 Toscana, C Italy
Grossglockner 73 C7 mountain W Austria
Grosskanizsa see Nagykanizsa
Gross-Karol see Carei
Grosskikinda see Kikinda
Grossmichel see Michalovce
Gross-Schlatten see Abrud
Grosswardein see Oradea
Groznyy 89 B8 Chechenskaya Respublika, SW Russia
Grudovo see Sredets
Grudziądz 76 C3 Ger. Graudenz. Kujawsko-pomorskie, C Poland
Grums 63 B6 Värmland, C Sweden
Grünberg/Grünberg in Schlesien see Zielona Góra
Grüneberg see Zielona Góra
Gruzinskaya SSR/Gruziya see Georgia
Gryazi 89 B6 Lipetskaya Oblast', W Russia
Gryfice 76 B2 Ger. Greifenberg, Greifenberg in Pommern. Zachodnio-pomorskie, NW Poland
Guabito 31 E4 Bocas del Toro, NW Panama
Guadalajara 28 D4 Jalisco, C Mexico
Guadalajara 71 E3 Ar. Wad Al-Hajarah; anc. Arriaca. Castilla-La Mancha, C Spain
Guadalcanal 122 C3 island C Solomon Islands
Guadalquivir 70 D4 river W Spain
Guadalupe 28 D3 Zacatecas, C Mexico
Guadalupe Peak 26 D3 mountain Texas, SW USA
Guadalupe River 27 G4 river SW USA
Guadarrama, Sierra de 71 E2 mountain range C Spain
Guadeloupe 33 H3 French overseas department E West Indies
Guadiana 70 C4 river Portugal/Spain
Guadix 71 E4 Andalucía, S Spain
Guaimaca 30 C2 Francisco Morazán, C Honduras
Guajira, Península de la 36 B1 peninsula N Colombia
Gualaco 30 D2 Olancho, C Honduras
Gualán 30 B2 Zacapa, C Guatemala
Gualdicciolo 74 D1 NW San Marino
Gualeguaychú 42 D4 Entre Ríos, E Argentina
Guam 122 B1 US unincorporated territory W Pacific Ocean
Guamúchil 28 C3 Sinaloa, C Mexico
Guanabacoa 32 B2 La Habana, W Cuba
Guanajuato 29 E4 Guanajuato, C Mexico
Guanare 36 C2 Portuguesa, N Venezuela
Guanare, Río 36 D2 river W Venezuela
Guangdong 106 C6 var. Guangdong Sheng, Kuang-tung, Kwangtung, Yue. province S China
Guangdong Sheng see Guangdong
Guangju see Gwangju
Guangxi see Guangxi Zhuangzu Zizhiqu
Guangxi Zhuangzu Zizhiqu 106 C6 var. Guangxi, Gui, Kuang-hsi, Kwangsi, Eng. Kwangsi Chuang Autonomous Region. autonomous region S China
Guangyuan 106 B5 var. Kuang-yuan, Kwangyuan. Sichuan, C China
Guangzhou 106 C6 var. Kuang-chou, Kwangchow, Eng. Canton. province capital Guangdong, S China
Guantánamo 32 D3 Guantánamo, SE Cuba
Guantánamo, Bahía de 32 D3 Eng. Guantanamo Bay. US military base SE Cuba
Guantanamo Bay see Guantánamo, Bahía de
Guaporé, Río 40 D3 var. Río Iténez. river Bolivia/Brazil
Guarda 70 C3 Guarda, N Portugal
Guarumal 31 F5 Veraguas, S Panama
Guasave 28 C3 Sinaloa, C Mexico
Guatemala 30 A2 off. Republic of Guatemala. country Central America
Guatemala Basin 13 B7 undersea basin E Pacific Ocean
Guatemala City see Ciudad de Guatemala
Guatemala, Republic of see Guatemala
Guaviare 34 B2 off. Comisaría Guaviare. province S Colombia
Guaviare, Comisaría see Guaviare
Guaviare, Río 36 D3 river E Colombia
Guayanas, Macizo de las see Guiana Highlands
Guayaquil 38 A2 var. Santiago de Guayaquil. Guayas, SW Ecuador
Guayaquil, Golfo de 38 A2 var. Gulf of Guayaquil. gulf SW Ecuador
Guayaquil, Gulf of see Guayaquil, Golfo de
Guaymas 28 B2 Sonora, NW Mexico
Gubadag 100 C2 Turkm. Tel'man; prev. Tel'mansk. Daşoguz Welaýaty, N Turkmenistan
Guben 72 D4 var. Wilhelm-Pieck-Stadt. Brandenburg, E Germany
Gudara see Ghŭdara
Gudauta 95 E1 NW Georgia
Guéret 68 C4 Creuse, C France
Guernsey 67 D8 British Crown Dependency Channel Islands, NW Europe
Guerrero Negro 28 A2 Baja California Sur, NW Mexico
Gui see Guangxi Zhuangzu Zizhiqu
Guiana see French Guiana
Guiana Highlands 40 D1 var. Macizo de las Guayanas. mountain range N South America
Guiba see Juba
Guidder see Guider
Guider 54 B4 var. Guidder. Nord, N Cameroon
Guidimouni 53 G3 Zinder, S Niger
Guildford 67 D7 SE England, United Kingdom
Guilin 106 C6 var. Kuei-lin, Kweilin. Guangxi Zhuangzu Zizhiqu, S China
Guimarães 70 B2 var. Guimaráes. Braga, N Portugal
Guimaráes see Guimarães
Guinea 52 C4 off. Republic of Guinea, var. Guinée; prev. French Guinea, People's Revolutionary Republic of Guinea. country W Africa
Guinea Basin 47 A5 undersea basin E Atlantic Ocean
Guinea-Bissau 52 B4 off. Republic of Guinea-Bissau, Fr. Guinée-Bissau, Port. Guiné-Bissau; prev. Portuguese Guinea. country W Africa
Guinea-Bissau, Republic of see Guinea-Bissau
Guinea, Gulf of 46 B4 Fr. Golfe de Guinée. gulf E Atlantic Ocean
Guinea, People's Revolutionary Republic of see Guinea
Guinea, Republic of see Guinea
Guiné-Bissau see Guinea-Bissau
Guinée see Guinea
Guinée-Bissau see Guinea-Bissau
Guinée, Golfe de see Guinea, Gulf of
Güiria 37 E1 Sucre, NE Venezuela
Guiyang 106 B6 var. Kuei-Yang, Kuei-yang, Kueyang, Kweiyang; prev. Kweichu. province capital Guizhou, S China
Guizhou 106 B6 Guangdong, SE China
Gujarāt 112 C4 var. Gujerat. cultural region W India
Gujerat see Gujarāt
Gujranwala 112 D2 Punjab, NE Pakistan
Gujrat 112 D2 Punjab, E Pakistan
Gulbarga see Kalaburagi
Gulbene 84 D3 Ger. Alt-Schwanenburg. Gulbene, NE Latvia
Gulf of Liaotung see Liaodong Wan
Gulfport 20 C3 Mississippi, S USA
Gulf, The see Persian Gulf
Gulistan see Guliston
Guliston 101 E2 Rus. Gulistan. Sirdaryo Viloyati, E Uzbekistan
Gulja see Yining
Gulkana 14 D3 Alaska, USA
Gulu 51 B6 N Uganda
Gulyantsi 82 C1 Pleven, N Bulgaria
Guma see Pishan
Gumbinnen see Gusev
Gumpolds see Humpolec
Gümülcine/Gümüljina see Komotiní
Gümüşane see Gümüşhane
Gümüşhane 95 E3 var. Gümüşane, Gumushkhane. Gümüşhane, NE Turkey (Türkiye)
Gumushkhane see Gümüşhane
Güney Doğu Toroslar 95 E4 mountain range SE Turkey (Türkiye)
Gunnbjørn Fjeld 60 D4 var. Gunnbjörns Bjerge. mountain C Greenland
Gunnbjörns Bjerge see Gunnbjørn Fjeld
Gunnedah 127 D6 New South Wales, SE Australia
Gunnison 22 C5 Colorado, C USA
Gurbansoltan Eje 100 C2 prev. Ýylanly, Rus. Il'yaly. Daşoguz Welaýaty, N Turkmenistan
Gurbantünggüt Shamo 104 B2 desert W China
Gurgan see Gorgān
Guri, Embalse de 37 E2 reservoir E Venezuela
Gurkfeld see Krško
Gurktaler Alpen 73 D7 mountain range S Austria
Gürün 94 D3 Sivas, C Turkey (Türkiye)
Gur'yev see Atyraū
Gusau 53 G4 Zamfara, NW Nigeria
Gusev 84 B4 Ger. Gumbinnen. Kaliningradskaya Oblast', W Russia
Gustavus 14 D4 Alaska, USA
Güstrow 72 C3 Mecklenburg-Vorpommern, NE Germany
Guta/Gúta see Kolárovo
Gütersloh 72 B4 Nordrhein-Westfalen, W Germany
Gutta see Kolárovo
Guttstadt see Dobre Miasto
Guwāhāti 113 G3 prev. Gauhāti. Assam, NE India
Guyana 37 F3 off. Co-operative Republic of Guyana; prev. British Guiana. country N South America
Guyana, Co-operative Republic of see Guyana
Guyane see French Guiana
Guymon 27 E1 Oklahoma, C USA
Güzelyurt Körfezi 80 C5 Gk. Kólpos Mórfu, Morphou. W Cyprus
Gvardeysk 84 A4 Ger. Tapaiu. Kaliningradskaya Oblast', W Russia
Gwadar 112 A3 var. Gwadur. Baluchistan, SW Pakistan
Gwadur see Gwādar
Gwalior 112 D3 Madhya Pradesh, C India
Gwanda 56 D3 Matabeleland South, SW Zimbabwe
Gwangju 107 E4 off. Gwangju Gwang-yeoksi, prev. Kwangju, var. Guangju, Kwangchu, Jap. Kōshū. SW South Korea
Gwangju Gwang-yeoksi see Gwangju
Gwy see Wye
Gyandzha see Gäncä
Gyangzê 104 C5 Xizang Zizhiqu, W China
Gyaring Co 104 C5 lake W China
Gyêgu see Yushu

Gyergyószentmiklós see Gheorgheni
Gyixong see Gonggar
Gympie 127 E5 Queensland, E Australia
Gyomaendrőd 77 D7 Békés, SE Hungary
Gyöngyös 77 D6 Heves, NE Hungary
Győr 77 C6 Ger. Raab, Lat. Arrabona. Győr-Moson-Sopron, NW Hungary
Gýtheio 83 B6 var. Githio; prev. Yíthion. Pelopónnisos, S Greece
Gyulafehérvár see Alba Iulia
Gyumri 95 F2 var. Giumri, Rus. Kumayri; prev. Aleksandropol', Leninakan. W Armenia
Gyzyrlabat see Serdar

H

Haabai see Ha'apai Group
Haacht 65 C6 Vlaams Brabant, C Belgium
Haaksbergen 64 E3 Overijssel, E Netherlands
Ha'apai Group 123 F4 var. Haabai. island group C Tonga
Haapsalu 84 D2 Ger. Hapsal. Läänemaa, W Estonia
Ha'Arava see 'Arabah, Wādī al
Haarlem 64 C3 prev. Harlem. Noord-Holland, W Netherlands
Haast 129 B6 West Coast, South Island, New Zealand
Hachijo-jima 109 D6 island Izu-shotō, SE Japan
Hachinohe 108 D3 Aomori, Honshū, C Japan
Hacıqabul 95 H3 prev. Qazimämmäd, Rus. Kazi Magomed. SE Azerbaijan
Haḍabat al Jilf al Kabīr 50 A2 var. Gilf Kebir Plateau. plateau SW Egypt
Hadama see Nazrēt
Hadejia 53 G4 Jigawa, N Nigeria
Hadejia 53 G3 river N Nigeria
Hadera 97 A6 var. Khadera; prev. Ḥadera. Haifa, C Israel
Ḥadera see Hadera
Hadhdhunmathi Atoll 110 A5 atoll S Maldives
Hadhramaut see Ḥaḍramawt
Ha Đông 114 D3 var. Hadong. Ha Tây, N Vietnam
Hadong see Ha Đông
Ḥaḍramawt 99 C6 Eng. Hadhramaut. mountain range S Yemen
Hadrianopolis see Edirne
Haerbin/Haerhpin/Ha-erh-pin see Harbin
Hafnia see Denmark
Hafnia see København
Hafren see Severn
Hafun, Ras see Xaafuun, Raas
Hagåtña 122 B1 , var. Agaña. dependent territory capital (Guam) NW Guam
Hagerstown 19 E4 Maryland, NE USA
Ha Giang 114 D3 Ha Giang, N Vietnam
Hagios Evstrátios see Ágios Efstrátios
HaGolan see Golan Heights
Hagondange 68 D3 Moselle, NE France
Haguenau 68 E3 Bas-Rhin, NE France
Haibowan see Wuhai
Haicheng 106 D3 Liaoning, NE China
Haida Gwaii 14 C5 prev. Queen Charlotte Islands, Fr. Îles de la Reine-Charlotte. island group British Columbia, SW Canada
Haidarabad see Hyderabad
Haifa see Hefa
Haifa, Bay of see Mifrats Hefa
Haifong see Hai Phong
Haikou 106 C7 var. Hai-k'ou, Hoihow, Fr. Hoï-Hao. province capital Hainan, S China
Hai-k'ou see Haikou
Ḥā'il 98 B4 Ḥā'il, NW Saudi Arabia
Hailuoto 62 D4 Swe. Karlö. island W Finland
Hainan 106 B7 var. Hainan Sheng, Qiong. province S China
Hainan Dao 106 C7 island S China
Hainan Sheng see Hainan
Hainasch see Ainaži
Haines 14 D4 Alaska, USA
Hainichen 72 D4 Sachsen, E Germany
Hai Phong 114 D3 var. Haifong, Haiphong. N Vietnam
Haiphong see Hai Phong
Haisyn 86 D3 prev. Haysyn, Rus. Gaysin. Vinnyts'ka Oblast', C Ukraine
Haiti 32 D3 off. Republic of Haiti. country C West Indies
Haiti, Republic of see Haiti
Haiya 50 C3 Red Sea, NE Sudan
Hajdúhadház 77 D6 Hajdú-Bihar, E Hungary
Hajîn 96 E3 var. Abū Ḥardān, Hajîne. Dayr az Zawr, E Syria
Hajîne see Hajîn
Hajnówka 76 E3 Ger. Hermhausen. Podlaskie, NE Poland
Hakodate 108 D3 Hokkaidō, NE Japan
Hal see Halle
Ḥalab 96 B2 Eng. Aleppo, Fr. Alep; anc. Beroea. Ḥalab, NW Syria
Hala'ib Triangle 50 C3 region Egypt/Sudan
Ḥalāniyāt, Juzur al 99 D6 var. Jazā'ir Bin Ghalfān, Eng. Kuria Muria Islands. island group S Oman
Halberstadt 72 C4 Sachsen-Anhalt, C Germany
Halden 63 B6 prev. Fredrikshald. Østfold, S Norway
Halfmoon Bay 129 A8 var. Oban. Stewart Island, Southland, New Zealand
Haliacmon see Aliákmonas
Halifax 17 F4 province capital Nova Scotia, SE Canada
Halkida see Chalkída
Halle 65 B6 Fr. Hal. Vlaams Brabant, C Belgium
Halle 72 C4 var. Halle an der Saale. Sachsen-Anhalt, C Germany
Halle an der Saale see Halle
Halle-Neustadt 72 C4 Sachsen-Anhalt, C Germany
Halley 132 B2 UK research station Antarctica
Hall Islands 120 B2 island group C Micronesia
Halls Creek 124 C3 Western Australia
Halmahera, Laut 117 F3 Eng. Halmahera Sea; sea E Indonesia
Halmahera, Pulau 117 F3 prev. Djailolo, Gilolo, Jailolo. island E Indonesia
Halmahera Sea see Halmahera, Laut
Halmstad 63 B7 Halland, S Sweden
Ha Long 114 E3 prev. Hông Gai; var. Hon Gai, Hongay. Quang Ninh, N Vietnam
Hälsingborg see Helsingborg
Hamada 109 B6 Shimane, Honshū, SW Japan
Hamadān 98 C3 anc. Ecbatana. Hamadān, W Iran
Ḥamāh 96 B3 var. Hama; anc. Epiphania, Bibl. Hamath. Ḥamāh, W Syria
Hamamatsu 109 D6 var. Hamamatu. Shizuoka, Honshū, S Japan
Hamamatu see Hamamatsu
Hamar 63 B5 prev. Storhammer. Hedmark, S Norway
Hamath see Ḥamāh
Hamburg 72 B3 Hamburg, N Germany
Hamd, Wadi al 98 A4 dry watercourse W Saudi Arabia
Hämeenlinna 63 D5 Swe. Tavastehus. Kanta-Häme, S Finland
HaMela h, Yam see Dead Sea
Hamersley Range 124 A4 mountain range Western Australia
Hamhŭng 107 E3 C North Korea
Hami 104 C3 var. Ha-mi, Uigh. Kumul, Qomul. Xinjiang Uygur Zizhiqu, NW China
Ha-mi see Hami
Hamilton 20 A5 dependent territory capital (Bermuda) C Bermuda
Hamilton 16 D5 Ontario, S Canada
Hamilton 128 D3 Waikato, North Island, New Zealand
Hamilton 66 C4 S Scotland, United Kingdom
Hamilton 20 C2 Alabama, S USA
Hamim, Wadi al 49 G2 river NE Libya
Hamīs Musait see Khamīs Mushayt
Hamm 72 B4 var. Hamm in Westfalen. Nordrhein-Westfalen, W Germany
Ḥammāmāt, Khalīj al see Hammamet, Golfe de
Hammamet, Golfe de 80 D3 Ar. Khalīj al Ḥammāmāt. gulf NE Tunisia
Ḥammār, Hawr al 98 C3 lake SE Iraq
Hamm in Westfalen see Hamm
Hampden 129 B7 Otago, South Island, New Zealand
Hampstead 67 A7 Maryland, USA
Ħamrun 80 B5 C Malta
Hāmūn, Daryācheh-ye see Şāberī, Hāmūn-e/Sīstān, Daryācheh-ye
Hamwih see Southampton
Hâncești see Hîncești
Hancewicze see Hantsavichy
Handan 106 C4 var. Han-tan. Hebei, E China
Haneda 108 A2 (Tōkyō) Tōkyō, Honshū, S Japan
HaNegev 97 A7 Eng. Negev. desert S Israel
Hanford 25 C6 California, W USA
Hangayn Nuruu 104 D2 mountain range C Mongolia
Hang-chou/Hangchow see Hangzhou
Hangö see Hanko
Hangzhou 106 D5 var. Hang-chou, Hangchow. province capital Zhejiang, SE China
Hania see Chaniá
Hanka, Lake see Khanka, Lake
Hanko 63 D6 Swe. Hangö. Uusimaa, SW Finland
Han-kou/Han-k'ou/Hankow see Wuhan
Hanmer Springs 129 C5 Canterbury, South Island, New Zealand
Hannibal 23 G4 Missouri, C USA
Hannover 72 B3 Eng. Hanover. Niedersachsen, NW Germany
Hanöbukten 63 B7 bay S Sweden
Ha Nôi 114 D3 Eng. Hanoi, Fr. Hanoï. country capital (Vietnam) N Vietnam
Hanover see Hannover
Han Shui 105 E4 river C China
Han-tan see Handan
Hantsavichy 85 B6 Pol. Hancewicze, Rus. Gantsevichi. Brestskaya Voblasts', SW Belarus
Hanyang see Wuhan
Hanzhong 106 B5 Shaanxi, C China
Hāora 113 F4 prev. Howrah. West Bengal, NE India
Haparanda 62 D4 Norrbotten, N Sweden
Hapsal see Haapsalu
Haradok 85 E5 Rus. Gorodok. Vitsyebskaya Voblasts', N Belarus
Haradzyets 85 B6 Rus. Gorodets. Brestskaya Voblasts', SW Belarus
Haramachi 108 D4 Fukushima, Honshū, E Japan
Harany 85 D5 Rus. Gorany. Vitsyebskaya Voblasts', N Belarus
Harare 56 D3 prev. Salisbury. country capital (Zimbabwe) Mashonaland East, NE Zimbabwe
Harbavichy 85 E6 Rus. Gorbovichi. Mahilyowskaya Voblasts', E Belarus
Harbel 52 C5 W Liberia
Harbin 107 E2 var. Haerbin, Ha-erh-pin, Kharbin; prev. Haerhpin, Pingkiang, Pinkiang. province capital Heilongjiang, NE China
Hardangerfjorden 63 A6 fjord S Norway
Hardangervidda 63 A6 plateau S Norway
Hardenberg 64 E3 Overijssel, E Netherlands
Harelbeke 65 A6 var. Harlebeke. West-Vlaanderen, W Belgium
Harem see Ḥārim
Haren 64 E2 Groningen, NE Netherlands
Hārer 51 D5 E Ethiopia
Hargeisa see Hargeysa
Hargeysa 51 D5 var. Hargeisa. Woqooyi Galbeed, NW Somalia
Hariana see Haryāna
Hari, Batang 116 B4 prev. Djambi. river Sumatera, W Indonesia
Ḥārim 96 B2 var. Harem. Idlib, W Syria
Harima-nada 109 B6 sea S Japan
Harī Rōd 101 E4 var. Harīrūd, Tedzhen, Turkm. Tejen. river Afghanistan/Iran
Harīrūd see Harī Rōd
Harlan 23 F3 Iowa, C USA
Harlebeke see Harelbeke
Harlem see Haarlem
Harlingen 64 D2 Fris. Harns. Fryslân, N Netherlands
Harlingen 27 G5 Texas, SW USA
Harlow 67 E6 E England, United Kingdom
Harney Basin 24 B4 basin Oregon, NW USA
Härnösand 63 C5 var. Hernösand. Västernorrland, C Sweden
Harns see Harlingen
Harper 52 D5 var. Cape Palmas. NE Liberia
Harricana 16 D3 river Québec, SE Canada
Harris 66 B3 physical region NW Scotland, United Kingdom
Harrisburg 19 E4 state capital Pennsylvania, NE USA
Harrisonburg 19 E4 Virginia, NE USA
Harrison, Cape 17 F2 headland Newfoundland and Labrador, E Canada
Harris Ridge see Lomonosov Ridge
Harrogate 67 D5 N England, United Kingdom
Hârşova 86 D5 prev. Hirşova. Constanța, SE Romania
Harstad 62 C2 Troms, N Norway
Hartford 19 G3 state capital Connecticut, NE USA
Hartlepool 67 D5 N England, United Kingdom
Harunabad see Eslāmābād
Har Us Gol 104 C2 lake Hovd, W Mongolia
Har Us Nuur 104 C2 lake NW Mongolia
Harwich 67 E6 E England, United Kingdom
Haryāna 112 D2 var. Hariana. cultural region N India
Hashemite Kingdom of Jordan see Jordan
Haskovo 82 D3 var. Khaskovo. Haskovo, S Bulgaria
Hasselt 65 C6 Limburg, NE Belgium
Hassetché see Al Ḥasakah
Hasta Colonia/Hasta Pompeia see Asti
Hastings 128 E4 Hawke's Bay, North Island, New Zealand
Hastings 67 E7 SE England, United Kingdom
Hastings 23 E4 Nebraska, C USA
Haţeg 86 B4 Ger. Wallenthal, Hung. Hátszeg; prev. Hatzeg, Hötzing. Hunedoara, SW Romania
Hátszeg see Haţeg
Hattem 64 D3 Gelderland, E Netherlands
Hatteras, Cape 21 G1 headland North Carolina, SE USA
Hatteras Plain 13 D6 abyssal plain Atlantic Ocean
Hattiesburg 20 C3 Mississippi, S USA
Hatton Bank see Hatton Ridge
Hatton Ridge 58 B2 var. Hatton Bank. undersea ridge N Atlantic Ocean
Hat Yai 115 C7 var. Ban Hat Yai. Songkhla, SW Thailand
Hatzeg see Haţeg
Hatzfeld see Jimbolia
Haugesund 63 A6 Rogaland, S Norway
Haukeligrend 63 A6 Telemark, S Norway
Haukivesi 63 E5 lake SE Finland
Hauraki Gulf 128 D2 var. Tīkapa Moana. gulf North Island, N New Zealand
Hauroko, Lake 129 A7 lake South Island, New Zealand
Haut Atlas 48 C2 Eng. High Atlas. mountain range C Morocco
Hautes Fagnes 65 D6 Ger. Hohes Venn. mountain range E Belgium
Hauts Plateaux 48 D2 plateau Algeria/Morocco
Hauzenberg 73 D6 Bayern, SE Germany
Havana 13 D6 Illinois, N USA
Havana see La Habana
Havant 67 D7 S England, United Kingdom
Havelock 21 F1 North Carolina, SE USA
Havelock North 128 E4 Hawke's Bay, North Island, New Zealand
Haverfordwest 67 C6 SW Wales, United Kingdom
Havířov 77 C5 Moravskoslezský Kraj, E Czechia (Czech Republic)
Havre 22 C1 Montana, NW USA
Havre see le Havre
Havre-St-Pierre 17 F3 Québec, E Canada
Hawaii 25 A8 off. State of Hawaii, also known as Aloha State, Paradise of the Pacific, var. Hawai'i. state USA, C Pacific Ocean
Hawai'i 25 B8 var. Hawaii. island Hawaiian Islands, USA, C Pacific Ocean
Hawaiian Islands 130 D2 prev. Sandwich Islands. island group Hawaii, USA
Hawaiian Ridge 130 H4 undersea ridge N Pacific Ocean
Hawea, Lake 129 B6 lake South Island, New Zealand
Hawera 128 D4 Taranaki, North Island, New Zealand
Hawick 66 C4 SE Scotland, United Kingdom
Hawke Bay 128 E4 bay North Island, New Zealand
Hawkeye State see Iowa
Hawlēr see Arbil
Hawthorne 25 C6 Nevada, W USA
Hay 127 C6 New South Wales, SE Australia
HaYarden see Jordan
Hayastan see Armenia
Hayastani Hanrapetut'yun see Armenia
Hayes 16 B2 river Manitoba, C Canada
Hay River 15 E4 Northwest Territories, W Canada
Hays 23 E5 Kansas, C USA
Haysyn see Haisyn
Hazar 100 B2 prev. Rus. Cheleken. Balkan Welaýaty, W Turkmenistan
Heard Island and McDonald Islands 119 B7 Australian external territory S Indian Ocean
Hearst 16 C4 Ontario, S Canada
Heart of Dixie see Alabama
Heathrow 67 A8 (London) SE England, United Kingdom
Hebei 106 C4 var. Hebei Sheng, Hopeh, Hopei, Ji; prev. Chihli. province E China
Hebei Sheng see Hebei
Hebron 97 A6 var. Al Khalīl, El Khalil, Heb. Hevron; anc. Kiriath-Arba. S West Bank, Middle East
Heemskerk 64 C3 Noord-Holland, W Netherlands
Heerde 64 D3 Gelderland, E Netherlands
Heerenveen 64 D2 Fris. It Hearrenfean. Fryslân, N Netherlands
Heerhugowaard 64 C2 Noord-Holland, NW Netherlands
Heerlen 65 D6 Limburg, SE Netherlands
Heerwegen see Polkowice
Hefa 97 A5 var. Haifa, hist. Caiffa, Caiphas; anc. Sycaminum. Haifa, N Israel
Ḥefa, Mifraz see Mifrats Hefa
Hefei 106 D5 var. Hofei, hist. Luchow. province capital Anhui, E China
Hegang 107 E2 Heilongjiang, NE China
Hei see Heilongjiang
Heide 72 B2 Schleswig-Holstein, N Germany
Heidelberg 73 B5 Baden-Württemberg, SW Germany
Heidenheim see Heidenheim an der Brenz
Heidenheim an der Brenz 73 B6 var. Heidenheim. Baden-Württemberg, S Germany
Hei-ho see Nagqu
Heilbronn 73 B6 Baden-Württemberg, SW Germany
Heiligenbeil see Mamonovo
Heilongjiang 106 D2 var. Hei, Heilongjiang Sheng, Hei-lung-chiang, Heilungkiang. province NE China
Heilong Jiang see Amur
Heilongjiang Sheng see Heilongjiang
Heiloo 64 C3 Noord-Holland, NW Netherlands
Heilsberg see Lidzbark Warmiński
Hei-lung-chiang/Heilungkiang see Heilongjiang
Heimdal 63 B5 Sør-Trøndelag, S Norway
Heinaste see Ainaži
Hekimhan 94 D3 Malatya, C Turkey (Türkiye)
Helena 22 B2 state capital Montana, NW USA
Helensville 128 D2 Auckland, North Island, New Zealand
Helgoland Bay see Helgoländer Bucht
Helgoländer Bucht 72 A2 var. Helgoland Bay, Heligoland Bight. bay NW Germany
Heligoland Bight see Helgoländer Bucht
Heliopolis see Baalbek
Hellas see Greece
Hellenic Republic see Greece
Hellevoetsluis 64 B4 Zuid-Holland, SW Netherlands
Hellín 71 E4 Castilla-La Mancha, C Spain
Helmand, Darya-ye see Helmand Rōd
Helmand Rōd 100 D5 var. Daryā-ye Helmand, Rūd-e Hīrmand, Helmand Rūd. river Afghanistan/Iran
Helmand Rūd see Helmand Rōd
Helmantica see Salamanca
Helmond 65 D5 Noord-Brabant, S Netherlands
Helsingborg 63 B7 prev. Hälsingborg. Skåne, S Sweden
Helsingfors see Helsinki
Helsinki 63 D6 Swe. Helsingfors. country capital (Finland) Uusimaa, S Finland
Heltau see Cisnădie
Helvetia see Switzerland
Henan 106 C5 var. Henan Sheng, Honan, Yu. province C China
Henderson 18 B5 Kentucky, S USA
Henderson 25 D7 Nevada, W USA
Henderson 27 H3 Texas, SW USA
Hendū Kosh see Hindu Kush
Hengchow see Hengyang
Hengduan Shan 106 A5 mountain range SW China
Hengelo 64 E3 Overijssel, E Netherlands
Hengnan see Hengyang
Hengyang 106 C6 var. Hengnan, Heng-yang; prev. Hengchow. Hunan, S China
Heng-yang see Hengyang
Henichesk 87 F4 Rus. Genichesk. Khersons'ka Oblast', S Ukraine
Hennebont 68 A3 Morbihan, NW France
Henrique de Carvalho see Saurimo
Henzada see Hinthada
Herakleion see Irákleio
Herāt 100 D4 var. Herat; anc. Aria. Herāt, W Afghanistan
Heredia 31 E4 Heredia, C Costa Rica
Hereford 27 E2 Texas, SW USA
Herford 72 B4 Nordrhein-Westfalen, NW Germany
Héristal see Herstal
Herk-de-Stad 65 C6 Limburg, NE Belgium
Herlen Gol/Herlen He see Kerulen
Hermannstadt see Sibiu
Hermansverk 63 A5 Sogn Og Fjordane, S Norway
Hermhausen see Hajnówka
Hermiston 24 C2 Oregon, NW USA
Hermon, Mount 97 B5 Ar. Jabal ash Shaykh. mountain S Syria
Hermosillo 28 B2 Sonora, NW Mexico
Hermoupolis see Ermoúpoli
Hernösand see Härnösand
Herrera del Duque 70 D3 Extremadura, W Spain
Herselt 65 C5 Antwerpen, C Belgium
Herstal 65 D6 Fr. Héristal. Liège, E Belgium
Herzogenbusch see 's-Hertogenbosch
Hesse see Hessen
Hessen 73 B5 Eng./Fr. Hesse. state C Germany
Hevron see Hebron
Hewlêr see Arbil
Heydebrech see Kędzierzyn-Kozle
Heydekrug see Šilutė
Heywood Islands 124 C3 island group Western Australia
Hibbing 23 F1 Minnesota, N USA
Hibernia see Ireland
Hidalgo del Parral 28 C2 var. Parral. Chihuahua, N Mexico
Hida-sanmyaku 109 C5 mountain range Honshū, S Japan
Hierosolyma see Jerusalem
Hierro 48 A3 var. Ferro. island Islas Canarias, Spain, NE Atlantic Ocean
High Atlas see Haut Atlas
High Plains see Great Plains
High Point 21 E1 North Carolina, SE USA
Hiiumaa 84 C2 Ger. Dagden, Swe. Dagö. island W Estonia
Hikurangi 128 D2 Northland, North Island, New Zealand
Hildesheim 72 B4 Niedersachsen, N Germany
Hilla see Al Ḥillah
Hillaby, Mount 33 G1 mountain N Barbados
Hill Bank 30 C1 Orange Walk, N Belize
Hillegom 64 C3 Zuid-Holland, W Netherlands
Hilo 25 B8 Hawaii, USA, C Pacific Ocean
Hilton Head Island 21 E2 South Carolina, SE USA
Hilversum 64 C3 Noord-Holland, C Netherlands
Himalaya/Himalaya Shan see Himalayas
Himalayas 113 E2 var. Himalaya, Chin. Himalaya Shan. mountain range S Asia
Himeji 109 C6 var. Himezi. Hyōgo, Honshū, SW Japan
Himezi see Himeji
Ḥimş 96 B4 var. Homs; anc. Emesa. Ḥimş, C Syria
Hîncești 86 D4 var. Hâncești; prev. Kotovsk. C Moldova
Hinchinbrook Island 126 D3 island Queensland, NE Australia
Hinds 129 C6 Canterbury, South Island, New Zealand
Hindu Kush 101 F4 Per. Hendū Kosh. mountain range Afghanistan/Pakistan
Hinesville 21 E3 Georgia, SE USA
Hinnøya 62 C3 Lapp. Iinnasuolu. island C Norway
Hinson Bay 20 A5 bay W Bermuda, W Atlantic Ocean
Hinthada 114 B4 var. Henzada. Ayeyarwady, SW Myanmar (Burma)
Hios see Chíos
Hirfanlı Barajı 94 C3 reservoir C Turkey (Türkiye)
Hīrmand, Rūd-e see Helmand Rōd
Hirosaki 108 D3 Aomori, Honshū, C Japan
Hiroshima 109 B6 var. Hirosima. Hiroshima, Honshū, SW Japan
Hirschberg/Hirschberg im Riesengebirge/Hirschberg in Schlesien see Jelenia Góra
Hirson 68 D3 Aisne, N France
Hîrşova see Hârşova
Hispalis see Sevilla
Hispana/Hispania see Spain
Hispaniola 34 B1 island Dominion Republic/Haiti
Hitachi 109 D5 var. Hitati. Ibaraki, Honshū, S Japan
Hitati see Hitachi
Hitra 62 A4 prev. Hitteren. island S Norway
Hitteren see Hitra
Hjälmaren 63 C6 Eng. Lake Hjalmar. lake C Sweden
Hjalmar, Lake see Hjälmaren
Hjørring 63 B7 Nordjylland, N Denmark
Hkakabo Razi 114 B1 mountain Myanmar (Burma)/China
Hlobyne 87 F2 Rus. Globino. Poltavs'ka Oblast', NE Ukraine
Hlukhiv 87 F1 Rus. Glukhov. Sums'ka Oblast', NE Ukraine
Hlybokaye 85 D5 Rus. Glubokoye. Vitsyebskaya Voblasts', N Belarus
Hoa Binh 114 D3 Hoa Binh, N Vietnam
Hoang Lien Son 114 D3 mountain range N Vietnam
Hobart 127 C8 prev. Hobarton, Hobart Town. state capital Tasmania, SE Australia
Hobarton/Hobart Town see Hobart
Hobbs 27 E3 New Mexico, SW USA
Hobro 63 A7 Nordjylland, N Denmark
Hô Chi Minh 115 E6 var. Ho Chi Minh City; prev. Saigon. S Vietnam
Ho Chi Minh City see Hô Chi Minh
Hodeidah see Al Ḥudaydah
Hódmezővásárhely 77 D7 Csongrád, SE Hungary
Hodna, Chott El 80 C4 var. Chott el-Hodna, Ar. Shatt al-Hodna. salt lake N Algeria
Hodna, Chott el-/Hodna, Shatt al- see Hodna, Chott El
Hodonín 77 C5 Ger. Göding. Jihomoravský Kraj, SE Czechia (Czech Republic)
Hoei see Huy
Hoey see Huy
Hof 73 C5 Bayern, SE Germany
Hofei see Hefei
Hōfu 109 B7 Yamaguchi, Honshū, SW Japan
Hofuf see Al Hufūf
Hogoley Islands see Chuuk Islands
Hohensalza see Inowrocław
Hohenstadt see Zábřeh
Hohes Venn see Hautes Fagnes
Hohe Tauern 73 C7 mountain range W Austria
Hohhot 105 F3 var. Huhehot, Huhohaote, Mong. Kukukhoto; prev. Kweisui, Kwesui. Nei Mongol Zizhiqu, N China
Hôi An 115 E5 prev. Faifo. Quang Nam-Đa Nẵng, C Vietnam
Hoï-Hao/Hoihow see Haikou
Hokianga Harbour 128 C2 inlet SE Tasman Sea
Hokitika 129 B5 West Coast, South Island, New Zealand
Hokkaido 108 C2 prev. Ezo, Yeso, Yezo. island NE Japan
Hola Prystan 87 E4 Rus. Golaya Pristan. Khersons'ka Oblast', S Ukraine
Holbrook 26 B2 Arizona, SW USA
Holetown 33 G1 prev. Jamestown. W Barbados
Holguín 32 C2 Holguín, SE Cuba
Hollabrunn 73 E6 Niederösterreich, NE Austria
Holland see Netherlands
Hollandia see Jayapura
Holly Springs 20 C1 Mississippi, S USA
Holman 15 E3 Victoria Island, Northwest Territories, N Canada
Holmsund 62 D4 Västerbotten, N Sweden
Holon 97 A6 var. Kholon; prev. Ḥolon. Tel Aviv, C Israel
Ḥolon see Holon
Holovanivsk 87 E3 Rus. Golovanevsk. Kirovohrads'ka Oblast', C Ukraine
Holstebro 63 A7 Ringkøbing, W Denmark
Holsteinborg/Holsteinsborg/Holstenborg/Holstensborg see Sisimiut
Holyhead 67 C5 Wel. Caer Gybi. NW Wales, United Kingdom
Hombori 53 E3 Mopti, S Mali
Homs see Al Khums, Libya
Homs see Ḥimş
Homyel' 85 D7 Rus. Gomel'. Homyel'skaya Voblasts', SE Belarus
Honan see Luoyang, China
Honan see Henan, China
Hondo 27 F4 Texas, SW USA
Hondo see Honshū
Honduras 30 C2 off. Republic of Honduras. country Central America
Honduras, Golfo de see Honduras, Gulf of
Honduras, Gulf of 30 C2 Sp. Golfo de Honduras. gulf W Caribbean Sea
Honduras, Republic of see Honduras
Hønefoss 63 B6 Buskerud, S Norway
Honey Lake 25 B5 lake California, W USA
Hon Gai see Ha Long
Hongay see Ha Long
Hông Gai see Ha Long
Hông Hà, Sông see Red River
Hong Kong 106 A1 Special administrative region Hong Kong, S China
Hong Kong Island 106 B2 island S China Asia
Honiara 122 C3 country capital (Solomon Islands) Guadalcanal, C Solomon Islands
Honjō 108 D4 var. Honzyô, Yurihonjō. Akita, Honshū, C Japan
Honolulu 25 A8 state capital O'ahu, Hawaii, USA, C Pacific Ocean
Honshu 109 E5 var. Hondo, Honsyû. island SW Japan
Honsyû see Honshū
Honte see Westerschelde
Honzyô see Honjō
Hoogeveen 64 E2 Drenthe, NE Netherlands
Hoogezand-Sappemeer 64 E2 Groningen, NE Netherlands
Hoorn 64 C2 Noord-Holland, NW Netherlands
Hoosier State see Indiana
Hopa 95 E2 Artvin, NE Turkey (Türkiye)
Hope 14 C3 British Columbia, SW Canada
Hopedale 17 F2 Newfoundland and Labrador, NE Canada
Hopeh/Hopei see Hebei
Hopkinsville 18 B5 Kentucky, S USA
Horasan 95 F3 Erzurum, NE Turkey (Türkiye)
Horizon Deep 130 D4 trench W Pacific Ocean

Horki *85 E6 Rus.* Gorki. Mahilyowskaya Voblasts', E Belarus
Horlivka *87 G3 Rom.* Adâncata, *Rus.* Gorlovka. Donets'ka Oblast', E Ukraine
Hormoz, Tangeh-ye *see* Hormuz, Strait of
Hormuz, Strait of *98 D4 var.* Strait of Ormuz, *Per.* Tangeh-ye Hormoz. *strait* Iran/Oman
Horn, Cape *see* Hornos, Cabo de
Hornos, Cabo de *43 C8 Eng.* Cape Horn. *headland* S Chile
Hornsby *126 E1* New South Wales, SE Australia
Horodnia *87 E1 prev.* Horodnya, *Rus.* Gorodnya. Chernihivs'ka Oblast', NE Ukraine
Horodnya *see* Horodnia
Horodok *86 B2 Pol.* Gródek Jagielloński, *Rus.* Gorodok, Gorodok Yagellonski. L'vivs'ka Oblast', NW Ukraine
Horodyshche *87 E2 Rus.* Gorodishche. Cherkas'ka Oblast', C Ukraine
Horoshiri-dake *108 D2 var.* Horosiri Dake. *mountain* Hokkaidō, N Japan
Horosiri Dake *see* Horoshiri-dake
Horsburgh Atoll *110 A4 var.* Goidhoo Atoll. *atoll* N Maldives
Horseshoe Bay *20 A5 bay* W Bermuda W Atlantic Ocean
Horseshoe Seamounts *58 A4 seamount range* E Atlantic Ocean
Horsham *127 B7* Victoria, SE Australia
Horst *65 D5* Limburg, SE Netherlands
Horten *63 B6* Vestfold, S Norway
Horyn' *85 B7 Rus.* Goryn. *river* NW Ukraine
Hosingen *65 D7* Diekirch, NE Luxembourg
Hospitalet *see* L'Hospitalet de Llobregat
Hotan *104 B4 var.* Khotan, *Chin.* Ho-t'ien. Xinjiang Uygur Zizhiqu, NW China
Ho-t'ien *see* Hotan
Hoting *62 C4* Jämtland, C Sweden
Hot Springs *20 B1* Arkansas, C USA
Hötzing *see* Haţeg
Houayxay *114 C3 var.* Ban Houayxay. Bokèo, N Laos
Houghton *18 B1* Michigan, N USA
Houilles *69 B5* Yvelines, Île-de-France, N France Europe
Houlton *19 H1* Maine, NE USA
Houma *20 B3* Louisiana, S USA
Houston *27 H4* Texas, SW USA
Hovd *104 C2 var.* Khovd, Kobdo; *prev.* Jirgalanta. Hovd, W Mongolia
Hove *67 E7* SE England, United Kingdom
Hoverla, Hora *86 C3 Rus.* Gora Goverla. *mountain* W Ukraine
Hovsgol, Lake *see* Hövsgöl Nuur
Hövsgöl Nuur *104 D1 var.* Lake Hovsgol. *lake* N Mongolia
Howa, Ouadi *see* Howar, Wādi
Howar, Wadi *50 A3 var.* Ouadi Howa. *river* Chad/Sudan
Howland Island *123 E2 US unincorporated territory* W Polynesia
Howrah *see* Hāora
Hoy *66 C2 island* N Scotland, United Kingdom
Hoyerswerda *72 D4 Lus.* Wojerecy. Sachsen, E Germany
Hpa-An *114 B4 var.* Pa-an. Kayin State, S Myanmar (Burma)
Hpyu *see* Phyu
Hradec Králové *77 B5 Ger.* Königgrätz. Královéhradecký Kraj, N Czechia (Czech Republic)
Hrandzichy *85 B5 Rus.* Grandichi. Hrodzyenskaya Voblasts', W Belarus
Hranice *77 C5 Ger.* Mährisch-Weisskirchen. Olomoucký Kraj, E Czechia (Czech Republic)
Hrebinka *87 E2 Rus.* Grebenka. Poltavs'ka Oblast', NE Ukraine
Hrodna *85 B5 Pol.* Grodno. Hrodzyenskaya Voblasts', W Belarus
Hrvatska/ Republika Hrvatska *see* Croatia
Hsia-men *see* Xiamen
Hsiang-t'an *see* Xiangtan
Hsi Chiang *see* Xi Jiang
Hsing-K'ai Hu *see* Khanka, Lake
Hsi-ning/Hsining *see* Xining
Hsinking *see* Changchun
Hsin-yang *see* Xinyang
Hsu-chou *see* Xuzhou
Htawei *see* Dawei
Huacho *38 C4* Lima, W Peru
Hua Hin *see* Ban Hua Hin
Huaihua *106 C5* Hunan, S China
Huailai *106 C3 var.* Shacheng. Hebei, E China
Huainan *106 D5 var.* Huai-nan, Hwainan. Anhui, E China
Huai-nan *see* Huainan
Huajuapan *29 F5 var.* Huajuapan de León. Oaxaca, SE Mexico
Huajuapan de León *see* Huajuapan
Hualapai Peak *26 A2 mountain* Arizona, SW USA
Huallaga, Río *38 C3 river* N Peru
Huambo *56 B2 Port.* Nova Lisboa. Huambo, C Angola
Huancavelica *38 D4* Huancavelica, SW Peru
Huancayo *38 D3* Junín, C Peru
Huang Hai *see* Yellow Sea
Huang He *106 C4 var.* Yellow River. *river* C China
Huangshi *106 C5 var.* Huang-shih, Hwangshih. Hubei, C China
Huang-shih *see* Huangshi
Huanta *38 D4* Ayacucho, C Peru
Huánuco *38 C3* Huánuco, C Peru
Huanuni *39 F4* Oruro, W Bolivia
Huaral *38 C4* Lima, W Peru
Huarás *see* Huaraz
Huaraz *38 C3 var.* Huarás. Ancash, W Peru
Huarmey *38 C3* Ancash, W Peru
Huatabampo *28 C2* Sonora, NW Mexico
Hubballi *102 B2 prev.* Hubli. Karnātaka, SW India
Hubli *see* Hubballi
Huddersfield *67 D5* N England, United Kingdom
Hudiksvall *63 C5* Gävleborg, C Sweden
Hudson Bay *15 G4 bay* NE Canada
Hudson, Détroit d' *see* Hudson Strait
Hudson Strait *15 H3 Fr.* Détroit d'Hudson. *strait* Northwest Territories/Québec, NE Canada
Hudur *see* Xuddur
Huê *114 E4* Th,a Thiên-Huê, C Vietnam
Huehuetenango *30 A2* Huehuetenango, W Guatemala
Huelva *70 C4 anc.* Onuba. Andalucía, SW Spain
Huesca *71 F2 anc.* Osca. Aragón, NE Spain
Huéscar *71 E4* Andalucía, S Spain
Hughenden *126 C3* Queensland, NE Australia
Hugo *27 G2* Oklahoma, C USA
Huhehot/Huhohaote *see* Hohhot
Huila Plateau *56 B2 plateau* S Angola
Huixtla *29 G5* Chiapas, SE Mexico
Hulingol *105 G2 prev.* Huolin Gol. Nei Mongol Zizhiqu, N China
Hull *16 D4* Québec, SE Canada
Hull *see* Kingston upon Hull
Hull Island *see* Orona
Hulst *65 B5* Zeeland, SW Netherlands
Hulun *see* Hulun Buir
Hulun Buir *105 F1 var.* Hailar; *prev.* Hulun. Nei Mongol Zizhiqu, N China
Hu-lun Ch'ih *see* Hulun Nur
Hulun Nur *105 F1 var.* Hu-lun Ch'ih; *prev.* Dalai Nor. *lake* NE China
Humaitá *40 D2* Amazonas, N Brazil
Humboldt River *25 C5 river* Nevada, W USA
Humphreys Peak *26 B1 mountain* Arizona, SW USA
Humpolec *77 B5 Ger.* Gumpolds, Humpoletz. Vysočina, C Czechia (Czech Republic)
Humpoletz *see* Humpolec
Hunan *106 C6 var.* Hunan Sheng, Xiang. *province* S China
Hunan Sheng *see* Hunan
Hunedoara *86 B4 Ger.* Eisenmarkt, *Hung.* Vajdahunyad. Hunedoara, SW Romania
Hünfeld *73 B5* Hessen, C Germany
Hungarian People's Republic *see* Hungary
Hungary *77 C6 off.* Hungary, *Ger.* Ungarn, *Hung.* Magyarország, *Rom.* Ungaria, *Croatian* Mađarska, *Ukr.* Uhorshchyna; *prev.* Hungarian People's Republic. *country* C Europe
Hungary, Plain of *see* Great Hungarian Plain
Hunjiang *see* Baishan
Hunter Island *127 B8 island* Tasmania, SE Australia
Huntington *18 D4* West Virginia, NE USA
Huntington Beach *25 B8* California, W USA
Huntly *128 D3* Waikato, North Island, New Zealand
Huntsville *20 D1* Alabama, S USA
Huntsville *27 G3* Texas, SW USA
Huolin Gol *see* Hulingol
Hurghada *see* Al Ghardaqah
Huron *23 E3* South Dakota, N USA
Huron, Lake *18 D2 lake* Canada/USA
Hurukawa *see* Furukawa
Hurunui *129 C5 river* South Island, New Zealand
Húsavík *61 E4* Nordhurland Eystra, NE Iceland
Husté *see* Khust
Husum *72 B2* Schleswig-Holstein, N Germany
Huszt *see* Khust
Hutchinson *23 E5* Kansas, C USA
Hutchinson Island *21 F4 island* Florida, SE USA
Huy *65 C6 Dut.* Hoei, Hoey. Liège, E Belgium
Huzi *see* Fuji
Hvannadalshnúkur *61 E5 volcano* S Iceland
Hvar *78 B4 It.* Lesina; *anc.* Pharus. *island* S Croatia
Hwainan *see* Huainan
Hwange *56 D3 prev.* Wankie. Matabeleland North, W Zimbabwe
Hwang-Hae *see* Yellow Sea
Hwangshih *see* Huangshi
Hyargas Nuur *104 C2 lake* NW Mongolia
Hyderābād *112 D5 var.* Haidarabad. *state capital* Telangana, C India
Hyderabad *112 B3 var.* Haidarabad. Sindh, SE Pakistan
Hyères *69 D6* Var, SE France
Hyères, Îles d' *69 D6 island group* S France
Hypanis *see* Kuban'
Hyrcania *see* Gorgān
Hyvinge *see* Hyvinkää
Hyvinkää *63 D5 Swe.* Hyvinge. Uusimaa, S Finland

I

Iader *see* Zadar
Ialomiţa *86 C5 river* SE Romania
Iaşi *86 D3 Ger.* Jassy. Iaşi, NE Romania
Ibadan *53 F5* Oyo, SW Nigeria
Ibagué *36 B3* Tolima, C Colombia
Ibar *78 D4 Alb.* Ibër. *river* C Serbia
Ibarra *38 B1 var.* San Miguel de Ibarra. Imbabura, N Ecuador
Ibër *see* Ibar
Iberia *see* Spain
Iberian Mountains *see* Ibérico, Sistema
Iberian Peninsula *58 B4 physical region* Portugal/Spain
Iberian Plain *58 B4 abyssal plain* E Atlantic Ocean
Ibérica, Cordillera *see* Ibérico, Sistema
Ibérico, Sistema *71 E2 var.* Cordillera Ibérica, *Eng.* Iberian Mountains. *mountain range* NE Spain
Ibiza *see* Eivissa
Ibo *see* Sassandra
Ica *38 D4* Ica, SW Peru
Icaria *see* Ikaría
Içá, Rio *see* Putumayo, Río
İçel *see* Mersin
Iceland *61 E4 off.* Republic of Iceland, *Dan.* Island, *Icel.* Ísland. *country* N Atlantic Ocean
Iceland Basin *58 B1 undersea basin* N Atlantic Ocean
Icelandic Plateau *see* Iceland Plateau
Iceland Plateau *133 B6 var.* Icelandic Plateau. *undersea plateau* S Greenland Sea
Iceland, Republic of *see* Iceland
Iconium *see* Konya
Iculisma *see* Angoulême
Idabel *27 H2* Oklahoma, C USA
Idaho *24 D3 off.* State of Idaho, *also known as* Gem of the Mountains, Gem State. *state* NW USA
Idaho Falls *24 E3* Idaho, NW USA
Idensalmi *see* Iisalmi
Idfū *50 B2 var.* Edfu. SE Egypt
Idi Amin, Lac *see* Edward, Lake
Idîni *52 B2* Trarza, W Mauritania
Idlib *96 B3* Idlib, NW Syria
Idre *63 B5* Dalarna, C Sweden
Iecava *84 C3* Bauska, S Latvia
Ieper *65 A6 Fr.* Ypres. West-Vlaanderen, W Belgium
Ierápetra *83 D8* Kríti, Greece, E Mediterranean Sea
Ierisós *see* Ierissós
Ierissós *82 C4 var.* Ierisós. Kentrikí Makedonía, N Greece
Iertra *see* Eritrea
Iferouâne *53 G2* Agadez, N Niger
Ifôghas, Adrar des *53 E2 var.* Adrar des Iforas. *mountain range* NE Mali
Iforas, Adrar des *see* Ifôghas, Adrar des
Igarka *92 D3* Krasnoyarskiy Kray, N Russia
Igaunija *see* Estonia
Iglau/Iglawa/Igława *see* Jihlava
Iglesias *75 A5* Sardegna, Italy, C Mediterranean Sea
Igloolik *15 G2* Nunavut, N Canada
Igoumenítsa *82 A4* Ípeiros, W Greece
Iguaçu, Rio *41 E4 Sp.* Río Iguazú. *river* Argentina/ Brazil
Iguaçu, Saltos do *41 E4 Sp.* Cataratas del Iguazú; *prev.* Victoria Falls. *waterfall* Argentina/Brazil
Iguala *29 E4 var.* Iguala de la Independencia. Guerrero, S Mexico
Iguala de la Independencia *see* Iguala
Iguazú, Cataratas del *see* Iguaçu, Saltos do
Iguazú, Río *see* Iguaçu, Rio
Iguid, Erg *see* Iguîdi, 'Erg
Iguîdi, 'Erg *48 C3 var.* Erg Iguid. *desert* Algeria/ Mauritania
Ihavandhippolhu Atoll *110 A3 var.* Ihavandiffulu Atoll. *atoll* N Maldives
Ihavandiffulu Atoll *see* Ihavandhippolhu Atoll
Ihosy *57 F4* Fianarantsoa, S Madagascar
Ihupuku, Motu *see* Campbell Island
Iinnasuolu *see* Hinnøya
Iisalmi *62 E4 var.* Idensalmi. Pohjois-Savo, C Finland
IJmuiden *64 C3* Noord-Holland, W Netherlands
IJssel *64 D3 var.* Yssel. *river* Netherlands
IJsselmeer *64 C2 prev.* Zuider Zee. *lake* N Netherlands
IJsselmuiden *64 D3* Overijssel, E Netherlands
Ijzer *65 A6 river* W Belgium
Ikaahuk *see* Sachs Harbour
Ikaluktutiak *see* Cambridge Bay
Ikaría *83 D6 var.* Kariot, Nicaria, Nikaria; *anc.* Icaria. *island* Dodekánisa, Greece, Aegean Sea
Ikela *55 D6* Equateur, C Dem. Rep. Congo
Iki *109 A7 island* SW Japan
Ilagan *117 E1* Luzon, N Philippines
Ilave *39 E4* Puno, S Peru
Iława *76 D3 Ger.* Deutsch-Eylau. Warmińsko-Mazurskie, NE Poland
Ilebo *55 C6 prev.* Port-Francqui. Kasai-Occidental, W Dem. Rep. Congo
Île-de-France *68 C3 cultural region* N France
Ilemi Triangle *51 B5 disputed region* Kenya/ South Sudan
Ilerda *see* Lleida
Ilfracombe *67 C7* SW England, United Kingdom
Ílhavo *70 B2* Aveiro, N Portugal
Iliamna Lake *14 C3 lake* Alaska, USA
Ilici *see* Elx
Iligan *117 E2 off.* Iligan City. Mindanao, S Philippines
Iligan City *see* Iligan
Illapel *42 B4* Coquimbo, C Chile
Illichivsk *see* Chornomorsk
Illicis *see* Elx
Illinois *18 A4 off.* State of Illinois, *also known as* Prairie State, Sucker State. *state* C USA
Illinois River *18 B4 river* Illinois, N USA
Illurco *see* Lorca
Illuro *see* Mataró
Ilo *39 E4* Moquegua, SW Peru
Iloilo *117 E2 off.* Iloilo City. Panay Island, C Philippines
Iloilo City *see* Iloilo
Ilorin *53 F4* Kwara, W Nigeria
Ilovlya *89 B6* Volgogradskaya Oblast', SW Russia
Iluh *see* Batman
Il'yaly *see* Gurbansoltan Eje
Imatra *63 E5* Etelä-Kariala, SE Finland
Imbros *see* Gökçeada
Imishli *see* İmişli
İmişli *95 H3 Rus.* Imishli. C Azerbaijan
Imola *74 C3* Emilia-Romagna, N Italy
Imperatriz *41 F2* Maranhão, NE Brazil
Imperia *74 A3* Liguria, NW Italy
Impfondo *55 C5* Likouala, NE Congo
Imphāl *113 H3 state capital* Manipur, NE India
Imroz Adası *see* Gökçeada
Inagua Islands *see* Little Inagua
Inagua Islands *see* Great Inagua
Inarijärvi *62 D2 Lapp.* Aanaarjävri, *Swe.* Enareträsk. *lake* N Finland
Inău *see* Ineu
Inawashiro-ko *109 D5 var.* Inawasiro Ko. *lake* Honshū, C Japan
Inawasiro Ko *see* Inawashiro-ko
İncesu *94 D3* Kayseri, Turkey (Türkiye)
Incheon *107 E4 off.* Incheon Gwang-yeoksi, *prev.* Inch'ŏn, *Jap.* Jinsen; *prev.* Chemulpo. NW South Korea
Incheon-Gwang-yeoksi *see* Incheon
Inch'ŏn *see* Incheon
Incudine, Monte *69 E7 mountain* Corse, France, C Mediterranean Sea
Indefatigable Island *see* Santa Cruz, Isla
Independence *23 F4* Missouri, C USA
Independence Fjord *61 E1 fjord* N Greenland
Independence Island *see* Malden Island
Independence Mountains *24 C4 mountain range* Nevada, W USA
India *102 B3 off.* Republic of India, *var.* Indian Union, Union of India, *Hind.* Bhārat. *country* S Asia
India *see* Inđija
Indiana *18 B4 off.* State of Indiana, *also known as* Hoosier State. *state* N USA
Indianapolis *18 C4 state capital* Indiana, N USA
Indian Church *30 C1* Orange Walk, N Belize
Indian Desert *see* Thar Desert
Indianola *23 F4* Iowa, C USA
Indian Union *see* India
India, Republic of *see* India
India, Union of *see* India
Indigirka *93 F2 river* NE Russia
Inđija *78 D3 Hung.* India; *prev.* Indjija. Vojvodina, N Serbia
Indira Point *110 G3 headland* Andaman and Nicobar Island, India, NE Indian Ocean
Indjija *see* Inđija
Indomed Fracture Zone *119 B6 tectonic feature* SW Indian Ocean
Indonesia *116 B4 off.* Republic of Indonesia, *Ind.* Republik Indonesia; *prev.* Dutch East Indies, Netherlands East Indies, United States of Indonesia. *country* SE Asia
Indonesian Borneo *see* Kalimantan
Indonesia, Republic of *see* Indonesia
Indonesia, Republik *see* Indonesia
Indonesia, United States of *see* Indonesia
Indore *112 D4* Madhya Pradesh, C India
Indreville *see* Châteauroux
Indus *112 C2 Chin.* Yindu He; *prev.* Yin-tu Ho. *river* S Asia
Indus Cone *see* Indus Fan
Indus Fan *90 C5 var.* Indus Cone. *undersea fan* N Arabian Sea
Indus, Mouths of the *112 B4 delta* S Pakistan
İnebolu *94 C2* Kastamonu, N Turkey (Türkiye)
Ineu *86 A4 Hung.* Borosjenő; *prev.* Inău. Arad, W Romania
Infiernillo, Presa del *29 E4 reservoir* S Mexico
Inglewood *24 D2* California, W USA
Ingolstadt *73 C6* Bayern, S Germany
Ingulets *see* Inhulets
Inguri *see* Enguri
Inhambane *57 E4* Inhambane, SE Mozambique
Inhulets *87 F3 Rus.* Ingulets. Dnipropetrovska Oblast, E Ukraine
I-ning *see* Yining
Inis *see* Ennis
Inis Ceithleann *see* Enniskillen
Inn *73 C6 river* C Europe
Innaanganeq *60 C1 var.* Kap York. *headland* NW Greenland
Inner Hebrides *66 B4 island group* W Scotland, United Kingdom
Inner Islands *57 H1 var.* Central Group. *island group* NE Seychelles
Innisfail *126 D3* Queensland, NE Australia
Inniskilling *see* Enniskillen
Innsbruch *see* Innsbruck
Innsbruck *73 C7 var.* Innsbruch. Tirol, W Austria
Inoucdjouac *see* Inukjuak
Inowrazlaw *see* Inowrocław
Inowrocław *76 C3 Ger.* Hohensalza; *prev.* Inowrazlaw. Kujawski-pomorskie, C Poland
I-n-Sakane, 'Erg *53 E2 desert* N Mali
I-n-Salah *48 D3 var.* In Salah. C Algeria
Insterburg *see* Chernyakhovsk
Insula *see* Lille
Inta *88 E3* Respublika Komi, NW Russia
Interamna *see* Teramo
Interamna Nahars *see* Terni
International Falls *23 F1* Minnesota, N USA
Inukjuak *16 D2 var.* Inoucdjouac; *prev.* Port Harrison. Québec, NE Canada
Inuuvik *see* Inuvik
Inuvik *14 D3 var.* Inuuvik. Northwest Territories, NW Canada
Invercargill *129 A7* Southland, South Island, New Zealand
Inverness *66 C3* N Scotland, United Kingdom
Investigator Ridge *119 D5 undersea ridge* E Indian Ocean
Investigator Strait *127 B7 strait* South Australia
Inyangani *56 D3 mountain* NE Zimbabwe
Ioánnina *82 A4 var.* Janina, Yannina. Ípeiros, W Greece
Iola *23 F5* Kansas, C USA
Ionia Basin *see* Ionian Basin
Ionian Basin *58 D5 var.* Ionia Basin. *undersea basin* Ionian Sea, C Mediterranean Sea
Ionian Islands *see* Iónia Nisiá
Ionian Sea *81 E3 Gk.* Iónio Pélagos, *It.* Mar Ionio. *sea* C Mediterranean Sea
Iónia Nisiá *83 A5 Eng.* Ionian Islands. *island group* W Greece
Ionio, Mar/Iónio Pélagos *see* Ionian Sea
Íos *83 D6 var.* Nio. *island* Kykládes, Greece, Aegean Sea
Ioulída *83 C6 prev.* Ioulís, Kykládes, Greece, Aegean Sea
Iowa *23 F3 off.* State of Iowa, *also known as* Hawkeye State. *state* C USA
Iowa City *23 G3* Iowa, C USA
Iowa Falls *23 G3* Iowa, C USA
Ipel' *77 C6 var.* Ipoly, *Ger.* Eipel. *river* Hungary/ Slovakia
Ipiales *36 A4* Nariño, SW Colombia
Ipoh *116 B3* Perak, Peninsular Malaysia
Ipoly *see* Ipel'
Ippy *54 C4* Ouaka, C Central African Republic
Ipswich *127 E5* Queensland, E Australia
Ipswich *67 E6 hist.* Gipeswic. E England, United Kingdom
Iqaluit *15 H3 prev.* Frobisher Bay. *province capital* Baffin Island, Nunavut, NE Canada
Iquique *42 B1* Tarapacá, N Chile
Iquitos *38 C1* Loreto, N Peru
Irákleio *83 D7 var.* Herakleion, *Eng.* Candia; *prev.* Iráklion. Kríti, Greece, E Mediterranean Sea
Iráklion *see* Irákleio
Iran *98 C3 off.* Islamic Republic of Iran; *prev.* Persia. *country* SW Asia
Iranian Plateau *98 D3 var.* Plateau of Iran. *plateau* N Iran
Iran, Islamic Republic of *see* Iran
Iran, Plateau of *see* Iranian Plateau
Irapuato *29 E4* Guanajuato, C Mexico
Iraq *98 B3 off.* Republic of Iraq, *Ar.* 'Irāq. *country* SW Asia
'Irāq *see* Iraq
Iraq, Republic of *see* Iraq
Irbid *97 B5* Irbid, N Jordan
Irbīl *see* Arbīl
Ireland *67 A5 off.* Ireland, *Ir.* Éire. *country* NW Europe
Ireland *58 C3 Lat.* Hibernia. *island* Ireland/ United Kingdom
Irian *see* New Guinea
Irian Barat *see* Papua
Irian Jaya *see* Papua
Irian, Teluk *see* Cenderawasih, Teluk
Iringa *51 C7* Iringa, C Tanzania
Iriomote-jima *108 A4 island* Sakishima-shotō, SW Japan
Iriona *30 D2* Colón, NE Honduras
Irish Sea *67 C5 Ir.* Muir Éireann. *sea* C British Isles
Irkutsk *93 E4* Irkutskaya Oblast', S Russia
Irminger Basin *see* Reykjanes Basin
Iroise *68 A3 sea* NW France
Iron Mountain *18 B2* Michigan, N USA
Ironwood *18 B1* Michigan, N USA
Irrawaddy *see* Ayeyarwady
Irrawaddy, Mouths of the *115 A5 delta* SW Myanmar (Burma)
Irtish *see* Irtysh
Irtysh *92 C4 var.* Irtish, *Kaz.* Ertis. *river* C Asia
Irun *71 E1 Cast.* Irún. País Vasco, N Spain
Irún *see* Irun
Iruña *see* Pamplona
Isabela, Isla *38 A5 var.* Albemarle Island. *island* Galápagos Islands, Ecuador, E Pacific Ocean
Isaccea *86 D4* Tulcea, E Romania
Isachsen *15 F1* Ellef Ringnes Island, Nunavut, N Canada
Ísafjördur *61 E4* Vestfirdhir, NW Iceland
Isbarta *see* Isparta
Isca Damnoniorum *see* Exeter
Ise *109 C6* Mie, Honshū, SW Japan
Iseghem *see* Izegem
Isère *69 D5 river* E France
Isernia *75 D5 var.* Æsernia. Molise, C Italy
Ise-wan *109 C6 bay* S Japan
Isfahan *see* Eşfahān
Isha Baydhabo *see* Baydhabo
Ishigaki-jima *108 A4 island* Sakishima-shotō, SW Japan
Ishikari-wan *108 C2 bay* Hokkaidō, NE Japan
Ishim *92 C4* Tyumenskaya Oblast', C Russia
Ishim *92 C4 Kaz.* Esil. *river* Kazakhstan/Russia
Ishinomaki *108 D4 var.* Isinomaki. Miyagi, Honshū, C Japan
Ishkashim *see* Ishkoshim
Ishkoshim *101 F3 Rus.* Ishkashim. S Tajikistan
Isinomaki *see* Ishinomaki
Isiro *55 E5* Orientale, NE Dem. Rep. Congo
Iskar *82 C2 var.* Iskŭr. *river* NW Bulgarias
İskenderun *94 D4 Eng.* Alexandretta. Hatay, S Turkey (Türkiye)
İskenderun Körfezi *96 A2 Eng.* Gulf of Alexandretta. *gulf* S Turkey (Türkiye)
Iskŭr *see* Iskar
Iskar, Yazovir *82 B2 prev.* Yazovir Stalin. *reservoir* W Bulgaria
Isla Cristina *70 C4* Andalucía, S Spain
Isla de León *see* San Fernando
Islamabad *112 C1 country capital* (Pakistan) Federal Capital Territory Islamabad, NE Pakistan
Island/Ísland *see* Iceland
Islay *66 B4 island* SW Scotland, United Kingdom
Isle *69 B5 river* W France
Isle of Man *67 B5 British Crown Dependency* NW Europe
Isles of Scilly *67 B8 island group* SW England, United Kingdom
Ismailia *see* Al Ismā'īlīya
Ismâ'ilîya *see* Al Ismā'īlīya
Ismid *see* İzmit
Isnā *50 B2 var.* Esna. SE Egypt
Isoka *56 D1* Northern, NE Zambia
İsparta *94 B4 var.* Isbarta. Isparta, SW Turkey (Türkiye)
İspir *95 E3* Erzurum, NE Turkey (Türkiye)
Israel *97 A7 off.* State of Israel, *var.* Medinat Israel, *Heb.* Yisrael, Yisra'el. *country* SW Asia
Israel, State of *see* Israel
Issa *see* Vis
Issiq Köl *see* Issyk-Kul', Ozero
Issoire *69 C5* Puy-de-Dôme, C France
Issyk-Kul' *see* Balykchy
Issyk-Kul', Ozero *101 G2 var.* Issiq Köl, *Kir.* Ysyk-Köl. *lake* E Kyrgyzstan
İstanbul *94 B2 Bul.* Tsarigrad, *Eng.* Istanbul, *prev.* Constantinople; *anc.* Byzantium. Istanbul, NW Turkey (Türkiye)
İstanbul Boğazi *94 B2 var.* Bosporus Thracius, *Eng.* Bosphorus, Bosporus, *Turk.* Karadeniz Boğazi. *strait* NW Turkey (Türkiye)
Istaravshan *101 E2 prev.* Ŭroteppa, *Rus.* Ura-Tyube. NW Tajikistan
Istarska Županija *see* Istra
Istra *78 A3 off.* Istarska Županija. *province* NW Croatia
Istra *78 A3 Eng.* Istria, *Ger.* Istrien. *cultural region* NW Croatia
Istria/Istrien *see* Istra
Itabuna *41 G3* Bahia, E Brazil
Itagüí *36 B3* Antioquia, W Colombia
Itaipú, Represa de *41 E4 reservoir* Brazil/Paraguay
Itaituba *41 E2* Pará, NE Brazil
Italia/Italiana, Republica/Italian Republic, The *see* Italy
Italian Somaliland *see* Somalia
Italy *74 C3 off.* The Italian Republic, *It.* Italia, Repubblica Italiana. *country* S Europe
Iténez, Río *see* Guaporé, Rio
Ithaca *19 E3* New York, NE USA
It Hearrenfean *see* Heerenveen
Itoigawa *109 C5* Niigata, Honshū, C Japan
Itseqqortoormiit *see* Ittoqqortoormiit
Ittoqqortoormiit *61 E3 var.* Itseqqortoormiit, *Dan.* Scoresbysund, *Eng.* Scoresby Sound. Tunu, C Greenland
Iturup, Ostrov *108 E1 island* Kuril'skiye Ostrova, SE Russia
Itzehoe *72 B2* Schleswig-Holstein, N Germany
Ivalo *62 D2 Lapp.* Avveel, Avvil. Lappi, N Finland
Ivanava *85 B7 Pol.* Janów, Janów Poleski, *Rus.* Ivanovo. Brestskaya Voblasts', SW Belarus
Ivangrad *see* Berane
Ivanhoe *127 C6* New South Wales, SE Australia
Ivano-Frankivsk *86 C2 Ger.* Stanislau, *Pol.* Stanisławów, *Rus.* Ivano-Frankovsk; *prev.* Stanislav. Ivano-Frankivs'ka Oblast', W Ukraine
Ivano-Frankovsk *see* Ivano-Frankivsk
Ivanovo *89 B5* Ivanovskaya Oblast', W Russia
Ivanovo *see* Ivanava
Ivantsevichi/Ivatsevichi *see* Ivatsevichy
Ivatsevichy *85 B6 Pol.* Iwacewicze, *Rus.* Ivantsevichi, Ivatsevichi. Brestskaya Voblasts', SW Belarus
Ivigtut *see* Ivittuut
Ivittuut *60 B4 var.* Ivigtut. Sermersooq, S Greenland
Iviza *see* Eivissa\Ibiza
Ivory Coast *52 D4 off.* République de la Côte d'Ivoire. *country* W Africa
Ivujivik *16 D1* Québec, NE Canada
Iwacewicze *see* Ivatsevichy
Iwaki *109 D5* Fukushima, Honshū, N Japan
Iwakuni *109 B7* Yamaguchi, Honshū, SW Japan
Iwanai *108 C2* Hokkaidō, NE Japan
Iwate *108 D3* Iwate, Honshū, N Japan
Ixtapa *29 E5* Guerrero, S Mexico
Ixtepec *29 F5* Oaxaca, SE Mexico
Iyo-nada *109 B7 sea* S Japan
Izabal, Lago de *30 B2 prev.* Golfo Dulce. *lake* E Guatemala
Izad Khvāst *98 D3* Fārs, C Iran
Iʿzāz *96 B2 var.* A'zāz. Ḥalab, NW Syria

Izegem *65 A6 prev.* Iseghem. West-Vlaanderen, W Belgium
Izhevsk *89 D5 prev.* Ustinov. Udmurtskaya Respublika, NW Russia
Iziaslav *86 C2 prev.* Izyaslav. Khmel'nyts'ka Oblast', W Ukraine
Izium *87 G2 prev.* Izyum. Kharkivs'ka Oblast', E Ukraine
Izmail *see* Izmayil
Izmayil *86 D4 Rus.* Izmail. Odes'ka Oblast', SW Ukraine
İzmir *94 A3 prev.* Smyrna. İzmir, W Turkey (Türkiye)
İzmit *94 B2 var.* Ismid; *anc.* Astacus. Kocaeli, NW Turkey (Türkiye)
İznik Gölü *94 B3 lake* NW Turkey (Türkiye)
Izu-hanto *109 D6 peninsula* Honshū, S Japan
Izu Shichito *see* Izu-shotō
Izu-shoto *109 D6 var.* Izu Shichito. *island group* S Japan
Izvor *82 B2* Pernik, W Bulgaria
Izyaslav *see* Iziaslav
Izyum *see* Izium

J

Jabal ash Shifa *98 A4 desert* NW Saudi Arabia
Jabalpur *113 E4 prev.* Jubbulpore. Madhya Pradesh, C India
Jabbūl, Sabkhat al *96 B2 sabkha* NW Syria
Jablah *96 A3 var.* Jeble, *Fr.* Djéblé. Al Lādhiqīyah, W Syria
Jaca *71 F1* Aragón, NE Spain
Jacaltenango *30 A2* Huehuetenango, W Guatemala
Jackson *20 B2 state capital* Mississippi, S USA
Jackson *23 H5* Missouri, C USA
Jackson *20 C1* Tennessee, S USA
Jackson Head *129 A6 headland* South Island, New Zealand
Jacksonville *21 E3* Florida, SE USA
Jacksonville *18 B4* Illinois, N USA
Jacksonville *21 F1* North Carolina, SE USA
Jacksonville *27 G3* Texas, SW USA
Jacmel *32 D3 var.* Jaquemel. S Haiti
Jacob *see* Nkayi
Jacobabad *112 B3* Sindh, SE Pakistan
Jadotville *see* Likasi
Jadransko More/Jadransko Morje *see* Adriatic Sea
Jaén *38 B2* Cajamarca, N Peru
Jaén *70 D4* Andalucía, SW Spain
Jaffna *110 D3* Northern Province, N Sri Lanka
Jagannath *see* Puri
Jagdalpur *113 E5* Chhattīsgarh, C India
Jagdaqi *105 G1* Nei Mongol Zizhiqu, N China
Jagodina *78 D4 prev.* Svetozarevo. Serbia, C Serbia
Jahra *see* Al Jahrā'
Jailolo *see* Halmahera, Pulau
Jaipur *112 D3 prev.* Jeypore. *state capital* Rājasthān, N India
Jaisalmer *112 C3* Rājasthān, NW India
Jajce *78 B3* Federacija Bosna I Hercegovina, W Bosnia and Herzegovina
Jakarta *116 C5 prev.* Djakarta, *Dut.* Batavia. *country capital* (Indonesia) Jawa, C Indonesia
Jakobstad *62 D4 Fin.* Pietarsaari. Österbotten, W Finland
Jakobstadt *see* Jēkabpils
Jalālābād *101 F4 var.* Jalalabad, Jelalabad. Nangarhār, E Afghanistan
Jalal-Abad *see* Dzhalal-Abad, Dzhalal-Abadskaya Oblast', Kyrgyzstan
Jalandhar *112 D2 prev.* Jullundur. Punjab, N India
Jalapa *30 D3* Nueva Segovia, NW Nicaragua
Jalpa *28 D4* Zacatecas, C Mexico
Jālū *49 G3 var.* Jūlā. NE Libya
Jaluit Atoll *122 D2 var.* Jālwōj. *atoll* Ralik Chain, S Marshall Islands
Jālwōj *see* Jaluit Atoll
Jamaame *51 D6 It.* Giamame; *prev.* Margherita. Jubbada Hoose, S Somalia
Jamaica *32 A4 country* W West Indies
Jamaica *34 A1 island* W West Indies
Jamaica Channel *32 D3 channel* Haiti/Jamaica
Jamālpur *113 F3* Bihār, NE India
Jambi *116 B4 var.* Telanaipura; *prev.* Djambi. Sumatera, W Indonesia
Jamdena *see* Yamdena, Pulau
James Bay *16 C3 bay* Ontario/Québec, E Canada
James River *23 E2 river* North Dakota/South Dakota, N USA
James River *19 E5 river* Virginia, NE USA
Jamestown *19 E3* New York, NE USA
Jamestown *23 E2* North Dakota, N USA
Jamestown *see* Holetown
Jammu *112 D2 prev.* Jummoo. *state capital* Jammu and Kashmir, NW India
Jammu and Kashmir *112 D1 disputed region* India/Pakistan
Jāmnagar *112 C4 prev.* Navanagar. Gujarāt, W India
Jamshedpur *113 F4* Jhārkhand, NE India
Jamuna *see* Brahmaputra
Janaúba *41 F3* Minas Gerais, SE Brazil
Janesville *18 B3* Wisconsin, N USA
Jang Bogo *132 C4 South Korean research station* Antarctica
Janina *see* Ioánnina
Janīn *97 E6* var. Jenin. N West Bank, Middle East
Janischken *see* Joniškis
Jankovac *see* Jánoshalma
Jan Mayen *61 F4 constituent part of Norway. island* N Atlantic Ocean
Jánoshalma *77 C7 Croatian* Jankovac. Bács-Kiskun, S Hungary
Janów *see* Ivanava, Belarus
Janow/Janów *see* Jonava, Lithuania
Janów Poleski *see* Ivanava
Japan *108 C4 var.* Nippon, *Jap.* Nihon. *country* E Asia
Japan, Sea of *108 A4 var.* East Sea, *Rus.* Yapanskoye More. *sea* NW Pacific Ocean
Japan Trench *103 F1 trench* NW Pacific Ocean
Japen *see* Yapen, Pulau
Japiim *40 C2 var.* Máncio Lima. Acre, W Brazil
Japurá, Rio *40 C2 var.* Río Caquetá, Yapurá. *river* Brazil/Colombia
Japurá, Rio *see* Caquetá, Río
Jaqué *31 G5* Darién, SE Panama
Jaquemel *see* Jacmel
Jarablos *see* Jarābulus
Jarābulus *96 C2 var.* Jarablos, Jerablus, *Fr.* Djérablous. Ḩalab, N Syria
Jarbah, Jazīrat *see* Djerba, Île de
Jardines de la Reina, Archipiélago de los *32 B2 island group* C Cuba
Jarīd, Shaṭṭ al *see* Jerid, Chott el
Jarocin *76 C4* Wielkopolskie, C Poland
Jaroslau *see* Jarosław
Jarosław *77 E5 Ger.* Jaroslau, *Rus.* Yaroslav. Podkarpackie, SE Poland
Jarqo'rg'on *101 E3 Rus.* Dzharkurgan. Surkhondaryo Viloyati, S Uzbekistan
Jarvis Island *123 G2 US unincorporated territory* C Pacific Ocean
Jashore *113 G4 prev.* Jessore. Khulna, W Bangladesh
Jasło *77 D5* Podkarpackie, SE Poland
Jastrzębie-Zdrój *77 C5* Śląskie, S Poland
Jataí *41 E3* Goiás, C Brazil
Jativa *see* Xàtiva
Jauf *see* Al Jawf
Jaunpiebalga *84 D3* Gulbene, NE Latvia
Jaunpur *113 E3* Uttar Pradesh, N India
Java *130 A3* South Dakota, N USA
Javalambre *71 E3 mountain* E Spain
Java Sea *116 D4 Ind.* Laut Jawa. *sea* W Indonesia
Java Trench *102 D5 var.* Sunda Trench. *trench* E Indian Ocean
Jawa, Laut *see* Java Sea
Jawhar *51 D6 var.* Jowhar, *It.* Giohar. Shabeellaha Dhexe, S Somalia
Jaworów *see* Yavoriv
Jaya, Puncak *117 G4 prev.* Puntjak Carstensz, Puntjak Sukarno. *mountain* Papua, E Indonesia
Jayapura *117 H4 var.* Djajapura, *Dut.* Hollandia; *prev.* Kotabaru, Sukarnapura. Papua, E Indonesia
Jay Dairen *see* Dalian
Jayhawker State *see* Kansas
Jaz Murian, Hamun-e *98 E4 lake* SE Iran
Jebba *53 F4* Kwara, W Nigeria
Jebel, Bahr el *see* White Nile
Jeble *see* Jablah
Jeddah *see* Jiddah
Jędrzejów *76 D4 Ger.* Endersdorf. Świętokrzyskie, C Poland
Jefferson City *23 G5 state capital* Missouri, C USA
Jega *53 F4* Kebbi, NW Nigeria
Jehol *see* Chengde
Jeju-do *107 E4 Jap.* Saishū; *prev.* Cheju-do, Quelpart. *island* S South Korea
Jeju Strait *107 E4 var.* Jeju-haehyŏp; *prev.* Cheju-Strait. *strait* S South Korea
Jeju-haehyŏp *see* Jeju Strait
Jēkabpils *84 D4 Ger.* Jakobstadt. Jēkabpils, S Latvia
Jelalabad *see* Jalālābād
Jelenia Góra *76 B4 Ger.* Hirschberg, Hirschberg im Riesengebirge, Hirschberg in Riesengebirge, Hirschberg in Schlesien. Dolnośląskie, SW Poland
Jelgava *84 C3 Ger.* Mitau. Jelgava, C Latvia
Jelondi *101 F3 prev.* Dzhelandy. SE Tajikistan
Jemappes *65 B6* Hainaut, S Belgium
Jember *116 D5 prev.* Djember. Jawa, C Indonesia
Jena *72 C4* Thüringen, C Germany
Jengish Chokusu *see* Tömür Feng
Jenin *see* Janīn
Jerablus *see* Jarābulus
Jerada *48 D2* NE Morocco
Jérémie *32 D3* SW Haiti
Jerez *see* Jeréz de la Frontera, Spain
Jerez de la Frontera *70 C5 var.* Jerez; *prev.* Xeres. Andalucía, SW Spain
Jerez de los Caballeros *70 C4* Extremadura, W Spain
Jericho *97 E7 var.* Arīḩā, Yeriẖo. C West Bank, Middle East
Jerid, Chott el *49 E2 var.* Shaṭṭ al Jarīd. *salt lake* SW Tunisia
Jersey *67 D8 British Crown Dependency* Channel Islands, NW Europe
Jerusalem *81 H4 Ar.* Al Quds, Al Quds ash Sharīf, *Heb.* Yerushalayim; *anc.* Hierosolyma. *country capital* (Israel - not internationally recognized) Jerusalem, NE Israel
Jesenice *73 D7 Ger.* Assling. NW Slovenia
Jesselton *see* Kota Kinabalu
Jessore *see* Jashore
Jesús María *42 C3* Córdoba, C Argentina
Jeypore *see* Jaipur, Rājasthān, India
Jhānsi *112 D3* Uttar Pradesh, N India
Jhārkhand *113 F4 cultural region* NE India
Jhelum *112 C2* Punjab, NE Pakistan
Ji *see* Hebei, China
Ji *see* Jilin, China
Jiangmen *106 C6* Guangdong, S China
Jiangsu *106 D4 var.* Chiang-su, Jiangsu Sheng, Kiangsu, Su. *province* E China
Jiangsu *see* Nanjing
Jiangsu Sheng *see* Jiangsu
Jiangxi *106 C6 var.* Chiang-hsi, Gan, Jiangxi Sheng, Kiangsi. *province* S China
Jiangxi Sheng *see* Jiangxi
Jiaxing *106 D5* Zhejiang, SE China
Jiayi *see* Chiayi
Jibuti *see* Djibouti
Jiddah *99 A5 Eng.* Jeddah. Makkah, W Saudi Arabia
Jih-k'a-tse *see* Xigazê
Jihlava *77 B5 Ger.* Iglau, *Pol.* Igława. Vysocina, S Czechia (Czech Republic)
Jilib *51 D6 It.* Gelib. Jubbada Dhexe, S Somalia
Jilin *106 E3 var.* Chi-lin, Girin, Kirin; *prev.* Yungki, Yunki. Jilin, NE China
Jilin *106 E3 var.* Chi-lin, Girin, Ji, Jilin Sheng, Kirin. *province* NE China
Jilin Sheng *see* Jilin
Jilong *see* Keelung
Jīma *51 C5 var.* Jimma, *It.* Gimma. Oromīya, C Ethiopia
Jimbolia *86 A4 Ger.* Hatzfeld, *Hung.* Zsombolya. Timiş, W Romania
Jiménez *28 D2* Chihuahua, N Mexico
Jimma *see* Jīma
Jimsar *104 C3* Xinjiang Uygur Zizhiqu, NW China
Jin *see* Shanxi
Jin *see* Tianjin Shi
Jinan *106 C4 var.* Chinan, Chi-nan, Tsinan. *province capital* Shandong, E China
Jingdezhen *106 C5* Jiangxi, S China
Jinghong *106 A6 var.* Yunjinghong. Yunnan, SW China
Jinhua *106 D5* Zhejiang, SE China
Jining *105 F3* Shandong, E China
Jinja *51 C6* S Uganda
Jinotega *30 D3* Jinotega, NW Nicaragua
Jinotepe *30 D3* Carazo, SW Nicaragua
Jinsen *see* Incheon
Jinzhong *106 C4 var.* Yuci. Shanxi, C China
Jinzhou *106 D3 var.* Chin-chou, Chinchow; *prev.* Chinhsien. Liaoning, NE China
Jirgalanta *see* Hovd
Jisr ash Shadadi *see* Ash Shadādah
Jiu *86 B5 Ger.* Schil, Schyl, *Hung.* Zsil, Zsily. *river* S Romania
Jiujiang *106 C5* Jiangxi, S China
Jixi *107 E2* Heilongjiang, NE China
Jizan *see* Jīzān
Jīzān *99 B6 var.* Qīzān, Jizan. Jīzān, SW Saudi Arabia
Jizzax *101 E2 Rus.* Dzhizak. Jizzax Viloyati, C Uzbekistan
João Belo *see* Xai-Xai
João Pessoa *41 G2 prev.* Paraíba. *state capital* Paraíba, E Brazil
Joazeiro *see* Juazeiro
Job'urg *see* Johannesburg
Jo-ch'iang *see* Ruoqiang
Jodhpur *112 C3* Rājasthān, NW India
Joensuu *63 E5* Pohjois-Karjala, SE Finland
Jōetsu *109 C5 var.* Zyôetu. Niigata, Honshū, C Japan
Jogjakarta *see* Yogyakarta
Johannesburg *56 D4 var.* Egoli, Erautini, Gauteng, *abbrev.* Job'urg. Gauteng, NE South Africa
Johannisburg *see* Pisz
John Day River *24 C3 river* Oregon, NW USA
John o'Groats *66 C2* N Scotland, United Kingdom
Johnston Atoll *121 E1 US unincorporated territory* C Pacific Ocean
Johor Baharu *see* Johor Bahru
Johor Bahru *116 B3 var.* Johor Baharu, Johore Bahru. Johor, Peninsular Malaysia
Johore Bahru *see* Johor Bahru
Johore Strait *116 A1 strait* Johor, Peninsular Malaysia, Malaysia/Singapore Asia Andaman Sea/ South China Sea
Joinvile *see* Joinville
Joinville *41 E4 var.* Joinvile. Santa Catarina, S Brazil
Jokkmokk *62 C3 Lapp.* Dálvvadis. Norrbotten, N Sweden
Jokyakarta *see* Yogyakarta
Joliet *18 B3* Illinois, N USA
Jonava *84 B4 Ger.* Janow, *Pol.* Janów. Kaunas, C Lithuania
Jonesboro *20 B1* Arkansas, C USA
Joniškis *84 C3 Ger.* Janischken. Šiaulai, N Lithuania
Jönköping *63 B7* Jönköping, S Sweden
Jonquière *17 E4* Québec, SE Canada
Joplin *23 F5* Missouri, C USA
Jordan *97 B6 off.* Hashemite Kingdom of Jordan, *Ar.* Al Mamlaka al Urdunīya al Hashemīyah, Al Urdunn; *prev.* Transjordan. *country* SW Asia
Jordan *97 B5 Ar.* Urdun, *Heb.* HaYarden. *river* SW Asia
Jorhāt *113 H3* Assam, NE India
Jos *53 G4* Plateau, C Nigeria
Joseph Bonaparte Gulf *124 D2 gulf* N Australia
Jos Plateau *53 G4 plateau* C Nigeria
Jotunheimen *63 A5 mountain range* S Norway
Joûnié *96 A4 var.* Junīyah. W Lebanon
Joure *64 D2 Fris.* De Jouwer. Fryslân, N Netherlands
Joutseno *63 E5* Etelä-Kariala, SE Finland
Jowhar *see* Jawhar
J.Storm Thurmond Reservoir *see* Clark Hill Lake
Juan Aldama *28 D3* Zacatecas, C Mexico
Juan de Fuca, Strait of *24 A1 strait* Canada/USA
Juan Fernández, Islas *35 A6 Eng.* Juan Fernandez Islands. *island group* W Chile
Juan Fernandez Islands *see* Juan Fernández, Islas
Juazeiro *41 G2 prev.* Joazeiro. Bahia, E Brazil
Juazeiro do Norte *41 G2* Ceará, E Brazil
Juba *51 B5 var.* Jūbā. *country capital* (South Sudan) Bahr el Gabel, S South Sudan
Juba *51 D6 Amh.* Genalē Wenz, *It.* Guiba, *Som.* Ganaane, Webi Jubba. *river* Ethiopia/Somalia
Jubba, Webi *see* Juba
Jubbulpore *see* Jabalpur
Júcar *71 E3 var.* Jucar. *river* C Spain
Juchitán *29 F5 var.* Juchitán de Zaragosa. Oaxaca, SE Mexico
Juchitán de Zaragosa *see* Juchitán
Judayyidat Hāmir *98 B3* Al Anbār, S Iraq
Judenburg *73 D7* Steiermark, C Austria
Juigalpa *30 D3* Chontales, S Nicaragua
Juiz de Fora *41 F4* Minas Gerais, SE Brazil
Jujuy *see* San Salvador de Jujuy
Jūlā *see* Jālū, Libya
Julia Beterrae *see* Béziers
Juliaca *39 E4* Puno, SE Peru
Juliana Top *37 G3 mountain* C Suriname
Julianehåb *see* Qaqortoq
Julio Briga *see* Bragança
Juliobriga *see* Logroño
Juliomagus *see* Angers
Jullundur *see* Jalandhar
Jumilla *71 E4* Murcia, SE Spain
Jummoo *see* Jammu
Jumna *see* Yamuna
Jumporn *see* Chumphon
Junction City *23 F4* Kansas, C USA
Juneau *14 D4 state capital* Alaska, USA
Junín *42 C4* Buenos Aires, E Argentina
Junīyah *see* Joûnié
Junkseylon *see* Phuket
Jur *51 B5 river* C Sudan
Jura *68 D4 cultural region* E France
Jura *73 A7 var.* Jura Mountains. *mountain range* France/Switzerland
Jura *66 B4 island* SW Scotland, United Kingdom
Jura Mountains *see* Jura
Jurbarkas *84 B4 Ger.* Georgenburg, Jurburg. Tauragė, W Lithuania
Jurburg *see* Jurbarkas
Jūrmala *84 C3* Rīga, C Latvia
Juruá, Rio *40 C2 var.* Río Yuruá. *river* Brazil/Peru
Juruena, Rio *40 D3 river* W Brazil
Jutiapa *30 B2* Jutiapa, S Guatemala
Juticalpa *30 D2* Olancho, C Honduras
Jutland *63 A7 Den.* Jylland. *peninsula* W Denmark
Juvavum *see* Salzburg
Juventud, Isla de la *32 A2 var.* Isla de Pinos, *Eng.* Isle of Youth; *prev.* The Isle of the Pines. *island* W Cuba
Južna Morava *79 E5 Ger.* Südliche Morava. *river* SE Serbia
Jwaneng *56 C4* Southern, S Botswana
Jylland *see* Jutland
Jyrgalan *see* Dzhergalan
Jyväskylä *63 D5* Keski-Suomi, C Finland

K

K2 *104 A4 Chin.* Qogir Feng, *Eng.* Mount Godwin Austen. *mountain* China/Pakistan
Kaafu Atoll *see* Male' Atoll
Kaaimanston *37 G3* Sipaliwini, N Suriname
Kaakhka *see* Kaka
Kaala *see* Caála
Kaamanen *62 D2 Lapp.* Gámas. Lappi, N Finland
Kaapstad *see* Cape Town
Kaaresuvanto *62 C3 Lapp.* Gárasavon. Lappi, N Finland
Kabale *51 B6* SW Uganda
Kabinda *55 D7* Kasai-Oriental, SE Dem. Rep. Congo
Kabinda *see* Cabinda
Kābol *see* Kābul
Kabompo *56 C2 river* W Zambia
Kābul *101 E4 prev.* Kābol, *Eng.* Kabul. *country capital* (Afghanistan) Kābul, E Afghanistan
Kabul *101 E4 var.* Daryā-ye Kābul. *river* Afghanistan/Pakistan
Kābul, Daryā-ye *see* Kabul
Kabwe *56 D2* Central, C Zambia
Kachchh, Gulf of *112 B4 var.* Gulf of Cutch, Gulf of Kutch. *gulf* W India
Kachchh, Rann of *112 B4 var.* Rann of Kachh, Rann of Kutch. *salt marsh* India/Pakistan
Kachh, Rann of *see* Kachchh, Rann of
Kadan Kyun *115 B5 prev.* King Island. *island* Myeik Archipelago, S Myanmar (Burma)
Kadapa *110 C2 prev.* Cuddapah. Andhra Pradesh, S India
Kadavu *123 E4 prev.* Kandavu. *island* S Fiji
Kadiivka *87 H3 prev.* Kadiyivka, *Rus.* Stakhanov. Luhans'ka Oblast', E Ukraine
Kadiyivka *see* Kadiivka
Kadoma *56 D3 prev.* Gatooma. Mashonaland West, C Zimbabwe
Kadugli *50 B4* Southern Kordofan, S Sudan
Kaduna *53 G4* Kaduna, C Nigeria
Kadzhi-Say *101 G2 Kir.* Kajisay. Issyk-Kul'skaya Oblast', NE Kyrgyzstan
Kaédi *52 C3* Gorgol, S Mauritania
Kaffa *see* Feodosiia
Kafue *56 D2* Lusaka, SE Zambia
Kafue *56 C2 river* C Zambia
Kaga Bandoro *54 C4 prev.* Fort-Crampel. Nana-Grébizi, C Central African Republic
Kagan *see* Kogon
Kâgheț *52 D1 var.* Karet. *physical region* N Mauritania
Kagi *see* Chiayi
Kagoshima *109 B8 var.* Kagosima. Kagoshima, Kyūshū, SW Japan
Kagoshima-wan *109 A8 bay* SW Japan
Kagosima *see* Kagoshima
Kagul *see* Cahul
Kharga *see* Al Khārjah
Kahmard, Darya-ye *101 E4 prev.* Darya-i-surkhab. *river* NE Afghanistan
Kahramanmaraş *94 D4 var.* Kahraman Maraş, Maraş, Marash. Kahramanmaraş, S Turkey (Türkiye)
Kaiapoi *129 C6* Canterbury, South Island, New Zealand
Kaifeng *106 C4* Henan, C China
Kai, Kepulauan *117 F4 prev.* Kei Islands. *island group* Maluku, SE Indonesia
Kaikohe *128 C2* Northland, North Island, New Zealand
Kaikoura *129 C5* Canterbury, South Island, New Zealand
Kaikoura Peninsula *129 C5 peninsula* South Island, New Zealand
Kainji Lake *see* Kainji Reservoir
Kainji Reservoir *53 F4 var.* Kainji Lake. *reservoir* W Nigeria
Kaipara Harbour *128 C2 harbour* North Island, New Zealand
Kairouan *49 E2 var.* Al Qayrawān. E Tunisia
Kaisaria *see* Kayseri
Kaiserslautern *73 A5* Rheinland-Pfalz, SW Germany
Kaišiadorys *85 B5* Kaunas, S Lithuania
Kaitaia *128 C2* Northland, North Island, New Zealand
Kajaani *62 E4 Swe.* Kajana. Kainuu, C Finland
Kajan *see* Kayan, Sungai
Kajana *see* Kajaani
Kajisay *see* Kadzhi-Say
Kaka *100 C2 Rus.* Kaakhka. Ahal Welaýaty, S Turkmenistan
Kake *14 D4* Kupreanof Island, Alaska, USA
Kakhovka *87 F4* Khersons'ka Oblast', S Ukraine
Kakhovske Vodoskhovyshche *87 F4 Rus.* Kakhovskoye Vodokhranilishche. *reservoir* SE Ukraine
Kakhovskoye Vodokhranilishche *see* Kakhovske Vodoskhovyshche
Kākināda *110 D1 prev.* Cocanada. Andhra Pradesh, E India
Kakshaal-Too, Khrebet *see* Kokshaal-Tau
Kaktovik *14 D2* Alaska, USA
Kalaallit Nunaat *see* Greenland
Kalaburagi *110 C1 prev.* Gulbarga. Karnātaka, C India
Kalahari Desert *56 B4 desert* Southern Africa
Kalaikhum *see* Qalaikhumb
Kalámai *see* Kalámata
Kalamariá *82 B4* Kentrikí Makedonía, N Greece
Kalamás *see* Thýamis
Kalámata *83 B6 prev.* Kalámai. Pelopónnisos, S Greece
Kalamazoo *18 C3* Michigan, N USA
Kalambaka *see* Kalampáka
Kálamos *83 C5* Attikí, C Greece
Kalampáka *82 B4 var.* Kalambaka. Thessalía, C Greece
Kalanchak *87 F4* Khersons'ka Oblast', S Ukraine
Kalarash *see* Călăraşi
Kalasin *114 D4 var.* Muang Kalasin. Kalasin, E Thailand
Kalat *112 B2 var.* Kelat, Khelat. Baluchistan, SW Pakistan
Kalāt *see* Qalāt
Kalbarri *125 A5* Western Australia
Kalecik *94 C3* Ankara, N Turkey (Türkiye)
Kalemie *55 E6 prev.* Albertville. Katanga, SE Dem. Rep. Congo
Kale Sultanie *see* Çanakkale
Kalgan *see* Zhangjiakou
Kalgoorlie *125 B6* Western Australia
Kalima *55 D6* Maniema, E Dem. Rep. Congo
Kalimantan *116 D4 Eng.* Indonesian Borneo. *geopolitical region* Borneo, C Indonesia
Kalinin *see* Tver'
Kaliningrad *84 A4* Kaliningradskaya Oblast', W Russia
Kaliningrad *see* Kaliningradskaya Oblast'
Kaliningradskaya Oblast' *84 B4 var.* Kaliningrad. *province and enclave* W Russia
Kalinkavichy *85 C7 Rus.* Kalinkovichi. Homyel'skaya Voblasts', SE Belarus
Kalinkovichi *see* Kalinkavichy
Kalisch/Kalish *see* Kalisz
Kalispell *22 B1* Montana, NW USA
Kalisz *76 C4 Ger.* Kalisch, *Rus.* Kalish; *anc.* Calisia. Wielkopolskie, C Poland
Kalix *62 D4* Norrbotten, N Sweden
Kalixälven *62 D3 river* N Sweden
Kallaste *84 E3 Ger.* Krasnogor. Tartumaa, SE Estonia
Kallavesi *63 E5 lake* SE Finland
Kalloní *83 D5* Lésvos, E Greece
Kalmar *63 C7 var.* Calmar. Kalmar, S Sweden
Kalmthout *65 C5* Antwerpen, N Belgium
Kalpáki *82 A4* Ípeiros, W Greece
Kalpeni Island *110 B3 island* Lakshadweep, India, N Indian Ocean
Kaltdorf *see* Pruszków
Kaluga *89 B5* Kaluzhskaya Oblast', W Russia
Kalush *86 C2 Pol.* Kałusz. Ivano-Frankivs'ka Oblast', W Ukraine
Kałusz *see* Kalush
Kalutara *110 D4* Western Province, SW Sri Lanka
Kalvarija *85 B5 Pol.* Kalwaria. Marijampolė, S Lithuania
Kalwaria *see* Kalvarija
Kalyān *112 C5* Mahārāshtra, W India
Kálymnos *83 D6 var.* Kálimnos. *island* Dodekánisa, Greece, Aegean Sea
Kama *88 D4 river* NW Russia
Kamarang *37 F3* W Guyana
Kambryk *see* Cambrai
Kamchatka *see* Kamchatka, Poluostrov
Kamchatka, Poluostrov *93 G3 Eng.* Kamchatka. *peninsula* E Russia
Kamenets-Podol'skiy *see* Kamianets-Podilskyi
Kamenka Dneprovskaya *see* Kamianka-Dniprovska
Kamenskoye *see* Kamianske
Kamensk-Shakhtinskiy *89 B6* Rostovskaya Oblast', SW Russia
Kamianets-Podilskyi *86 C3 prev.* Kam"yanets'-Podil's'kyy, *Rus.* Kamenets-Podol'skiy. Khmel'nyts'ka Oblast', W Ukraine
Kamianka-Dniprovska *87 F3 prev.* Kam"yanka-Dniprovs'ka, *Rus.* Kamenka Dneprovskaya. Zaporiz'ka Oblast', SE Ukraine
Kamianske *87 F3 prev.* Kam"yans'ke, Kamenskoye; *Rus.* Dniprodzerzhyns'k. Dnipropetrovska Oblast, E Ukraine
Kamina *55 D7* Katanga, S Dem. Rep. Congo
Kamishli *see* Al Qāmishlī
Kamloops *15 E5* British Columbia, SW Canada
Kammu Seamount *130 C2 guyot* N Pacific Ocean
Kampala *51 B6 country capital* (Uganda) S Uganda
Kampong Cham *115 D6 Khmer.* Kâmpóng Cham. Kampong Cham, C Cambodia
Kâmpóng Cham *see* Kampong Cham
Kampong Chhang *115 D6 Khmer.* Kâmpóng Chhnăng. Kampong Chhang, C Cambodia
Kâmpóng Chhnăng *see* Kampong Chhang
Kampong Speu *115 D6 Khmer.* Kâmpóng Spoe. Kampong Speu, S Cambodia
Kâmpóng Spoe *see* Kampong Speu
Kampong Thom *115 D5 Khmer.* Kâmpóng Thum, *prev.* Trâpeăng Vêng. Kampong Thom, C Cambodia
Kâmpóng Thum *see* Kampong Thom
Kâmpóng Trâbêk *115 D5 prev.* Phumĭ Kâmpóng Trâbêk, Phum Kompong Trabek. Kampong Thom, C Cambodia
Kampot *115 D6 Khmer.* Kâmpôt. Kampot, SW Cambodia
Kâmpôt *see* Kampot
Kampuchea *see* Cambodia
Kampuchea, Democratic *see* Cambodia
Kampuchea, People's Democratic Republic of *see* Cambodia
Kam"yanets'-Podil's'kyy *see* Kamianets-Podilskyi
Kam"yanka-Dniprovs'ka *see* Kamianka-Dniprovska
Kam"yans'ke *see* Kamianske
Kamyshin *89 B6* Volgogradskaya Oblast', SW Russia
Kanaky *see* New Caledonia
Kananga *55 D6 prev.* Luluabourg. Kasai-Occidental, S Dem. Rep. Congo
Kanara *see* Karnātaka
Kanash *89 C5* Chuvashskaya Respublika, W Russia
Kanazawa *109 C5* Ishikawa, Honshū, SW Japan
Kanbe *114 B4* Yangon, SW Myanmar (Burma)
Kānchīpuram *110 C2 prev.* Conjeeveram. Tamil Nādu, SE India
Kandahār *101 E5 Per.* Qandahār. Kandahār, S Afghanistan
Kandalakša *see* Kandalaksha
Kandalaksha *88 B2 var.* Kandalakša, *Fin.* Kantalahti. Murmanskaya Oblast', NW Russia
Kandangan *116 D4* Borneo, C Indonesia
Kandau *see* Kandava
Kandava *84 C3 Ger.* Kandau. Tukums, W Latvia
Kandavu *see* Kadavu
Kandi *53 F4* N Benin
Kandy *110 D3* Central Province, C Sri Lanka
Kane Fracture Zone *44 B4 fracture zone* NW Atlantic Ocean
Kāne'ohe *25 A8 var.* Kaneohe. O'ahu, Hawaii, USA, C Pacific Ocean
Kanestron, Akrotírio *see* Palioúri, Akrotírio
Kanëv *see* Kaniv
Kanevskoye Vodokhranilishche *see* Kanivske Vodoskhovyshche
Kangān *see* Bandar-e Kangān
Kangaroo Island *127 A7 island* South Australia
Kangertittivaq *61 E4 Dan.* Scoresby Sund. *fjord* E Greenland

Kangikajik 61 E4 var. Kap Brewster. headland E Greenland
Kaniv 87 E2 Rus. Kanëv. Cherkas'ka Oblast', C Ukraine
Kanivske Vodoskhovyshche 87 E2 prev. Kanivs'ke Vodoskhovyshche. Rus. Kanevskoye Vodokhranilishche. reservoir C Ukraine
Kanivs'ke Vodoskhovyshche see Kanivske Vodoskhovyshche
Kanjiža 78 D2 Ger. Altkanischa, Hung. Magyarkanizsa, Ókanizsa; prev. Stara Kanjiža. Vojvodina, N Serbia
Kankaanpää 63 D5 Satakunta, SW Finland
Kankakee 18 B3 Illinois, N USA
Kankan 52 D4 E Guinea
Kannur 110 B2 var. Cannanore. Kerala, SW India
Kano 53 G4 Kano, N Nigeria
Kānpur 113 E3 Eng. Cawnpore. Uttar Pradesh, N India
Kansas 23 F5 off. State of Kansas, also known as Jayhawker State, Sunflower State. state C USA
Kansas City 23 F4 Kansas, C USA
Kansas City 23 F4 Missouri, C USA
Kansas River 23 F5 river Kansas, C USA
Kansk 93 E4 Krasnoyarskiy Kray, S Russia
Kansu see Gansu
Kantalahti see Kandalaksha
Kántanos 83 C7 Kríti, Greece, E Mediterranean Sea
Kantemirovka 89 B6 Voronezhskaya Oblast', W Russia
Kantipur see Kathmandu
Kanton 123 F3 var. Abariringa, Canton Island; prev. Mary Island. atoll Phoenix Islands, C Kiribati
Kanye 56 C4 Southern, SE Botswana
Kaohsiung 106 D6 var. Gaoxiong, Jap. Takao, Takow. S Taiwan
Kaolack 52 B3 var. Kaolak. W Senegal
Kaolak see Kaolack
Kaolan see Lanzhou
Kaoma 56 C2 Western, W Zambia
Kapelle 65 B5 Zeeland, SW Netherlands
Kapellen 65 C5 Antwerpen, N Belgium
Kapka, Massif du 54 C2 mountain range E Chad
Kaplangky, Plato see Gaplaňgyr Platosy
Kapoeas see Kapuas, Sungai
Kapoeta 51 C5 E Equatoria, SE South Sudan
Kaposvár 77 C7 Somogy, SW Hungary
Kappeln 72 B2 Schleswig-Holstein, N Germany
Kaproncza see Koprivnica
Kapstad see Cape Town
Kapsukas see Marijampolė
Kaptsevichy 85 C7 Rus. Koptsevichi. Homyel'skaya Voblasts', SE Belarus
Kapuas, Sungai 116 C4 prev. Kapoeas. river Borneo, C Indonesia
Kapuskasing 16 C4 Ontario, S Canada
Kapyl' 85 C6 Rus. Kopyl'. Minskaya Voblasts', C Belarus
Kara-Balta 101 F2 Chuyskaya Oblast', N Kyrgyzstan
Karabil', Vozvyshennost' see Garabil Belentligi
Kara-Bogaz-Gol, Zaliv see Garabogaz Aylagy
Karabük 94 C2 Karabük, NW Turkey (Türkiye)
Karachi 112 B3 Sindh, SE Pakistan
Karácsonkő see Piatra-Neamţ
Karadeniz see Black Sea
Karadeniz Boğazi see İstanbul Boğazı
Karaferiye see Véroia
Karagandy/Karaganda see Qaraghandy
Karaginskiy, Ostrov 93 H2 island E Russia
Karagumskiy Kanal see Garagum Kanaly
Karak see Al Karak
Kara-Kala see Magtymguly
Karakax see Moyu
Karakılısse see Ağrı
Karakol 101 G2 prev. Przheval'sk. Issyk-Kul'skaya Oblast', NE Kyrgyzstan
Karakol 101 G2 var. Karakolka. Issyk-Kul'skaya Oblast', NE Kyrgyzstan
Karakolka see Karakol
Karakoram Range 112 D1 mountain range C Asia
Karaköse see Ağrı
Karakul' see Qarokŭl, Tajikistan
Kara Kum see Garagum
Kara Kum Canal/Karakumskiy Kanal see Garagum Kanaly
Karakumy, Peski see Garagum
Karamai see Karamay
Karaman 94 C4 Karaman, S Turkey (Türkiye)
Karamay 104 B2 var. Karamai, Kelamayi; prev. Chin. K'o-la-ma-i. Xinjiang Uygur Zizhiqu, NW China
Karamea Bight 129 B5 gulf South Island, New Zealand
Karapelit 82 E1 Rom. Stejarul. Dobrich, NE Bulgaria
Kara-Say 101 G2 Issyk-Kul'skaya Oblast', NE Kyrgyzstan
Karasburg 56 B4 Karas, S Namibia
Kara Sea see Karskoye More
Kara Strait see Karskiye Vorota, Proliv
Karatau 92 C5 Kaz. Qarataū. Zhambyl, S Kazakhstan
Karavás 83 B7 Kýthira, S Greece
Karbalā' 98 B3 var. Kerbala, Kerbela. Karbalā', S Iraq
Kardeljevo see Ploče
Kardhítsa see Kardítsa
Kardítsa 83 B5 var. Kardhítsa. Thessalía, C Greece
Kärdla 84 C2 Ger. Kertel. Hiiumaa, W Estonia
Kardzhali 82 D3 var. Kirdzhali, Kŭrdzhali. Kardzhali, S Bulgaria
Karet see Kâgheţ
Kargı 94 C2 Çorum, N Turkey (Türkiye)
Kargilik see Yecheng
Kariba 56 D2 Mashonaland West, N Zimbabwe
Kariba, Lake 56 D3 reservoir Zambia/Zimbabwe
Karibib 56 B3 Erongo, C Namibia
Kariega 56 C5 prev. Uitenhage. Eastern Cape, S South Africa
Karies see Karyés
Karigasniemi 62 D2 Lapp. Garegegasnjárga. Lappi, N Finland
Karimata, Selat 116 C4 strait W Indonesia
Karīmnagar 112 D5 Telangana, C India
Karin 50 D4 Sahil, N Somalia
Kariot see Ikaría
Káristos see Kárystos
Karkinitska Zatoka 87 E4 Rus. Karkinitskiy Zaliv. gulf S Ukraine
Karkinitskiy Zaliv see Karkinitska Zatoka
Karkūk see Kirkūk
Karleby see Kokkola
Karl-Marx-Stadt see Chemnitz
Karlö see Hailuoto
Karlovac 78 B3 Ger. Karlstadt, Hung. Károlyváros. Karlovac, C Croatia
Karlovy Vary 77 A5 Ger. Karlsbad; prev. Eng. Carlsbad. Karlovarský Kraj, W Czechia (Czech Republic)
Karlsbad see Karlovy Vary
Karlsburg see Alba Iulia
Karlskrona 63 C7 Blekinge, S Sweden
Karlsruhe 73 B6 var. Carlsruhe. Baden-Württemberg, SW Germany
Karlstad 63 B6 Värmland, C Sweden
Karlstadt see Karlovac
Karnāl 112 D2 Haryāna, N India
Karnātaka 110 C1 var. Kanara; prev. Maisur, Mysore. cultural region W India
Karnobat 82 D2 Burgas, E Bulgaria
Karnul see Kurnool
Karol see Carei
Károly-Fehérvár see Alba Iulia
Károlyváros see Karlovac
Karpaten see Carpathian Mountains
Kárpathos 83 E7 Kárpathos, SE Greece
Kárpathos 83 E7 It. Scarpanto; anc. Carpathos, Carpathus. island SE Greece
Karpaty see Carpathian Mountains
Karpenísi 83 B5 prev. Karpenísion. Stereá Ellás, C Greece
Karpenísion see Karpenísi
Karpilovka see Aktsyabrski
Kars 95 F2 var. Qars. Kars, NE Turkey (Türkiye)
Karsau see Kārsava
Kārsava 84 D4 Ger. Karsau; prev. Rus. Korsovka. Ludza, E Latvia
Karshi see Qarshi, Uzbekistan
Karskiye Vorota, Proliv 88 E2 Eng. Kara Strait. strait N Russia
Karskoye More 92 D2 Eng. Kara Sea. sea Arctic Ocean
Karyés 82 C4 var. Karies. Ágion Óros, N Greece
Kárystos 83 C6 var. Káristos. Évvoia, C Greece
Kasai 55 C6 var. Cassai, Kassai. river Angola/Dem. Rep. Congo
Kasaji 55 D7 Katanga, S Dem. Rep. Congo
Kasama 56 D1 Northern, N Zambia
Kasan see Koson
Kāsaragod 110 B2 Kerala, SW India
Kaschau see Košice
Kāshān 98 C3 Eşfahān, C Iran
Kashgar see Kashi
Kashi 104 A3 Chin. Kaxgar, K'o-shih, Uigh. Kashgar. Xinjiang Uygur Zizhiqu, NW China
Kasi see Vārānasi
Kasongo 55 D6 Maniema, E Dem. Rep. Congo
Kasongo-Lunda 55 C7 Bandundu, SW Dem. Rep. Congo
Kásos 83 D7 island S Greece
Kaspiy Mangy Oypaty see Caspian Depression
Kaspiysk 89 B8 Respublika Dagestan, SW Russia
Kaspiyskoye More/Kaspiy Tengizi see Caspian Sea
Kassa see Košice
Kassai see Kasai
Kassala 50 C4 Kassala, E Sudan
Kassel 72 B4 prev. Cassel. Hessen, C Germany
Kasserine 49 E2 var. Al Qaşrayn. W Tunisia
Kastamonu 94 C2 var. Castamoni, Kastamuni. Kastamonu, N Turkey (Türkiye)
Kastamuni see Kastamonu
Kastaneá 82 B4 Kentrikí Makedonía, N Greece
Kastélli see Kíssamos
Kastoriá 82 B4 Dytikí Makedonía, N Greece
Kástro 83 C6 Sífnos, Kykládes, Greece, Aegean Sea
Kastsyukovichy 85 E7 Rus. Kostyukovichi. Mahilyowskaya Voblasts', E Belarus
Kastsyukowka 85 D7 Rus. Kostyukovka. Homyel'skaya Voblasts', SE Belarus
Kasulu 51 B7 Kigoma, W Tanzania
Kasumiga-ura 109 D5 lake Honshū, S Japan
Katahdin, Mount 19 G1 mountain Maine, NE USA
Katalla 14 C3 Alaska, USA
Katana see Qaţanā
Katanning 125 B7 Western Australia
Katawaz see Zarghūn Shahr
Katchall Island 111 F3 island Nicobar Islands, India, NE Indian Ocean
Kateríni 82 B4 Kentrikí Makedonía, N Greece
Katha 114 B2 Sagaing, N Myanmar (Burma)
Katherine 126 A2 Northern Territory, N Australia
Kathmandu 102 C3 prev. Kantipur. country capital (Nepal) Central, C Nepal
Katikati 128 D3 Bay of Plenty, North Island, New Zealand
Katima Mulilo 56 C3 Caprivi, NE Namibia
Katiola 52 D4 C Ivory Coast
Kati Thanda see Eyre, Lake
Kā Tiritiri o Te Moana see Southern Alps
Káto Achaïa 83 B5 var. Kato Ahaia, Káto Akhaía. Dytikí Ellás, S Greece
Kato Ahaia/Káto Akhaía see Káto Achaïa
Katoúna 83 A5 Dytikí Ellás, C Greece
Katowice 77 C5 Ger. Kattowitz. Śląskie, S Poland
Katsina 53 G3 Katsina, N Nigeria
Kattakurgan see Kattaqo'rg'on
Kattaqo'rg'on 101 E2 Rus. Kattakurgan. Samarqand Viloyati, C Uzbekistan
Kattavía 83 E7 Ródos, Dodekánisa, Greece, Aegean Sea
Kattegat 63 B7 Dan. Kattegatt. strait N Europe
Kattegatt see Kattegat
Kattowitz see Katowice
Kaua'i 25 A7 var. Kauai. island Hawaiian Islands, Hawaii, USA, C Pacific Ocean
Kauai see Kaua'i
Kauen see Kaunas
Kaufbeuren 73 C6 Bayern, S Germany
Kaunas 84 B4 Ger. Kauen, Pol. Kowno; prev. Rus. Kovno. Kaunas, C Lithuania
Kavadar see Kavadarci
Kavadarci 79 E6 Turk. Kavadar. C Macedonia
Kavaja see Kavajë
Kavajë 79 C6 It. Cavaia, Kavaja. Tiranë, W Albania
Kavakli see Topolovgrad
Kavála 82 C3 prev. Kavála. Anatolikí Makedonía kai Thráki, NE Greece
Kāvali 110 D2 Andhra Pradesh, E India
Kavála see Kavála
Kavango see Cubango/Okavango
Kavaratti Island 110 A3 island Lakshadweep, Lakshadweep, SW India Asia N Indian Ocean
Kavarna 82 E2 Dobrich, NE Bulgaria
Kavengo see Cubango/Okavango
Kavir, Dasht-e 98 D3 var. Great Salt Desert. salt pan N Iran
Kavkaz see Caucasus
Kawagoe 109 D5 Saitama, Honshū, S Japan
Kawasaki 108 A2 Kanagawa, Honshū, S Japan
Kawerau 128 E3 Bay of Plenty, North Island, New Zealand
Kaxgar see Kashi
Kaya 53 E3 C Burkina
Kayan 114 B4 Yangon, SW Burma (Myanmar)
Kayan, Sungai 116 D3 prev. Kajan. river Borneo, C Indonesia
Kayes 52 C3 Kayes, W Mali
Kayseri 94 D3 var. Kaisaria; anc. Caesarea Mazaca, Mazaca. Kayseri, C Turkey (Türkiye)
Kazach'ye 93 F2 Respublika Sakha (Yakutiya), NE Russia
Kazakh Soviet Socialist Republic see Kazakhstan
Kazakhskiy Melkosopochnik see Saryarqa
Kazakhstan 92 B4 off. Republic of Kazakhstan, var. Kazakstan, Kaz. Qazaqstan, Qazaqstan Respūblikasy; prev. Kazakh Soviet Socialist Republic, Rus. Respublika Kazakhstan. country C Asia
Kazakhstan, Republic of see Kazakhstan
Kazakhstan, Respublika see Kazakhstan
Kazakh Steppe 92 C4 var. Kirghiz Steppe. uplands C Kazakhstan
Kazakstan see Kazakhstan
Kazan' 89 C5 Respublika Tatarstan, W Russia
Kazandzhik see Bereket
Kazanlak 82 D2 var. Kazanlŭk, Kazanlik. Stara Zagora, C Bulgaria
Kazanlik see Kazanlak
Kazanlŭk see Kazanlak
Kazatin see Koziatyn
Kazbegi see Kazbek
Kazbek 95 F1 var. Kazbegi, Geor. Mqinvartsveri. mountain N Georgia
Kāzerūn 98 D4 Fārs, S Iran
Kazi Magomed see Hacıqabul
Kazvin see Qazvīn
Kéa 83 C6 var. Tziá, Kéos; anc. Ceos. island Kykládes, Greece, Aegean Sea
Kea, Mauna 25 B8 mountain Hawaii, USA
Kéamu see Aneityum
Kearney 23 E4 Nebraska, C USA
Keban Barajı 95 E3 reservoir C Turkey (Türkiye)
Kebkabiya 50 A4 Northern Darfur, W Sudan
Kebnekaise 62 C3 mountain N Sweden
Kecskemét 77 D7 Bács-Kiskun, C Hungary
Kediri 116 D5 Jawa, C Indonesia
Kędzierzyn-Kozle 77 C5 Ger. Heydebrech. Opolskie, S Poland
Keelung 106 D6 var. Chilung, Jilong; Jap. Kirun, Kirun'; prev. Sp. Santissima Trinidad. N Taiwan
Keetmanshoop 56 B4 Karas, S Namibia
Kefallinía see Kefalloniá
Kefalloniá 83 A5 var. Kefallinía. island Iónia Nisiá, Greece, C Mediterranean Sea
Kefe see Feodosiia
Kegel see Keila
Kehl 73 A6 Baden-Württemberg, SW Germany
Kei Islands see Kai, Kepulauan
Keijō see Seoul
Keila 84 D2 Ger. Kegel. Harjumaa, NW Estonia
Keïta 53 F3 Tahoua, C Niger
Keith 127 B7 South Australia
Këk-Art 101 G2 prev. Alaykel', Alay-Kuu. Oshskaya Oblast', SW Kyrgyzstan
Kékes 77 C6 mountain N Hungary
Kelamayi see Karamay
Kelang see Klang
Kelat see Kālat
Kelifskiy Uzboy see Kelif Uzboýy
Kelif Uzboýy 100 D3 Rus. Kelifskiy Uzboy. salt marsh E Turkmenistan
Kelkit Çayı 95 E3 river N Turkey (Türkiye)
Kelmė 84 B4 Šiaulai, C Lithuania
Kélo 54 B4 Tandjilé, SW Chad
Kelowna 15 E5 British Columbia, SW Canada
Kelso 24 B2 Washington, NW USA
Keltsy see Kielce
Keluang 116 B3 var. Kluang. Johor, Peninsular Malaysia, Malaysia
Kem' 88 B3 Respublika Kareliya, NW Russia
Kemah 95 E3 Erzincan, E Turkey (Türkiye)
Kemaman see Cukai
Kemerovo 92 D4 prev. Shcheglovsk. Kemerovskaya Oblast', C Russia
Kemi 62 D4 Lappi, NW Finland
Kemijärvi 62 D3 Swe. Kemiträsk. Lappi, N Finland
Kemijoki 62 D3 river NW Finland
Kemin 101 G2 prev. Bystrovka. Chuyskaya Oblast', N Kyrgyzstan
Kemins Island see Nikumaroro
Kemiträsk see Kemijärvi
Kemmuna 80 A5 var. Comino. island C Malta
Kempele 62 D4 Pohjois-Pohjanmaa, C Finland
Kempten 73 B7 Bayern, S Germany
Kendal 67 D5 NW England, United Kingdom
Kendari 117 E4 Sulawesi, C Indonesia
Kenedy 27 G4 Texas, SW USA
Kenema 52 C4 SE Sierra Leone
Këneurgench see Köneürgenç
Kenge 55 C6 Bandundu, SW Dem. Rep. Congo
Kengtung 114 C3 prev. Keng Tung. Shan State, E Myanmar (Burma)
Keng Tung see Kengtung
Kénitra 48 C2 prev. Port-Lyautey. NW Morocco
Kennett 23 H5 Missouri, C USA
Kennewick 24 C2 Washington, NW USA
Kenora 16 A3 Ontario, S Canada
Kenosha 18 B3 Wisconsin, N USA
Kentau 92 B5 Türkistan/Turkestan, S Kazakhstan
Kentucky 18 C5 off. Commonwealth of Kentucky, also known as Bluegrass State. state C USA
Kentucky Lake 18 B5 reservoir Kentucky/Tennessee, S USA
Kentung see Keng Tung
Kenya 51 C6 off. Republic of Kenya. country E Africa
Kenya, Mount see Kirinyaga
Kenya, Republic of see Kenya
Keokuk 23 G4 Iowa, C USA
Kéos see Keá
Kępno 76 C4 Wielkopolskie, C Poland
Keppel Island see Niuatoputapu
Kerak see Al Karak
Kerala 110 C2 cultural region S India
Kerasunt see Giresun
Keratéa 83 C6 var. Keratea. Attikí, C Greece
Keratea see Keratéa
Kerbala/Kerbela see Karbalā'
Kerch 87 G5 Rus. Kerch'. Respublika Krym, SE Ukraine
Kerch' see Kerch
Kerchens'ka Protska/Kerchenskiy Proliv see Kerch Strait
Kerch Strait 87 G4 var. Bosporus Cimmerius, Enikale Strait, Rus. Kerchenskiy Proliv, Ukr. Kerchens'ka Protska. strait Black Sea/Sea of Azov
Keremitlik see Lyulyakovo
Kerguelen 119 C7 island C French Southern and Antarctic Lands
Kerguelen Plateau 119 C7 undersea plateau S Indian Ocean
Kerí 83 A6 Zákynthos, Iónia Nisiá, Greece, C Mediterranean Sea
Kerikeri 128 D2 Northland, North Island, New Zealand
Kerkenah, Îles de 80 D4 var. Kerkennah Islands, Ar. Juzur Qarqannah. island group E Tunisia
Kerkennah Islands see Kerkenah, Îles de
Kerki see Atamyrat
Kérkira see Kérkyra
Kerkrade 65 D6 Limburg, SE Netherlands
Kerkuk see Kirkūk
Kérkyra 82 A4 var. Kérkira, Eng. Corfu. Kérkyra, Iónia Nisiá, Greece, C Mediterranean Sea
Kermadec Islands 130 C4 island group New Zealand, SW Pacific Ocean
Kermadec Trench 121 E4 trench SW Pacific Ocean
Kermān 98 D3 var. Kirman; anc. Carmana. Kermān, C Iran
Kermānshāh 98 C3 var. Qahremānshahr; prev. Bākhtarān. Kermānshāhān, W Iran
Kerrville 27 F4 Texas, SW USA
Kertel see Kärdla
Kerulen 105 E2 Chin. Herlen He, Mong. Herlen Gol. river China/Mongolia
Kerýneia see Girne
Kesennuma 108 D4 Miyagi, Honshū, C Japan
Keszthely 77 C7 Zala, SW Hungary
Ketchikan 14 D4 Revillagigedo Island, Alaska, USA
Kętrzyn 76 D2 Ger. Rastenburg. Warmińsko-Mazurskie, NE Poland
Kettering 67 D6 C England, United Kingdom
Kettering 18 C4 Ohio, N USA
Keupriya see Primorsko
Keuruu 63 D5 Keski-Suomi, C Finland
Keweenaw Peninsula 18 B1 peninsula Michigan, N USA
Key Largo 21 F5 Key Largo, Florida, SE USA
Keystone State see Pennsylvania
Key West 21 E5 Florida Keys, Florida, SE USA
Kezdivásárhely see Târgu Secuiesc
Khabarovsk 93 G4 Khabarovskiy Kray, SE Russia
Khachmas see Xaçmaz
Khadera see Hadera
Khairpur 112 B3 Sindh, SE Pakistan
Khalij as Suways 50 B2 var. Suez, Gulf of. gulf NE Egypt
Khalkidhikí see Chalkidikí
Khalkís see Chalkída
Khambhat, Gulf of 112 C4 Eng. Gulf of Cambay. gulf W India
Khamīs Mushayt 99 B6 var. Hamïs Musait. 'Asīr, SW Saudi Arabia
Khānābād 101 E3 Kunduz, NE Afghanistan
Khān al Baghdādī see Al Baghdādī
Khandwa 112 D4 Madhya Pradesh, C India
Khanh Hung see Soc Trăng
Khaniá see Chaniá
Khanka, Lake 107 E2 var. Hsing-K'ai Hu, Lake Hanka, Chin. Xingkai Hu, Rus. Ozero Khanka. lake China/Russia
Khanka, Ozero see Khanka, Lake
Khankendi see Xankändi
Khanthabouli 114 D4 prev. Savannakhét. Savannakhét, S Laos
Khanty-Mansiysk 92 C3 prev. Ostyako-Voguls'k. Khanty-Mansiyskiy Avtonomnyy Okrug-Yugra, C Russia
Khān Yūnis 97 A7 var. Khān Yūnus. S Gaza Strip
Khān Yūnus see Khān Yūnis
Khanzi see Ghanzi
Kharagpur 113 F4 West Bengal, NE India
Kharbin see Harbin
Kharkiv 87 G2 Rus. Khar'kov. Kharkivs'ka Oblast', NE Ukraine
Khar'kov see Kharkiv
Kharmanli 82 D3 Haskovo, S Bulgaria
Khartoum 50 B4 var. El Khartûm, Khartum. country capital (Sudan) Khartoum, C Sudan
Khartum see Khartoum
Khasavyurt 89 B8 Respublika Dagestan, SW Russia
Khash, Dasht-e 100 D5 Eng. Khash Desert. desert SW Afghanistan
Khash Desert see Khāsh, Dasht-e
Khashim Al Qirba/Khashm al Qirbah see Khashm el Girba
Khashm el Girba 50 C4 var. Khashim Al Qirba, Khashm al Qirbah. Kassala, E Sudan
Khaskovo see Haskovo
Khaybar, Kowtal-e see Khyber Pass
Khaydarkan 101 F2 var. Khaydarken. Batkenskaya Oblast', SW Kyrgyzstan
Khaydarken see Khaydarkan
Khazar, Baḩr-e/Khazar, Daryā-ye see Caspian Sea
Khelat see Kālat
Kherson 87 E4 Khersons'ka Oblast', S Ukraine
Kheta 93 E2 river N Russia
Khíos see Chíos
Khiva/Khiwa see Xiva
Khmel'nitskiy see Khmel 'nyts'kyy
Khmelnytskyi 86 C2 prev. Khmel'nyts'kyy, Rus. Khmel'nitskiy; prev. Proskurov. Khmel'nyts'ka Oblast', W Ukraine
Khmel 'nyts'kyy see Khmelnytskyi
Khodasy 85 E6 Rus. Khodosy. Mahilyowskaya Voblasts', E Belarus
Khodoriv 86 C2 Pol. Chodorów, Rus. Khodorov. L'vivs'ka Oblast', NW Ukraine
Khodorov see Khodoriv
Khodosy see Khodasy
Khodzhent see Khujand
Khoi see Khvoy
Khojend see Khujand
Khokand see Qo'qon
Kholm see Khulm
Kholm see Chełm
Kholon see Holon
Khoms see Al Khums
Khong Sedone see Muang Khôngxédôn
Khon Kaen 114 D4 var. Muang Khon Kaen. Khon Kaen, E Thailand
Khor 93 G4 Khabarovskiy Kray, SE Russia
Khorat see Nakhon Ratchasima
Khorog see Khorugh
Khorugh 101 F3 Rus. Khorog. S Tajikistan
Khōst 101 F4 prev. Khowst. Khōst, E Afghanistan
Khotan see Hotan
Khouribga 48 B2 C Morocco
Khovd see Hovd
Khowst see Khōst
Khoy see Khvoy
Khoyniki 85 D8 Homyel'skaya Voblasts', SE Belarus
Khrustalnyi 87 H3 prev. Khrustal'nyy, Krindachevka. Rus. Krasnyi Luch. Luhans'ka Oblast', E Ukraine
Khrustal'nyy see Khrustalnyi
Khudzhand see Khujand
Khujand 101 E2 var. Khodzhent, Khojend, Rus. Khudzhand; prev. Leninabad, Taj. Leninobod. N Tajikistan
Khulm 101 E3 var. Kholm, Tashqurghan. Balkh, N Afghanistan
Khulna 113 G4 Khulna, SW Bangladesh
Khums see Al Khums
Khust 86 B3 var. Husté, Cz. Chust, Hung. Huszt. Zakarpats'ka Oblast', W Ukraine
Khvoy 98 C2 var. Khoi, Khoy. Āžarbāyjān-e Bākhtarī, NW Iran
Khyber Pass 112 C1 var. Kowtal-e Khaybar. pass Afghanistan/Pakistan
Kiangmai see Chiang Mai
Kiang-ning see Nanjing
Kiangsi see Jiangxi
Kiangsu see Jiangsu
Kiáto 83 B6 prev. Kiáton. Pelopónnisos, S Greece
Kiáton see Kiáto
Kiayi see Chiayi
Kibangou 55 B6 Niari, SW Congo
Kibombo 55 D6 Maniema, E Dem. Rep. Congo
Kıbrıs/Kıbrıs Cumhuriyeti see Cyprus
Kičevo 79 D6 SW Macedonia
Kidderminster 67 D6 C England, United Kingdom
Kiel 72 B2 Schleswig-Holstein, N Germany
Kielce 76 D4 Rus. Keltsy. Świętokrzyskie, C Poland
Kieler Bucht 72 B2 bay N Germany
Kiev see Kyiv
Kiffa 52 C3 Assaba, S Mauritania
Kigali 51 B6 country capital (Rwanda) C Rwanda
Kigoma 51 B7 Kigoma, W Tanzania
Kihnu 84 C2 var. Kihnu Saar, Ger. Kühnö. island SW Estonia
Kihnu Saar see Kihnu
Kii-suido 109 C7 strait S Japan
Kikinda 78 D3 Ger. Grosskikinda, Hung. Nagykikinda; prev. Velika Kikinda. Vojvodina, N Serbia
Kikládhes see Kykládes
Kikwit 55 C6 Bandundu, W Dem. Rep. Congo
Kilien Mountains see Qilian Shan
Kilimane see Quelimane
Kilimanjaro 47 E5 region E Tanzania
Kilimanjaro 51 C7 var. Uhuru Peak. volcano NE Tanzania
Kilingi-Nõmme 84 D3 Ger. Kurkund. Pärnumaa, SW Estonia
Kilis 94 D4 Kilis, S Turkey (Türkiye)
Kiliya 86 D4 Rom. Chilia-Nouă. Odes'ka Oblast', SW Ukraine
Kilkenny 67 B6 Ir. Cill Chainnigh. Kilkenny, S Ireland
Kilkís 82 B3 Kentrikí Makedonía, N Greece
Killarney 67 A6 Ir. Cill Airne. Kerry, SW Ireland
Killeen 27 G3 Texas, SW USA
Kilmain see Quelimane
Kilmarnock 66 C4 W Scotland, United Kingdom
Kilwa see Kilwa Kivinje
Kilwa Kivinje 51 C7 var. Kilwa. Lindi, SE Tanzania
Kimberley 56 C4 Northern Cape, C South Africa
Kimberley Plateau 124 C3 plateau Western Australia
Kimch'aek 107 E3 prev. Sŏngjin. E North Korea
Kími see Kými
Kinabalu, Gunung 116 D3 mountain East Malaysia
Kindersley 15 F5 Saskatchewan, S Canada
Kindia 52 C4 Guinée-Maritime, SW Guinea
Kindley Field 20 A4 air base E Bermuda
Kindu 55 D6 prev. Kindu-Port-Empain. Maniema, C Dem. Rep. Congo
Kindu-Port-Empain see Kindu
Kineshma 89 C5 Ivanovskaya Oblast', W Russia
King Abdullah Economic City 99 A5 W Saudi Arabia
King Charles Islands see Kong Karls Land
King Christian IX Land see Kong Christian IX Land
King Frederik VI Coast see Kong Frederik VI Kyst
King Frederik VIII Land see Kong Frederik VIII Land
King Island 127 B8 island Tasmania, SE Australia
King Island see Kadan Kyun
Kingissepp see Kuressaare
Kingman 26 A1 Arizona, SW USA
Kingman Reef 123 E2 US territory C Pacific Ocean
Kingsford Smith 126 E2 (Sydney) New South Wales, SE Australia
King's Lynn 67 E6 var. Bishop's Lynn, Kings Lynn, Lynn, Lynn Regis. E England, United Kingdom
Kings Lynn see King's Lynn
King Sound 124 B3 sound Western Australia
Kingsport 21 E1 Tennessee, S USA
Kingston 32 B5 country capital (Jamaica) E Jamaica
Kingston 16 D5 Ontario, SE Canada
Kingston 19 F3 New York, NE USA
Kingston upon Hull 67 D5 var. Hull. E England, United Kingdom
Kingston upon Thames 67 A8 SE England, United Kingdom
Kingstown 33 H4 country capital (Saint Vincent and the Grenadines) Saint Vincent, Saint Vincent and the Grenadines
Kingstown see Dún Laoghaire
Kingsville 27 G5 Texas, SW USA
King William Island 15 F3 island Nunavut, N Canada

Kinneret, Yam 97 B5 var. Chinnereth, Sea of Bahr Tabariya, Sea of Galilee, Lake Tiberias, Ar. Bahrat Tabariya. lake N Israel
Kinrooi 65 D5 Limburg, NE Belgium
Kinshasa 55 B6 prev. Léopoldville. country capital (Dem. Rep. Congo) Kinshasa, W Dem. Rep. Congo
Kintyre 66 B4 peninsula W Scotland, United Kingdom
Kinyeti 51 B5 mountain S South Sudan
Kiparissía see Kyparissía
Kipili 51 B7 Rukwa, W Tanzania
Kipushi 55 D8 Katanga, SE Dem. Rep. Congo
Kirdzhali see Kardzhali
Kirghizia see Kyrgyzstan
Kirghiz Range 101 F2 Rus. Kirgizskiy Khrebet; prev. Alexander Range. mountain range Kazakhstan/Kyrgyzstan
Kirghiz SSR see Kyrgyzstan
Kirghiz Steppe see Kazakh Steppe
Kirgizskaya SSR see Kyrgyzstan
Kirgizskiy Khrebet see Kirghiz Range
Kiriath-Arba see Hebron
Kiribati 123 F2 off. Republic of Kiribati. country C Pacific Ocean
Kiribati, Republic of see Kiribati
Kırıkhan 94 D4 Hatay, S Turkey (Türkiye)
Kırıkkale 94 C3 province C Turkey (Türkiye)
Kirin see Jilin
Kirinyaga 51 C6 prev. Mount Kenya. volcano C Kenya
Kirishi 88 B4 var. Kirisi. Leningradskaya Oblast', NW Russia
Kirisi see Kirishi
Kiritimati 123 G2 prev. Christmas Island. atoll Line Islands, E Kiribati
Kirkenes 62 E2 Fin. Kirkkoniemi. Finnmark, N Norway
Kirk-Kilissa see Kırklareli
Kirkkoniemi see Kirkenes
Kirkland Lake 16 D4 Ontario, S Canada
Kırklareli 94 A2 prev. Kirk-Kilissa. Kırklareli, NW Turkey (Türkiye)
Kirkpatrick, Mount 132 B3 mountain Antarctica
Kirksville 23 G4 Missouri, C USA
Kirkūk 98 B3 var. Karkūk, Kerkuk. At Ta'mīn, N Iraq
Kirkwall 66 C2 NE Scotland, United Kingdom
Kirkwood 23 G4 Missouri, C USA
Kir Moab/Kir of Moab see Al Karak
Kirov 89 C5 prev. Vyatka. Kirovskaya Oblast', NW Russia
Kirovabad see Gäncä
Kirovakan see Vanadzor
Kirovo-Chepetsk 89 D5 Kirovskaya Oblast', NW Russia
Kirovohrad see Kropyvnytskyi
Kirovo see Kropyvnytskyi
Kirthar Range 112 B3 mountain range S Pakistan
Kiruna 62 C3 Lapp. Giron. Norrbotten, N Sweden
Kirun/Kirun' see Keelung
Kisalföld see Little Alföld
Kisangani 55 D5 prev. Stanleyville. Orientale, NE Dem. Rep. Congo
Kishinev see Chişinău
Kislovodsk 89 B7 Stavropol'skiy Kray, SW Russia
Kismaayo 51 D6 var. Chisimayu, Kismayu, It. Chisimaio. Jubbada Hoose, S Somalia
Kismayu see Kismaayo
Kíssamos 83 C7 prev. Kastélli. Kríti, Greece, E Mediterranean Sea
Kissidougou 52 C4 Guinée-Forestière, S Guinea
Kissimmee, Lake 21 E4 lake Florida, SE USA
Kistna see Krishna
Kisumu 51 C6 prev. Port Florence. Nyanza, W Kenya
Kisvárda 77 E6 Ger. Kleinwardein. Szabolcs-Szatmár-Bereg, E Hungary
Kita 52 D3 Kayes, W Mali
Kitab see Kitob
Kitakyūshū 109 A7 var. Kitakyûsyû. Fukuoka, Kyūshū, SW Japan
Kitakyûsyû see Kitakyūshū
Kitami 108 D2 Hokkaidō, NE Japan
Kitchener 16 C5 Ontario, S Canada
Kitega see Gitega
Kíthnos see Kýthnos
Kitimat 14 D4 British Columbia, SW Canada
Kitinen 62 D3 river N Finland
Kitob 101 E3 Rus. Kitab. Qashqadaryo Viloyati, S Uzbekistan
Kitwe 56 D2 var. Kitwe-Nkana. Copperbelt, C Zambia
Kitwe-Nkana see Kitwe
Kitzbüheler Alpen 73 C7 mountain range W Austria
Kivalina 14 C2 Alaska, USA
Kivalo 62 D3 ridge C Finland
Kivertsi 86 C1 Pol. Kiwerce, Rus. Kivertsy. Volyns'ka Oblast', NW Ukraine
Kivertsy see Kivertsi
Kivu, Lac see Kivu, Lake
Kivu, Lake 55 E6 Fr. Lac Kivu. lake Rwanda/ Dem. Rep. Congo
Kiwerce see Kivertsi
Kiyev see Kyiv
Kiyevskoye Vodokhranilishche see Kyivske Vodoskhovyshche
Kızıl Irmak 94 C3 river C Turkey (Türkiye)
Kizil Kum see Kyzylkum Desert
Kizyl-Arvat see Serdar
Kjølen see Kölen
Kladno 77 A5 Středočeský, NW Czechia (Czech Republic)
Klagenfurt 73 D7 Slvn. Celovec. Kärnten, S Austria
Klaipėda 84 B3 Ger. Memel. Klaipeda, NW Lithuania
Klamath Falls 24 B4 Oregon, NW USA
Klamath Mountains 24 A4 mountain range California/Oregon, W USA
Klang 116 B3 var. Kelang; prev. Port Swettenham. Selangor, Peninsular Malaysia
Klarälven 63 B6 river Norway/Sweden
Klatovy 77 A5 Ger. Klattau. Plzeňský Kraj, W Czechia (Czech Republic)
Klattau see Klatovy
Klausenburg see Cluj-Napoca
Klazienaveen 64 E2 Drenthe, NE Netherlands
Kleines Ungarisches Tiefland see Little Alföld
Klein Karas 56 B4 Karas, S Namibia
Kleinwardein see Kisvárda
Kleisoúra 83 A5 Ípeiros, W Greece
Klerksdorp 56 D4 North-West, N South Africa
Klimavichy 85 E7 Rus. Klimovichi. Mahilyowskaya Voblasts', E Belarus
Klimovichi see Klimavichy
Klintsy 89 A5 Bryanskaya Oblast', W Russia
Klisura 82 C2 Plovdiv, C Bulgaria
Ključ 78 B3 Federacija Bosna I Hercegovina, NW Bosnia and Herzegovina
Kłobuck 76 C4 Śląskie, S Poland
Klosters 73 B7 Graubünden, SE Switzerland
Kluang see Keluang
Kluczbork 76 C4 Ger. Kreuzburg, Kreuzburg in Oberschlesien. Opolskie, S Poland
Klyuchevskaya Sopka, Vulkan 93 H3 volcano E Russia
Knin 78 B4 Šibenik-Knin, S Croatia
Knjaževac 78 E4 Serbia, E Serbia
Knokke-Heist 65 A5 West-Vlaanderen, NW Belgium
Knoxville 20 D1 Tennessee, S USA
Knud Rasmussen Land 60 D1 physical region N Greenland
Kobdo see Hovd
Kōbe 109 C6 Hyōgo, Honshū, SW Japan
København 63 B7 Eng. Copenhagen; anc. Hafnia. country capital (Denmark) Sjælland, København, E Denmark
Kobenni 52 D3 Hodh el Gharbi, S Mauritania
Koblenz 73 A5 prev. Coblenz, Fr. Coblence; anc. Confluentes. Rheinland-Pfalz, W Germany
Kobrin see Kobryn
Kobryn 85 A6 Rus. Kobrin. Brestskaya Voblasts', SW Belarus
Kobuleti 95 F2 prev. K'obulet'i. W Georgia
K'obulet'i see Kobuleti
Kočani 79 E6 NE Macedonia
Kočevje 73 D8 Ger. Gottschee. S Slovenia
Koch Bihār 113 G3 West Bengal, NE India
Kochchi see Kochi
Kochi 110 C3 var. Cochin, Kochchi. Kerela SW India
Kōchi 109 B7 var. Kôti. Kōchi, Shikoku, SW Japan
Kochiu see Gejiu
Kodiak 14 C3 Kodiak Island, Alaska, USA
Kodiak Island 14 C3 island Alaska, USA
Koedoes see Kudus
Koeln see Köln
Koepang see Kupang
Ko-erh-mu see Golmud
Koetai see Mahakam, Sungai
Koetaradja see Bandaaceh
Kōfu 109 D5 var. Kôhu. Yamanashi, Honshū, S Japan
Kogarah 126 E2 New South Wales, E Australia
Kogon 100 D2 Rus. Kagan. Buxoro Viloyati, C Uzbekistan
Kōhalom see Rupea
Kohīma 113 H3 state capital Nāgāland, E India
Kohtla-Järve 84 E2 Ida-Virumaa, NE Estonia
Kôhu see Kōfu
Kokand see Qo'qon
Kokchetav see Kökshetaū
Kokkola 62 D4 Swe. Karleby; prev. Swe. Gamlakarleby. Österbotten, W Finland
Koko 53 F4 Kebbi, W Nigeria
Kokomo 18 C4 Indiana, N USA
Koko Nor see Qinghai, China
Koko Nor see Qinghai Hu, China
Kokrines 14 C2 Alaska, USA
Kokshaal-Tau 101 G2 Rus. Khrebet Kakshaal-Too. mountain range China/Kyrgyzstan
Kokshetau see Kökshetaū
Kökshetaū 92 C4 var. Kokshetau; prev. Kokchetav. Kokshetau, N Kazakhstan
Koksijde 65 A5 West-Vlaanderen, W Belgium
Koksoak 16 D2 river Québec, E Canada
Kokstad 56 D5 KwaZulu/Natal, E South Africa
Kolaka 117 E4 Sulawesi, C Indonesia
K'o-la-ma-i see Karamay
Kola Peninsula see Kol'skiy Poluostrov
Kolari 62 D3 Lappi, NW Finland
Kolárovo 77 C6 Ger. Gutta; prev. Guta, Hung. Gúta. Nitriansky Kraj, SW Slovakia
Kolberg see Kołobrzeg
Kolda 52 C3 S Senegal
Kolding 63 A7 Vejle, C Denmark
Kölen 59 E1 Nor. Kjølen. mountain range Norway/Sweden
Kolguyev, Ostrov 88 C2 island NW Russia
Kolhāpur 110 B1 Mahārāshtra, SW India
Kolhumadulu 110 A5 var. Thaa Atoll. atoll S Maldives
Kolín 77 B5 Ger. Kolin. Střední Čechy, C Czechia (Czech Republic)
Kolka 84 C2 Talsi, NW Latvia
Kolkasrags 84 C2 prev. Eng. Cape Domesnes. headland NW Latvia
Kolkāta 113 G4 prev. Calcutta. West Bengal, N India
Kollam 110 C3 var. Quilon. Kerala, SW India
Kolmar see Colmar
Köln 72 A4 var. Koeln, Eng./Fr. Cologne, prev. Cöln; anc. Colonia Agrippina, Oppidum Ubiorum. Nordrhein-Westfalen, W Germany
Koło 76 C3 Wielkopolskie, C Poland
Kołobrzeg 76 B2 Ger. Kolberg. Zachodnio-pomorskie, NW Poland
Kolokani 52 D3 Koulikoro, W Mali
Kolomea see Kolomyia
Kolomna 89 B5 Moskovskaya Oblast', W Russia
Kolomyya 86 C3 prev. Kolomyya, Ger. Kolomea. Ivano-Frankivs'ka Oblast', W Ukraine
Kolomyya see Kolomyia
Kolosjoki see Nikel'
Kolozsvár see Cluj-Napoca
Kolpa 78 A2 Ger. Kulpa, Croatian Kupa. river Croatia/Slovenia
Kolpino 88 B4 Leningradskaya Oblast', NW Russia
Kólpos Mórfu see Güzelyurt Körfezi
Kol'skiy Poluostrov 88 C2 Eng. Kola Peninsula. peninsula NW Russia
Kolwezi 55 D7 Katanga, S Dem. Rep. Congo
Kolyma 93 G2 river NE Russia
Komatsu 109 C5 var. Komatu. Ishikawa, Honshū, SW Japan
Komatu see Komatsu
Kommunizm, Qullai see Ismoili Somoní, Qullai
Komoé 53 E4 var. Komoé Fleuve. river E Ivory Coast
Komoé Fleuve see Komoé
Komotau see Chomutov
Komotiní 82 D3 var. Gümüljina, Turk. Gümülcine. Anatolikí Makedonía kai Thráki, NE Greece
Kompong Som see Sihanoukville
Komrat see Comrat
Komsomolets, Ostrov 93 E1 island Severnaya Zemlya, N Russia
Komsomol'sk-na-Amure 93 G4 Khabarovskiy Kray, SE Russia
Kondolovo 82 E3 Burgas, E Bulgaria
Kondopoga 88 B3 Respublika Kareliya, NW Russia
Kondoz see Kunduz
Köneürgenç 100 C2 var. Köneürgench, Rus. Këneurgench; prev. Kunya-Urgench. Daşoguz Welaýaty, N Turkmenistan
Kong Christian IX Land 60 D4 Eng. King Christian IX Land. physical region SE Greenland
Kong Frederik IX Land 60 C3 physical region SW Greenland
Kong Frederik VIII Land 61 E2 Eng. King Frederik VIII Land. physical region NE Greenland
Kong Frederik VI Kyst 60 C4 Eng. King Frederik VI Coast. physical region SE Greenland
Kong Karls Land 61 G2 Eng. King Charles Islands. island group SE Svalbard
Kongo see Congo (river)
Kongolo 55 D6 Katanga, E Dem. Rep. Congo
Kongor 51 B5 Jonglei, E South Sudan
Kong Oscar Fjord 61 E3 fjord E Greenland
Kongsberg 63 B6 Buskerud, S Norway
Kông, Tônle 115 E5 var. Xê Kong. river Cambodia/Laos
Kong, Xê see Kông, Tônle
Königgrätz see Hradec Králové
Königshütte see Chorzów
Konin 76 C3 Ger. Kuhnau. Weilkopolskie, C Poland
Koninkrijk der Nederlanden see Netherlands
Konispol 79 C7 var. Konispoli. Vlorë, S Albania
Konispoli see Konispol
Kónitsa 82 A4 Ípeiros, W Greece
Konitz see Chojnice
Konjic 78 C4 Federacija Bosna I Hercegovina, S Bosnia and Herzegovina
Konosha 88 C4 Arkhangel'skaya Oblast', NW Russia
Konotop 87 F1 Sums'ka Oblast', NE Ukraine
Konstantinovka see Kostiantynivka
Konstanz 73 B7 var. Constanz, Eng. Constance, hist. Kostnitz; anc. Constantia. Baden-Württemberg, S Germany
Konstanza see Constanţa
Konya 94 C4 var. Konieh, prev. Konia; anc. Iconium. Konya, C Turkey (Türkiye)
Kopaonik 79 D5 mountain range S Serbia
Kopar see Koper
Koper 73 D8 It. Capodistria; prev. Kopar. SW Slovenia
Köpetdag Gershi 100 C3 mountain range Iran/ Turkmenistan
Köpetdag Gershi/Kopetdag, Khrebet see Koppeh Dāgh
Koppeh Dāgh 98 D2 Rus. Khrebet Kopetdag, Turkm. Köpetdag Gershi. mountain range Iran/ Turkmenistan
Kopreinitz see Koprivnica
Koprivnica 78 B2 Ger. Kopreinitz, Hung. Kaproncza. Koprivnica-Križevci, N Croatia
Köprülü see Veles
Koptsevichi see Kaptsevichy
Kopyl' see Kapyl'
Korat see Nakhon Ratchasima
Korat Plateau 114 D4 plateau E Thailand
Korba 113 E4 Chhattīsgarh, C India
Korça see Korçë
Korçë 79 D6 var. Korça, Gk. Korytsa, It. Corriza; prev. Koritsa. Korçë, SE Albania
Korčula 78 B4 It. Curzola; anc. Corcyra Nigra. island S Croatia
Korea Bay 105 G3 bay China/North Korea
Korea, Democratic People's Republic of see North Korea
Korea, Republic of see South Korea
Korea Strait 109 A7 Jap. Chōsen-kaikyō, Kor. Taehan-haehyŏp. channel Japan/South Korea
Korhogo 52 D4 N Ivory Coast
Korinthos 83 B6 anc. Corinthus Eng. Corinth. Pelopónnisos, S Greece
Korinthiakós Kólpos 83 B5 Eng. Gulf of Corinth; anc. Corinthiacus Sinus. gulf C Greece
Koritsa see Korçë
Kōriyama 109 D5 Fukushima, Honshū, C Japan
Korla 104 C3 Chin. K'u-erh-lo. Xinjiang Uygur Zizhiqu, NW China
Körmend 77 B7 Vas, W Hungary
Koróni 83 B6 Pelopónnisos, S Greece
Koror 122 A2 (Palau) Oreor, N Palau
Kőrös see Križevci
Korosten' 86 D1 Zhytomyrs'ka Oblast', NW Ukraine
Koro Toro 54 C2 Borkou-Ennedi-Tibesti, N Chad
Korsovka see Kārsava
Kortrijk 65 A6 Fr. Courtrai. West-Vlaanderen, W Belgium
Koryak Range 93 H2 var. Koryakskiy Khrebet, Eng. Koryak Range. mountain range NE Russia
Koryak Range see Koryakskoye Nagor'ye
Koryakskiy Khrebet see Koryakskoye Nagor'ye
Koryazhma 88 C4 Arkhangel'skaya Oblast', NW Russia
Korytsa see Korçë
Kos 83 E6 Kos, Dodekánisa, Greece, Aegean Sea
Kos 83 E6 It. Coo; anc. Cos. island Dodekánisa, Greece, Aegean Sea
Ko-saki 109 A7 headland Nagasaki, Tsushima, SW Japan
Kościan 76 B4 Ger. Kosten. Wielkopolskie, C Poland
Kościerzyna 76 C2 Pomorskie, NW Poland
Kosciusko, Mount see Kosciuszko, Mount
Kosciuszko, Mount 127 C7 prev. Mount Kosciusko. mountain New South Wales, SE Australia
K'o-shih see Kashi
Koshikijima-retto 109 A8 var. Kosikizima Rettô. island group SW Japan
Kōshū see Gwangju
Košice 77 D6 Ger. Kaschau, Hung. Kassa. Košický Kraj, E Slovakia
Kosikizima Rettô see Koshikijima-rettō
Köslin see Koszalin
Koson 101 E3 Rus. Kasan. Qashqadaryo Viloyati, S Uzbekistan
Kosovo 79 D5 prev. Autonomous Province of Kosovo and Metohija. country (not fully recognised) SE Europe
Kosovo and Metohija, Autonomous Province of see Kosovo
Kosovo Polje see Fushë Kosovë
Kosovska Mitrovica see Mitrovicë
Kosrae 122 C2 prev. Kusaie. island Caroline Islands, E Micronesia
Kossou, Lac de 52 D5 lake C Ivory Coast
Kostanay see Qostanay
Kosten see Kościan
Kostenets 82 C2 prev. Georgi Dimitrov. Sofiya, W Bulgaria
Kostiantynivka 87 G3 prev. Kostyantynivka, Rus. Konstantinovka. Donets'ka Oblast', SE Ukraine
Kostnitz see Konstanz
Kostroma 88 B4 Kostromskaya Oblast', NW Russia
Kostyantynivka see Kostiantynivka
Kostyukovichi see Kastsyukovichy
Kostyukovka see Kastsyukowka
Koszalin 76 B2 Ger. Köslin. Zachodnio-pomorskie, NW Poland
Kota 112 D3 prev. Kotah. Rājasthān, N India
Kota Baharu see Kota Bharu
Kota Bahru see Kota Bharu
Kotabaru see Jayapura
Kota Bharu 116 B3 var. Kota Baharu, Kota Bahru. Kelantan, Peninsular Malaysia
Kotaboemi see Kotabumi
Kotabumi 116 B4 prev. Kotaboemi. Sumatera, W Indonesia
Kotah see Kota
Kota Kinabalu 116 D3 prev. Jesselton. Sabah, East Malaysia
Kotel'nyy, Ostrov 93 E2 island Novosibirskiye Ostrova, N Russia
Kotka 63 E5 Kymenlaakso, S Finland
Kotlas 88 C4 Arkhangel'skaya Oblast', NW Russia
Kotonu see Cotonou
Kotor 79 C5 It. Cattaro. SW Montenegro
Kotovsk see Podilsk
Kotovsk see Hînceşti
Kottbus see Cottbus
Kotto 54 D4 river Central African Republic/ Dem. Rep. Congo
Kotuy 93 E2 river N Russia
Koudougou 53 E4 C Burkina
Koulamoutou 55 B6 Ogooué-Lolo, C Gabon
Koulikoro 52 D3 Koulikoro, SW Mali
Koumra 54 C4 Moyen-Chari, S Chad
Kourou 37 H3 N French Guiana
Kousseir see Al Quşayr
Kousséri 54 B3 prev. Fort-Foureau. Extrême-Nord, NE Cameroon
Koutiala 52 D4 Sikasso, S Mali
Kouvola 63 E5 Kymenlaakso, S Finland
Kovel 86 C1 Pol. Kowel. Volyns'ka Oblast', NW Ukraine
Kovno see Kaunas
Koweit see Kuwait
Kowel see Kovel
Kowloon 106 A2 Hong Kong, S China
Kowno see Kaunas
Kozáni 82 B4 Dytikí Makedonía, N Greece
Kozara 78 B3 mountain range NW Bosnia and Herzegovina
Kozarska Dubica see Bosanska Dubica
Kozhikode 110 C2 var. Calicut. Kerala, SW India
Koziatyn 86 D2 prev. Kozyatyn, Rus. Kazatin. Vinnyts'ka Oblast', C Ukraine
Kozu-shima 109 D6 island E Japan
Kozyatyn see Koziatyn
Kpalimé 53 E5 var. Palimé. SW Togo
Krâchéh see Kratie
Kragujevac 78 D4 Serbia, C Serbia
Krainburg see Kranj
Kra, Isthmus of 115 B6 isthmus Malaysia/ Thailand
Krakau see Kraków
Kraków 77 D5 Eng. Cracow, Ger. Krakau; anc. Cracovia. Małopolskie, S Poland
Krâlănh 115 D5 Siĕmréab, NW Cambodia
Kralendijk 33 E5 dependent territory capital (Bonaire) Lesser Antilles, S Caribbean Sea
Kraljevo 78 D4 prev. Rankovićevo. Serbia, C Serbia
Kramatorsk 87 G3 Donets'ka Oblast', SE Ukraine
Kramfors 63 C5 Västernorrland, C Sweden
Kranéa see Kraniá
Kraniá 82 B4 var. Kranéa. Dytikí Makedonía, N Greece
Kranj 73 D7 Ger. Krainburg. NW Slovenia
Kranz see Zelenogradsk
Krāslava 84 D4 Krāslava, SE Latvia
Krasnaye 85 C5 Rus. Krasnoye. Minskaya Voblasts', C Belarus
Krasnoarmeysk 89 C6 Saratovskaya Oblast', W Russia
Krasnodar 89 A7 prev. Ekaterinodar, Yekaterinodar. Krasnodarskiy Kray, SW Russia
Krasnodon see Sorokyne
Krasnogor see Kallaste
Krasnogvardeyskoye see Kurman
Krasnohvardiyske see Kurman
Krasnokamensk 93 F4 Chitinskaya Oblast', S Russia
Krasnokamsk 89 D5 Permskaya Oblast', W Russia
Krasnoperekopsk see Yany Kapu
Krasnostav see Krasnystaw
Krasnovodsk see Türkmenbaşy
Krasnovodskiy Zaliv see Türkmenbasy Aylagy
Krasnowodsk Aylagy see Türkmenbasy Aylagy
Krasnoyarsk 92 D4 Krasnoyarskiy Kray, S Russia
Krasnoye see Krasnaye
Krasnystaw 76 E4 Rus. Krasnostav. Lubelskie, SE Poland
Krasnyy Kut 89 C6 Saratovskaya Oblast', W Russia
Krasnyi Luch see Khrustalnyi
Kratie 115 D6 Khmer. Krâchéh. Kratie, E Cambodia
Krâvanh, Chuŏr Phnum 115 C6 Eng. Cardamom Mountains, Fr. Chaîne des Cardamomes. mountain range W Cambodia
Krefeld 72 A4 Nordrhein-Westfalen, W Germany
Kreisstadt see Krosno Odrzańskie
Kremenchug see Kremenchuk
Kremenchugskoye Vodokhranilishche/ Kremenchuk Reservoir see Kremenchutske Vodoskhovyshche
Kremenchuk 87 F2 Rus. Kremenchug. Poltavs'ka Oblast', NE Ukraine
Kremenchuk Reservoir see Kremenchutske Vodoskhovyshche
Kremenchutske Vodoskhovyshche 87 F2 Eng. Kremenchuk Reservoir, Rus. Kremenchugskoye Vodokhranilishche. reservoir C Ukraine
Kremenets 86 C2 Pol. Krzemieniec. Ternopil's'ka Oblast', W Ukraine
Kremennaya see Kreminna
Kreminna 87 G2 Rus. Kremennaya. Luhans'ka Oblast', E Ukraine
Kresena see Kresna
Kresna 82 C3 var. Kresena. Blagoevgrad, SW Bulgaria
Kretikon Delagos see Kritikó Pélagos
Kretinga 84 B3 Ger. Krottingen. Klaipėda, NW Lithuania
Kreutz see Cristuru Secuiesc
Kreuz see Križevci, Croatia
Kreuz see Risti, Estonia
Kreuzburg/Kreuzburg in Oberschlesien see Kluczbork
Krichëv see Krychaw
Krievija see Russia
Krindachevka see Khrustal'nyy
Krishna 110 C1 prev. Kistna. river C India
Krishnagiri 110 C2 Tamil Nādu, SE India
Kristiania see Oslo
Kristiansand 63 A6 var. Christiansand. Vest-Agder, S Norway
Kristianstad 63 B7 Skåne, S Sweden
Kristiansund 62 A4 var. Christiansund. Møre og Romsdal, S Norway
Kríti 83 C7 Eng. Crete. island Greece, Aegean Sea
Kritikó Pélagos 83 D7 var. Kretikon Delagos, Eng. Sea of Crete; anc. Mare Creticum. sea Greece, Aegean Sea
Krivoy Rog see Kryvyi Rih
Križevci 78 B2 Ger. Kreuz, Hung. Kőrös. Varaždin, NE Croatia
Krk 78 A3 It. Veglia; anc. Curieta. island NW Croatia
Kroatien see Croatia
Krolevets 87 F1 Sums'ka Oblast', NE Ukraine
Królewska Huta see Chorzów
Kronach 73 C5 Bayern, E Germany
Kronstadt see Braşov
Kroonstad 56 D4 Free State, C South Africa
Kropotkin 89 A7 Krasnodarskiy Kray, SW Russia
Kropyvnytskyi 87 E3 prev. Kropyvnyts'kyy, Rus. Kirovohrad; prev. Kirovo, Yelizavetgrad, Zinov'yevsk. Kirovohrads'ka Oblast', C Ukraine
Kropyvnyts'kyy see Kropyvnytskyi
Krosno 77 D5 Ger. Krossen. Podkarpackie, SE Poland
Krosno Odrzańskie 76 B3 Ger. Crossen, Kreisstadt. Lubuskie, W Poland
Krossen see Krosno
Krottingen see Kretinga
Krško 73 E8 Ger. Gurkfeld; prev. Videm-Krško. E Slovenia
Krugloye see Kruhlaye
Kruhlaye 85 D6 Rus. Krugloye. Mahilyowskaya Voblasts', E Belarus
Kruja see Krujë
Krujë 79 C6 var. Kruja, It. Croia. Durrës, C Albania
Krummau see Český Krumlov
Krung Thep 115 C5 var. Krung Thep Mahanakhon, Eng. Bangkok. country capital (Thailand) Bangkok, C Thailand
Krung Thep, Ao 115 C5 var. Bight of Bangkok. bay S Thailand
Krung Thep Mahanakhon see Krung Thep
Krupki 85 D6 Minskaya Voblasts', C Belarus
Krušné Hory see Erzgebirge
Krychaw 85 E7 Rus. Krichëv. Mahilyowskaya Voblasts', E Belarus
Krym 87 F5 prev. Kryms'kyy Pivostriv, Eng. Crimea. (Ukrainian territory annexed by Russia since 2014). peninsula S Ukraine
Krymski Hory 87 F5 mountain range S Ukraine
Kryms'kyy Pivostriv see Krym
Krynica 77 D5 Ger. Tannenhof. Małopolskie, S Poland
Kryve Ozero 87 E3 Odes'ka Oblast', SW Ukraine
Kryvyi Rih 87 F3 prev. Kryvyy Rih, Rus. Krivoy Rog. Dnipropetrovska Oblast, SE Ukraine
Kryvyy Rih see Kryvyi Rih
Krzemieniec see Kremenets
Ksar al Kabir see Ksar-el-Kebir
Ksar al Soule see Er-Rachidia
Ksar-el-Kebir 48 C2 var. Alcázar, Ksar al Kabir, Ksar-el-Kébir, Ar. Al-Kasr al-Kebir, Al-Qsar al-Kbir, Sp. Alcazarquivir. NW Morocco
Ksar-el-Kébir see Ksar-el-Kebir
Kuala Dungun see Dungun
Kuala Lumpur 116 B3 country capital (Malaysia) Kuala Lumpur, Peninsular Malaysia
Kuala Terengganu 116 B3 var. Kuala Trengganu. Terengganu, Peninsular Malaysia
Kualatungkal 116 B4 Sumatera, W Indonesia
Kuang-chou see Guangzhou
Kuang-hsi see Guangxi Zhuangzu Zizhiqu
Kuang-tung see Guangdong
Kuang-yuan see Guangyuan
Kuantan 116 B3 Pahang, Peninsular Malaysia
Kuba see Quba
Kuban' 87 G5 var. Hypanis. river SW Russia
Kubango see Cubango/Okavango
Kuching 116 C3 prev. Sarawak. Sarawak, East Malaysia
Kūchnay Darwāshān 100 D5 prev. Kūchnay Darweyshān. Helmand, S Afghanistan
Kūchnay Darweyshān see Kūchnay Darwāshān
Kuçova see Kuçovë
Kuçovë 79 C6 var. Kuçova; prev. Qyteti Stalin. Berat, C Albania
Kudara see Ghūdara
Kudus 116 C5 prev. Koedoes. Jawa, C Indonesia
Kuei-lin see Guilin
Kuei-Yang/Kuei-yang see Guiyang
K'u-erh-lo see Korla
Kueyang see Guiyang
Kugaaruk 15 G3 prev. Pelly Bay. Nunavut, N Canada
Kugluktuk 31 E3 var. Qurlurtuuq; prev. Coppermine. Nunavut, NW Canada
Kuhmo 62 E4 Kainuu, E Finland
Kuhnau see Konin
Kühnö see Kihnu
Kuibyshev see Kuybyshevskoye Vodokhranilishche
Kuito see Cuito
Kuji 108 D3 var. Kuzi. Iwate, Honshū, C Japan

Kukës 79 D5 var. Kukësi. Kukës, NE Albania
Kukësi see Kukës
Kukong see Shaoguan
Kukukhoto see Hohhot
Kula Kangri 113 G3 var. Kulhakangri. mountain Bhutan/China
Kuldīga 84 B3 Ger. Goldingen. Kuldīga, W Latvia
Kuldja see Yining
Kulhakangri see Kula Kangri
Kullorsuaq 60 D2 var. Kuvdlorssuak. Avannaata, C Greenland
Kulm see Chełmno
Kulmsee see Chełmża
Kŭlob 101 F3 Rus. Kulyab. SW Tajikistan
Kulpa see Kolpa
Kulu 94 C3 Konya, W Turkey (Türkiye)
Kulunda Steppe 92 C4 Kaz. Qulyndy Zhazyghy, Rus. Kulundinskaya Ravnina. grassland Kazakhstan/Russia
Kulundinskaya Ravnina see Kulunda Steppe
Kulyab see Kŭlob
Kum see Qom
Kuma 89 B7 river SW Russia
Kumamoto 109 A7 Kumamoto, Kyūshū, SW Japan
Kumanova see Kumanovo
Kumanovo 79 E5 Turk. Kumanova. N Macedonia
Kumasi 53 E5 prev. Coomassie. C Ghana
Kumayri see Gyumri
Kumba 55 A5 Sud-Ouest, W Cameroon
Kumertau 89 D6 Respublika Bashkortostan, W Russia
Kumillā see Cumilla
Kumo 53 G4 Gombe, E Nigeria
Kumon Range 114 B2 mountain range N Myanmar (Burma)
Kumul see Hami
Kunashiri see Kunashir, Ostrov
Kunashir, Ostrov 108 E1 var. Kunashiri. island Kuril'skiye Ostrova, SE Russia
Kunda 84 E2 Lääne-Virumaa, NE Estonia
Kunduz 101 E3 prev. Kondoz. NE Afghanistan
Kunene 47 C6 var. Kunene. river Angola/Namibia
Kunene see Cunene
Kungsbacka 63 B7 Halland, S Sweden
Kungur 89 D5 Permskaya Oblast', NW Russia
Kunlun Mountains see Kunlun Shan
Kunlun Shan 104 B4 Eng. Kunlun Mountains. mountain range NW China
Kunming 106 B6 var. K'un-ming; prev. Yunnan. province capital Yunnan, SW China
K'un-ming see Kunming
Kununurra 124 D3 Western Australia
Kunya-Urgench see Köneürgenç
Kuopio 63 E5 Pohjois-Savo, C Finland
Kupa see Kolpa
Kupang 117 E5 prev. Koepang. Timor, C Indonesia
Kupiansk 87 G2 prev. Kup"yans'k, Rus. Kupyansk. Kharkivs'ka Oblast', E Ukraine
Kup"yans'k see Kupiansk
Kupyansk see Kupiansk
Kür see Kura
Kura 95 H3 Az. Kür, Geor. Mtkvari, Turk. Kura Nehri. river SW Asia
Kura Nehri see Kura
Kurashiki 109 B6 var. Kurasiki. Okayama, Honshū, SW Japan
Kurasiki see Kurashiki
Kurdistan 95 F4 cultural region SW Asia
Kŭrdzhali see Kardzhali
Kure 109 B7 Hiroshima, Honshū, SW Japan
Küre Dağları 94 C2 mountain range N Turkey (Türkiye)
Kuressaare 84 C2 Ger. Arensburg; prev. Kingissepp. Saaremaa, W Estonia
Kureyka 90 D2 river N Russia
Kurgan-Tyube see Bokhtar
Kuria Muria Islands see Ḥalāniyāt, Juzur al
Kuril'skiye Ostrova 93 H4 Eng. Kuril Islands. island group SE Russia
Kuril Islands see Kuril'skiye Ostrova
Kuril-Kamchatka Depression see Kuril-Kamchatka Trench
Kuril-Kamchatka Trench 91 F3 var. Kuril-Kamchatka Depression. trench NW Pacific Ocean
Kuril'sk 108 E1 Jap. Shana. Kuril'skiye Ostrova, Sakhalinskaya Oblats', SE Russia
Ku-ring-gai 126 E1 New South Wales, E Australia
Kurisches Haff see Courland Lagoon
Kurkund see Kilingi-Nõmme
Kurman 87 F4 prev. Krasnohvardiiske, Rus. Krasnogvardeyskoye. Respublika Krym, S Ukraine
Kurnool 110 C1 var. Karnul. Andhra Pradesh, S India
Kursk 89 A6 Kurskaya Oblast', W Russia
Kurskiy Zaliv see Courland Lagoon
Kuršumlija 79 D5 Serbia, S Serbia
Kurtbunar see Tervel
Kurtitsch/Kürtös see Curtici
Kuruktag 104 C3 mountain range NW China
Kurume 109 A7 Fukuoka, Kyūshū, SW Japan
Kurupukari 37 F3 C Guyana
Kusaie see Kosrae
Kushiro 108 D2 var. Kusiro. Hokkaidō, NE Japan
Kushka see Serhetabat
Kusiro see Kushiro
Kuskokwim Mountains 14 C3 mountain range Alaska, USA
Kustanay see Qostanay
Küstence/Küstendje see Constanța
Kütahya 94 B3 prev. Kutaia. Kütahya, W Turkey (Türkiye)
Kutai see Mahakam, Sungai
Kutaisi 95 F2 prev. K'ut'aisi 95 F2 W Georgia W Georgia
K'ut'aisi see Kutaisi
Küt al 'Amārah see Al Küt
Kut al Imara see Al Küt
Kutaradja/Kutaraja see Bandaaceh
Kutch, Gulf of see Kachchh, Gulf of
Kutch, Rann of see Kachchh, Rann of
Kutina 78 B3 Sisak-Moslavina, NE Croatia
Kutno 76 C3 Łódzkie, C Poland
Kuujjuaq 17 E2 prev. Fort-Chimo. Québec, E Canada
Kuusamo 62 E3 Pohjois-Pohjanmaa, E Finland
Kuvango see Cubango
Kuvdlorssuak see Kullorsuaq
Kuwait 98 C4 off. State of Kuwait, var. Dawlat al Kuwait, Koweit, Kuweit. country SW Asia
Kuwait see Al Kuwayt
Kuwait City see Al Kuwayt
Kuwait, Dawlat al see Kuwait
Kuwait, State of see Kuwait
Kuwajleen see Kwajalein Atoll
Kuwayt 98 C3 Maysān, E Iraq
Kuweit see Kuwait
Kuybyshev see Samara
Kuybyshev Reservoir see Kuybyshevskoye Vodokhranilishche
Kuybyshevskoye Vodokhranilishche 89 C5 var. Kuibyshev, Eng. Kuybyshev Reservoir. reservoir W Russia
Kuytun 104 B2 Xinjiang Uygur Zizhiqu, NW China
Kuzi see Kuji
Kuznetsk 89 B6 Penzenskaya Oblast', W Russia
Kuźnica 76 E2 Białystok, NE Poland Europe
Kvaløya 62 C2 island N Norway
Kvarnbergsvattnet 62 B4 var. Frostviken. lake N Sweden
Kvarner 78 A3 var. Carnaro, It. Quarnero. gulf W Croatia
Kvitøya 61 G1 island NE Svalbard
Kwajalein Atoll 122 C1 var. Kuwajleen. atoll Ralik Chain, C Marshall Islands
Kwando see Cuando
Kwangchow see Guangzhou
Kwangchu see Gwangju
Kwangju see Gwangju
Kwango 55 C7 Port. Cuango. river Angola/Dem. Rep. Congo
Kwangsi/Kwangsi Chuang Autonomous Region see Guangxi Zhuangzu Zizhiqu
Kwangtung see Guangdong
Kwangyuan see Guangyuan
Kwanza see Cuanza
Kweichu see Guiyang
Kweilin see Guilin
Kweisui see Hohhot
Kweiyang see Guiyang
Kwekwe 56 D3 prev. Que Que. Midlands, C Zimbabwe
Kwesui see Hohhot
Kwidzyń 76 C2 Ger. Marienwerder. Pomorskie, N Poland
Kwigillingok 14 C3 Alaska, USA
Kwilu 55 C6 river W Dem. Rep. Congo
Kwito see Cuito
Kyabé 54 C4 Moyen-Chari, S Chad
Kyaikkami 115 B5 prev. Amherst. Mon State, S Myanmar (Burma)
Kyaiklat 114 B4 Ayeyarwady, SW Myanmar (Burma)
Kyaikto 114 B4 Mon State, S Myanmar (Burma)
Kyakhta 93 E5 Respublika Buryatiya, S Russia
Kyaukse 114 B3 Mandalay, C Myanmar (Burma)
Kyiv 87 E2 var. Kyyiv, Kiev, Rus. Kiyev. country capital (Ukraine) Kyivska Oblast, N Ukraine
Kyiv Reservoir see Kyivske Vodoskhovyshche
Kyivske Vodoskhovyshche 87 E1 var. Kyyivs'ke Vodoskhovyshche, Eng. Kyiv Reservoir, Rus. Kiyevskoye Vodokhranilishche. reservoir N Ukraine
Kyjov 77 C5 Ger. Gaya. Jihomoravský Kraj, SE Czechia (Czech Republic)
Kykládes 83 D6 var. Kikládhes, Eng. Cyclades. island group SE Greece
Kými 83 C5 prev. Kími. Évvoia, C Greece
Kyŏngsŏng see Seoul
Kyōto 109 C6 Kyōto, Honshū, SW Japan
Kyparissía 83 B6 var. Kiparissía. Pelopónnisos, S Greece
Kypriakí Dimokratía see Cyprus
Kýpros see Cyprus
Kyrá Panagía 83 C5 island Vóreies Sporádes, Greece, Aegean Sea
Kyrenia see Girne
Kyrgyz Republic see Kyrgyzstan
Kyrgyzstan 101 F2 off. Kyrgyz Republic, var. Kirghizia; prev. Kirgizskaya SSR, Kirghiz SSR, Republic of Kyrgyzstan. country C Asia
Kyrgyzstan, Republic of see Kyrgyzstan
Kythira 83 C7 var. Kíthira, It. Cerigo, Lat. Cythera. island S Greece
Kýthnos 83 C6 Kńythnos, Kykládes, Greece, Aegean Sea
Kythnos 83 C6 var. Kíthnos, Thermiá, It. Termia; anc. Cythnos. island Kykládes, Greece, Aegean Sea
Kythréa see Değirmenlik
Kyushu 109 B7 var. Kyûsyû. island SW Japan
Kyushu-Palau Ridge 103 F3 var. Kyusyu-Palau Ridge. undersea ridge W Pacific Ocean
Kyustendil 82 B2 anc. Pautalia. Kyustendil, W Bulgaria
Kyûsyû see Kyūshū
Kyusyu-Palau Ridge see Kyushu-Palau Ridge
Kyyiv see Kyiv
Kyyivs'ke Vodoskhovyshche see Kyivske Vodoskhovyshche
Kyzyl 92 D4 Respublika Tyva, C Russia
Kyzyl Kum see Kyzylkum Desert
Kyzylkum, Peski see Kyzylkum Desert
Kyzylkum Desert 100 D2 var. Kizil Kum, Kyzyl Kum, Qizil Qum, Uzb. Qizilqum, Kaz. Qyzylqum, Rus. Peski Kyzylkum. desert Kazakhstan/Uzbekistan
Kyzylrabot see Qizilrabot
Kyzyl-Suu 101 G2 prev. Pokrovka. Issyk-Kul'skaya Oblast', NE Kyrgyzstan
Kzylorda see Qyzylorda
Kzyl-Orda see Qyzylorda

L

Laaland see Lolland
La Algaba 70 C4 Andalucía, S Spain
Laarne 65 B5 Oost-Vlaanderen, NW Belgium
La Asunción 37 E1 Nueva Esparta, NE Venezuela
Laatokka see Ladozhskoye, Ozero
Laâyoune 48 B3 var. Aaiún. country capital (Western Sahara) NW Western Sahara
La Banda Oriental see Uruguay
la Baule-Escoublac 68 A4 Loire-Atlantique, NW France
Labé 52 C4 NW Guinea
Labe see Elbe
Laborca see Laborec
Laborec 77 E5 Hung. Laborca. river E Slovakia
Labrador 17 F2 cultural region Newfoundland and Labrador, SW Canada
Labrador Basin 12 E3 var. Labrador Sea Basin. undersea basin Labrador Sea
Labrador Sea 60 A4 sea NW Atlantic Ocean
Labrador Sea Basin see Labrador Basin
Labudalin see Ergun
Labutta 115 A5 Ayeyarwady, SW Myanmar (Burma)
Laç 79 C6 var. Laci. Lezhë, C Albania
La Calera 42 B4 Valparaíso, C Chile
La Carolina 70 D4 Andalucía, S Spain
Laccadive Islands 110 A3 Eng. Laccadive Islands. island group India, N Indian Ocean
Laccadive Islands/Laccadive Minicoy and Amindivi Islands, the see Lakshadweep
La Ceiba 30 D2 Atlántida, N Honduras
Lachanás 82 B3 Kentrikí Makedonía, N Greece
La Chaux-de-Fonds 73 A7 Neuchâtel, W Switzerland
Lachlan River 127 C6 river New South Wales, SE Australia
Laci see Laç
la Ciotat 69 D6 anc. Citharista. Bouches-du-Rhône, SE France
Lacobriga see Lagos
La Concepción 31 E5 var. Concepción. Chiriquí, W Panama
La Concepción 36 C1 Zulia, NW Venezuela
La Condamine 69 C8 W Monaco
Laconia 19 G2 New Hampshire, NE USA
La Crosse 18 A2 Wisconsin, N USA
La Cruz 30 D4 Guanacaste, NW Costa Rica
Ladoga, Lake see Ladozhskoye, Ozero
Ladozhskoye, Ozero 88 B3 Eng. Lake Ladoga, Fin. Laatokka. lake NW Russia
Ladysmith 18 B2 Wisconsin, N USA
Lae 122 B3 Morobe, W Papua New Guinea
La Esperanza 30 C2 Intibucá, SW Honduras
Lafayette 18 C4 Indiana, N USA
Lafayette 20 B3 Louisiana, S USA
La Fé 32 A2 Pinar del Río, W Cuba
Lafia 53 G4 Nassarawa, C Nigeria
la Flèche 68 B4 Sarthe, NW France
Lagdo, Lac de 54 B4 lake N Cameroon
Laghouat 48 D2 N Algeria
Lagos 53 F5 Lagos, SW Nigeria
Lagos 70 B5 anc. Lacobriga. Faro, S Portugal
Lagos de Moreno 29 E4 Jalisco, SW Mexico
Lagouira 48 A4 SW Western Sahara
La Grande 24 C3 Oregon, NW USA
La Guaira 44 B4 Distrito Federal, N Venezuela
Lagunas 42 B1 Tarapacá, N Chile
Lagunillas 39 G4 Santa Cruz, SE Bolivia
La Habana 32 B2 var. Havana. country capital (Cuba) Ciudad de La Habana, W Cuba
Lahat 116 B4 Sumatera, W Indonesia
La Haye see 's-Gravenhage
Laholm 63 B7 Halland, S Sweden
Lahore 112 D2 Punjab, NE Pakistan
Lahr 73 A6 Baden-Württemberg, S Germany
Lahti 63 D5 Swe. Lahtis. Päijät-Häme, S Finland
Lahtis see Lahti
Laï 54 B4 prev. Behagle, De Behagle. Tandjilé, S Chad
Laibach see Ljubljana
Lai Châu 114 D3 Lai Châu, N Vietnam
Laila see Laylā
La Junta 22 D5 Colorado, C USA
Lake Charles 20 A3 Louisiana, S USA
Lake City 21 E3 Florida, SE USA
Lake District 67 C5 physical region NW England, United Kingdom
Lake Havasu City 26 A2 Arizona, SW USA
Lake Jackson 27 H4 Texas, SW USA
Lakeland 21 E4 Florida, SE USA
Lakeside 25 C8 California, W USA
Lake State see Michigan
Lakewood 22 D4 Colorado, C USA
Lakhnau see Lucknow
Lakonikós Kólpos 83 B7 gulf S Greece
Lakselv 62 D2 Lapp. Leavdnja. Finnmark, N Norway
la Laon see Laon
Lalibela 50 C4 Āmara, Ethiopia
La Libertad 30 B1 Petén, N Guatemala
La Ligua 42 B4 Valparaíso, C Chile
Lalín 70 C1 Galicia, NW Spain
Lalitpur 113 F3 Central, C Nepal
La Louvière 65 B6 Hainaut, S Belgium
la Maddalena 74 A4 Sardegna, Italy, C Mediterranean Sea
la Manche see English Channel
Lamar 22 D5 Colorado, C USA
La Marmora, Punta 75 A5 mountain Sardegna, Italy, C Mediterranean Sea
La Massana 69 A8 La Massana, W Andorra Europe
Lambaréné 55 A6 Moyen-Ogooué, W Gabon
Lamego 70 C2 Viseu, N Portugal
Lamesa 27 E3 Texas, SW USA
Lamezia Terme 75 D6 Calabria, SE Italy
Lamía 83 B5 Stereá Ellás, C Greece
Lamoni 23 F4 Iowa, C USA
Lampang 114 C4 var. Muang Lampang. Lampang, NW Thailand
Lámpeia 83 B6 Dytikí Ellás, S Greece
Lanbi Kyun 115 B6 prev. Sullivan Island. island Myeik Archipelago, S Myanmar (Burma)
Lancang Jiang see Mekong
Lancaster 67 D5 NW England, United Kingdom
Lancaster 25 C7 California, W USA
Lancaster 19 F4 Pennsylvania, NE USA
Lancaster Sound 15 F2 sound Nunavut, N Canada
Lan-chou/Lan-chow/Lanchow see Lanzhou
Landao see Lantau Island
Landen 65 C6 Vlaams Brabant, C Belgium
Lander 22 C3 Wyoming, C USA
Landerneau 68 A3 Finistère, NW France
Landes 69 B5 cultural region SW France
Land of Enchantment see New Mexico
The Land of Opportunity see Arkansas
Land of Steady Habits see Connecticut
Land of the Midnight Sun see Alaska
Landsberg see Gorzów Wielkopolski, Lubuskie, Poland
Landsberg an der Warthe see Gorzów Wielkopolski
Land's End 67 B8 headland SW England, United Kingdom
Landshut 73 C6 Bayern, SE Germany
Langar 101 E2 Rus. Lyangar. Navoiy Viloyati, C Uzbekistan
Langfang 106 D4 Hebei, E China
Langkawi, Pulau 115 B7 island Peninsular Malaysia
Langres 68 D4 Haute-Marne, N France
Langsa 116 A3 Sumatera, W Indonesia
Lang Shan 105 E3 mountain range N China
Lang Sơn 114 D3 var. Langson. Lang Sơn, N Vietnam
Langson see Lang Sơn
Lang Suan 115 B6 Chumphon, SW Thailand
Languedoc 69 C6 cultural region S France
Länkäran 95 H3 Rus. Lenkoran'. S Azerbaijan
Lansing 18 C3 state capital Michigan, N USA
Lanta, Ko 115 B7 island S Thailand
Lantau Island 106 A2 Cant. Tai Yue Shan, Chin. Landao. island Hong Kong, S China
Lan-ts'ang Chiang see Mekong
Lantung, Gulf of see Liaodong Wan
Lanzarote 48 B3 island Islas Canarias, Spain, NE Atlantic Ocean
Lanzhou 106 B4 var. Lan-chou, Lanchow, Lan-chow; prev. Kaolan. province capital Gansu, C China
Lao Cai 114 D3 Lao Cai, N Vietnam
Laodicea/Laodicea ad Mare see Al Lādhiqīyah
Laoet see Laut, Pulau
Laojunmiao 106 A3 prev. Yumen. Gansu, N China
Laon 68 D3 var. la Laon; anc. Laudunum. Aisne, N France
Lao People's Democratic Republic see Laos
La Orchila, Isla 36 D1 island N Venezuela
La Oroya 38 C3 Junín, C Peru
Laos 114 D4 off. Lao People's Democratic Republic. country SE Asia
La Palma 31 G5 Darién, SE Panama
La Palma 48 A3 island Islas Canarias, Spain, NE Atlantic Ocean
La Paz 39 F4 var. La Paz de Ayacucho. country capital (Bolivia-seat of government) La Paz, W Bolivia
La Paz 28 B3 Baja California Sur, NW Mexico
La Paz, Bahía de 28 B3 bay NW Mexico
La Paz de Ayacucho see La Paz
La Pérouse Strait 108 D1 Jap. Sōya-kaikyō, Rus. Proliv Laperuza. strait Japan/Russia
Laperuza, Proliv see La Pérouse Strait
Lápithos see Lapta
Lapland 62 D3 Fin. Lappi, Swe. Lappland. cultural region N Europe
La Plata 42 D4 Buenos Aires, E Argentina
La Plata see Sucre
La Pola 70 D1 var. Pola de Lena. Asturias, N Spain
Lappeenranta 63 E5 Swe. Villmanstrand. Etelä-Karjala, SE Finland
Lappi/Lappland see Lapland
Lappo see Lapua
Lapta 80 C5 Gk. Lápithos. NW Cyprus
Laptev Sea see Laptevykh, More
Laptevykh, More 93 E2 Eng. Laptev Sea. sea Arctic Ocean
Lapua 63 D5 Swe. Lappo. Etelä-Pohjanmaa, W Finland
Lapurdum see Bayonne
Łapy 76 E3 Podlaskie, NE Poland
La Quiaca 42 C2 Jujuy, N Argentina
L'Aquila 74 C4 var. Aquila, Aquila degli Abruzzi. Abruzzo, C Italy
Laracha 70 B1 Galicia, NW Spain
Laramie 22 C4 Wyoming, C USA
Laramie Mountains 22 C3 mountain range Wyoming, C USA
Laredo 71 E1 Cantabria, N Spain
Laredo 27 F5 Texas, SW USA
La Réunion see Réunion
Largeau see Faya
Largo 21 E4 Florida, SE USA
Largo, Cayo 32 B2 island W Cuba
Lario see Como, Lago di
La Rioja 42 C3 La Rioja, NW Argentina
La Rioja 71 E2 autonomous community N Spain
Lárisa 82 B4 var. Larissa. Thessalía, C Greece
Larissa see Lárisa
Larkana 112 B3 var. Larkhana. Sindh, SE Pakistan
Larkhana see Lārkāna
Larnaca see Lárnaka
Lárnaka 80 C5 var. Larnaca, Larnax. SE Cyprus
Larnax see Lárnaka
la Rochelle 68 B4 anc. Rupella. Charente-Maritime, W France
la Roche-sur-Yon 68 B4 prev. Bourbon Vendée, Napoléon-Vendée. Vendée, NW France
La Roda 71 E3 Castilla-La Mancha, C Spain
La Romana 33 E3 E Dominican Republic
Larvotto 69 C8 N Monaco Europe
La-sa see Lhasa
Las Cabezas de San Juan 70 C5 Andalucía, S Spain
Las Cruces 26 D3 New Mexico, SW USA
La See d'Urgel see La Seu d'Urgell
La Serena 42 B3 Coquimbo, C Chile
La Seu d'Urgell 71 G1 prev. La See d'Urgel, Seo de Urgel. Cataluña, NE Spain
la Seyne-sur-Mer 69 D6 Var, SE France
Lashio 114 B3 Shan State, E Myanmar (Burma)
Lashkar Gāh 100 D5 var. Lash-Kar-Gar'. Helmand, S Afghanistan
Lash-Kar-Gar' see Lashkar Gāh
La Sila 75 D6 mountain range SW Italy
La Sirena 30 D3 Región Autónoma Atlántico Sur, E Nicaragua
Łask 76 C4 Łódzkie, C Poland
Las Lomitas 42 D2 Formosa, N Argentina
La Solana 71 E4 Castilla-La Mancha, C Spain
Las Palmas 48 A3 var. Las Palmas de Gran Canaria. Gran Canaria, Islas Canarias, Spain, NE Atlantic Ocean
Las Palmas de Gran Canaria see Las Palmas
La Spezia 74 B3 Liguria, NW Italy
Lassa see Lhasa
Las Tablas 31 F5 Los Santos, S Panama
Last Frontier, The see Alaska
Las Tunas 32 C2 var. Victoria de las Tunas. Las Tunas, E Cuba
La Suisse see Switzerland
Las Vegas 25 D7 Nevada, W USA
Latacunga 38 B1 Cotopaxi, C Ecuador
Latakia see Al Lādhiqīyah
la Teste 69 B5 Gironde, SW France
Latina 75 C5 prev. Littoria. Lazio, C Italy
La Tortuga, Isla 37 E1 var. Isla Tortuga. island N Venezuela
La Tuque 17 E4 Québec, SE Canada
Latvia 84 C3 off. Republic of Latvia, Ger. Lettland, Latv. Latvija, Latvijas Republika; prev. Latvian SSR, Rus. Latviyskaya SSR. country NE Europe
Latvian SSR/Latvija/Latvijas Republika/ Latviyskaya SSR see Latvia
Latvia, Republic of see Latvia
Laudunum see Laon
Laudus see St-Lô
Lauenburg/Lauenburg in Pommern see Lębork
Lau Group 123 E4 island group E Fiji
Lauis see Lugano
Launceston 127 C8 Tasmania, SE Australia
La Unión 30 C2 Olancho, C Honduras
La Unión 71 F4 Murcia, SE Spain
Laurel 20 C3 Mississippi, S USA
Laurel 22 C2 Montana, NW USA
Laurentian Highlands see Laurentian Mountains
Laurentian Mountains 17 E3 var. Laurentian Highlands, Fr. Les Laurentides. plateau Newfoundland and Labrador/Québec, Canada
Laurentides, Les see Laurentian Mountains
Lauria 75 D6 Basilicata, S Italy
Laurinburg 21 F1 North Carolina, SE USA
Lauru see Choiseul
Lausanne 73 A7 It. Losanna. Vaud, SW Switzerland
Laut, Pulau 116 D4 prev. Laoet. island Borneo, C Indonesia
Laval 16 D4 Québec, SE Canada
Laval 68 B3 Mayenne, NW France
La Vall d'Uixó 71 F3 var. Vall D'Uxó. Comunitat Valenciana, E Spain
La Vega 33 E3 var. Concepción de la Vega. C Dominican Republic
La Vila Joíosa 71 F4 var. Villajoyosa. Comunitat Valenciana, E Spain
Lávrio 83 C6 prev. Lávrion. Attikí, C Greece
Lávrion see Lávrio
Lawrence 19 G3 Massachusetts, NE USA
Lawrenceburg 20 C1 Tennessee, S USA
Lawton 27 F2 Oklahoma, C USA
La Yarada 39 E4 Tacna, SW Peru
Laylā 99 C5 var. Laila. Ar Riyāḍ, C Saudi Arabia
Lazarev Sea 132 B1 sea Antarctica
Lázaro Cárdenas 29 E5 Michoacán, SW Mexico
Leal see Lihula
Leamhcán see Lucan
Leamington 16 C5 Ontario, S Canada
Leavdnja see Lakselv
Lebak 117 E3 Mindanao, S Philippines
Lebanese Republic see Lebanon
Lebanon 23 G5 Missouri, C USA
Lebanon 19 G2 New Hampshire, NE USA
Lebanon 24 B3 Oregon, NW USA
Lebanon 96 A4 off. Lebanese Republic, Ar. Lubnān, Fr. Liban. country SW Asia
Lebanon, Mount see Liban, Jebel
Lebap 100 D2 Lebapskiy Velayat, NE Turkmenistan
Lebedin see Lebedyn
Lebedyn 87 F2 Rus. Lebedin. Sums'ka Oblast', NE Ukraine
Lębork 76 C2 var. Lębórk, Ger. Lauenburg, Lauenburg in Pommern. Pomorskie, N Poland
Lebrija 70 C5 Andalucía, S Spain
Lebu 43 A5 Bío Bío, C Chile
le Cannet 69 D6 Alpes-Maritimes, SE France
Le Cap see Cap-Haïtien
Lecce 75 E6 Puglia, SE Italy
Lechainá 83 B6 var. Lehena, Lekhainá. Dytikí Ellás, S Greece
Ledo Salinarius see Lons-le-Saunier
Leduc 15 E5 Alberta, SW Canada
Leech Lake 23 F2 lake Minnesota, N USA
Leeds 67 D5 N England, United Kingdom
Leek 64 E2 Groningen, NE Netherlands
Leer 72 A3 Niedersachsen, NW Germany
Leeuwarden 64 D1 Fris. Ljouwert. Fryslân, N Netherlands
Leeuwin, Cape 120 A5 headland Western Australia
Leeward Islands 33 G3 island group E West Indies
Leeward Islands see Sotavento, Ilhas de
Lefkáda 83 A5 prev. Levkás. Lefkáda, Iónia Nisiá, Greece, C Mediterranean Sea
Lefkáda 83 A5 It. Santa Maura, prev. Levkás; anc. Leucas. island Iónia Nisiá, Greece, C Mediterranean Sea
Lefká Óri 83 C7 mountain range Kríti, Greece, E Mediterranean Sea
Lefkímmi 83 A5 var. Levkímmi. Kérkyra, Iónia Nisiá, Greece, C Mediterranean Sea
Lefkosía/Lefkoşa see Nicosia
Legaceaster see Chester
Legaspi see Legazpi City
Legazpi City 117 E2 var. Legaspi. C Philippines
Leghorn see Livorno
Legnica 76 B4 Ger. Liegnitz. Dolnośląskie, SW Poland
le Havre 68 B3 Eng. Havre; prev. le Havre-de-Grâce. Seine-Maritime, N France
le Havre-de-Grâce see le Havre
Lehena see Lechainá
Leicester 67 D6 Lat. Batae Coritanorum. C England, United Kingdom
Leiden 64 B3 prev. Leyden; anc. Lugdunum Batavorum. Zuid-Holland, W Netherlands
Leie 68 D2 Fr. Lys. river Belgium/France
Leinster 67 B6 Ir. Cúige Laighean. cultural region E Ireland
Leipsic see Leipzig
Leipsoí 83 E6 island Dodekánisa, Greece, Aegean Sea
Leipzig 72 C4 Pol. Lipsk, hist. Leipsic; anc. Lipsia. Sachsen, E Germany
Leiria 70 B3 anc. Collipo. Leiria, C Portugal
Leirvik 63 A6 Hordaland, S Norway
Lek 64 C4 river SW Netherlands
Lekhainá see Lechainá
Lekhchevo see Boychinovtsi
Leksand 63 C5 Dalarna, C Sweden
Lel'chitsy see Lyel'chytsy
le Léman see Geneva, Lake
Lelystad 64 D3 Flevoland, C Netherlands
Léman, Lac see Geneva, Lake
le Mans 68 B3 Sarthe, NW France
Lemberg see Lviv
Lemesós 80 C5 var. Limassol. SW Cyprus
Lemhi Range 24 D3 mountain range Idaho, C USA North America
Lemnos see Límnos
Lemovices see Limoges
Lena 93 F3 river NE Russia
Lena Tablemount 119 B7 seamount S Indian Ocean
Len Dao 106 C8 island S Spratly Islands
Lengshuitan see Yongzhou
Leninabad see Khujand
Leninakan see Gyumri
Lenine see Yedy Kuiu
Leningrad see Sankt-Peterburg
Lenino see Yedy Kuiu
Leninobod see Khujand
Leninogor see Ridder/Ridder

Leninogorsk see Ridder/Ridder
Leninpol' 101 F2 Talasskaya Oblast', NW Kyrgyzstan
Lenin-Turkmenski see Türkmenabat
Lenkoran' see Länkäran
Lenti 77 B7 Zala, SW Hungary
Lentia see Linz
Leoben 73 E7 Steiermark, C Austria
León 29 E4 var. León de los Aldamas. Guanajuato, C Mexico
León 30 C3 León, NW Nicaragua
León 70 D1 Castilla y León, NW Spain
León de los Aldamas see León
Leonídi see Leonídio
Leonídio 83 B6 var. Leonídi. Pelopónnisos, S Greece
Léopold II, Lac see Mai-Ndombe, Lac
Léopoldville see Kinshasa
Lepe 70 C4 Andalucía, S Spain
Lepel' see Lyepyel'
le Portel 68 C2 Pas-de-Calais, N France
Le Puglie see Puglia
le Puy 69 C5 prev. le Puy-en-Velay, hist. Anicium, Podium Anicensis. Haute-Loire, C France
le Puy-en-Velay see le Puy
Léré 54 B4 Mayo-Kébbi, SW Chad
Lérida see Lleida
Lerma 70 D2 Castilla y León, N Spain
Lernayin Gharabagh see Nagornyy-Karabakh
Leros 83 D6 island Dodekánisa, Greece, Aegean Sea
Lerwick 66 D1 NE Scotland, United Kingdom
Lesbos see Lésvos
Les Cayes see Cayes
Les Gonaïves see Gonaïves
Leshan 106 B5 Sichuan, C China
les Herbiers 68 B4 Vendée, NW France
Lesh/Leshi see Lezhë
Lesina see Hvar
Leskovac 79 E5 Serbia, SE Serbia
Lesnoy 92 C3 Sverdlovskaya Oblast', C Russia
Lesotho 56 D4 off. Kingdom of Lesotho; prev. Basutoland. country S Africa
Lesotho, Kingdom of see Lesotho
les Sables-d'Olonne 68 B4 Vendée, NW France
Lesser Antarctica see West Antarctica
Lesser Antilles 33 G4 island group E West Indies
Lesser Caucasus 95 F2 Rus. Malyy Kavkaz. mountain range SW Asia
Lesser Khingan Range see Xiao Hinggan Ling
Lesser Sunda Islands 117 E5 Eng. Lesser Sunda Islands. island group C Indonesia
Lesser Sunda Islands see Nusa Tenggara
Lésvos 94 A3 anc. Lesbos. island E Greece
Leszno 76 B4 Ger. Lissa. Wielkopolskie, C Poland
Lethbridge 15 E5 Alberta, SW Canada
Lethem 37 F3 S Guyana
Leti, Kepulauan 117 F5 island group E Indonesia
Letpadan 114 B4 Bago, SW Myanmar (Burma)
Letsôk-aw Kyun 115 B6 var. Letsutan Island; prev. Domel Island. island Myeik Archipelago, S Myanmar (Burma)
Letsutan Island see Letsôk-aw Kyun
Lettland see Latvia
Lëtzebuerg see Luxembourg
Leucas see Lefkáda
Leuven 65 C6 Fr. Louvain, Ger. Löwen. Vlaams Brabant, C Belgium
Leuze see Leuze-en-Hainaut
Leuze-en-Hainaut 65 B6 var. Leuze. Hainaut, SW Belgium
Léva see Levice
Levanger 62 B4 Nord-Trøndelag, C Norway
Levelland 27 E2 Texas, SW USA
Leverkusen 72 A4 Nordrhein-Westfalen, W Germany
Levice 77 C6 Ger. Lewentz, Hung. Léva, Lewenz. Nitriansky Kraj, SW Slovakia
Levin 128 D4 Manawatu-Wanganui, North Island, New Zealand
Levkás see Lefkáda
Levkímmi see Lefkímmi
Lewentz/Lewenz see Levice
Lewis, Isle of 66 B2 island NW Scotland, United Kingdom
Lewis Range 22 B1 mountain range Montana, NW USA
Lewiston 24 C2 Idaho, NW USA
Lewiston 19 G2 Maine, NE USA
Lewistown 22 C1 Montana, NW USA
Lexington 18 C5 Kentucky, S USA
Lexington 23 E4 Nebraska, C USA
Leyden see Leiden
Leyte 117 F2 island C Philippines
Leżajsk 77 E5 Podkarpackie, SE Poland
Lezha see Lezhë
Lezhë 79 C6 var. Lezha; prev. Lesh, Leshi. Lezhë, NW Albania
Lhasa 104 C5 var. La-sa, Lassa. Xizang Zizhiqu, W China
Lhaviyani Atoll see Faadhippolhu Atoll
Lhazê 104 C5 var. Quxar. Xizang Zizhiqu, China E Asia
L'Hospitalet de Llobregat 71 G2 var. Hospitalet. Cataluña, NE Spain
Liancourt Rocks 109 A5 Jap. Takeshima, Kor. Dokdo. island group Japan/South Korea
Lianyungang 106 D4 var. Xinpu. Jiangsu, E China
Liao see Liaoning
Liaodong Wan 105 G3 Eng. Gulf of Lantung, Gulf of Liaotung. gulf NE China
Liao He 103 E1 river NE China
Liaoning 106 D3 var. Liao, Liaoning Sheng, Shengking, hist. Fengtien, Shenking. province NE China
Liaoning Sheng see Liaoning
Liaoyuan 107 E3 var. Dongliao, Shuang-liao, Jap. Chengchiatun. Jilin, NE China
Liard see Fort Liard
Liban see Lebanon
Liban, Jebel 96 B4 Ar. Jabal al Gharbt, Jabal Lubnān, Eng. Mount Lebanon. mountain range C Lebanon
Libau see Liepāja
Libby 22 A1 Montana, NW USA
Liberal 23 E5 Kansas, C USA
Liberalitas Julia see Évora
Liberec 76 B4 Ger. Reichenberg. Liberecký Kraj, N Czechia (Czech Republic)
Liberia 30 D4 Guanacaste, NW Costa Rica
Liberia 52 C5 off. Republic of Liberia. country W Africa
Liberia, Republic of see Liberia
Libian Desert see Libyan Desert
Libīyah, Aş Şahrā' al see Libyan Desert
Libiyā, Dawlat see Libya
Libourne 69 B5 Gironde, SW France
Libreville 55 A5 country capital (Gabon) Estuaire, NW Gabon
Libya 49 F3 off. State of Libya, Ar. Dawlat Libiyā; prev. Libyan Arab Republic, Great Socialist People's Libyan Arab Jamahiriya. country N Africa
Libya, State of see Libya
Libyan Arab Republic see Libya
Libyan Desert 49 H4 var. Libian Desert, Ar. Aş Şahrā' al Libīyah. desert N Africa
Libyan Plateau 81 F4 var. Aḍ Ḍiffah. plateau Egypt/Libya
Lichtenfels 73 C5 Bayern, SE Germany
Lichtenvoorde 64 E4 Gelderland, E Netherlands
Lichuan 106 C5 Hubei, C China
Lida 85 B5 Hrodzyenskaya Voblasts', W Belarus
Lidhorikíon see Lidoríki
Lidköping 63 B6 Västra Götaland, S Sweden
Lidokhorikion see Lidoríki
Lidoríki 83 B5 prev. Lidhorikíon, Lidokhorikion. Stereá Ellás, C Greece
Lidzbark Warmiński 76 D2 Ger. Heilsberg. Olsztyn, N Poland
Liechtenstein 72 D1 off. Principality of Liechtenstein. country C Europe
Liechtenstein, Principality of see Liechtenstein
Liège 65 D6 Dut. Luik, Ger. Lüttich. Liège, E Belgium
Liegnitz see Legnica
Lienz 73 D7 Tirol, W Austria
Liepāja 84 B3 Ger. Libau. Liepāja, W Latvia
Lietuva see Lithuania
Lievenhof see Livāni
Liezen 73 D7 Steiermark, C Austria
Liffey 67 B6 river Ireland
Lifou 122 D5 island Îles Loyauté, E New Caledonia
Liger see Loire
Ligure, Appennino 74 A2 Eng. Ligurian Mountains. mountain range NW Italy
Ligure, Mar see Ligurian Sea
Ligurian Mountains see Ligure, Appennino
Ligurian Sea 74 A3 Fr. Mer Ligurienne, It. Mar Ligure. sea N Mediterranean Sea
Ligurienne, Mer see Ligurian Sea
Lihu'e 25 A7 var. Lihue. Kaua'i, Hawaii, USA
Lihue see Lihu'e
Lihula 84 D2 Ger. Leal. Läänemaa, W Estonia
Liivi Laht see Riga, Gulf of
Likasi 55 D7 prev. Jadotville. Shaba, SE Dem. Rep. Congo
Likenes 63 A6 Vest-Agder, S Norway
Lille 68 C2 var. l'Isle, Dut. Rijssel, Flem. Ryssel, prev. Lisle; anc. Insula. Nord, N France
Lillehammer 63 B5 Oppland, S Norway
Lillestrøm 63 B6 Akershus, S Norway
Lilongwe 57 E2 country capital (Malawi) Central, W Malawi
Lilybaeum see Marsala
Lima 38 C4 country capital (Peru) Lima, W Peru
Limanowa 77 D5 Małopolskie, S Poland
Limassol see Lemesós
Limerick 67 A6 Ir. Luimneach. Limerick, SW Ireland
Limín Vathéos see Sámos
Límnos 81 F3 anc. Lemnos. island E Greece
Limoges 69 C5 anc. Augustoritum Lemovicensium, Lemovices. Haute-Vienne, C France
Limón 31 E4 var. Puerto Limón. Limón, E Costa Rica
Limón 30 D2 Colón, NE Honduras
Limonum see Poitiers
Limousin 69 C5 cultural region C France
Limoux 69 C6 Aude, S France
Limpopo 56 D3 var. Crocodile. river S Africa
Linares 42 B4 Maule, C Chile
Linares 29 E3 Nuevo León, NE Mexico
Linares 70 D4 Andalucía, S Spain
Lincoln 67 D5 anc. Lindum, Lindum Colonia. E England, United Kingdom
Lincoln 19 H2 Maine, NE USA
Lincoln 23 F4 state capital Nebraska, C USA
Lincoln Sea 12 D2 sea Arctic Ocean
Linden 37 F3 E Guyana
Líndhos see Líndos
Lindi 51 D8 Lindi, SE Tanzania
Líndos 83 E7 var. Líndhos. Ródos, Dodekánisa, Greece, Aegean Sea
Lindum/Lindum Colonia see Lincoln
Line Islands 123 G3 island group E Kiribati
Lingeh see Bandar-e Lengeh
Lingen 72 A3 var. Lingen an der Ems. Niedersachsen, NW Germany
Lingen an der Ems see Lingen
Lingga, Kepulauan 116 B4 island group W Indonesia
Linköping 63 C6 Östergötland, S Sweden
Linz 73 D6 anc. Lentia. Oberösterreich, N Austria
Lion, Golfe du 69 C7 Eng. Gulf of Lion, Gulf of Lions; anc. Sinus Gallicus. gulf S France
Lion, Gulf of/Lions, Gulf of see Lion, Golfe du
Liozno see Lyozna
Lipari 75 D6 island Isole Eolie, S Italy
Lipari Islands/Lipari, Isole see Eolie, Isole
Lipetsk 89 B5 Lipetskaya Oblast', W Russia
Lipno 76 C3 Kujawsko-pomorskie, C Poland
Lipova 86 A4 Hung. Lippa. Arad, W Romania
Lipovets see Lypovets
Lippa see Lipova
Lipsia/Lipsk see Leipzig
Lira 51 B6 N Uganda
Lisala 55 C5 Equateur, N Dem. Rep. Congo
Lisboa 70 B4 Eng. Lisbon; anc. Felicitas Julia, Olisipo. country capital (Portugal) Lisboa, W Portugal
Lisbon see Lisboa
Lisichansk see Lysychansk
Lisieux 68 B3 anc. Noviomagus. Calvados, N France
Liski 89 B6 prev. Georgiu-Dezh. Voronezhskaya Oblast', W Russia
Lisle/l'Isle see Lille
Lismore 127 E5 New South Wales, SE Australia
Lissa see Vis, Croatia
Lissa see Leszno, Poland
Lisse 64 C3 Zuid-Holland, W Netherlands
Litang 106 A5 var. Gaocheng. Sichuan, C China
Lītani, Nahr el 97 B5 var. Nahr al Litant. river C Lebanon
Litant, Nahr al see Lītani, Nahr el
Litauen see Lithuania
Lithgow 127 D6 New South Wales, SE Australia
Lithuania 84 B4 off. Republic of Lithuania, Ger. Litauen, Lith. Lietuva, Pol. Litwa, Rus. Litva; prev. Lithuanian SSR, Rus. Litovskaya SSR. country NE Europe
Lithuanian SSR see Lithuania
Lithuania, Republic of see Lithuania
Litóchoro 82 B4 var. Litohoro, Litókhoron. Kentrikí Makedonía, N Greece
Litohoro/Litókhoron see Litóchoro
Litovskaya SSR see Lithuania
Little Alföld 77 C6 Ger. Kleines Ungarisches Tiefland, Hung. Kisalföld, Slvk. Podunajská Rovina. plain Hungary/Slovakia
Little Andaman 111 F2 island Andaman Islands, India, NE Indian Ocean
Little Barrier Island see Te Hauturu-o-Toi
Little Bay 71 H5 bay Alboran Sea, Mediterranean Sea
Little Cayman 32 B3 island E Cayman Islands
Little Falls 23 F2 Minnesota, N USA
Littlefield 27 E2 Texas, SW USA
Little Inagua 32 D2 var. Inagua Islands. island S The Bahamas
Little Minch, The 66 B3 strait NW Scotland, United Kingdom
Little Missouri River 22 D2 river NW USA
Little Nicobar 111 G3 island Nicobar Islands, India, NE Indian Ocean
Little Rhody see Rhode Island
Little Rock 20 B1 state capital Arkansas, C USA
Little Saint Bernard Pass 69 D5 Fr. Col du Petit St-Bernard, It. Colle del Piccolo San Bernardo. pass France/Italy
Little Sound 20 A5 bay Bermuda, NW Atlantic Ocean
Littleton 22 D4 Colorado, C USA
Littoria see Latina
Litva/Litwa see Lithuania
Liubotyn 87 G2 prev. Lyubotyn, Rus. Lyubotin. Kharkivs'ka Oblast', E Ukraine
Liu-chou/Liuchow see Liuzhou
Liuzhou 106 C6 var. Liu-chou, Liuchow. Guangxi Zhuangzu Zizhiqu, S China
Livanátai see Livanátes
Livanátes 83 B5 prev. Livanátai. Stereá Ellás, C Greece
Līvāni 84 D4 Ger. Lievenhof. Preiļi, SE Latvia
Liverpool 17 F5 Nova Scotia, SE Canada
Liverpool 67 C5 NW England, United Kingdom
Livingston 22 B2 Montana, NW USA
Livingston 27 H3 Texas, SW USA
Livingstone 56 C3 var. Maramba. Southern, S Zambia
Livingstone Mountains 129 A7 mountain range South Island, New Zealand
Livno 78 B4 Federicija Bosna I Hercegovina, SW Bosnia and Herzegovina
Livojoki 62 D4 river C Finland
Livonia 18 D3 Michigan, N USA
Livorno 74 B3 Eng. Leghorn. Toscana, C Italy
Lixian Jiang see Black River
Lixoúri 83 A5 prev. Lixoúrion. Kefallinía, Iónia Nisiá, Greece, C Mediterranean Sea
Lixoúrion see Lixoúri
Lizarra see Estella
Ljouwert see Leeuwarden
Ljubelj see Loibl Pass
Ljubljana 73 D7 Ger. Laibach, It. Lubiana; anc. Aemona, Emona. country capital (Slovenia) C Slovenia
Ljungby 63 B7 Kronoberg, S Sweden
Ljusdal 63 C5 Gävleborg, C Sweden
Ljusnan 63 C5 river C Sweden
Llanelli 67 C6 prev. Llanelly. SW Wales, United Kingdom
Llanelly see Llanelli
Llanes 70 D1 Asturias, N Spain
Llanos 36 D2 physical region Colombia/Venezuela
Lleida 71 F2 Cast. Lérida; anc. Ilerda. Cataluña, NE Spain
Llucmajor 71 G3 Mallorca, Spain, W Mediterranean Sea
Loaita Island 106 C8 island W Spratly Islands
Loanda see Luanda
Lobamba 56 D4 country capital (Eswatini- royal and legislative) NW Eswatini
Lobatse 56 C4 var. Lobatsi. Kgatleng, SE Botswana
Lobatsi see Lobatse
Löbau 72 D4 Sachsen, E Germany
Lobito 56 B2 Benguela, W Angola
Lob Nor see Lop Nur
Lobositz see Lovosice
Loburi see Lop Buri
Locarno 73 B8 Ger. Luggarus. Ticino, S Switzerland
Lochem 64 E3 Gelderland, E Netherlands
Lockport 19 E3 New York, NE USA
Lodja 55 D6 Kasai-Oriental, C Dem. Rep. Congo
Lodwar 51 C6 Rift Valley, NW Kenya
Łódź 76 D4 Rus. Lodz. Łódź, C Poland
Loei 114 C4 var. Loey, Muang Loei. Loei, C Thailand
Loey see Loei
Lofoten 62 B3 var. Lofoten Islands. island group C Norway
Lofoten Islands see Lofoten
Logan 22 B3 Utah, W USA
Logan, Mount 14 D3 mountain Yukon, W Canada
Logroño 71 E1 anc. Vareia, Lat. Juliobriga. La Rioja, N Spain
Loibl Pass 73 D7 Ger. Loiblpass, Slvn. Ljubelj. pass Austria/Slovenia
Loiblpass see Loibl Pass
Loikaw 114 B4 Kayah State, C Myanmar (Burma)
Loire 68 B4 var. Liger. river C France
Loja 38 B2 Loja, S Ecuador
Lokitaung 51 C5 Rift Valley, NW Kenya
Lokoja 53 G4 Kogi, C Nigeria
Loksa 84 E2 Ger. Loxa. Harjumaa, NW Estonia
Lolland 63 B8 prev. Laaland. island S Denmark
Lom 82 C1 prev. Lom-Palanka. Montana, NW Bulgaria
Lomami 55 D6 river C Dem. Rep. Congo
Lomas 38 D4 Arequipa, SW Peru
Lomas de Zamora 42 D4 Buenos Aires, E Argentina
Lombamba 56 D4 country capital (Eswatini - royal and legislative) NW Eswatini
Lombardia 74 B2 Eng. Lombardy. region N Italy
Lombardy see Lombardia
Lombok, Pulau 116 D5 island Nusa Tenggara, C Indonesia
Lomé 53 F5 country capital (Togo) S Togo
Lomela 55 D6 Kasai-Oriental, C Dem. Rep. Congo
Lommel 65 C5 Limburg, N Belgium
Lomond, Loch 66 B4 lake C Scotland, United Kingdom
Lomonosov Ridge 133 B3 var. Harris Ridge, Rus. Khrebet Homonsova. undersea ridge Arctic Ocean
Lomonsova, Khrebet see Lomonosov Ridge
Lom-Palanka see Lom
Lompoc 25 B7 California, W USA
Lom Sak 114 C4 var. Muang Lom Sak. Phetchabun, C Thailand
Łomża 76 D3 Rus. Lomzha. Podlaskie, NE Poland
Lomzha see Łomża
Loncoche 43 B5 Araucanía, C Chile
Londinium see London
London 67 A7 anc. Augusta, Lat. Londinium. country capital (United Kingdom) SE England, United Kingdom
London 16 C5 Ontario, S Canada
London 18 C5 Kentucky, S USA
Londonderry 66 B4 var. Derry, Ir. Doire. NW Northern Ireland, United Kingdom
Londonderry, Cape 124 C2 cape Western Australia
Londrina 41 E4 Paraná, S Brazil
Lone Star State see Texas
Long Bay 21 F2 bay W Jamaica
Long Beach 25 C7 California, W USA
Longford 67 B5 Ir. An Longfort. Longford, C Ireland
Long Island 32 D2 island C The Bahamas
Long Island 19 G4 island New York, NE USA
Longlac 16 C3 Ontario, S Canada
Longmont 22 D4 Colorado, C USA
Longreach 126 C4 Queensland, E Australia
Long Strait 93 G1 Eng. Long Strait. strait NE Russia
Long Strait see Longa, Proliv
Longview 27 H3 Texas, SW USA
Longview 24 B2 Washington, NW USA
Long Xuyên 115 D6 var. Longxuyen. An Giang, S Vietnam
Longxuyen see Long Xuyên
Longyan 106 D6 Fujian, SE China
Longyearbyen 61 G2 dependent territory capital (Svalbard) Spitsbergen, W Svalbard
Lons-le-Saunier 68 D4 anc. Ledo Salinarius. Jura, E France
Lop Buri 115 C5 var. Loburi. Lop Buri, C Thailand
Lop Nor see Lop Nur
Lop Nur 104 C3 var. Lob Nor, Lop Nor, Lo-pu Po. seasonal lake NW China
Loppersum 64 E1 Groningen, NE Netherlands
Lo-pu Po see Lop Nur
Lorca 71 E4 Ar. Lurka; anc. Eliocroca, Lat. Illurco. Murcia, S Spain
Lord Howe Island 120 C4 island E Australia
Lord Howe Rise 120 C4 undersea rise SW Pacific Ocean
Loreto 28 B3 Baja California Sur, NW Mexico
Lorient 68 A3 prev. l'Orient. Morbihan, NW France
l'Orient see Lorient
Lorn, Firth of 66 B4 inlet W Scotland, United Kingdom
Lörrach 73 A7 Baden-Württemberg, S Germany
Lorraine 68 D3 cultural region NE France
Los Alamos 26 C1 New Mexico, SW USA
Los Amates 30 B2 Izabal, E Guatemala
Los Ángeles 43 B5 Bío Bío, C Chile
Los Angeles 25 C7 California, W USA
Losanna see Lausanne
Lošinj 78 A3 Ger. Lussin, It. Lussino. island W Croatia
Loslau see Wodzisław Śląski
Los Mochis 28 C3 Sinaloa, C Mexico
Losonc/Losontz see Lučenec
Los Roques, Islas 36 D1 island group N Venezuela
Lot 69 B5 cultural region S France
Lot 69 B5 river S France
Lotagipi Swamp 51 C5 wetland Kenya/ South Sudan
Lötzen see Giżycko
Loualaba see Lualaba
Louangnamtha 114 C3 var. Luong Nam Tha. Louang Namtha, N Laos
Louangphabang 102 D3 var. Louangphrabang, Luang Prabang. Louangphabang, N Laos
Louangphrabang see Louangphabang
Loubomo see Dolisie
Loudéac 68 A3 Côtes d'Armor, NW France
Loudi 106 C5 Hunan, S China
Louga 52 B3 NW Senegal
Louisiade Archipelago 122 B4 island group SE Papua New Guinea
Louisiana 20 A2 off. State of Louisiana, also known as Creole State, Pelican State. state S USA
Louisville 18 C5 Kentucky, S USA
Louisville Ridge 121 E4 undersea ridge S Pacific Ocean
Loup River 23 E4 river Nebraska, C USA
Lourdes 69 B6 Hautes-Pyrénées, S France
Lourenço Marques see Maputo
Louth 67 E5 E England, United Kingdom
Loutrá 82 C4 Kentrikí Makedonía, N Greece
Louvain see Leuven
Louvain-la Neuve 65 C6 Walloon Brabant, C Belgium
Louviers 68 C3 Eure, N France
Lovech 82 C2 Lovech, N Bulgaria
Loveland 22 D4 Colorado, C USA
Lovosice 76 A4 Ger. Lobositz. Ústecký Kraj, NW Czechia (Czech Republic)
Lóvua 56 C1 Moxico, E Angola
Lowell 19 G3 Massachusetts, NE USA
Löwen see Leuven
Lower California see Baja California
Lower Hutt 129 D5 Wellington, North Island, New Zealand
Lower Lough Erne 67 A5 lake SW Northern Ireland, United Kingdom
Lower Red Lake 23 F1 lake Minnesota, N USA
Lower Rhine see Neder Rijn
Lower Tunguska see Nizhnyaya Tunguska
Lowestoft 67 E6 E England, United Kingdom
Loxa see Loksa
Lo-yang see Luoyang
Loyauté, Îles 122 D5 island group S New Caledonia
Loyev see Loyew
Loyew 85 D8 Rus. Loyev. Homyel'skaya Voblasts', SE Belarus
Loznica 78 C3 Serbia, W Serbia
Lu see Shandong, China
Lualaba 55 D6 Fr. Loualaba. river SE Dem. Rep. Congo
Luanda 56 A1 var. Loanda, Port. São Paulo de Loanda. country capital (Angola) Luanda, NW Angola
Luang Prabang see Louangphabang
Luang, Thale 115 C7 lagoon S Thailand
Luangua, Rio see Luangwa
Luangwa 51 B8 var. Aruângua, Rio Luangua. river Mozambique/Zambia
Luanshya 56 D2 Copperbelt, C Zambia
Luarca 70 C1 Asturias, N Spain
Lubaczów 77 E5 var. Lúbaczów. Podkarpackie, SE Poland
Lubnān see Lebanon
L'uban' 76 B4 Leningradskaya Oblast', Russia
Lubānas Ezers see Lubāns
Lubango 56 B2 Port. Sá da Bandeira. Huíla, SW Angola
Lubāns 84 D4 var. Lubānas Ezers. lake E Latvia
Lubao 55 D6 Kasai-Oriental, C Dem. Rep. Congo
Lübben 72 D4 Brandenburg, E Germany
Lübbenau 72 D4 Brandenburg, E Germany
Lubbock 27 E2 Texas, SW USA
Lübeck 72 C2 Schleswig-Holstein, N Germany
Lubelska, Wyżyna 76 E4 plateau SE Poland
Lüben see Lubin
Lubiana see Ljubljana
Lubin 76 B4 Ger. Lüben. Dolnośląskie, SW Poland
Lublin 76 E4 Rus. Lyublin. Lubelskie, E Poland
Lubliniec 76 C4 Śląskie, S Poland
Lubnān, Jabal see Liban, Jebel
Lubny 87 F2 Poltavs'ka Oblast', NE Ukraine
Lubsko 76 B4 Ger. Sommerfeld. Lubuskie, W Poland
Lubumbashi 55 E8 prev. Élisabethville. Shaba, SE Dem. Rep. Congo
Lubutu 55 D6 Maniema, E Dem. Rep. Congo
Luca see Lucca
Lucan 67 B5 Ir. Leamhcán. Dublin, E Ireland
Lucanian Mountains see Lucano, Appennino
Lucano, Appennino 75 D5 Eng. Lucanian Mountains. mountain range S Italy
Lucapa 56 C1 var. Lukapa. Lunda Norte, NE Angola
Lucca 74 B3 anc. Luca. Toscana, C Italy
Lucea 32 A4 W Jamaica
Lucena 117 E1 off. Lucena City. Luzon, N Philippines
Lucena 70 D4 Andalucía, S Spain
Lucena City see Lucena
Lučenec 77 D6 Ger. Losontz, Hung. Losonc. Banskobystrický Kraj, C Slovakia
Lucentum see Alicante
Lucerna/Lucerne see Luzern
Luchow see Hefei
Łuck see Lutsk
Lucknow 113 E3 var. Lakhnau. state capital Uttar Pradesh, N India
Lüda see Dalian
Luda Kamchia 82 D2 river E Bulgaria
Ludasch see Luduş
Lüderitz 56 B4 prev. Angra Pequena. Karas, SW Namibia
Ludhiāna 112 D2 Punjab, N India
Ludington 18 C2 Michigan, N USA
Ludsan see Ludza
Luduş 86 B4 Ger. Ludasch, Hung. Marosludas. Mureş, C Romania
Ludvika 63 C6 Dalarna, C Sweden
Ludwigsburg 73 B6 Baden-Württemberg, SW Germany
Ludwigsfelde 72 D3 Brandenburg, NE Germany
Ludwigshafen 73 B5 var. Ludwigshafen am Rhein. Rheinland-Pfalz, W Germany
Ludwigshafen am Rhein see Ludwigshafen
Ludwigslust 72 C3 Mecklenburg-Vorpommern, N Germany
Ludza 84 D4 Ger. Ludsan. Ludza, E Latvia
Luebo 55 C6 Kasai-Occidental, SW Dem. Rep. Congo
Luena 56 C2 var. Lwena, Port. Luso. Moxico, E Angola
Lufira 55 E7 river SE Dem. Rep. Congo
Lufkin 27 H3 Texas, SW USA
Luga 88 A4 Leningradskaya Oblast', NW Russia
Lugano 73 B8 Ger. Lauis. Ticino, S Switzerland
Lugdunum see Lyon
Lugdunum Batavorum see Leiden
Lugenda, Rio 57 E2 river N Mozambique
Luggarus see Locarno
Lugh Ganana see Luuq
Lugo 70 C1 anc. Lugus Augusti. Galicia, NW Spain
Lugoj 86 A4 Ger. Lugosch, Hung. Lugos. Timiş, W Romania
Lugos/Lugosch see Lugoj
Lugus Augusti see Lugo
Luguvallium/Luguvallum see Carlisle
Luhansk 87 H3 Voroshilovgrad. Luhans'ka Oblast', E Ukraine
Luimneach see Limerick
Lukapa see Lucapa
Lukenie 55 C6 river C Dem. Rep. Congo
Lukovit 82 C2 Lovech, N Bulgaria
Łuków 76 E4 Ger. Bogendorf. Lubelskie, E Poland
Lukuga 55 D7 river SE Dem. Rep. Congo
Luleå 62 D4 Norrbotten, N Sweden
Luleälven 62 C3 river N Sweden
Lulonga 55 C5 river NW Dem. Rep. Congo
Lulua 55 D7 river S Dem. Rep. Congo
Luluabourg see Kananga
Lumber State see Maine
Lumbo 57 F2 Nampula, NE Mozambique
Lumsden 129 A7 Southland, South Island, New Zealand
Lund 63 B7 Skåne, S Sweden
Lüneburg 72 C3 Niedersachsen, N Germany
Lunga, Isola see Dugi Otok
Lungkiang see Qiqihar
Lungué-Bungo 56 C2 var. Lungwebungu. river Angola/Zambia
Lungwebungu see Lungué-Bungo
Luninets see Luninyets
Łuniniec see Luninyets
Luninyets 85 B7 Pol. Łuniniec, Rus. Luninets. Brestskaya Voblasts', SW Belarus
Lunteren 64 D4 Gelderland, C Netherlands

Luong Nam Tha see Louangnamtha
Luoyang 106 C4 var. Honan, Lo-yang. Henan, C China
Lupatia see Altamura
Lúrio 57 F2 Nampula, NE Mozambique
Lúrio, Rio 57 E2 river NE Mozambique
Lurka see Lorca
Lusaka 56 D2 country capital (Zambia) Lusaka, SE Zambia
Lushnja see Lushnjë
Lushnjë 79 C6 var. Lushnja. Fier, C Albania
Luso see Luena
Lussin/Lussino see Lošinj
Lūt, Baḩrat/Lut, Bahret see Dead Sea
Lut, Dasht-e 98 D3 var. Kavīr-e Lūt. desert E Iran
Lutetia/Lutetia Parisiorum see Paris
Lūt, Kavīr-e see Lūt, Dasht-e
Luton 67 D6 E England, United Kingdom
Łutselk'e 15 F4 prev. Snowdrift. Northwest Territories, W Canada
Lutsk 86 C1 Pol. Łuck. Volyns'ka Oblast', NW Ukraine
Lüttich see Liège
Lutzow-Holm Bay 132 C2 var. Lutzow-Holm Bay. bay Antarctica
Lutzow-Holm Bay see Lützow Holmbukta
Luuq 51 D6 It. Lugh Ganana. Gedo, SW Somalia
Luvua 55 D7 river SE Dem. Rep. Congo
Luwego 51 C8 river S Tanzania
Luxembourg 65 D8 country capital (Luxembourg) Luxembourg, S Luxembourg
Luxembourg 65 D8 off. Grand Duchy of Luxembourg, var. Lëtzebuerg, Luxemburg. country NW Europe
Luxemburg see Luxembourg
Luxor see Al Uqşur
Luza 88 C4 Kirovskaya Oblast', NW Russia
Luz, Costa de la 70 C5 coastal region SW Spain
Luzern 73 B7 Fr. Lucerne, It. Lucerna. Luzern, C Switzerland
Luzon 117 E1 island N Philippines
Luzon Strait 103 E3 strait Philippines/Taiwan
Lviv 86 B2 Ger. Lemberg, Pol. Lwów, Rus. L'vov. L'vivs'ka Oblast', W Ukraine
L'vov see Lviv
Lwena see Luena
Lwów see Lviv
Lyakhavichy 85 B6 Rus. Lyakhovichi. Brestskaya Voblasts', SW Belarus
Lyakhovichi see Lyakhavichy
Lyallpur see Faisalābād
Lyangar see Langar
Lyck see Ełk
Lycksele 62 C4 Västerbotten, N Sweden
Lycopolis see Asyūţ
Lyel'chytsy 85 C7 Rus. Lel'chitsy. Homyel'skaya Voblasts', SE Belarus
Lyepyel' 85 D5 Rus. Lepel'. Vitsyebskaya Voblasts', N Belarus
Lyme Bay 67 C7 bay S England, United Kingdom
Lynchburg 19 E5 Virginia, NE USA
Lynn see King's Lynn
Lynn Lake 15 F4 Manitoba, C Canada
Lynn Regis see King's Lynn
Lyon 69 D5 Eng. Lyons; anc. Lugdunum. Rhône, E France
Lyons see Lyon
Lyozna 85 E6 Rus. Liozno. Vitsyebskaya Voblasts', NE Belarus
Lypovets 86 D2 Rus. Lipovets. Vinnyts'ka Oblast', C Ukraine
Lys see Leie
Lysychansk 87 H3 Rus. Lisichansk. Luhans'ka Oblast', E Ukraine
Lyttelton 129 C6 South Island, New Zealand
Lyublin see Lublin
Lyubotin see Liubotyn
Lyubotyn see Liubotyn
Lyulyakovo 82 E2 prev. Keremitlik. Burgas, E Bulgaria
Lyusina 85 B6 Rus. Lyusino. Brestskaya Voblasts', SW Belarus
Lyusino see Lyusina

M

Maale 110 B4 var. Male'. country capital (Maldives) Male' Atoll, C Maldive
Ma'ān 97 B7 Ma'ān, SW Jordan
Maardu 84 D2 Ger. Maart. Harjumaa, NW Estonia
Ma'aret-en-Nu'man see Ma'arrat an Nu'mān
Ma'arrat an Nu'mān 96 B3 var. Ma'aret-en-Nu'man, Fr. Maarret enn Naamâne. Idlib, NW Syria
Maarret enn Naamâne see Ma'arrat an Nu'mān
Maart see Maardu
Maas see Meuse
Maaseik 65 D5 prev. Maeseyck. Limburg, NE Belgium
Maastricht 65 D6 var. Maestricht; anc. Traiectum ad Mosam, Traiectum Tungorum. Limburg, SE Netherlands
Macao 107 C6 Port. Macau. Special administrative region Guangdong, SE China
Macapá 41 E1 state capital Amapá, N Brazil
Macarsca see Makarska
Macassar see Makassar
Macău see Makó, Hungary
Macau see Macao
MacCluer Gulf see Berau, Teluk
Macdonnell Ranges 124 D4 mountain range Northern Territory, C Australia
Macedonia see North Macedonia
Maceió 41 G3 state capital Alagoas, E Brazil
Machachi 38 B1 Pichincha, C Ecuador
Machala 38 B2 El Oro, SW Ecuador
Machanga 57 E3 Sofala, E Mozambique
Machilīpatnam 110 D1 var. Bandar Masulipatnam. Andhra Pradesh, E India
Machiques 36 C2 Zulia, NW Venezuela
Macías Nguema Biyogo see Bioco, Isla de
Măcin 86 D5 Tulcea, SE Romania
Mackay 126 D4 Queensland, NE Australia
Mackay, Lake 124 C4 salt lake Northern Territory/Western Australia
Mackenzie 15 E3 river Northwest Territories, NW Canada
Mackenzie Bay 132 D3 bay Antarctica
Mackenzie Mountains 14 D3 mountain range Northwest Territories, NW Canada
Macleod, Lake 124 A4 lake Western Australia
Macomb 18 A4 Illinois, N USA
Macomer 75 A5 Sardegna, Italy, C Mediterranean Sea
Mâcon 69 D5 anc. Matisco, Matisco Ædourum. Saône-et-Loire, C France
Macon 20 D2 Georgia, SE USA
Macon 23 G4 Missouri, C USA
Macquarie Ridge 132 C5 undersea ridge SW Pacific Ocean
Macuspana 29 G4 Tabasco, SE Mexico
Ma'dabā 97 B6 var. Mādabā, Madeba; anc. Medeba. Ma'dabā, NW Jordan
Mādabā see Ma'dabā
Madagascar 57 F3 off. Republic of Madagascar, Malg. Madagasikara; prev. Democratic Republic of Madagascar, Malagasy Republic. country W Indian Ocean
Madagascar 57 F3 island W Indian Ocean
Madagascar Basin 47 E7 undersea basin W Indian Ocean
Madagascar, Democratic Republic of see Madagascar
Madagascar, Republic of see Madagascar
Madagascar Plateau 47 E7 var. Madagascar Ridge, Madagascar Rise, Rus. Madagaskarskiy Khrebet. undersea plateau W Indian Ocean
Madagascar Rise/Madagascar Ridge see Madagascar Plateau
Madagasikara see Madagascar
Madagaskarskiy Khrebet see Madagascar Plateau
Madang 122 B3 Madang, N Papua New Guinea
Madanīyīn see Médenine
Mađarska see Hungary
Made 64 C4 Noord-Brabant, S Netherlands
Madeba see Ma'dabā
Madeira 48 A2 var. Ilha de Madeira. island Madeira, Portugal, NE Atlantic Ocean
Madeira, Ilha de see Madeira
Madeira Plain 44 C3 abyssal plain E Atlantic Ocean
Madeira, Rio 40 D2 var. Río Madera. river Bolivia/Brazil
Madeleine, Îles de la 17 F4 Eng. Magdalen Islands. island group Québec, E Canada
Madera 25 B6 California, W USA
Madera, Río see Madeira, Rio
Madhya Pradesh 113 E4 prev. Central Provinces and Berar. cultural region C India
Madīnat ath Thawrah see Ath Thawrah
Madioen see Madiun
Madison 23 F3 South Dakota, N USA
Madison 18 B3 state capital Wisconsin, N USA
Madiun 116 D5 prev. Madioen. Jawa, C Indonesia
Madoera see Madura
Madona 84 D4 Ger. Modohn. Madona, E Latvia
Madras see Chennai
Madras see Tamil Nādu
Madre de Dios, Río 39 E3 river Bolivia/Peru
Madre del Sur, Sierra 29 E5 mountain range S Mexico
Madre, Laguna 29 F3 lagoon NE Mexico
Madre, Laguna 27 G5 lagoon Texas, SW USA
Madre Occidental, Sierra 28 C3 var. Western Sierra Madre. mountain range C Mexico
Madre Oriental, Sierra 29 E3 var. Eastern Sierra Madre. mountain range C Mexico
Madre, Sierra 30 B2 var. Sierra de Soconusco. mountain range Guatemala/Mexico
Madrid 70 D3 country capital (Spain) Madrid, C Spain
Madura see Madurai
Madurai 110 C3 prev. Madura, Mathurai. Tamil Nādu, S India
Madura, Pulau 116 D5 prev. Madoera. island C Indonesia
Maebashi 109 D5 var. Maebasi, Mayebashi. Gunma, Honshū, S Japan
Maebasi see Maebashi
Mae Nam Khong see Mekong
Mae Nam Nan 114 C4 river NW Thailand
Mae Nam Yom 114 C4 river W Thailand
Maeseyck see Maaseik
Maestricht see Maastricht
Maéwo 122 D4 prev. Aurora. island C Vanuatu
Mafia 51 D7 island E Tanzania
Mafraq/Muḩāfaẓat al Mafraq see Al Mafraq
Magadan 93 G3 Magadanskaya Oblast', E Russia
Magallanes see Punta Arenas
Magallanes, Estrecho de see Magellan, Strait of
Magangué 36 B2 Bolívar, N Colombia
Magdalena 39 F3 Beni, N Bolivia
Magdalena 28 B1 Sonora, NW Mexico
Magdalena, Isla 28 B3 island NW Mexico
Magdalena, Río 36 B2 river C Colombia
Magdalen Islands see Madeleine, Îles de la
Magdeburg 72 C4 Sachsen-Anhalt, C Germany
Magelang 116 C5 Jawa, C Indonesia
Magellan, Strait of 43 B8 Sp. Estrecho de Magallanes. strait Argentina/Chile
Magerøy see Magerøya
Magerøya 62 D1 var. Magerøy, Lapp. Máhkarávju. island N Norway
Maggiore, Lago see Maggiore, Lake
Maggiore, Lake 74 B1 It. Lago Maggiore. lake Italy/Switzerland
Maglaj 78 C3 Federacija Bosna I Hercegovina, N Bosnia and Herzegovina
Maglie 75 E6 Puglia, SE Italy
Magna 22 B4 Utah, W USA
Magnesia see Manisa
Magnitogorsk 92 B4 Chelyabinskaya Oblast', C Russia
Magnolia State see Mississippi
Magta' Laḩjar 52 C3 var. Magta Lahjar, Magta' Lahjar, Magtá Lahjar. Brakna, SW Mauritania
Magtymguly 100 C3 prev. Garrygala; Rus. Kara-gala. W Turkmenistan
Magway 114 A3 var. Magwe. Magway, W Myanmar (Burma)
Magyar-Becse see Bečej
Magyarkanizsa see Kanjiža
Magyarország see Hungary
Mahajanga 57 F2 var. Majunga. Mahajanga, NW Madagascar
Mahakam, Sungai 116 D4 var. Koetai, Kutai. river Borneo, C Indonesia
Mahalapye 56 D3 var. Mahalatswe. Central, SE Botswana
Mahalatswe see Mahalapye
Māhān 98 D3 Kermān, E Iran
Mahanādi 113 F4 river E India
Mahārāshtra 112 D5 cultural region W India
Mahbés see El Mahbas
Mahbūbnagar 112 D5 Telangana, C India
Mahdia 49 F2 var. Al Mahdīyah, Mehdia. NE Tunisia
Mahé 57 H1 island Inner Islands, NE Seychelles
Mahia Peninsula 128 E4 peninsula North Island, New Zealand
Mahilyow 85 D6 Rus. Mogilëv. Mahilyowskaya Voblasts', E Belarus
Máhkarávju see Magerøya
Mahmūd-e 'Erāqī see Maḩmūd-e Rāqī
Maḩmūd-e Rāqī 101 E4 var. Mahmūd-e 'Erāqī. Kāpīsā, NE Afghanistan
Mahón see Maó
Mähren see Moravia
Mährisch-Weisskirchen see Hranice
Maicao 36 C1 La Guajira, N Colombia
Mai Ceu/Mai Chio see Maych'ew
Maīdān Shahr 101 E4 prev. Meydān Shahr. Vardak, E Afghanistan
Maidstone 67 E7 SE England, United Kingdom
Maiduguri 53 H4 Borno, NE Nigeria
Mailand see Milano
Maīmanah 100 D3 var. Meymaneh, Maymana. Fāryāb, NW Afghanistan
Main 73 B5 river C Germany
Mai-Ndombe, Lac 55 C6 prev. Lac Léopold II. lake W Dem. Rep. Congo
Maine 19 G2 off. State of Maine, also known as Lumber State, Pine Tree State. state NE USA
Maine 68 B3 cultural region NW France
Maine, Gulf of 19 H2 gulf NE USA
Mainland 66 C2 island N Scotland, United Kingdom
Mainland 66 D1 island NE Scotland, United Kingdom
Mainz 73 B5 Fr. Mayence. Rheinland-Pfalz, SW Germany
Maio 52 A3 var. Mayo. island Ilhas de Sotavento, SE Cape Verde (Cabo Verde)
Maisur see Mysūru, India
Maisur see Karnātaka, India
Maitri 132 C2 Indian research station Antarctica
Maizhokunggar 104 C5 Xizang Zizhiqu, W China
Majorca see Mallorca
Mājro see Majuro Atoll
Majunga see Mahajanga
Majuro Atoll 122 D2 var. Mājro. atoll and capital (Marshall Islands) Ratak Chain, SE Marshall Islands
Makale see Mek'elē
Makarov Basin 133 B3 undersea basin Arctic Ocean
Makarska 78 B4 It. Macarsca. Split-Dalmacija, SE Croatia
Makasar see Makassar
Makasar, Selat see Makassar Straits
Makassar 117 E4 var. Macassar, Makasar; prev. Ujungpandang. Sulawesi, C Indonesia
Makassar Straits 116 D4 Ind. Makasar Selat. strait C Indonesia
Makay 57 F3 var. Massif du Makay. mountain range SW Madagascar
Makay, Massif du see Makay
Makeni 52 C4 C Sierra Leone
Makeyevka see Makiivka
Makhachkala 92 A4 prev. Petrovsk-Port. Respublika Dagestan, SW Russia
Makiivka 87 G3 prev. Makiyivka, Rus. Makeyevka; prev. Dmitriyevsk. Donets'ka Oblast', E Ukraine
Makin 122 D2 prev. Pitt Island. atoll Tungaru, W Kiribati
Makira see San Cristobal
Makiyivka see Makiivka
Makkah 99 A5 Eng. Mecca. Makkah, W Saudi Arabia
Makkovik 17 F2 Newfoundland and Labrador, NE Canada
Makó 77 D7 Rom. Macău. Csongrád, SE Hungary
Makoua 55 B5 Cuvette, C Congo
Makran Coast 98 E4 coastal region SE Iran
Makrany 85 A6 Rus. Mokrany. Brestskaya Voblasts', SW Belarus
Mākū 98 B2 Āžarbāyjān-e Gharbī, NW Iran
Makurdi 53 G4 Benue, C Nigeria
Mala see Malaita, Solomon Islands
Malabār Coast 110 B3 coast SW India
Malabo 55 A5 prev. Santa Isabel. country capital (Equatorial Guinea) Isla de Bioco, NW Equatorial Guinea
Malaca see Málaga
Malacca, Strait of 116 B3 Ind. Selat Malaka. strait Indonesia/Malaysia
Malacka see Malacky
Malacky 77 C6 Hung. Malacka. Bratislavský Kraj, W Slovakia
Maladzyechna 85 C5 Pol. Molodeczno, Rus. Molodechno. Minskaya Voblasts', C Belarus
Málaga 70 D5 anc. Malaca. Andalucía, S Spain
Malagarasi River 51 B7 river W Tanzania Africa
Malagasy Republic see Madagascar
Malaita 122 C3 var. Mala. island N Solomon Islands
Malakal 51 B5 Upper Nile, NE South Sudan
Malakula see Malekula
Malang 116 D5 Jawa, C Indonesia
Malange see Malanje
Malanje 56 B1 var. Malange. Malanje, NW Angola
Mälaren 63 C6 lake C Sweden
Malatya 95 E4 anc. Melitene. Malatya, SE Turkey (Türkiye)
Mala Vyska 87 E3 Rus. Malaya Viska. Kirovohrads'ka Oblast', S Ukraine
Malawi 57 E1 off. Republic of Malawi; prev. Nyasaland, Nyasaland Protectorate. country S Africa
Malawi, Lake see Nyasa, Lake
Malawi, Republic of see Malawi
Malaya Viska see Mala Vyska
Malay Peninsula 102 D4 peninsula Malaysia/Thailand
Malaysia 116 B3 off. Malaysia, var. Federation of Malaysia; prev. the separate territories of Federation of Malaya, Sarawak and Sabah (North Borneo) and Singapore. country SE Asia
Malaysia, Federation of see Malaysia
Malbork 76 C2 Ger. Marienburg, Marienburg in Westpreussen. Pomorskie, N Poland
Malchin 72 C3 Mecklenburg-Vorpommern, N Germany
Malden 23 H5 Missouri, C USA
Malden Island 123 G3 prev. Independence Island. atoll E Kiribati
Maldives 110 A4 off. Republic of Maldives. country N Indian Ocean
Maldives, Republic of see Maldives
Male' see Maale
Male' Atoll 110 B4 var. Kaafu Atoll. atoll C Maldives
Malekula 122 D4 var. Malakula; prev. Mallicolo. island W Vanuatu
Malesína 83 C5 Stereá Ellás, E Greece
Malheur Lake 24 C3 lake Oregon, NW USA
Mali 53 E3 off. Republic of Mali, Fr. République du Mali; prev. French Sudan, Sudanese Republic. country W Africa
Malik, Wadi al see Milk, Wadi el
Mali Kyun 115 B5 var. Tavoy Island. island Myeik Archipelago, S Myanmar (Burma)
Malin see Malyn
Malindi 51 D7 Coast, SE Kenya
Malines see Mechelen
Mali, Republic of see Mali
Mali, République du see Mali
Malkiye see Al Mālikīyah
Malko Tarnovo 82 E3 var. Malko Tŭrnovo. Burgas, E Bulgaria
Malko Tŭrnovo see Malko Tarnovo
Mallaig 66 B3 N Scotland, United Kingdom
Mallawī 50 B2 var. Mallawi. C Egypt
Mallawi see Mallawī
Mallicolo see Malekula
Mallorca 71 G3 Eng. Majorca; anc. Baleares Major. island Islas Baleares, Spain, W Mediterranean Sea
Malmberget 62 C3 Lapp. Malmivaara. Norrbotten, N Sweden
Malmédy 65 D6 Liège, E Belgium
Malmivaara see Malmberget
Malmö 63 B7 Skåne, S Sweden
Maloelap see Maloelap Atoll
Maloelap Atoll 122 D1 var. Maļoeļap. atoll E Marshall Islands
Małopolska, Wyżyna 76 D4 plateau S Poland
Malozemel'skaya Tundra 88 D3 physical region NW Russia
Malta 84 D4 Rēzekne, SE Latvia
Malta 22 C1 Montana, NW USA
Malta 75 C8 off. Republic of Malta. country C Mediterranean Sea
Malta 75 C8 island Malta, C Mediterranean Sea
Malta, Canale di see Malta Channel
Malta Channel 75 C8 It. Canale di Malta. strait Italy/Malta
Malta, Republic of see Malta
Maluku 117 F4 Dut. Molukken, Eng. Moluccas; prev. Spice Islands. island group E Indonesia
Maluku, Laut see Molucca Sea
Malung 63 B6 Dalarna, C Sweden
Malventum see Benevento
Malvina, Isla Gran see West Falkland
Malvinas, Islas see Falkland Islands
Malyn 86 D2 Rus. Malin. Zhytomyrs'ka Oblast', N Ukraine
Malyy Kavkaz see Lesser Caucasus
Mamberamo, Sungai 117 H4 river Papua, E Indonesia
Mambij see Manbij
Mamonovo 84 A4 Ger. Heiligenbeil. Kaliningradskaya Oblast', W Russia
Mamoré, Rio 39 F3 river Bolivia/Brazil
Mamou 52 C4 W Guinea
Mamoudzou 57 F2 dependent territory capital (Mayotte) C Mayotte
Mamuno 56 C3 Ghanzi, W Botswana
Manacor 71 G3 Mallorca, Spain, W Mediterranean Sea
Manado 117 F3 prev. Menado. Sulawesi, C Indonesia
Managua 30 D3 country capital (Nicaragua) Managua, W Nicaragua
Managua, Lake 30 C3 var. Xolotlán. lake W Nicaragua
Manakara 57 G4 Fianarantsoa, SE Madagascar
Manama see Al Manāmah
Mananjary 57 G3 Fianarantsoa, SE Madagascar
Manáos see Manaus
Manapouri, Lake 129 A7 lake South Island, New Zealand
Manar see Mannar
Manas, Gora 101 E2 mountain Kyrgyzstan/Uzbekistan
Manaung Island 114 A4 prev. Cheduba Island. island W Myanmar (Burma)
Manaus 40 D2 prev. Manáos. state capital Amazonas, NW Brazil
Manavgat 94 B4 Antalya, SW Turkey (Türkiye)
Manawatāwhi 128 C1 var. Three Kings Islands. island group N New Zealand
Manbij 96 C2 var. Mambij, Fr. Membidj. Ḩalab, N Syria
Manchester 67 D5 Lat. Mancunium. NW England, United Kingdom
Manchester 19 G3 New Hampshire, NE USA
Man-chou-li see Manzhouli
Manchurian Plain 103 E1 plain NE China
Máncio Lima see Japiim
Mancunium see Manchester
Mand see Mand, Rūd-e
Mandalay 114 B3 Mandalay, C Myanmar (Burma)
Mandan 23 E2 North Dakota, N USA
Mandeville 32 B5 C Jamaica
Mándra 83 C6 Attikí, C Greece
Mand, Rud-e 98 D4 var. Mand. river S Iran
Mandurah 125 A6 Western Australia
Manduria 75 E5 Puglia, SE Italy
Mandya 110 C2 Karnātaka, C India
Manfredonia 75 D5 Puglia, SE Italy
Mangai 55 C6 Bandundu, W Dem. Rep. Congo
Mangaia 123 G5 island group S Cook Islands
Mangalia 86 D5 anc. Callatis. Constanța, SE Romania
Mangalmé 54 C3 Guéra, SE Chad
Mangalore see Mangalūru
Mangalūru 110 B2 prev. Mangalore. Karnātaka, W India
Mangaung see Bloemfontein
Mango 53 E4 var. Sansanné-Mango. N Togo
Mangoky 57 F3 river W Madagascar
Manhattan 23 F4 Kansas, C USA
Manicouagan, Réservoir 16 D3 lake Québec, E Canada
Manihiki 123 G4 atoll N Cook Islands
Manihiki Plateau 121 E3 undersea plateau C Pacific Ocean
Maniitsoq 60 C3 var. Manîtsoq, Dan. Sukkertoppen. Qeqqata, S Greenland
Manila 117 E1 off. City of Manila; Fil. Maynila. country capital (Philippines) Luzon, N Philippines
Manila, City of see Manila
Manisa 94 A3 var. Manissa, prev. Saruhan; anc. Magnesia. Manisa, W Turkey (Türkiye)
Manissa see Manisa
Manitoba 15 F5 province S Canada
Manitoba, Lake 15 F5 lake Manitoba, S Canada
Manitoulin Island 16 C4 island Ontario, S Canada
Manîtsoq see Maniitsoq
Manizales 36 B3 Caldas, W Colombia
Manjimup 125 A7 Western Australia
Mankato 23 F3 Minnesota, N USA
Manlleu 71 G2 Cataluña, NE Spain
Manly 126 E1 Iowa, C USA
Manmād 112 C5 Mahārāshtra, W India
Mannar 110 C3 var. Manar. Northern Province, NW Sri Lanka
Mannar, Gulf of 110 C3 gulf India/Sri Lanka
Mannheim 73 B5 Baden-Württemberg, SW Germany
Manokwari 117 G4 New Guinea, E Indonesia
Manono 55 E7 Shaba, SE Dem. Rep. Congo
Manosque 69 D6 Alpes-de-Haute-Provence, SE France
Manra 123 F3 prev. Sydney Island. atoll Phoenix Islands, C Kiribati
Mansa 56 D2 prev. Fort Rosebery. Luapula, N Zambia
Mansel Island 15 G3 island Nunavut, NE Canada
Mansfield 18 D4 Ohio, N USA
Manta 38 A2 Manabí, W Ecuador
Manteca 25 B6 California, W USA
Mantoue see Mantova
Mantova 74 B2 Eng. Mantua, Fr. Mantoue. Lombardia, NW Italy
Mantua see Mantova
Manuae 123 G4 island S Cook Islands
Manukau see Manurewa
Manurewa 128 D3 var. Manukau. Auckland, North Island, New Zealand
Manzanares 71 E3 Castilla-La Mancha, C Spain
Manzanillo 32 C3 Granma, E Cuba
Manzanillo 28 D4 Colima, SW Mexico
Manzhouli 105 F1 var. Man-chou-li. Nei Mongol Zizhiqu, N China
Mao 54 B3 Kanem, W Chad
Maó 71 H3 Cast. Mahón, Eng. Port Mahon; anc. Portus Magonis. Menorca, Spain, W Mediterranean Sea
Maoke, Pegunungan 117 H4 Dut. Sneeuw-gebergte, Eng. Snow Mountains. mountain range Papua, E Indonesia
Maoming 106 C6 Guangdong, S China
Mapmaker Seamounts 103 H2 seamount range N Pacific Ocean
Maputo 56 D4 prev. Lourenço Marques. country capital (Mozambique) Maputo, S Mozambique
Marabá 41 F2 Pará, NE Brazil
Maracaibo 36 C1 Zulia, NW Venezuela
Maracaibo, Gulf of see Venezuela, Golfo de
Maracaibo, Lago de 36 C2 var. Lake Maracaibo. inlet NW Venezuela
Maracaibo, Lake see Maracaibo, Lago de
Maracay 36 D2 Aragua, N Venezuela
Marada see Marādah
Marādah 49 G3 var. Marada. N Libya
Maradi 53 G3 Maradi, S Niger
Maragha see Marāgheh
Marāgheh 98 C2 var. Maragha. Āžarbāyjān-e Khāvarī, NW Iran
Marajó, Baía de 41 F1 bay N Brazil
Marajó, Ilha de 41 E1 island N Brazil
Marakesh see Marrakech
Maramba see Livingstone
Maranhão 41 F2 off. Estado do Maranhão. state/region E Brazil
Maranhão, Estado do see Maranhão
Marañón, Río 38 B2 river N Peru
Marathon 16 C4 Ontario, S Canada
Marathón see Marathónas
Marathónas 83 C5 prev. Marathón. Attikí, C Greece
Marbella 70 D5 Andalucía, S Spain
Marble Bar 124 B4 Western Australia
Marburg see Marburg an der Lahn, Germany
Marburg see Maribor, Slovenia
Marburg an der Lahn 72 B4 hist. Marburg. Hessen, W Germany
March see Morava
Marche 74 C3 Eng. Marches. region C Italy
Marche 69 C5 cultural region C France
Marche-en-Famenne 65 C7 Luxembourg, SE Belgium
Marchena, Isla 38 B5 var. Bindloe Island. island Galápagos Islands, Ecuador, E Pacific Ocean
Marches see Marche
Mar Chiquita, Laguna 42 C3 lake C Argentina
Marcounda see Markounda
Mardan 112 C1 North-West Frontier Province, N Pakistan
Mar del Plata 43 D5 Buenos Aires, E Argentina
Mardin 95 E4 Mardin, SE Turkey (Türkiye)
Maré 122 D5 island Îles Loyauté, E New Caledonia
Marea Neagră see Black Sea
Mareeba 126 D3 Queensland, NE Australia
Marek see Dupnitsa
Marganets see Marhanets
Margarita, Isla de 37 E1 island N Venezuela
Margate 67 E7 prev. Margate. SE England, United Kingdom
Margherita see Jamaame
Margherita, Lake 51 C5 Eng. Lake Margherita, It. Abbaia. lake SW Ethiopia
Margherita, Lake see Ābaya Hāyk'
Marghita 86 B3 Hung. Margitta. Bihor, NW Romania
Margitta see Marghita
Mārgō, Dasht-e 100 D5 desert SW Afghanistan
Marhanets 87 F3 Rus. Marganets. Dnipropetrovska Oblast, E Ukraine
María Cleofas, Isla 28 C4 island C Mexico
Maria Island 127 C8 island Tasmania, SE Australia
María Madre, Isla 28 C4 island C Mexico
María Magdalena, Isla 28 C4 island C Mexico
Mariana Trench 103 G4 trench W Pacific Ocean
Mariánské Lázně 77 A5 Ger. Marienbad. Karlovarský Kraj, W Czechia (Czech Republic)
Marías, Islas 28 C4 island group C Mexico
Maria-Theresiopel see Subotica
Maribor 73 E7 Ger. Marburg. NE Slovenia
Marica see Maritsa

Maridi *51 B5* W Equatoria, S South Sudan
Marie Byrd Land *132 A3 physical region* Antarctica
Marie-Galante *33 G4 var.* Ceyre to the Caribs. *island* SE Guadeloupe
Marienbad *see* Mariánské Lázně
Marienburg *see* Alūksne, Latvia
Marienburg *see* Malbork, Poland
Marienburg in Westpreussen *see* Malbork
Marienhausen *see* Viļaka
Mariental *56 B4* Hardap, SW Namibia
Marienwerder *see* Kwidzyń
Mariestad *63 B6* Västra Götaland, S Sweden
Marietta *20 D2* Georgia, SE USA
Marijampolė *84 B4 prev.* Kapsukas. Marijampolė, S Lithuania
Marília *41 E4* São Paulo, S Brazil
Marín *70 B1* Galicia, NW Spain
Mar'ina Gorka *see* Mar''ina Horka
Mar''ina Horka *85 C6 Rus.* Mar'ina Gorka. Minskaya Voblasts', C Belarus
Maringá *41 E4* Paraná, S Brazil
Marion *23 G3* Iowa, C USA
Marion *18 D4* Ohio, N USA
Marion, Lake *21 E2 reservoir* South Carolina, SE USA
Mariscal Estigarribia *42 D2* Boquerón, NW Paraguay
Maritsa *82 D3 var.* Marica, *Gk.* Évros, *Turk.* Meriç; *anc.* Hebrus. *river* SW Europe
Maritzburg *see* Pietermaritzburg
Mariupol *87 G4 prev.* Zhdanov. Donets'ka Oblast', SE Ukraine
Marka *51 D6 var.* Merca. Shabeellaha Hoose, S Somalia
Markham, Mount *132 B4 mountain* Antarctica
Markounda *54 C4 var.* Marcounda. Ouham, NW Central African Republic
Marktredwitz *73 C5* Bayern, E Germany
Marlborough *126 D4* Queensland, E Australia
Marmanda *see* Marmande
Marmande *69 B5 anc.* Marmanda. Lot-et-Garonne, SW France
Sea of Marmara *94 A2 Eng.* Sea of Marmara. *sea* NW Turkey (Türkiye)
Marmara, Sea of *see* Marmara Denizi
Marmaris *94 A4* Muğla, SW Turkey (Türkiye)
Marne *68 C3 cultural region* N France
Marne *68 D3 river* N France
Maro *54 C4* Moyen-Chari, S Chad
Maroantsetra *57 G2* Toamasina, NE Madagascar
Maromokotro *57 G2 mountain* N Madagascar
Maroni *37 G3 Dut.* Marowijne. *river* French Guiana/Suriname
Marosheviz *see* Toplița
Marosludas *see* Luduș
Marosvásárhely *see* Târgu Mureș
Marotiri *121 F4 var.* Îlots de Bass, Morotiri. *island group* Îles Australes, SW French Polynesia
Maroua *54 B3* Extrême-Nord, N Cameroon
Marowijne *see* Maroni
Marquesas Fracture Zone *131 E3 fracture zone* E Pacific Ocean
Marquette *18 B1* Michigan, N USA
Marrakech *48 C2 var.* Marakesh, *Eng.* Marrakesh; *prev.* Morocco. W Morocco
Marrakesh *see* Marrakech
Marrawah *127 C8* Tasmania, SE Australia
Marree *127 B5* South Australia
Marsá al Burayqah *49 G3 var.* Al Burayqah. N Libya
Marsabit *51 C6* Eastern, N Kenya
Marsala *75 B7 anc.* Lilybaeum. Sicilia, Italy, C Mediterranean Sea
Marsberg *72 B4* Nordrhein-Westfalen, W Germany
Marseille *69 D6 Eng.* Marseilles; *anc.* Massilia. Bouches-du-Rhône, SE France
Marseilles *see* Marseille
Marshall *23 F2* Minnesota, N USA
Marshall *27 H2* Texas, SW USA
Marshall Islands *122 C1 off.* Republic of the Marshall Islands. *country* W Pacific Ocean
Marshall Islands, Republic of the *see* Marshall Islands
Marshall Seamounts *103 H3 seamount range* SW Pacific Ocean
Marsh Harbour *32 C1* Great Abaco, W The Bahamas
Martaban *see* Mottama
Martha's Vineyard *19 G3 island* Massachusetts, NE USA
Martigues *69 D6* Bouches-du-Rhône, SE France
Martin *77 C5 Ger.* Sankt Martin, *Hung.* Turócszentmárton; *prev.* Turčianský Svätý Martin. Žilinský Kraj, N Slovakia
Martinique *33 G4 French overseas department* E West Indies
Martinique Channel *see* Martinique Passage
Martinique Passage *33 G4 var.* Dominica Channel, Martinique Channel. *channel* Dominica/Martinique
Marton *128 D4* Manawatu-Wanganui, North Island, New Zealand
Martos *70 D4* Andalucía, S Spain
Marungu *55 E7 mountain range* SE Dem. Rep. Congo
Mary *100 D3 prev.* Merv. Mary Welaýaty, S Turkmenistan
Maryborough *127 D4* Queensland, E Australia
Maryborough *see* Port Laoise
Mary Island *see* Kanton
Maryland *19 E5 off.* State of Maryland, *also known as* America in Miniature, Cockade State, Free State, Old Line State. *state* NE USA
Maryland, State of *see* Maryland
Maryville *23 F4* Missouri, C USA
Maryville *20 D1* Tennessee, S USA
Masai Steppe *51 C7 grassland* NW Tanzania
Masaka *51 B6* SW Uganda
Masallı *95 H3 Rus.* Masally. S Azerbaijan
Masally *see* Masallı
Masasi *51 C8* Mtwara, SE Tanzania
Masawa/Massawa *see* Mits'iwa
Masaya *30 D3* Masaya, W Nicaragua
Mascarene Basin *119 B5 undersea basin* W Indian Ocean
Mascarene Islands *57 H4 island group* W Indian Ocean
Mascarene Plain *119 B5 abyssal plain* W Indian Ocean
Mascarene Plateau *119 B5 undersea plateau* W Indian Ocean
Maseru *56 D4 country capital* (Lesotho) W Lesotho
Mashhad *98 E2 var.* Meshed. Khorāsān-Razavī, NE Iran
Masindi *51 B6* W Uganda
Masīra *see* Maşīrah, Jazīrat
Masira, Gulf of *see* Maşīrah, Khalīj
Maşīrah, Jazīrat *99 E5 var.* Masīra. *island* E Oman
Maşīrah, Khalīj *99 E5 var.* Gulf of Masira. *bay* E Oman
Masis *see* Büyük Ağrı Dağı
Maskat *see* Masqaţ
Maskav *101 E3 Rus.* Moskovskiy; *prev.* Chubek, Moskva. SW Tajikistan
Mason City *23 F3* Iowa, C USA
Masqaţ *99 E5 var.* Maskat, *Eng.* Muscat. *country capital* (Oman) NE Oman
Massa *74 B3* Toscana, C Italy
Massachusetts *19 G3 off.* Commonwealth of Massachusetts, *also known as* Bay State, Old Bay State, Old Colony State. *state* NE USA
Massenya *54 B3* Chari-Baguirmi, SW Chad
Massif Central *69 C5 plateau* C France
Massilia *see* Marseille
Massoukou *see* Franceville
Mastanli *see* Momchilgrad
Masterton *129 D5* Wellington, North Island, New Zealand
Masty *85 B5 Rus.* Mosty. Hrodzyenskaya Voblasts', W Belarus
Masuda *109 B6* Shimane, Honshū, SW Japan
Masuku *see* Franceville
Masvingo *56 D3 prev.* Fort Victoria, Nyanda, Victoria. Masvingo, SE Zimbabwe
Maşyāf *96 B3 Fr.* Misiaf. Ḥamāh, C Syria
Mata-Au *see* Clutha River
Matadi *55 B6* Bas-Congo, W Dem. Rep. Congo
Matagalpa *30 D3* Matagalpa, C Nicaragua
Matale *110 D3* Central Province, C Sri Lanka
Matam *52 C3* NE Senegal
Matamata *128 D3* Waikato, North Island, New Zealand
Matamoros *28 D3* Coahuila, NE Mexico
Matamoros *29 E2* Tamaulipas, C Mexico
Matane *17 E4* Québec, SE Canada
Matanzas *32 B2* Matanzas, NW Cuba
Matara *110 D4* Southern Province, S Sri Lanka
Mataram *116 D5* Pulau Lombok, C Indonesia
Mataró *71 G2 anc.* Illuro. Cataluña, E Spain
Mataura *129 B7* Southland, South Island, New Zealand
Mataura *129 B7 river* South Island, New Zealand
Mata Uta *see* Matā'utu
Matā'utu *123 E4 var.* Mata Uta. *dependent territory capital* (Wallis and Futuna) Île Uvea, Wallis and Futuna
Matera *75 E5* Basilicata, S Italy
Mathurai *see* Madurai
Matianus *see* Orūmīyeh, Daryācheh-ye
Matías Romero *29 F5* Oaxaca, SE Mexico
Matisco/Matisco Ædourum *see* Mâcon
Mato Grosso *41 E3 off.* Estado de Mato Grosso; *prev.* Matto Grosso. *state/region* W Brazil
Mato Grosso do Sul *41 E4 off.* Estado de Mato Grosso do Sul. *state/region* S Brazil
Mato Grosso do Sul, Estado de *see* Mato Grosso do Sul
Mato Grosso, Estado de *see* Mato Grosso
Mato Grosso, Planalto de *34 C4 plateau* C Brazil
Matosinhos *70 B2 prev.* Matozinhos. Porto, NW Portugal
Matozinhos *see* Matosinhos
Matsue *109 B6 var.* Matsuye, Matue. Shimane, Honshū, SW Japan
Matsumoto *109 C5 var.* Matumoto. Nagano, Honshū, S Japan
Matsuyama *109 B7 var.* Matuyama. Ehime, Shikoku, SW Japan
Matsuye *see* Matsue
Matterhorn *73 A8 It.* Monte Cervino. *mountain* Italy/Switzerland
Matthews Ridge *37 F2* N Guyana
Matthew Town *32 D2* Great Inagua, S The Bahamas
Matto Grosso *see* Mato Grosso
Matucana *38 C4* Lima, W Peru
Matue *see* Matsue
Matumoto *see* Matsumoto
Maturín *37 E2* Monagas, NE Venezuela
Matuyama *see* Matsuyama
Mau *113 E3 var.* Maunāth Bhanjan. Uttar Pradesh, N India
Maui *25 B8 island* Hawaii, USA, C Pacific Ocean
Maun *56 C3* North-West, C Botswana
Maunāth Bhanjan *see* Mau
Mauren *72 E1* NE Liechtenstein Europe
Maurice *see* Mauritius
Mauritania *52 C2 off.* Islamic Republic of Mauritania, *Ar.* Mūrītānīyah. *country* W Africa
Mauritania, Islamic Republic of *see* Mauritania
Mauritius *57 H3 off.* Republic of Mauritius, *Fr.* Maurice. *country* W Indian Ocean
Mauritius *119 B5 island* W Indian Ocean
Mauritius, Republic of *see* Mauritius
Mawlamyaing *see* Mawlamyine
Mawlamyine *114 B4 var.* Mawlamyaing, Moulmein. Mon State, S Myanmar (Burma)
Mawson *132 D2 Australian research station* Antarctica
Mayadin *see* Al Mayādīn
Mayaguana *32 D2 island* SE The Bahamas
Mayaguana Passage *32 D2 passage* SE The Bahamas
Mayagüez *33 F3* W Puerto Rico
Mayamey *98 D2* Semnān, N Iran
Maya Mountains *30 B2 Sp.* Montañas Mayas. *mountain range* Belize/Guatemala
Mayas, Montañas *see* Maya Mountains
Maych'ew *50 C4 var.* Mai Chio, *It.* Mai Ceu. Tigray, N Ethiopia
Mayebashi *see* Maebashi
Mayence *see* Mainz
Mayfield *129 B6* Canterbury, South Island, New Zealand
Maykop *89 A7* Respublika Adygeya, SW Russia
Maymana *see* Maimanah
Maymyo *see* Pyin-Oo-Lwin
Maynila *see* Manila
Mayo *see* Maio
Mayor Island *128 D3 island* NE New Zealand
Mayor Pablo Lagerenza *see* Capitán Pablo Lagerenza
Mayotte *57 F2 French overseas department* E Africa
May Pen *32 B5* C Jamaica
Mayyit, Al Baḩr al *see* Dead Sea
Mazabuka *56 D2* Southern, S Zambia
Mazaca *see* Kayseri
Mazagan *see* El-Jadida
Mazār-e Sharīf *101 E3 var.* Mazār-i Sharif. Balkh, N Afghanistan
Mazār-i Sharif *see* Mazār-e Sharīf
Mazatlán *28 C3* Sinaloa, C Mexico
Mažeikiai *84 B3* Telšiai, NW Lithuania
Mazirbe *84 C2* Talsi, NW Latvia
Mazra'a *see* Al Mazra'ah
Mazury *76 D3 physical region* NE Poland
Mazyr *85 C7 Rus.* Mozyr'. Homyel'skaya Voblasts', SE Belarus
Mbabane *56 D4 country capital* (Eswatini - administrative) NW Eswatini
Mbacké *see* Mbaké
Mbaïki *55 C5 var.* M'Baiki. Lobaye, SW Central African Republic
M'Baiki *see* Mbaïki
Mbaké *52 B3 var.* Mbacké. W Senegal
Mbala *56 D1 prev.* Abercorn. Northern, NE Zambia
Mbale *51 C6* E Uganda
Mbandaka *55 C5 prev.* Coquilhatville. Equateur, NW Dem. Rep. Congo
M'Banza Congo *56 B1 var.* Mbanza Congo; *prev.* São Salvador, São Salvador do Congo. Dem. Rep. Congo, NW Angola
Mbanza-Ngungu *55 B6* Bas-Congo, W Dem. Rep. Congo
Mbarara *51 B6* SW Uganda
Mbé *54 B4* Nord, N Cameroon
Mbeya *51 C7* Mbeya, SW Tanzania
Mbomou/M'Bomu/Mbomu *see* Bomu
Mbour *52 B3* W Senegal
Mbuji-Mayi *55 D7 prev.* Bakwanga. Kasai-Oriental, S Dem. Rep. Congo
McAlester *27 G2* Oklahoma, C USA
McAllen *27 G5* Texas, SW USA
McCamey *27 E3* Texas, SW USA
M'Clintock Channel *15 F2 channel* Nunavut, N Canada
McComb *20 B3* Mississippi, S USA
McCook *23 E4* Nebraska, C USA
McKean Island *123 E3 island* Phoenix Islands, C Kiribati
McKinley, Mount *see* Denali
McKinley Park *see* Denali Park
McMinnville *24 B3* Oregon, NW USA
McMurdo *132 B4 US research station* Antarctica
McPherson *23 E5* Kansas, C USA
McPherson *see* Fort McPherson
Mdantsane *56 D5* Eastern Cape, SE South Africa
Mead, Lake *25 D6 reservoir* Arizona/Nevada, W USA
Mecca *see* Makkah
Mechelen *65 C5 Eng.* Mechlin, *Fr.* Malines. Antwerpen, C Belgium
Mechlin *see* Mechelen
Mecklenburger Bucht *72 C2 bay* N Germany
Mecsek *77 C7 mountain range* SW Hungary
Medan *116 B3* Sumatera, E Indonesia
Medeba *see* Ma'dabā
Medellín *36 B3* Antioquia, NW Colombia
Médenine *49 F2 var.* Madanīyīn. SE Tunisia
Medeshamstede *see* Peterborough
Medford *24 B4* Oregon, NW USA
Medgidia *86 D5* Constanța, SE Romania
Medgyes *see* Mediaș
Mediaș *86 B4 Ger.* Mediasch, *Hung.* Medgyes. Sibiu, C Romania
Mediasch *see* Mediaș
Medicine Hat *15 F5* Alberta, SW Canada
Medina *see* Al Madīnah
Medinaceli *71 E2* Castilla y León, N Spain
Medina del Campo *70 D2* Castilla y León, N Spain
Medinat Israel *see* Israel
Mediolanum *see* Saintes, France
Mediolanum *see* Milano, Italy
Mediomatrica *see* Metz
Mediterranean Sea *80 D3 Fr.* Mer Méditerranée. *sea* Africa/Asia/Europe
Méditerranée, Mer *see* Mediterranean Sea
Médoc *69 B5 cultural region* SW France
Medvezh'yegorsk *88 B3* Respublika Kareliya, NW Russia
Meekatharra *125 B5* Western Australia
Meemu Atoll *see* Mulakatholhu
Meerssen *65 D6 var.* Mersen. Limburg, SE Netherlands
Meerut *112 D2* Uttar Pradesh, N India
Megáli Préspa, Límni *see* Prespa, Lake
Meghālaya *91 G3 cultural region* NE India
Mehdia *see* Mahdia
Meheso *see* Mī'ēso
Me Hka *see* Nmai Hka
Mehriz *98 D3* Yazd, C Iran
Mehtar Lām *101 F4 var.* Mehtarlām, Meterlam, Methariam, Metharlam. Laghmān, E Afghanistan
Mehtarlām *see* Mehtar Lām
Meiktila *114 B3* Mandalay, C Myanmar (Burma)
Méjico *see* Mexico
Mejillones *42 B2* Antofagasta, N Chile
Mek'elē *50 C4 var.* Makale. Tigray, N Ethiopia
Mékhé *52 B3* NW Senegal
Mekong *102 D3 var.* Lan-ts'ang Chiang, *Cam.* Mékôngk, *Chin.* Lancang Jiang, *Lao.* Mènam Khong, *Th.* Mae Nam Khong, *Tib.* Dza Chu, *Vtn.* Sông Tiên Giang. *river* SE Asia
Mékôngk *see* Mekong
Mekong, Mouths of the *115 E6 delta* S Vietnam
Melaka *116 B3 var.* Malacca. Melaka, Peninsular Malaysia
Melaka, Selat *see* Malacca, Strait of
Melanesia *122 D3 island group* W Pacific Ocean
Melanesian Basin *120 C2 undersea basin* W Pacific Ocean
Melbourne *127 C7 state capital* Victoria, SE Australia
Melbourne *21 E4* Florida, SE USA
Meleda *see* Mljet
Melghir, Chott *49 E2 var.* Chott Melrhir. *salt lake* E Algeria
Melilla *58 B5 anc.* Rusaddir, Russadir. Melilla, Spain, N Africa
Melilla *48 D2 autonomous city of Spain* Spain, N Africa
Melita *15 F5* Manitoba, S Canada
Melita *see* Mljet
Melitene *see* Malatya
Melitopol *87 F4* Zaporiz'ka Oblast', SE Ukraine
Melle *65 B5* Oost-Vlaanderen, NW Belgium
Mellerud *63 B6* Västra Götaland, S Sweden
Mellieha *80 B5* E Malta
Mellizo Sur, Cerro *43 A7 mountain* S Chile
Melo *42 E4* Cerro Largo, NE Uruguay
Melodunum *see* Melun
Melrhir, Chott *see* Melghir, Chott
Melsungen *72 B4* Hessen, C Germany
Melun *68 C3 anc.* Melodunum. Seine-et-Marne, N France
Melville Bay/Melville Bugt *see* Qimusseriarsuaq
Melville Island *124 D2 island* Northern Territory, N Australia
Melville Island *15 E2 island* Parry Islands, Northwest Territories, NW Canada
Melville, Lake *17 F2 lake* Newfoundland and Labrador, E Canada
Melville Peninsula *15 G3 peninsula* Nunavut, NE Canada
Melville Sound *see* Viscount Melville Sound
Membidj *see* Manbij
Memel *see* Neman, NE Europe
Memel *see* Klaipeda, Lithuania
Memmingen *73 B6* Bayern, S Germany
Memphis *20 C1* Tennessee, S USA
Menaam *see* Menaldum
Ménaka *53 F3* Goa, E Mali
Menaldum *64 D1 Fris.* Menaam. Fryslân, N Netherlands
Mènam Khong *see* Mekong
Mendaña Fracture Zone *131 F4 fracture zone* E Pacific Ocean
Mende *69 C5 anc.* Mimatum. Lozère, S France
Mendeleyev Ridge *133 B2 undersea ridge* Arctic Ocean
Mendocino Fracture Zone *130 D2 fracture zone* NE Pacific Ocean
Mendoza *42 B4* Mendoza, W Argentina
Menemen *94 A3* İzmir, W Turkey (Türkiye)
Menengiyn Tal *105 F2 plain* E Mongolia
Menongue *56 B2 var.* Vila Serpa Pinto, *Port.* Serpa Pinto. Cuando Cubango, C Angola
Menorca *71 H3 Eng.* Minorca; *anc.* Balearis Minor. *island* Islas Baleares, Spain, W Mediterranean Sea
Mentawai, Kepulauan *116 A4 island group* W Indonesia
Meppel *64 D2* Drenthe, NE Netherlands
Meran *see* Merano
Merano *74 C1 Ger.* Meran. Trentino-Alto Adige, N Italy
Merca *see* Marka
Mercedes *42 D3* Corrientes, NE Argentina
Mercedes *42 D4* Soriano, SW Uruguay
Meredith, Lake *27 E1 reservoir* Texas, SW USA
Merefa *87 G2* Kharkivs'ka Oblast', E Ukraine
Mergate *see* Margate
Mergui *see* Myeik
Mergui Archipelago *see* Myeik Archipelago
Mérida *29 H3* Yucatán, SW Mexico
Mérida *70 C4 anc.* Augusta Emerita. Extremadura, W Spain
Mérida *36 C2* Mérida, W Venezuela
Meridian *20 C2* Mississippi, S USA
Mérignac *69 B5* Gironde, SW France
Merín, Laguna *see* Mirim Lagoon
Merkulovichi *see* Myerkulavichy
Merowe *50 B3* Northern, N Sudan
Merredin *125 B6* Western Australia
Mersen *see* Meerssen
Mersey *67 D5 river* NW England, United Kingdom
Mersin *94 C4 var.* İçel. İçel, S Turkey (Türkiye)
Mērsrags *84 C3* Talsi, NW Latvia
Meru *51 C6* Eastern, C Kenya
Merv *see* Mary
Merzifon *94 D2* Amasya, N Turkey (Türkiye)
Merzig *73 A5* Saarland, SW Germany
Mesa *26 B2* Arizona, SW USA
Meseritz *see* Międzyrzecz
Meshed *see* Mashhad
Mesopotamia *35 C5 var.* Mesopotamia Argentina. *physical region* NE Argentina
Mesopotamia Argentina *see* Mesopotamia
Messalo, Rio *57 E2 var.* Mualo. *river* NE Mozambique
Messana/Messene *see* Messina
Messina *see* Musina
Messina, Strait of *see* Messina, Stretto di
Messina, Stretto di *75 D7 Eng.* Strait of Messina. *strait* SW Italy
Messíni *83 B6* Pelopónnisos, S Greece
Mesta *see* Néstos
Mestghanem *see* Mostaganem
Mestia *95 F1 var.* Mestiya. N Georgia
Mestiya *see* Mestia
Mestre *74 C2* Veneto, NE Italy
Metairie *20 B3* Louisiana, S USA
Metán *42 C2* Salta, N Argentina
Metapán *30 B2* Santa Ana, NW El Salvador
Meta, Río *36 D3 river* Colombia/Venezuela
Meterlam *see* Mehtar Lām
Methariam/Metharlam *see* Mehtar Lām
Metis *see* Metz
Metković *78 B4* Dubrovnik-Neretva, SE Croatia
Métsovo *82 B4 prev.* Métsovon. Ípeiros, C Greece
Métsovon *see* Métsovo
Metz *68 D3 anc.* Divodurum Mediomatricum, Mediomatrica, Metis. Moselle, NE France
Meulaboh *116 A3* Sumatera, W Indonesia
Meuse *65 C6 Dut.* Maas. *river* W Europe
Mexcala, Río *see* Balsas, Río
Mexicali *28 A1* Baja California Norte, NW Mexico
Mexicanos, Estados Unidos *see* Mexico
Mexico *23 G4* Missouri, C USA
Mexico *28 C3 off.* United Mexican States, *var.* Méjico, México, *Sp.* Estados Unidos Mexicanos. *country* N Central America
México *see* Mexico
Mexico City *see* México, Ciudad de
México, Ciudad de *29 E4 var.* Mexico City. *country capital* (Mexico) México, C Mexico
México, Golfo de *see* Mexico, Gulf of
Mexico, Gulf of *29 F2 Sp.* Golfo de México. *gulf* W Atlantic Ocean
Meyadine *see* Al Mayādīn
Meydān Shahr *see* Maīdān Shahr
Meymaneh *see* Maimanah
Mezen' *88 D3 river* NW Russia
Mezőtúr *77 D7* Jász-Nagykun-Szolnok, E Hungary
Mġarr *80 A5* Gozo, N Malta
Miadzioł Nowy *see* Myadzyel
Miahuatlán *29 F5 var.* Miahuatlán de Porfirio Díaz. Oaxaca, SE Mexico
Miahuatlán de Porfirio Díaz *see* Miahuatlán
Miami *21 F5* Florida, SE USA
Miami *27 G1* Oklahoma, C USA
Miami Beach *21 F5* Florida, SE USA
Mīāneh *98 C2 var.* Miyāneh. Āżarbāyjān-e Sharqī, NW Iran
Mianyang *106 B5* Sichuan, C China
Miastko *76 C2 Ger.* Rummelsburg in Pommern. Pomorskie, N Poland
Mi Chai *see* Nong Khai
Michalovce *77 E5 Ger.* Grossmichel, *Hung.* Nagymihály. Košický Kraj, E Slovakia
Michigan *18 C1 off.* State of Michigan, *also known as* Great Lakes State, Lake State, Wolverine State. *state* N USA
Michigan, Lake *18 C2 lake* N USA
Michurin *see* Tsarevo
Michurinsk *89 B5* Tambovskaya Oblast', W Russia
Micoud *33 F2* SE Saint Lucia
Micronesia *122 B1 off.* Federated States of Micronesia. *country* W Pacific Ocean
Micronesia *122 C1 island group* W Pacific Ocean
Micronesia, Federated States of *see* Micronesia
Mid-Atlantic Cordillera *see* Mid-Atlantic Ridge
Mid-Atlantic Ridge *44 C3 var.* Mid-Atlantic Cordillera, Mid-Atlantic Rise, Mid-Atlantic Swell. *undersea ridge* Atlantic Ocean
Mid-Atlantic Rise/Mid-Atlantic Swell *see* Mid-Atlantic Ridge
Middelburg *65 B5* Zeeland, SW Netherlands
Middelharnis *64 B4* Zuid-Holland, SW Netherlands
Middelkerke *65 A5* West-Vlaanderen, W Belgium
Middle America Trench *13 B7 trench* E Pacific Ocean
Middle Andaman *111 F2 island* Andaman Islands, India, NE Indian Ocean
Middle Atlas *48 C2 Eng.* Middle Atlas. *mountain range* N Morocco
Middle Atlas *see* Moyen Atlas
Middleburg Island *see* 'Eua
Middle Congo *see* Congo (Republic of)
Middlesboro *18 C5* Kentucky, S USA
Middlesbrough *67 D5* N England, United Kingdom
Middletown *19 F4* New Jersey, NE USA
Middletown *19 F3* New York, NE USA
Mid-Indian Basin *119 C5 undersea basin* N Indian Ocean
Mid-Indian Ridge *119 C5 var.* Central Indian Ridge. *undersea ridge* C Indian Ocean
Midland *16 D5* Ontario, S Canada
Midland *18 C3* Michigan, N USA
Midland *27 E3* Texas, SW USA
Mid-Pacific Mountains *130 C2 var.* Mid-Pacific Seamounts. *seamount range* NW Pacific Ocean
Mid-Pacific Seamounts *see* Mid-Pacific Mountains
Midway Islands *130 D2 US unincorporated territory* C Pacific Ocean
Miechów *77 D5* Małopolskie, S Poland
Międzyrzec Podlaski *76 E3* Lubelskie, E Poland
Międzyrzecz *76 B3 Ger.* Meseritz. Lubuskie, W Poland
Mielec *77 D5* Podkarpackie, SE Poland
Miercurea-Ciuc *86 C4 Ger.* Szeklerburg, *Hung.* Csíkszereda. Harghita, C Romania
Mieres del Camín *70 D1 var.* Mieres del Camino. Asturias, NW Spain
Mieres del Camino *see* Mieres del Camín
Mī'ēso *51 D5 var.* Meheso, Miesso. Oromīya, C Ethiopia
Miesso *see* Mī'ēso
Mifrats Hefa *97 A5 Eng.* Bay of Haifa; *prev.* Mifraz Ḥefa. *bay* N Israel
Miguel Asua *28 D3 var.* Miguel Auza. Zacatecas, C Mexico
Miguel Auza *see* Miguel Asua
Mijdrecht *64 C3* Utrecht, C Netherlands
Mikashevichi *see* Mikashevichy
Mikashevichy *85 C7 Pol.* Mikaszewicze, *Rus.* Mikashevichi. Brestskaya Voblasts', SW Belarus
Mikaszewicze *see* Mikashevichy
Mikhaylovgrad *see* Montana
Mikhaylovka *89 B6* Volgogradskaya Oblast', SW Russia
Míkonos *see* Mýkonos
Mikre *82 C2* Lovech, N Bulgaria
Mikun' *88 D4* Respublika Komi, NW Russia
Mikuni-sanmyaku *109 D5 mountain range* Honshū, N Japan Asia
Mikura-jima *109 D6 island* E Japan
Milagro *38 B2* Guayas, SW Ecuador
Milan *see* Milano
Milange *57 E2* Zambézia, NE Mozambique
Milano *74 B2 Eng.* Milan, *Ger.* Mailand; *anc.* Mediolanum. Lombardia, N Italy
Milas *94 A4* Muğla, SW Turkey (Türkiye)
Milashavichy *85 C7 Rus.* Milashevichi. Homyel'skaya Voblasts', SE Belarus
Milashevichi *see* Milashavichy
Mildura *127 C6* Victoria, SE Australia
Mile *see* Mili Atoll
Miles *127 D5* Queensland, E Australia
Miles City *22 C2* Montana, NW USA
Milford *see* Milford Haven
Milford Haven *67 C6 prev.* Milford. SW Wales, United Kingdom
Milford Sound *129 A6 var.* Piopiotahi. Southland, South Island, New Zealand
Milford Sound *129 A6 inlet* South Island, New Zealand
Milh, Bahr al *see* Razāzah, Buḩayrat ar
Mili Atoll *122 D2 var.* Mile. *atoll* Ratak Chain, SE Marshall Islands
Mil'kovo *93 H3* Kamchatskaya Oblast', E Russia
Milk River *15 E5* Alberta, SW Canada
Milk River *22 C1 river* Montana, NW USA
Milk, Wadi el *66 B4 var.* Wadi al Malik. *river* C Sudan
Milledgeville *21 E2* Georgia, SE USA
Mille Lacs Lake *23 F2 lake* Minnesota, N USA
Millennium Island *121 F3 prev.* Caroline Island, Thornton Island. *atoll* Line Islands, E Kiribati
Millerovo *89 B6* Rostovskaya Oblast', SW Russia
Mílos *83 C7 island* Kykládes, Greece, Aegean Sea
Milton *129 B7* Otago, South Island, New Zealand
Milton Keynes *67 D6* SE England, United Kingdom
Milwaukee *18 B3* Wisconsin, N USA
Mimatum *see* Mende
Min *see* Fujian
Mīnā' as Sulṭān Qābūs *118 B3* NE Oman

Minas Gerais 41 F3 off. Estado de Minas Gerais. state/region E Brazil
Minas Gerais, Estado de see Minas Gerais
Minatitlán 29 F4 Veracruz-Llave, E Mexico
Minbu 114 A3 Magway, W Myanmar (Burma)
Minch, The 66 B3 var. North Minch. strait NW Scotland, United Kingdom
Mindanao 117 F2 island S Philippines
Mindanao Sea see Bohol Sea
Mindelheim 73 C6 Bayern, S Germany
Mindello see Mindelo
Mindelo 52 A2 var. Mindello; prev. Porto Grande. São Vicente, N Cape Verde (Cabo Verde)
Minden 72 B4 anc. Minthun. Nordrhein-Westfalen, NW Germany
Mindoro 117 E2 island N Philippines
Mindoro Strait 117 E2 strait W Philippines
Mineral Wells 27 F2 Texas, SW USA
Mingäçevir 95 G2 Rus. Mingechaur, Mingechevir. C Azerbaijan
Mingechaur/Mingechevir see Mingäçevir
Mingora see Saidu Sharif
Minho 70 B2 former province N Portugal
Minho 70 B2 Sp. Miño. river Portugal/Spain
Minho, Rio see Miño
Minicoy Island 110 B3 island SW India
Minius see Miño
Minna 53 G4 Niger, C Nigeria
Minneapolis 23 F2 Minnesota, N USA
Minnesota 23 F2 off. State of Minnesota, also known as Gopher State, New England of the West, North Star State. state N USA
Miño 70 B2 var. Mino, Minius, Port. Rio Minho. river Portugal/Spain
Miño see Minho, Rio
Minorca see Menorca
Minot 23 E1 North Dakota, N USA
Minsk 85 C6 country capital (Belarus) Minskaya Voblasts', C Belarus
Minskaya Wzvyshsha 85 C6 mountain range C Belarus
Mińsk Mazowiecki 76 D3 var. Nowo-Minsk. Mazowieckie, C Poland
Minthun see Minden
Minto, Lac 16 D2 lake Québec, C Canada
Minya see Al Minyā
Miraflores 28 C3 Baja California Sur, NW Mexico
Miranda de Ebro 71 E1 La Rioja, N Spain
Mirgorod see Myrhorod
Miri 116 D3 Sarawak, East Malaysia
Mirim Lagoon 41 E5 var. Lake Mirim, Sp. Laguna Merín. lagoon Brazil/Uruguay
Mirim, Lake see Mirim Lagoon
Mírina see Mýrina
Mirjāveh 98 E4 Sīstān va Balūchestān, SE Iran
Mirny 132 C3 Russian research station Antarctica
Mirnyy 93 F3 Respublika Sakha (Yakutiya), NE Russia
Mirpur Khas 112 B3 Sindh, SE Pakistan
Mirtóo Pélagos 83 C6 Eng. Myrtoan Sea; anc. Myrtoum Mare. sea S Greece
Misiaf see Maşyāf
Miskito Coast see Mosquito Coast
Miskitos, Cayos 31 E2 island group NE Nicaragua
Miskolc 77 D6 Borsod-Abaúj-Zemplén, NE Hungary
Misool, Pulau 117 F4 island Maluku, E Indonesia
Mişr see Egypt
Mişrātah 49 F2 var. Misurata. NW Libya
Mission 27 G5 Texas, SW USA
Mississippi 20 B2 off. State of Mississippi, also known as Bayou State, Magnolia State. state SE USA
Mississippi Delta 20 B4 delta Louisiana, S USA
Mississippi River 13 C6 river C USA
Missoula 22 B1 Montana, NW USA
Missouri 23 F5 off. State of Missouri, also known as Bullion State, Show Me State. state C USA
Missouri River 23 E3 river C USA
Mistassini, Lac 16 D3 lake Québec, SE Canada
Mistelbach an der Zaya 73 E6 Niederösterreich, NE Austria
Misti, Volcán 39 E4 volcano S Peru
Misurata see Mişrātah
Mitau see Jelgava
Mitchell 127 D5 Queensland, E Australia
Mitchell 23 E3 South Dakota, N USA
Mitchell, Mount 21 E1 mountain North Carolina, SE USA
Mitchell River 126 C2 river Queensland, NE Australia
Mi Tho see My Tho
Mitilíni see Mytilíni
Mito 109 D5 Ibaraki, Honshū, S Japan
Mitrovica see Mitrovicë
Mitrovica/Mitrowitz see Sremska Mitrovica, Serbia
Mitrovicë 79 D5 Serb. Mitrovica, prev. Kosovska Mitrovica, Titova Mitrovica. N Kosovo
Mits'iwa 50 C4 var. Masawa, Massawa. E Eritrea
Mitspe Ramon 97 A7 prev. Mizpe Ramon. Southern, S Israel
Mittelstadt see Baia Sprie
Mitú 36 C4 Vaupés, SE Colombia
Mitumba, Chaîne des/Mitumba Range see Mitumba, Monts
Mitumba Monts 55 E7 var. Chaîne des Mitumba, Mitumba Range. mountain range E Dem. Rep. Congo
Miueru Wantipa, Lake 55 E7 lake N Zambia
Miyake-jima 109 D6 island Sakishima-shotō, SW Japan
Miyako 108 D4 Iwate, Honshū, C Japan
Miyakonojō 109 B8 var. Miyakonzyô. Miyazaki, Kyūshū, SW Japan
Miyakonzyô see Miyakonojō
Miyāneh see Miāneh
Miyazaki 109 B8 Miyazaki, Kyūshū, SW Japan
Mizia 82 C1 var. Miziya. Vratsa, NW Bulgaria
Mizil 86 C5 Prahova, SE Romania
Miziya see Mjøs
Mizpe Ramon see Mitspe Ramon
Mjøsa 63 B6 var. Mjøsen. lake S Norway
Mjøsen see Mjøsa
Mladenovac 78 D4 Serbia, C Serbia
Mława 76 D3 Mazowieckie, C Poland
Mljet 79 B5 It. Meleda; anc. Melita. island S Croatia
Mmabatho 56 C4 North-West, N South Africa
Moab 22 B5 Utah, W USA
Moa Island 126 C1 island Queensland, NE Australia
Moanda 55 B6 var. Mouanda. Haut-Ogooué, SE Gabon
Moba 55 E7 Katanga, E Dem. Rep. Congo
Mobay see Montego Bay
Mobaye 55 C5 Basse-Kotto, S Central African Republic
Moberly 23 G4 Missouri, C USA
Mobile 20 C3 Alabama, S USA
Mobutu Sese Seko, Lac see Albert, Lake
Moçâmedes see Namibe
Mochudi 56 C4 Kgatleng, SE Botswana
Mocímboa da Praia 57 F2 var. Vila de Mocímboa da Praia. Cabo Delgado, N Mozambique
Môco 56 B2 var. Morro de Môco. mountain W Angola
Mocoa 36 A4 Putumayo, SW Colombia
Môco, Morro de see Môco
Mocuba 57 E3 Zambézia, NE Mozambique
Modena 74 B3 anc. Mutina. Emilia-Romagna, N Italy
Modesto 25 B6 California, W USA
Modica 75 C7 anc. Motyca. Sicilia, Italy, C Mediterranean Sea
Modimolle 56 D4 prev. Nylstroom. Limpopo, NE South Africa
Modohn see Madona
Modriča 78 C3 Republika Srpska, N Bosnia and Herzegovina
Moe 127 C7 Victoria, SE Australia
Möen see Møn, Denmark
Moero, Lac see Mweru, Lake
Moeskroen see Mouscron
Mogadiscio/Mogadishu see Muqdisho
Mogador see Essaouira
Mogilëv see Mahilyow
Mogilev-Podol'skiy see Mohyliv-Podilskyi
Mogilno 76 C3 Kujawsko-pomorskie, C Poland
Moḩammadābād-e Rīgān 98 E4 Kermān, SE Iran
Mohammedia 48 C2 prev. Fédala. NW Morocco
Mohave, Lake 25 D7 reservoir Arizona/Nevada, W USA
Mohawk River 19 F3 river New York, NE USA
Mohéli see Mwali
Mohns Ridge 61 F3 undersea ridge Greenland Sea/Norwegian Sea
Moho 39 E4 Puno, SE Peru
Mohoro 51 C7 Pwani, E Tanzania
Mohua see Golden Bay
Mohyliv-Podilskyi 86 D3 prev. Mohyliv-Podil's'kyy, Rus. Mogilev-Podol'skiy. Vinnyts'ka Oblast', C Ukrain
Mohyliv-Podil's'kyy see Mohyliv-Podilskyi
Moili see Mwali
Mo i Rana 62 C3 Nordland, C Norway
Mõisaküla 84 D3 Ger. Moiseküll. Viljandimaa, S Estonia
Moiseküll see Mõisaküla
Moissac 69 B6 Tarn-et-Garonne, S France
Mojácar 71 E5 Andalucía, S Spain
Mojave Desert 25 D7 plain California, W USA
Mokrany see Makrany
Moktama see Mottama
Mol 65 C5 prev. Moll. Antwerpen, N Belgium
Moldavia see Moldova
Moldavian SSR/Moldavskaya SSR see Moldova
Molde 63 A5 Møre og Romsdal, S Norway
Moldotau, Khrebet see Moldo-Too, Khrebet
Moldo-Too, Khrebet 101 G2 prev. Khrebet Moldotau. mountain range C Kyrgyzstan
Moldova 86 D3 off. Republic of Moldova, var. Moldavia; prev. Moldavian SSR, Rus. Moldavskaya SSR. country SE Europe
Moldova Nouă 86 A4 Ger. Neumoldowa, Hung. Újmoldova. Caraş-Severin, SW Romania
Moldova, Republic of see Moldova
Moldoveanul see Vârful Moldoveanu
Molfetta 75 E5 Puglia, SE Italy
Moll see Mol
Mollendo 39 E4 Arequipa, SW Peru
Mölndal 63 B7 Västra Götaland, S Sweden
Molochansk 87 G4 Zaporiz'ka Oblast', SE Ukraine
Molodechno/Molodeczno see Maladzyechna
Moloka'i 25 B8 var. Molokai. island Hawaiian Islands, Hawaii, USA
Molokai Fracture Zone 131 E2 tectonic feature NE Pacific Ocean
Molopo 56 C4 seasonal river Botswana/South Africa
Mólos 83 B5 Stereá Ellás, C Greece
Molotov see Severodvinsk, Arkhangel'skaya Oblast', Russia
Molotov see Perm', Permskaya Oblast', Russia
Moluccas see Maluku
Molucca Sea 117 F4 Ind. Laut Maluku. sea E Indonesia
Molukken see Maluku
Mombasa 51 D7 Coast, SE Kenya
Mombetsu see Monbetsu
Momchilgrad 82 D3 prev. Mastanli. Kardzhali, S Bulgaria
Møn 63 B8 prev. Möen. island SE Denmark
Mona, Canal de la see Mona Passage
Monaco 69 C7 var. Monaco-Ville; anc. Monoecus. country capital (Monaco) S Monaco
Monaco 69 E6 off. Principality of Monaco. country W Europe
Monaco see München
Monaco, Port de 69 C8 bay S Monaco W Mediterranean Sea
Monaco, Principality of see Monaco
Monaco-Ville see Monaco
Monahans 27 E3 Texas, SW USA
Mona, Isla 33 E3 island W Puerto Rico
Mona Passage 33 E3 Sp. Canal de la Mona. channel Dominican Republic/Puerto Rico
Monastir see Bitola
Monbetsu 108 D2 var. Mombetsu, Monbetu. Hokkaidō, NE Japan
Monbetu see Monbetsu
Moncalieri 74 A2 Piemonte, NW Italy
Monchegorsk 88 C2 Murmanskaya Oblast', NW Russia
Monclova 28 D2 Coahuila, NE Mexico
Moncton 17 F4 New Brunswick, SE Canada
Mondovì 74 A2 Piemonte, NW Italy
Monfalcone 74 D2 Friuli-Venezia Giulia, NE Italy
Monforte de Lemos 70 C1 Galicia, NW Spain
Mongo 54 C3 Guéra, C Chad
Mongolia 104 C2 Mong. Mongol Uls. country E Asia
Mongolia, Plateau of 102 D1 plateau E Mongolia
Mongol Uls see Mongolia
Mongora see Saidu Sharif
Mongos, Chaîne des see Bongo, Massif des
Mongu 56 C2 Western, W Zambia
Monkchester see Newcastle upon Tyne
Monkey Bay 57 E2 Southern, SE Malawi
Monkey River see Monkey River Town
Monkey River Town 30 C2 var. Monkey River. Toledo, SE Belize
Monoecus see Monaco
Mono Lake 25 C6 lake California, W USA
Monostor see Beli Manastir
Monóvar see Monòver
Monòver 71 F4 var. Monóvar. Comunitat Valenciana, E Spain
Monroe 20 B2 Louisiana, S USA
Monrovia 52 C5 country capital (Liberia) W Liberia
Mons 65 B6 Dut. Bergen. Hainaut, S Belgium
Monselice 74 C2 Veneto, NE Italy
Montana 82 C2 prev. Ferdinand, Mikhaylovgrad. Montana, NW Bulgaria
Montana 22 B1 off. State of Montana, also known as Mountain State, Treasure State. state NW USA
Montargis 68 C4 Loiret, C France
Montauban 69 B6 Tarn-et-Garonne, S France
Montbéliard 68 D4 Doubs, E France
Mont Cenis, Col du 69 D5 pass E France
Mont-de-Marsan 69 B6 Landes, SW France
Monteagudo 39 G4 Chuquisaca, S Bolivia
Montecarlo 69 C8 Misiones, NE Argentina
Monte Caseros 42 D3 Corrientes, NE Argentina
Monte Cristi 32 D3 var. San Fernando de Monte Cristi. NW Dominican Republic
Monte Croce Carnico, Passo di see Plöcken Pass
Montegiardino 74 E2 SE San Marino
Montego Bay 32 A4 var. Mobay. W Jamaica
Montélimar 69 D5 anc. Acunum Acusio, Montilium Adhemari. Drôme, E France
Montemorelos 29 E3 Nuevo León, NE Mexico
Montenegro 79 C5 Serb. Crna Gora. country SW Europe
Monte Patria 42 B3 Coquimbo, N Chile
Monterey 25 B6 California, W USA
Monterey see Monterrey
Monterey Bay 25 A6 bay California, W USA
Montería 36 B2 Córdoba, NW Colombia
Montero 39 G4 Santa Cruz, C Bolivia
Monterrey 29 E3 var. Monterey. Nuevo León, NE Mexico
Montes Claros 41 F3 Minas Gerais, SE Brazil
Montevideo 42 D4 country capital (Uruguay) Montevideo, S Uruguay
Montevideo 23 F2 Minnesota, N USA
Montgenèvre, Col de 69 D5 pass France/Italy
Montgomery 20 D2 state capital Alabama, S USA
Montgomery see Sahiwal
Monthey 73 A7 Valais, SW Switzerland
Montilium Adhemari see Montélimar
Montluçon 68 C4 Allier, C France
Montoro 70 D4 Andalucía, S Spain
Montpelier 19 G2 state capital Vermont, NE USA
Montpellier 69 C6 Hérault, S France
Montréal 17 E4 Eng. Montreal. Québec, SE Canada
Montrose 66 D3 E Scotland, United Kingdom
Montrose 22 C5 Colorado, C USA
Montserrat 33 G3 var. Emerald Isle. UK Overseas Territory E West Indies
Monywa 114 B3 Sagaing, C Myanmar (Burma)
Monza 74 B2 Lombardia, N Italy
Monze 56 D2 Southern, S Zambia
Monzón 71 F2 Aragón, NE Spain
Moonie 127 D5 Queensland, E Australia
Moon-Sund see Väinameri
Moora 125 A6 Western Australia
Moore 27 G1 Oklahoma, C USA
Moore, Lake 125 B6 lake Western Australia
Moorhead 23 F2 Minnesota, N USA
Moose 16 C3 river Ontario, S Canada
Moosehead Lake 19 G1 lake Maine, NE USA
Moosonee 16 C3 Ontario, SE Canada
Mopti 53 E3 Mopti, C Mali
Moquegua 39 E4 Moquegua, SE Peru
Mora 63 C5 Dalarna, C Sweden
Morales 30 C2 Izabal, E Guatemala
Morant Bay 32 B5 E Jamaica
Moratalla 71 E4 Murcia, SE Spain
Morava 77 C5 var. March. river C Europe
Morava see Moravia, Czechia (Czech Republic)
Morava see Velika Morava, Serbia
Moravia 77 B5 Cz. Morava, Ger. Mähren. cultural region E Czechia (Czech Republic)
Moray Firth 66 C3 inlet N Scotland, United Kingdom
Morea see Pelopónnisos
Moreau River 22 D2 river South Dakota, N USA
Moree 127 D5 New South Wales, SE Australia
Morelia 29 E4 Michoacán, S Mexico
Morena, Sierra 70 C4 mountain range S Spain
Moreni 86 C5 Dâmboviţa, S Romania
Morgan City 20 B3 Louisiana, S USA
Morghab, Darya-ye 100 D3 Rus. Murgab, Murghab, Turkm. Murgap, Murgap Deryasy. river Afghanistan/Turkmenistan
Morioka 108 D4 Iwate, Honshū, C Japan
Morlaix 68 A3 Finistère, NW France
Mormon State see Utah
Mornington Abyssal Plain 45 A7 abyssal plain SE Pacific Ocean
Mornington Island 126 B2 island Wellesley Islands, Queensland, N Australia
Morocco 48 B3 off. Kingdom of Morocco, Ar. Al Maghrib. country N Africa
Morocco see Marrakech
Morocco, Kingdom of see Morocco
Morogoro 51 C7 Morogoro, E Tanzania
Moro Gulf 117 E3 gulf S Philippines
Morón 32 C2 Ciego de Ávila, C Cuba
Mörön 104 D2 Hövsgöl, N Mongolia
Morondava 57 F3 Toliara, W Madagascar
Moroni 57 F2 country capital (Comoros) Grande Comore, NW Comoros
Morotai, Pulau 117 F3 island Maluku, E Indonesia
Morotiri see Marotiri
Morphou see Güzelyurt
Morrinsville 128 D3 Waikato, North Island, New Zealand
Morris 23 F2 Minnesota, N USA
Morris Jesup, Kap 61 E1 headland N Greenland
Morvan 68 D4 physical region C France
Moscow 24 C2 Idaho, NW USA
Moscow see Moskva
Mosel 73 A5 Fr. Moselle. river W Europe
Mosel see Moselle
Moselle 65 E8 Ger. Mosel. river W Europe
Moselle see Mosel
Mosgiel 129 B7 Otago, South Island, New Zealand
Moshi 51 C7 Kilimanjaro, NE Tanzania
Mosjøen 62 B4 Nordland, C Norway
Moskovskiy see Maskav
Moskva 89 B5 Eng. Moscow. country capital (Russia) Gorod Moskva, W Russia
Moskva see Maskav
Moson and Magyaróvár see Mosonmagyaróvár
Mosonmagyaróvár 77 C6 Ger. Wieselburg-Ungarisch-Altenburg; prev. Moson and Magyaróvár, Ger. Wieselburg and Ungarisch-Altenburg. Győr-Moson-Sopron, NW Hungary
Mosquito Coast 31 E3 var. Miskito Coast, Eng. Mosquito Coast. coastal region E Nicaragua
Mosquito Coast see La Mosquitia
Mosquito Gulf 31 F4 Eng. Mosquito Gulf. gulf N Panama
Mosquito Gulf see Mosquitos, Golfo de los
Moss 63 B6 Østfold, S Norway
Mossâmedes see Namibe
Mosselbaai 56 C5 var. Mosselbai, Eng. Mossel Bay. Western Cape, SW South Africa
Mosselbai/Mossel Bay see Mosselbaai
Mossendjo 55 B6 Niari, SW Congo
Mossoró 41 G2 Rio Grande do Norte, NE Brazil
Most 76 A4 Ger. Brüx. Ústecký Kraj, NW Czechia (Czech Republic)
Mosta 80 B5 var. Musta. C Malta
Mostaganem 48 D2 var. Mestghanem. NW Algeria
Mostar 78 C4 Federacija Bosna I Hercegovina, S Bosnia and Herzegovina
Mosty see Masty
Mosul see Al Mawşil
Mota del Cuervo 71 E3 Castilla-La Mancha, C Spain
Motagua, Río 30 B2 river Guatemala/Honduras
Mother of Presidents/Mother of States see Virginia
Motril 70 D5 Andalucía, S Spain
Motru 86 B4 Gorj, SW Romania
Mottama 114 B4 prev. Martaban; var. Moktama. Mon State, S Myanmar (Burma)
Motueka 129 C5 Tasman, South Island, New Zealand
Motul 29 H3 var. Motul de Felipe Carrillo Puerto. Yucatán, SE Mexico
Motul de Felipe Carrillo Puerto see Motul
Motyca see Modica
Mouanda see Moanda
Mouhoun see Black Volta
Mouila 55 A6 Ngounié, C Gabon
Moukden see Shenyang
Mould Bay 15 E2 Prince Patrick Island, Northwest Territories, N Canada
Moulins 68 C4 Allier, C France
Moulmein see Mawlamyine
Moundou 54 B4 Logone-Occidental, SW Chad
Moŭng Roessei 115 D5 Battambang, W Cambodia
Moun Hou see Black Volta
Mountain Home 20 B1 Arkansas, C USA
Mountain State see Montana
Mountain State see West Virginia
Mount Cook see Aoraki (mountain)
Mount Cook see Aoraki (populated place)
Mount Desert Island 19 H2 island Maine, NE USA
Mount Gambier 127 B7 South Australia
Mount Isa 126 B3 Queensland, C Australia
Mount Magnet 125 B5 Western Australia
Mount Pleasant 23 G4 Iowa, C USA
Mount Pleasant 18 C3 Michigan, N USA
Mount Vernon 18 B5 Illinois, N USA
Mount Vernon 24 B1 Washington, NW USA
Mourdi, Dépression du 54 C2 desert lowland Chad/Sudan
Mouscron 65 A6 Dut. Moeskroen. Hainaut, W Belgium
Mouse River see Souris River
Moussoro 54 B3 Kanem, W Chad
Moyen-Congo see Congo (Republic of)
Mo'ynoq 100 C1 Rus. Muynak. Qoraqalpog'iston Respublikasi, NW Uzbekistan
Moyobamba 38 B2 San Martín, NW Peru
Moyu 104 B3 var. Karakax. Xinjiang Uygur Zizhiqu, NW China
Moyynkum, Peski 101 F1 Kaz. Moyynqum. desert S Kazakhstan
Moyynqum see Moyynkum, Peski
Mozambika, Lakandranon' i see Mozambique Channel
Mozambique 57 E3 off. Republic of Mozambique; prev. People's Republic of Mozambique, Portuguese East Africa. country S Africa
Mozambique Basin see Natal Basin
Mozambique, Canal de see Mozambique Channel
Mozambique Channel 57 E3 Fr. Canal de Mozambique, Mal. Lakandranon' i Mozambika. strait W Indian Ocean
Mozambique, People's Republic of see Mozambique
Mozambique Plateau 47 D7 var. Mozambique Rise. undersea plateau SW Indian Ocean
Mozambique, Republic of see Mozambique
Mozambique Rise see Mozambique Plateau
Mozyr' see Mazyr
Mpama 55 B6 river C Congo
Mpika 56 D2 Northern, NE Zambia
Mqinvartsveri see Kazbek
Mrągowo 76 D2 Ger. Sensburg. Warmińsko-Mazurskie, NE Poland
Mthatha 56 D5 prev. Umtata. Eastern Cape, SE South Africa
Mtkvari see Kura
Mtwara 51 D8 Mtwara, SE Tanzania
Mualo see Messalo, Rio
Muang Chiang Rai see Chiang Rai
Muang Kalasin see Kalasin
Muang Khammouan see Thakhèk
Muang Không 115 D5 Champasak, S Laos
Muang Khôngxédôn 115 D5 var. Khong Sedone. Salavan, S Laos
Muang Khon Kaen see Khon Kaen
Muang Lampang see Lampang
Muang Loei see Loei
Muang Lom Sak see Lom Sak
Muang Nakhon Sawan see Nakhon Sawan
Muang Namo 114 C3 Oudômxai, N Laos
Muang Nan see Nan
Muang Phalan 114 D4 var. Muang Phalane. Savannakhét, S Laos
Muang Phalane see Muang Phalan
Muang Phayao see Phayao
Muang Phitsanulok see Phitsanulok
Muang Phrae see Phrae
Muang Roi Et see Roi Et
Muang Sakon Nakhon see Sakon Nakhon
Muang Samut Prakan see Samut Prakan
Muang Sing 114 C3 Louang Namtha, N Laos
Muang Ubon see Ubon Ratchathani
Muang Xaignabouri see Xaignabouli
Muar 116 B3 var. Bandar Maharani. Johor, Peninsular Malaysia
Mucojo 57 F2 Cabo Delgado, N Mozambique
Mudanjiang 107 E3 var. Mu-tan-chiang. Heilongjiang, NE China
Mudon 115 B5 Mon State, S Myanmar (Burma)
Muenchen see München
Muenster see Münster
Mufulira 56 D2 Copperbelt, C Zambia
Mughla see Muğla
Muğla 94 A4 var. Mughla. Muğla, SW Turkey (Türkiye)
Mūḩ, Sabkhat al 96 C3 lake C Syria
Muhu Väin see Väinameri
Muisne 38 A1 Esmeraldas, NW Ecuador
Mukacheve see Mukachevo
Mukachevo 86 B3 prev. Mukacheve, Hung. Munkács. Zakarpats'ka Oblast', W Ukraine
Mukalla see Al Mukallā
Mukden see Shenyang
Mula 71 E4 Murcia, SE Spain
Mulakatholhu 110 B4 var. Meemu Atoll, Mulaku Atoll. atoll C Maldives
Mulaku Atoll see Mulakatholhu
Muleshoe 27 E2 Texas, SW USA
Mulhacén 71 E5 var. Cerro de Mulhacén. mountain S Spain
Mulhacén, Cerro de see Mulhacén
Mülhausen see Mulhouse
Mülheim 73 A6 var. Mulheim an der Ruhr. Nordrhein-Westfalen, W Germany
Mulheim an der Ruhr see Mülheim
Mulhouse 68 E4 Ger. Mülhausen. Haut-Rhin, NE France
Müller-gerbergte see Muller, Pegunungan
Muller, Pegunungan 116 D4 Dut. Müller-gerbergte. mountain range Borneo, C Indonesia
Mull, Isle of 66 B4 island W Scotland, United Kingdom
Mulongo 55 D7 Katanga, SE Dem. Rep. Congo
Multan 112 C2 Punjab, E Pakistan
Mumbai 112 C5 prev. Bombay. state capital Mahārāshtra, W India
Munamägi see Suur Munamägi
Münchberg 73 C5 Bayern, E Germany
München 73 C6 var. Muenchen, Eng. Munich, It. Monaco. Bayern, SE Germany
Muncie 18 C4 Indiana, N USA
Mungbere 55 E5 Orientale, NE Dem. Rep. Congo
Mu Nggava see Rennell
Munich see München
Munkács see Mukachevo
Münster 72 A4 var. Muenster, Münster in Westfalen. Nordrhein-Westfalen, W Germany
Munster 67 A6 Ir. Cúige Mumhan. cultural region S Ireland
Münster in Westfalen see Münster
Muong Xiang Ngeun 114 C4 var. Xieng Ngeun. Louangphabang, N Laos
Muonio 62 D3 Lappi, N Finland
Muonioälv/Muoniojoki see Muonionjoki
Muonionjoki 62 D3 var. Muoniojoki, Swe. Muonioälv. river Finland/Sweden
Muqāţ 97 C5 Al Mafraq, E Jordan
Muqdisho 51 D6 Eng. Mogadishu, It. Mogadiscio. country capital (Somalia) Banaadir, S Somalia
Mur 73 E7 Croatian Mura. river C Europe
Mura see Mur
Muradiye 95 F3 Van, E Turkey (Türkiye)
Murapara see Murupara
Murata 74 E2 S San Marino
Murchison River 125 A5 river Western Australia
Murcia 71 F4 Murcia, SE Spain
Murcia 71 E4 autonomous community SE Spain
Mureş 86 A4 river Hungary/Romania
Murfreesboro 20 D1 Tennessee, S USA
Murgab see Morghāb, Daryā-yeb
Murgap 100 D3 Mary Welaýaty, S Turkmenistan
Murgap see Morghāb, Daryā-ye
Murgap Deryasy see Morghāb, Daryā-ye
Murghab see Morghāb, Daryā-ye
Murghob 101 F3 Rus. Murgab. SE Tajikistan
Murgon 127 E5 Queensland, E Australia
Mūrītānīyah see Mauritania
Müritz 72 C3 var. Müritzee. lake NE Germany
Müritzee see Müritz
Murmansk 88 C2 Murmanskaya Oblast', NW Russia
Murmashi 88 C2 Murmanskaya Oblast', NW Russia
Murom 89 B5 Vladimirskaya Oblast', W Russia
Muroran 108 D3 Hokkaidō, NE Japan
Muros 70 B1 Galicia, NW Spain
Murray Fracture Zone 131 E2 fracture zone NE Pacific Ocean
Murray Range see Murray Ridge
Murray Ridge 90 C5 var. Murray Range. undersea ridge N Arabian Sea
Murray River 127 B6 river SE Australia
Murrumbidgee River 127 C6 river New South Wales, SE Australia
Murska Sobota 73 E7 Ger. Olsnitz. NE Slovenia
Murupara 128 E3 var. Murapara. Bay of Plenty, North Island, New Zealand
Murviedro see Sagunt
Murwāra 113 E4 Madhya Pradesh, N India
Murwillumbah 127 E5 New South Wales, SE Australia
Murzuq, Edeyin see Murzuq, Idhān
Murzuq, Idhan 49 F4 var. Edeyin Murzuq. desert SW Libya
Mürzzuschlag 73 E7 Steiermark, E Austria
Muş 95 F3 var. Mush. Muş, E Turkey (Türkiye)
Musa, Gebel 50 C2 var. Gebel Mûsa. mountain NE Egypt
Mûsa, Gebel see Mūsá, Jabal
Musala 82 B3 mountain W Bulgaria
Muscat see Masqaţ
Muscat and Oman see Oman
Muscatine 23 G3 Iowa, C USA
Musgrave Ranges 125 D5 mountain range South Australia
Musina 75 D7 var. Messana, Messene; anc. Zancle. Sicilia, Italy, C Mediterranean Sea

Musina *56 D3* prev. Messina. Limpopo, NE South Africa
Muskegon *18 C3* Michigan, N USA
Muskogean see Tallahassee
Muskogee *27 G1* Oklahoma, C USA
Musoma *51 C6* Mara, N Tanzania
Musta see Mosta
Mustafa-Pasha see Svilengrad
Musters, Lago *43 B6* lake S Argentina
Muswellbrook *127 D6* New South Wales, SE Australia
Mut *94 C4* İçel, S Turkey (Türkiye)
Mu-tan-chiang see Mudanjiang
Mutare *56 D3* var. Mutari; prev. Umtali. Manicaland, E Zimbabwe
Mutari see Mutare
Mutina see Modena
Mutsu-wan *108 D3* bay N Japan
Muttonbird Islands see Titi
Mu Us Shadi *105 E3* var. Ordos Desert; prev. Mu Us Shamo. desert N China
Mu Us Shamo see Mu Us Shadi
Muy Muy *30 D3* Matagalpa, C Nicaragua
Muynak see Mo'ynoq
Mužlja *78 D3* Hung. Felsőmuzslya; prev. Gornja Mužlja. Vojvodina, N Serbia
Mwali *57 F2* var. Moili, Fr. Mohéli. island S Comoros
Mwanza *51 B6* Mwanza, NW Tanzania
Mweka *55 C6* Kasai-Occidental, C Dem. Rep. Congo
Mwene-Ditu *55 D7* Kasai-Oriental, S Dem. Rep. Congo
Mweru, Lake *55 D7* var. Lac Moero. lake Dem. Rep. Congo/Zambia
Myadel' see Myadzyel
Myadzyel *85 C5* Pol. Miadzioł Nowy, Rus. Myadel'. Minskaya Voblasts', N Belarus
Myanaung *114 B4* Ayeyarwady, SW Myanmar (Burma)
Myanmar *114 A3* off. Republic of the Union of Myanmar; prev. Union of Myanmar, var. Burma. country SE Asia
Myanmar, Republic of the Union of see Myanmar
Myanmar, Union of see Myanmar
Myaungmya *114 A4* Ayeyarwady, SW Myanmar (Burma)
Myaydo see Aunglan
Myeik *115 B6* var. Mergui. Tanintharyi, S Myanmar (Burma)
Myeik Archipelago *115 B6* prev. Mergui Archipelago. island group S Myanmar (Burma)
Myerkulavichy *85 D7* Rus. Merkulovichi. Homyel'skaya Voblasts', SE Belarus
Myingyan *114 B3* Mandalay, C Myanmar (Burma)
Myitkyina *114 B2* Kachin State, N Myanmar (Burma)
Mykolaiv *87 E4* prev. Mykolayiv, Rus. Nikolayev. Mykolayivs'ka Oblast', S Ukraine
Mykolayiv see Mykolaiv
Mykonos *83 D6* var. Míkonos. island Kykládes, Greece, Aegean Sea
Myrhorod *87 F2* Rus. Mirgorod. Poltavs'ka Oblast', NE Ukraine
Mýrina *82 D4* var. Mírina. Límnos, SE Greece
Myrtle Beach *21 F2* South Carolina, SE USA
Myrtoan Sea see Mirtóo Pélagos
Mýrtos *83 D8* Kríti, Greece, E Mediterranean Sea
Myrtoum Mare see Mirtóo Pélagos
Myślibórz *76 B3* Zachodnio-pomorskie, NW Poland
Mysore see Mysūru
Mysore see Karnātaka
Mysūru *110 C2* prev. Mysore, var. Maisur. Karnātaka, W India
My Tho *115 E6* var. Mi Tho. Tiên Giang, S Vietnam
Mytilene see Mytilíni
Mytilíni *83 D5* var. Mitilíni; anc. Mytilene. Lésvos, E Greece
Mzuzu *57 E2* Northern, N Malawi

N

Naberezhnyye Chelny *89 D5* prev. Brezhnev. Respublika Tatarstan, W Russia
Nāblus *97 A6* var. Nābulus, Heb. Shekhem; anc. Neapolis, Bibl. Shechem. N West Bank, Middle East
Nābulus see Nablus
Nacala *57 F2* Nampula, NE Mozambique
Na-Ch'ii see Nagqu
Nada see Danzhou
Nadi *123 E4* prev. Nandi. Viti Levu, W Fiji
Nadur *80 A5* Gozo, N Malta
Nadvirna *86 C3* Pol. Nadwórna, Rus. Nadvornaya. Ivano-Frankivs'ka Oblast', W Ukraine
Nadvoitsy *88 B3* Respublika Kareliya, NW Russia
Nadvornaya/Nadwórna see Nadvirna
Nadym *92 C3* Yamalo-Nenetskiy Avtonomnyy Okrug, N Russia
Náfpaktos *83 B5* var. Návpaktos. Dytikí Ellás, C Greece
Náfplio *83 B6* prev. Návplion. Pelopónnisos, S Greece
Naga *117 E2* off. Naga City; prev. Nueva Caceres. Luzon, N Philippines
Naga City see Naga
Nagano *109 C5* Nagano, Honshū, S Japan
Nagaoka *109 D5* Niigata, Honshū, C Japan
Nagara Pathom see Nakhon Pathom
Nagara Sridharmaraj see Nakhon Si Thammarat
Nagara Svarga see Nakhon Sawan
Nagasaki *109 A7* Nagasaki, Kyūshū, SW Japan
Nagato *109 A7* Yamaguchi, Honshū, SW Japan
Nāgercoil *110 C3* Tamil Nādu, SE India
Nagorno-Karabakh see Nagornyy-Karabakh
Nagorno-Karabakhskaya Avtonomnaya Oblast see Nagornyy-Karabakh
Nagornyy Karabakh *95 G3* prev. Nagorno-Karabakh, Nagorno-Karabakhskaya Avtonomnaya Oblast, Arm. Lernayin Gharabagh, Az. Dağlıq Qarabağ. former autonomous region SW Azerbaijan
Nagoya *109 C6* Aichi, Honshū, SW Japan
Nāgpur *112 D4* Mahārāshtra, C India
Nagqu *104 C5* Chin. Na-Ch'ii; prev. Hei-ho. Xizang Zizhiqu, W China
Nagybánya see Baia Mare
Nagybecskerek see Zrenjanin
Nagydisznód see Cisnădie
Nagyenyed see Aiud
Nagykálló *77 E6* Szabolcs-Szatmár-Bereg, E Hungary
Nagykanizsa *77 C7* Ger. Grosskanizsa. Zala, SW Hungary
Nagykároly see Carei
Nagykikinda see Kikinda
Nagykőrös *77 D7* Pest, C Hungary
Nagy-Mihály see Michalovce
Nagysurány see Šurany
Nagyszalonta see Salonta
Nagyszeben see Sibiu
Nagyszentmiklós see Sânnicolau Mare
Nagyszöllös see Vynohradiv
Nagyszombat see Trnava
Nagytapolcsány see Topoľčany
Nagyvárad see Oradea
Naha *108 A3* Okinawa, Okinawa, SW Japan
Nahariya *97 A5* prev. Nahariyya. Northern, N Israel
Nahariyya see Nahariya
Nahuel Huapí, Lago *43 B5* lake W Argentina
Nain *17 F2* Newfoundland and Labrador, NE Canada
Nā'īn *98 D3* Eşfahān, C Iran
Nairobi *47 E5* country capital (Kenya) Nairobi Area, S Kenya
Nairobi *51 C6* Nairobi Area, S Kenya
Naissus see Niš
Najaf see An Najaf
Najima see Fukuoka
Najin see Rajin
Najrān *99 B6* var. Abā as Su'ūd. Najrān, S Saudi Arabia
Nakambé see White Volta
Nakamura *109 B7* var. Shimanto. Kōchi, Shikoku, SW Japan
Nakatsugawa *109 C6* var. Nakatugawa. Gifu, Honshū, SW Japan
Nakatugawa see Nakatsugawa
Nakhichevan' see Naxçıvan
Nakhodka *93 G5* Primorskiy Kray, SE Russia
Nakhon Pathom *115 C5* var. Nagara Pathom, Nakorn Pathom. Nakhon Pathom, W Thailand
Nakhon Ratchasima *115 C5* var. Khorat, Korat. Nakhon Ratchasima, E Thailand
Nakhon Sawan *115 C5* var. Muang Nakhon Sawan, Nagara Svarga. Nakhon Sawan, W Thailand
Nakhon Si Thammarat *115 C7* var. Nagara Sridharmaraj, Nakhon Sithamnaraj. Nakhon Si Thammarat, SW Thailand
Nakhon Sithamnaraj see Nakhon Si Thammarat
Nakorn Pathom see Nakhon Pathom
Nakuru *51 C6* Rift Valley, SW Kenya
Nal'chik *89 B8* Kabardino-Balkarskaya Respublika, SW Russia
Nālūt *49 F2* NW Libya
Namakan Lake *18 A1* lake Canada/USA
Namangan *101 F2* Namangan Viloyati, E Uzbekistan
Nambala *56 D2* Central, C Zambia
Nam Co *104 C5* lake W China
Nam Định *114 D3* Nam Ha, N Vietnam
Namib Desert *56 B3* desert W Namibia
Namibe *56 A2* Port. Moçâmedes, Mossâmedes. Namibe, SW Angola
Namibia *56 B3* off. Republic of Namibia, var. South West Africa, Afr. Suidwes-Afrika, Ger. Deutsch-Südwestafrika; prev. German Southwest Africa, South-West Africa. country S Africa
Namibia, Republic of see Namibia
Namnetes see Nantes
Namo see Namu Atoll
Nam Ou *114 C3* river N Laos
Nampa *24 D3* Idaho, NW USA
Nampula *57 E2* Nampula, NE Mozambique
Namsos *62 B4* Nord-Trøndelag, C Norway
Nam Tha *114 C4* river N Laos
Namu Atoll *122 D2* var. Namo. atoll Ralik Chain, C Marshall Islands
Namur *65 C6* Dut. Namen. Namur, SE Belgium
Namyit Island *106 C8* island S Spratly Islands
Nan *114 C4* var. Muang Nan. Nan, NW Thailand
Nanaimo *14 D5* Vancouver Island, British Columbia, SW Canada
Nanchang *106 C5* var. Nan-ch'ang, Nanch'ang-hsien. province capital Jiangxi, S China
Nan-ch'ang see Nanchang
Nanch'ang-hsien see Nanchang
Nan-ching see Nanjing
Nancy *68 D3* Meurthe-et-Moselle, NE France
Nandaime *30 D3* Granada, SW Nicaragua
Nānded *112 D5* Mahārāshtra, C India
Nandi see Nadi
Nándorhgy see Oţelu Roşu
Nandyāl *110 C1* Andhra Pradesh, E India
Naniwa see Ōsaka
Nanjing *106 D5* var. Nan-ching, Nanking; prev. Chianning, Chian-ning, Kiang-ning, Jiangsu. province capital Jiangsu, E China
Nanking see Nanjing
Nanning *106 B6* var. Nan-ning; prev. Yung-ning. Guangxi Zhuangzu Zizhiqu, S China
Nan-ning see Nanning
Nanortalik *60 C5* Kujalleq, S Greenland
Nanpan Jiang *114 D2* river S China
Nanping *106 D6* var. Nan-p'ing; prev. Yenping. Fujian, SE China
Nan-p'ing see Nanping
Nansei-shotō *108 A2* Eng. Ryukyu Islands. island group SW Japan
Nansei Syotō Trench see Ryukyu Trench
Nansen Basin *133 C4* undersea basin Arctic Ocean
Nansen Cordillera *133 B3* var. Arctic Mid Oceanic Ridge, Nansen Ridge. seamount range Arctic Ocean
Nansen Ridge see Nansen Cordillera
Nanterre *68 D1* Hauts-de-Seine, N France
Nantes *68 B4* Bret. Naoned; anc. Condivincum, Namnetes. Loire-Atlantique, NW France
Nantucket Island *19 G3* island Massachusetts, NE USA
Nanumaga *123 E3* var. Nanumanga. atoll NW Tuvalu
Nanumanga see Nanumaga
Nanumea Atoll *123 E3* atoll NW Tuvalu
Nanyang *106 C5* var. Nan-yang. Henan, C China
Nan-yang see Nanyang
Naoned see Nantes
Napa *25 B6* California, W USA
Napier *128 E4* Hawke's Bay, North Island, New Zealand
Naples *21 E5* Florida, SE USA
Naples see Napoli
Napo *34 A3* province NE Ecuador
Napoléon-Vendée see la Roche-sur-Yon
Napoli *75 C5* Eng. Naples, Ger. Neapel; anc. Neapolis. Campania, S Italy
Napo, Río *38 C1* river Ecuador/Peru
Naracoorte *127 B7* South Australia
Naradhivas see Narathiwat
Narathiwat *115 C7* var. Naradhivas. Narathiwat, SW Thailand
Narbada see Narmada
Narbo Martius see Narbonne
Narbonne *69 C6* anc. Narbo Martius. Aude, S France
Narborough Island see Fernandina, Isla
Nares Abyssal Plain see Nares Plain
Nares Plain *13 E6* var. Nares Abyssal Plain. abyssal plain NW Atlantic Ocean
Nares Stræde see Nares Strait
Nares Strait *60 D1* Dan. Nares Stræde. strait Canada/Greenland
Narew *76 E3* river E Poland
Narmada *102 B3* var. Narbada. river C India
Narova see Narva
Narovlya see Narowlya
Narowlya *85 C8* Rus. Narovlya. Homyel'skaya Voblasts', SE Belarus
Närpes *63 D5* Fin. Närpiö. Österbotten, W Finland
Närpiö see Närpes
Narrabri *127 D6* New South Wales, SE Australia
Narrogin *125 B6* Western Australia
Narva *84 E2* Ida-Virumaa, NE Estonia
Narva *84 E2* prev. Narova. river Estonia/Russia
Narva Bay *84 E2* Est. Narva Laht, Ger. Narwa-Bucht, Rus. Narvskiy Zaliv. bay Estonia/Russia
Narva Laht see Narva Bay
Narva Reservoir *84 E2* Est. Narva Veehoidla, Rus. Narvskoye Vodokhranilishche. reservoir Estonia/Russia
Narva Veehoidla see Narva Reservoir
Narvik *62 C3* Nordland, C Norway
Narvskiy Zaliv see Narva Bay
Narvskoye Vodokhranilishche see Narva Reservoir
Narwa-Bucht see Narva Bay
Nar'yan-Mar *88 D3* prev. Beloshchel'ye, Dzerzhinskiy. Nenetskiy Avtonomnyy Okrug, NW Russia
Naryn *101 G2* Narynskaya Oblast', C Kyrgyzstan
Nassau *32 C1* country capital (The Bahamas) New Providence, N The Bahamas
Năsăud *86 B3* Ger. Nussdorf, Hung. Naszód. Bistriţa-Năsăud, N Romania
Nase see Naze
Nāshik *112 C5* prev. Nāsik. Mahārāshtra, W India
Nashua *19 G3* New Hampshire, NE USA
Nashville *20 C1* state capital Tennessee, S USA
Näsijärvi *63 D5* lake SW Finland
Nāsik see Nāshik
Nasir, Buhayrat/Nâşir, Buḥeiret see Nasser, Lake
Nâsiri see Ahvāz
Nasiriya see An Nāşirīyah
Nasser, Lake *50 B3* var. Buhayrat Nasir, Buḩayrat Nāşir, Buḥeiret Nâşir. lake Egypt/Sudan
Naszód see Năsăud
Nata *56 C3* Central, NE Botswana
Natal *41 G2* state capital Rio Grande do Norte, E Brazil
Natal Basin *119 A6* var. Mozambique Basin. undersea basin W Indian Ocean
Natanya see Netanya
Natchez *20 B3* Mississippi, S USA
Natchitoches *20 A2* Louisiana, S USA
Nathanya see Netanya
Natitingou *53 F4* NW Benin
Natsrat see Natzrat
Natuna Islands see Natuna, Kepulauan
Natuna, Kepulauan *102 D4* var. Natuna Islands. island group W Indonesia
Naturaliste Plateau *119 E6* undersea plateau E Indian Ocean
Natzrat *97 A5* var. Natsrat, Ar. En Nazira, Eng. Nazareth; prev. Nazerat. Northern, N Israel
Naugard see Nowogard
Naujaat *15 G3* prev. Repulse Bay, Nunavut, N Canada
Naujamiestis *84 C4* Panevėžys, C Lithuania
Nauru *122 D2* off. Republic of Nauru; prev. Pleasant Island. country W Pacific Ocean
Nauru, Republic of see Nauru
Nauta *38 C2* Loreto, N Peru
Navahrudak *85 C6* Pol. Nowogródek, Rus. Novogrudok. Hrodzyenskaya Voblasts', W Belarus
Navanagar see Jāmnagar
Navapolatsk *85 D5* Rus. Novopolotsk. Vitsyebskaya Voblasts', N Belarus
Navarra *71 E2* Eng./Fr. Navarre. autonomous community N Spain
Navarre see Navarra
Navassa Island *32 C3* US unincorporated territory C West Indies
Navoi see Navoiy
Navoiy *101 E2* Rus. Navoi. Navoiy Viloyati, C Uzbekistan
Navojoa *28 C2* Sonora, NW Mexico
Navolat see Navolato
Navolato *28 C3* var. Navolat. Sinaloa, C Mexico
Návpaktos see Náfpaktos
Návplion see Náfplio
Nawabashah see Nawabshah
Nawabshah *112 B3* var. Nawabashah. Sindh, S Pakistan
Naxçıvan *95 G3* Rus. Nakhichevan'. SW Azerbaijan
Náxos *83 D6* var. Naxos. Náxos, Kykládes, Greece, Aegean Sea
Náxos *83 D6* island Kykládes, Greece, Aegean Sea
Nayoro *108 D2* Hokkaidō, NE Japan
Nay Pyi Taw *114 B4* country capital Myanmar (Burma)) Mandalay, C Myanmar (Burma)
Nazareth see Natzrat
Nazca *38 D4* Ica, S Peru
Nazca Ridge *35 A5* undersea ridge E Pacific Ocean
Naze *108 B3* var. Nase. Kagoshima, Amami-ōshima, SW Japan
Nazerat see Natzrat
Nazilli *94 A4* Aydın, SW Turkey (Türkiye)
Nazrēt *51 C5* var. Adama, Hadama. Oromīya, C Ethiopia
N'Dalatando *56 B1* Port. Salazar, Vila Salazar. Cuanza Norte, NW Angola
Ndélé *54 C4* Bamingui-Bangoran, N Central African Republic
Ndendé *55 B6* Ngounié, S Gabon
Ndindi *55 A6* Nyanga, S Gabon
N'Djaména *54 B3* var. Ndjamena; prev. Fort-Lamy. country capital (Chad) Chari-Baguirmi, W Chad
Ndjamena see N'Djaména
Ndjolé *55 A5* Moyen-Ogooué, W Gabon
Ndola *56 D2* Copperbelt, C Zambia
Ndzouani see Anjouan
Neagh, Lough *67 B5* lake E Northern Ireland, United Kingdom
Néa Moudaniá *82 C4* var. Néa Moudhaniá. Kentrikí Makedonía, N Greece
Néa Moudhaniá see Néa Moudaniá
Neapel see Napoli
Neápoli *82 B4* prev. Neápolis. Dytikí Makedonía, N Greece
Neápoli *83 D8* Kríti, Greece, E Mediterranean Sea
Neápoli *83 C7* Pelopónnisos, S Greece
Neápolis see Neápoli, Greece
Neapolis see Napoli, Italy
Neapolis see Nablus, West Bank
Near Islands *14 A2* island group Aleutian Islands, Alaska, USA
Néa Zíchni *82 C3* var. Néa Zíkhni; prev. Néa Zíkhna. Kentrikí Makedonía, NE Greece
Néa Zíkhna/Néa Zíkhni see Néa Zíchni
Nebaj *30 B2* Quiché, W Guatemala
Nebitdag see Balkanabat
Neblina, Pico da *40 C1* mountain NW Brazil
Nebraska *22 D4* off. State of Nebraska, also known as Blackwater State, Cornhusker State, Tree Planters State. state C USA
Nebraska City *23 F4* Nebraska, C USA
Neches River *27 H3* river Texas, SW USA
Neckar *73 B6* river SW Germany
Necochea *43 D5* Buenos Aires, E Argentina
Nederland see Netherlands
Neder Rijn *64 D4* Eng. Lower Rhine. river C Netherlands
Nederweert *65 D5* Limburg, SE Netherlands
Neede *64 E3* Gelderland, E Netherlands
Neerpelt *65 D5* Limburg, NE Belgium
Neftekamsk *89 D5* Respublika Bashkortostan, W Russia
Neftezavodsk see Seýdi
Negara Brunei Darussalam see Brunei
Negēlē *51 D5* var. Negelli, It. Neghelli. Oromīya, C Ethiopia
Negelli see Negēlē
Negev see HaNegev
Neghelli see Negēlē
Negomane *57 E2* var. Negomano. Cabo Delgado, N Mozambique
Negomano see Negomane
Negombo *110 C3* Western Province, SW Sri Lanka
Negotin *78 E4* Serbia, E Serbia
Negra, Punta *38 A3* headland NW Peru
Negreşti see Negreşti-Oaş
Negreşti-Oaş *86 B3* Hung. Avasfelsőfalu; prev. Negreşti. Satu Mare, NE Romania
Negro, Río *43 C5* river E Argentina
Negro, Rio *40 D1* river N South America
Negro, Río *42 D4* river Brazil/Uruguay
Negros *117 E2* island C Philippines
Nehbandān *98 E3* Khorāsān, E Iran
Neijiang *106 B5* Sichuan, C China
Neiva *36 B3* Huila, S Colombia
Nellore *110 D2* Andhra Pradesh, E India
Nelson *129 C5* Nelson, South Island, New Zealand
Nelson *15 G4* river Manitoba, C Canada
Néma *52 D3* Hodh ech Chargui, SE Mauritania
Neman *84 B4* Ger. Ragnit. Kaliningradskaya Oblast', W Russia
Neman *84 A4* Bel. Nyoman, Ger. Memel, Lith. Nemunas, Pol. Niemen. river NE Europe
Nemausus see Nîmes
Neméa *83 B6* Pelopónnisos, S Greece
Nemetocenna see Arras
Nemours *68 C3* Seine-et-Marne, N France
Nemunas see Neman
Nemuro *108 E2* Hokkaidō, NE Japan
Neochóri *83 B5* Dytikí Ellás, C Greece
Nepal *113 E3* off. Nepal. country S Asia
Nepal see Nepal
Nereta *84 C4* Aizkraukle, S Latvia
Neretva *78 C4* river Bosnia and Herzegovina/Croatia
Neris *85 C5* Bel. Viliya, Pol. Wilia; prev. Pol. Wilja. river Belarus/Lithuania
Neris see Viliya
Nerva *70 C4* Andalucía, S Spain
Neryungri *93 F4* Respublika Sakha (Yakutiya), NE Russia
Neskaupstaður *61 E5* Austurland, E Iceland
Ness, Loch *66 C3* lake N Scotland, United Kingdom
Nesterov see Zhovkva
Néstos *82 C3* Bul. Mesta, Turk. Kara Su. river Bulgaria/Greece
Nesvizh see Nyasvizh
Netanya *97 A6* var. Natanya, Nathanya. Central, C Israel
Netherlands *64 C3* off. Kingdom of the Netherlands, var. Holland, Dut. Koninkrijk der Nederlanden, Nederland. country NW Europe
Netherlands East Indies see Indonesia
Netherlands Guiana see Suriname
Netherlands, Kingdom of the see Netherlands
Netherlands New Guinea see Papua
Nettilling Lake *15 G3* lake Baffin Island, Nunavut, N Canada
Netze see Noteć
Neu Amerika see Puławy
Neubrandenburg *72 D3* Mecklenburg-Vorpommern, NE Germany
Neuchâtel *73 A7* Ger. Neuenburg. Neuchâtel, W Switzerland
Neuchâtel, Lac de *73 A7* Ger. Neuenburger See. lake W Switzerland
Neuenburger See see Neuchâtel, Lac de
Neufchâteau *65 D8* Luxembourg, SE Belgium
Neugradisk see Nova Gradiška
Neuhof see Zgierz
Neukuhren see Pionerskiy
Neumarkt see Târgu Secuiesc, Covasna, Romania
Neumarkt see Târgu Mureş
Neumayer III *132 A2* German research station Antarctica
Neumoldowa see Moldova Nouă
Neumünster *72 B2* Schleswig-Holstein, N Germany
Neunkirchen *73 A5* Saarland, SW Germany
Neuquén *43 B5* Neuquén, SE Argentina
Neuruppin *72 C3* Brandenburg, NE Germany
Neusalz an der Oder see Nowa Sól
Neu Sandec see Nowy Sącz
Neusatz see Novi Sad
Neusiedler See *73 E6* Hung. Fertő. lake Austria/Hungary
Neusohl see Banská Bystrica
Neustadt see Baia Mare, Maramureş, Romania
Neustadt an der Haardt see Neustadt an der Weinstrasse
Neustadt an der Weinstrasse *73 B5* prev. Neustadt an der Haardt, hist. Niewenstat; anc. Nova Civitas. Rheinland-Pfalz, SW Germany
Neustadtl see Novo mesto
Neustettin see Szczecinek
Neustrelitz *72 D3* Mecklenburg-Vorpommern, NE Germany
Neutra see Nitra
Neu-Ulm *73 B6* Bayern, S Germany
Neuwied *73 A5* Rheinland-Pfalz, W Germany
Neuzen see Terneuzen
Nevada *25 C5* off. State of Nevada, also known as Battle Born State, Sagebrush State, Silver State. state W USA
Nevada, Sierra *70 D5* mountain range S Spain
Nevada, Sierra *25 C6* mountain range W USA
Nevers *68 C4* anc. Noviodunum. Nièvre, C France
Neves *54 E2* São Tomé, S Sao Tome and Principe, Africa
Nevinnomyssk *89 B7* Stavropol'skiy Kray, SW Russia
Nevşehir *94 C3* var. Nevshehr. Nevşehir, C Turkey (Türkiye)
Newala *51 C8* Mtwara, SE Tanzania
New Albany *18 C5* Indiana, N USA
New Amsterdam *37 G3* E Guyana
Newark *19 F4* New Jersey, NE USA
New Bedford *19 G3* Massachusetts, NE USA
Newberg *24 B3* Oregon, NW USA
New Bern *21 F1* North Carolina, SE USA
New Braunfels *27 G4* Texas, SW USA
Newbridge *67 B6* Ir. An Droichead Nua. Kildare, C Ireland
New Britain *122 B3* island E Papua New Guinea
New Brunswick *17 E4* Fr. Nouveau-Brunswick. province SE Canada
New Caledonia *122 D4* var. Kanaky, Fr. Nouvelle-Calédonie. French self-governing territory of special status SW Pacific Ocean
New Caledonia *122 C5* island SW Pacific Ocean
New Caledonia Basin *120 C4* undersea basin W Pacific Ocean
Newcastle *127 D6* New South Wales, SE Australia
Newcastle see Newcastle upon Tyne
Newcastle upon Tyne *66 D4* var. Newcastle, hist. Monkchester, Lat. Pons Aelii. NE England, United Kingdom
New Delhi *112 D3* country capital (India) Delhi, N India
New England of the West see Minnesota
Newfoundland *17 G3* Fr. Terre-Neuve. island Newfoundland and Labrador, SE Canada
Newfoundland and Labrador *17 F2* Fr. Terre Neuve. province E Canada
Newfoundland Basin *44 B3* undersea feature NW Atlantic Ocean
New Georgia Islands *122 C3* island group NW Solomon Islands
New Glasgow *17 F4* Nova Scotia, SE Canada
New Goa see Panaji
New Guinea *122 A3* Dut. Nieuw Guinea, Ind. Irian. island Indonesia/Papua New Guinea
New Hampshire *19 F2* off. State of New Hampshire, also known as Granite State. state NE USA
New Haven *19 G3* Connecticut, NE USA
New Hebrides see Vanuatu
New Iberia *20 B3* Louisiana, S USA
New Ireland *122 C3* island NE Papua New Guinea
New Jersey *19 F4* off. State of New Jersey, also known as The Garden State. state NE USA
Newman *124 B4* Western Australia
Newmarket *67 E6* E England, United Kingdom
New Mexico *26 C2* off. State of New Mexico, also known as Land of Enchantment, Sunshine State. state SW USA
New Orleans *20 B3* Louisiana, S USA
New Plymouth *128 C4* Taranaki, North Island, New Zealand
Newport *67 C7* SE Wales, United Kingdom
Newport *18 C4* Kentucky, S USA
Newport *19 G2* Vermont, NE USA
Newport News *19 F5* Virginia, NE USA
New Providence *32 C1* island N The Bahamas
Newquay *67 C7* SW England, United Kingdom
Newry *67 B5* Ir. An tIúr. SE Northern Ireland, United Kingdom
New Sarum see Salisbury
New Siberian Islands see Novosibirskiye Ostrova
New South Wales *127 C6* state SE Australia
Newton *23 G3* Iowa, C USA
Newton *23 F5* Kansas, C USA
Newtownabbey *67 B5* Ir. Baile na Mainistreach. E Northern Ireland, United Kingdom
New Ulm *23 F2* Minnesota, N USA
New York *19 F4* New York, NE USA
New York *19 F3* state NE USA
New Zealand *128 A4* var. Aotearoa. country SW Pacific Ocean
Neyveli *110 C2* Tamil Nādu, SE India
Nezhin see Nizhyn
Ngangze Co *104 B5* lake W China
Ngaoundéré *54 B4* var. N'Gaoundéré. Adamaoua, N Cameroon
N'Gaoundéré see Ngaoundéré
Ngazidja *57 F2* Fr. Grande-Comore. island NW Comoros
Ngerulmud *56 D4* country capital (Palau) NW Pacific Ocean
N'Giva *56 B3* var. Ondjiva, Port. Vila Pereira de Eça. Cunene, S Angola
Ngo *55 B6* Plateaux, SE Congo
Ngoko *55 B5* river Cameroon/Congo
Ngourti *53 H3* Diffa, E Niger
Nguigmi *53 H3* var. N'Guigmi. Diffa, SE Niger
N'Guigmi see Nguigmi
N'Gunza see Sumbe
Nguru *53 G3* Yobe, NE Nigeria
Nha Trang *115 E6* Khanh Hoa, S Vietnam

Niagara Falls 16 *D5* Ontario, S Canada
Niagara Falls 19 *E3* New York, NE USA
Niagara Falls 18 *D3* *waterfall* Canada/USA
Niamey 53 *F3* *country capital* (Niger) Niamey, SW Niger
Niangay, Lac 53 *E3* *lake* E Mali
Nia-Nia 55 *E5* Orientale, NE Dem. Rep. Congo
Nias, Pulau 116 *A3* *island* W Indonesia
Nicaea *see* Nice
Nicaragua 30 *D3* *off.* Republic of Nicaragua. *country* Central America
Nicaragua, Lago de 30 *D4* *var.* Cocibolca, Gran Lago, *Eng.* Lake Nicaragua. *lake* S Nicaragua
Nicaragua, Lake *see* Nicaragua, Lago de
Nicaragua, Republic of *see* Nicaragua
Nicaria *see* Ikaría
Nice 69 *D6* *It.* Nizza; *anc.* Nicaea. Alpes-Maritimes, SE France
Nicephorium *see* Ar Raqqah
Nicholas II Land *see* Severnaya Zemlya
Nicholls Town 32 *C1* Andros Island, NW The Bahamas
Nicobar Islands 102 *B4* *island group* India, E Indian Ocean
Nicosia 80 *C5* *Gk.* Lefkosía, *Turk.* Lefkoşa. *country capital* (Cyprus) C Cyprus
Nicoya 30 *D4* Guanacaste, W Costa Rica
Nicoya, Golfo de 30 *D5* *gulf* W Costa Rica
Nicoya, Península de 30 *D4* *peninsula* NW Costa Rica
Nida 84 *A3* *Ger.* Nidden. Klaipėda, SW Lithuania
Nidaros *see* Trondheim
Nidden *see* Nida
Nidzica 76 *D3* *Ger.* Niedenburg. Warmińsko-Mazurskie, NE Poland
Niedenburg *see* Nidzica
Niedere Tauern 77 *A6* *mountain range* C Austria
Niemen *see* Neman
Nieśwież *see* Nyasvizh
Nieuw Amsterdam 37 *G3* Commewijne, NE Suriname
Nieuw-Bergen 64 *D4* Limburg, SE Netherlands
Nieuwegein 64 *C4* Utrecht, C Netherlands
Nieuw Guinea *see* New Guinea
Nieuw Nickerie 37 *G3* Nickerie, NW Suriname
Niewenstat *see* Neustadt an der Weinstrasse
Niğde 94 *C4* Niğde, C Turkey (Türkiye)
Niger 53 *F3* *off.* Republic of Niger. *country* W Africa
Niger 53 *F4* *river* W Africa
Nigeria 53 *F4* *off.* Federal Republic of Nigeria. *country* W Africa
Nigeria, Federal Republic of *see* Nigeria
Niger, Mouths of the 53 *F5* *delta* S Nigeria
Niger, Republic of *see* Niger
Nihon *see* Japan
Niigata 109 *D5* Niigata, Honshū, C Japan
Niihama 109 *B7* Ehime, Shikoku, SW Japan
Ni'ihau 25 *A7* *var.* Niihau. *island* Hawaii, USA, C Pacific Ocean
Nii-jima 109 *D6* *island* E Japan
Nijkerk 64 *D3* Gelderland, C Netherlands
Nijlen 65 *C5* Antwerpen, N Belgium
Nijmegen 64 *D4* *Ger.* Nimwegen; *anc.* Noviomagus. Gelderland, SE Netherlands
Nikaria *see* Ikaría
Nikel' 88 *C2* *Finn.* Kolosjoki. Murmanskaya Oblast', NW Russia
Nikiniki 117 *E5* Timor, S Indonesia
Niklasmarkt *see* Gheorgheni
Nikolainkaupunki *see* Vaasa
Nikolayev *see* Mykolayiv
Nikol'sk *see* Ussuriysk
Nikol'sk-Ussuriyskiy *see* Ussuriysk
Nikopol 87 *F3* Dnipropetrovska Oblast, SE Ukraine
Nikšić 79 *C5* C Montenegro
Nikumaroro 123 *E3* ; *prev.* Gardner Island. *atoll* Phoenix Islands, C Kiribati
Nikunau 123 *E3* *var.* Nukunau; *prev.* Byron Island. *atoll* Tungaru, W Kiribati
Nile 50 *B2* *former province* NW Uganda
Nile 46 *D3* *Ar.* Nahr an Nil. *river* N Africa
Nile Delta 50 *B1* *delta* N Egypt
Nīl, Nahr an *see* Nile
Nîmes 69 *C6* *anc.* Nemausus, Nismes. Gard, S France
Nimwegen *see* Nijmegen
Nine Degree Channel 110 *B3* *channel* India/ Maldives
Ninetyeast Ridge 119 *D5* *undersea feature* E Indian Ocean
Ninety Mile Beach *see* Te-Oneroa-a-Tōhē
Ningbo 106 *D5* *var.* Ning-po, Yin-hsien; *prev.* Ninghsien. Zhejiang, SE China
Ning-hsia *see* Ningxia
Ninghsien *see* Ningbo
Ning-po *see* Ningbo
Ningsia/Ningsia Hui/Ningsia Hui Autonomous Region *see* Ningxia
Ningxia 106 *B4* *off.* Ningxia Huizu Zizhiqu, *var.* Ning-hsia, Ningsia, *Eng.* Ningsia Hui, Ningsia Hui Autonomous Region. *autonomous region* N China
Ningxia Huizu Zizhiqu *see* Ningxia
Nio *see* Íos
Niobrara River 23 *E3* *river* Nebraska/ Wyoming, C USA
Nioro 52 *D3* *var.* Nioro du Sahel. Kayes, W Mali
Nioro du Sahel *see* Nioro
Niort 68 *B4* Deux-Sèvres, W France
Nipigon 16 *B4* Ontario, S Canada
Nipigon, Lake 16 *B3* *lake* Ontario, S Canada
Nippon *see* Japan
Niš 79 *E5* *Eng.* Nish, *Ger.* Nisch; *anc.* Naissus. Serbia, SE Serbia
Nişab 98 *B4* Al Ḥudūd ash Shamālīyah, N Saudi Arabia
Nisch/Nish *see* Niš
Nisibin *see* Nusaybin
Nisiros *see* Nísyros
Nisko 76 *E4* Podkrapackie, SE Poland
Nismes *see* Nîmes
Nistru *see* Dniester
Nísyros 83 *E7* *var.* Nisiros. *island* Dodekánisa, Greece, Aegean Sea
Nitra 77 *C6* *Ger.* Neutra, *Hung.* Nyitra. Nitriansky Kraj, SW Slovakia
Nitra 77 *C6* *Ger.* Neutra, *Hung.* Nyitra. *river* W Slovakia
Niuatobutabu *see* Niuatoputapu
Niuatoputapu 123 *E4* *var.* Niuatobutabu; *prev.* Keppel Island. *island* N Tonga
Niue 123 *F4* *self-governing territory in free association with New Zealand* S Pacific Ocean
Niulakita 123 *E3* *var.* Nurakita. *atoll* S Tuvalu
Niutao 123 *E3* *atoll* NW Tuvalu
Nivernais 68 *C4* *cultural region* C France
Nizāmābād 112 *D5* Telangana, C India
Nizhnegorskiy *see* Nyzhnohirskyi
Nizhnekamsk 89 *C5* Respublika Tatarstan, W Russia
Nizhnevartovsk 92 *D3* Khanty-Mansiyskiy Avtonomnyy Okrug-Yugra, C Russia
Nizhniy Novgorod 89 *C5* *prev.* Gor'kiy. Nizhegorodskaya Oblast', W Russia
Nizhniy Odes 88 *D4* Respublika Komi, NW Russia
Nizhnyaya Tunguska 93 *E3* *Eng.* Lower Tunguska. *river* N Russia
Nizhyn 87 *E1* *Rus.* Nezhin. Chernihivs'ka Oblast', NE Ukraine
Nizza *see* Nice
Njombe 51 *C8* Iringa, S Tanzania
Nkayi 55 *B6* *prev.* Jacob. Bouenza, S Congo
Nkongsamba 54 *A4* *var.* N'Kongsamba. Littoral, W Cameroon
N'Kongsamba *see* Nkongsamba
Nmai Hka 114 *B2* *var.* Me Hka. *river* N Myanmar (Burma)
Nobeoka 109 *B7* Miyazaki, Kyūshū, SW Japan
Noboribetsu 108 *D3* *var.* Noboribetu. Hokkaidō, NE Japan
Noboribetu *see* Noboribetsu
Nogales 28 *B1* Sonora, NW Mexico
Nogales 26 *B3* Arizona, SW USA
Nogal Valley *see* Dooxo Nugaaleed
Noire, Rivi`ere *see* Black River
Nokia 63 *D5* Pirkanmaa, W Finland
Nokou 54 *B3* Kanem, W Chad
Nola 55 *B5* Sangha-Mbaéré, SW Central African Republic
Nolinsk 89 *C5* Kirovskaya Oblast', NW Russia
Nongkaya *see* Nong Khai
Nong Khai 114 *C4* *var.* Mi Chai, Nongkaya. Nong Khai, E Thailand
Nonouti 122 *D2* *prev.* Sydenham Island. *atoll* Tungaru, W Kiribati
Noord-Beveland 64 *B4* *var.* North Beveland. *island* SW Netherlands
Noordwijk aan Zee 64 *C3* Zuid-Holland, W Netherlands
Noordzee *see* North Sea
Nora 63 *C6* Örebro, C Sweden
Norak 101 *E3* *Rus.* Nurek. W Tajikistan
Nordaustlandet 61 *G1* *island* NE Svalbard
Norden 72 *A3* Niedersachsen, NW Germany
Norderstedt 72 *B3* Schleswig-Holstein, N Germany
Nordfriesische Inseln *see* North Frisian Islands
Nordhausen 72 *C4* Thüringen, C Germany
Nordhorn 72 *A3* Niedersachsen, NW Germany
Nord, Mer du *see* North Sea
Nord-Ouest, Territoires du *see* Northwest Territories
Nordsee/Nordsjøen/Nordsøen *see* North Sea
Norfolk 23 *E3* Nebraska, C USA
Norfolk 19 *F5* Virginia, NE USA
Norfolk Island 120 *D4* *Australian self-governing territory* SW Pacific Ocean
Norfolk Ridge 120 *D4* *undersea feature* W Pacific Ocean
Norge *see* Norway
Norias 27 *G5* Texas, SW USA
Noril'sk 92 *D3* Taymyrskiy (Dolgano-Nenetskiy) Avtonomnyy Okrug, N Russia
Norman 27 *G1* Oklahoma, C USA
Normandes, Îles *see* Channel Islands
Normandie 68 *B3* *Eng.* Normandy. *cultural region* N France
Normandy *see* Normandie
Normanton 126 *C3* Queensland, NE Australia
Norrköping 63 *C6* Östergötland, S Sweden
Norrtälje 63 *C6* Stockholm, C Sweden
Norseman 125 *B6* Western Australia
Norske Havet *see* Norwegian Sea
North Albanian Alps 79 *C5* *Alb.* Bjeshkët e Namuna, *Croatian* Prokletije. *mountain range* SE Europe
Northallerton 67 *D5* N England, United Kingdom
Northam 125 *A6* Western Australia
North America 12 *continent*
Northampton 67 *D6* C England, United Kingdom
North Andaman 111 *F2* *island* Andaman Islands, India, NE Indian Ocean
North Australian Basin 119 *E5* *Fr.* Bassin Nord de l' Australie. *undersea feature* E Indian Ocean
North Bay 16 *D4* Ontario, S Canada
North Beveland *see* Noord-Beveland
North Borneo *see* Sabah
North Cape 128 *C1* *headland* North Island, New Zealand
North Cape 62 *D1* *Eng.* North Cape. *headland* N Norway
North Cape *see* Nordkapp
North Carolina 21 *E1* *off.* State of North Carolina, *also known as* Old North State, Tar Heel State, Turpentine State. *state* SE USA
North Channel 18 *D2* *lake channel* Canada/USA
North Charleston 21 *F2* South Carolina, SE USA
North Dakota 22 *D2* *off.* State of North Dakota, *also known as* Flickertail State, Peace Garden State, Sioux State. *state* N USA
North Devon Island *see* Devon Island
North East Frontier Agency/North East Frontier Agency of Assam *see* Arunāchal Pradesh
Northeast Providence Channel 32 *C1* *channel* N The Bahamas
Northeim 72 *B4* Niedersachsen, C Germany
Northern Cook Islands 123 *F4* *island group* N Cook Islands
Northern Dvina *see* Severnaya Dvina
Northern Ireland 66 *B4* *var.* The Six Counties. *cultural region* Northern Ireland, United Kingdom
Northern Mariana Islands 120 *B1* *US commonwealth territory* W Pacific Ocean
Northern Rhodesia *see* Zambia
Northern Sporades *see* Vóreies Sporádes
Northern Territory 122 *A5* *territory* N Australia
North European Plain 59 *E3* *plain* N Europe
Northfield 23 *F2* Minnesota, N USA
North Fiji Basin 120 *D3* *undersea feature* N Coral Sea
North Frisian Islands 72 *B2* *var.* Nordfriesische Inseln. *island group* N Germany
North Huvadhu Atoll 110 *B5* *var.* Gaafu Alifu Atoll. *atoll* S Maldives
North Island 128 *B2* *var.* Te Ika-a-Māui. *island* N New Zealand
North Korea 107 *E3* *off.* Democratic People's Republic of Korea, *Kor.* Chosŏn-minjujuŭi-inmin-kanghwaguk. *country* E Asia
North Little Rock 20 *B1* Arkansas, C USA
North Macedonia 79 *D6* *off.* Republic of North Macedonia, *Mac.* Severna Makedonija, *prev.* Macedonia. *country* SE Europe
North Macedonia, Republic of *see* North Macedonia
North Minch *see* Minch, The
North Mole 71 *G4* *harbour wall* NW Gibraltar, Europe
North Platte 23 *E4* Nebraska, C USA
North Platte River 22 *D4* *river* C USA
North Pole 133 *B3* *pole* Arctic Ocean
North Saskatchewan 15 *F5* *river* Alberta/ Saskatchewan, S Canada
North Sea 58 *D3* *Dan.* Nordsøen, *Dut.* Noordzee, *Fr.* Mer du Nord, *Ger.* Nordsee, *Nor.* Nordsjøen; *prev.* German Ocean, *Lat.* Mare Germanicum. *sea* NW Europe
North Siberian Lowland 93 *E2* *var.* North Siberian Plain, *Eng.* North Siberian Lowland. *lowlands* N Russia
North Siberian Lowland/North Siberian Plain *see* Severo-Sibirskaya Nizmennost'
North Star State *see* Minnesota
North Taranaki Bight 128 *C3* *gulf* North Island, New Zealand
North Uist 66 *B3* *island* NW Scotland, United Kingdom
Northwest Atlantic Mid-Ocean Canyon 12 *E4* *undersea feature* N Atlantic Ocean
North West Highlands 66 *C3* *mountain range* N Scotland, United Kingdom
Northwest Pacific Basin 91 *G4* *undersea feature* NW Pacific Ocean
Northwest Territories 15 *E3* *Fr.* Territoires du Nord-Ouest. *territory* NW Canada
Northwind Plain 133 *B2* *undersea feature* Arctic Ocean
Norton Sound 14 *C2* *inlet* Alaska, USA
Norway 63 *A5* *off.* Kingdom of Norway, *Nor.* Norge. *country* N Europe
Norway, Kingdom of *see* Norway
Norwegian Basin 61 *F4* *undersea feature* NW Norwegian Sea
Norwegian Sea 61 *F4* *var.* Norske Havet. *sea* NE Atlantic Ocean
Norwich 67 *E6* E England, United Kingdom
Nösen *see* Bistrița
Noshiro 108 *D4* *var.* Nosiro; *prev.* Noshirominato. Akita, Honshū, C Japan
Noshirominato/Nosiro *see* Noshiro
Nosivka 87 *E1* *Rus.* Nosovka. Chernihivs'ka Oblast', NE Ukraine
Nosop 56 *C4* *river* E Namibia
Nosovka *see* Nosivka
Noṣratābād 98 *E3* Sīstān va Balūchestān, E Iran
Noteć 76 *C3* *Ger.* Netze. *river* NW Poland
Nóties Sporádes *see* Dodekánisa
Nottingham 67 *D6* C England, United Kingdom
Nouâdhibou 52 *B2* *prev.* Port-Étienne. Dakhlet Nouâdhibou, W Mauritania
Nouakchott 52 *B2* *country capital* (Mauritania) Nouakchott District, SW Mauritania
Nouméa 122 *C5* *dependent territory capital* (New Caledonia) Province Sud, S New Caledonia
Nouveau-Brunswick *see* New Brunswick
Nouvelle-Calédonie *see* New Caledonia
Nouvelle Écosse *see* Nova Scotia
Nova Civitas *see* Neustadt an der Weinstrasse
Nova Gorica 73 *D8* W Slovenia
Nova Gradiška 78 *C3* *Ger.* Neugradisk, *Hung.* Újgradiska. Brod-Posavina, NE Croatia
Nova Iguaçu 41 *F4* Rio de Janeiro, SE Brazil
Nova Kakhovka 87 *F4* S Ukraine
Nova Lisboa *see* Huambo
Novara 74 *B2* *anc.* Novaria. Piemonte, NW Italy
Novaria *see* Novara
Nova Scotia 17 *F4* *Fr.* Nouvelle Écosse. *province* SE Canada
Nova Scotia 13 *E5* *physical region* SE Canada
Novaya Sibir', Ostrov 93 *F1* *island* Novosibirskiye Ostrova, NE Russia
Novaya Zemlya 88 *D1* *island group* N Russia
Novaya Zemlya Trough *see* East Novaya Zemlya Trough
Novgorod *see* Velikiy Novgorod
Novi Grad *see* Bosanski Novi
Novi Iskar 82 *C2* *var.* Novi Iksŭr. Sofia-Grad, W Bulgaria
Novi Iksŭr *see* Novi Iskar
Noviodunum *see* Nevers, Nièvre, France
Noviomagus *see* Lisieux, Calvados, France
Noviomagus *see* Nijmegen, Netherlands
Novi Pazar 79 *D5* *Turk.* Yenipazar. Serbia, S Serbia
Novi Sad 78 *D3* *Ger.* Neusatz, *Hung.* Újvidék. Vojvodina, N Serbia
Novoazovsk 87 *G4* Donets'ka Oblast', E Ukraine
Novocheboksarsk 89 *C5* Chuvashskaya Respublika, W Russia
Novocherkassk 89 *B7* Rostovskaya Oblast', SW Russia
Novodvinsk 88 *C3* Arkhangel'skaya Oblast', NW Russia
Novograd-Volynskiy *see* Novohrad-Volynskyi
Novogrudok *see* Navahrudak
Novohrad-Volynskyi 86 *D2* *prev.* Novohrad-Volyns'kyy, *Rus.* Novograd-Volynskiy. Zhytomyrs'ka Oblast', N Ukrain
Novohrad-Volyns'kyy *see* Novohrad-Volynskyi
Novokazalinsk *see* Ayteke Bi
Novokuznetsk 92 *D4* *prev.* Stalinsk. Kemerovskaya Oblast', S Russia
Novolazarevskaya 132 *C2* *Russian research station* Antarctica
Novo mesto 73 *E8* *Ger.* Rudolfswert; *prev. Ger.* Neustadtl. SE Slovenia
Novomoskovsk 89 *B5* Tul'skaya Oblast', W Russia
Novomoskovsk 87 *F3* Dnipropetrovska Oblast, E Ukraine
Novopolotsk *see* Navapolatsk
Novoradomsk *see* Radomsko
Novo Redondo *see* Sumbe
Novorossiysk 89 *A7* Krasnodarskiy Kray, SW Russia
Novoshakhtinsk 89 *B6* Rostovskaya Oblast', SW Russia
Novosibirsk 92 *D4* Novosibirskaya Oblast', C Russia
Novosibirskiye Ostrova 93 *F1* *Eng.* New Siberian Islands. *island group* N Russia
Novotroitsk 89 *D6* Orenburgskaya Oblast', W Russia
Novotroitskoye *see* Novotroyitske
Novotroyitske 87 *F4* *prev.* Novotroyits'ke, *Rus.* Novotroitskoye. Khersons'ka Oblast', S Ukraine
Novotroyits'ke *see* Novotroyitske
Novo-Urgench *see* Urganch
Novovolynsk 86 *C1* Volyns'ka Oblast', NW Ukraine
Novy Dvor 85 *B6* *Rus.* Novyy Dvor. Hrodzyenskaya Voblasts', W Belarus
Novyy Bug *see* Novyy Buh
Novyy Buh 87 *E3* *Rus.* Novyy Bug. Mykolayivs'ka Oblast', S Ukraine
Novyy Dvor *see* Novy Dvor
Novyy Margilan *see* Farg'ona
Novyy Uzen' *see* Zhangaözen
Nowa Sól 76 *B4* *var.* Nowasól, *Ger.* Neusalz an der Oder. Lubuskie, W Poland
Nowasól *see* Nowa Sól
Nowogard 76 *B2* *var.* Nowógard, *Ger.* Naugard. Zachodnio-pomorskie, NW Poland
Nowogródek *see* Navahrudak
Nowo-Minsk *see* Mińsk Mazowiecki
Nowy Dwór Mazowiecki 76 *D3* Mazowieckie, C Poland
Nowy Sącz 77 *D5* *Ger.* Neu Sandec. Małopolskie, S Poland
Nowy Tomyśl 76 *B3* *var.* Nowy Tomysl. Wielkopolskie, C Poland
Nowy Tomysl *see* Nowy Tomyśl
Noyon 68 *C3* Oise, N France
Nsanje 57 *E3* Southern, S Malawi
Nsawam 53 *E5* SE Ghana
Ntomba, Lac 55 *C6* *var.* Lac Tumba. *lake* NW Dem. Rep. Congo
Nubian Desert 50 *B3* *desert* NE Sudan
Nu Chiang *see* Salween
Nueva Caceres *see* Naga
Nueva Gerona 32 *B2* Isla de la Juventud, S Cuba
Nueva Rosita 28 *D2* Coahuila, NE Mexico
Nuevitas 32 *C2* Camagüey, E Cuba
Nuevo, Bajo 31 *G1* *island* NW Colombia South America
Nuevo Casas Grandes 28 *C1* Chihuahua, N Mexico
Nuevo, Golfo 43 *C6* *gulf* S Argentina
Nuevo Laredo 29 *E2* Tamaulipas, NE Mexico
Nui Atoll 123 *E3* *atoll* W Tuvalu
Nu Jiang *see* Salween
Nûk *see* Nuuk
Nukha *see* Şǎki
Nuku'alofa 123 *E5* *country capital* (Tonga) Tongatapu, S Tonga
Nukufetau Atoll 123 *E3* *atoll* C Tuvalu
Nukulaelae Atoll 123 *E3* *var.* Nukulailai. *atoll* E Tuvalu
Nukulailai *see* Nukulaelae Atoll
Nukunau *see* Nikunau
Nukunonu Atoll 123 *E3* *island* C Tokelau
Nukus 100 *C2* Qoraqalpog'iston Respublikasi, W Uzbekistan
Nullarbor Plain 125 *C6* *plateau* South Australia/ Western Australia
Nunap Isua 60 *C5* *var.* Uummannarsuaq, *Dan.* Kap Farvel, *Eng.* Cape Farewell. *cape* S Greenland
Nunavut 15 *F3* *territory* N Canada
Nuneaton 67 *D6* C England, United Kingdom
Nunivak Island 14 *B2* *island* Alaska, USA
Nunspeet 64 *D3* Gelderland, E Netherlands
Nuoro 75 *A5* Sardegna, Italy, C Mediterranean Sea
Nuquí 36 *A3* Chocó, W Colombia
Nurakita *see* Niulakita
Nurata *see* Nurota
Nurek *see* Norak
Nuremberg *see* Nürnberg
Nurmes 62 *E4* Pohjois-Karjala, E Finland
Nürnberg 73 *C5* *Eng.* Nuremberg. Bayern, S Germany
Nurota 101 *E2* *Rus.* Nurata. Navoiy Viloyati, C Uzbekistan
Nur-Sultan 92 *C4* *prev.* Astana, Akmola, Akmolinsk, Tselinograd, Aqmola. *country capital* (Kazakhstan) Akmola, N Kazakhstan
Nusaybin 95 *F4* *var.* Nisibin. Manisa, SE Turkey (Türkiye)
Nussdorf *see* Năsăud
Nutmeg State *see* Connecticut
Nuuk 60 *C4* *var.* Nûk, *Dan.* Godthaab, Godthåb. *dependent territory capital* (Greenland) Sermersooq, SW Greenland
Nyagan' 92 *C3* Khanty-Mansiyskiy Avtonomnyy Okrug-Yugra, N Russia
Nyainqentanglha Shan 104 *C5* *mountain range* W China
Nyala 50 *A4* Southern Darfur, W Sudan
Nyamapanda 56 *D3* Mashonaland East, NE Zimbabwe
Nyamtumbo 51 *C8* Ruvuma, S Tanzania
Nyanda *see* Masvingo
Nyandoma 88 *C4* Arkhangel'skaya Oblast', NW Russia
Nyantakara 51 *B7* Kagera, NW Tanzania
Nyasa, Lake 57 *E2* *var.* Lake Malawi; *prev.* Lago Nyassa. *lake* E Africa
Nyasaland/Nyasaland Protectorate *see* Malawi
Nyassa, Lago *see* Nyasa, Lake
Nyasvizh 85 *C6* *Pol.* Nieśwież, *Rus.* Nesvizh. Minskaya Voblasts', C Belarus
Nyaunglebin 114 *B4* Bago, SW Myanmar (Burma)
Nyeri 51 *C6* Central, C Kenya
Nyíregyháza 77 *D6* Szabolcs-Szatmár-Bereg, NE Hungary
Nyitra *see* Nitra
Nykøbing 63 *B8* Storstrøm, SE Denmark
Nyköping 63 *C6* Södermanland, S Sweden
Nylstroom *see* Modimolle
Nyngan 127 *D6* New South Wales, SE Australia
Nyoman *see* Neman
Nyurba 93 *F3* Respublika Sakha (Yakutiya), NE Russia
Nyzhnohirskyi 87 *F4* *prev.* Nyzhn'ohirs'kyy, *Rus.* Nizhnegorskiy. Respublika Krym, S Ukraine
Nyzhn'ohirs'kyy *see* Nyzhnohirskyi
NZ *see* New Zealand
Nzega 51 *C7* Tabora, C Tanzania
Nzérékoré 52 *D4* SE Guinea
Nzwani *see* Anjouan

O

Oa'hu 25 *A7* *var.* Oahu. *island* Hawaiian Islands, Hawaii, USA
Oak Harbor 24 *B1* Washington, NW USA
Oakland 25 *B6* California, W USA
Oamaru 129 *B7* Otago, South Island, New Zealand
Oaxaca 29 *F5* *var.* Oaxaca de Juárez; *prev.* Antequera. Oaxaca, SE Mexico
Oaxaca de Juárez *see* Oaxaca
Ob' 90 *C2* *river* C Russia
Obal' 85 *D5* *Rus.* Obol'. Vitsyebskaya Voblasts', N Belarus
Oban 66 *C4* W Scotland, United Kingdom
Oban *see* Halfmoon Bay
Obando *see* Puerto Inírida
Obdorsk *see* Salekhard
Óbecse *see* Bečej
Obeliai 84 *C4* Panevėžys, NE Lithuania
Oberhollabrunn *see* Tulln
Ob', Gulf of *see* Obskaya Guba
Obidovichi *see* Abidavichy
Obihiro 108 *D2* Hokkaidō, NE Japan
Obo 54 *D4* Haut-Mbomou, E Central African Republic
Obock 50 *D4* E Djibouti
Obol' *see* Obal'
Oborniki 76 *C3* Wielkolpolskie, W Poland
Obrovo *see* Abrova
Obskaya Guba 92 *D3* *Eng.* Gulf of Ob. *gulf* N Russia
Ob' Tablemount 119 *B7* *undersea feature* S Indian Ocean
Ocala 21 *E4* Florida, SE USA
Ocaña 36 *B2* Norte de Santander, N Colombia
Ocaña 70 *D3* Castilla-La Mancha, C Spain
O Carballiño 70 *C1* *Cast.* Carballiño. Galicia, NW Spain
Occidental, Cordillera 36 *B2* *mountain range* W South America
Occidental, Cordillera 39 *E4* *mountain range* Bolivia/Chile
Ocean Falls 14 *D5* British Columbia, SW Canada
Ocean Island *see* Banaba
Oceanside 25 *C8* California, W USA
Ocean State *see* Rhode Island
Ochakiv 87 *E4* *Rus.* Ochakov. Mykolayivs'ka Oblast', S Ukraine
Ochakov *see* Ochakiv
Ochamchire 95 *E2* *prev.* Och'amch'ire. W Georgia
Och'amch'ire *see* Ochamchire
Ocho Rios 32 *B4* C Jamaica
Ochrida *see* Ohrid
Ochrida, Lake *see* Ohrid, Lake
Ocotal 30 *D3* Nueva Segovia, NW Nicaragua
Ocozocuautla 29 *G5* Chiapas, SE Mexico
October Revolution Island *see* Oktyabr'skoy Revolyutsii, Ostrov
Ocú 31 *F5* Herrera, S Panama
Ōdate 108 *D3* Akita, Honshū, C Japan
Oddur *see* Xuddur
Ödemiş 94 *A4* İzmir, SW Turkey (Türkiye)
Ödenburg *see* Sopron
Odenpäh *see* Otepää
Odense 63 *B7* Fyn, C Denmark
Oder 76 *B3* *Cz./Pol.* Odra. *river* C Europe
Oderhaff *see* Szczeciński, Zalew
Odesa 87 *E4* *Rus.* Odessa. Odes'ka Oblast', SW Ukraine
Odessa 27 *E3* Texas, SW USA
Odessa *see* Odesa
Odessus *see* Varna
Odienné 52 *D4* NW Ivory Coast
Odisha 113 *F4* *prev.* Orissa. *cultural region* NE India
Ŏdŏngk 115 *D6* Kampong Speu, S Cambodia
Odoorn 64 *E2* Drenthe, NE Netherlands
Odra *see* Oder
Oesel *see* Saaremaa
Of 95 *E2* Trabzon, NE Turkey (Türkiye)
Ofanto 75 *D5* *river* S Italy
Offenbach 73 *B5* *var.* Offenbach am Main. Hessen, W Germany
Offenbach am Main *see* Offenbach
Offenburg 73 *B6* Baden-Württemberg, SW Germany
Ogaadeen *see* Ogaden
Ogaden 51 *D5* *Som.* Ogaadeen. *plateau* Ethiopia/ Somalia
Ōgaki 109 *C6* Gifu, Honshū, SW Japan
Ogallala 22 *D4* Nebraska, C USA
Ogbomosho 53 *F4* *var.* Ogmoboso. Oyo, W Nigeria
Ogden 22 *B4* Utah, W USA
Ogdensburg 19 *F2* New York, NE USA
Ogmoboso *see* Ogbomosho
Ogulin 78 *A3* Karlovac, NW Croatia
Ohio 18 *C4* *off.* State of Ohio, *also known as* Buckeye State. *state* N USA
Ohio River 18 *C4* *river* N USA
Ohlau *see* Oława
Ohri *see* Ohrid
Ohrid 79 *D6* *Turk.* Ochrida, Ohri. SW North Macedonia
Ohrid, Lake 79 *D6* *var.* Lake Ochrida, *Alb.* Liqeni i Ohrit, *Mac.* Ohridsko Ezero. *lake* Albania/ North Macedonia
Ohridsko Ezero/Ohrit, Liqeni i *see* Ohrid, Lake
Ohura 128 *D3* Manawatu-Wanganui, North Island, New Zealand
Oirschot 65 *C5* Noord-Brabant, S Netherlands
Oise 68 *C3* *river* N France
Oistins 33 *G2* S Barbados
Ōita 109 *B7* Ōita, Kyūshū, SW Japan
Ojinaga 28 *D2* Chihuahua, N Mexico
Ojos del Salado, Cerro 42 *B3* *mountain* W Argentina
Okaihau 128 *C2* Northland, North Island, New Zealand
Ókanizsa *see* Kanjiža
Okara 112 *C2* Punjab, E Pakistan
Okavanggo *see* Cubango/Okavango
Okavango *see* Cubango
Okavango Delta 56 *C3* *wetland* N Botswana
Okayama 109 *B6* Okayama, Honshū, SW Japan
Okazaki 109 *C6* Aichi, Honshū, C Japan
Okeechobee, Lake 21 *E4* *lake* Florida, SE USA
Okefenokee Swamp 21 *E3* *wetland* Georgia, SE USA

Okhotsk 93 G3 Khabarovskiy Kray, E Russia
Okhotsk, Sea of 91 F3 sea NW Pacific Ocean
Okhtyrka 87 F2 Rus. Akhtyrka. Sums'ka Oblast', NE Ukraine
Oki-guntō see Oki-shotō
Okinawa 108 A3 island SW Japan
Okinawa-shoto 108 A3 island group Nansei-shotō, SW Japan Asia
Oki-shoto 109 B6 var. Oki-guntō. island group SW Japan
Oklahoma 27 F2 off. State of Oklahoma, also known as The Sooner State. state C USA
Oklahoma City 27 G1 state capital Oklahoma, C USA
Okmulgee 27 G1 Oklahoma, C USA
Oko, Wadi 50 C3 river NE Sudan
Oktyabr'skiy 89 D6 Volgogradskaya Oblast', SW Russia
Oktyabr'skiy see Aktsyabrski
Oktyabr'skoy Revolyutsii, Ostrov 93 E2 Eng. October Revolution Island. island Severnaya Zemlya, N Russia
Okulovka 88 B4 var. Okulovka. Novgorodskaya Oblast', W Russia
Okulovka see Okulovka
Okushiri-to 108 C3 var. Okusiri Tô. island NE Japan
Okusiri Tô see Okushiri-tō
Oláh-Toplicza see Toplița
Öland 63 C7 island S Sweden
Olavarría 43 D5 Buenos Aires, E Argentina
Oława 76 C4 Ger. Ohlau. Dolnośląskie, SW Poland
Olbia 75 A5 prev. Terranova Pausania. Sardegna, Italy, C Mediterranean Sea
Old Bay State/Old Colony State see Massachusetts
Old Dominion see Virginia
Oldebroek 64 D3 Gelderland, E Netherlands
Oldenburg 72 B3 Niedersachsen, NW Germany
Oldenburg 72 C2 var. Oldenburg in Holstein. Schleswig-Holstein, N Germany
Oldenburg in Holstein see Oldenburg
Oldenzaal 64 E3 Overijssel, E Netherlands
Old Harbour 32 B5 C Jamaica
Old Line State see Maryland
Old North State see North Carolina
Olëkma 93 F4 river C Russia
Olëkminsk 93 F3 Respublika Sakha (Yakutiya), NE Russia
Oleksandriia 87 F3 prev. Oleksandriya, Rus. Aleksandriya. Kirovohrads'ka Oblast', C Ukraine
Oleksandrivka 87 E3 Rus. Aleksandrovka. Kirovohrads'ka Oblast', C Ukraine
Oleksandriya see Oleksandriia
Olenegorsk 88 C2 Murmanskaya Oblast', NW Russia
Olenëk 93 E3 Respublika Sakha (Yakutiya), NE Russia
Olenëk 93 E3 river NE Russia
Oléron, Île d' 69 A5 island W France
Oleshky 87 E4 Rus. Tsiurupynsk. Khersons'ka Oblast', S Ukraine
Olevsk 86 D1 Zhytomyrs'ka Oblast', N Ukraine
Ölgiy 104 C2 Bayan-Ölgiy, W Mongolia
Olhão 70 B5 Faro, S Portugal
Olifa 56 B3 Kunene, NW Namibia
Ólimbos see Ólympos
Olimpo see Fuerte Olimpo
Olisipo see Lisboa
Olita see Alytus
Oliva 71 F4 Comunitat Valenciana, E Spain
Olivet 68 C4 Loiret, C France
Olmaliq 101 E2 Rus. Almalyk. Toshkent Viloyati, E Uzbekistan
Olmütz see Olomouc
Olomouc 77 C5 Ger. Olmütz, Pol. Ołomuniec. Olomoucký Kraj, E Czechia (Czech Republic)
Ołomuniec see Olomouc
Olonets 88 B3 Respublika Kareliya, NW Russia
Olovyannaya 93 F4 Chitinskaya Oblast', S Russia
Olpe 72 B4 Nordrhein-Westfalen, W Germany
Olsnitz see Murska Sobota
Olsztyn 76 D2 Ger. Allenstein. Warmińsko-Mazurskie, N Poland
Olt 86 B5 var. Oltul, Ger. Alt. river S Romania
Oltenița 86 C5 prev. Eng. Oltenitsa; anc. Constantiola. Călărași, SE Romania
Oltenitsa see Oltenița
Oltul see Olt
Olvera 70 D5 Andalucía, S Spain
Olviopol see Pervomaisk
Olympia 24 B2 state capital Washington, NW USA
Olympic Mountains 24 A2 mountain range Washington, NW USA
Olympus, Mount 82 B4 var. Ólimbos, Eng. Mount Olympus. mountain N Greece
Omagh 67 B5 Ir. An Ómaigh. W Northern Ireland, United Kingdom
Omaha 23 F4 Nebraska, C USA
Oman 99 D6 off. Sultanate of Oman, Ar. Salṭanat 'Umān; prev. Muscat and Oman. country SW Asia
Oman, Gulf of 98 E4 Ar. Khalij 'Umān. gulf N Arabian Sea
Oman, Sultanate of see Oman
Omboué 55 A6 Ogooué-Maritime, W Gabon
Omdurman 50 B4 var. Umm Durmān. Khartoum, C Sudan
Ometepe, Isla de 30 D4 island S Nicaragua
Ommen 64 E3 Overijssel, E Netherlands
Omsk 92 C4 Omskaya Oblast', C Russia
Ōmuta 109 A7 Fukuoka, Kyūshū, SW Japan
Onda 71 F3 Comunitat Valenciana, E Spain
Ondjiva see N'Giva
Öndörhaan 105 E2 var. Undur Khan; prev. Tsetsen Khan. Hentiy, E Mongolia
Onega 88 C3 Arkhangel'skaya Oblast', NW Russia
Onega 88 B4 river NW Russia
Onega, Lake see Onezhskoye Ozero
Onex 73 A7 Genève, SW Switzerland
Onezhskoye Ozero 88 B4 Eng. Lake Onega. lake NW Russia
Ongole 110 D1 Andhra Pradesh, E India
Onitsha 53 G5 Anambra, S Nigeria
Onon Gol 105 E2 river N Mongolia
Onslow 124 A4 Western Australia
Onslow Bay 21 F1 bay North Carolina, E USA
Ontario 16 B3 province S Canada
Ontario, Lake 19 E3 lake Canada/USA
Onteniente see Ontinyent
Ontinyent 71 F4 var. Onteniente. Comunitat Valenciana, E Spain
Ontong Java Rise 103 H4 undersea feature W Pacific Ocean
Onuba see Huelva
Oodeypore see Udaipur
Oos-Londen see East London
Oostakker 65 B5 Oost-Vlaanderen, NW Belgium
Oostburg 65 B5 Zeeland, SW Netherlands
Oostende 65 A5 Eng. Ostend, Fr. Ostende. West-Vlaanderen, NW Belgium
Oosterbeek 64 D4 Gelderland, SE Netherlands
Oosterhout 64 C4 Noord-Brabant, S Netherlands
Opatija 78 A2 It. Abbazia. Primorje-Gorski Kotar, NW Croatia
Opava 77 C5 Ger. Troppau. Moravskoslezský Kraj, E Czechia (Czech Republic)
Ópazova see Stara Pazova
Opelika 20 D2 Alabama, S USA
Opelousas 20 B3 Louisiana, S USA
Ophiusa see Formentera
Opmeer 64 C2 Noord-Holland, NW Netherlands
Opochka 88 A4 Pskovskaya Oblast', W Russia
Opole 76 C4 Ger. Oppeln. Opolskie, S Poland
Oporto see Porto
Opotiki 128 E3 Bay of Plenty, North Island, New Zealand
Oppeln see Opole
Oppidum Ubiorum see Köln
Oqtosh 101 E2 Rus. Aktash. Samarqand Viloyati, C Uzbekistan
Oradea 86 B3 prev. Oradea Mare, Ger. Grosswardein, Hung. Nagyvárad. Bihor, NW Romania
Oradea Mare see Oradea
Orahovac see Rahovec
Oral 92 B3 var. Ural'sk. Zapadnyy Kazakhstan, NW Kazakhstan
Oran 48 D2 var. Ouahran, Wahran. NW Algeria
Orange 127 D6 New South Wales, SE Australia
Orange 69 D6 anc. Arausio. Vaucluse, SE France
Orangeburg 21 E2 South Carolina, SE USA
Orange Cone see Orange Fan
Orange Fan 47 C7 var. Orange Cone. undersea feature SW Indian Ocean
Orange Mouth/Orangemund see Oranjemund
Orange River 56 B4 Afr. Oranjerivier. river S Africa
Orange Walk 30 C1 Orange Walk, N Belize
Oranienburg 72 D3 Brandenburg, NE Germany
Oranjemund 56 B4 var. Orangemund; prev. Orange Mouth. Karas, SW Namibia
Oranjerivier see Orange River
Oranjestad 33 E5 dependent territory capital (Aruba) Lesser Antilles, S Caribbean Sea
Orantes see Orontes
Orany see Varėna
Orașul Stalin see Brașov
Oravicabánya see Oravița
Oravița 86 A4 Ger. Orawitza, Hung. Oravicabánya. Caraș-Severin, SW Romania
Orawitza see Oravița
Orbetello 74 B4 Toscana, C Italy
Orcadas 132 A1 Argentinian research station South Orkney Islands, Antarctica
Orchard Homes 22 B1 Montana, NW USA
Ordino 69 A8 Ordino, NW Andorra Europe
Ordos Desert see Mu Us Shadi
Ordu 94 D2 anc. Cotyora. Ordu, N Turkey (Türkiye)
Ordzhonikidze see Pokrov, Ukraine
Ordzhonikidze see Vladikavkaz, Russia
Ordzhonikidze see Yenakiieve
Orealla 37 G3 E Guyana
Örebro 63 C6 Örebro, C Sweden
Oregon 24 B3 off. State of Oregon, also known as Beaver State, Sunset State, Valentine State, Webfoot State. state NW USA
Oregon City 24 B3 Oregon, NW USA
Oregon, State of see Oregon
Orekhov see Orikhiv
Orël 89 B5 Orlovskaya Oblast', W Russia
Orem 22 B4 Utah, W USA
Ore Mountains see Erzgebirge/Krušné Hory
Orenburg 89 D6 prev. Chkalov. Orenburgskaya Oblast', W Russia
Orense see Ourense
Orestiáda 82 D3 prev. Orestiás. Anatolikí Makedonía kai Thráki, NE Greece
Orestiás see Orestiáda
Organ Peak 26 D3 mountain New Mexico, SW USA
Orgeyev see Orhei
Orhei 86 D3 var. Orheiu, Rus. Orgeyev. N Moldova
Orheiu see Orhei
Oriental, Cordillera 38 D3 mountain range Bolivia/Peru
Oriental, Cordillera 36 B3 mountain range C Colombia
Orihuela 71 F4 Comunitat Valenciana, E Spain
Orikhiv 87 G3 Rus. Orekhov. Zaporiz'ka Oblast', SE Ukraine
Orinoco, Río 37 E2 river Colombia/Venezuela
Orinoquía 36 C3 region NE Colombia
Orissa see Odisha
Orissaar see Orissaare
Orissaare 84 C2 Ger. Orissaar. Saaremaa, W Estonia
Oristano 75 A5 Sardegna, Italy, C Mediterranean Sea
Orito 36 A4 Putumayo, SW Colombia
Orizaba, Volcán Pico de 13 C7 var. Citlaltépetl. mountain S Mexico
Orkney see Orkney Islands
Orkney Islands 66 C2 var. Orkney, Orkneys. island group N Scotland, United Kingdom
Orkneys see Orkney Islands
Orlando 21 E4 Florida, SE USA
Orléanais 68 C4 cultural region C France
Orléans 68 C4 anc. Aurelianum. Loiret, C France
Orléansville see Chlef
Orly 68 E2 (Paris) Essonne, N France
Orlya 85 B6 Hrodzyenskaya Voblasts', W Belarus
Ormsö see Vormsi
Ormuz, Strait of see Hormuz, Strait of
Örnsköldsvik 63 C5 Västernorrland, C Sweden
Orolaunum see Arlon
Orol Dengizi see Aral Sea
Oromocto 17 F4 New Brunswick, SE Canada
Orona 123 F3 prev. Hull Island. atoll Phoenix Islands, C Kiribati
Orontes 94 D5 var. Orantes, Nahr El Aassi , Nahr Al 'Āṣī , Asi Nehri . river Lebanon/Syria/Turkey
Oropeza see Cochabamba
Oroseirá Rodópis see Rhodope Mountains
Orpington 67 B8 United Kingdom
Orschowa see Orșova
Orsha 85 E6 Vitsyebskaya Voblasts', NE Belarus
Orsk 92 B4 Orenburgskaya Oblast', W Russia
Orșova 86 A4 Ger. Orschowa, Hung. Orsova. Mehedinți, SW Romania
Ortelsburg see Szczytno
Orthez 69 B6 Pyrénées-Atlantiques, SW France
Ortona 74 D4 Abruzzo, C Italy
Oruba see Aruba
Orūmīyeh, Daryācheh-ye 99 C2 var. Matianus, Sha Hi, Urumi Yeh, Eng. Lake Urmia; prev. Daryācheh-ye Reẓā'īyeh. lake NW Iran
Oruro 39 F4 Oruro, W Bolivia
Oryokko see Yalu
Oss 64 D4 Noord-Brabant, S Netherlands
Ōsaka 109 C6 hist. Naniwa. Ōsaka, Honshū, SW Japan
Ōsaki see Furukawa
Osa, Península de 31 E5 peninsula S Costa Rica
Osborn Plateau 119 D5 undersea feature E Indian Ocean
Osca see Huesca
Ösel see Saaremaa
Osh 101 F2 Oshskaya Oblast', SW Kyrgyzstan
Oshawa 16 D5 Ontario, SE Canada
Oshikango 56 B3 Ohangwena, N Namibia
O-shima 109 D6 island S Japan
Oshkosh 18 B2 Wisconsin, N USA
Oshmyany see Ashmyany
Osiek see Osijek
Osijek 78 C3 prev. Osiek, Osjek, Ger. Esseg, Hung. Eszék. Osijek-Baranja, E Croatia
Osipenko see Berdiansk
Osipovichi see Asipovichy
Osjek see Osijek
Oskaloosa 23 G4 Iowa, C USA
Oskarshamn 63 C7 Kalmar, S Sweden
Öskemen 92 D5 var. Ust'-Kamenogorsk. Vostochnyy Kazakhstan, E Kazakhstan
Oskol see Oskil
Oskil 87 G2 Rus. Oskol. river Russia/Ukraine
Oslo 63 B6 prev. Christiania, Kristiania. country capital (Norway) Oslo, S Norway
Osmaniye 94 D4 Osmaniye, S Turkey (Türkiye)
Osnabrück 72 A3 Niedersachsen, NW Germany
Osogov Mountains 82 B3 var. Osogovske Planine, Osogovski Planina, Mac. Osogovski Planini. mountain range Bulgaria/North Macedonia
Osogovske Planine/Osogovski Planina/ Osogovski Planini see Osogov Mountains
Oșorhei see Târgu Mureș
Osorno 43 B5 Los Lagos, C Chile
Ossa, Serra d' 70 C4 mountain range SE Portugal
Ossora 93 H2 Koryakskiy Avtonomnyy Okrug, E Russia
Ostee see Baltic Sea
Ostend/Ostende see Oostende
Oster 87 E1 Chernihivs'ka Oblast', N Ukraine
Östermyra see Seinäjoki
Osterode/Osterode in Ostpreussen see Ostróda
Österreich see Austria
Östersund 63 C5 Jämtland, C Sweden
Ostia Aterni see Pescara
Ostiglia 74 C2 Lombardia, N Italy
Ostrava 77 C5 Moravskoslezský Kraj, E Czechia (Czech Republic)
Ostróda 76 D3 Ger. Osterode, Osterode in Ostpreussen. Warmińsko-Mazurskie, NE Poland
Ostrołęka 76 D3 Ger. Wiesenhof, Rus. Ostrolenka. Mazowieckie, C Poland
Ostrolenka see Ostrołęka
Ostrov 88 A4 Latv. Austrava. Pskovskaya Oblast', W Russia
Ostrovets see Ostrowiec Świętokrzyski
Ostrovnoy 88 C2 Murmanskaya Oblast', NW Russia
Ostrów see Ostrów Wielkopolski
Ostrowiec see Ostrowiec Świętokrzyski
Ostrowiec Świętokrzyski 76 D4 var. Ostrowiec, Rus. Ostrovets. Świętokrzyskie, C Poland
Ostrów Mazowiecka 76 D3 var. Ostrów Mazowiecki. Mazowieckie, NE Poland
Ostrów Mazowiecki see Ostrów Mazowiecka
Ostrowo see Ostrów Wielkopolski
Ostrów Wielkopolski 76 C4 var. Ostrów, Ger. Ostrowo. Wielkopolskie, C Poland
Ostyako-Voguls'k see Khanty-Mansiysk
Osum see Osumit, Lumi i
Osumi-shoto 109 A8 island group Kagoshima, Nansei-shotō, SW Japan Asia East China Sea Pacific Ocean
Osumit, Lumi i 79 D7 var. Osum. river SE Albania
Osuna 70 D4 Andalucía, S Spain
Oswego 19 F2 New York, NE USA
Otago Peninsula 129 B7 peninsula South Island, New Zealand
Otaki 128 D4 Wellington, North Island, New Zealand
Otaru 108 C2 Hokkaidō, NE Japan
Otavalo 38 B1 Imbabura, N Ecuador
Otavi 56 B3 Otjozondjupa, N Namibia
Oțelu Roșu 86 B4 Ger. Ferdinandsberg, Hung. Nándorhgy. Caras-Severin, SW Romania
Otepää 84 D3 Ger. Odenpäh. Valgamaa, SE Estonia
Oti 53 E4 river N Togo
Otira 129 C6 West Coast, South Island, New Zealand
Otjiwarongo 56 B3 Otjozondjupa, N Namibia
Otorohanga 128 D3 Waikato, North Island, New Zealand
Otranto, Canale d' see Otranto, Strait of
Otranto, Strait of 79 C6 It. Canale d'Otranto. strait Albania/Italy
Otrokovice 77 C5 Ger. Otrokowitz. Zlínský Kraj, E Czechia (Czech Republic)
Otrokowitz see Otrokovice
Ōtsu 109 C6 var. Ôtu. Shiga, Honshū, SW Japan
Ottawa 16 D5 country capital (Canada) Ontario, SE Canada
Ottawa 18 B3 Illinois, N USA
Ottawa 23 F5 Kansas, C USA
Ottawa 19 E2 Fr. Outaouais. river Ontario/Québec, SE Canada
Ottawa Islands 16 C1 island group Nunavut, C Canada
Ottignies 65 C6 Wallon Brabant, C Belgium
Ottumwa 23 G4 Iowa, C USA
Ôtu see Ōtsu
Ouachita Mountains 20 A1 mountain range Arkansas/Oklahoma, C USA
Ouachita River 20 B2 river Arkansas/Louisiana, C USA
Ouagadougou 53 E4 var. Wagadugu. country capital (Burkina) C Burkina
Ouahigouya 53 E3 NW Burkina
Ouahran see Oran
Oualâta 52 D3 var. Oualata. Hodh ech Chargui, SE Mauritania
Ouanary 37 H3 E French Guiana
Ouanda Djallé 54 D4 Vakaga, NE Central African Republic
Ouarâne 52 D2 desert C Mauritania
Ouargla 49 E2 var. Wargla. NE Algeria
Ouarzazate 48 C3 S Morocco
Oubangui 55 C5 Fr. Oubangui. river C Africa
Oubangui see Ubangi
Oubangui-Chari see Central African Republic
Oubangui-Chari, Territoire de l' see Central African Republic
Oudjda see Oujda
Ouessant, Île d' 68 A3 Eng. Ushant. island NW France
Ouésso 55 B5 Sangha, NW Congo
Oujda 48 D2 Ar. Oudjda, Ujda. NE Morocco
Oujeft 52 C2 Adrar, C Mauritania
Oulu 62 D4 Swe. Uleåborg. Pohjois-Pohjanmaa, C Finland
Oulujärvi 62 D4 Swe. Uleträsk. lake C Finland
Oulujoki 62 D4 Swe. Uleälv. river C Finland
Ounasjoki 62 D3 river N Finland
Ounianga Kébir 54 C2 Borkou-Ennedi-Tibesti, N Chad
Oup see Auob
Oupeye 65 D6 Liège, E Belgium
Our 65 D6 river NW Europe
Ourense 70 C1 Cast. Orense, Lat. Aurium. Galicia, NW Spain
Ourique 70 B4 Beja, S Portugal
Ours, Grand Lac de l' see Great Bear Lake
Ourthe 65 D7 river E Belgium
Ouse 67 D5 river N England, United Kingdom
Outaouais see Ottawa
Outer Hebrides 66 B3 var. Western Isles. island group NW Scotland, United Kingdom
Outer Islands 57 G1 island group SW Seychelles Africa W Indian Ocean
Outes 70 B1 Galicia, NW Spain
Ouvéa 122 D5 island Îles Loyauté, NE New Caledonia
Ouyen 127 C6 Victoria, SE Australia
Ovalle 42 B3 Coquimbo, N Chile
Ovar 70 B2 Aveiro, N Portugal
Overflakkee 64 B4 island SW Netherlands
Overijse 65 C6 Vlaams Brabant, C Belgium
Oviedo 70 C1 var. Uviéu, anc. Asturias. Asturias, NW Spain
Ovilava see Wels
Ovruch 86 D1 Zhytomyrs'ka Oblast', N Ukraine
Owando 55 B5 prev. Fort Rousset. Cuvette, C Congo
Owase 109 C6 Mie, Honshū, SW Japan
Owatonna 23 F3 Minnesota, N USA
Owen Fracture Zone 118 B4 tectonic feature W Arabian Sea
Owen, Mount 129 C5 mountain South Island, New Zealand
Owensboro 18 B5 Kentucky, S USA
Owen Stanley Range 122 B3 mountain range S Papua New Guinea
Owerri 53 G5 Imo, S Nigeria
Owo 53 F5 Ondo, SW Nigeria
Owyhee River 24 C4 river Idaho/Oregon, NW USA
Oxford 129 C6 Canterbury, South Island, New Zealand
Oxford 67 D6 Lat. Oxonia. S England, United Kingdom
Oxkutzcab 29 H4 Yucatán, SE Mexico
Oxnard 25 B7 California, W USA
Oxonia see Oxford
Oxus see Amu Darya
Oyama 109 D5 Tochigi, Honshū, S Japan
Oyem 55 B5 Woleu-Ntem, N Gabon
Oyo 55 B6 Cuvette, C Congo
Oyo 53 F4 Oyo, W Nigeria
Ozark 20 D3 Alabama, S USA
Ozark Plateau 23 G5 plain Arkansas/Missouri, C USA
Ozarks, Lake of the 23 F5 reservoir Missouri, C USA
Ozbourn Seamount 130 D4 undersea feature W Pacific Ocean
Ózd 77 D6 Borsod-Abaúj-Zemplén, NE Hungary
Ozieri 75 A5 Sardegna, Italy, C Mediterranean Sea

P

Paamiut 60 B4 var. Pâmiut, Dan. Frederikshåb. S Greenland
Pa-an see Hpa-An
Pabianice 76 C4 Łódzski, Poland
Pabna 113 G4 Rajshahi, W Bangladesh
Pacaraima, Sierra/Pacaraim, Serra see Pakaraima Mountains
Pachuca 29 E4 var. Pachuca de Soto. Hidalgo, C Mexico
Pachuca de Soto see Pachuca
Pacific-Antarctic Ridge 132 B5 undersea feature S Pacific Ocean
Pacific Ocean 130 D3 ocean
Padalung see Phatthalung
Padang 116 B4 Sumatera, W Indonesia
Paderborn 72 B4 Nordrhein-Westfalen, NW Germany
Padma see Brahmaputra
Padma see Ganges
Padova 74 C2 Eng. Padua; anc. Patavium. Veneto, NE Italy
Padre Island 27 G5 island Texas, SW USA
Padua see Padova
Paducah 18 B5 Kentucky, S USA
Paeroa 128 D3 Waikato, North Island, New Zealand
Páfos 80 C5 var. Paphos. W Cyprus
Pag 78 A3 It. Pago. island Zadar, C Croatia
Page 26 B1 Arizona, SW USA
Pago see Pag
Pago Pago 123 F4 dependent territory capital (American Samoa) Tutuila, W American Samoa
Pahiatua 128 D4 Manawatu-Wanganui, North Island, New Zealand
Pahsien see Chongqing
Paide 84 D2 Ger. Weissenstein. Järvamaa, N Estonia
Paihia 128 D2 Northland, North Island, New Zealand
Päijänne 63 D5 lake S Finland
Paine, Cerro 43 A7 mountain S Chile
Painted Desert 26 B1 desert Arizona, SW USA
Paisance see Piacenza
Paisley 66 C4 W Scotland, United Kingdom
País Vasco 71 E1 cultural region N Spain
Paita 38 B3 Piura, NW Peru
Pakanbaru see Pekanbaru
Pakaraima Mountains 37 E3 var. Serra Pacaraim, Sierra Pacaraima. mountain range N South America
Pakistan 112 A2 off. Islamic Republic of Pakistan, var. Islami Jamhuriya e Pakistan. country S Asia
Pakistan, Islamic Republic of see Pakistan
Pakistan, Islami Jamhuriya e see Pakistan
Paknam see Samut Prakan
Pakokku 114 A3 Magway, C Myanmar (Burma)
Pak Phanang 115 C7 var. Ban Pak Phanang. Nakhon Si Thammarat, SW Thailand
Pakruojis 84 C4 Šiaulai, N Lithuania
Paks 77 C7 Tolna, S Hungary
Pakxé see Pakxé
Pakxé 115 D5 var. Paksé. Champasak, S Laos
Palafrugell 71 G2 Cataluña, NE Spain
Palagruža 79 B5 It. Pelagosa. island SW Croatia
Palaiá Epídavros 83 C6 Pelopónnisos, S Greece
Palaiseau 68 D2 Essonne, N France
Palamós 71 G2 Cataluña, NE Spain
Palamuse 84 E2 Ger. Sankt-Bartholomäi. Jõgevamaa, E Estonia
Palanka see Bačka Palanka
Pālanpur 112 C4 Gujarāt, W India
Palantia see Palencia
Palapye 56 D3 Central, SE Botswana
Palau 122 A2 var. Belau. country W Pacific Ocean
Palawan 117 E2 island W Philippines
Palawan Passage 116 D2 passage W Philippines
Paldiski 84 D2 prev. Baltiski, Eng. Baltic Port, Ger. Baltischport. Harjumaa, NW Estonia
Palembang 116 B4 Sumatera, W Indonesia
Palencia 70 D2 anc. Palantia, Pallantia. Castilla y León, NW Spain
Palerme see Palermo
Palermo 75 C7 Fr. Palerme; anc. Panhormus, Panormus. Sicilia, Italy, C Mediterranean Sea
Palestine, State of see West Bank; Gaza Strip
Pāli 112 C3 Rājasthān, N India
Palikir 122 C2 country capital (Micronesia) Pohnpei, E Micronesia
Palimé see Kpalimé
Palioúri, Akrotírio 82 C4 var. Akrotírio Kanestron. headland N Greece
Palk Strait 110 C3 strait India/Sri Lanka
Pallantia see Palencia
Palliser, Cape 129 D5 headland North Island, New Zealand
Palma 71 G3 var. Palma de Mallorca. Mallorca, Spain, W Mediterranean Sea
Palma del Río 70 D4 Andalucía, S Spain
Palma de Mallorca see Palma
Palmar Sur 31 E5 Puntarenas, SE Costa Rica
Palma Soriano 32 C3 Santiago de Cuba, E Cuba
Palm Beach 126 E1 New South Wales, E Australia
Palmer 132 A2 US research station Antarctica
Palmer Land 132 A3 physical region Antarctica
Palmerston 123 F4 island S Cook Islands
Palmerston see Darwin
Palmerston North 128 D4 Manawatu-Wanganui, North Island, New Zealand
Palmetto State, The see South Carolina
Palmi 75 D7 Calabria, SW Italy
Palmira 36 B3 Valle del Cauca, W Colombia
Palm Springs 25 D7 California, W USA
Palmyra see Tadmur
Palmyra Atoll 123 G2 US incorporated territory C Pacific Ocean
Palo Alto 25 B6 California, W USA
Paloe see Denpasar, Bali, C Indonesia
Paloe see Palu
Palu 117 E4 prev. Paloe. Sulawesi, C Indonesia
Pamiers 69 B6 Ariège, S France
Pamir 101 F3 var. Daryā-ye Pāmīr, Taj. Dar"yoi Pomir. river Afghanistan/Tajikistan
Pāmīr, Daryā-ye see Pamir
Pamir/Pāmīr, Daryā-ye see Pamirs
Pamirs 101 F3 Pash. Daryā-ye Pāmīr, Rus. Pamir. mountain range C Asia
Pâmiut see Paamiut
Pamlico Sound 21 G1 sound North Carolina, SE USA
Pampa 27 E1 Texas, SW USA
Pampa Aullagas, Lago see Poopó, Lago
Pampas 42 C4 plain C Argentina
Pampeluna see Pamplona
Pamplona 36 C2 Norte de Santander, N Colombia
Pamplona 71 E1 Basq. Iruña, prev. Pampeluna; anc. Pompaelo. Navarra, N Spain
Panaji 110 B1 var. Pangim, Panjim, New Goa. state capital Goa, W India
Panamá 31 G4 var. Ciudad de Panama, Eng. Panama City. country capital (Panama) Panamá, C Panama
Panama 31 G5 off. Republic of Panama. country Central America
Panama Basin 13 C8 undersea feature E Pacific Ocean
Panama Canal 31 F4 canal E Panama
Panama City 20 D3 Florida, SE USA
Panama City see Panamá
Panamá, Golfo de 31 G5 var. Gulf of Panama. gulf S Panama
Panama, Gulf of see Panamá, Golfo de
Panama, Isthmus of see Panama, Istmo de
Panama, Istmo de 31 G4 Eng. Isthmus of Panama; prev. Isthmus of Darien. isthmus E Panama
Panama, Republic of see Panama
Panay Island 117 E2 island C Philippines
Pančevo 78 D3 Ger. Pantschowa, Hung. Pancsova. Vojvodina, N Serbia
Pancsova see Pančevo
Paneas see Bāniyās
Panevėžys 84 C4 Panevėžys, C Lithuania
Pangim see Panaji
Pangkalpinang 116 C4 Pulau Bangka, W Indonesia
Pang-Nga see Phang-Nga
Panhormus see Palermo
Panjim see Panaji
Panopolis see Akhmīm
Pánormos 83 C7 Kríti, Greece, E Mediterranean Sea

Panormus *see* Palermo
Pantanal *41 E3* var. Pantanalmato-Grossense. swamp SW Brazil
Pantanalmato-Grossense *see* Pantanal
Pantelleria, Isola di *75 B7* island SW Italy
Pantschowa *see* Pančevo
Pánuco *29 E3* Veracruz-Llave, E Mexico
Pao-chi/Paoki *see* Baoji
Paola *80 B5* E Malta
Pao-shan *see* Baoshan
Pao-t'ou/Paotow *see* Baotou
Papagayo, Golfo de *30 C4* gulf NW Costa Rica
Papakura *128 D3* Auckland, North Island, New Zealand
Papantla *29 F4* var. Papantla de Olarte. Veracruz-Llave, E Mexico
Papantla de Olarte *see* Papantla
Papeete *123 H4* dependent territory capital (French Polynesia) Tahiti, W French Polynesia
Paphos *see* Páfos
Papilė *84 B3* Šiaulai, NW Lithuania
Papillion *23 F4* Nebraska, C USA
Papua *117 H4* var. Irian Barat, West Irian, West New Guinea, West Papua; prev. Dutch New Guinea, Irian Jaya, Netherlands New Guinea. province E Indonesia
Papua and New Guinea, Territory of *see* Papua New Guinea
Papua, Gulf of *122 B3* gulf S Papua New Guinea
Papua New Guinea *122 B3* off. Independent State of Papua New Guinea; prev. Territory of Papua and New Guinea. country NW Melanesia
Papua New Guinea, Independent State of *see* Papua New Guinea
Papuk *78 C3* mountain range NE Croatia
Pará *41 E2* off. Estado do Pará. state/region NE Brazil
Pará *see* Belém
Paracel Islands *103 E3* disputed territory SE Asia
Paraćin *78 D4* Serbia, C Serbia
Paradise of the Pacific *see* Hawaii
Pará, Estado do *see* Pará
Paragua, Río *37 E3* river SE Venezuela
Paraguay *42 C2* country C South America
Paraguay *42 D2* var. Río Paraguay. river C South America
Paraguay, Río *see* Paraguay
Parahiba/Parahyba *see* Paraíba
Paraíba *41 G2* off. Estado da Paraíba; prev. Parahiba, Parahyba. state/region E Brazil
Paraíba *see* João Pessoa
Paraíba, Estado da *see* Paraíba
Parakou *53 F4* C Benin
Paramaribo *37 G3* country capital (Suriname) Paramaribo, N Suriname
Paramushir, Ostrov *93 H3* island SE Russia
Paraná *41 E4* Entre Ríos, E Argentina
Paraná *41 E5* off. Estado do Paraná. state/region S Brazil
Paraná *35 C5* var. Alto Paraná. river C South America
Paraná, Estado do *see* Paraná
Paranésti *82 C3* var. Paranéstio. Anatolikí Makedonía kai Thráki, NE Greece
Paranéstio *see* Paranésti
Paraparaumu *129 D5* Wellington, North Island, New Zealand
Parchim *72 C3* Mecklenburg-Vorpommern, N Germany
Parczew *76 E4* Lubelskie, E Poland
Pardubice *77 B5* Ger. Pardubitz. Pardubický Kraj, C Czechia (Czech Republic)
Pardubitz *see* Pardubice
Parechcha *85 B5* Pol. Porzecze, Rus. Porech'ye. Hrodzyenskaya Voblasts', W Belarus
Parecis, Chapada dos *40 D3* var. Serra dos Parecis. mountain range W Brazil
Parecis, Serra dos *see* Parecis, Chapada dos
Parenzo *see* Poreč
Parepare *117 E4* Sulawesi, C Indonesia
Párga *83 A5* Ípeiros, W Greece
Paria, Golfo de *see* Paria, Gulf of
Paria, Gulf of *37 E1* var. Golfo de Paria. gulf Trinidad and Tobago/Venezuela
Parika *37 F2* NE Guyana
Parikiá *83 D6* Kykládes, Greece, Aegean Sea
Paris *68 D1* anc. Lutetia, Lutetia Parisiorum, Parisii. country capital (France) Paris, N France
Paris *27 G2* Texas, SW USA
Parisii *see* Paris
Parkersburg *18 D4* West Virginia, NE USA
Parkes *127 D6* New South Wales, SE Australia
Parkhar *see* Farkhor
Parma *74 B2* Emilia-Romagna, N Italy
Parnahyba *see* Parnaíba
Parnaíba *41 F2* var. Parnahyba. Piauí, E Brazil
Pärnu *84 D2* Ger. Pernau, Latv. Pērnava; prev. Rus. Pernov. Pärnumaa, SW Estonia
Pärnu *84 D2* var. Parnu Jõgi, Ger. Pernau. river SW Estonia
Pärnu-Jaagupi *84 D2* Ger. Sankt-Jakobi. Pärnumaa, SW Estonia
Parnu Jõgi *see* Pärnu
Pärnu Laht *84 D2* Ger. Pernauer Bucht. bay SW Estonia
Paroikiá *see* Páros
Paropamisus Range *see* Safēd Kōh, Silsilah-ye
Páros *83 D6* var. Paroikiá, Greece
Páros *83 D6* island Kykládes, Greece, Aegean Sea
Parral *42 B4* Maule, C Chile
Parral *see* Hidalgo del Parral
Parramatta *126 D1* New South Wales, SE Australia
Parras *28 D3* var. Parras de la Fuente. Coahuila, NE Mexico
Parras de la Fuente *see* Parras
Parsons *23 F5* Kansas, C USA
Pasadena *25 C7* California, W USA
Pasadena *27 H4* Texas, SW USA
Paşcani *86 C3* Hung. Páskán. Iaşi, NE Romania
Pasco *24 C2* Washington, NW USA
Pascua, Isla de *131 F4* var. Rapa Nui, Easter Island. island E Pacific Ocean
Pasewalk *72 D3* Mecklenburg-Vorpommern, NE Germany
Pashkeni *see* Bolyarovo
Pasinler *95 F3* Erzurum, NE Turkey (Türkiye)
Páskán *see* Paşcani
Pasłęk *76 D2* Ger. Preußisch Holland. Warmińsko-Mazurskie, NE Poland
Pasni *112 A3* Baluchistan, SW Pakistan
Paso de Indios *43 B6* Chubut, S Argentina
Passarowitz *see* Požarevac
Passau *73 D6* Bayern, SE Germany
Passo Fundo *41 E5* Rio Grande do Sul, S Brazil
Pastavy *85 C5* Pol. Postawy, Rus. Postovy. Vitsyebskaya Voblasts', NW Belarus
Pastaza, Río *38 B2* river Ecuador/Peru
Pasto *36 A4* Nariño, SW Colombia
Pasvalys *84 C4* Panevėžys, N Lithuania
Patagonia *35 B7* physical region Argentina/Chile
Patalung *see* Phatthalung
Patani *see* Pattani
Patavium *see* Padova
Patea *128 D4* Taranaki, North Island, New Zealand
Paterson *19 F3* New Jersey, NE USA
Pathein *114 A4* var. Bassein. Ayeyarwady, SW Myanmar (Burma)
Pátmos *83 D6* island Dodekánisa, Greece, Aegean Sea
Patna *113 F3* var. Azimabad. state capital Bihār, N India
Patnos *95 F3* Ağrı, E Turkey (Türkiye)
Patos, Lagoa dos *41 E5* lagoon S Brazil
Pátra *83 B5* Eng. Patras; prev. Pátrai. Dytikí Elláa, S Greece
Pátrai/Patras *see* Pátra
Pattani *115 C7* var. Patani. Pattani, SW Thailand
Pattaya *115 C5* Chon Buri, S Thailand
Patuca, Río *30 D2* river E Honduras
Pau *69 B6* Pyrénées-Atlantiques, SW France
Paulatuk *15 E3* Northwest Territories, NW Canada
Paungde *114 B4* Bago, C Myanmar (Burma)
Pautalia *see* Kyustendil
Pavia *74 B2* anc. Ticinum. Lombardia, N Italy
Pāvilosta *84 B3* Liepāja, W Latvia
Pavlikeni *82 D2* Veliko Tarnovo, N Bulgaria
Pavlodar *92 C4* Pavlodar, NE Kazakhstan
Pavlograd *see* Pavlohrad
Pavlohrad *87 G3* Rus. Pavlograd. Dnipropetrovska Oblast, E Ukraine
Pawai, Pulau *116 A2* island SW Singapore Asia
Pawn *114 B3* river C Myanmar (Burma)
Pax Augusta *see* Badajoz
Pax Julia *see* Beja
Paxoí *83 A5* island Iónia Nisiá, Greece, C Mediterranean Sea
Payo Obispo *see* Chetumal
Paysandú *42 D4* Paysandú, W Uruguay
Pazar *95 E2* Rize, NE Turkey (Türkiye)
Pazardzhik *82 C3* prev. Tatar Pazardzhik. Pazardzhik, SW Bulgaria
Peace Garden State *see* North Dakota
Peach State *see* Georgia
Pearl Islands *31 G5* Eng. Pearl Islands. island group SE Panama
Pearl Islands *see* Perlas, Archipiélago de las
Pearl Lagoon *see* Perlas, Laguna de
Pearl River *20 B3* river Louisiana/Mississippi, S USA
Pearsall *27 F4* Texas, SW USA
Peawanuk *16 C2* Ontario, S Canada
Peć *see* Pejë
Pechora *88 D3* Respublika Komi, NW Russia
Pechora *88 D3* river NW Russia
Pechora Sea *see* Pechorskoye More
Pechorskoye More *88 D2* Eng. Pechora Sea. sea NW Russia
Pecos *27 E3* Texas, SW USA
Pecos River *27 E3* river New Mexico/Texas, SW USA
Pécs *77 C7* Ger. Fünfkirchen, Lat. Sopianae. Baranya, SW Hungary
Pedra Lume *52 A3* Sal, NE Cape Verde (Cabo Verde)
Pedro Cays *32 C3* island group Greater Antilles, S Jamaica North America N Caribbean Sea W Atlantic Ocean
Pedro Juan Caballero *42 D2* Amambay, E Paraguay
Peer *65 D5* Limburg, NE Belgium
Pegasus Bay *129 C6* bay South Island, New Zealand
Pegu *see* Bago
Pehuajó *42 C4* Buenos Aires, E Argentina
Pei-ching *see* Beijing/Beijing Shi
Peine *72 B3* Niedersachsen, C Germany
Pei-p'ing *see* Beijing/Beijing Shi
Peipsi Järv/Peipus-See *see* Peipus, Lake
Peipus, Lake *84 E3* Est. Peipsi Järv, Ger. Peipus-See, Rus. Chudskoye Ozero. lake Estonia/Russia
Peiraiás *83 C6* prev. Piraiévs, Eng. Piraeus. Attikí, C Greece
Pejë *79 D5* Serb. Peć. W Kosovo
Pèk *see* Phônsaven
Pekalongan *116 C4* Jawa, C Indonesia
Pekanbaru *116 B3* var. Pakanbaru. Sumatera, W Indonesia
Pekin *18 B4* Illinois, N USA
Peking *see* Beijing/Beijing Shi
Pelagie *75 B8* island group SW Italy
Pelagosa *see* Palagruža
Pelican State *see* Louisiana
Pelly Bay *see* Kugaaruk
Pélmonostor *see* Beli Manastir
Peloponnese *83 B6* var. Morea, Eng. Peloponnese; anc. Peloponnesus. peninsula S Greece
Peloponnese/Peloponnesus *see* Pelopónnisos
Pematangsiantar *116 B3* Sumatera, W Indonesia
Pemba *57 F2* prev. Port Amelia, Porto Amélia. Cabo Delgado, NE Mozambique
Pemba *51 D7* island E Tanzania
Pembroke *16 D4* Ontario, SE Canada
Penang *see* Pinang, Pulau, Peninsular Malaysia
Penang *see* George Town
Penas, Golfo de *43 A7* gulf S Chile
Penderma *see* Bandırma
Pendleton *24 C3* Oregon, NW USA
Pend Oreille, Lake *24 D2* lake Idaho, NW USA
Peneius *see* Pineiós
Peng-pu *see* Bengbu
Penibético, Sistema *see* Béticos, Sistemas
Peniche *70 B3* Leiria, W Portugal
Peninsular State *see* Florida
Penninae, Alpes/Pennine, Alpi *see* Pennine Alps
Pennine Alps *73 A8* Fr. Alpes Pennines, It. Alpi Pennine, Lat. Alpes Penninae. mountain range Italy/Switzerland
Pennine Chain *see* Pennines
Pennines *67 D5* var. Pennine Chain. mountain range N England, United Kingdom
Pennines, Alpes *see* Pennine Alps
Pennsylvania *19 E4* off. Commonwealth of Pennsylvania, also known as Keystone State. state NE USA
Penobscot River *19 G2* river Maine, NE USA
Penong *127 A6* South Australia
Penonomé *31 F5* Coclé, C Panama
Penrhyn *123 G3* atoll N Cook Islands
Penrhyn Basin *121 F3* undersea feature C Pacific Ocean
Penrith *126 D1* New South Wales, SE Australia
Penrith *67 D5* NW England, United Kingdom
Pensacola *20 C3* Florida, SE USA
Pentecost *122 D4* Fr. Pentecôte. island C Vanuatu
Pentecôte *see* Pentecost
Penza *89 C6* Penzenskaya Oblast', W Russia
Penzance *67 C7* SW England, United Kingdom
Peoria *18 B4* Illinois, N USA
Perchtoldsdorf *73 E6* Niederösterreich, NE Austria
Percival Lakes *124 C4* lakes Western Australia
Perdido, Monte *71 F1* mountain NE Spain
Perece Vela Basin *see* West Mariana Basin
Pereira *36 B3* Risaralda, W Colombia
Peremyshl *see* Przemyśl
Pergamino *42 C4* Buenos Aires, E Argentina
Périgueux *69 C5* anc. Vesuna. Dordogne, SW France
Perito Moreno *43 B6* Santa Cruz, S Argentina
Perlas, Laguna de *31 E3* Eng. Pearl Lagoon. lagoon E Nicaragua
Perleberg *72 C3* Brandenburg, N Germany
Perlepe *see* Prilep
Perm' *92 C3* prev. Molotov. Permskaya Oblast', NW Russia
Pernambuco *41 G2* off. Estado de Pernambuco. state/region E Brazil
Pernambuco *see* Recife
Pernambuco Abyssal Plain *see* Pernambuco Plain
Pernambuco, Estado de *see* Pernambuco
Pernambuco Plain *45 C5* var. Pernambuco Abyssal Plain. undersea feature E Atlantic Ocean
Pernau *see* Pärnu
Pernauer Bucht *see* Pärnu Laht
Pērnava *see* Pärnu
Pernik *82 B2* prev. Dimitrovo. Pernik, W Bulgaria
Pernov *see* Pärnu
Perote *29 F4* Veracruz-Llave, E Mexico
Pérouse *see* Perugia
Perovsk *see* Qyzylorda
Perpignan *69 C6* Pyrénées-Orientales, S France
Perryton *27 F1* Texas, SW USA
Perryville *23 H5* Missouri, C USA
Persia *see* Iran
Persian Gulf *98 C4* var. Gulf, The, Ar. Khalīj al 'Arabī, Per. Khalīj-e Fars. gulf SW Asia
Perth *125 A6* state capital Western Australia
Perth *66 C4* C Scotland, United Kingdom
Perth Basin *119 E6* undersea feature SE Indian Ocean
Peru *38 C3* off. Republic of Peru. country W South America
Peru *see* Beru
Peru Basin *45 A5* undersea feature E Pacific Ocean
Peru-Chile Trench *34 A4* undersea feature E Pacific Ocean
Perugia *74 C4* Fr. Pérouse; anc. Perusia. Umbria, C Italy
Perugia, Lake of *see* Trasimeno, Lago
Peru, Republic of *see* Peru
Perusia *see* Perugia
Péruwelz *65 B6* Hainaut, SW Belgium
Pervomaisk *87 E3* prev. Pervomays'k, Olviopol. Mykolayivs'ka Oblast', S Ukraine
Pervomays'k *see* Pervomaisk
Pervyy Kuril'skiy Proliv *93 H3* strait E Russia
Pesaro *74 C3* anc. Pisaurum. Marche, C Italy
Pescara *74 D4* anc. Aternum, Ostia Aterni. Abruzzo, C Italy
Peshawar *112 C1* North-West Frontier Province, N Pakistan
Peshkopi *79 C6* var. Peshkopia, Peshkopija. Dibër, NE Albania
Peshkopia/Peshkopija *see* Peshkopi
Pessac *69 B5* Gironde, SW France
Petach-Tikva *see* Petah Tikva
Petah Tikva *97 A6* var. Petach-Tikva, Petah Tiqva, Petakh Tikva; prev. Petaẖ Tiqwa. Tel Aviv, C Israel
Petaẖ Tiqwa *see* Petah Tikva
Petakh Tikva/Petah Tiqva *see* Petah Tikva
Pétange *65 D8* Luxembourg, SW Luxembourg
Petchaburi *see* Phetchaburi
Peterborough *127 B6* South Australia
Peterborough *16 D5* Ontario, SE Canada
Peterborough *67 E6* prev. Medeshamstede. E England, United Kingdom
Peterhead *66 D3* NE Scotland, United Kingdom
Peter I Øy *132 A3* Norwegian dependency Antarctica
Petermann Bjerg *61 E3* mountain C Greenland
Petersburg *19 E5* Virginia, NE USA
Peters Mine *37 F3* var. Peter's Mine. N Guyana
Petit St-Bernard, Col du *see* Little Saint Bernard Pass
Peto *29 H4* Yucatán, SE Mexico
Petoskey *18 C2* Michigan, N USA
Petra *see* Wādī Mūsā
Petrich *82 C3* Blagoevgrad, SW Bulgaria
Petrikau *see* Piotrków Trybunalski
Petrikov *see* Pyetrykaw
Petrinja *78 B3* Sisak-Moslavina, C Croatia
Petroaleksandrovsk *see* To'rtkok'l
Petrodvorets *88 A4* Fin. Pietarhovi. Leningradskaya Oblast', NW Russia
Petrograd *see* Sankt-Peterburg
Petrokov *see* Piotrków Trybunalski
Petropavl *see* Petropavlovsk
Petropavlovsk *92 C4* Kaz. Petropavl. Severnyy Kazakhstan, N Kazakhstan
Petropavlovsk-Kamchatskiy *93 H3* Kamchatskaya Oblast', E Russia
Petroşani *86 B4* var. Petroşeni, Ger. Petroschen, Hung. Petrozsény. Hunedoara, W Romania
Petroschen/Petroşeni *see* Petroşani
Petroskoi *see* Petrozavodsk
Petrovgrad *see* Zrenjanin
Petrovsk-Port *see* Makhachkala
Petrozavodsk *92 B2* Fin. Petroskoi. Respublika Kareliya, NW Russia
Petrozsény *see* Petroşani
Pettau *see* Ptuj
Pevek *93 G1* Chukotskiy Avtonomnyy Okrug, NE Russia
Pezinok *77 C6* Ger. Bösing, Hung. Bazin. Bratislavský Kraj, W Slovakia
Pforzheim *73 B6* Baden-Württemberg, SW Germany
Pfungstadt *73 B5* Hessen, W Germany
Phangan, Ko *115 C6* island SW Thailand
Phang-Nga *115 B7* var. Pang-Nga, Phangnga. Phangnga, SW Thailand
Phangnga *see* Phang-Nga
Phan Rang/Phanrang *see* Phan Rang-Thap Cham
Phan Rang-Thap Cham *115 E6* var. Phanrang, Phan Rang, Phan Rang Thap Cham. Ninh Thuận, S Vietnam
Phan Thiêt *115 E6* Binh Thuận, S Vietnam
Pharnacia *see* Giresun
Pharus *see* Hvar
Phatthalung *115 C7* var. Padalung, Patalung. Phatthalung, SW Thailand
Phayao *114 C4* var. Muang Phayao. Phayao, NW Thailand
Phenix City *20 D2* Alabama, S USA
Phet Buri *see* Phetchaburi
Phetchaburi *115 C5* var. Bejraburi, Petchaburi, Phet Buri. Phetchaburi, SW Thailand
Philadelphia *19 F4* Pennsylvania, NE USA
Philadelphia *see* 'Ammān
Philippine Basin *103 F3* undersea feature W Pacific Ocean
Philippine Islands *117 E1* island group W Pacific Ocean
Philippines *117 E1* off. Republic of the Philippines. country SE Asia
Philippine Sea *103 F3* sea W Pacific Ocean
Philippines, Republic of the *see* Philippines
Philippine Trench *120 A1* undersea feature W Philippine Sea
Philippopolis *see* Plovdiv
Phitsanulok *114 C4* var. Bisnulok, Muang Phitsanulok, Pitsanulok. Phitsanulok, C Thailand
Phlórina *see* Flórina
Phnom Penh *115 D6* Khmer. Phnum Pénh. country capital (Cambodia) Phnom Penh, S Cambodia
Phnum Pénh *see* Phnom Penh
Phoenix *26 B2* state capital Arizona, SW USA
Phoenix Islands *123 E3* island group C Kiribati
Phôngsali *114 C3* var. Phong Saly. Phôngsali, N Laos
Phong Saly *see* Phôngsali
Phônsaven *114 D4* var. Xieng Khouang; prev. Pèk, Xiangkhoang. Xiangkhoang, N Laos
Phrae *114 C4* var. Muang Phrae, Prae. Phrae, NW Thailand
Phra Nakhon Si Ayutthaya *see* Ayutthaya
Phra Thong, Ko *115 B6* island SW Thailand
Phuket *115 B7* var. Bhuket, Puket, Mal. Ujung Salang; prev. Junkseylon, Salang. Phuket, SW Thailand
Phuket, Ko *115 B7* island SW Thailand
Phumĭ Kâmpóng Trâbêk *see* Kâmpóng Trâbêk
Phumĭ Sâmraông *see* Samraong
Phum Kompong Trabek *see* Kâmpóng Trâbêk
Phum Samrong *see* Samraong
Phu Vinh *see* Tra Vinh
Phyu *114 B4* var. Hpyu, Pyu. Bago, C Myanmar (Burma)
Piacenza *74 B2* Fr. Paisance; anc. Placentia. Emilia-Romagna, N Italy
Piatra-Neamţ *86 C4* Hung. Karácsonkő. Neamţ, NE Romania
Piatykhatky *87 F3* prev. P"yatykhatky, Rus. Pyatikhatki. Dnipropetrovska Oblast, E Ukraine
Piauhy *see* Piauí
Piauí *41 F2* off. Estado do Piauí; prev. Piauhy. state/region E Brazil
Piauí, Estado do *see* Piauí
Picardie *68 C3* Eng. Picardy. cultural region N France
Picardy *see* Picardie
Piccolo San Bernardo, Colle di *see* Little Saint Bernard Pass
Pichilemu *42 B4* Libertador, C Chile
Pico *70 A5* var. Ilha do Pico. island Azores, Portugal, NE Atlantic Ocean
Pico, Ilha do *see* Pico
Picos *41 F2* Piauí, E Brazil
Picton *129 C5* Marlborough, South Island, New Zealand
Piedmont *see* Piemonte
Piedras Negras *29 E2* var. Ciudad Porfirio Díaz. Coahuila, NE Mexico
Pielavesi *62 D4* lake C Finland
Pielinen *62 E4* var. Pielisjärvi. lake E Finland
Pielisjärvi *see* Pielinen
Piemonte *74 A2* Eng. Piedmont. region NW Italy
Pierre *23 E3* state capital South Dakota, N USA
Piešt'any *77 C6* Ger. Pistyan, Hung. Pöstyén. Tranavský Kraj, W Slovakia
Pietarhovi *see* Petrodvorets
Pietari *see* Sankt-Peterburg
Pietarsaari *see* Jakobstad
Pietermaritzburg *56 C5* var. Maritzburg. KwaZulu/Natal, E South Africa
Pietersburg *see* Polokwane
Pigs, Bay of *see* Cochinos, Bahía de
Pihkva Järv *see* Pskov, Lake
Pijijiapan *29 G5* Chiapas, SE Mexico
Pikes Peak *22 C5* mountain Colorado, C USA
Pikeville *18 D5* Kentucky, S USA
Pikinni *see* Bikini Atoll
Piła *76 B3* Ger. Schneidemühl. Wielkopolskie, C Poland
Pilar *42 D3* var. Villa del Pilar. Ñeembucú, S Paraguay
Pilcomayo, Rio *35 C5* river C South America
Pilos *see* Pýlos
Pilsen *see* Plzeň
Pilzno *see* Plzeň
Pinang *see* Pinang, Pulau, Peninsular Malaysia
Pinang *see* George Town
Pinang, Pulau *116 B3* var. Penang, Pinang; prev. Prince of Wales Island. island Peninsular Malaysia
Pinar del Río *32 A2* Pinar del Río, W Cuba
Píndhos/Píndhos Óros *see* Píndos
Píndos *82 A4* var. Píndhos Óros, Eng. Pindus Mountains; prev. Píndhos. mountain range C Greece
Pindus Mountains *see* Píndos
Pine Bluff *20 B2* Arkansas, C USA
Pine Creek *124 D2* Northern Territory, N Australia
Pinega *88 C3* river NW Russia
Pineiós *82 B4* var. Piniós; anc. Peneius. river C Greece
Pineland *27 H3* Texas, SW USA
Pínes, Akrotírio *82 C4* var. Akrotírio Pínnes. headland N Greece
Pines, The Isle of the *see* Juventud, Isla de la
Pine Tree State *see* Maine
Pingdingshan *106 C4* Henan, C China
Pingkiang *see* Harbin
Ping, Mae Nam *114 B4* river W Thailand
Piniós *see* Pineiós
Pinkiang *see* Harbin
Pínnes, Akrotírio *see* Pínes, Akrotírio
Pinos, Isla de *see* Juventud, Isla de la
Pinotepa Nacional *29 F5* var. Santiago Pinotepa Nacional. Oaxaca, SE Mexico
Pinsk *85 B7* Pol. Pińsk. Brestskaya Voblasts', SW Belarus
Pinta, Isla *38 A5* var. Abingdon. island Galápagos Islands, Ecuador, E Pacific Ocean
Piombino *74 B3* Toscana, C Italy
Pioneer Mountains *24 D3* mountain range Montana, N USA North America
Pionerskiy *84 A4* Ger. Neukuhren. Kaliningradskaya Oblast', W Russia
Piopiotahi *see* Milford Sound
Piotrków Trybunalski *76 D4* Ger. Petrikau, Rus. Petrokov. Łodzkie, C Poland
Piraeus/Piraiévs *see* Peiraías
Pírgos *see* Pýrgos
Pirineos *see* Pyrenees
Piripiri *41 F2* Piauí, E Brazil
Pirna *72 D4* Sachsen, E Germany
Pirot *79 E5* Serbia, SE Serbia
Piryatin *see* Pyriatyn
Pisa *74 B3* var. Pisae. Toscana, C Italy
Pisae *see* Pisa
Pisaurum *see* Pesaro
Pisco *38 D4* Ica, SW Peru
Písek *77 A5* Budějovický Kraj, S Czechia (Czech Republic)
Pishan *104 A3* var. Guma. Xinjiang Uygur Zizhiqu, NW China
Pishpek *see* Bishkek
Pistoia *74 B3* anc. Pistoria, Pistoriæ. Toscana, C Italy
Pistoria/Pistoriæ *see* Pistoia
Pistyan *see* Piešt'any
Pisz *76 D3* Ger. Johannisburg. Warmińsko-Mazurskie, NE Poland
Pita *52 C4* NW Guinea
Pitalito *36 B4* Huila, S Colombia
Pitcairn Group of Islands *see* Pitcairn, Henderson, Ducie & Oeno Islands
Pitcairn Island *121 G4* island Pitcairn, Henderson, Ducie & Oeno Islands
Pitcairn, Henderson, Ducie & Oeno Islands *121 G4* var. Pitcairn Group of Islands. UK Overseas Territory C Pacific Ocean
Piteå *62 D4* Norrbotten, N Sweden
Piteşti *86 B5* Argeş, S Romania
Pitsanulok *see* Phitsanulok
Pitt Island *see* Makin
Pittsburg *23 F5* Kansas, C USA
Pittsburgh *19 E4* Pennsylvania, NE USA
Pittsfield *19 F3* Massachusetts, NE USA
Piura *38 B2* Piura, NW Peru
Pivdennyi Buh *87 E3* prev. Pivdennyy Buh, Rus. Yuzhnyy Bug. river S Ukraine
Pivdennyy Buh *see* Pivdennyi Buh
Placentia *see* Piacenza
Placetas *32 B2* Villa Clara, C Cuba
Plainview *27 E2* Texas, SW USA
Pláka *83 C7* Kykládes, Greece, Aegean Sea
Planeta Rica *36 B2* Córdoba, NW Colombia
Planken *72 E1* C Liechtenstein Europe
Plano *27 G2* Texas, SW USA
Plasencia *70 C3* Extremadura, W Spain
Plate, River *42 D4* var. River Plate. estuary Argentina/Uruguay
Plate, River *see* Plata, Río de la
Platinum *14 C3* Alaska, USA
Plattensee *see* Balaton
Platte River *23 E4* river Nebraska, C USA
Plattsburgh *19 F2* New York, NE USA
Plauen *73 C5* var. Plauen im Vogtland. Sachsen, E Germany
Plauen im Vogtland *see* Plauen
Pļaviņas *84 D4* Ger. Stockmannshof. Aizkraukle, S Latvia
Plây Cu *115 E5* var. Pleiku. Gia Lai, C Vietnam
Pleasant Island *see* Nauru
Pleiku *see* Plây Cu
Plenty, Bay of *128 E3* bay North Island, New Zealand
Plérin *68 A3* Côtes d'Armor, NW France
Plesetsk *88 C3* Arkhangel'skaya Oblast', NW Russia
Pleshchenitsy *see* Plyeshchanitsy
Pleskau *see* Pskov
Pleskauer See *see* Pskov, Lake
Pleskava *see* Pskov
Pleszew *76 C4* Wielkopolskie, C Poland
Pleven *82 C2* prev. Plevna. Pleven, N Bulgaria
Plevlja/Plevlje *see* Pljevlja
Plevna *see* Pleven
Pljevlja *78 C4* prev. Plevlja, Plevlje. N Montenegro
Plocce *see* Ploče
Ploče *78 B4* It. Plocce; prev. Kardeljevo. Dubrovnik-Neretva, SE Croatia
Płock *76 D3* Ger. Plozk. Mazowieckie, C Poland
Plöcken Pass *73 C7* Ger. Plöckenpass, It. Passo di Monte Croce Carnico. pass SW Austria
Plöckenpass *see* Plöcken Pass
Ploeşti *see* Ploieşti
Ploieşti *86 C5* prev. Ploeşti. Prahova, SE Romania
Plomári *83 D5* prev. Plomárion. Lésvos, E Greece
Plomárion *see* Plomári
Płońsk *76 D3* Mazowieckie, C Poland
Plovdiv *82 C3* prev. Eumolpias; anc. Evmolpia, Philippopolis, Lat. Trimontium. Plovdiv, C Bulgaria
Plozk *see* Płock
Plungė *84 B3* Telšiai, W Lithuania
Plyeshchanitsy *85 D5* Rus. Pleshchenitsy. Minskaya Voblasts', N Belarus
Plymouth *67 C7* SW England, United Kingdom
Plzeň *77 A5* Ger. Pilsen, Pol. Pilzno. Plzeňský Kraj, W Czechia (Czech Republic)
Po *58 D4* river N Italy
Pobedy, Pik *see* Tömür Feng
Po, Bocche del *see* Po, Foci del

Pocahontas 20 B1 Arkansas, C USA
Pocatello 24 E4 Idaho, NW USA
Pochinok 89 A5 Smolenskaya Oblast', W Russia
Pocking 73 D6 Bayern, SE Germany
Poděbrady 77 B5 Ger. Podiebrad. Středočeský Kraj, C Czechia (Czech Republic)
Podgorica 79 C5 prev. Titograd. country capital (Montenegro) S Montenegro
Podiebrad see Poděbrady
Podilsk 86 D3 prev. Kotovsk. Odes'ka Oblast', SW Ukraine
Podilska Vysochyna 86 D3 prev. Podil's'ka Vysochina. plateau W Ukraine
Podil's'ka Vysochina see Podilska Vysochyna
Podium Anicensis see le Puy
Podol'sk 89 B5 Moskovskaya Oblast', W Russia
Podravska Slatina see Slatina
Podujevë 79 D5 Serb. Podujevo. N Kosovo
Podujevo see Podujevë
Podunajská Rovina see Little Alföld
Poetovio see Ptuj
Pogradec 79 D6 var. Pogradeci. Korçë, SE Albania
Pogradeci see Pogradec
Pohjanlahti see Bothnia, Gulf of
Pohnpei 122 C2 prev. Ponape Ascension Island. island E Micronesia
Poictiers see Poitiers
Poinsett, Cape 132 D4 headland Antarctica
Point de Galle see Galle
Pointe-à-Pitre 33 G3 Grande Terre, C Guadeloupe
Pointe-Noire 55 B6 Kouilou, S Congo
Point Lay 14 C2 Alaska, USA
Poitiers 68 B4 prev. Poictiers; anc. Limonum. Vienne, W France
Poitou 68 B4 cultural region W France
Pokharā 113 E3 Western, C Nepal
Pokrov 87 F3 Rus. Ordzhonikidze. Dnipropetrovska Oblast, E Ukraine
Pokrovka see Kyzyl-Suu
Pokrovske 87 G3 Rus. Pokrovskoye. Dnipropetrovska Oblast, E Ukraine
Pokrovskoye see Pokrovske
Pola see Pula
Pola de Lena see La Pola
Poland 76 B4 off. Republic of Poland, var. Polish Republic, Pol. Polska, Rzeczpospolita Polska; prev. Pol. Polska Rzeczpospolita Ludowa, The Polish People's Republic. country C Europe
Poland, Republic of see Poland
Polatlı 94 C3 Ankara, C Turkey (Türkiye)
Polatsk 85 D5 Rus. Polotsk. Vitsyebskaya Voblasts', N Belarus
Pol-e Khomrī see Pul-e Khumrī
Poli see Pólis
Polikastro/Políkastron see Polýkastro
Polikrayshte see Dolna Oryahovitsa
Pólis 80 C5 var. Poli. W Cyprus
Polish People's Republic, The see Poland
Polish Republic see Poland
Polkowice 76 B4 Ger. Heerwegen. Dolnośląskie, W Poland
Pollença 71 G3 Mallorca, Spain, W Mediterranean Sea
Pologi see Polohy
Polohy 87 G3 Rus. Pologi. Zaporiz'ka Oblast', SE Ukraine
Polokwane 56 D4 prev. Pietersburg. Limpopo, NE South Africa
Polonne 86 D2 Rus. Polonnoye. Khmel'nyts'ka Oblast', NW Ukraine
Polonnoye see Polonne
Polotsk see Polatsk
Polska/Polska, Rzeczpospolita/Polska Rzeczpospolita Ludowa see Poland
Polski Trambesh 82 D2 prev. Polsko Kosovo, var. Polski Trŭmbesh. Ruse, N Bulgari
Polski Trŭmbesh see Polski Trambesh
Polsko Kosovo see Polski Trambesh
Poltava 87 F2 Poltavs'ka Oblast', NE Ukraine
Poltoratsk see Aşgabat
Põlva 84 E3 Ger. Pölwe. Põlvamaa, SE Estonia
Pölwe see Põlva
Polyarnyy 88 C2 Murmanskaya Oblast', NW Russia
Polýkastro 82 B3 var. Polikastro; prev. Políkastron. Kentrikí Makedonía, N Greece
Polynesia 121 F4 island group C Pacific Ocean
Pomeranian Bay 72 D2 Ger. Pommersche Bucht, Pol. Zatoka Pomorska. bay Germany/Poland
Pomir, Dar"yoi see Pamir/Pāmīr, Daryā-ye
Pommersche Bucht see Pomeranian Bay
Pomorska, Zatoka see Pomeranian Bay
Pomorskiy Proliv 88 D2 strait NW Russia
Po, Mouth of the 74 C2 var. Bocche del Po. river NE Italy
Pompaelo see Pamplona
Pompano Beach 21 F5 Florida, SE USA
Ponape Ascension Island see Pohnpei
Ponca City 27 G1 Oklahoma, C USA
Ponce 33 F3 C Puerto Rico
Pondicherry see Puducherry
Pondichéry see Puducherry
Ponferrada 70 C1 Castilla y León, NW Spain
Poniatowa 76 E4 Lubelskie, E Poland
Pons Aelii see Newcastle upon Tyne
Pons Vetus see Pontevedra
Ponta Delgada 70 B5 São Miguel, Azores, Portugal, NE Atlantic Ocean
Ponta Grossa 41 E4 Paraná, S Brazil
Pontarlier 68 D4 Doubs, E France
Ponteareas 70 B2 Galicia, NW Spain
Ponte da Barca 70 B2 Viana do Castelo, N Portugal
Pontevedra 70 B1 anc. Pons Vetus. Galicia, NW Spain
Pontiac 18 D3 Michigan, N USA
Pontianak 116 C4 Borneo, C Indonesia
Pontisarae see Pontoise
Pontivy 68 A3 Morbihan, NW France
Pontoise 68 C3 anc. Briva Isarae, Cergy-Pontoise, Pontisarae. Val-d'Oise, N France
Ponziane Island 75 C5 island C Italy
Poole 67 D7 S England, United Kingdom
Poona see Pune
Poopó, Lago 39 F4 var. Lago Pampa Aullagas. lake W Bolivia
Popayán 36 B4 Cauca, SW Colombia
Poperinge 65 A6 West-Vlaanderen, W Belgium
Poplar Bluff 23 G5 Missouri, C USA
Popocatépetl 29 E4 volcano S Mexico
Popper see Poprad
Poprad 77 D5 Ger. Deutschendorf, Hung. Poprád. Prešovský Kraj, E Slovakia
Poprád 77 D5 Ger. Popper, Hung. Poprád. river Poland/Slovakia
Porbandar 112 B4 Gujarāt, W India
Porcupine Plain 58 B3 undersea feature E Atlantic Ocean
Pordenone 74 C2 anc. Portenau. Friuli-Venezia Giulia, NE Italy
Poreč 78 A2 It. Parenzo. Istra, NW Croatia
Porech'ye see Parechcha
Pori 63 D5 Swe. Björneborg. Satakunta, SW Finland
Porirua 129 D5 Wellington, North Island, New Zealand
Porkhov 88 A4 Pskovskaya Oblast', W Russia
Porlamar 37 E1 Nueva Esparta, NE Venezuela
Póros 83 C6 Póros, S Greece
Póros 83 A5 Kefallinía, Iónia Nisiá, Greece, C Mediterranean Sea
Pors see Porsangenfijorden
Porsangenfijorden 62 D2 Lapp. Pors. fjord N Norway
Porsgrunn 63 B6 Telemark, S Norway
Portachuelo 39 G4 Santa Cruz, C Bolivia
Portadown 67 B5 Ir. Port An Dúnáin. S Northern Ireland, United Kingdom
Portalegre 70 C3 anc. Ammaia, Amoea. Portalegre, E Portugal
Port Alexander 14 D4 Baranof Island, Alaska, USA
Port Alfred 56 D5 Eastern Cape, S South Africa
Port Amelia see Pemba
Port An Dúnáin see Portadown
Port Angeles 24 B1 Washington, NW USA
Port Antonio 32 B5 NE Jamaica
Port Arthur 27 H4 Texas, SW USA
Port Augusta 127 B6 South Australia
Port-au-Prince 32 D3 var. Pòtoprens. country capital (Haiti) C Haiti
Port Blair 111 F2 Andaman and Nicobar Islands, SE India
Port Charlotte 21 E4 Florida, SE USA
Port Darwin see Darwin
Port d'Envalira 69 B8 E Andorra Europe
Port Douglas 126 D3 Queensland, NE Australia
Port Elizabeth see Gqeberha
Portenau see Pordenone
Porterville 25 C7 California, W USA
Port-Étienne see Nouâdhibou
Port Florence see Kisumu
Port-Francqui see Ilebo
Port-Gentil 55 A6 Ogooué-Maritime, W Gabon
Port Harcourt 53 G5 Rivers, S Nigeria
Port Hardy 14 D5 Vancouver Island, British Columbia, SW Canada
Port Harrison see Inukjuak
Port Hedland 124 B4 Western Australia
Port Huron 18 D3 Michigan, N USA
Portimão 70 B4 var. Vila Nova de Portimão. Faro, S Portugal
Port Jackson 126 E1 harbour New South Wales, E Australia
Portland 127 B7 Victoria, SE Australia
Portland 19 G2 Maine, NE USA
Portland 24 B3 Oregon, NW USA
Portland 27 G4 Texas, SW USA
Portland Bight 32 B5 bay S Jamaica
Portlaoighise see Port Laoise
Port Laoise 67 B6 var. Portlaoise, Ir. Portlaoighise; prev. Maryborough. C Ireland
Portlaoise see Port Laoise
Port Lavaca 27 G4 Texas, SW USA
Port Lincoln 127 A6 South Australia
Port Louis 57 H3 country capital (Mauritius) NW Mauritius
Port-Lyautey see Kénitra
Port Macquarie 127 E6 New South Wales, SE Australia
Port Mahon see Mahón
Portmore 32 B5 C Jamaica
Port Moresby 122 B3 country capital (Papua New Guinea) Central/National Capital District, SW Papua New Guinea
Port Natal see Durban
Porto 70 B2 Eng. Oporto; anc. Portus Cale. Porto, NW Portugal
Porto Alegre 41 E5 var. Pôrto Alegre. state capital Rio Grande do Sul, S Brazil
Porto Alegre 54 E2 São Tomé, S Sao Tome and Principe, Africa
Porto Alexandre see Tombua
Porto Amélia see Pemba
Porto Bello see Portobelo
Portobelo 31 G4 var. Porto Bello, Puerto Bello. Colón, N Panama
Port O'Connor 27 G4 Texas, SW USA
Porto Edda see Sarandë
Portoferraio 74 B4 Toscana, C Italy
Port of Spain 33 H5 country capital (Trinidad and Tobago) Trinidad, Trinidad and Tobago
Porto Grande see Mindelo
Portogruaro 74 C2 Veneto, NE Italy
Porto-Novo 53 F5 country capital (Benin - official) S Benin
Porto Rico see Puerto Rico
Porto Santo 48 A2 var. Ilha do Porto Santo. island Madeira, Portugal, NE Atlantic Ocean
Porto Santo, Ilha do see Porto Santo
Porto Torres 75 A5 Sardegna, Italy, C Mediterranean Sea
Porto Velho 40 D2 var. Velho. state capital Rondônia, W Brazil
Portoviejo 38 A2 var. Puertoviejo. Manabí, W Ecuador
Port Pirie 127 B6 South Australia
Port Rex see East London
Port Said see Būr Saʻīd
Portsmouth 67 D7 S England, United Kingdom
Portsmouth 19 G3 New Hampshire, NE USA
Portsmouth 18 D4 Ohio, N USA
Portsmouth 19 F5 Virginia, NE USA
Port Stanley see Stanley
Port Sudan 50 C3 Red Sea, NE Sudan
Port Swettenham see Klang/Pelabuhan Klang
Port Talbot 67 C7 S Wales, United Kingdom
Portugal 70 B3 off. Portuguese Republic. country SW Europe
Portuguese East Africa see Mozambique
Portuguese Guinea see Guinea-Bissau
Portuguese Republic see Portugal
Portuguese Timor see East Timor
Portuguese West Africa see Angola
Portus Cale see Porto
Portus Magnus see Almería
Portus Magonis see Mahón
Port-Vila 122 D4 var. Vila. country capital (Vanuatu) Éfaté, C Vanuatu
Porvenir 39 E3 Pando, NW Bolivia
Porvenir 43 B8 Magallanes, S Chile
Porvoo 63 E6 Swe. Borgå. Uusimaa, S Finland
Porzecze see Parechcha
Posadas 42 D3 Misiones, NE Argentina
Posad-Pokrovske 87 E4 Khersonska Oblast, S Ukraine
Poschega see Požega
Posen see Poznań
Posnania see Poznań
Postavy/Postawy see Pastavy
Posterholt 65 D5 Limburg, SE Netherlands
Postojna 73 D8 Ger. Adelsberg, It. Postumia. SW Slovenia
Postumia see Postojna
Pöstyén see Piešťany
Potamós 83 C7 Antikýthira, S Greece
Potentia see Potenza
Potenza 75 D5 anc. Potentia. Basilicata, S Italy
Poti 95 F2 prev. P'ot'i. W Georgia
P'ot'i see Poti
Potiskum 53 G4 Yobe, NE Nigeria
Potomac River 19 E5 river NE USA
Pòtoprens see Port-au-Prince
Potosí 39 F4 Potosí, S Bolivia
Potsdam 72 D3 Brandenburg, NE Germany
Potwar Plateau 112 C2 plateau NE Pakistan
Poŭthĭsăt see Pursat
Po, Valle del see Po Valley
Po Valley 74 C2 It. Valle del Po. valley N Italy
Považská Bystrica 77 C5 Ger. Waagbistritz, Hung. Vágbeszterce. Trenčiansky Kraj, W Slovakia
Poverty Bay see Tūranganui-a-Kiwa
Póvoa de Varzim 70 B2 Porto, NW Portugal
Powder River 22 D2 river Montana/Wyoming, NW USA
Powell 22 C2 Wyoming, C USA
Powell, Lake 22 B5 lake Utah, W USA
Požarevac 78 D4 Ger. Passarowitz. Serbia, NE Serbia
Poza Rica 29 F4 var. Poza Rica de Hidalgo. Veracruz-Llave, E Mexico
Poza Rica de Hidalgo see Poza Rica
Požega 78 D4 prev. Slavonska Požega, Ger. Poschega, Hung. Pozsega. Požega-Slavonija, NE Croatia
Požega 78 D4 Serbia
Poznań 76 C3 Ger. Posen, Posnania. Wielkolpolskie, C Poland
Pozoblanco 70 D4 Andalucía, S Spain
Pozsega see Požega
Pozsony see Bratislava
Pozzallo 75 C8 Sicilia, Italy, C Mediterranean Sea
Prachatice 77 A5 Ger. Prachatitz. Jihočeský Kraj, S Czechia (Czech Republic)
Prachatitz see Prachatice
Prado del Ganso see Goose Green
Prae see Phrae
Prag/Praga/Prague see Praha
Praha 77 A5 Eng. Prague, Ger. Prag, Pol. Praga. country capital (Czechia (Czech Republic)) Středočeský Kraj, NW Czechia (Czech Republic)
Praia 52 A3 country capital (Cape Verde) Santiago, S Cape Verde (Cabo Verde)
Prairie State see Illinois
Prathet Thai see Thailand
Prato 74 B3 Toscana, C Italy
Pratt 23 E5 Kansas, C USA
Prattville 20 D2 Alabama, S USA
Pravda see Glavinitsa
Pravia 70 C1 Asturias, N Spain
Prayagraj 113 E3 prev. Allahābād. Uttar Pradesh, N India
Preăh Seihânŭ see Sihanoukville
Preny see Prienai
Prenzlau 72 D3 Brandenburg, NE Germany
Prerau see Přerov
Přerov 77 C5 Ger. Prerau. Olomoucký Kraj, E Czechia (Czech Republic)
Preschau see Prešov
Prescott 26 B2 Arizona, SW USA
Preševo 79 D5 Serbia, SE Serbia
Presidente Epitácio 41 E4 São Paulo, S Brazil
Presidente Stroessner see Ciudad del Este
Prešov 77 D5 var. Preschau, Ger. Eperies, Hung. Eperjes. Prešovský Kraj, E Slovakia
Prespa, Lake 79 D6 Alb. Liqen i Prespës, Gk. Límni Megáli Préspa, Limni Prespa, Mac. Prespansko Ezero, Serb. Prespansko Jezero. lake SE Europe
Prespa, Limni/Prespansko Ezero/Prespansko Jezero/Prespës, Liqen i see Prespa, Lake
Presque Isle 19 H1 Maine, NE USA
Pressburg see Bratislava
Preston 67 D5 NW England, United Kingdom
Prestwick 66 C4 W Scotland, United Kingdom
Pretoria 56 D4 var. Epitoli. country capital (South Africa-administrative capital) Gauteng, NE South Africa
Preussisch Eylau see Bagrationovsk
Preußisch Holland see Pasłęk
Preussisch-Stargard see Starogard Gdański
Préveza 83 A5 Ípeiros, W Greece
Pribilof Islands 14 A3 island group Alaska, USA
Priboj 78 C4 Serbia, W Serbia
Price 22 B4 Utah, W USA
Prichard 20 C3 Alabama, S USA
Priekulė 84 B3 Ger. Prökuls. Klaipėda, W Lithuania
Prienai 85 B5 Pol. Preny. Kaunas, S Lithuania
Prieska 56 C4 Northern Cape, C South Africa
Prijedor 78 B3 Republika Srpska, NW Bosnia and Herzegovina
Prijepolje 78 D4 Serbia, W Serbia
Prikaspiyskaya Nizmennost' see Caspian Depression
Prilep 79 D6 Turk. Perlepe. S North Macedonia
Priluki see Pryluky
Primorsk 84 A4 Ger. Fischhausen. Kaliningradskaya Oblast', W Russia
Primorsko 82 E2 prev. Keupriya. Burgas, E Bulgaria
Primorskoye see Prymorsk
Prince Albert 15 F5 Saskatchewan, S Canada
Prince Edward Island 17 F4 Fr. Île-du Prince-Édouard. province SE Canada
Prince Edward Islands 47 E8 island group S South Africa
Prince George 15 E5 British Columbia, SW Canada
Prince of Wales Island 126 B1 island Queensland, E Australia
Prince of Wales Island 15 F2 island Queen Elizabeth Islands, Nunavut, NW Canada
Prince of Wales Island see Pinang, Pulau
Prince Patrick Island 15 E2 island Parry Islands, Northwest Territories, NW Canada
Prince Rupert 14 D4 British Columbia, SW Canada
Prince's Island see Príncipe
Princess Charlotte Bay 126 C2 bay Queensland, NE Australia
Princess Elizabeth Land 132 C3 physical region Antarctica
Príncipe 55 A5 var. Príncipe Island, Eng. Prince's Island. island N Sao Tome and Principe
Príncipe Island see Príncipe
Prinzapolka 31 E3 Región Autónoma Atlántico Norte, NE Nicaragua
Pripet see Pripyat
Pripiat see Pripyat
Pripyat 85 C7 var. Pripet, Bel. Prypyats', Ukr. Prypiat. river Belarus/Ukraine
Pripyat Marshes 85 B7 Bel. Prypyatskiya baloty, Ukr. Prypiatski bolota. wetland Belarus/Ukraine
Prishtinë 79 D5 Eng. Pristina, Serb. Priština. country capital (Kosovo) C Kosovo
Pristina/Priština see Prishtinë
Privas 69 D5 Ardèche, E France
Privolzhskaya Vozvyshennost' 59 G3 var. Volga Uplands. mountain range W Russia
Prizren 79 D5 S Kosovo
Probolinggo 116 D5 Jawa, C Indonesia
Probstberg see Wyszków
Progreso 29 H3 Yucatán, SE Mexico
Prokhladnyy 89 B8 Kabardino-Balkarskaya Respublika, SW Russia
Prokletije see North Albanian Alps
Prökuls see Priekulė
Prokuplje 79 D5 Serbia, SE Serbia
Prome see Pyay
Promyshlennyy 88 E3 Respublika Komi, NW Russia
Prościejów see Prostějov
Proskurov see Khmelnytskyi
Prossnitz see Prostějov
Prostějov 77 C5 Ger. Prossnitz, Pol. Prościejów. Olomoucký Kraj, E Czechia (Czech Republic)
Provence 69 D6 cultural region SE France
Providence 19 G3 state capital Rhode Island, NE USA
Providence see Fort Providence
Providencia, Isla de 31 F3 island NW Colombia, Caribbean Sea
Provideniya 133 B1 Chukotskiy Avtonomnyy Okrug, NE Russia
Provo 22 B4 Utah, W USA
Prudhoe Bay 14 D2 Alaska, USA
Prusa see Bursa
Pruszków 76 D3 Ger. Kaltdorf. Mazowieckie, C Poland
Prut 86 D4 Ger. Pruth. river E Europe
Pruth see Prut
Pružana see Pruzhany
Pruzhany 85 B6 Pol. Prużana. Brestskaya Voblasts', SW Belarus
Prychornomor'ska Nyzovyna see Black Sea Lowland
Prydniprovska Nyzovyna/Prydnyaprowskaya Nizina see Dnieper Lowland
Prydz Bay 132 D3 bay Antarctica
Pryluky 87 E2 Rus. Priluki. Chernihivs'ka Oblast', NE Ukraine
Prymorsk 87 G4 prev. Primorskoye. Zaporiz'ka Oblast', SE Ukraine
Prypiatski bolota see Pripyat Marshes
Pryp"yat'/Prypyats' see Pripet
Prypyatskiya baloty see Pripyat Marshes
Przemyśl 77 E5 Rus. Peremyshl. Podkarpackie, C Poland
Przheval'sk see Karakol
Psará 83 D5 island E Greece
Psel 87 F2 Rus. Psël. river Russia/Ukraine
Psël see Psel
Pskov 92 B2 Ger. Pleskau, Latv. Pleskava. Pskovskaya Oblast', W Russia
Pskov, Lake 84 E3 Est. Pihkva Järv, Ger. Pleskauer See, Rus. Pskovskoye Ozero. lake Estonia/Russia
Pskovskoye Ozero see Pskov, Lake
Ptich' see Ptsich
Ptsich 85 C7 Rus. Ptich'. Homyel'skaya Voblasts', SE Belarus
Ptsich 85 C7 Rus. Ptich'. river SE Belarus
Ptuj 73 E7 Ger. Pettau; anc. Poetovio. NE Slovenia
Pucallpa 38 C3 Ucayali, C Peru
Puck 76 C2 Pomorskie, N Poland
Pudasjärvi 62 D4 Pohjois-Pohjanmaa, C Finland
Puducheri see Puducherry
Puducherry 110 C2 prev. Pondicherry; var. Puduchcheri, Fr. Pondichéry. Pondicherry, SE India
Puebla 29 F4 var. Puebla de Zaragoza. Puebla, S Mexico
Puebla de Zaragoza see Puebla
Pueblo 22 D5 Colorado, C USA
Puerto Acosta 39 E4 La Paz, W Bolivia
Puerto Aisén 43 B6 Aisén, S Chile
Puerto Ángel 29 F5 Oaxaca, SE Mexico
Puerto Argentino see Stanley
Puerto Ayacucho 36 D3 Amazonas, SW Venezuela
Puerto Baquerizo Moreno 38 B5 var. Baquerizo Moreno. Galápagos Islands, Ecuador, E Pacific Ocean
Puerto Barrios 30 C2 Izabal, E Guatemala
Puerto Bello see Portobelo
Puerto Berrío 36 B2 Antioquia, C Colombia
Puerto Cabello 36 D1 Carabobo, N Venezuela
Puerto Cabezas 31 E2 var. Bilwi. Región Autónoma Atlántico Norte, NE Nicaragua
Puerto Carreño 36 D3 Vichada, E Colombia
Puerto Cortés 30 C2 Cortés, NW Honduras
Puerto Cumarebo 36 C1 Falcón, N Venezuela
Puerto Deseado 43 C7 Santa Cruz, SE Argentina
Puerto Escondido 29 F5 Oaxaca, SE Mexico
Puerto Francisco de Orellana 38 B1 var. Coca. NE Ecuador
Puerto Gallegos see Río Gallegos
Puerto Inírida 36 D3 var. Obando. Guainía, E Colombia
Puerto La Cruz 37 E1 Anzoátegui, NE Venezuela
Puerto Lempira 31 E2 Gracias a Dios, E Honduras
Puerto Limón see Limón
Puertollano 70 D4 Castilla-La Mancha, C Spain
Puerto López 36 C1 La Guajira, N Colombia
Puerto Maldonado 39 E3 Madre de Dios, E Peru
Puerto México see Coatzacoalcos
Puerto Montt 43 B5 Los Lagos, C Chile
Puerto Natales 43 B7 Magallanes, S Chile
Puerto Obaldía 31 H5 Kuna Yala, NE Panama
Puerto Plata 33 E3 var. San Felipe de Puerto Plata. N Dominican Republic
Puerto Presidente Stroessner see Ciudad del Este
Puerto Princesa 117 E2 off. Puerto Princesa City. Palawan, W Philippines
Puerto Princesa City see Puerto Princesa
Puerto Príncipe see Camagüey
Puerto Rico 33 F3 off. Commonwealth of Puerto Rico; prev. Porto Rico. US commonwealth territory C West Indies
Puerto Rico 34 B1 island C West Indies
Puerto Rico, Commonwealth of see Puerto Rico
Puerto Rico Trench 34 B1 trench NE Caribbean Sea
Puerto San José see San José
Puerto San Julián 43 B7 var. San Julián. Santa Cruz, SE Argentina
Puerto Suárez 39 H4 Santa Cruz, E Bolivia
Puerto Vallarta 28 D4 Jalisco, SW Mexico
Puerto Varas 43 B5 Los Lagos, C Chile
Puerto Viejo 31 E4 Heredia, NE Costa Rica
Puertoviejo see Portoviejo
Puget Sound 24 B1 sound Washington, NW USA
Puglia 75 E5 var. Le Puglie, Eng. Apulia. region SE Italy
Puhiwaero see South West Cape
Pukaki, Lake 129 B6 lake South Island, New Zealand
Pukekohe 128 D3 Auckland, North Island, New Zealand
Puket see Phuket
Pukhavichy 85 C6 Rus. Pukhovichi. Minskaya Voblasts', C Belarus
Pukhovichi see Pukhavichy
Pula 78 A3 It. Pola; prev. Pulj. Istra, NW Croatia
Pulaski 18 D5 Virginia, NE USA
Puławy 76 D4 Ger. Neu Amerika. Lubelskie, E Poland
Pul-e Khumrī 101 E4 prev. Pol-e Khomrī, var. Pul-i-Khumri. Baghlān, NE Afghanistan
Pul-i-Khumri see Pul-e Khumrī
Pulj see Pula
Pullman 24 C2 Washington, NW USA
Pułtusk 76 D3 Mazowieckie, C Poland
Puná, Isla 38 A2 island SW Ecuador
Pune 112 C5 prev. Poona. Mahārāshtra, W India
Punjab 112 C2 prev. West Punjab, Western Punjab. province E Pakistan
Puno 39 E4 Puno, SE Peru
Punta Alta 43 C5 Buenos Aires, E Argentina
Punta Arenas 43 B8 prev. Magallanes. Magallanes, S Chile
Punta Gorda 30 C2 Toledo, SE Belize
Punta Gorda 31 E4 Región Autónoma Atlántico Sur, SE Nicaragua
Puntarenas 30 D4 Puntarenas, W Costa Rica
Punto Fijo 36 C1 Falcón, N Venezuela
Pupuya, Nevado 39 E4 mountain W Bolivia
Puri 113 F5 var. Jagannath. Odisha, E India
Puriramya see Buriram
Purmerend 64 C3 Noord-Holland, C Netherlands
Pursat 115 D6 Khmer. Poŭthĭsăt. Pursat, W Cambodia
Purus, Rio 40 C2 var. Río Purús. river Brazil/Peru
Pusan see Busan
Pushkino see Biläsuvar
Püspökladány 77 D6 Hajdú-Bihar, E Hungary
Putorana, Gory/Putorana Mountains see Putorana, Plato
Putorana Mountains 93 E3 var. Gory Putorana, Eng. Putorana Mountains. mountain range N Russia
Putrajaya 116 B3 administrative capital (Malaysia) Kuala Lumpur, Peninsular Malaysia
Puttalam 110 C3 North Western Province, W Sri Lanka
Puttgarden 72 C2 Schleswig-Holstein, N Germany
Putumayo, Río 36 B5 var. Içá, Rio. river NW South America
Puurmani 84 D2 Ger. Talkhof. Jõgevamaa, E Estonia
Pyatigorsk 89 B7 Stavropol'skiy Kray, SW Russia
Pyatikhatki see Piatykhatky
P"yatykhatky see Piatykhatky
Pyay 114 B4 var. Prome, Pye. Bago, C Myanmar (Burma)
Pye see Pyay
Pyetrykaw 85 C7 Rus. Petrikov. Homyel'skaya Voblasts', SE Belarus
Pyin-Oo-Lwin 114 B3 var. Maymyo. Mandalay, C Myanmar (Burma)
Pýlos 83 B6 var. Pilos. Pelopónnisos, S Greece
P'yŏngyang 107 E3 var. P'yŏngyang-jikhalsi, Eng. Pyongyang. country capital (North Korea) SW North Korea
P'yŏngyang-jikhalsi see P'yŏngyang
Pyramid Lake 25 C5 lake Nevada, W USA
Pyrenaei Montes see Pyrenees
Pyrenees 80 B2 Fr. Pyrénées, Sp. Pirineos; anc. Pyrenaei Montes. mountain range SW Europe
Pýrgos 83 B6 var. Pírgos. Dytikí Ellás, S Greece
Pyriatyn 87 E2 prev. Pyryatyn, Rus. Piryatin. Poltavs'ka Oblast', NE Ukraine
Pyritz see Pyrzyce
Pyryatyn see Pyriatyn
Pyrzyce 76 B3 Ger. Pyritz. Zachodnio-pomorskie, NW Poland
Pyu see Phyu
Pyuntaza 114 B4 Bago, SW Myanmar (Burma)

Q

Qāʻ al Jafr 97 C7 lake S Jordan
Qaanaaq 60 D1 var. Qânâq, Dan. Thule. Avannaata, N Greenland
Qabāṭiyah 97 E6 N West Bank, Middle East
Qābis see Gabès
Qābis, Khalīj see Gabès, Golfe de
Qacentina see Constantine
Qafṣah see Gafsa
Qagan Us see Dulan
Qahremānshahr see Kermānshāh
Qaidam Pendi 104 C4 basin C China
Qal'aikhum see Qalaikhumb
Qalaikhumb 101 F3 prev. Qal'aikhum, Rus. Kalaikhum. S Tajikistan

Qalāt 101 E5 Per. Kalāt. Zābol, S Afghanistan
Qal'at Bīshah 99 B5 'Asīr, SW Saudi Arabia
Qalqiliya see Qalqīlyah
Qalqīlyah 97 D6 var. Qalqiliya. Central, W West Bank, Middle East
Qamdo 104 D5 Xizang Zizhiqu, W China
Qamishly see Al Qāmishlī
Qânâq see Qaanaaq
Qaqortoq 60 C4 Dan. Julianehåb. Kujalleq, S Greenland
Qaraghandy 92 C4 var. Karagandy; prev. Karaganda. Karagandy, C Kazakhstan
Qara Qum see Garagum
Qarataū see Karatau, Zhambyl, Kazakhstan
Qarkilik see Ruoqiang
Qarokŭl 101 F3 Rus. Karakul'. E Tajikistan
Qarqannah, Juzur see Kerkenah, Îles de
Qars see Kars
Qarshi 101 E3 Rus. Karshi; prev. Bek-Budi. Qashqadaryo Viloyati, S Uzbekistan
Qasigianguit see Qasigiannguit
Qasigiannguit 60 C3 var. Qasigianguit, Dan. Christianshåb. Qeqertalika, C Greenland
Qaşr al Farāfirah 50 B2 var. Qasr Farâfra. W Egypt
Qasr Farâfra see Qaşr al Farāfirah
Qaţanā 97 B5 var. Katana. Dimashq, S Syria
Qatar 98 C4 off. State of Qatar, Ar. Dawlat Qaţar. country SW Asia
Qatar, State of see Qatar
Qattara Depression see Qaţţārah, Munkhafaḑ al
Qaţţâra, Monkhafad el see Qaţţārah, Munkhafaḑ al
Qaţţārah, Munkhafaḑ al 50 A1 var. Monkhafad el Qaţţâra, Eng. Qattara Depression. desert NW Egypt
Qausuittuq see Resolute
Qazaqstan/Qazaqstan Respūblikasy see Kazakhstan
Qazimämmäd see Hacıqabul
Qazris see Cáceres
Qazvīn 98 C2 var. Kazvin. Qazvīn, N Iran
Qena see Qinā
Qeqertarssuaq see Qeqertarsuaq
Qeqertarsuaq 60 C3 var. Qeqertarssuaq, Dan. Godhavn. Qeqertalik, S Greenland
Qeqertarsuaq 60 C3 island W Greenland
Qeqertarsuup Tunua 60 C3 Dan. Disko Bugt. inlet W Greenland
Qerveh see Qorveh
Qeshm 98 D4 var. Jazīreh-ye Qeshm, Qeshm Island. island S Iran
Qeshm Island/Qeshm, Jazīreh-ye see Qeshm
Qilian Shan 104 D3 var. Kilien Mountains. mountain range N China
Qimusseriarsuaq 60 C2 Dan. Melville Bugt, Eng. Melville Bay. bay NW Greenland
Qinā 50 B2 var. Qena; anc. Caene, Caenepolis. E Egypt
Qing see Qinghai
Qingdao 106 D4 var. Ching-Tao, Ch'ing-tao, Tsingtao, Tsintao, Ger. Tsingtau. Shandong, E China
Qinghai 104 C4 var. Chinghai, Koko Nor, Qing, Qinghai Sheng, Tsinghai. province C China
Qinghai Hu 104 D4 var. Ch'ing Hai, Tsing Hai, Mong. Koko Nor. lake C China
Qinghai Sheng see Qinghai
Qingzang Gaoyuan 104 B4 var. Xizang Gaoyuan, Eng. Plateau of Tibet. plateau W China
Qinhuangdao 106 D3 Hebei, E China
Qinzhou 106 B6 Guangxi Zhuangzu Zizhiqu, S China
Qiong see Hainan
Qiqihar 106 D2 var. Ch'i-ch'i-ha-erh, Tsitsihar; prev. Lungkiang. Heilongjiang, NE China
Qira 104 B4 Xinjiang Uygur Zizhiqu, NW China
Qita Ghazzah see Gaza Strip
Qitai 104 C3 Xinjiang Uygur Zizhiqu, NW China
Qīzān see Jīzān
Qizil Orda see Qyzylorda
Qizil Qum/Qizilqum see Kyzylkum Desert
Qizilrabot 101 G3 Rus. Kyzylrabot. SE Tajikistan
Qogir Feng see K2
Qom 98 C3 var. Kum, Qum. Qom, N Iran
Qomolangma Feng see Everest, Mount
Qomul see Hami
Qo'qon 101 F2 var. Khokand, Rus. Kokand. Farg'ona Viloyati, E Uzbekistan
Qorveh 98 C3 var. Qerveh, Qurveh. Kordestān, W Iran
Qostanay 92 C4 var. Kostanay, Kustanay. Kostanay, N Kazakhstan
Qoubaïyât 96 B4 var. Al Qubayyāt. N Lebanon
Qoussantîna see Constantine
Quang Ngai 115 E5 var. Quangngai, Quang Nghia. Quang Ngai, C Vietnam
Quangngai see Quang Ngai
Quang Nghia see Quang Ngai
Quan Long see Ca Mau
Quanzhou 106 D6 var. Ch'uan-chou, Tsinkiang; prev. Chin-chiang. Fujian, SE China
Quanzhou 106 C6 Guangxi Zhuangzu Zizhiqu, S China
Qu'Appelle 15 F5 river Saskatchewan, S Canada
Quarles, Pegunungan 117 E4 mountain range Sulawesi, C Indonesia
Quarnero see Kvarner
Quartu Sant' Elena 75 A6 Sardegna, Italy, C Mediterranean Sea
Quba 95 H2 Rus. Kuba. N Azerbaijan
Qubba see Ba'qūbah
Québec 17 E4 var. Quebec. province capital Québec, SE Canada
Québec 16 D3 var. Quebec. province SE Canada
Queen Charlotte Islands see Haida Gwaii
Queen Charlotte Sound 14 C5 sea area British Columbia, W Canada
Queen Elizabeth Islands 15 E1 Fr. Îles de la Reine-Élisabeth. island group Nunavut, N Canada
Queensland 126 B4 state N Australia
Queenstown 129 B7 Otago, South Island, New Zealand
Queenstown 56 D5 Eastern Cape, S South Africa
Quelimane 57 E3 var. Kilimane, Kilmain, Quilimane. Zambézia, NE Mozambique
Quelpart see Jeju-do
Quepos 31 E4 Puntarenas, S Costa Rica
Que Que see Kwekwe
Quera see Chur
Querétaro 29 E4 Querétaro de Arteaga, C Mexico
Quesada 31 E4 var. Ciudad Quesada, San Carlos. Alajuela, N Costa Rica
Quetta 112 B2 Baluchisan, SW Pakistan
Quetzalcoalco see Coatzacoalcos
Quezaltenango 30 A2 var. Quetzaltenango. Quezaltenango, W Guatemala
Quibdó 36 A3 Chocó, W Colombia
Quilimane see Quelimane
Quillabamba 38 D3 Cusco, C Peru
Quilon see Kollam
Quimper 68 A3 anc. Quimper Corentin. Finistère, NW France
Quimper Corentin see Quimper
Quimperlé 68 A3 Finistère, NW France
Quincy 18 A4 Illinois, N USA
Qui Nhon/Quinhon see Quy Nhơn
Quissico 57 E4 Inhambane, S Mozambique
Quito 38 B1 country capital (Ecuador) Pichincha, N Ecuador
Qulyndy Zhazyghy see Kulunda Steppe
Qum see Qom
Qurein see Al Kuwayt
Qŭrghonteppa see Bokhtar
Qurlurtuuq see Kugluktuk
Qurveh see Qorveh
Quşayr see Al Quşayr
Quxar see Lhazê
Quy Nhơn 115 E5 var. Quinhon, Qui Nhon. Binh Đinh, C Vietnam
Qyteti Stalin see Kuçovë
Qyzylorda 92 B5 var. Kzylorda, Kzyl-Orda, Qizil Orda; prev. Perovsk. Kyzylorda, S Kazakhstan
Qyzylqum see Kyzylkum Desert

R

Raab 78 B1 Hung. Rába. river Austria/Hungary
Raab see Rába
Raab see Győr
Raahe 62 D4 Swe. Brahestad. Pohjois-Pohjanmaa, W Finland
Raalte 64 D3 Overijssel, E Netherlands
Raamsdonksveer 64 C4 Noord-Brabant, S Netherlands
Raasiku 84 D2 Ger. Rasik. Harjumaa, NW Estonia
Rába 77 B7 Ger. Raab. river Austria/Hungary
Rába see Raab
Rabat 48 C2 var. al Dar al Baida. country capital (Morocco) NW Morocco
Rabat 80 B5 W Malta
Rabat see Victoria
Rabbah Ammon/Rabbath Ammon see 'Ammān
Rabinal 30 B2 Baja Verapaz, C Guatemala
Rabka 77 D5 Małopolskie, S Poland
Râbniţa see Rîbniţa
Rabyânah, Ramlat 49 G4 var. Rebiana Sand Sea, Şaḩrā' Rabyānah. desert SE Libya
Rabyānah, Şaḩrā' see Rabyânah, Ramlat
Răcari see Durankulak
Race, Cape 17 H3 headland Newfoundland, Newfoundland and Labrador, E Canada
Rach Gia 115 D6 Kiên Giang, S Vietnam
Rach Gia, Vinh 115 D6 bay S Vietnam
Racine 18 B3 Wisconsin, N USA
Rácz-Becse see Bečej
Rădăuţi 86 C3 Ger. Radautz, Hung. Rádóc. Suceava, N Romania
Radautz see Rădăuţi
Rádeyilikóé see Fort Good Hope
Rádóc see Rădăuţi
Radom 76 D4 Mazowieckie, C Poland
Radomsko 76 D4 Rus. Novoradomsk. Łódzkie, C Poland
Radomyshl 86 D2 Zhytomyrs'ka Oblast', N Ukraine
Radoviš 79 E6 prev. Radovište. E North Macedonia
Radovište see Radoviš
Radviliškis 84 B4 Šiaulai, N Lithuania
Radzyń Podlaski 76 E4 Lubelskie, E Poland
Rae-Edzo see Edzo
Raetihi 128 D4 Manawatu-Wanganui, North Island, New Zealand
Rafa see Rafah
Rafaela 42 C3 Santa Fe, E Argentina
Rafah 97 A7 var. Rafa, Rafaḩ, Heb. Rafiaḩ, Raphiah. SW Gaza Strip
Rafḩah 98 B4 Al Ḩudūd ash Shamālīyah, N Saudi Arabia
Rafiaḩ see Rafah
Raga 51 A5 W Bahr el Ghazal, W South Sudan
Ragged Island Range 32 C2 island group S The Bahamas
Ragnit see Neman
Ragusa 75 C7 Sicilia, Italy, C Mediterranean Sea
Ragusa see Dubrovnik
Rahachow 85 D7 Rus. Rogachëv. Homyel'skaya Voblasts', SE Belarus
Rahaeng see Tak
Rahaţ, Ḩarrat 99 B5 lava flow W Saudi Arabia
Rahimyar Khan 112 C3 Punjab, SE Pakistan
Rahovec 79 D5 Serb. Orahovac. W Kosovo
Raiatea 123 G4 island Îles Sous le Vent, W French Polynesia
Rāichūr 110 C1 Karnātaka, C India
Raidestos see Tekirdağ
Rainier, Mount 12 A4 volcano Washington, NW USA
Rainy Lake 16 A4 lake Canada/USA
Raipur 113 E4 Chhattīsgarh, C India
Rājahmundry 113 E5 Andhra Pradesh, E India
Rajang see Rajang, Batang
Rajang, Batang 116 D3 var. Rajang. river East Malaysia
Rājapālaiyam 110 C3 Tamil Nādu, SE India
Rājasthān 112 C3 cultural region NW India
Rajin 107 E3 prev. Najin. NE North Korea
Rājkot 112 C4 Gujarāt, W India
Rāj Nāndgaon 113 E4 Chhattīsgarh, C India
Rajshahi 113 G3 prev. Rampur Boalia. Rajshahi, W Bangladesh
Rakahanga 123 F3 atoll N Cook Islands
Rakaia 129 B6 river South Island, New Zealand
Rakiura see Stewart Island
Rakka see Ar Raqqah
Rakke 84 E2 Lääne-Virumaa, NE Estonia
Rakvere 84 E2 Ger. Wesenberg. Lääne-Virumaa, N Estonia
Raleigh 21 F1 state capital North Carolina, SE USA
Ralik Chain 122 D1 island group Ralik Chain, W Marshall Islands
Ramadi see Ar Ramādī
Ramallah 97 E7 C West Bank, Middle East
Râmnicul-Sărat see Râmnicu Sărat
Râmnicu Sărat 86 C4 prev. Râmnicul-Sărat, Rîmnicu-Sărat. Buzău, E Romania
Râmnicu Vâlcea 86 B4 prev. Rîmnicu Vîlcea. Vâlcea, C Romania
Rampur Boalia see Rajshahi
Ramree Island 114 A4 island W Myanmar (Burma)
Ramtha see Ar Ramthā
Rancagua 42 B4 Libertador, C Chile
Rānchi 113 F4 Jhārkhand, N India
Randers 63 B7 Århus, C Denmark
Råneš see Ringvassøya
Rangiora 129 C6 Canterbury, South Island, New Zealand
Rangitikei 128 D4 river North Island, New Zealand
Rangitoto ki te Tonga 128 C4 var. D'Urville Island. island C New Zealand
Rangoon see Yangon
Rangpur 113 G3 Rajshahi, N Bangladesh
Rankin Inlet 15 G3 Nunavut, C Canada
Rankovićevo see Kraljevo
Ranong 115 B6 Ranong, SW Thailand
Rapa Nui see Pascua, Isla de
Raphiah see Rafah
Rapid City 22 D3 South Dakota, N USA
Räpina 84 E3 Ger. Rappin. Põlvamaa, SE Estonia
Rapla 84 D2 Ger. Rappel. Raplamaa, NW Estonia
Rappel see Rapla
Rappin see Räpina
Rarotonga 123 G5 island S Cook Islands, C Pacific Ocean
Ras al'Ain see Ra's al 'Ayn
Ra's al 'Ayn 96 D1 var. Ras al'Ain. Al Ḩasakah, N Syria
Ra's an Naqb 97 B7 Ma'ān, S Jordan
Raseiniai 84 B4 Kaunas, C Lithuania
Rasht 98 C2 var. Resht. Gīlān, NW Iran
Rasik see Raasiku
Râşnov 86 C4 prev. Rîşno, Rozsnyó, Hung. Barcarozsnyó. Braşov, C Romania
Rastenburg see Kętrzyn
Ratak Chain 122 D1 island group Ratak Chain, E Marshall Islands
Ratān 63 C5 Jämtland, C Sweden
Rat Buri see Ratchaburi
Ratchaburi 115 C5 var. Rat Buri. Ratchaburi, W Thailand
Ratisbon/Ratisbona/Ratisbonne see Regensburg
Rat Islands 14 A2 island group Aleutian Islands, Alaska, USA
Ratlām 112 D4 prev. Rutlam. Madhya Pradesh, C India
Ratnapura 110 D4 Sabaragamuwa Province, S Sri Lanka
Raton 26 D1 New Mexico, SW USA
Rättvik 63 C5 Dalarna, C Sweden
Raudhatain see Ar Rawḑatayn
Raufarhöfn 61 E4 Nordhurland Eystra, NE Iceland
Raukawa see Cook Strait
Raukumara Range 128 E3 mountain range North Island, New Zealand
Rāulakela see Rāurkela
Rauma 63 D5 Swe. Raumo. Satakunta, SW Finland
Raumo see Rauma
Rāurkela 113 F4 var. Rāulakela, Rourkela. Odisha, E India
Ravenna 74 C3 Emilia-Romagna, N Italy
Ravi 112 C2 river India/Pakistan
Rawalpindi 112 C1 Punjab, NE Pakistan
Rawa Mazowiecka 76 D4 Łódzkie, C Poland
Rawicz 76 C4 Ger. Rawitsch. Wielkopolskie, C Poland
Rawitsch see Rawicz
Rawlins 22 C3 Wyoming, C USA
Rawson 43 C6 Chubut, SE Argentina
Rayak 96 B4 var. Rayaq, Riyāq. E Lebanon
Rayaq see Rayak
Rayong 115 C5 Rayong, S Thailand
Razazah, Buhayrat ar 98 B3 var. Baḩr al Milḩ. lake C Iraq
Razdolnoye see Rozdolne
Razelm, Lacul see Razim, Lacul
Razgrad 82 D2 Razgrad, N Bulgaria
Razim, Lacul 86 D5 prev. Lacul Razelm. lagoon NW Black Sea
Reading 67 D7 S England, United Kingdom
Reading 19 F4 Pennsylvania, NE USA
Realicó 42 C4 La Pampa, C Argentina
Reăng Kesei 115 D5 Battambang, W Cambodia
Greater Antarctica see East Antarctica
Rebecca, Lake 125 C6 lake Western Australia
Rebiana Sand Sea see Rabyânah, Ramlat
Rebun-to 108 C2 island NE Japan
Rechitsa see Rechytsa
Rechytsa 85 D7 Rus. Rechitsa. Brestskaya Voblasts', SW Belarus
Recife 41 G2 prev. Pernambuco. state capital Pernambuco, E Brazil
Recklinghausen 72 A4 Nordrhein-Westfalen, W Germany
Recogne 65 C7 Luxembourg, SE Belgium
Reconquista 42 D3 Santa Fe, C Argentina
Red Deer 15 E5 Alberta, SW Canada
Redding 25 B5 California, W USA
Redon 68 B4 Ille-et-Vilaine, NW France
Red River 23 E1 river Canada/USA
Red River 114 C2 var. Yuan, Chin. Yuan Jiang, Vtn. Sông Hông Hà. river China/Vietnam
Red River 13 C6 river S USA
Red River 20 B3 river Louisiana, S USA
Red Sea 50 C3 var. Sinus Arabicus. sea Africa/Asia
Red Wing 23 G2 Minnesota, N USA
Reefton 129 C5 West Coast, South Island, New Zealand
Reese River 25 C5 river Nevada, W USA
Refahiye 95 E3 Erzincan, C Turkey (Türkiye)
Regensburg 73 C6 Eng. Ratisbon, Fr. Ratisbonne, hist. Ratisbona; anc. Castra Regina, Reginum. Bayern, SE Germany
Regenstauf 73 C6 Bayern, SE Germany
Rēgestān see Rīgistān
Reggane 48 D3 C Algeria
Reggio see Reggio nell'Emilia
Reggio Calabria see Reggio di Calabria
Reggio di Calabria 75 D7 var. Reggio Calabria, Gk. Rhegion; anc. Regium, Rhegium. Calabria, SW Italy
Reggio Emilia see Reggio nell'Emilia
Reggio nell'Emilia 74 B2 var. Reggio Emilia, abbrev. Reggio; anc. Regium Lepidum. Emilia Romagna, N Italy
Reghin 86 C4 Ger. Sächsisch-Reen, Hung. Szászrégen; prev. Reghinul Săsesc, Ger. Sächsisch-Regen. Mureş, C Romania
Reghinul Săsesc see Reghin
Regina 15 F5 province capital Saskatchewan, S Canada
Reginum see Regensburg
Rēgīstān 100 D5 var. Rigestān, Rēgestān. desert region S Afghanistan
Regium see Reggio di Calabria
Regium Lepidum see Reggio nell'Emilia
Rehoboth 56 B3 Hardap, C Namibia
Rehovot 97 A6 ; prev. Reḩovot. Central, C Israel
Reḩovot see Rehovot
Reichenau see Bogatynia, Poland
Reichenberg see Liberec
Reid 125 D6 Western Australia
Reikjavik see Reykjavík
Ré, Île de 68 A4 island W France
Reims 68 D3 Eng. Rheims; anc. Durocortorum, Remi. Marne, N France
Reindeer Lake 15 F4 lake Manitoba/Saskatchewan, C Canada
Reine-Charlotte, Îles de la see Haida Gwaii
Reine-Élisabeth, Îles de la see Queen Elizabeth Islands
Reinga, Cape 128 C1 var. Te Rerenga Wairua. headland North Island, New Zealand
Reinosa 70 D1 Cantabria, N Spain
Reka see Rijeka
Rekhovot see Rehovot
Reliance 15 F4 Northwest Territories, C Canada
Remi see Reims
Rendina see Rentína
Rendsburg 72 B2 Schleswig-Holstein, N Germany
Rengat 116 B4 Sumatera, W Indonesia
Reni 86 D4 Odes'ka Oblast', SW Ukraine
Rennell 122 C4 var. Mu Nggava. island S Solomon Islands
Rennes 68 B3 Bret. Roazon; anc. Condate. Ille-et-Vilaine, NW France
Reno 25 C5 Nevada, W USA
Renqiu 106 C4 Hebei, E China
Rentína 83 B5 var. Rendina. Thessalía, C Greece
Reps see Rupea
Repulse Bay see Naujaat
Reschitza see Reşiţa
Resht see Rasht
Resicabánya see Reşiţa
Resistencia 42 D3 Chaco, NE Argentina
Reşiţa 86 A4 Ger. Reschitza, Hung. Resicabánya. Caraş-Severin, W Romania
Resolute 15 F2 Inuit Qausuittuq. Cornwallis Island, Nunavut, N Canada
Resolution Island 17 E1 island Nunavut, NE Canada
Resolution Island 129 A7 island SW New Zealand
Réunion 57 H4 off. La Réunion. French overseas department W Indian Ocean
Réunion 119 B5 island W Indian Ocean
Reus 71 F2 Cataluña, E Spain
Reutlingen 73 B6 Baden-Württemberg, S Germany
Reuver 65 D5 Limburg, SE Netherlands
Reval/Revel see Tallinn
Revillagigedo Island 28 B5 island Alexander Archipelago, Alaska, USA
Rexburg 24 E3 Idaho, NW USA
Reyes 39 F3 Beni, NW Bolivia
Rey, Isla del 31 G5 island Archipiélago de las Perlas, SE Panama
Reykjanes Basin 60 C5 var. Irminger Basin. undersea basin N Atlantic Ocean
Reykjanes Ridge 58 A1 undersea ridge N Atlantic Ocean
Reykjavík 61 E5 var. Reikjavik. country capital (Iceland) Höfudhborgarsvaedhi, W Iceland
Reynosa 29 E2 Tamaulipas, C Mexico
Reẕā'īyeh, Daryācheh-ye see Orūmīyeh, Daryācheh-ye
Rezé 68 A4 Loire-Atlantique, NW France
Rēzekne 84 D4 Ger. Rositten; prev. Rus. Rezhitsa. Rēzekne, SE Latvia
Rezhitsa see Rēzekne
Rezovo 82 E3 Turk. Rezve. Burgas, E Bulgaria
Rezve see Rezovo
Rhaedestus see Tekirdağ
Rhegion/Rhegium see Reggio di Calabria
Rheims see Reims
Rhein see Rhine
Rheine 72 A3 var. Rheine in Westfalen. Nordrhein-Westfalen, NW Germany
Rheine in Westfalen see Rheine
Rheinisches Schiefergebirge 73 A5 var. Rhine State Uplands, Eng. Rhenish Slate Mountains. mountain range W Germany
Rhenish Slate Mountains see Rheinisches Schiefergebirge
Rhin see Rhine
Rhine 58 D4 Dut. Rijn, Fr. Rhin, Ger. Rhein. river W Europe
Rhinelander 18 B2 Wisconsin, N USA
Rhine State Uplands see Rheinisches Schiefergebirge
Rho 74 B2 Lombardia, N Italy
Rhode Island 19 G3 off. State of Rhode Island and Providence Plantations, also known as Little Rhody, Ocean State. state NE USA
Rhodes 83 E7 var. Ródhos, Eng. Rhodes, It. Rodi; anc. Rhodos. island Dodekánisa, Greece, Aegean Sea
Rhodes see Ródos
Rhodesia see Zimbabwe
Rhodope Mountains 82 C3 var. Rodópi Óri, Bul. Rodopi Planina, Rodopi, Gk. Oroseirá Rodópis, Turk. Dospad Dagh. mountain range Bulgaria/Greece
Rhône 58 C4 river France/Switzerland
Rhum 66 B3 var. Rum. island W Scotland, United Kingdom
Ribble 67 D5 river NW England, United Kingdom
Ribeira see Santa Uxía de Ribeira
Ribeirão Preto 41 F4 São Paulo, S Brazil
Riberalta 39 F2 Beni, N Bolivia
Rîbniţa 86 D3 var. Râbniţa, Rus. Rybnitsa. NE Moldova
Rice Lake 18 A2 Wisconsin, N USA
Richard Toll 52 B3 N Senegal
Richfield 22 B4 Utah, W USA
Richland 24 C2 Washington, NW USA
Richmond 129 C5 Tasman, South Island, New Zealand
Richmond 18 C5 Kentucky, S USA
Richmond 19 E5 state capital Virginia, NE USA
Richmond Range 129 C5 mountain range South Island, New Zealand
Ricobayo, Embalse de 70 C2 reservoir NW Spain
Ricomagus see Riom
Ridder/Ridder 92 D4 prev. Leninogor, Leninogorsk. Vostochnyy Kazakhstan, E Kazakhstan
Ridgecrest 25 C7 California, W USA
Ried see Ried im Innkreis
Ried im Innkreis 73 D6 var. Ried. Oberösterreich, NW Austria
Riemst 65 D6 Limburg, NE Belgium
Riesa 72 D4 Sachsen, E Germany
Rift Valley see Great Rift Valley
Rīga 84 C3 Eng. Riga. country capital (Latvia) Rīga, C Latvia
Rigaer Bucht see Riga, Gulf of
Riga, Gulf of 84 C3 Est. Liivi Laht, Ger. Rigaer Bucht, Latv. Rīgas Jūras Līcis, Rus. Rizhskiy Zaliv; prev. Est. Riia Laht. gulf Estonia/Latvia
Rīgas Jūras Līcis see Riga, Gulf of
Rigestān see Rēgīstān
Riia Laht see Riga, Gulf of
Riihimäki 63 D5 Kanta-Häme, S Finland
Rijeka 78 A2 Ger. Sankt Veit am Flaum, It. Fiume, Slvn. Reka; anc. Tarsatica. Primorje-Gorski Kotar, NW Croatia
Rijn see Rhine
Rijssel see Lille
Rijssen 64 E3 Overijssel, E Netherlands
Rimah, Wadi ar 98 B4 var. Wādī ar Rummah. dry watercourse C Saudi Arabia
Rimini 74 C3 anc. Ariminum. Emilia-Romagna, N Italy
Rîmnicu-Sărat see Râmnicu Sărat
Rîmnicu Vîlcea see Râmnicu Vâlcea
Rimouski 17 E4 Québec, SE Canada
Ringebu 63 B5 Oppland, S Norway
Ringen see Rõngu
Ringkøbing Fjord 63 A7 fjord W Denmark
Ringvassøya 62 C2 Lapp. Ráneš. island N Norway
Rio see Rio de Janeiro
Riobamba 38 B1 Chimborazo, C Ecuador
Rio Branco 34 B3 state capital Acre, W Brazil
Rio Branco, Território de see Roraima
Río Bravo 29 E2 Tamaulipas, C Mexico
Río Cuarto 42 C4 Córdoba, C Argentina
Rio de Janeiro 41 F4 var. Rio. state capital Rio de Janeiro, SE Brazil
Río Gallegos 43 B7 var. Gallegos, Puerto Gallegos. Santa Cruz, S Argentina
Rio Grande 41 E5 var. São Pedro do Rio Grande do Sul. Rio Grande do Sul, S Brazil
Río Grande 28 D3 Zacatecas, C Mexico
Rio Grande do Norte 41 G2 off. Estado do Rio Grande do Norte. state/region E Brazil
Rio Grande do Norte, Estado do see Rio Grande do Norte
Rio Grande do Sul 41 E5 off. Estado do Rio Grande do Sul. state/region S Brazil
Rio Grande do Sul, Estado do see Rio Grande do Sul
Rio Grande Plateau see Rio Grande Rise
Rio Grande Rise 35 E6 var. Rio Grande Plateau. undersea plateau SW Atlantic Ocean
Ríohacha 36 B1 La Guajira, N Colombia
Río Lagartos 29 H3 Yucatán, SE Mexico
Riom 69 C5 anc. Ricomagus. Puy-de-Dôme, C France
Río Verde 29 E4 var. Rioverde. San Luis Potosí, C Mexico
Rioverde see Río Verde
Ripoll 71 G2 Cataluña, NE Spain
Rishiri-to 108 C2 var. Risiri Tô. island NE Japan
Risiri Tô see Rishiri-tō
Rîşno see Râşnov
Risti 84 D2 Ger. Kreuz. Läänemaa, W Estonia
Rivas 30 D4 Rivas, SW Nicaragua
Rivera 42 D3 Rivera, NE Uruguay
River Falls 18 A2 Wisconsin, N USA
Riverside 25 C7 California, W USA
Riverton 129 A7 var. Aparima. Southland, South Island, New Zealand
Riverton 22 C3 Wyoming, C USA
Rivière-du-Loup 17 E4 Québec, SE Canada
Rivne 86 C2 Pol. Równe, Rus. Rovno. Rivnens'ka Oblast', NW Ukraine
Rivoli 74 A2 Piemonte, NW Italy
Riyadh/Riyāḑ, Minţaqat ar see Ar Riyāḑ
Riyāq see Rayak
Rize 95 E2 Rize, NE Turkey (Türkiye)
Rizhao 106 D4 Shandong, E China
Rizhskiy Zaliv see Riga, Gulf of
Rkîz 52 C3 Trarza, W Mauritania
Road Town 33 F3 dependent territory capital (British Virgin Islands) Tortola, C British Virgin Islands
Roanne 69 C5 anc. Rodunma. Loire, E France
Roanoke 19 E5 Virginia, NE USA
Roanoke River 21 F1 river North Carolina/Virginia, SE USA
Roatán 30 C2 var. Coxen Hole, Coxin Hole. Islas de la Bahía, N Honduras
Roat Kampuchea see Cambodia
Roazon see Rennes
Robbie Ridge 121 E3 undersea ridge W Pacific Ocean
Robert Williams see Caála
Robinson Range 125 B5 mountain range Western Australia
Robson, Mount 15 E5 mountain British Columbia, SW Canada
Robstown 27 G4 Texas, SW USA
Roca Partida, Isla 28 B5 island W Mexico
Rocas, Atol das 41 G2 island E Brazil
Rochefort 65 C7 Namur, SE Belgium
Rochefort 68 B4 var. Rochefort sur Mer. Charente-Maritime, W France
Rochefort sur Mer see Rochefort
Rochester 23 G3 Minnesota, N USA
Rochester 19 G2 New Hampshire, NE USA
Rocheuses, Montagnes/Rockies see Rocky Mountains
Rockall Bank 58 B2 undersea bank N Atlantic Ocean
Rockall Trough 58 B2 trough N Atlantic Ocean
Rockdale 126 E2 Texas, SW USA
Rockford 18 B3 Illinois, N USA
Rockhampton 126 D4 Queensland, E Australia
Rock Hill 21 E1 South Carolina, SE USA
Rockingham 125 A6 Western Australia
Rock Island 18 B3 Illinois, N USA

Rock Sound *32 C1* Eleuthera Island, C The Bahamas
Rock Springs *22 C3* Wyoming, C USA
Rockstone *37 F3* C Guyana
Rocky Mount *21 F1* North Carolina, SE USA
Rocky Mountains *12 B4* var. Rockies, Fr. Montagnes Rocheuses. mountain range Canada/USA
Roden *64 E2* Drenthe, NE Netherlands
Rodez *69 C5* anc. Segodunum. Aveyron, S France
Ródhos/Rodi see Ródos
Rodopi/ Rodópi Óri see Rhodope Mountains
Rodopi Planina see Rhodope Mountains
Rodosto see Tekirdağ
Rodunma see Roanne
Roermond *65 D5* Limburg, SE Netherlands
Roeselare *65 A6* Fr. Roulers; prev. Rousselaere. West-Vlaanderen, W Belgium
Rogachëv see Rahachow
Rogatica *78 C4* Republika Srpska, SE Bosnia and Herzegovina
Rogers *20 A1* Arkansas, C USA
Roger Simpson Island see Abemama
Roi Ed see Roi Et
Roi Et *115 D5* var. Muang Roi Et, Roi Ed. Roi Et, E Thailand
Roja *84 C2* Talsi, NW Latvia
Rokiškis *84 C4* Panevėžys, NE Lithuania
Rokycany *77 A5* Ger. Rokytzan. Plzeňský Kraj, W Czechia (Czech Republic)
Rokytzan see Rokycany
Rôlas, Ilha das *54 E2* island S Sao Tome and Principe, Africa, E Atlantic Ocean
Rolla *23 G5* Missouri, C USA
Röm see Rømø
Roma *74 C4* Eng. Rome. country capital (Italy) Lazio, C Italy
Roma *127 D5* Queensland, E Australia
Roman *82 C2* Vratsa, NW Bulgaria
Roman *86 C4* Hung. Romanvásár. Neamţ, NE Romania
Romania *86 B4* Bul. Rumŭniya, Ger. Rumänien, Hung. Románia, Rom. România, Croatian Rumunjska, Ukr. Rumuniya, prev. Republica Socialistă România, Roumania, Rumania, Socialist Republic of Romania, prev.Rom. Romînia. country SE Europe
România, Republica Socialistă see Romania
Romania, Socialist Republic of see Romania
Románvásár see Roman
Rome *20 D2* Georgia, SE USA
Rome see Roma
Romînia see Romania
Romny *87 F2* Sums'ka Oblast', NE Ukraine
Rømø *63 A7* Ger. Röm. island SW Denmark
Roncador, Serra do *34 D4* mountain range C Brazil
Ronda *70 D5* Andalucía, S Spain
Rondônia *40 D3* off. Estado de Rondônia; prev. Território de Rondônia. state/region W Brazil
Rondônia, Estado de see Rondônia
Rondônia, Território de see Rondônia
Rondonópolis *41 E3* Mato Grosso, W Brazil
Rongelap Atoll *122 D1* var. Rōnlap. atoll Ralik Chain, NW Marshall Islands
Rõngu *84 D3* Ger. Ringen. Tartumaa, SE Estonia
Rōnlap see Rongelap Atoll
Rønne *63 B8* Bornholm, E Denmark
Ronne Ice Shelf *132 A3* ice shelf Antarctica
Roosendaal *65 C5* Noord-Brabant, S Netherlands
Roosevelt Island *132 B4* island Antarctica
Roraima *40 D1* off. Estado de Roraima; prev. Território de Rio Branco, Território de Roraima. state/region N Brazil
Roraima, Estado de see Roraima
Roraima, Mount *37 E3* mountain N South America
Roraima, Território de see Roraima
Røros *63 B5* Sør-Trøndelag, S Norway
Ross *129 B6* West Coast, South Island, New Zealand
Rosa, Lake *32 D2* lake Great Inagua, S The Bahamas
Rosario *42 D4* Santa Fe, C Argentina
Rosario *42 D2* San Pedro, C Paraguay
Rosario see Rosarito
Rosarito *28 A1* var. Rosario. Baja California Norte, NW Mexico
Roscianum see Rossano
Roscommon *18 C2* Michigan, N USA
Roseau *33 G4* prev. Charlotte Town. country capital (Dominica) SW Dominica
Roseburg *24 B4* Oregon, NW USA
Rosenau see Rožňava
Rosenberg *27 G4* Texas, SW USA
Rosenberg see Ružomberok, Slovakia
Rosengarten *72 B3* Niedersachsen, N Germany
Rosenheim *73 C6* Bayern, S Germany
Rosia *71 H5* W Gibraltar Europe
Rosia Bay *71 H5* bay SW Gibraltar Europe W Mediterranean Sea Atlantic Ocean
Roşiori de Vede *86 B5* Teleorman, S Romania
Rositten see Rēzekne
Roslavl' *89 A5* Smolenskaya Oblast', W Russia
Rosmalen *64 C4* Noord-Brabant, S Netherlands
Rossano *75 E6* anc. Roscianum. Calabria, SW Italy
Ross Ice Shelf *132 B4* ice shelf Antarctica
Rossiyskaya Federatsiya see Russia
Rosso *52 B3* Trarza, SW Mauritania
Rossosh' *89 B6* Voronezhskaya Oblast', W Russia
Ross Sea *132 B4* sea Antarctica
Rostak see Ar Rustāq
Rostock *72 C2* Mecklenburg-Vorpommern, NE Germany
Rostov see Rostov-na-Donu
Rostov-na-Donu *89 B7* var. Rostov, Eng. Rostov-on-Don. Rostovskaya Oblast', SW Russia
Rostov-on-Don see Rostov-na-Donu
Roswell *26 D2* New Mexico, SW USA
Rota *122 B1* island S Northern Mariana Islands
Rotcher Island see Tamana
Rothera *132 A2* UK research station Antarctica
Rotomagus see Rouen
Rotorua *128 D3* Bay of Plenty, North Island, New Zealand
Rotorua, Lake *128 D3* lake North Island, New Zealand
Rotterdam *64 C4* Zuid-Holland, SW Netherlands
Rottweil *73 B6* Baden-Württemberg, S Germany
Rotuma *123 E4* island NW Fiji Oceania, S Pacific Ocean
Roubaix *68 C2* Nord, N France
Rouen *68 C3* anc. Rotomagus. Seine-Maritime, N France
Roulers see Roeselare
Roumania see Romania
Round Rock *27 G3* Texas, SW USA
Rourkela see Rāurkela
Rousselaere see Roeselare
Roussillon *69 C6* cultural region S France
Rouyn-Noranda *16 D4* Québec, SE Canada
Rovaniemi *62 D3* Lappi, N Finland
Rovigno see Rovinj
Rovigo *74 C2* Veneto, NE Italy
Rovinj *78 A3* It. Rovigno. Istra, NW Croatia
Rovno see Rivne
Rovuma, Rio *57 F2* var. Ruvuma. river Mozambique/Tanzania
Rovuma, Rio see Ruvuma
Równe see Rivne
Roxas City *117 E2* Panay Island, C Philippines
Royale, Isle *18 B1* island Michigan, N USA
Royan *69 B5* Charente-Maritime, W France
Rozdolne *87 F4* Rus. Razdolnoye. Respublika Krym, S Ukraine
Rožňava *77 D6* Ger. Rosenau, Hung. Rozsnyó. Košický Kraj, E Slovakia
Rózsahegy see Ružomberok
Rozsnyó see Râşnov, Romania
Rozsnyó see Rožňava, Slovakia
Ruanda see Rwanda
Ruapehu, Mount *128 D4* volcano North Island, New Zealand
Ruapuke Island *129 B8* island SW New Zealand
Ruatoria *128 E3* Gisborne, North Island, New Zealand
Ruawai *128 D2* Northland, North Island, New Zealand
Rubezhnoye see Rubizhne
Rubizhne *87 H3* Rus. Rubezhnoye. Luhans'ka Oblast', E Ukraine
Ruby Mountains *25 D5* mountain range Nevada, W USA
Rucava *84 B3* Liepāja, SW Latvia
Rudensk see Rudzyensk
Rūdiškės *85 B5* Vilnius, S Lithuania
Rudnik see Dolni Chiflik
Rudny see Rudnyy
Rudnyy *92 C4* var. Rudny. Kostanay, N Kazakhstan
Rudolf, Lake see Turkana, Lake
Rudolfswert see Novo mesto
Rudzyensk *85 C6* Rus. Rudensk. Minskaya Voblasts', C Belarus
Rufiji *51 C7* river E Tanzania
Rufino *42 C4* Santa Fe, C Argentina
Rugāji *84 D4* Balvi, E Latvia
Rügen *72 D2* headland NE Germany
Ruggell *72 E1* N Liechtenstein Europe
Ruhja see Rūjiena
Ruhnu *84 C2* var. Ruhnu Saar, Swe. Runö. island SW Estonia
Ruhnu Saar see Ruhnu
Rujen see Rūjiena
Rūjiena *84 D3* Est. Ruhja, Ger. Rujen. Valmiera, N Latvia
Rukwa, Lake *51 B7* lake SE Tanzania
Rum see Rhum
Ruma *78 D3* Vojvodina, N Serbia
Rumadiya see Ar Ramādī
Rumania/Rumänien see Romania
Rumbek *51 B5* El Buhayrat, C South Sudan
Rum Cay *32 D2* island C The Bahamas
Rumia *76 C2* Pomorskie, N Poland
Rummah, Wādī ar see Rimah, Wādī ar
Rummelsburg in Pommern see Miastko
Rumuniya/Rumŭniya/Rumunjska see Romania
Runanga *129 B5* West Coast, South Island, New Zealand
Rundu *56 C3* var. Runtu. Okavango, NE Namibia
Runö see Ruhnu
Runtu see Rundu
Ruoqiang *104 C3* var. Jo-ch'iang, Uigh. Charkhlik, Charkhliq, Qarkilik. Xinjiang Uygur Zizhiqu, NW China
Rupea *86 C4* Ger. Reps, Hung. Kőhalom; prev. Cohalm. Braşov, C Romania
Rupel *65 B5* river N Belgium
Rupella see la Rochelle
Rupert, Rivière de *16 D3* river Québec, C Canada
Rusaddir see Melilla
Ruschuk/Rusçuk see Ruse
Ruse *82 D1* var. Ruschuk, Rustchuk, Turk. Rusçuk. Ruse, N Bulgaria
Russadir see Melilla
Russellville *20 A1* Arkansas, C USA
Russia *90 D2* off. Russian Federation, Latv. Krievija, Rus. Rossiyskaya Federatsiya. country Asia/Europe
Russian America see Alaska
Russian Federation see Russia
Rustaq see Ar Rustāq
Rustavi *95 G2* prev. Rust'avi. SE Georgia
Rust'avi see Rustavi
Rustchuk see Ruse
Ruston *20 B2* Louisiana, S USA
Rutanzige, Lake see Edward, Lake
Rutba see Ar Ruţbah
Rutlam see Ratlām
Rutland *19 F2* Vermont, NE USA
Rutog *104 A4* var. Rutög, Rutok. Xizang Zizhiqu, W China
Rutok see Rutog
Ruvuma see Rovuma, Rio
Ruwenzori *55 E5* mountain range Dem. Rep. Congo/Uganda
Ruzhany *85 B6* Brestskaya Voblasts', SW Belarus
Ružomberok *77 C5* Ger. Rosenberg, Hung. Rózsahegy. Žilinský Kraj, N Slovakia
Rwanda *51 B6* off. Republic of Rwanda; prev. Ruanda. country C Africa
Rwanda, Republic of see Rwanda
Ryazan' *89 B5* Ryazanskaya Oblast', W Russia
Rybach'ye see Balykchy
Rybinsk *88 B4* prev. Andropov. Yaroslavskaya Oblast', W Russia
Rybnik *77 C5* Śląskie, S Poland
Rybnitsa see Rîbniţa
Ryde *126 E1* United Kingdom
Ryki *76 D4* Lubelskie, E Poland
Rykovo see Yenakiieve
Rypin *76 C3* Kujawsko-pomorskie, C Poland
Ryssel see Lille
Rysy *77 C5* mountain S Poland
Ryukyu Islands see Nansei-shotō
Ryukyu Trench *103 F3* var. Nansei Syotō Trench. trench S East China Sea
Rzeszów *77 E5* Podkarpackie, SE Poland
Rzhev *88 B4* Tverskaya Oblast', W Russia

S

Saale *72 C4* river C Germany
Saalfeld *73 C5* var. Saalfeld an der Saale. Thüringen, C Germany
Saalfeld an der Saale see Saalfeld
Saarbrücken *73 A6* Fr. Sarrebruck. Saarland, SW Germany
Sääre *84 C2* var. Sjar. Saaremaa, W Estonia
Saare see Saaremaa
Saaremaa *84 C2* Ger. Oesel, Ösel; prev. Saare. island W Estonia
Saariselkä *62 D2* Lapp. Suoločielgi. Lappi, N Finland
Sab' Ābār *96 C4* var. Sab'a Biyar, Sa'b Bi'ār. Ḩimş, C Syria
Sab'a Biyar see Sab' Ābār
Šabac *78 D3* Serbia, W Serbia
Sabadell *71 G2* Cataluña, E Spain
Sabah *116 D3* prev. British North Borneo, North Borneo. state East Malaysia
Sabanalarga *36 B1* Atlántico, N Colombia
Sabaneta *36 C1* Falcón, N Venezuela
Sabaria see Szombathely
Sab'atayn, Ramlat as *99 C6* desert C Yemen
Sabaya *39 F4* Oruro, S Bolivia
Sa'b Bi'ār see Sab' Ābār
Sá da Bandeira see Lubango
Şa'dah *99 B6* NW Yemen
Sado *109 C5* var. Sadoga-shima. island C Japan
Sadoga-shima see Sado
Saena Julia see Siena
Safad see Tsefat
Şafāqis see Sfax
Şafāshahr *98 D3* var. Deh Bīd. Fārs, C Iran
Safed see Tsefat
Safēd Kōh, Silsilah-ye *100 D4* prev. Selseleh-ye Safīd Kūh, Eng. Paropamisus Range. mountain range W Afghanistan
Säffle *63 B6* Värmland, C Sweden
Safford *26 C3* Arizona, SW USA
Safi *48 B2* W Morocco
Safid Kuh, Selseleh-ye see Safēd Kōh, Silsilah-ye
Sagaing *114 B3* Sagaing, C Myanmar (Burma)
Sagami-nada *109 D6* inlet SW Japan
Sagan see Żagań
Sāgar *112 D4* prev. Saugor. Madhya Pradesh, C India
Sagarmāthā see Everest, Mount
Sagebrush State see Nevada
Saghez see Saqqez
Saginaw *18 C3* Michigan, N USA
Saginaw Bay *18 D2* lake bay Michigan, N USA
Sagua la Grande *32 B2* Villa Clara, C Cuba
Sagunto/Saguntum see Sagunt
Sagunt *71 F3* var. Sagunto, Ar. Murviedro; anc. Saguntum. Comunitat Valenciana, E Spain
Sahara *46 B3* desert Libya/Algeria
Sahara el Gharbîya see Şaḩrā' al Gharbīyah
Saharan Atlas see Atlas Saharien
Sahel *52 D3* physical region C Africa
Sāḩilīyah, Jibāl as *96 B3* mountain range NW Syria
Sahiwal *112 C2* prev. Montgomery. Punjab, E Pakistan
Şaḩrā' al Gharbīyah *50 B2* var. Sahara el Gharbîya, Eng. Western Desert. desert C Egypt
Şaḩrā' ash Sharqīyah *81 H5* Eng. Arabian Desert, Eastern Desert. desert E Egypt
Saïda *97 A5* var. Şaydā, Sayida; anc. Sidon. W Lebanon
Sa'īdābād see Sīrjān
Saidpur *113 G3* var. Syedpur. Rajshahi, NW Bangladesh
Saidu Sharif *112 C1* var. Mingora, Mongora. North-West Frontier Province, N Pakistan
Saigon see Hô Chi Minh
Saimaa *63 E5* lake SE Finland
St Albans *67 E6* anc. Verulamium. E England, United Kingdom
Saint Albans *18 D5* West Virginia, NE USA
St Andrews *66 C4* E Scotland, United Kingdom
Saint Anna Trough see Svyataya Anna Trough
St. Ann's Bay *32 B4* C Jamaica
St. Anthony *17 G3* Newfoundland and Labrador, SE Canada
Saint Augustine *21 E3* Florida, SE USA
St Austell *67 C7* SW England, United Kingdom
St Barthélemy *33 G3* French overseas collectivity, E West Indies
St. Botolph's Town see Boston
St-Brieuc *68 A3* Côtes d'Armor, NW France
St. Catharines *16 D5* Ontario, S Canada
St-Chamond *69 D5* Loire, E France
Saint Christopher and Nevis, Federation of see Saint Kitts and Nevis
Saint Christopher-Nevis see Saint Kitts and Nevis
Saint Clair, Lake *18 D3* var. Lac à L'Eau Claire. lake Canada/USA
St-Claude *69 D5* anc. Condate. Jura, E France
Saint Cloud *23 F2* Minnesota, N USA
Saint Croix *33 F3* island S Virgin Islands (US)
Saint Croix River *18 A2* river Minnesota/Wisconsin, N USA
St David's Island *20 B5* island E Bermuda
St-Denis *57 G4* dependent territory capital (Réunion) NW Réunion
St-Dié *68 E4* Vosges, NE France
St-Egrève *69 D5* Isère, E France
Sainte Marie, Cap see Vohimena, Tanjona
Saintes *69 B5* anc. Mediolanum. Charente-Maritime, W France
St-Étienne *69 D5* Loire, E France
St-Flour *69 C5* Cantal, C France
St-Gall/Saint Gall/St. Gallen see Sankt Gallen
St-Gaudens *69 B6* Haute-Garonne, S France
Saint George *127 D5* Queensland, E Australia
St George *20 B4* N Bermuda
Saint George *22 A5* Utah, W USA
St. George's *33 G5* country capital (Grenada) SW Grenada
St-Georges *17 E4* Québec, SE Canada
St-Georges *37 H3* E French Guiana
Saint George's Channel *67 B6* channel Ireland/Wales, United Kingdom
St George's Island *20 B4* island E Bermuda
St Helena see St Helena, Ascension and Tristan da Cunha
St Helena, Ascension and Tristan da Cunha *47 A6* UK Overseas territory C Atlantic Ocean
St Helier *67 D8* dependent territory capital (Jersey) S Jersey, Channel Islands
St.Iago de la Vega see Spanish Town
Saint Ignace *18 C2* Michigan, N USA
St-Jean, Lac *17 E4* lake Québec, SE Canada
Saint Joe River *24 D2* river Idaho, NW USA North America
St. John *17 F4* New Brunswick, SE Canada
Saint-John see Saint John
Saint John *19 H1* Fr. Saint-John. river Canada/USA
St John's *33 G3* country capital (Antigua and Barbuda) Antigua, Antigua and Barbuda
St. John's *17 H3* province capital Newfoundland and Labrador, E Canada
Saint Joseph *23 F4* Missouri, C USA
St Julian's see San Giljan
St Kilda *66 A3* island NW Scotland, United Kingdom
Saint Kitts and Nevis *33 F3* off. Federation of Saint Christopher and Nevis, var. Saint Christopher-Nevis. country E West Indies
St-Laurent see St-Laurent-du-Maroni
St-Laurent-du-Maroni *37 H3* var. St-Laurent. NW French Guiana
St-Laurent, Fleuve see St. Lawrence
St. Lawrence *17 E4* Fr. Fleuve St-Laurent. river Canada/USA
St. Lawrence, Gulf of *17 F3* gulf NW Atlantic Ocean
Saint Lawrence Island *14 B2* island Alaska, USA
St-Lô *68 B3* anc. Briovera, Laudus. Manche, N France
St-Louis *68 E4* Haut-Rhin, NE France
Saint Louis *52 B3* NW Senegal
Saint Louis *23 G4* Missouri, C USA
Saint Lucia *33 E1* country SE West Indies
Saint Lucia Channel *33 H4* channel Martinique/Saint Lucia
St-Malo *68 B3* Ille-et-Vilaine, NW France
St-Malo, Golfe de *68 A3* gulf NW France
Saint Martin see Sint Maarten
St Martin *33 G3* French overseas collectivity. E West Indies
St.Matthew's Island see Zadetkyi Kyun
St.Matthias Group *122 B3* island group NE Papua New Guinea
St. Moritz *73 B7* Ger. Sankt Moritz, Rmsch. San Murezzan. Graubünden, SE Switzerland
St-Nazaire *68 A4* Loire-Atlantique, NW France
Saint Nicholas see São Nicolau
Saint-Nicolas see Sint-Niklaas
St-Omer *68 C2* Pas-de-Calais, N France
Saint Paul *23 F2* state capital Minnesota, N USA
St-Paul, Île *119 C6* var. St.Paul Island. island Île St-Paul, NE French Southern and Antarctic Lands Antarctica Indian Ocean
St.Paul Island see St-Paul, Île
St Peter Port *67 D8* dependent territory capital (Guernsey) C Guernsey, Channel Islands
Saint Petersburg *21 E4* Florida, SE USA
Saint Petersburg see Sankt-Peterburg
St-Pierre and Miquelon *17 G4* Fr. Îles St-Pierre et Miquelon. French overseas collectivity NE North America
St-Quentin *68 C3* Aisne, N France
Saint Thomas see São Tomé, Sao Tome and Principe
Saint Thomas see Charlotte Amalie, Virgin Islands (US)
Saint Ubes see Setúbal
Saint Vincent *33 G4* island N Saint Vincent and the Grenadines
Saint Vincent see São Vicente
Saint Vincent and the Grenadines *33 H4* country SE West Indies
Saint Vincent, Cape see São Vicente, Cabo de
Saint Vincent Passage *33 H4* passage Saint Lucia/Saint Vincent and the Grenadines
Saint Yves see Setúbal
Saipan *120 B1* island/country capital (Northern Mariana Islands) S Northern Mariana Islands
Saishū see Jeju-do
Sajama, Nevado *39 F4* mountain W Bolivia
Sajószentpéter *77 D6* Borsod-Abaúj-Zemplén, NE Hungary
Sakākah *98 B4* Al Jawf, NW Saudi Arabia
Sakakawea, Lake *22 D1* reservoir North Dakota, N USA
Sakartvelo see Georgia
Sakata *108 D4* Yamagata, Honshū, C Japan
Sakhalin *93 G4* var. Sakhalin. island SE Russia
Sakhalin see Sakhalin, Ostrov
Sakhon Nakhon see Sakon Nakhon
Şäki *95 G2* Rus. Sheki; prev. Nukha. NW Azerbaijan
Saki see Saky
Sakishima-shoto *108 A3* var. Sakisima Syotô. island group SW Japan
Sakisima Syotô see Sakishima-shotō
Sakiz see Saqqez
Sakiz-Adasi see Chíos
Sakon Nakhon *114 D4* var. Muang Sakon Nakhon, Sakhon Nakhon. Sakon Nakhon, E Thailand
Saky *87 F5* Rus. Saki. Respublika Krym, S Ukraine
Sal *52 A3* island Ilhas de Barlavento, NE Cape Verde (Cabo Verde)
Sala *63 C6* Västmanland, C Sweden
Salacgrīva *84 C3* Est. Salatsi. Limbaži, N Latvia
Sala Consilina *75 D5* Campania, S Italy
Salado, Río *40 D5* river E Argentina
Salado, Río *42 C3* river C Argentina
Şalālah *99 D6* SW Oman
Salamá *30 B2* Baja Verapaz, C Guatemala
Salamanca *42 B4* Coquimbo, C Chile
Salamanca *70 D2* anc. Helmantica, Salmantica. Castilla y León, NW Spain
Salamīyah *96 B3* var. As Salamīyah. Ḩamāh, W Syria
Salang see Phuket
Salantai *84 B3* Klaipėda, NW Lithuania
Salatsi see Salacgrīva
Salavan *115 D5* var. Saravan, Saravane. Salavan, S Laos
Salavat *89 D6* Respublika Bashkortostan, W Russia
Sala y Gomez *131 F4* island Chile, E Pacific Ocean
Sala y Gomez Fracture Zone see Sala y Gomez Ridge
Sala y Gomez Ridge *131 G4* var. Sala y Gomez Fracture Zone. fracture zone SE Pacific Ocean
Salazar see N'Dalatando
Šalčininkai *85 C5* Vilnius, SE Lithuania
Salduba see Zaragoza
Saldus *84 B3* Ger. Frauenburg. Saldus, W Latvia
Sale *127 C7* Victoria, SE Australia
Salé *48 C2* NW Morocco
Salekhard *92 D3* prev. Obdorsk. Yamalo-Nenetskiy Avtonomnyy Okrug, N Russia
Salem *110 C2* Tamil Nādu, SE India
Salem *24 B3* state capital Oregon, NW USA
Salerno *75 D5* anc. Salernum. Campania, S Italy
Salerno, Gulf of *75 C5* Eng. Gulf of Salerno. gulf S Italy
Salerno, Gulf of see Salerno, Golfo di
Salernum see Salerno
Salfit *97 E6* C West Bank, Middle East
Salihorsk *85 C7* Rus. Soligorsk. Minskaya Voblasts', S Belarus
Salima *57 E2* Central, C Malawi
Salina *23 E5* Kansas, C USA
Salina Cruz *29 F5* Oaxaca, SE Mexico
Salinas *38 A2* Guayas, W Ecuador
Salinas *25 B6* California, W USA
Salisbury *67 D7* var. New Sarum. S England, United Kingdom
Salisbury see Harare
Sállan see Sørøya
Salliq see Coral Harbour
Sallyana see Şalyān
Salmantica see Salamanca
Salmon River *24 D3* river Idaho, NW USA
Salmon River Mountains *24 D3* mountain range Idaho, NW USA
Salo *63 D6* Länsi-Suomi, SW Finland
Salon-de-Provence *69 D6* Bouches-du-Rhône, SE France
Salonica/Salonika see Thessaloníki
Salonta *86 A3* Hung. Nagyszalonta. Bihor, W Romania
Sal'sk *89 B7* Rostovskaya Oblast', SW Russia
Salt see As Salţ
Salta *42 C2* Salta, NW Argentina
Saltash *67 C7* SW England, United Kingdom
Saltillo *29 E3* Coahuila, NE Mexico
Salt Lake City *22 B4* state capital Utah, W USA
Salto *42 D4* Salto, N Uruguay
Salton Sea *25 D8* lake California, W USA
Salvador *41 G3* prev. São Salvador. state capital Bahia, E Brazil
Salween *102 C2* Bur. Thanlwin, Chin. Nu Chiang, Nu Jiang. river SE Asia
Şalyān *113 E3* var. Sallyana. Mid Western, W Nepal
Salzburg *73 D6* anc. Juvavum. Salzburg, N Austria
Salzgitter *72 C4* prev. Watenstedt-Salzgitter. Niedersachsen, C Germany
Salzwedel *72 C3* Sachsen-Anhalt, N Germany
Šamac see Bosanski Šamac
Samakhixai see Attapu
Samalayuca *28 C1* Chihuahua, N Mexico
Samar *117 F2* island C Philippines
Samara *92 B3* prev. Kuybyshev. Samarskaya Oblast', W Russia
Samarang see Semarang
Samarinda *116 D4* Borneo, C Indonesia
Samarkand see Samarqand
Samarkandski/Samarkandskoye see Temirtau
Samarobriva see Amiens
Samarqand *101 E2* Rus. Samarkand. Samarqand Viloyati, C Uzbekistan
Samawa see As Samāwah
Şamaxı *95 H2* E Azerbaijan
Sambalpur *113 F4* Odisha, E India
Sambava *57 G2* Antsiranana, NE Madagascar
Sambir *86 B2* Rus. Sambor. L'vivs'ka Oblast', NW Ukraine
Sambor see Sambir
Sambre *68 D2* river Belgium/France
Samfya *56 D2* Luapula, N Zambia
Saminatal *72 E2* valley Austria/Liechtenstein, Europe
Samnān see Semnān
Sam Neua see Xam Nua
Samoa *123 E4* off. Independent State of Western Samoa, var. Sāmoa; prev. Western Samoa. country W Polynesia
Sāmoa see Samoa
Samoa Basin *121 E3* undersea basin W Pacific Ocean
Samobor *78 A2* Zagreb, N Croatia
Sámos *83 E6* prev. Limín Vathéos. Sámos, Dodekánisa, Greece, Aegean Sea
Sámos *83 D6* island Dodekánisa, Greece, Aegean Sea
Samothrace see Samothráki
Samothráki *82 D4* Samothráki, NE Greece
Samothraki *82 C4* anc. Samothrace. island NE Greece
Sampit *116 C4* Borneo, C Indonesia
Samraong *115 D5* prev. Phumĭ Sâmraông, Phum Samrong. Oddar Meanchey, NW Cambodia
Samsun *94 D2* anc. Amisus. Samsun, N Turkey (Türkiye)
Samt'redia *95 F2* W Georgia
Samui, Ko *115 C6* island SW Thailand
Samut Prakan *115 C5* var. Muang Samut Prakan, Paknam. Samut Prakan, C Thailand

San 52 D3 Ségou, C Mali
San 77 E5 river SE Poland
Şan'ā' 99 B6 Eng. Sanaa. country capital (Yemen) W Yemen
Sana 78 B3 river NW Bosnia and Herzegovina
Sanaa see Şan'ā'
Sanae 132 B2 South African research station Antarctica
Sanaga 55 B5 river C Cameroon
San Ambrosio, Isla 35 A5 Eng. San Ambrosio Island. island W Chile
San Ambrosio Island see San Ambrosio, Isla
Sanandaj 98 C3 prev. Sinneh. Kordestān, W Iran
San Andrés, Isla de 31 F3 island NW Colombia, Caribbean Sea
San Andrés Tuxtla 29 F4 var. Tuxtla. Veracruz-Llave, E Mexico
San Angelo 27 F3 Texas, SW USA
San Antonio 30 B2 Toledo, S Belize
San Antonio 42 B4 Valparaíso, C Chile
San Antonio 27 F4 Texas, SW USA
San Antonio Oeste 43 C5 Río Negro, E Argentina
San Antonio River 27 G4 river Texas, SW USA
Sanāw 99 C6 var. Sanaw. NE Yemen
San Benedicto, Isla 28 B4 island W Mexico
San Benito 30 B1 Petén, N Guatemala
San Benito 27 G5 Texas, SW USA
San Bernardino 25 C7 California, W USA
San Blas 28 C3 Sinaloa, C Mexico
San Blas, Cape 20 D3 headland Florida, SE USA
San Blas, Cordillera de 31 G4 mountain range NE Panama
San Carlos 30 D4 Río San Juan, S Nicaragua
San Carlos 26 B2 Arizona, SW USA
San Carlos see Quesada, Costa Rica
San Carlos de Ancud see Ancud
San Carlos de Bariloche 43 B5 Río Negro, SW Argentina
San Carlos del Zulia 36 C2 Zulia, W Venezuela
San Clemente Island 25 B8 island Channel Islands, California, W USA
San Cristóbal 36 C2 Táchira, W Venezuela
San Cristóbal 122 C4 var. Makira. island SE Solomon Islands
San Cristóbal see San Cristóbal de Las Casas
San Cristóbal de Las Casas 29 G5 var. San Cristóbal. Chiapas, SE Mexico
San Cristóbal, Isla 38 B5 var. Chatham Island. island Galápagos Islands, Ecuador, E Pacific Ocean
Sancti Spíritus 32 B2 Sancti Spíritus, C Cuba
Sandakan 116 D3 Sabah, East Malaysia
Sandalwood Island see Sumba, Pulau
Sandanski 82 C3 prev. Sveti Vrach. Blagoevgrad, SW Bulgaria
Sanday 66 D2 island NE Scotland, United Kingdom
Sanders 26 C2 Arizona, SW USA
Sand Hills 22 D3 mountain range Nebraska, C USA
San Diego 25 C8 California, W USA
Sandnes 63 A6 Rogaland, S Norway
Sandomierz 76 D4 Rus. Sandomir. Świętokrzyskie, C Poland
Sandomir see Sandomierz
Sandoway see Thandwe
Sandpoint 24 C1 Idaho, NW USA
Sand Springs 27 G1 Oklahoma, C USA
Sandusky 18 D3 Ohio, N USA
Sandvika 63 A6 Akershus, S Norway
Sandviken 63 C6 Gävleborg, C Sweden
Sandwich Island see Efate
Sandwich Islands see Hawaiian Islands
Sandy Bay 71 H5 Saskatchewan, C Canada
Sandy City 22 B4 Utah, W USA
Sandy Lake 16 B3 lake Ontario, C Canada
San Esteban 30 D2 Olancho, C Honduras
San Eugenio/San Eugenio del Cuareim see Artigas
San Felipe 36 D1 Yaracuy, NW Venezuela
San Felipe de Puerto Plata see Puerto Plata
San Félix, Isla 35 A5 Eng. San Felix Island. island W Chile
San Felix Island see San Félix, Isla
San Fernando 70 C5 prev. Isla de León. Andalucía, S Spain
San Fernando 33 H5 Trinidad, Trinidad and Tobago
San Fernando 24 D1 California, W USA
San Fernando 36 D2 var. San Fernando de Apure. Apure, C Venezuela
San Fernando de Apure see San Fernando
San Fernando del Valle de Catamarca 42 C3 var. Catamarca. Catamarca, NW Argentina
San Fernando de Monte Cristi see Monte Cristi
San Francisco 25 B6 California, W USA
San Francisco del Oro 28 C2 Chihuahua, N Mexico
San Francisco de Macorís 33 E3 C Dominican Republic
San Fructuoso see Tacuarembó
San Gabriel 38 B1 Carchi, N Ecuador
San Gabriel Mountains 24 E1 mountain range California, USA
Sangihe, Kepulauan see Sangir, Kepulauan
San Giljan 80 B5 var. St Julian's. N Malta
Sangir, Kepulauan 117 F3 var. Kepulauan Sangihe. island group N Indonesia
Sāngli 110 B1 Mahārāshtra, W India
Sangmélima 55 B5 Sud, S Cameroon
Sangre de Cristo Mountains 26 D1 mountain range Colorado/New Mexico, C USA
San Ignacio 30 B1 prev. Cayo, El Cayo. Cayo, W Belize
San Ignacio 39 F3 Beni, N Bolivia
San Ignacio 28 B2 Baja California Sur, NW Mexico
San Joaquin Valley 25 B7 valley California, W USA
San Jorge, Golfo 43 C6 var. Gulf of San Jorge. gulf S Argentina
San Jorge, Gulf of see San Jorge, Golfo
San José 31 E4 country capital (Costa Rica) San José, C Costa Rica
San José 39 G3 var. San José de Chiquitos. Santa Cruz, E Bolivia
San José 30 B3 var. Puerto San José. Escuintla, S Guatemala
San Jose 25 B6 California, W USA
San José see San José del Guaviare, Colombia
San José de Chiquitos see San José
San José de Cúcuta see Cúcuta
San José del Guaviare 36 C4 var. San José. Guaviare, S Colombia
San Juan 42 B4 San Juan, W Argentina
San Juan 33 F3 dependent territory capital (Puerto Rico) NE Puerto Rico
San Juan see San Juan de los Morros
San Juan Bautista 42 D3 Misiones, S Paraguay
San Juan Bautista see Villahermosa
San Juan Bautista Tuxtepec see Tuxtepec
San Juan de Alicante see Sant Joan d'Alacant
San Juan del Norte 31 E4 var. Greytown. Río San Juan, SE Nicaragua
San Juan de los Morros 36 D2 var. San Juan. Guárico, N Venezuela
San Juanito, Isla 28 C4 island C Mexico
San Juan Mountains 26 D1 mountain range Colorado, C USA
San Juan, Río 31 E4 river Costa Rica/Nicaragua
San Juan River 26 C1 river Colorado/Utah, W USA
San Julián see Puerto San Julián
Sankt-Bartholomäi see Palamuse
Sankt Gallen 73 B7 var. St. Gallen, Eng. Saint Gall, Fr. St-Gall. Sankt Gallen, NE Switzerland
Sankt-Georgen see Sfântu Gheorghe
Sankt-Jakobi see Pärnu-Jaagupi, Pärnumaa, Estonia
Sankt Martin see Martin
Sankt Moritz see St. Moritz
Sankt-Peterburg 88 B4 prev. Leningrad, Petrograd, Eng. Saint Petersburg, Fin. Pietari. Leningradskaya Oblast', NW Russia
Sankt Pölten 73 E6 Niederösterreich, N Austria
Sankt Veit am Flaum see Rijeka
Sankuru 55 D6 river C Dem. Rep. Congo
Şanlıurfa 95 E4 prev. Sanli Urfa, Urfa; anc. Edessa. Şanlıurfa, S Turkey (Türkiye)
Sanli Urfa see Şanlıurfa
San Lorenzo 39 G5 Tarija, S Bolivia
San Lorenzo 38 A1 Esmeraldas, N Ecuador
San Lorenzo, Isla 38 C4 island W Peru
Sanlúcar de Barrameda 70 C5 Andalucía, S Spain
San Luis 42 C4 San Luis, C Argentina
San Luis 30 B2 Petén, NE Guatemala
San Luis see San Luis Río Colorado
San Luis Obispo 25 B7 California, W USA
San Luis Potosí 29 E3 San Luis Potosí, C Mexico
San Luis Río Colorado 28 A1 var. San Luis Río Colorado. Sonora, NW Mexico
San Marcos 30 A2 San Marcos, W Guatemala
San Marcos 27 G4 Texas, SW USA
San Marcos de Arica see Arica
San Marino 74 E1 country capital (San Marino) C San Marino
San Marino 74 D1 off. Republic of San Marino. country S Europe
San Marino, Republic of see San Marino
San Martín 132 A2 Argentinian research station Antarctica
San Mateo 37 E2 Anzoátegui, NE Venezuela
San Matías 39 H3 Santa Cruz, E Bolivia
San Matías, Gulf of 43 C5 var. Gulf of San Matías. gulf E Argentina
San Matías, Gulf of see San Matías, Golfo
Sanmenxia 106 C4 var. Shan Xian. Henan, C China
Sânmiclăuş Mare see Sânnicolau Mare
San Miguel 30 C3 San Miguel, SE El Salvador
San Miguel 28 D2 Coahuila, N Mexico
San Miguel de Ibarra see Ibarra
San Miguel de Tucumán 42 C3 var. Tucumán. Tucumán, N Argentina
San Miguelito 31 G4 Panamá, C Panama
San Miguel, Río 39 G3 river E Bolivia
San Murezzan see St. Moritz
Sannār see Sennar
Sânnicolaul-Mare see Sânnicolau Mare
Sânnicolau Mare 86 A4 var. Sânnicolaul-Mare, Hung. Nagyszentmiklós; prev. Sânmiclăuş Mare, Sînnicolau Mare. Timiş, W Romania
Sanok 77 E5 Podkarpackie, SE Poland
San Pablo 39 F5 Potosí, S Bolivia
San Pedro 30 C1 Corozal, NE Belize
San-Pédro 52 D5 S Ivory Coast
San Pedro 28 D3 var. San Pedro de las Colonias. Coahuila, NE Mexico
San Pedro de la Cueva 28 C2 Sonora, NW Mexico
San Pedro de las Colonias see San Pedro
San Pedro de Lloc 38 B3 La Libertad, NW Peru
San Pedro Mártir, Sierra 28 A1 mountain range NW Mexico
San Pedro Sula 30 C2 Cortés, NW Honduras
San Rafael 42 B4 Mendoza, W Argentina
San Rafael Mountains 25 C7 mountain range California, W USA
San Ramón de la Nueva Orán 42 C2 Salta, N Argentina
San Remo 74 A3 Liguria, NW Italy
San Salvador 30 B3 country capital (El Salvador) San Salvador, SW El Salvador
San Salvador 32 D2 prev. Watlings Island. island E The Bahamas
San Salvador de Jujuy 42 C2 var. Jujuy. Jujuy, N Argentina
San Salvador, Isla 38 A4 island Ecuador
Sansanné-Mango see Mango
San Sebastián 71 E1 País Vasco, N Spain see also Donostia
Sansepolcro 74 C3 Toscana, C Italy
San Severo 75 D5 Puglia, SE Italy
Santa Ana 39 F3 Beni, N Bolivia
Santa Ana 30 B3 Santa Ana, NW El Salvador
Santa Ana 24 D2 California, W USA
Santa Ana de Coro see Coro
Santa Ana Mountains 24 E2 mountain range California, W USA
Santa Barbara 28 C2 Chihuahua, N Mexico
Santa Barbara 25 C7 California, W USA
Santa Catalina de Armada 70 B1 var. Santa Comba. Galicia, NW Spain
Santa Catalina Island 25 B8 island Channel Islands, California, W USA
Santa Catarina 41 E5 off. Estado de Santa Catarina. state/region S Brazil
Santa Catarina, Estado de see Santa Catarina
Santa Clara 32 B2 Villa Clara, C Cuba
Santa Clarita 24 D1 California, USA
Santa Comba see Santa Cataliña de Armada
Santa Cruz 54 E2 São Tomé, S Sao Tome and Principe, Africa
Santa Cruz 25 B6 California, W USA
Santa Cruz 39 G4 department E Bolivia
Santa Cruz Barillas see Barillas
Santa Cruz del Quiché 30 B2 Quiché, W Guatemala
Santa Cruz de Tenerife 48 A3 Tenerife, Islas Canarias, Spain, NE Atlantic Ocean
Santa Cruz, Isla 38 B5 var. Indefatigable Island, Isla Chávez. island Galápagos Islands, Ecuador, E Pacific Ocean
Santa Cruz Islands 122 D3 island group E Solomon Islands
Santa Cruz, Río 43 B7 river S Argentina
Santa Elena 30 B1 Cayo, W Belize
Santa Fe 42 C4 Santa Fe, C Argentina
Santa Fe 26 D1 state capital New Mexico, SW USA
Santa Fe see Bogotá
Santa Fe de Bogotá see Bogotá
Santa Genoveva 28 B3 mountain NW Mexico
Santa Isabel 122 C3 var. Bughotu. island N Solomon Islands
Santa Isabel see Malabo
Santa Lucia Range 25 B7 mountain range California, W USA
Santa Margarita, Isla 28 B3 island NW Mexico
Santa Maria 41 E5 Rio Grande do Sul, S Brazil
Santa Maria 25 B7 California, USA
Santa Maria 70 A5 island Azores, Portugal, NE Atlantic Ocean
Santa María del Buen Aire see Buenos Aires
Santa María, Isla 38 A5 var. Isla Floreana, Charles Island. island Galápagos Islands, Ecuador, E Pacific Ocean
Santa Marta 36 B1 Magdalena, N Colombia
Santa Maura see Lefkáda
Santa Monica 24 D1 California, W USA
Santana 54 E2 São Tomé, C Sao Tome and Principe, Africa
Santander 70 D1 Cantabria, N Spain
Santarém 41 E2 Pará, N Brazil
Santarém 70 B3 anc. Scalabis. Santarém, W Portugal
Santa Rosa 42 C4 La Pampa, C Argentina
Santa Rosa see Santa Rosa de Copán
Santa Rosa de Copán 30 C2 var. Santa Rosa. Copán, W Honduras
Santa Rosa Island 25 B8 island California, W USA
Santa Uxía de Ribeira 70 B1 var. Ribeira NW Spain
Sant Carles de la Rápida see Sant Carles de la Ràpita
Sant Carles de la Ràpita 71 F3 var. Sant Carles de la Rápida. Cataluña, NE Spain
Santiago 42 B4 var. Gran Santiago. country capital (Chile) Santiago, C Chile
Santiago 33 E3 var. Santiago de los Caballeros. N Dominican Republic
Santiago 31 F5 Veraguas, S Panama
Santiago see Santiago de Compostela
Santiago 52 A3 var. São Tiago. island Ilhas de Sotavento, S Cape Verde (Cabo Verde)
Santiago see Santiago de Cuba, Cuba
Santiago de Compostela 70 B1 var. Santiago, Eng. Compostella; anc. Campus Stellae. Galicia, NW Spain
Santiago de Cuba 32 C3 var. Santiago. Santiago de Cuba, E Cuba
Santiago de Guayaquil see Guayaquil
Santiago del Estero 42 C3 Santiago del Estero, C Argentina
Santiago de los Caballeros see Santiago, Dominican Republic
Santiago de los Caballeros see Ciudad de Guatemala, Guatemala
Santiago Pinotepa Nacional see Pinotepa Nacional
Santiago, Río 38 B2 river N Peru
Santi Quaranta see Sarandë
Santissima Trinidad see Keelung
Sant Joan d'Alacant 71 F4 Cast. San Juan de Alicante. Comunitat Valenciana, E Spain
Sant Julià de Lòria 69 A8 Sant Julià de Lòria, SW Andorra Europe
Santo see Espiritu Santo
Santo Antão 52 A2 island Ilhas de Barlavento, N Cape Verde (Cabo Verde)
Santo António 54 E1 Príncipe, N Sao Tome and Principe, Africa
Santo Domingo 33 D3 prev. Ciudad Trujillo. country capital (Dominican Republic) SE Dominican Republic
Santo Domingo de los Colorados 38 B1 Pichincha, NW Ecuador
Santo Domingo Tehuantepec see Tehuantepec
Santorini see Thíra
Santorini 83 D7 island Kykládes, Greece, Aegean Sea
San Tomé de Guayana see Ciudad Guayana
Santos 41 F4 São Paulo, S Brazil
Santos Plateau 35 D5 undersea plateau SW Atlantic Ocean
Santo Tomé 42 D3 Corrientes, NE Argentina
Santo Tomé de Guayana see Ciudad Guayana
San Valentín, Cerro 43 A6 mountain S Chile
San Vicente 30 C3 San Vicente, C El Salvador
São Francisco, Rio 41 F3 river E Brazil
Sao Hill 51 C7 Iringa, S Tanzania
São João da Madeira 70 B2 Aveiro, N Portugal
São Jorge 70 A5 island Azores, Portugal, NE Atlantic Ocean
São Luís 41 F2 state capital Maranhão, NE Brazil
São Mandol see São Manuel, Rio
São Manuel, Rio 41 E3 var. São Mandol, Teles Pirés. river C Brazil
São Marcos, Baía de 41 F1 bay N Brazil
São Miguel 70 A5 island Azores, Portugal, NE Atlantic Ocean
Saona, Isla 33 E3 island SE Dominican Republic
Saône 69 D5 river E France
São Nicolau 52 A3 Eng. Saint Nicholas. island Ilhas de Barlavento, N Cape Verde (Cabo Verde)
São Paulo 41 E4 state capital São Paulo, S Brazil
São Paulo 41 E4 off. Estado de São Paulo. state/region S Brazil
São Paulo de Loanda see Luanda
São Paulo, Estado de see São Paulo
São Pedro do Rio Grande do Sul see Rio Grande
São Roque, Cabo de 41 G2 headland E Brazil
São Salvador see Salvador, Brazil
São Salvador/São Salvador do Congo see M'Banza Congo, Angola
São Tiago see Santiago
São Tomé 55 A5 country capital (Sao Tome and Principe) São Tomé, S Sao Tome and Principe
São Tomé 54 E2 Eng. Saint Thomas. island S Sao Tome and Principe
Sao Tome and Principe 54 D1 off. Democratic Republic of Sao Tome and Principe. country E Atlantic Ocean
Sao Tome and Principe, Democratic Republic of see Sao Tome and Principe
São Tomé, Pico de 54 D2 mountain São Tomé, C Sao Tome and Principe, Africa
São Vicente 52 A3 Eng. Saint Vincent. island Ilhas de Barlavento, N Cape Verde (Cabo Verde)
São Vicente, Cabo de 70 B5 Eng. Cape Saint Vincent, Port. Cabode São Vicente. cape S Portugal
São Vicente, Cabo de see São Vicente, Cabo de
Sápai see Sápes
Sapele 53 F5 Delta, S Nigeria
Sápes 82 D3 var. Sápai. Anatolikí Makedonía kai Thráki, NE Greece
Sapir 97 B7 prev. Sappir. Southern, S Israel
Sa Pobla 71 G3 Mallorca, Spain, W Mediterranean Sea
Sappir see Sapir
Sapporo 108 D2 Hokkaidō, NE Japan
Sapri 75 D6 Campania, S Italy
Sapulpa 27 G1 Oklahoma, C USA
Saqqez 98 C2 var. Saghez. Sakiz. Saqqiz. Kordestān, NW Iran
Saqqiz see Saqqez
Sara Buri 115 C5 var. Saraburi. Saraburi, C Thailand
Saraburi see Sara Buri
Saragossa see Zaragoza
Saragt see Sarahs
Saraguro 38 B2 Loja, S Ecuador
Sarahs 100 D3 var. Saragt, Rus. Serakhs. Ahal Welaýaty, S Turkmenistan
Sarajevo 78 C4 country capital (Bosnia and Herzegovina) Federacija Bosna I Hercegovina, SE Bosnia and Herzegovina
Sarakhs 98 E2 Khorāsān-Razavī, NE Iran
Saraktash 89 D6 Orenburgskaya Oblast', W Russia
Saran' 92 C4 Kaz. Saran. Karagandy, C Kazakhstan
Saranda see Sarandë
Sarandë 79 C7 var. Saranda, It. Porto Edda; prev. Santi Quaranta. Vlorë, S Albania
Saransk 89 C5 Respublika Mordoviya, W Russia
Sarasota 21 E4 Florida, SE USA
Saratov 92 B3 Saratovskaya Oblast', W Russia
Saravan/Saravane see Salavan
Sarawak 116 D3 state East Malaysia
Sarawak see Kuching
Sarcelles 68 D1 Val-d'Oise, Île-de-France, N France Europe
Sardegna 75 A5 Eng. Sardinia. island Italy, C Mediterranean Sea
Sardinia see Sardegna
Sarera, Teluk see Cenderawasih, Teluk
Sargasso Sea 44 B4 sea W Atlantic Ocean
Sargodha 112 C2 Punjab, NE Pakistan
Sarh 54 C4 prev. Fort-Archambault. Moyen-Chari, S Chad
Sārī 98 D2 var. Sari, Sāri. Māzandarān, N Iran
Saría 83 E7 island SE Greece
Sarıkamış 95 F3 Kars, NE Turkey (Türkiye)
Sarikol Range 101 G3 Rus. Sarykol'skiy Khrebet. mountain range China/Tajikistan
Sark 67 D8 Fr. Sercq. island Channel Islands
Şarkışla 94 D3 Sivas, C Turkey (Türkiye)
Sarmiento 43 B6 Chubut, S Argentina
Sarnia 16 C5 Ontario, S Canada
Sarny 86 C1 Rivnens'ka Oblast', NW Ukraine
Sarochyna 85 D5 Rus. Sorochino. Vitsyebskaya Voblasts', N Belarus
Sarov 89 C5 prev. Sarova. Respublika Mordoviya, SW Russia
Sarova see Sarov
Sarpsborg 63 B6 Østfold, S Norway
Sarrebruck see Saarbrücken
Sartène 69 E7 Corse, France, C Mediterranean Sea
Sarthe 68 B4 river N France
Sárti 82 C4 Kentrikí Makedonía, N Greece
Saruhan see Manisa
Saryarqa 92 C4 var. Kazakhskiy Melkosopochnik. uplands C Kazakhstan
Sarykol'skiy Khrebet see Sarikol Range
Sary-Tash 101 F2 Oshskaya Oblast', SW Kyrgyzstan
Saryyesik-Atyrau, Peski 101 G1 desert E Kazakhstan
Sasebo 109 A7 Nagasaki, Kyūshū, SW Japan
Saskatchewan 15 F5 province SW Canada
Saskatchewan 15 F5 river Manitoba/Saskatchewan, C Canada
Saskatoon 15 F5 Saskatchewan, S Canada
Sasovo 89 B5 Ryazanskaya Oblast', W Russia
Sassandra 52 D5 S Ivory Coast
Sassandra 52 D5 var. Ibo, Sassandra Fleuve. river S Ivory Coast
Sassandra Fleuve see Sassandra
Sassari 75 A5 Sardegna, Italy, C Mediterranean Sea
Sassenheim 64 C3 Zuid-Holland, W Netherlands
Sassnitz 72 D2 Mecklenburg-Vorpommern, NE Germany
Sathmar see Satu Mare
Sátoraljaújhely 77 D6 Borsod-Abaúj-Zemplén, NE Hungary
Satpura Range 112 D4 mountain range C India
Satsuma-Sendai 109 A8 Kagoshima, Kyūshū, SW Japan
Satsunan-shoto 108 A3 island group Nansei-shotō, SW Japan Asia
Sattanen 62 D3 Lappi, NE Finland
Satu Mare 86 B3 Ger. Sathmar, Hung. Szatmárrnémeti. Satu Mare, NW Romania
Sau see Sava
Saudi Arabia 99 B5 off. Kingdom of Saudi Arabia, Al 'Arabīyah as Su'ūdīyah, Ar. Al Mamlakah al 'Arabīyah as Su'ūdīyah. country SW Asia
Saudi Arabia, Kingdom of see Saudi Arabia
Sauer see Sûre
Saugor see Sāgar
Saulkrasti 84 C3 Rīga, C Latvia
Sault Sainte Marie 16 C4 Michigan, N USA
Sault Sainte Marie 18 C1 Michigan, N USA
Sault Ste. Marie 16 C4 Ontario, S Canada
Saumur 68 B4 Maine-et-Loire, NW France
Saurimo 56 C1 Port. Henrique de Carvalho, Vila Henrique de Carvalho. Lunda Sul, NE Angola
Sava 85 E6 Mahilyowskaya Voblasts', E Belarus
Sava 78 B3 Eng. Save, Ger. Sau, Hung. Száva. river SE Europe
Savá 30 D2 Colón, N Honduras
Savai'i 123 E4 island NW Samoa
Savannah 21 E2 Georgia, SE USA
Savannah River 21 E2 river Georgia/South Carolina, SE USA
Savannakhét see Khanthabouli
Savanna-La-Mar 32 A5 W Jamaica
Savaria see SzombathelySave see Sava
Save, Rio 57 E3 var. Sabi. river Mozambique/Zimbabwe
Saverne 68 E3 var. Zabern; anc. Tres Tabernae. Bas-Rhin, NE France
Savigliano 74 A2 Piemonte, NW Italy
Savigsivik see Savissivik
Savinski see Savinskiy
Savinskiy 88 C3 var. Savinski. Arkhangel'skaya Oblast', NW Russia
Savissivik 60 D1 var. Savigsivik. Avannaata, N Greenland
Savoie 69 D5 cultural region E France
Savona 74 A2 Liguria, NW Italy
Savu Sea 117 E5 Ind. Laut Sawu. sea S Indonesia
Sawakin see Suakin
Sawdirī see Sodiri
Sawhāj see Sūhāj
Şawqirah 99 D6 var. Suqrah. S Oman
Sawu, Laut see Savu Sea
Saxe see Sachsen
Saxony see Sachsen
Sayaboury see Xaignabouli
Sayanskiy Khrebet 90 D3 mountain range S Russia
Saýat 100 D3 Rus. Sayat. Lebap Welaýaty, E Turkmenistan
Sayaxché 30 B2 Petén, N Guatemala
Şaydā/Sayida see Saïda
Sayḩūt 99 D6 E Yemen
Saynshand 105 E2 Dornogovī, SE Mongolia
Sayre 19 E3 Pennsylvania, NE USA
Say'ūn 99 C6 var. Saywūn. C Yemen
Saywūn see Say'ūn
Scalabis see Santarém
Scandinavia 44 D2 geophysical region NW Europe
Scarborough 67 D5 N England, United Kingdom
Scarpanto see Kárpathos
Scebeli see Shebeli
Schaan 72 E1 W Liechtenstein Europe
Schaerbeek 65 C6 Brussels, C Belgium
Schaffhausen 73 B7 Fr. Schaffhouse. Schaffhausen, N Switzerland
Schaffhouse see Schaffhausen
Schagen 64 C2 Noord-Holland, NW Netherlands
Schaulen see Šiauliai
Schebschi Mountains see Shebshi Mountains
Scheessel 72 B3 Niedersachsen, NW Germany
Schefferville 17 E2 Québec, E Canada
Schelde see Scheldt
Scheldt 65 B5 Dut. Schelde, Fr. Escaut. river W Europe
Schell Creek Range 25 D5 mountain range Nevada, W USA
Schenectady 19 F3 New York, NE USA
Schertz 27 G4 Texas, SW USA
Schiermonnikoog 64 D1 Fris. Skiermûntseach. island Waddeneilanden, N Netherlands
Schijndel 64 D4 Noord-Brabant, S Netherlands
Schil see Jiu
Schiltigheim 68 E3 Bas-Rhin, NE France
Schivelbein see Świdwin
Schleswig 72 B2 Schleswig-Holstein, N Germany
Schleswig-Holstein 72 B2 state N Germany
Schlettstadt see Sélestat
Schlochau see Człuchów
Schneekoppe see Sněžka
Schneidemühl see Piła
Schoden see Skuodas
Schönebeck 72 C4 Sachsen-Anhalt, C Germany
Schönlanke see Trzcianka
Schooten see Schoten
Schoten 65 C5 var. Schooten. Antwerpen, N Belgium
Schouwen 64 B4 island SW Netherlands
Schwabenalb see Schwäbische Alb
Schwäbische Alb 73 B6 var. Schwabenalb, Eng. Swabian Jura. mountain range S Germany
Schwandorf 73 C5 Bayern, SE Germany
Schwaz 73 C7 Tirol, W Austria
Schweidnitz see Świdnica
Schweinfurt 73 B5 Bayern, SE Germany
Schweiz see Switzerland
Schwerin 72 C3 Mecklenburg-Vorpommern, N Germany
Schwertberg see Świecie
Schwiebus see Świebodzin
Schwyz 73 B7 var. Schwiz. Schwyz, C Switzerland
Schyl see Jiu
Scio see Chíos
Scoresby Sound/Scoresbysund see Ittoqqortoormiit
Scoresby Sund see Kangertittivaq
Scotia Sea 35 C8 sea SW Atlantic Ocean
Scotland 66 C3 cultural region Scotland, United Kingdom
Scott Base 132 B4 NZ research station Antarctica
Scott Island 132 B5 island Antarctica
Scottsbluff 22 D3 Nebraska, C USA
Scottsboro 20 D1 Alabama, S USA
Scottsdale 26 B2 Arizona, SW USA
Scranton 19 F3 Pennsylvania, NE USA
Scrobesbyrig' see Shrewsbury
Scupi see Skopje
Scutari see Shkodër
Scutari, Lake 79 C5 Alb. Liqeni i Shkodrës, Croatian Skadarsko Jezero. lake Albania/Montenegro
Scyros see Skýros
Searcy 20 B1 Arkansas, C USA
Seattle 24 B2 Washington, NW USA
Sébaco 30 D3 Matagalpa, W Nicaragua
Sebaste/Sebastia see Sivas
Sebastián Vizcaíno, Bahía 28 A2 bay NW Mexico
Sebastopol see Sevastopol
Sebenico see Šibenik
Sechura, Bahía de 38 A3 bay NW Peru
Secunderābād 112 D5 var. Sikandarabad. Telangana, C India
Sedan 68 D3 Ardennes, N France
Seddon 129 D5 Marlborough, South Island, New Zealand
Seddonville 129 C5 West Coast, South Island, New Zealand
Sédhiou 52 B3 SW Senegal
Sedlez see Siedlce
Sedona 26 B2 Arizona, SW USA
Sedunum see Sion

Seeland see Sjælland
Seenu Atoll see Addu Atoll
Seesen 72 B4 Niedersachsen, C Germany
Segestica see Sisak
Segezha 88 B3 Respublika Kareliya, NW Russia
Seghedin see Szeged
Segna see Senj
Segodunum see Rodez
Ségou 52 D3 var. Segu. Ségou, C Mali
Segovia 70 D2 Castilla y León, C Spain
Segoviao Wangkí see Coco, Río
Segu see Ségou
Séguédine 53 H2 Agadez, NE Niger
Seguin 27 G4 Texas, SW USA
Segura 71 E4 river S Spain
Seinäjoki 63 D5 Swe. Östermyra. Etelä-Pohjanmaa, W Finland
Seine 68 D1 river N France
Seine, Baie de la 68 B3 bay N France
Sejong 106 E4 var. Sejong City. administrative capital (South Korea) C South Korea
Sejong City see Sejong
Sekondi see Sekondi-Takoradi
Sekondi-Takoradi 53 E5 var. Sekondi. S Ghana
Selânik see Thessaloníki
Selenga 105 E1 Mong. Selenge Mörön. river Mongolia/Russia
Selenge Mörön see Selenga
Sélestat 68 E4 Ger. Schlettstadt. Bas-Rhin, NE France
Seleucia see Silifke
Selfoss 61 E5 Sudhurland, SW Iceland
Sélibabi 52 C3 var. Sélibaby. Guidimaka, S Mauritania
Sélibaby see Sélibabi
Selma 25 C6 California, W USA
Selway River 24 D2 river Idaho, NW USA
Selwyn Range 126 B3 mountain range Queensland, C Australia
Selzaete see Zelzate
Semarang 116 C5 var. Samarang. Jawa, C Indonesia
Sembé 55 B5 Sangha, NW Congo
Semendria see Smederevo
Semey 92 D4 prev. Semipalatinsk. Vostochnyy Kazakhstan, E Kazakhstan
Semechevo see Syemyezhava
Seminole 27 E3 Texas, SW USA
Seminole, Lake 20 D3 reservoir Florida/Georgia, SE USA
Semipalatinsk see Semey
Semnān 98 D3 var. Samnān. Semnān, N Iran
Semois 65 C8 river SE Belgium
Sendai 108 D4 Miyagi, Honshū, C Japan
Sendai-wan 108 D4 bay E Japan
Senec 77 C6 Ger. Wartberg, Hung. Szenc; prev. Szempcz. Bratislavský Kraj, W Slovakia
Senegal 52 B3 off. Republic of Senegal, Fr. Sénégal. country W Africa
Senegal 52 C3 Fr. Sénégal. river W Africa
Senegal, Republic of see Senegal
Senftenberg 72 D4 Brandenburg, E Germany
Senia see Senj
Senica 77 C6 Ger. Senitz, Hung. Szenice. Trnavský Kraj, W Slovakia
Seniça see Sjenica
Senitz see Senica
Senj 78 A3 Ger. Zengg, It. Segna; anc. Senia. Lika-Senj, NW Croatia
Senja 62 C2 prev. Senjen. island N Norway
Senjen see Senja
Senkaku Islands see Senkaku-shoto
Senkaku-shoto 108 A3 var. Senkaku Islands, Chin. Diaoyutai Lieyu, Diaoyutai Qundao. disputed island group SW Japan
Senlis 68 C3 Oise, N France
Sennar 50 C4 var. Sannār. Sinnar, C Sudan
Senones see Sens
Sens 68 C3 anc. Agendicum, Senones. Yonne, C France
Sensburg see Mrągowo
Sên, Stœng 115 D5 river C Cambodia
Senta 78 D3 Hung. Zenta. Vojvodina, N Serbia
Seo de Urgel see La See d'Urgel
Seoul 107 E4 off. Seoul Teukbyeolsi, prev. Sŏul, Jap. Keijō; prev. Kyŏngsŏng. country capital (South Korea) NW South Korea
Seoul Teukbyeolsi see Seoul
Şepşi-Sângeorz/Sepsiszentgyőrgy see Sfântu Gheorghe
Sept-Îles 17 E3 Québec, SE Canada
Seraing 65 D6 Liège, E Belgium
Serakhs see Sarahs
Seram, Laut 117 F4 Eng. Ceram Sea. sea E Indonesia
Pulau Seram 117 F4 var. Serang, Eng. Ceram. island Maluku, E Indonesia
Serang 116 C5 Jawa, C Indonesia
Serang see Seram, Pulau
Serasan, Selat 116 C3 strait Indonesia/Malaysia
Serbia 78 D4 off. Republic of Serbia; prev. Federal Republic of Serbia, Yugoslavia, Serb. Srbija, Republika Srbija. country SE Europe
Serbia, Federal Republic of see Serbia
Serbia, Republic of see Serbia
Sercq see Sark
Serdar 100 C2 prev. Rus. Gyzyrlabat, Kizyl-Arvat. Balkan Welaýaty, W Turkmenistan
Serdica see Sofia
Serdobol' see Sortavala
Serenje 56 D2 Central, E Zambia
Seres see Sérres
Seret/Sereth see Siret
Serhetabat 100 D4 prev. Rus. Gushgy, Kushka. Mary Welaýaty, S Turkmenistan
Sérifos 83 C6 anc. Seriphos. island Kykládes, Greece, Aegean Sea
Seriphos see Sérifos
Serov 92 C3 Sverdlovskaya Oblast', C Russia
Serowe 56 D3 Central, SE Botswana
Serpa Pinto see Menongue
Serpent's Mouth, The 37 F2 Sp. Boca de la Serpiente. strait Trinidad and Tobago/Venezuela
Serpiente, Boca de la see Serpent's Mouth, The
Serpukhov 89 B5 Moskovskaya Oblast', W Russia
Sérrai see Sérres
Serrana, Cayo de 31 F2 island group NW Colombia South America
Serranilla, Cayo de 31 F2 island group NW Colombia South America Caribbean Sea
Serravalle 74 E1 N San Marino
Sérres 82 C3 var. Seres; prev. Sérrai. Kentrikí Makedonía, NE Greece
Sesdlets see Siedlce
Sesto San Giovanni 74 B2 Lombardia, N Italy
Sesvete 78 B2 Zagreb, N Croatia
Setabis see Xàtiva
Sète 69 C6 prev. Cette. Hérault, S France
Setesdal 63 A6 valley S Norway
Sétif 49 E2 var. Stif. N Algeria
Setté Cama 55 A6 Ogooué-Maritime, SW Gabon
Setúbal 70 B4 Eng. Saint Ubes, Saint Yves. Setúbal, W Portugal
Setúbal, Baía de 70 B4 bay W Portugal
Seul, Lac 16 B3 lake Ontario, S Canada
Sevan 95 G2 C Armenia
Sevana Lich 95 G3 Eng. Lake Sevan, Rus. Ozero Sevan. lake E Armenia
Sevan, Lake/Sevan, Ozero see Sevana Lich
Sevastopol 87 F5 Eng. Sebastopol. Misto Sevastopol, S Ukraine
Severn 16 B2 river Ontario, S Canada
Severn 67 D6 Wel. Hafren. river England/Wales, United Kingdom
Severna Makedonija see North Macedonia
Severnaya Dvina 88 C4 var. Northern Dvina. river NW Russia
Severnaya Zemlya 93 E2 var. Nicholas II Land. island group N Russia
Severnyy 88 E3 Respublika Komi, NW Russia
Severodonetsk see Sievierodonetsk
Severodvinsk 88 C3 prev. Molotov, Sudostroy. Arkhangel'skaya Oblast', NW Russia
Severomorsk 88 C2 Murmanskaya Oblast', NW Russia
Seversk 92 D4 Tomskaya Oblast', C Russia
Sevier Lake 22 A4 lake Utah, W USA
Sevilla 70 C4 Eng. Seville; anc. Hispalis. Andalucía, SW Spain
Seville see Sevilla
Sevlievo 82 D2 Gabrovo, N Bulgaria
Sevluš/Sevlyush see Vynohradiv
Seward's Folly see Alaska
Seychelles 57 G1 off. Republic of Seychelles. country W Indian Ocean
Seychelles, Republic of see Seychelles
Seyðisfjörður 61 E5 Austurland, E Iceland
Seýdi 100 D2 Rus. Seydi; prev. Neftezavodsk. Lebap Welaýaty, E Turkmenistan
Seyhan see Adana
Sfákia see Chóra Sfakíon
Sfântu Gheorghe 86 C4 Ger. Sankt-Georgen, Hung. Sepsiszentgyőrgy; prev. Şepşi-Sângeorz, Sfîntu Gheorghe. Covasna, C Romania
Sfax 49 F2 Ar. Şafāqis. E Tunisia
Sfîntu Gheorghe see Sfântu Gheorghe
's-Gravenhage 64 B4 var. Den Haag, Eng. The Hague, Fr. La Haye. country capital (Netherlands-seat of government) Zuid-Holland, W Netherlands
's-Gravenzande 64 B4 Zuid-Holland, W Netherlands
Shaan/Shaanxi Sheng see Shaanxi
Shaanxi 106 B5 var. Shaan, Shaanxi Sheng, Shan-hsi, Shenshi, Shensi. province C China
Shabani see Zvishavane
Shabeelle, Webi see Shebeli
Shache 104 A3 var. Yarkant. Xinjiang Uygur Zizhiqu, NW China
Shacheng see Huailai
Shackleton Ice Shelf 132 D3 ice shelf Antarctica
Shaddādī see Ash Shadādah
Shāhābād see Eslāmābād
Sha Hi see Orūmīyeh, Daryācheh-ye
Shahjahanabad see Delhi
Shahr-e Kord 98 C3 var. Shahr Kord. Chahār Maḩall va Bakhtīārī, C Iran
Shahr Kord see Shahr-e Kord
Shāhrūd 98 D2 prev. Emāmrūd, Emāmshahr. Semnān, N Iran
Shalkar see Shalqar
Shalqar 92 B4 var. Shalkar, Chelkar. Aktyubinsk, W Kazakhstan
Shām, Bādiyat ash see Syrian Desert
Shana see Kuril'sk
Shandi see Shendi
Shandong 106 D4 var. Lu, Shandong Sheng, Shantung. province E China
Shandong Sheng see Shandong
Shanghai 106 D5 var. Shang-hai. Shanghai Shi, E China
Shangrao 106 D5 Jiangxi, S China
Shan-hsi see Shaanxi, China
Shan-hsi see Shanxi, China
Shannon 67 A6 Ir. An tSionainn. river W Ireland
Shan Plateau 114 B3 plateau E Myanmar (Burma)
Shansi see Shanxi
Shantar Islands see Shantarskiye Ostrova
Shantarskiye Ostrova 93 G3 Eng. Shantar Islands. island group E Russia
Shantou 106 D6 var. Shan-t'ou, Swatow. Guangdong, S China
Shan-t'ou see Shantou
Shantung see Shandong
Shanxi 106 C4 var. Jin, Shan-hsi, Shansi, Shanxi Sheng. province C China
Shan Xian see Sanmenxia
Shanxi Sheng see Shanxi
Shaoguan 106 C6 var. Shao-kuan, Cant. Kukong; prev. Ch'u-chiang. Guangdong, S China
Shao-kuan see Shaoguan
Shaqrā' 98 B4 Ar Riyāḑ, C Saudi Arabia
Shaqrā see Shuqrah
Shar 92 D5 var. Charsk. Vostochnyy Kazakhstan, E Kazakhstan
Shari 108 D2 Hokkaidō, NE Japan
Shari see Chari
Sharjah see Ash Shāriqah
Shark Bay 125 A5 bay Western Australia
Sharqi, Al Jabal ash/Sharqi, Jebel esh see Anti-Lebanon
Shashe 56 D3 var. Shashi. river Botswana/Zimbabwe
Shashi see Shashe
Shatskiy Rise 103 G1 undersea rise N Pacific Ocean
Shawnee 27 G1 Oklahoma, C USA
Shaykh, Jabal ash see Hermon, Mount
Shchadryn 85 D7 Rus. Shchedrin. Homyel'skaya Voblasts', SE Belarus
Shchedrin see Shchadryn
Shcheglovsk see Kemerovo
Shchëkino 89 B5 Tul'skaya Oblast', W Russia
Shchors see Snovsk
Shchuchin see Shchuchyn
Shchuchinsk 92 C4 prev. Shchuchye. Akmola, N Kazakhstan
Shchuchye see Shchuchinsk
Shchuchyn 85 B5 Pol. Szczuczyn Nowogródzki, Rus. Shchuchin. Hrodzyenskaya Voblasts', W Belarus
Shebekino 89 A6 Belgorodskaya Oblast', W Russia
Shebelē Wenz, Wabē see Shebeli
Shebeli 51 D5 Amh. Wabē Shebelē Wenz, It. Scebeli, Som. Webi Shabeelle. river Ethiopia/Somalia
Sheberghān see Shibirghān
Sheboygan 18 B2 Wisconsin, N USA
Shebshi Mountains 54 A4 var. Schebschi Mountains. mountain range E Nigeria
Shechem see Nablus
Shedadi see Ash Shadādah
Sheffield 67 D5 N England, United Kingdom
Shekhem see Nablus
Sheki see Şäki
Shelby 22 B1 Montana, NW USA
Sheldon 23 F3 Iowa, C USA
Shelekhov Gulf see Shelikhova, Zaliv
Shelikhova, Zaliv 93 G2 Eng. Shelekhov Gulf. gulf E Russia
Shendi 50 C4 var. Shandī. River Nile, NE Sudan
Shengking see Liaoning
Shenking see Liaoning
Shenshi/Shensi see Shaanxi
Shenyang 106 D3 Chin. Shen-yang, Eng. Moukden, Mukden; prev. Fengtien. province capital Liaoning, NE China
Shen-yang see Shenyang
Shepetivka 86 D2 Rus. Shepetovka. Khmel'nyts'ka Oblast', NW Ukraine
Shepetovka see Shepetivka
Shepparton 127 C7 Victoria, SE Australia
Sherbrooke 17 E4 Québec, SE Canada
Shereik 50 C3 River Nile, N Sudan
Sheridan 22 C2 Wyoming, C USA
Sherman 27 G2 Texas, SW USA
's-Hertogenbosch 64 C4 Fr. Bois-le-Duc, Ger. Herzogenbusch. Noord-Brabant, S Netherlands
Shetland Islands 66 D1 island group NE Scotland, United Kingdom
Shevchenko see Aqtaū
Shiberghān/Shiberghan see Shibirghān
Shibirghān 101 E3 var. Sheberghān, Shiberghan, Shiberghān. Jowzjān, N Afghanistan
Shibetsu 108 D2 var. Sibetu. Hokkaidō, NE Japan
Shibh Jazirat Sīnā' 50 C2 var. Sinai Peninsula, Ar. Shibh Jazirat Sīnā', Sīnā'. physical region NE Egypt
Shibushi-wan 109 B8 bay SW Japan
Shigatse see Xigazê
Shih-chia-chuang/Shihmen see Shijiazhuang
Shihezi 104 C2 Xinjiang Uygur Zizhiqu, NW China
Shiichi see Shyichy
Shijiazhuang 106 C4 var. Shih-chia-chuang; prev. Shihmen. province capital Hebei, E China
Shikarpur 112 B3 Sindh, S Pakistan
Shikoku 109 C7 var. Sikoku. island SW Japan
Shikoku Basin 103 F2 var. Sikoku Basin. undersea basin N Philippine Sea
Shikotan, Ostrov 108 E2 Jap. Shikotan-tō. island NE Russia
Shikotan-tō see Shikotan, Ostrov
Shilabo 51 D5 Sumalē, E Ethiopia
Shiliguri 113 F3 prev. Siliguri. West Bengal, NE India
Shilka 93 F4 river S Russia
Shillong 113 G3 state capital Meghālaya, NE India
Shimanto see Nakamura
Shimbir Berris see Shimbiris
Shimbiris 50 E4 var. Shimbir Berris. mountain N Somalia
Shimoga see Shivamogga
Shimonoseki 109 A7 var. Simonoseki, hist. Akamagaseki, Bakan. Yamaguchi, Honshū, SW Japan
Shinano-gawa 109 C5 var. Sinano Gawa. river Honshū, C Japan
Shīnḑanḑ 100 D4 prev. Shīndand. Herāt, W Afghanistan
Shīndand see Shīnḑanḑ
Shingū 109 C6 var. Singû. Wakayama, Honshū, SW Japan
Shinjō 108 D4 var. Sinzyô. Yamagata, Honshū, C Japan
Shinyanga 51 C7 Shinyanga, NW Tanzania
Shiprock 26 C1 New Mexico, SW USA
Shīrāz 98 D4 var. Shīrāz. Fārs, S Iran
Shishchitsy see Shyshchytsy
Shivamogga 110 C2 prev. Shimoga. Karnātaka, W India
Shivpuri 112 D3 Madhya Pradesh, C India
Shizugawa 108 D4 Miyagi, Honshū, NE Japan
Shizuoka 109 D6 var. Sizuoka. Shizuoka, Honshū, S Japan
Shklov see Shklow
Shklow 85 D6 Rus. Shklov. Mahilyowskaya Voblasts', E Belarus
Shkodër 79 C5 var. Shkodra, It. Scutari, Croatian Skadar. Shkodër, NW Albania
Shkodra see Shkodër
Shkodrës, Liqeni i see Scutari, Lake
Shkumbinit, Lumi i 79 C6 var. Shkumbî, Shkumbin. river C Albania
Shkumbî/Shkumbin see Shkumbinit, Lumi i
Sholāpur see Solāpur
Shostka 87 F1 Sums'ka Oblast', NE Ukraine
Show Low 26 B2 Arizona, SW USA
Show Me State see Missouri
Shpola 87 E3 Cherkas'ka Oblast', N Ukraine
Shqipëria/Shqipërisë, Republika e see Albania
Shreveport 20 A2 Louisiana, S USA
Shrewsbury 67 D6 hist. Scrobesbyrig'. W England, United Kingdom
Shu 92 C5 Kaz. Shū. Zhambyl, SE Kazakhstan
Shuang-liao see Liaoyuan
Shū, Kazakhstan see Shu
Shumagin Islands 14 B3 island group Alaska, USA
Shumen 82 D2 Shumen, NE Bulgaria
Shumilina 85 E5 Rus. Shumilino. Vitsyebskaya Voblasts', NE Belarus
Shumilino see Shumilina
Shunsen see Chuncheon
Shuqrah 99 B7 var. Shaqrā. SW Yemen
Shwebo 114 B3 Sagaing, C Myanmar (Burma)
Shyichy 85 C7 Rus. Shiichi. Homyel'skaya Voblasts', SE Belarus
Shymkent 92 B5 prev. Chimkent. Türkistan/Turkestan, S Kazakhstan
Shyshchytsy 85 C6 Rus. Shishchitsy. Minskaya Voblasts', C Belarus
Siam see Thailand
Siam, Gulf of see Thailand, Gulf of
Sian see Xi'an
Siang see Brahmaputra
Siangtan see Xiangtan
Šiauliai 84 B4 Ger. Schaulen. Šiauliai, N Lithuania
Siazan' see Siyäzän
Sibay 89 D6 Respublika Bashkortostan, W Russia
Šibenik 78 B4 It. Sebenico. Šibenik-Knin, S Croatia
Siberia see Sibir'
Siberoet see Siberut, Pulau
Siberut, Pulau 116 A4 prev. Siberoet. island Kepulauan Mentawai, W Indonesia
Sibi 112 B2 Baluchistān, SW Pakistan
Sibir' 93 E3 var. Siberia. physical region NE Russia
Sibiti 55 B6 Lékoumou, S Congo
Sibiu 86 B4 Ger. Hermannstadt, Hung. Nagyszeben. Sibiu, C Romania
Sibolga 116 B3 Sumatera, W Indonesia
Sibu 116 D3 Sarawak, East Malaysia
Sibut 54 C4 prev. Fort-Sibut. Kémo, S Central African Republic
Sibuyan Sea 117 E2 sea W Pacific Ocean
Sichon 115 C6 var. Ban Sichon, Si Chon. Nakhon Si Thammarat, SW Thailand
Si Chon see Sichon
Sichuan 106 B5 var. Chuan, Sichuan Sheng, Ssu-ch'uan, Szechuan, Szechwan. province C China
Sichuan Pendi 106 B5 basin C China
Sichuan Sheng see Sichuan
Sicilian Channel see Sicily, Strait of
Sicily 75 C7 Eng. Sicily; anc. Trinacria. island Italy, C Mediterranean Sea
Sicily, Strait of 75 B7 var. Sicilian Channel. strait C Mediterranean Sea
Sicuani 39 E4 Cusco, S Peru
Sidári 82 A4 Kérkyra, Iónia Nisiá, Greece, C Mediterranean Sea
Sidas 116 C4 Borneo, C Indonesia
Siderno 75 D7 Calabria, SW Italy
Sidhirókastron see Sidirókastro
Sîdi Barrâni 50 A1 NW Egypt
Sidi Bel Abbès 48 D2 var. Sidi bel Abbès, Sidi-Bel-Abbès. NW Algeria
Sidirókastro 82 C3 prev. Sidhirókastron. Kentrikí Makedonía, NE Greece
Sidley, Mount 132 B4 mountain Antarctica
Sidney 22 D1 Montana, NW USA
Sidney 22 D4 Nebraska, C USA
Sidney 18 C4 Ohio, N USA
Sidon see Saïda
Sidra see Surt
Sidra/Sidra, Gulf of see Surt, Khalīj, N Libya
Siebenbürgen see Transylvania
Siedlce 76 E3 Ger. Sedlez, Rus. Sesdlets. Mazowieckie, C Poland
Siegen 72 B4 Nordrhein-Westfalen, W Germany
Siemiatycze 76 E3 Podlaskie, NE Poland
Siena 74 B3 Fr. Sienne; anc. Saena Julia. Toscana, C Italy
Sienne see Siena
Sieradz 76 C4 Sieradz, C Poland
Sierpc 76 D3 Mazowieckie, C Poland
Sierra Leone 52 C4 off. Republic of Sierra Leone. country W Africa
Sierra Leone Basin 44 C4 undersea basin E Atlantic Ocean
Sierra Leone, Republic of see Sierra Leone
Sierra Leone Ridge see Sierra Leone Rise
Sierra Leone Rise 44 C4 var. Sierra Leone Ridge, Sierra Leone Schwelle. undersea rise E Atlantic Ocean
Sierra Leone Schwelle see Sierra Leone Rise
Sierra Vista 26 B3 Arizona, SW USA
Sievierodonetsk 87 H3 prev. Syeverodonets'k, Rus. Severodonetsk. Luhans'ka Oblast', E Ukraine
Sífnos 83 C6 anc. Siphnos. island Kykládes, Greece, Aegean Sea
Sigli 116 A3 Sumatera, W Indonesia
Siglufjörður 61 E4 Nordhurland Vestra, N Iceland
Signal Peak 26 A2 mountain Arizona, SW USA
Signan see Xi'an
Siguatepeque 30 C2 Comayagua, W Honduras
Siguiri 52 D4 NE Guinea
Sihanoukville 115 D6 Khmer. Preăh Seihânŭ; prev. Kompong Som. Sihanoukville, SW Cambodia
Siilinjärvi 62 E4 Pohjois-Savo, C Finland
Siirt 95 F4 var. Sert; anc. Tigranocerta. Siirt, SE Turkey (Türkiye)
Sikandarabad see Secunderābād
Sikasso 52 D4 Sikasso, S Mali
Sikeston 23 H5 Missouri, C USA
Sikhote-Alin', Khrebet 93 G4 mountain range SE Russia
Siking see Xi'an
Siklós 77 C7 Baranya, SW Hungary
Sikoku see Shikoku
Sikoku Basin see Shikoku Basin
Šilalė 84 B4 Tauragė, W Lithuania
Silchar 113 G3 Assam, NE India
Silesia 76 B4 physical region SW Poland
Silifke 94 C4 anc. Seleucia. İçel, S Turkey (Türkiye)
Siliguri see Shiliguri
Siling Co 104 C5 lake W China
Silinhot see Xilinhot
Silistra 82 E1 var. Silistria; anc. Durostorum. Silistra, NE Bulgaria
Silistria see Silistra
Sillamäe 84 E2 Ger. Sillamäggi. Ida-Virumaa, NE Estonia
Sillamäggi see Sillamäe
Sillein see Žilina
Šilutė 84 B4 Ger. Heydekrug. Klaipėda, W Lithuania
Silvan 95 E4 Dıyarbakır, SE Turkey (Türkiye)
Silva Porto see Kuito
Silver State see Colorado
Silver State see Nevada
Simanichy 85 C7 Rus. Simonichi. Homyel'skaya Voblasts', SE Belarus
Simav 94 B3 Kütahya, W Turkey (Türkiye)
Simav Çayı 94 A3 river NW Turkey (Türkiye)
Simbirsk see Ul'yanovsk
Simeto 75 C7 river Sicilia, Italy, C Mediterranean Sea
Simeulue, Pulau 116 A3 island NW Indonesia
Simferopol 87 F5 Respublika Krym, S Ukraine
Simitla 82 C3 Blagoevgrad, SW Bulgaria
Şimlăul Silvaniei/Şimleul Silvaniei see Şimleu Silvaniei
Şimleu Silvaniei 86 B3 Hung. Szilágysomlyó; prev. Şimlăul Silvaniei, Şimleul Silvaniei. Sălaj, NW Romania
Simonichi see Simanichy
Simonoseki see Shimonoseki
Simpelveld 65 D6 Limburg, SE Netherlands
Simplon Pass 73 B8 pass S Switzerland
Simpson see Fort Simpson
Simpson Desert 126 B4 desert Northern Territory/South Australia
Sīnā' see Sīnā', Shibh Jazirat
Sīnā', Shibh Jazirat 50 C2 var. Sinai Peninsula, Sinai; Ar. Sīnā'. physical region NE Egypt
Sinai/Sinai Peninsula see Sīnā', Shibh Jazirat
Sinaia 86 C4 Prahova, SE Romania
Sinano Gawa see Shinano-gawa
Sincelejo 36 B2 Sucre, NW Colombia
Sind see Sindh
Sindelfingen 73 B6 Baden-Württemberg, SW Germany
Sindh 112 B3 var. Sind. province SE Pakistan
Sindi 84 D2 Ger. Zintenhof. Pärnumaa, SW Estonia
Sines 70 B4 Setúbal, S Portugal
Singan see Xi'an
Singapore 116 B3 country capital (Singapore) S Singapore
Singapore 116 A1 off. Republic of Singapore. country SE Asia
Singapore, Republic of see Singapore
Singen 73 B6 Baden-Württemberg, S Germany
Singida 51 C7 Singida, C Tanzania
Singkang 117 E4 Sulawesi, C Indonesia
Singkawang 116 C3 Borneo, C Indonesia
Singora see Songkhla
Singû see Shingū
Sining see Xining
Siniscola 75 A5 Sardegna, Italy, C Mediterranean Sea
Sinj 78 B4 Split-Dalmacija, SE Croatia
Sinkiang/Sinkiang Uighur Autonomous Region see Xinjiang Uygur Zizhiqu
Sinnamarie see Sinnamary
Sinnamary 37 H3 var. Sinnamarie. N French Guiana
Sinneh see Sanandaj
Sinnicolau Mare see Sânnicolau Mare
Sinoe, Lacul see Sinoie, Lacul
Sinoie, Lacul 86 D5 prev. Lacul Sinoe. lagoon SE Romania
Sinop 94 D2 anc. Sinope. Sinop, N Turkey (Türkiye)
Sinope see Sinop
Sinsheim 73 B6 Baden-Württemberg, SW Germany
Sint Maarten 33 G3 Eng. Saint Martin. Self-governing country of the Netherlands. E West Indies
Sint-Michielsgestel 64 C4 Noord-Brabant, S Netherlands
Sîn-Miclăuş see Gheorgheni
Sint-Niklaas 65 B5 Fr. Saint-Nicolas. Oost-Vlaanderen, N Belgium
Sint-Pieters-Leeuw 65 B6 Vlaams Brabant, C Belgium
Sintra 70 B3 prev. Cintra. Lisboa, W Portugal
Sinujiif 51 E5 Nugaal, NE Somalia
Sinus Aelaniticus see Aqaba, Gulf of
Sinus Gallicus see Lion, Golfe du
Sinyang see Xinyang
Sinzyô see Shinjō
Sion 73 A7 Ger. Sitten; anc. Sedunum. Valais, SW Switzerland
Sioux City 23 F3 Iowa, C USA
Sioux Falls 23 F3 South Dakota, N USA
Sioux State see North Dakota
Siphnos see Sífnos
Siping 106 D3 var. Ssu-p'ing, Szeping; prev. Ssu-p'ing-chieh. Jilin, NE China
Siple, Mount 132 A4 mountain Siple Island, Antarctica
Siquirres 31 E4 Limón, E Costa Rica
Siracusa 75 D7 Eng. Syracuse. Sicilia, Italy, C Mediterranean Sea
Sir Edward Pellew Group 126 B2 island group Northern Territory, NE Australia
Siret 86 C3 var. Siretul, Ger. Sereth, Rus. Seret. river Romania/Ukraine
Siretul see Siret
Siria see Syria
Sirikit Reservoir 114 C4 lake N Thailand
Sīrjān 98 D4 prev. Sa'īdābād. Kermān, S Iran
Sirna see Sýrna
Şırnak 95 F4 Şırnak, SE Turkey (Türkiye)
Síros see Sýros
Sirte see Surt
Sirte, Gulf of see Surt, Khalīj
Sirti, Gulf of see Surt, Khalīj
Şirvan 95 H3 prev. Äli-Bayramı. SE Azerbaijan
Sisak 78 B3 var. Siscia, Ger. Sissek, Hung. Sziszek; anc. Segestica. Sisak-Moslavina, C Croatia
Siscia see Sisak
Sisimiut 60 C3 var. Holsteinborg, Holsteinsborg, Holstenborg, Holstensborg. Qeqqata, S Greenland
Sissek see Sisak
Sīstān, Daryācheh-ye see Şāberī, Hāmūn-e
Sitaş Cristuru see Cristuru Secuiesc
Siteía 83 D8 var. Sitía. Kríti, Greece, E Mediterranean Sea
Sitges 71 G2 Cataluña, NE Spain
Sitía see Siteía
Sittang see Sittaung
Sittard 65 D5 Limburg, SE Netherlands
Sittaung 114 B4 var. Sittoung, Sittang. river S Myanmar (Burma)
Sitten see Sion
Sittoung see Sittaung
Sittwe 114 A3 var. Akyab. Rakhine State, W Myanmar (Burma)
Siuna 30 D3 Región Autónoma Atlántico Norte, NE Nicaragua
Siut see Asyūţ

Sivas 94 D3 anc. Sebastia, Sebaste. Sivas, C Turkey (Türkiye)
Siverek 95 E4 Şanlıurfa, S Turkey (Türkiye)
Severskiy Donets see Siverskyi Donets
Siverskyi Donets 87 G2 var. Donets, Rus. Severskiy Donets. river E Ukraine
Siwa see Sīwah
Sīwah 50 A2 var. Siwa. NW Egypt
Six Counties, The see Northern Ireland
Six-Fours-les-Plages 69 D6 Var, SE France
Siyäzän 95 H2 Rus. Siazan'. NE Azerbaijan
Sjar see Sääre
Sjenica 79 D5 Turk. Seniça. Serbia, SW Serbia
Skadar see Shkodër
Skadarsko Jezero see Scutari, Lake
Skagerak see Skagerrak
Skagerrak 63 A6 var. Skagerak. channel N Europe
Skagit River 24 B1 river Washington, NW USA
Skalka 62 C3 lake N Sweden
Skarżysko-Kamienna 76 D4 Świętokrzyskie, C Poland
Skaudvilė 84 B4 Tauragė, SW Lithuania
Skegness 67 E6 E England, United Kingdom
Skellefteå 62 D4 Västerbotten, N Sweden
Skellefteälven 62 C4 river N Sweden
Ski 63 B6 Akershus, S Norway
Skiáthos 83 C5 Skiáthos, Vóreies Sporádes, Greece, Aegean Sea
Skidal' 85 B5 Rus. Skidel'. Hrodzyenskaya Voblasts', W Belarus
Skidel' see Skidal'
Skiermûntseach see Schiermonnikoog
Skierniewice 76 D3 Łódzkie, C Poland
Skiftet 84 C1 strait Finland Atlantic Ocean Baltic Sea Gulf of Bothnia/Gulf of Finland
Skíros see Skýros
Skópelos 83 C5 Skópelos, Vóreies Sporádes, Greece, Aegean Sea
Skopje 79 D6 var. Üsküb, Turk. Üsküp, prev. Skoplje; anc. Scupi. country capital (North Macedonia) N North Macedonia
Skoplje see Skopje
Skovorodino 93 F4 Amurskaya Oblast', SE Russia
Skudnesfjorden 63 A6 fjord S Norway
Skuodas 84 B3 Ger. Schoden, Pol. Szkudy. Klaipėda, NW Lithuania
Skye, Isle of 66 B3 island NW Scotland, United Kingdom
Skylge see Terschelling
Skýros 83 C5 var. Skíros. Skýros, Vóreies Sporádes, Greece, Aegean Sea
Skýros 83 C5 var. Skíros; anc. Scyros. island Vóreies Sporádes, Greece, Aegean Sea
Slagelse 63 B7 Vestsjælland, E Denmark
Slatina 78 C3 Hung. Szlatina; prev. Podravska Slatina. Virovitica-Podravina, NE Croatia
Slatina 86 B5 Olt, S Romania
Slavgorod see Slawharad
Slavonski Brod 78 C3 Ger. Brod, Hung. Bród; prev. Brod, Brod na Savi. Brod-Posavina, NE Croatia
Slavuta 86 C2 Khmel'nyts'ka Oblast', NW Ukraine
Slavyansk see Sloviansk
Slawharad 85 E7 Rus. Slavgorod. Mahilyowskaya Voblasts', E Belarus
Sławno 76 C2 Zachodnio-pomorskie, NW Poland
Slêmanî see As Sulaymānīyah
Sliema 80 B5 N Malta
Sligo 67 A5 Ir. Sligeach. Sligo, NW Ireland
Sliven 82 D2 var. Slivno. Sliven, C Bulgaria
Slivnitsa 82 B2 Sofiya, W Bulgaria
Slivno see Sliven
Slobozia 86 C5 Ialomiţa, SE Romania
Slonim 85 B6 Pol. Słonim. Hrodzyenskaya Voblasts', W Belarus
Słonim see Slonim
Slovakia 77 C6 off. Slovak Republic, Ger. Slowakei, Hung. Szlovákia, Slvk. Slovensko, Slovenská Republika. country C Europe
Slovak Ore Mountains see Slovenské rudohorie
Slovak Republic see Slovakia
Slovenia 73 D8 off. Republic of Slovenia, Ger. Slowenien, Slvn. Slovenija. country SE Europe
Slovenia, Republic of see Slovenia
Slovenija see Slovenia
Slovenská Republika see Slovakia
Slovenské rudohorie 77 D6 Eng. Slovak Ore Mountains, Ger. Slowakisches Erzgebirge, Ungarisches Erzgebirge. mountain range C Slovakia
Slovensko see Slovakia
Sloviansk 87 G3 prev. Slov''yans'k, Rus. Slavyansk. Donets'ka Oblast', E Ukraine
Slov''yans'k see Sloviansk
Slowakei see Slovakia
Slowakisches Erzgebirge see Slovenské rudohorie
Slowenien see Slovenia
Słubice 76 B3 Ger. Frankfurt. Lubuskie, W Poland
Sluch 86 D1 river NW Ukraine
Słupsk 76 C2 Ger. Stolp. Pomorskie, N Poland
Slutsk 85 C6 Minskaya Voblasts', S Belarus
Smallwood Reservoir 17 F2 lake Newfoundland and Labrador, S Canada
Smara 48 B3 var. Es Semara. N Western Sahara
Smarhon' 85 C5 Pol. Smorgonie, Rus. Smorgon'. Hrodzyenskaya Voblasts', W Belarus
Smederevo 78 D4 Ger. Semendria. Serbia, N Serbia
Smederevska Palanka 78 D4 Serbia, C Serbia
Smela see Smila
Smila 87 E2 Rus. Smela. Cherkas'ka Oblast', C Ukraine
Smilten see Smiltene
Smiltene 84 D3 Ger. Smilten. Valka, N Latvia
Smøla 62 A4 island W Norway
Smolensk 89 A5 Smolenskaya Oblast', W Russia
Smorgon'/Smorgonie see Smarhon'
Smyrna see İzmir
Snake 12 B4 river Yukon, NW Canada
Snake River 24 C3 river NW USA
Snake River Plain 24 D4 plain Idaho, NW USA
Sneek 64 D2 var. Snits. Fryslân, N Netherlands
Sneeuw-gebergte see Maoke, Pegunungan
Sněžka 76 B4 Ger. Schneekoppe, Pol. Śnieżka. mountain N Czechia (Czech Republic) /Poland
Śniardwy, Jezioro 76 D2 Ger. Spirdingsee. lake NE Poland
Sniečkus see Visaginas
Śnieżka see Sněžka
Snina 77 E5 Hung. Szinna. Prešovský Kraj, E Slovakia
Snits see Sneek
Snovsk 87 E1 Rus. Shchors. Chernihivs'ka Oblast', N Ukraine
Snowdonia 67 C6 mountain range NW Wales, United Kingdom
Snowdrift see Łutselk'e
Snow Mountains see Maoke, Pegunungan
Snyder 27 F3 Texas, SW USA
Sobradinho, Barragem de see Sobradinho, Represa de
Sobradinho, Represa de 41 F2 var. Barragem de Sobradinho. reservoir E Brazil
Sochi 89 A7 Krasnodarskiy Kray, SW Russia
Société, Îles de la/Society Islands see Société, Archipel de la
Society Islands 123 G4 var. Archipel de Tahiti, Îles de la Société, Eng. Society Islands. island group W French Polynesia
Soconusco, Sierra de see Madre, Sierra
Socorro 26 D2 New Mexico, SW USA
Socorro, Isla 28 B5 island W Mexico
Socotra see Suquṭrā
Soc Trăng 115 D6 var. Khanh Hung. Soc Trăng, S Vietnam
Socuéllamos 71 E3 Castilla-La Mancha, C Spain
Sodankylä 62 D3 Lappi, N Finland
Sodari see Sodiri
Söderhamn 63 C5 Gävleborg, C Sweden
Södertälje 63 C6 Stockholm, C Sweden
Sodiri 50 B4 var. Sawdirī, Sodari. Northern Kordofan, C Sudan
Soekaboemi see Sukabumi
Soemba see Sumba, Pulau
Soengaipenoeh see Sungaipenuh
Soerabaja see Surabaya
Soerakarta see Surakarta
Sofia 82 C2 var. Sophia, Sofiya, Lat. Serdica. country capital (Bulgaria) Sofia-Grad, W Bulgaria
Sofiya see Sofia
Sogamoso 36 B3 Boyacá, C Colombia
Sognefjorden 63 A5 fjord NE North Sea
Sohag see Sūhāj
Sohar see Ṣuḥār
Sohm Plain 44 B3 abyssal plain NW Atlantic Ocean
Sohrau see Żory
Sokal 86 C2 L'vivs'ka Oblast', NW Ukraine
Söke 94 A4 Aydın, SW Turkey (Türkiye)
Sokhumi 95 E1 Rus. Sukhumi. NW Georgia
Sokodé 53 F4 C Togo
Sokol 88 C4 Vologodskaya Oblast', NW Russia
Sokółka 76 E3 Podlaskie, NE Poland
Sokolov 77 A5 Ger. Falkenau an der Eger; prev. Falknov nad Ohří. Karlovarský Kraj, W Czechia (Czech Republic)
Sokone 52 B3 W Senegal
Sokoto 53 F3 Sokoto, NW Nigeria
Sokoto 53 F4 river NW Nigeria
Sokotra see Suquṭrā
Solāpur 102 B3 var. Sholāpur. Mahārāshtra, W India
Solca 86 C3 Ger. Solka. Suceava, N Romania
Sol, Costa del 70 D5 coastal region S Spain
Soldeu 69 B7 NE Andorra Europe
Solec Kujawski 76 C3 Kujawsko-pomorskie, C Poland
Soledad 36 B1 Atlántico, N Colombia
Isla Soledad see East Falkland
Soligorsk see Salihorsk
Solikamsk 92 C3 Permskaya Oblast', NW Russia
Sol'-Iletsk 89 D6 Orenburgskaya Oblast', W Russia
Solingen 72 A4 Nordrhein-Westfalen, W Germany
Solka see Solca
Sollentuna 63 C6 Stockholm, C Sweden
Solo see Surakarta
Solok 116 B4 Sumatera, W Indonesia
Solomon Islands 122 C3 prev. British Solomon Islands Protectorate. country W Solomon Islands N Melanesia W Pacific Ocean
Solomon Islands 122 C3 island group Papua New Guinea/Solomon Islands
Solomon Sea 122 B3 sea W Pacific Ocean
Soltau 72 B3 Niedersachsen, NW Germany
Sol'tsy 88 A4 Novgorodskaya Oblast', W Russia
Solun see Thessaloníki
Solwezi 56 D2 North Western, NW Zambia
Sōma 108 D4 Fukushima, Honshū, C Japan
Somalia 51 D5 off. Somali Democratic Republic, Som. Jamhuuriyadda Federaalka Soomaaliya, Soomaaliya; prev. Italian Somaliland, Somaliland Protectorate. country E Africa
Somali Basin 47 E5 undersea basin W Indian Ocean
Somali Democratic Republic see Somalia
Somaliland 51 D5 disputed territory N Somalia
Somaliland Protectorate see Somalia
Sombor 78 C3 Hung. Zombor. Vojvodina, NW Serbia
Someren 65 D5 Noord-Brabant, SE Netherlands
Somerset 20 A5 var. Somerset Village. W Bermuda
Somerset 18 C5 Kentucky, S USA
Somerset Island 20 A5 island W Bermuda
Somerset Island 15 F2 island Queen Elizabeth Islands, Nunavut, NW Canada
Somerset Village see Somerset
Somers Islands see Bermuda
Somerton 26 A2 Arizona, SW USA
Someş 86 B3 river Hungary/Romania Europe
Somme 68 C2 river N France
Sommerfeld see Lubsko
Somotillo 30 C3 Chinandega, NW Nicaragua
Somoto 30 D3 Madriz, NW Nicaragua
Songea 51 C8 Ruvuma, S Tanzania
Sŏngjin see Kimch'aek
Songkhla 115 C7 var. Songkla, Mal. Singora. Songkhla, SW Thailand
Songkla see Songkhla
Sonoran Desert 26 A3 var. Desierto de Altar. desert Mexico/USA
Sonsonate 30 B3 Sonsonate, W El Salvador
Soochow see Suzhou
Soomaaliya/Soomaaliya, Jamhuuriyadda Federaalka see Somalia
Soome Laht see Finland, Gulf of
Sop Hao 114 D3 Houaphan, N Laos
Sophia see Sofia
Sopianae see Pécs
Sopot 76 C2 Ger. Zoppot. Pomorskie, N Poland
Sopron 77 B6 Ger. Ödenburg. Győr-Moson-Sopron, NW Hungary
Sorau/Sorau in der Niederlausitz see Żary
Sorgues 69 D6 Vaucluse, SE France
Sorgun 94 D3 Yozgat, C Turkey (Türkiye)
Soria 71 E2 Castilla y León, N Spain
Soroca 86 D3 Rus. Soroki. N Moldova
Sorochino see Sarochyna
Soroki see Soroca
Sorokyne 87 H3 Rus. Krasnodon. Luhans'ka Oblast', E Ukraine
Sorong 117 F4 Papua, E Indonesia
Sørøy see Sørøya
Sørøya 62 C2 var. Sørøy, Lapp. Sállan. island N Norway
Sortavala 88 B3 prev. Serdobol'. Respublika Kareliya, NW Russia
Sotavento, Ilhas de 52 A3 var. Leeward Islands. island group S Cape Verde (Cabo Verde)
Sotkamo 62 E4 Kainuu, C Finland
Souanké 55 B5 Sangha, NW Congo
Soueida see As Suwaydā'
Souflí 82 D3 prev. Souflíon. Anatolikí Makedonía kai Thráki, NE Greece
Souflíon see Souflí
Soufrière 33 F2 W Saint Lucia
Soukhné see As Sukhnah
Sŏul see Seoul
Soûr 97 A5 var. Şūr; anc. Tyre. SW Lebanon
Souris River 23 E1 var. Mouse River. river Canada/USA
Soúrpi 83 B5 Thessalía, C Greece
Sousse 49 F2 var. Sūsah. NE Tunisia
South Africa 56 C4 off. Republic of South Africa, Afr. Suid-Afrika. country S Africa
South Africa, Republic of see South Africa
South America 34 continent
Southampton 67 D7 hist. Hamwih, Lat. Clausentum. S England, United Kingdom
Southampton Island 15 G3 island Nunavut, NE Canada
South Andaman 111 F2 island Andaman Islands, India, NE Indian Ocean
South Australia 127 A5 state S Australia
South Australian Basin 120 B5 undersea basin SW Indian Ocean
South Bend 18 C3 Indiana, N USA
South Beveland see Zuid-Beveland
South Bruny Island 127 C8 island Tasmania, SE Australia
South Carolina 21 E2 off. State of South Carolina, also known as The Palmetto State. state SE USA
South Carpathians see Carpaţii Meridionali
South China Basin 103 E4 undersea basin SE South China Sea
South China Sea 103 E4 sea SE Asia
South Dakota 22 D2 off. State of South Dakota, also known as The Coyote State, Sunshine State. state N USA
Southeast Indian Ridge 119 D7 undersea ridge Indian Ocean/Pacific Ocean
Southeast Pacific Basin 131 E5 var. Belling Hausen Mulde. undersea basin SE Pacific Ocean
South East Point 127 C7 headland Victoria, S Australia
Southend-on-Sea 67 E6 E England, United Kingdom
Southern Alps 129 B6 var. Kā Tiritiri o Te Moana. mountain range South Island, New Zealand
Southern Cook Islands 123 F4 island group S Cook Islands
Southern Cross 125 B6 Western Australia
Southern Indian Lake 15 F4 lake Manitoba, C Canada
Southern Ocean 45 B7 ocean Atlantic Ocean/Indian Ocean/Pacific Ocean
Southern Uplands 66 C4 mountain range S Scotland, United Kingdom
South Fiji Basin 120 D4 undersea basin S Pacific Ocean
South Geomagnetic Pole 132 B3 pole Antarctica
South Georgia 35 D8 island South Georgia and the South Sandwich Islands, SW Atlantic Ocean
South Georgia and the South Sandwich Islands 35 D8 UK Overseas Territory SW Atlantic Ocean
South Goulburn Island 124 E2 island Northern Territory, N Australia
South Huvadhu Atoll 110 A5 atoll S Maldives
South Indian Basin 119 D7 undersea basin Indian Ocean/Pacific Ocean
South Island 129 C6 var. Te Waipounamu. island S New Zealand
South Korea 107 E4 off. Republic of Korea, Kor. Taehan Min'guk. country E Asia
South Lake Tahoe 25 C5 California, W USA
South Orkney Islands 132 A2 island group Antarctica
South Ossetia 95 F2 former autonomous region SW Georgia
South Pacific Basin see Southwest Pacific Basin
South Platte River 22 D4 river Colorado/Nebraska, C USA
South Pole 132 B3 pole Antarctica
South Sandwich Islands 35 D8 island group SW Atlantic Ocean
South Sandwich Trench 35 E8 trench SW Atlantic Ocean
South Shetland Islands 132 A2 island group Antarctica
South Shields 66 D4 NE England, United Kingdom
South Sioux City 23 F3 Nebraska, C USA
South Sudan 50 B5 off. Republic of South Sudan, country N Africa
South Taranaki Bight 128 C4 bight SE Tasman Sea
South Tasmania Plateau see Tasman Plateau
South Uist 66 B3 island NW Scotland, United Kingdom
South-West Africa/South West Africa see Namibia
South West Cape 129 A8 var. Puhiwaero. headland Stewart Island, New Zealand
Southwest Indian Ocean Ridge see Southwest Indian Ridge
Southwest Indian Ridge 119 B6 var. Southwest Indian Ocean Ridge. undersea ridge SW Indian Ocean
Southwest Pacific Basin 121 E4 var. South Pacific Basin. undersea basin SE Pacific Ocean
Sovereign Base Area 80 C5 uk military installation S Cyprus
Soweto 56 D4 Gauteng, NE South Africa
Sōya-kaikyō see La Pérouse Strait
Spain 70 D3 off. Kingdom of Spain, Sp. España; anc. Hispania, Iberia, Lat. Hispana. country SW Europe
Spain, Kingdom of see Spain
Spalato see Split
Spanish Town 32 B5 hist. St.Iago de la Vega. C Jamaica
Sparks 25 C5 Nevada, W USA
Sparta see Spárti
Spartanburg 21 E1 South Carolina, SE USA
Spárti 83 B6 Eng. Sparta. Pelopónnisos, S Greece
Spearfish 22 D2 South Dakota, N USA
Speightstown 33 G1 NW Barbados
Spencer 23 F3 Iowa, C USA
Spencer Gulf 127 B6 gulf South Australia
Spey 66 C3 river NE Scotland, United Kingdom
Spice Islands see Maluku
Spiess Seamount 45 C7 seamount S Atlantic Ocean
Spijkenisse 64 B4 Zuid-Holland, SW Netherlands
Spíli 83 C8 Kríti, Greece, E Mediterranean Sea
Spīn Bōldak 101 E5 prev. Spīn Būldak. Kandahār, S Afghanistan
Spīn Būldak see Spīn Bōldak
Spirdingsee see Śniardwy, Jezioro
Spitsbergen 61 F2 island NW Svalbard
Split 78 B4 It. Spalato. Split-Dalmacija, S Croatia
Špoģi 84 D4 Daugvapils, SE Latvia
Spokane 24 C2 Washington, NW USA
Spratly Islands 116 B2 disputed territory SE Asia
Spree 72 D4 river E Germany
Springbok 56 B5 NE South Africa
Springfield 18 B4 state capital Illinois, N USA
Springfield 19 G3 Massachusetts, NE USA
Springfield 23 G5 Missouri, C USA
Springfield 18 C4 Ohio, N USA
Springfield 24 B3 Oregon, NW USA
Spring Garden 37 F2 NE Guyana
Spring Hill 21 E4 Florida, SE USA
Springs Junction 129 C5 West Coast, South Island, New Zealand
Springsure 126 D4 Queensland, E Australia
Sprottau see Szprotawa
Spruce Knob 19 E4 mountain West Virginia, NE USA
Srbija/Republika Srbija see Serbia
Srbinje see Foča
Srbobran 78 D3 var. Bácsszenttamás, Hung. Szenttamás. Vojvodina, N Serbia
Srebrenica 78 C4 Republika Srpska, E Bosnia and Herzegovina
Sredets 82 D2 prev. Syulemeshlii. Stara Zagora, C Bulgaria
Sredets 82 E2 prev. Grudovo. Burgas, E Bulgaria
Srednerusskaya Vozvyshennost' 87 G1 Eng. Central Russian Upland. mountain range W Russia
Sremska Mitrovica 78 C3 prev. Mitrovica, Ger. Mitrowitz. Vojvodina, NW Serbia
Srepok, Sông see Srêpôk, Tônle
Srêpôk, Tônle 115 E5 var. Sông Srepok. river Cambodia/Vietnam
Sri Aman 116 C3 Sarawak, East Malaysia
Sri Jayawardenapura Kotte 110 D3 administrative capital (Sri Lanka) Western Province, W Sri Lanka
Srīkākulam 113 F5 Andhra Pradesh, E India
Sri Lanka 110 D3 off. Democratic Socialist Republic of Sri Lanka; prev. Ceylon. country S Asia
Sri Lanka, Democratic Socialist Republic of see Sri Lanka
Srinagarind Reservoir 115 C5 lake W Thailand
Srpska, Republika 78 B3 republic Bosnia and Herzegovina
Ssu-ch'uan see Sichuan
Ssu-p'ing/Ssu-p'ing-chieh see Siping
Stabroek 65 B5 Antwerpen, N Belgium
Stade 72 B3 Niedersachsen, NW Germany
Stadskanaal 64 E2 Groningen, NE Netherlands
Stafford 67 D6 C England, United Kingdom
Staicele 84 D3 Limbaži, N Latvia
Ştaierdorf-Anina see Anina
Stájerlakanina see Anina
Stakhanov see Kadiivka
Stalin see Varna
Stalinabad see Dushanbe
Stalingrad see Volgograd
Stalino see Donets'k
Stalinobod see Dushanbe
Stalinsk see Novokuznetsk
Stalinski Zaliv see Varnenski Zaliv
Stalin, Yazovir see Iskar, Yazovir
Stalowa Wola 76 E4 Podkarpackie, SE Poland
Stamford 19 F3 Connecticut, NE USA
Stampalia see Astypálaia
Stanislau see Ivano-Frankivsk
Stanislav see Ivano-Frankivsk
Stanisławów see Ivano-Frankivsk
Stanke Dimitrov see Dupnitsa
Stanley 43 D7 var. Port Stanley, Puerto Argentino. dependent territory capital (Falkland Islands) East Falkland, Falkland Islands
Stanleyville see Kisangani
Stann Creek see Dangriga
Stanovoy Khrebet 91 E3 mountain range SE Russia
Stanthorpe 127 D5 Queensland, E Australia
Staphorst 64 D2 Overijssel, E Netherlands
Starachowice 76 D4 Świętokrzyskie, C Poland
Stara Kanjiža see Kanjiža
Stara Pazova 78 D3 Ger. Altpasua, Hung. Ópazova. Vojvodina, N Serbia
Stara Planina see Balkan Mountains
Stara Zagora 82 D2 Lat. Augusta Trajana. Stara Zagora, C Bulgaria
Starbuck Island 123 G3 prev. Volunteer Island. island E Kiribati
Stargard in Pommern see Stargard Szczeciński
Stargard Szczeciński 76 B3 Ger. Stargard in Pommern. Zachodnio-pomorskie, NW Poland
Stari Bečej see Bečej
Starobel'sk see Starobilsk
Starobilsk 87 H2 Rus. Starobel'sk. Luhans'ka Oblast', E Ukraine
Starobin 85 C7 var. Starobyn. Minskaya Voblasts', S Belarus
Starobyn see Starobin
Starogard Gdański 76 C2 Ger. Preussisch-Stargard. Pomorskie, N Poland
Starokonstantinov see Starokostiantyniv
Starokostiantyniv 86 D2 prev. Starokostyantyniv, Rus. Starokonstantinov. Khmel'nyts'ka Oblast', NW Ukraine
Starokostyantyniv see Starokostiantyniv
Starominskaya 89 A7 Krasnodarskiy Kray, SW Russia
Staryya Darohi 85 C6 Rus. Staryye Dorogi. Minskaya Voblasts', S Belarus
Staryye Dorogi see Staryya Darohi
Staryy Oskol 89 B6 Belgorodskaya Oblast', W Russia
State College 19 E4 Pennsylvania, NE USA
Staten Island see Estados, Isla de los
Statesboro 21 E2 Georgia, SE USA
States, The see United States of America
Station Nord 61 F1 N Greenland
Staunton 19 E5 Virginia, NE USA
Stavanger 63 A6 Rogaland, S Norway
Stavers Island see Vostok Island
Stavropol' 89 B7 prev. Voroshilovsk. Stavropol'skiy Kray, SW Russia
Stavropol' see Tol'yatti
Steamboat Springs 22 C4 Colorado, C USA
Steenwijk 64 D2 Overijssel, N Netherlands
Steier see Steyr
Steierdorf/Steierdorf-Anina see Anina
Steinamanger see Szombathely
Steinkjer 62 B4 Nord-Trøndelag, C Norway
Stejarul see Karapelit
Stendal 72 C3 Sachsen-Anhalt, C Germany
Stepanakert see Xankändi
Stephenville 27 F3 Texas, SW USA
Sterling 22 D4 Colorado, C USA
Sterling 18 B3 Illinois, N USA
Sterlitamak 92 B3 Respublika Bashkortostan, W Russia
Stettin see Szczecin
Stettiner Haff see Szczeciński, Zalew
Stevenage 67 E6 E England, United Kingdom
Stevens Point 18 B2 Wisconsin, N USA
Stewart Island 129 A8 var. Rakiura. island S New Zealand
Steyerlak-Anina see Anina
Steyr 73 D6 var. Steier. Oberösterreich, N Austria
St. Helena Bay 56 B5 bay SW South Africa
Stif see Sétif
Stillwater 27 G1 Oklahoma, C USA
Štip 79 E6 E North Macedonia
Stirling 66 C4 C Scotland, United Kingdom
Stjørdalshalsen 62 B4 Nord-Trøndelag, C Norway
St-Maur-des-Fossés 68 E2 Val-de-Marne, Île-de-France, N France Europe
Stockach 73 B6 Baden-Württemberg, S Germany
Stockholm 63 C6 country capital (Sweden) Stockholm, C Sweden
Stockmannshof see Pļaviņas
Stockton 25 B6 California, W USA
Stockton Plateau 27 E4 plain Texas, SW USA
Stœng Trêng see Stung Treng
Stoke see Stoke-on-Trent
Stoke-on-Trent 67 D6 var. Stoke. C England, United Kingdom
Stolbce see Stowbtsy
Stolbtsy see Stowbtsy
Stolp see Słupsk
Stolpmünde see Ustka
Stómio 82 B4 Thessalía, C Greece
Store Bælt see Storebælt
Storebelt see Storebælt
Støren 63 B5 Sør-Trøndelag, S Norway
Storfjorden 61 G2 fjord S Norway
Storhammer see Hamar
Stornoway 66 B2 NW Scotland, United Kingdom
Storsjön 63 B5 lake C Sweden
Storuman 62 C4 Västerbotten, N Sweden
Storuman 62 C4 lake N Sweden
Stowbtsy 85 C6 Pol. Stolbce, Rus. Stolbtsy. Minskaya Voblasts', C Belarus
Strabane 67 B5 Ir. An Srath Bán. W Northern Ireland, United Kingdom
Strakonice 77 A5 Ger. Strakonitz. Jihočeský Kraj, S Czechia (Czech Republic)
Strakonitz see Strakonice
Stralsund 72 D2 Mecklenburg-Vorpommern, NE Germany
Stranraer 67 C5 S Scotland, United Kingdom
Strasbourg 68 E3 Ger. Strassburg; anc. Argentoratum. Bas-Rhin, NE France
Strășeni 86 D3 var. Strasheny. C Moldova
Strasheny see Strășeni
Strassburg see Strasbourg, France
Strassburg see Aiud, Romania
Stratford 128 D4 Taranaki, North Island, New Zealand
Strathfield 126 E2 New South Wales, E Australia
Straubing 73 C6 Bayern, SE Germany
Strehaia 86 B5 Mehedinţi, SW Romania
Strelka 92 D4 Krasnoyarskiy Kray, C Russia
Strigonium see Esztergom
Strofilia see Strofyliá
Strofyliá 83 C5 var. Strofilia. Évvoia, C Greece
Stromboli 75 D6 island Isole Eolie, S Italy
Stromeferry 66 C3 N Scotland, United Kingdom
Strömstad 63 B6 Västra Götaland, S Sweden
Strömsund 62 C4 Jämtland, C Sweden
Struga 79 D6 SW North Macedonia
Struma see Strymónas
Strumica 79 E6 E North Macedonia
Strumyani 82 C3 Blagoevgrad, SW Bulgaria
Strymónas 82 C3 Bul. Struma. river Bulgaria/Greece
Stryi 86 B2 prev. Stryy. L'vivs'ka Oblast', NW Ukraine
Stryy see Stryi
Studholme 129 B6 Canterbury, South Island, New Zealand
Stuhlweissenberg see Székesfehérvár
Stung Treng 115 D5 Khmer. Stœng Trêng. Stung Treng, NE Cambodia
Sturgis 22 D3 South Dakota, N USA
Stuttgart 73 B6 Baden-Württemberg, SW Germany
Stykkishólmur 61 E4 Vesturland, W Iceland
Styr 86 C1 Rus. Styr'. river Belarus/Ukraine
Su see Jiangsu
Suakin 50 C3 var. Sawakin. Red Sea, NE Sudan
Subačius 84 C4 Panevėžys, NE Lithuania
Subaykhān 96 E3 Dayr az Zawr, E Syria
Subotica 78 D2 Ger. Maria-Theresiopel, Hung. Szabadka. Vojvodina, N Serbia
Suceava 86 C3 Ger. Suczawa, Hung. Szucsava. Suceava, NE Romania
Su-chou see Suzhou
Suchow see Suzhou, Jiangsu, China
Suchow see Xuzhou, Jiangsu, China
Sucker State see Illinois

Sucre 39 F4 *hist.* Chuquisaca, La Plata. *country capital* (Bolivia-legal capital) Chuquisaca, S Bolivia
Suczawa *see* Suceava
Sudak 87 F5 S Ukraine
Sudan 50 A4 *off.* Republic of Sudan, *Ar.* Jumhuriyat as-Sudan; *prev.* Anglo-Egyptian Sudan. *country* N Africa
Sudanese Republic *see* Mali
Sudan, Jumhuriyat as- *see* Sudan
Sudan, Republic of *see* Sudan
Sudbury 16 C4 Ontario, S Canada
Sudd 51 B5 *swamp region* N South Sudan
Sudeten 76 B4 *var.* Sudetes, Sudetic Mountains, *Cz./Pol.* Sudety. *mountain range* Czechia (Czech Republic) /Poland
Sudetes/Sudetic Mountains/Sudety *see* Sudeten
Südkarpaten *see* Carpaţii Meridionali
Südliche Morava *see* Južna Morava
Sudong, Pulau 116 A2 *island* SW Singapore Asia
Sudostroy *see* Severodvinsk
Sue 51 B5 *river* S South Sudan
Sueca 71 F3 Comunitat Valenciana, E Spain
Sue Wood Bay 20 B5 *bay* W Bermuda North America W Atlantic Ocean
Suez *see* As Suways
Suez Canal *see* Suways, Qanāt as
Suez, Gulf of *see* Khalīj as Suways
Suğla Gölü 94 C4 *lake* SW Turkey (Türkiye)
Sūhāj 50 B2 *var.* Sawhāj, Suliag; *Eng.* Sohag. C Egypt
Şuḩār 99 D5 *var.* Sohar. N Oman
Sühbaatar 105 E1 Selenge, N Mongolia
Suhl 73 C5 Thüringen, C Germany
Suicheng *see* Suixi
Suid-Afrika *see* South Africa
Suidwes-Afrika *see* Namibia
Suixi 106 C6 *var.* Suicheng. Guangdong, S China
Sujawal 112 B3 Sindh, SE Pakistan
Sukabumi 116 C5 *prev.* Soekaboemi. Jawa, C Indonesia
Sukagawa 109 D5 Fukushima, Honshū, C Japan
Sukarnapura *see* Jayapura
Sukarno, Puntjak *see* Jaya, Puncak
Sukhne *see* As Sukhnah
Sukhona 88 C4 *var.* Tot'ma. *river* NW Russia
Sukhumi *see* Sokhumi
Sukkertoppen *see* Maniitsoq
Sukkur 112 B3 Sindh, SE Pakistan
Sukumo 109 B7 Kōchi, Shikoku, SW Japan
Sulaimaniya *see* As Sulaymānīyah
Sulaiman Range 112 C2 *mountain range* C Pakistan
Sula, Kepulauan 117 E4 *island group* C Indonesia
Sulawesi 117 E4 *Eng.* Celebes. *island* C Indonesia
Sulawesi, Laut *see* Celebes Sea
Sulechów 76 B3 *Ger.* Züllichau. Lubuskie, W Poland
Suliag *see* Sūhāj
Sullana 38 B2 Piura, NW Peru
Sullivan Island *see* Lanbi Kyun
Sulphur Springs 27 G2 Texas, SW USA
Sultānābād *see* Arāk
Sulu Archipelago 117 E3 *island group* SW Philippines
Sülüktü *see* Sulyukta
Sulu, Laut *see* Sulu Sea
Sulu Sea 117 E2 *var.* Laut Sulu. *sea* SW Philippines
Sulyukta 101 E2 *Kir.* Sülüktü. Batkenskaya Oblast', SW Kyrgyzstan
Sumatera 115 B8 *Eng.* Sumatra. *island* W Indonesia
Sumatra *see* Sumatera
Šumava *see* Bohemian Forest
Sumba, Pulau 117 E5 *Eng.* Sandalwood Island; *prev.* Soemba. *island* Nusa Tenggara, C Indonesia
Sumba, Selat 117 E5 *strait* Nusa Tenggara, S Indonesia
Sumbawanga 51 B7 Rukwa, W Tanzania
Sumbe 56 B2 *var.* N'Gunza, *Port.* Novo Redondo. Cuanza Sul, W Angola
Sumeih 51 B5 Southern Darfur, S Sudan
Sumgait *see* Sumqayıt, Azerbaijan
Summer Lake 24 B4 *lake* Oregon, NW USA
Summit 71 H5 Alaska, USA
Sumqayıt 95 H2 *Rus.* Sumgait. E Azerbaijan
Sumy 87 F2 Sums'ka Oblast', NE Ukraine
Sunbury 127 C7 Victoria, SE Australia
Sunda Islands *see* Greater Sunda Islands
Sunda, Selat 116 B5 *strait* Jawa/Sumatera, SW Indonesia
Sunda Trench *see* Java Trench
Sunderland 66 D4 *var.* Wearmouth. NE England, United Kingdom
Sundsvall 63 C5 Västernorrland, C Sweden
Sunflower State *see* Kansas
Sungaipenuh 116 B4 *prev.* Soengaipenoeh. Sumatera, W Indonesia
Sunnyvale 25 A6 California, W USA
Sunset State *see* Oregon
Sunshine State *see* Florida
Sunshine State *see* New Mexico
Sunshine State *see* South Dakota
Suntar 93 F3 Respublika Sakha (Yakutiya), NE Russia
Sunyani 53 E5 W Ghana
Suoločielgi *see* Saariselkä
Suomenlahti *see* Finland, Gulf of
Suomen Tasavalta/Suomi *see* Finland
Suomussalmi 62 E4 Kainuu, E Finland
Suong 115 D6 Tbong Khmum, C Cambodia
Suoyarvi 88 B3 Respublika Kareliya, NW Russia
Supe 38 C3 Lima, W Peru
Supérieur, Lac *see* Superior, Lake
Superior 18 A1 Wisconsin, N USA
Superior, Lake 18 B1 *Fr.* Lac Supérieur. *lake* Canada/USA
Suqrah *see* Şawqirah
Suquṭrá 99 C7 *var.* Sokotra, *Eng.* Socotra. *island* SE Yemen
Şūr 99 E5 NE Oman
Şūr *see* Soûr
Surabaja *see* Surabaya
Surabaya 116 D5 *prev.* Surabaja, Soerabaja. Jawa, C Indonesia
Surakarta 116 C5 *Eng.* Solo; *prev.* Soerakarta. Jawa, S Indonesia
Šurany 77 C6 *Hung.* Nagysurány. Nitriansky Kraj, SW Slovakia
Sūrat 112 C4 Gujarāt, W India
Suratdhani *see* Surat Thani
Surat Thani 115 C6 *var.* Suratdhani. Surat Thani, SW Thailand
Surazh 85 E5 Vitsyebskaya Voblasts', NE Belarus
Surdulica 79 E5 Serbia, SE Serbia
Sûre 65 D7 *var.* Sauer. *river* W Europe
Surendranagar 112 C4 Gujarāt, W India
Surfers Paradise 127 E5 Queensland, E Australia
Surgut 92 D3 Khanty-Mansiyskiy Avtonomnyy Okrug-Yugra, C Russia
Surin 115 D5 Surin, E Thailand
Surinam *see* Suriname
Suriname 37 G3 *off.* Republic of Suriname, *var.* Surinam; *prev.* Dutch Guiana, Netherlands Guiana. *country* N South America
Suriname, Republic of *see* Suriname
Sūrīyah/Sūrīyah, Al-Jumhūrīyah al-'Arabīyah as- *see* Syria
Surkhab, Darya-i- *see* Kahmard, Daryā-ye
Surkhob 101 F3 *river* C Tajikistan
Surt 49 G2 *var.* Sidra, Sirte. N Libya
Surt, Khalīj 49 F2 *var.* Gulf of Sirte, Gulf of Sidra, Gulf of Sirti, Sidra. *gulf* N Libya
Surtsey 61 E5 *island* S Iceland
Suruga-wan 109 D6 *bay* SE Japan
Susa 74 A2 Piemonte, NE Italy
Sūsah *see* Sousse
Susanville 25 B5 California, W USA
Susitna 14 C3 Alaska, USA
Susteren 65 D5 Limburg, SE Netherlands
Susuman 93 G3 Magadanskaya Oblast', E Russia
Sutlej 112 C2 *river* India/Pakistan
Suur Munamägi 84 D3 *var.* Munamägi, *Ger.* Eier-Berg. *mountain* SE Estonia
Suur Väin 84 C2 *Ger.* Grosser Sund. *strait* W Estonia
Suva 123 E4 *country capital* (Fiji) Viti Levu, W Fiji
Suvalkai/Suvalki *see* Suwałki
Suvorovo 82 E2 *prev.* Vetrino. Varna, E Bulgaria
Suwałki 76 E2 *Lith.* Suvalkai, *Rus.* Suvalki. Podlaskie, NE Poland
Şuwār *see* Aş Şuwār
Suways, Qanāt as 50 B1 *Eng.* Suez Canal. *canal* NE Egypt
Suweida *see* As Suwaydā'
Suzhou 106 D5 *var.* Soochow, Su-chou, Suchow; *prev.* Wuhsien. Jiangsu, E China
Svalbard 61 E1 *constituent part of Norway. island group* Arctic Ocean
Svartisen 62 C3 *glacier* C Norway
Svay Rieng 115 D6 *Khmer.* Svay Riĕng. Svay Rieng, S Cambodia
Svay Riĕng *see* Svay Rieng
Sveg 63 B5 Jämtland, C Sweden
Svenstavik 63 C5 Jämtland, C Sweden
Sverdlovsk *see* Yekaterinburg
Sverige *see* Sweden
Sveti Vrach *see* Sandanski
Svetlogorsk *see* Svietlahorsk
Svetlograd 89 B7 Stavropol'skiy Kray, SW Russia
Svetlovodsk *see* Svitlovodsk
Svetozarevo *see* Jagodina
Svilengrad 82 D3 *prev.* Mustafa-Pasha. Haskovo, S Bulgaria
Svitlovodsk 87 F3 *Rus.* Svetlovodsk. Kirovohrads'ka Oblast', C Ukraine
Svizzera *see* Switzerland
Svobodnyy 93 G4 Amurskaya Oblast', SE Russia
Svyataya Anna Trough 133 C4 *var.* Saint Anna Trough. *trough* N Kara Sea
Svyetlahorsk 85 D7 *Rus.* Svetlogorsk. Homyel'skaya Voblasts', SE Belarus
Swabian Jura *see* Schwäbische Alb
Swakopmund 56 B3 Erongo, W Namibia
Swan Islands 31 E1 *island group* NE Honduras North America
Swansea 67 C7 *Wel.* Abertawe. S Wales, United Kingdom
Swarzędz 76 C3 Poznań, W Poland
Swatow *see* Shantou
Swaziland *see* Eswatini
Sweden 62 B4 *off.* Kingdom of Sweden, *Swe.* Sverige. *country* N Europe
Sweden, Kingdom of *see* Sweden
Sweetwater 27 F3 Texas, SW USA
Świdnica 76 B4 *Ger.* Schweidnitz. Wałbrzych, SW Poland
Świdwin 76 B2 *Ger.* Schivelbein. Zachodnio-pomorskie, NW Poland
Świebodzice 76 B4 *Ger.* Freiburg in Schlesien, Swiebodzice. Walbrzych, SW Poland
Świebodzin 76 B3 *Ger.* Schwiebus. Lubuskie, W Poland
Świecie 76 C3 *Ger.* Schwertberg. Kujawsko-pomorskie, C Poland
Swindon 67 D7 S England, United Kingdom
Swinemünde *see* Świnoujście
Świnoujście 76 B2 *Ger.* Swinemünde. Zachodnio-pomorskie, NW Poland
Swiss Confederation *see* Switzerland
Switzerland 73 A7 *off.* Swiss Confederation, *Fr.* La Suisse, *Ger.* Schweiz, *It.* Svizzera; *anc.* Helvetia. *country* C Europe
Sycaminum *see* Hefa
Sydenham Island *see* Nonouti
Sydney 126 D1 *state capital* New South Wales, SE Australia
Sydney 17 G4 Cape Breton Island, Nova Scotia, SE Canada
Sydney Island *see* Manra
Syedpur *see* Saidpur
Syemyezhava 85 C6 *Rus.* Semechevo. Minskaya Voblasts', C Belarus
Syene *see* Aswān
Syeverodonets'k *see* Sievierodonetsk
Syktyvkar 88 D4 *prev.* Ust'-Sysol'sk. Respublika Komi, NW Russia
Sylhet 113 G3 Sylhet, NE Bangladesh
Synelnykove 87 G3 Dnipropetrovska Oblast, E Ukraine
Syowa 132 C2 *Japanese research station* Antarctica
Syracuse 19 E3 New York, NE USA
Syracuse *see* Siracusa
Syrdar'ya 92 B4 Sirdaryo Viloyati, E Uzbekistan
Syria 96 B3 *off.* Syrian Arab Republic, *var.* Siria, Syrie, *Ar.* Al-Jumhūrīyah al-'Arabīyah as-Sūrīyah, Sūrīyah. *country* SW Asia
Syrian Arab Republic *see* Syria
Syrian Desert 97 D5 *Ar.* Al Hamad, Bādiyat ash Shām. *desert* SW Asia
Syrie *see* Syria
Sýrna 83 E7 *var.* Sirna. *island* Kykládes, Greece, Aegean Sea
Sýros 83 C6 *var.* Síros. *island* Kykládes, Greece, Aegean Sea
Syulemeshlii *see* Sredets
Syvash, Zaliv *see* Syvash, Zatoka
Syvash, Zatoka 87 F4 *Rus.* Zaliv Syvash. *inlet* S Ukraine
Syzran' 89 C6 Samarskaya Oblast', W Russia
Szabadka *see* Subotica
Szamotuły 76 B3 Poznań, W Poland
Szászrégen *see* Reghin
Szatmárrnémeti *see* Satu Mare
Száva *see* Sava
Szczecin 76 B3 *Eng./Ger.* Stettin. Zachodnio-pomorskie, NW Poland
Szczecinek 76 B2 *Ger.* Neustettin. Zachodnio-pomorskie, NW Poland
Szczeciński, Zalew 76 A2 *var.* Stettiner Haff, *Ger.* Oderhaff. *bay* Germany/Poland
Szczuczyn Nowogródzki *see* Shchuchyn
Szczytno 76 D3 *Ger.* Ortelsburg. Warmińsko-Mazurskie, NE Poland
Szechuan/Szechwan *see* Sichuan
Szeged 77 D7 *Ger.* Szegedin, *Rom.* Seghedin. Csongrád, SE Hungary
Szegedin *see* Szeged
Székelykeresztúr *see* Cristuru Secuiesc
Székesfehérvár 77 C6 *Ger.* Stuhlweissenberg; *anc.* Alba Regia. Fejér, W Hungary
Szeklerburg *see* Miercurea-Ciuc
Szekler Neumarkt *see* Târgu Secuiesc
Szekszárd 77 C7 Tolna, S Hungary
Szempcz/Szenc *see* Senec
Szenice *see* Senica
Szenttamás *see* Srbobran
Szeping *see* Siping
Szilágysomlyó *see* Şimleu Silvaniei
Szinna *see* Snina
Sziszek *see* Sisak
Szitás-Keresztúr *see* Cristuru Secuiesc
Szkudy *see* Skuodas
Szlatina *see* Slatina
Szlovákia *see* Slovakia
Szolnok 77 D6 Jász-Nagykun-Szolnok, C Hungary
Szombathely 77 B6 *Ger.* Steinamanger; *anc.* Sabaria, Savaria. Vas, W Hungary
Szprotawa 76 B4 *Ger.* Sprottau. Lubuskie, W Poland
Sztálinváros *see* Dunaújváros
Szucsava *see* Suceava

T

Tabariya, Bahrat *see* Kinneret, Yam
Table Rock Lake 27 G1 *reservoir* Arkansas/ Missouri, C USA
Tábor 77 B5 Jihočeský Kraj, S Czechia (Czech Republic)
Tabora 51 B7 Tabora, W Tanzania
Tabrīz 98 C2 *var.* Tebriz; *anc.* Tauris. Āžarbāyjān-e Sharqī, NW Iran
Tabuaeran 123 G2 *prev.* Fanning Island. *atoll* Line Islands, E Kiribati
Tabūk 98 A4 Tabūk, NW Saudi Arabia
Täby 63 C6 Stockholm, C Sweden
Tachau *see* Tachov
Tachov 77 A5 *Ger.* Tachau. Plveňský Kraj, W Czechia (Czech Republic)
Tacloban 117 F2 *off.* Tacloban City. Leyte, C Philippines
Tacloban City *see* Tacloban
Tacna 39 E4 Tacna, SE Peru
Tacoma 24 B2 Washington, NW USA
Tacuarembó 42 D4 *prev.* San Fructuoso. Tacuarembó, C Uruguay
Tademaït, Plateau du 48 D3 *plateau* C Algeria
Tadmor *see* Tadmur
Tadmur 96 C3 *var.* Tamar, *Gk.* Palmyra, *Bibl.* Tadmor. Ḩimş, C Syria
Tādpatri 110 C2 Andhra Pradesh, E India
Tadzhikistan *see* Tajikistan
Taegu *see* Daegu
Taehan-haehyŏp *see* Korea Strait
Taehan Min'guk *see* South Korea
Taejŏn *see* Daejeon
Tafassâsset, Ténéré du 53 G2 *desert* N Niger
Tafila/Ţafilah, Muḩāfaẓat aţ *see* Aţ Ţafilah
Taganrog 89 A7 Rostovskaya Oblast', SW Russia
Taganrog, Gulf of 87 G4 *Rus.* Taganrogskiy Zaliv, *Ukr.* Tahanroz'ka Zatoka. *gulf* Russia/Ukraine
Taganrogskiy Zaliv *see* Taganrog, Gulf of
Taguatinga 41 F3 Tocantins, C Brazil
Tagus 70 C3 *Port.* Rio Tejo, *Sp.* Río Tajo. *river* Portugal/Spain
Tagus Plain 58 A4 *abyssal plain* E Atlantic Ocean
Tahanroz'ka Zatoka *see* Taganrog, Gulf of
Tahat 49 E4 *mountain* SE Algeria
Tahiti 123 H4 *island* Îles du Vent, W French Polynesia
Tahiti, Archipel de *see* Société, Archipel de la
Tahlequah 27 G1 Oklahoma, C USA
Tahoe, Lake 25 B5 *lake* California/Nevada, W USA
Tahoua 53 F3 Tahoua, W Niger
Taibei *see* Taipei
Taichū *see* Taichung
Taichung 106 D6 *Jap.* Taichū; *var.* Taizhong, Taiwan. C Taiwan
Taiden *see* Daejeon
Taieri 129 B7 *river* South Island, New Zealand
Taihape 128 D4 Manawatu-Wanganui, North Island, New Zealand
Taihoku *see* Taibei
Taikyū *see* Daegu
Tailem Bend 127 B7 South Australia
T'ainan *see* Tainan
Tainan 106 D6 *prev.* T'ainan, Dainan. S Taiwan
Taipei 106 D6 *var.* Taibei, T'aipei; *Jap.* Taihoku; *prev.* Daihoku. *capital* (Taiwan) N Taiwan
Taiping 116 B3 Perak, Peninsular Malaysia
Taitetimu 129 A7 *var.* Caswell Sound. *sound* South Island, New Zealand
Taiwan 106 D6 *off.* Republic of China, *var.* Formosa, Formo'sa. *country* E Asia
Taiwan *see* Taichung
T'aiwan Haihsia/Taiwan Haixia *see* Taiwan Strait
Taiwan Strait 106 D6 *var.* Formosa Strait, *Chin.* T'aiwan Haihsia, Taiwan Haixia. *strait* China/ Taiwan
Taiyuan 106 C4 *var.* T'ai-yuan, T'ai-yüan; *prev.* Yangku. *province capital* Shanxi, C China
T'ai-yuan/T'ai-yüan *see* Taiyuan
Taizhong *see* Tiachung
Ta'izz 99 B7 SW Yemen
Tajikistan 101 E3 *off.* Republic of Tajikistan, *Rus.* Tadzhikistan, *Taj.* Jumhurii Tojikiston; *prev.* Tajik S.S.R. *country* C Asia
Tajikistan, Republic of *see* Tajikistan
Tajik S.S.R *see* Tajikistan
Tajo, Río *see* Tagus
Tak 114 C4 *var.* Rahaeng. Tak, W Thailand
Takao *see* Kaohsiung
Takaoka 109 C5 Toyama, Honshū, SW Japan
Takapuna 128 D2 Auckland, North Island, New Zealand
Takeshima *see* Liancourt Rocks
Takhiatash *see* Taxiatosh
Takhtakupyr *see* Taxtako'pir
Takikawa 108 D2 Hokkaidō, NE Japan
Takla Makan Desert *see* Taklimakan Shamo
Taklimakan Shamo 104 B3 *Eng.* Takla Makan Desert. *desert* NW China
Takow *see* Kaohsiung
Takutea 123 G4 *island* S Cook Islands
Talabriga *see* Aveiro, Portugal
Talabriga *see* Talavera de la Reina, Spain
Talachyn 85 D6 *Rus.* Tolochin. Vitsyebskaya Voblasts', NE Belarus
Talamanca, Cordillera de 31 E5 *mountain range* S Costa Rica
Talara 38 B2 Piura, NW Peru
Talas 101 F2 Talasskaya Oblast', NW Kyrgyzstan
Talaud, Kepulauan 117 F3 *island group* E Indonesia
Talavera de la Reina 70 D3 *anc.* Caesarobriga, Talabriga. Castilla-La Mancha, C Spain
Talca 42 B4 Maule, C Chile
Talcahuano 43 B5 Bío Bío, C Chile
Taldykorgan *see* Taldyqorghan
Taldy-Kurgan *see* Taldyqorghan
Taldyqorghan 92 C5 *var.* Taldykorgan; *prev.* Taldy-Kurgan. Taldykorgan, SE Kazakhstan
Ta-lien *see* Dalian
Taliq-an *see* Tāluqān
Tal'ka 85 C6 Minskaya Voblasts', C Belarus
Talkhof *see* Puurmani
Tall Abyaḑ 96 C2 *var.* At Tall al Abyaḑ, Tell Abyad, *Fr.* Tell Abiad. Ar Raqqah, N Syria
Tallahassee 20 D3 *prev.* Muskogean. *state capital* Florida, SE USA
Tallin *see* Tallinn
Tallinn 84 D2 *Ger.* Reval, *Rus.* Tallin; *prev.* Revel. *country capital* (Estonia) Harjumaa, NW Estonia
Tall Kalakh 96 B4 *var.* Tell Kalakh. Ḩimş, C Syria
Tallulah 20 B2 Louisiana, S USA
Talnakh 92 D3 Taymyrskiy (Dolgano-Nenetskiy) Avtonomnyy Okrug, N Russia
Talne 87 E3 *Rus.* Tal'noye. Cherkas'ka Oblast', C Ukraine
Tal'noye *see* Talne
Taloga 27 F1 Oklahoma, C USA
Tāloqān *see* Tāluqān
Tāluqān 101 E3 *var.* Tāloqān, Taliq-an. Takhār, NE Afghanistan
Talsen *see* Talsi
Talsi 84 C3 *Ger.* Talsen. Talsi, NW Latvia
Taltal 42 B2 Antofagasta, N Chile
Talvik 62 D2 Finnmark, N Norway
Tamabo, Banjaran 116 D3 *mountain range* East Malaysia
Tamale 53 E4 C Ghana
Tamana 123 E3 *prev.* Rotcher Island. *atoll* Tungaru, W Kiribati
Tamanrasset 49 E4 *var.* Tamenghest. S Algeria
Tamar 67 C7 *river* SW England, United Kingdom
Tamar *see* Tadmur
Tamatave *see* Toamasina
Tamazunchale 29 E4 San Luis Potosí, C Mexico
Tambacounda 52 C3 SE Senegal
Tambov 89 B6 Tambovskaya Oblast', W Russia
Tambura 51 B5 W Equatoria, SW South Sudan
Tamchaket *see* Tâmchekkeţ
Tâmchekkeţ 52 C3 *var.* Tamchaket. Hodh el Gharbi, S Mauritania
Tamenghest *see* Tamanrasset
Tamil Nādu 110 C3 *prev.* Madras. *cultural region* SE India
Tam Ky 115 E5 Quang Nam-ƒa Năng, C Vietnam
Tammerfors *see* Tampere
Tampa 21 E4 Florida, SE USA
Tampa Bay 21 E4 *bay* Florida, SE USA
Tampere 63 D5 *Swe.* Tammerfors. Pirkanmaa, W Finland
Tampico 29 E3 Tamaulipas, C Mexico
Tamworth 127 D6 New South Wales, SE Australia
Tanabe 109 C7 Wakayama, Honshū, SW Japan
Tana Bru 62 D2 Finnmark, N Norway
T'ana Hāyk' 50 C4 *var.* Lake Tana. *lake* NW Ethiopia
Tanais *see* Don
Tana, Lake *see* T'ana Hāyk'
Tanami Desert 124 D3 *desert* Northern Territory, N Australia
Tananarive *see* Antananarivo
Ţăndărei 86 D5 Ialomiţa, SE Romania
Tandil 43 D5 Buenos Aires, E Argentina
Tandjoengkarang *see* Bandar Lampung
Tanega-shima 109 B8 *island* Nansei-shotō, SW Japan
Tanen Taunggyi *see* Tane Range
Tane Range 114 B4 *Bur.* Tanen Taunggyi. *mountain range* W Thailand
Tanezrouft 48 D4 *desert* Algeria/Mali
Ţanf, Jabal aţ 96 D4 *mountain* SE Syria
Tanga 51 C7 Tanga, E Tanzania
Tanganyika and Zanzibar *see* Tanzania
Tanganyika, Lake 51 B7 *lake* E Africa
Tanger 48 C2 *var.* Tangiers, Tangier, *Fr./Ger.* Tangerk, *Sp.* Tánger; *anc.* Tingis. NW Morocco
Tangerk *see* Tanger
Tanggula Shan 104 C4 *mountain* W China
Tangier *see* Tanger
Tangiers *see* Tanger
Tangra Yumco 104 B5 *var.* Tangro Tso. *lake* W China
Tangro Tso *see* Tangra Yumco
Tangshan 106 D3 *var.* T'ang-shan. Hebei, E China
T'ang-shan *see* Tangshan
Tanimbar, Kepulauan 117 F5 *island group* Maluku, E Indonesia
Tanintharyi 115 B6 *prev.* Tenasserim. S Myanmar (Burma)
Tanjungkarang/Tanjungkarang-Telukbetung *see* Bandar Lampung
Tanna 122 D4 *island* S Vanuatu
Tannenhof *see* Krynica
Tan-Tan 48 B3 SW Morocco
Tan-tung *see* Dandong
Tanzania 51 C7 *off.* United Republic of Tanzania, *Swa.* Jamhuri ya Muungano wa Tanzania; *prev.* German East Africa, Tanganyika and Zanzibar. *country* E Africa
Tanzania, Jamhuri ya Muungano wa *see* Tanzania
Tanzania, United Republic of *see* Tanzania
Taoudenit *see* Taoudenni
Taoudenni *see* Tombouctou
Tapa 84 E2 *Ger.* Taps. Lääne-Virumaa, NE Estonia
Tapachula 29 G5 Chiapas, SE Mexico
Tapaiu *see* Gvardeysk
Tapajós, Rio 41 E2 *var.* Tapajóz. *river* NW Brazil
Tapajóz *see* Tapajós, Rio
Taps *see* Tapa
Ţarābulus 49 F2 *var.* Ţarābulus al Gharb, *Eng.* Tripoli. *country capital* (Libya) NW Libya
Ţarābulus al Gharb *see* Ţarābulus
Tarābulus/Ţarābulus ash Shām *see* Tripoli
Taraclia 86 D4 *Rus.* Tarakilya. S Moldova
Tarakilya *see* Taraclia
Taranaki, Mount 128 C4 *var.* Egmont. *volcano* North Island, New Zealand
Tarancón 71 E3 Castilla-La Mancha, C Spain
Taranto 75 E5 *var.* Tarentum. Puglia, SE Italy
Taranto, Gulf of 75 E6 *Eng.* Gulf of Taranto. *gulf* S Italy
Taranto, Gulf of *see* Taranto, Golfo di
Tarapoto 38 C2 San Martín, N Peru
Tarare 69 D5 Rhône, E France
Tarascon 69 D6 Bouches-du-Rhône, SE France
Tarawa 122 D2 *atoll and capital* (Kiribati) Tungaru, W Kiribati
Taraz 92 C5 *prev.* Aulie Ata, Auliye-Ata, Dzhambul, Zhambyl. Zhambyl, S Kazakhstan
Tarazona 71 E2 Aragón, NE Spain
Tarbes 69 B6 *anc.* Bigorra. Hautes-Pyrénées, S France
Tarcoola 127 A6 South Australia
Taree 127 D6 New South Wales, SE Australia
Tarentum *see* Taranto
Târgovişte 86 C5 *prev.* Tîrgovişte. Dâmboviţa, S Romania
Targu Jiu 86 B4 *prev.* Tîrgu Jiu. Gorj, W Romania
Târgul-Neamţ *see* Târgu-Neamţ
Târgul-Săcuiesc *see* Târgu Secuiesc
Târgu Mureş 86 B4 *prev.* Oşorhei, Tîrgu Mures, *Ger.* Neumarkt, *Hung.* Marosvásárhely. Mureş, C Romania
Târgu-Neamţ 86 C3 *var.* Târgul-Neamţ; *prev.* Tîrgu-Neamţ. Neamţ, NE Romania
Târgu Ocna 86 C4 *Hung.* Aknavásár; *prev.* Tîrgu Ocna. Bacău, E Romania
Târgu Secuiesc 86 C4 *Ger.* Neumarkt, Szekler Neumarkt, *Hung.* Kezdivásárhely; *prev.* Chezdi-Oşorheiu, Târgul-Săcuiesc, Tîrgu Secuiesc. Covasna, E Romania
Tar Heel State *see* North Carolina
Tarija 39 G5 Tarija, S Bolivia
Tarīm 99 C6 C Yemen
Tarim Basin *see* Tarim Pendi
Tarim Pendi 102 C2 *Eng.* Tarim Basin. *basin* NW China
Tarim He 104 B3 *river* NW China
Tarma 38 C3 Junín, C Peru
Tarn 69 C6 *cultural region* S France
Tarn 69 C6 *river* S France
Tarnobrzeg 76 D4 Podkarpackie, SE Poland
Tarnopol *see* Ternopil
Tarnów 77 D5 Małopolskie, S Poland
Tarraco *see* Tarragona
Tarragona 71 G2 *anc.* Tarraco. Cataluña, E Spain
Tarrasa *see* Terrassa
Tàrrega 71 F2 *var.* Tarrega. Cataluña, NE Spain
Tarsatica *see* Rijeka
Tarsus 94 C4 İçel, S Turkey (Türkiye)
Tartous/Tartouss *see* Ţarţūs
Tartu 84 D3 *Ger.* Dorpat; *prev. Rus.* Yurev, Yury'ev. Tartumaa, SE Estonia
Ţarţūs 96 A3 *off.* Muḩāfaẓat Ţarţūs, *var.* Tartous, Tartus. *governorate* W Syria
Ţarţūs, Muḩāfaẓat *see* Ţarţūs
Ta Ru Tao, Ko 115 B7 *island* S Thailand Asia
Tarvisio 74 D2 Friuli-Venezia Giulia, NE Italy
Tarvisium *see* Treviso
Tashauz *see* Daşoguz
Tashi Chho Dzong *see* Thimphu
Tashkent *see* Toshkent
Tash-Kömür *see* Tash-Kumyr
Tash-Kumyr 101 F2 *Kir.* Tash-Kömür. Dzhalal-Abadskaya Oblast', W Kyrgyzstan
Tashqurghan *see* Khulm
Tasiilaq 60 D4 *var.* Angmagssalik. Semersooq, S Greenland
Tasikmalaja *see* Tasikmalaya
Tasikmalaya 116 C5 *prev.* Tasikmalaja. Jawa, C Indonesia
Tasman Basin 120 C5 *var.* East Australian Basin. *undersea basin* S Tasman Sea
Tasman Bay 129 C5 *var.* Te Tai-o-Aorere. *inlet* South Island, New Zealand
Tasmania 127 B8 *prev.* Van Diemen's Land. *state* SE Australia
Tasmania 130 B4 *island* SE Australia
Tasman Plateau 120 C5 *var.* South Tasmania Plateau. *undersea plateau* SW Tasman Sea
Tasman Sea 120 C5 *sea* SW Pacific Ocean
Tassili-n-Ajjer 49 E4 *plateau* E Algeria
Tatabánya 77 C6 Komárom-Esztergom, NW Hungary
Tatar Pazardzhik *see* Pazardzhik
Tathlīth 99 B5 'Asīr, S Saudi Arabia
Tatra Mountains 77 D5 *Ger.* Tatra, *Hung.* Tátra, *Pol./Slvk.* Tatry. *mountain range* Poland/Slovakia
Tatra/Tátra *see* Tatra Mountains
Tatry *see* Tatra Mountains
Ta-t'ung/Tatung *see* Datong
Tatvan 95 F3 Bitlis, SE Turkey (Türkiye)
Ta'ū 123 F4 *var.* Tau. *island* Manua Islands, E American Samoa
Taukum, Peski 101 G1 *desert* SE Kazakhstan
Taumarunui 128 D4 Manawatu-Wanganui, North Island, New Zealand
Taungdwingyi 114 B3 Magway, C Myanmar (Burma)
Taunggyi 114 B3 Shan State, C Myanmar (Burma)
Taungoo 114 B4 Bago, C Myanmar (Burma)
Taunton 67 C7 SW England, United Kingdom

Taupo *128 D3* Waikato, North Island, New Zealand
Taupo, Lake *128 D3 lake* North Island, New Zealand
Tauragė *84 B4 Ger.* Tauroggen. Tauragė, SW Lithuania
Tauranga *128 D3* Bay of Plenty, North Island, New Zealand
Tauris *see* Tabrīz
Tauroggen *see* Tauragė
Taurus Mountains *see* Toros Dağları
Tavas *94 B4* Denizli, SW Turkey (Türkiye)
Tavastehus *see* Hämeenlinna
Tavira *70 C5* Faro, S Portugal
Tavoy *see* Dawei
Tavoy Island *see* Mali Kyun
Ta Waewae Bay *129 A7 bay* South Island, New Zealand
Tawakoni, Lake *27 G2 reservoir* Texas, SW USA
Tawau *116 D3* Sabah, East Malaysia
Ṭawkar *see* Tokar
Tawzar *see* Tozeur
Taxco *29 E4 var.* Taxco de Alarcón. Guerrero, S Mexico
Taxco de Alarcón *see* Taxco
Taxiatosh *100 C2 Rus.* Takhiatash. Qoraqalpog'iston Respublikasi, W Uzbekistan
Taxtako'pir *100 D1 Rus.* Takhtakupyr. Qoraqalpog'iston Respublikasi, NW Uzbekistan
Tay *66 C3 river* C Scotland, United Kingdom
Taylor *27 G3* Texas, SW USA
Taymā' *98 A4* Tabūk, NW Saudi Arabia
Taymyr, Ozero *93 E2 lake* N Russia
Taymyr, Poluostrov *93 E2 peninsula* N Russia
Taz *92 D3 river* N Russia
Tbilisi *95 G2 var.* T'bilisi, *Eng.* Tiflis. *country capital* (Georgia) SE Georgia
T'bilisi *see* **Tbilisi**
Tchad *see* Chad
Tchad, Lac *see* Chad, Lake
Tchien *see* Zwedru
Tchongking *see* Chongqing
Tczew *76 C2 Ger.* Dirschau. Pomorskie, N Poland
Te Anau *129 A7* Southland, South Island, New Zealand
Te Anau, Lake *129 A7 lake* South Island, New Zealand
Teapa *29 G4* Tabasco, SE Mexico
Teate *see* Chieti
Tebingtinggi *116 B3* Sumatera, N Indonesia
Tebriz *see* Tabrīz
Techirghiol *86 D5* Constanța, SE Romania
Tecomán *28 D4* Colima, SW Mexico
Tecpan *29 E5 var.* Tecpan de Galeana. Guerrero, S Mexico
Tecpan de Galeana *see* Tecpan
Tecuci *86 C4* Galați, E Romania
Tedzhen *see* Harī Rōd
Tees *67 D5 river* N England, United Kingdom
Tefé *40 D2* Amazonas, N Brazil
Tegal *116 C4* Jawa, C Indonesia
Tegelen *65 D5* Limburg, SE Netherlands
Tegucigalpa *30 C3 country capital* (Honduras) Francisco Morazán, SW Honduras
Te Hauturu-o-Toi *128 D2 var.* Little Barrier Island. *island* N New Zealand
Teheran *see* Tehrān
Te Houhou *129 A7 var.* Gerorge Sound. *sound* South Island, New Zealand
Tehrān *98 C3 var.* Teheran. *country capital* (Iran) Tehrān, N Iran
Tehuacán *29 F4* Puebla, S Mexico
Tehuantepec *29 F5 var.* Santo Domingo Tehuantepec. Oaxaca, SE Mexico
Tehuantepec, Golfo de *29 F5 var.* Gulf of Tehuantepec. *gulf* S Mexico
Tehuantepec, Gulf of *see* Tehuantepec, Golfo de
Tehuantepec, Isthmus of *see* Tehuantepec, Istmo de
Tehuantepec, Istmo de *29 F5 var.* Isthmus of Tehuantepec. *isthmus* SE Mexico
Te Ika-a-Māui *see* North Island
Tejen *100 C3 Rus.* Tedzhen. Ahal Welaýaty, S Turkmenistan
Tejen *see* Harī Rōd
Tejo, Rio *see* Tagus
Te Kao *128 C1* Northland, North Island, New Zealand
Tekax *29 H4 var.* Tekax de Álvaro Obregón. Yucatán, SE Mexico
Tekax de Álvaro Obregón *see* Tekax
Tekeli *92 C5* Almaty, SE Kazakhstan
Tekirdağ *94 A2 It.* Rodosto; *anc.* Bisanthe, Raidestos, Rhaedestus. Tekirdağ, NW Turkey (Türkiye)
Te Kuiti *128 D3* Waikato, North Island, New Zealand
Tela *30 C2* Atlántida, NW Honduras
Telanaipura *see* Jambi
Telangana *112 D5 cultural region* SE India
Tel Aviv-Jaffa *see* Tel Aviv-Yafo
Tel Aviv-Yafo *97 A6 var.* Tel Aviv-Jaffa. Tel Aviv, C Israel
Teles Pirés *see* São Manuel, Rio
Telish *82 C2 prev.* Azizie. Pleven, N Bulgaria
Tell Abiad/Tell Abyad *see* Tall Abyaḍ
Tell Kalakh *see* Tall Kalakh
Tell Shedadi *see* Ash Shadādah
Tel'man/Tel'mansk *see* Gubadag
Teloekbetoeng *see* Bandar Lampung
Telo Martius *see* Toulon
Telschen *see* Telšiai
Telšiai *84 B3 Ger.* Telschen. Telšiai, NW Lithuania
Telukbetung *see* Bandar Lampung
Temerin *78 D3* Vojvodina, N Serbia
Temeschburg/Temeschwar *see* Timișoara
Temesvár/Temeswar *see* Timișoara
Temirtau *92 C4 prev.* Samarkandski, Samarkandskoye. Karagandy, C Kazakhstan
Tempio Pausania *75 A5* Sardegna, Italy, C Mediterranean Sea
Temple *27 G3* Texas, SW USA
Temuco *43 B5* Araucanía, C Chile
Temuka *129 B6* Canterbury, South Island, New Zealand
Tenasserim *see* Tanintharyi
Ténenkou *52 D3* Mopti, C Mali
Ténéré *53 G3 physical region* C Niger
Tenerife *48 A3 island* Islas Canarias, Spain, NE Atlantic Ocean
Tengger Shamo *105 E3 desert* N China
Tengréla *52 D4 var.* Tingréla. N Ivory Coast
Tenkodogo *53 E4* S Burkina
Tennant Creek *126 A3* Northern Territory, C Australia
Tennessee *20 C1 off.* State of Tennessee, *also known as* The Volunteer State. *state* SE USA
Tennessee River *20 C1 river* S USA
Tenos *see* Tínos
Te-Oneroa-a-Tōhē *128 C1 var.* Ninety Mile Beach. *beach* North Island, New Zealand
Tepelena *see* Tepelenë
Tepelenë *79 C7 var.* Tepelena, *It.* Tepeleni. Gjirokastër, S Albania
Tepeleni *see* Tepelenë
Tepic *28 D4* Nayarit, C Mexico
Teplice *76 A4 Ger.* Teplitz; *prev.* Teplice-Šanov, Teplitz-Schönau. Ústecký Kraj, NW Czechia (Czech Republic)
Teplice-Šanov/Teplitz/Teplitz-Schönau *see* Teplice
Tequila *28 D4* Jalisco, SW Mexico
Teraina *123 G2 prev.* Washington Island. *atoll* Line Islands, E Kiribati
Teramo *74 C4 anc.* Interamna. Abruzzi, C Italy
Tercan *95 E3* Erzincan, NE Turkey (Türkiye)
Terceira *70 A5 var.* Ilha Terceira. *island* Azores, Portugal, NE Atlantic Ocean
Terceira, Ilha *see* Terceira
Terekhovka *see* Tsyerakhowka
Te Rerenga Wairua *see* Reinga, Cape
Teresina *41 F2 var.* Therezina. *state capital* Piauí, NE Brazil
Termez *see* Termiz
Termia *see* Kýthnos
Términos, Laguna de *29 G4 lagoon* SE Mexico
Termiz *101 E3 Rus.* Termez. Surkhondaryo Viloyati, S Uzbekistan
Termoli *74 D4* Molise, C Italy
Terneuzen *65 B5 var.* Neuzen. Zeeland, SW Netherlands
Terni *74 C4 anc.* Interamna Nahars. Umbria, C Italy
Ternopil *86 C2 Pol.* Tarnopol, *Rus.* Ternopol'. Ternopil's'ka Oblast', W Ukraine
Ternopol' *see* Ternopil
Terracina *75 C5* Lazio, C Italy
Terranova di Sicilia *see* Gela
Terranova Pausania *see* Olbia
Terrassa *71 G2 Cast.* Tarrasa. Cataluña, E Spain
Terre Adélie *132 C4 physical region* Antarctica
Terre Haute *18 B4* Indiana, N USA
Terre Neuve *see* Newfoundland and Labrador
Terschelling *64 C1 Fris.* Skylge. *island* Waddeneilanden, N Netherlands
Teruel *71 F3 anc.* Turba. Aragón, E Spain
Tervel *82 E1 prev.* Kurtbunar, *Rom.* Curtbunar. Dobrich, NE Bulgaria
Tervueren *see* Tervuren
Tervuren *65 C6 var.* Tervueren. Vlaams Brabant, C Belgium
Teseney *50 C4 var.* Tessenei. W Eritrea
Tessalit *53 E2* Kidal, NE Mali
Tessaoua *53 G3* Maradi, S Niger
Tessenderlo *65 C5* Limburg, NE Belgium
Tessenei *see* Teseney
Testigos, Islas los *37 E1 island group* N Venezuela
Te Tai-o-Aorere *see* Tasman Bay
Tete *57 E2* Tete, NW Mozambique
Teterow *72 C3* Mecklenburg-Vorpommern, NE Germany
Tétouan *48 C2 var.* Tetouan, Tetuán. N Morocco
Tetovo *79 D5* Razgrad, N Bulgaria
Tetschen *see* Děčín
Tetuán *see* Tétouan
Teverya *see* Tverya
Te Waewae Bay *129 A7 bay* South Island, New Zealand
Te Waipounamu *see* South Island
Texarkana *20 A2* Arkansas, C USA
Texarkana *27 H2* Texas, SW USA
Texas *27 F3 off.* State of Texas, *also known as* Lone Star State. *state* S USA
Texas City *27 H4* Texas, SW USA
Texel *64 C2 island* Waddeneilanden, NW Netherlands
Texoma, Lake *27 G2 reservoir* Oklahoma/Texas, C USA
Teziutlán *29 F4* Puebla, S Mexico
Thaa Atoll *see* Kolhumadulu
Thai, Ao *see* Thailand, Gulf of
Thai Binh *114 D3* Thai Binh, N Vietnam
Thailand *115 C5 off.* Kingdom of Thailand, *Th.* Prathet Thai; *prev.* Siam. *country* SE Asia
Thailand, Gulf of *115 C6 var.* Gulf of Siam, *Th.* Ao Thai, *Vtn.* Vinh Thai Lan. *gulf* SE Asia
Thailand, Kingdom of *see* Thailand
Thai Lan, Vinh *see* Thailand, Gulf of
Thai Nguyên *114 D3* Bắc Thai, N Vietnam
Thakhèk *114 D4 var.* Muang Khammouan. Khammouan, C Laos
Thamarīd *see* Thamarīt
Thamarīt *99 D6 var.* Thamarīd, Thumrayt. SW Oman
Thames *128 D3* Waikato, North Island, New Zealand
Thames *67 B8 river* S England, United Kingdom
Thandwe *114 A4 var.* Sandoway. Rakhine State, W Myanmar (Burma)
Thanh Hoa *114 D3* Thanh Hoa, N Vietnam
Thanintari Taungdan *see* Bilauktaung Range
Thanlwin *see* Salween
Thar Desert *112 C3 var.* Great Indian Desert, Indian Desert. *desert* India/Pakistan
Tharthar, Buhayrat ath *98 B3 lake* C Iraq
Thásos *82 C4* Thásos, E Greece
Thásos *82 C4 island* E Greece
Thaton *114 B4* Mon State, S Myanmar (Burma)
Thayet *114 A4 var.* Thayetmyo. Magway, C Myanmar (Burma)
Thayetmyo *see* Thayet
The Crane *33 H2 var.* Crane. S Barbados
The Dalles *24 B3* Oregon, NW USA
The Flatts Village *see* Flatts Village
The Hague *see* 's-Gravenhage
Theodosia *see* Feodosiia
The Pas *15 F5* Manitoba, C Canada
Therezina *see* Teresina
Thérma *83 D6* Ikaría, Dodekánisa, Greece, Aegean Sea
Thermaic Gulf/Thermaicus Sinus *see* Thermaïkós Kólpos
Thermaïkós Kólpos *82 B4 Eng.* Thermaic Gulf; *anc.* Thermaicus Sinus. *gulf* N Greece
Thermiá *see* Kýthnos
Thérmo *83 B5* Dytikí Ellás, C Greece
The Rock *71 H4* New South Wales, SE Australia
The Sooner State *see* Oklahoma
Thessaloníki *82 C3 Eng.* Salonica, Salonika, *Croatian* Solun, *Turk.* Selânik. Kentrikí Makedonía, N Greece
The Valley *33 G3 dependent territory capital* (Anguilla) E Anguilla
The Village *27 G1* Oklahoma, C USA
The Volunteer State *see* Tennessee
Thiamis *see* Thýamis
Thian Shan *see* Tien Shan
Thibet *see* Xizang Zizhiqu
Thief River Falls *23 F1* Minnesota, N USA
Thienen *see* Tienen
Thiers *69 C5* Puy-de-Dôme, C France
Thiès *52 B3* W Senegal
Thikombia *see* Cikobia
Thimbu *see* Thimphu
Thimphu *113 G3 var.* Thimbu; *prev.* Tashi Chho Dzong. *country capital* (Bhutan) W Bhutan
Thionville *68 D3 Ger.* Diedenhofen. Moselle, NE France
Thíra *83 D7 var.* Santorini. Kykládes, Greece, Aegean Sea
Thiruvananthapuram *110 C3 var.* Trivandrum, Tiruvantapuram. *state capital* Kerala, SW India
Thitu Island *106 C8 island* NW Spratly Islands
Tholen *64 B4 island* SW Netherlands
Thomasville *20 D3* Georgia, SE USA
Thompson *15 F4* Manitoba, C Canada
Thonon-les-Bains *69 D5* Haute-Savoie, E France
Thoothukudi *see* Tuticorin
Thorenburg *see* Turda
Thorlákshöfn *61 E5* Sudhurland, SW Iceland
Thorn *see* Toruń
Thornton Island *see* Millennium Island
Thorshavn *see* Tórshavn
Thospitis *see* Van Gölü
Thouars *68 B4* Deux-Sèvres, W France
Thoune *see* Thun
Thracian Sea *82 D4 Gk.* Thrakikó Pélagos; *anc.* Thracium Mare. *sea* Greece/Turkey (Türkiye)
Thracium Mare/Thrakikó Pélagos *see* Thracian Sea
Three Gorges Reservoir *107 C5 reservoir* C China
Three Kings Islands *see* Manawatāwhi
Thrissur *110 C3 var.* Trichūr. Kerala, SW India
Thuin *65 B7* Hainaut, S Belgium
Thule *see* Qaanaaq
Thumrayt *see* Thamarīt
Thun *73 A7 Fr.* Thoune. Bern, W Switzerland
Thunder Bay *16 B4* Ontario, S Canada
Thuner See *73 A7 lake* C Switzerland
Thung Song *115 C7 var.* Cha Mai. Nakhon Si Thammarat, SW Thailand
Thurso *66 C2* N Scotland, United Kingdom
Thýamis *82 A4 var.* Kalamás; *prev.* Thiamis. *river* W Greece
Tianjin *106 D4 var.* Tientsin. Tianjin Shi, E China
Tianjin *see* Tianjin Shi
Tianjin Shi *106 D4 var.* Jin, Tianjin, T'ien-ching, Tientsin. *municipality* E China
Tian Shan *see* Tien Shan
Tianshui *106 B4* Gansu, C China
Tiba *see* Chiba
Tiber *74 C4 Eng.* Tiber. *river* C Italy
Tiber *see* Tevere, Italy
Tiber *see* Tivoli, Italy
Tiberias *see* Tverya
Tiberias, Lake *see* Yam Kinneret
Tibesti *54 C2 var.* Tibesti Massif, *Ar.* Tibisti. *mountain range* N Africa
Tibesti Massif *see* Tibesti
Tibet *see* Xizang Zizhiqu
Tibetan Autonomous Region *see* Xizang Zizhiqu
Tibet, Plateau of *see* Qingzang Gaoyuan
Tibisti *see* Tibesti
Tibnī *see* At Tibnī
Tiburón, Isla *28 B2 var.* Isla del Tiburón. *island* NW Mexico
Tiburón, Isla del *see* Tiburón, Isla
Tichau *see* Tychy
Tîchît *52 D2 var.* Tichitt. Tagant, C Mauritania
Tichitt *see* Tîchît
Ticinum *see* Pavia
Ticul *29 H3* Yucatán, SE Mexico
Tidjikdja *see* Tidjikja
Tidjikja *52 C2 var.* Tidjikdja; *prev.* Fort-Cappolani. Tagant, C Mauritania
T'ien-ching *see* Tianjin Shi
Tienen *65 C6 var.* Thienen, *Fr.* Tirlemont. Vlaams Brabant, C Belgium
Tiên Giang, Sông *see* Mekong
Tien Shan *104 B3 Chin.* Thian Shan, Tian Shan, T'ien Shan, *Rus.* Tyan'-Shan'. *mountain range* C Asia
Tientsin *see* Tianjin
Tierp *63 C6* Uppsala, C Sweden
Tierra del Fuego *43 B8 island* Argentina/Chile
Tiflis *see* T'bilisi
Tifton *20 D3* Georgia, SE USA
Tifu *117 F4* Pulau Buru, E Indonesia
Tighina *see* Bender
Tigranocerta *see* Siirt
Tigris *98 B2 Ar.* Dijlah, *Turk.* Dicle. *river* Iraq/Syria/Turkey (Türkiye)
Tiguentourine *49 E3* E Algeria
Ti-hua/Tihwa *see* Ürümqi
Tijuana *28 A1* Baja California Norte, NW Mexico
Tikapa Moana *see* Hauraki Gulf
Tikhoretsk *89 A7* Krasnodarskiy Kray, SW Russia
Tikhvin *88 B4* Leningradskaya Oblast', NW Russia
Tikiarjuaq *see* Whale Cove
Tiki Basin *121 G3 undersea basin* S Pacific Ocean
Tikinsso *52 C4 river* C Guinea
Tiksi *93 F2* Respublika Sakha (Yakutiya), NE Russia
Tilburg *64 C4* Noord-Brabant, S Netherlands
Tilimsen *see* Tlemcen
Tilio Martius *see* Toulon
Tillabéri *53 F3 var.* Tillabéry. Tillabéri, W Niger
Tillabéry *see* Tillabéri
Tílos *83 E7 island* Dodekánisa, Greece, Aegean Sea
Timan Ridge *see* Timanskiy Kryazh
Timanskiy Kryazh *88 D3 Eng.* Timan Ridge. *ridge* NW Russia
Timaru *129 B6* Canterbury, South Island, New Zealand
Timbaki/Timbákion *see* Tympáki
Timbedgha *52 D3 var.* Timbédra. Hodh ech Chargui, SE Mauritania
Timbédra *see* Timbedgha
Timbuktu *see* Tombouctou
Timiș *86 A4 county* SW Romania
Timișoara *86 A4 Ger.* Temeschwar, Temeswar, *Hung.* Temesvár; *prev.* Temeschburg. Timiș, W Romania
Timmins *16 C4* Ontario, S Canada
Timor *103 F5 island* Nusa Tenggara, C Indonesia
Timor-Leste *see* East Timor
Timor Lorosa'e *see* East Timor
Timor Sea *103 F5 sea* E Indian Ocean
Timor Timur *see* East Timor
Timor Trench *see* Timor Trough
Timor Trough *103 F5 var.* Timor Trench. *trough* NE Timor Sea
Timrå *63 C5* Västernorrland, C Sweden
Tindouf *48 C3* W Algeria
Tineo *70 C1* Asturias, N Spain
Tingis *see* Tanger
Tingo María *38 C3* Huánuco, C Peru
Tingréla *see* Tengréla
Tinhosa Grande *54 E2 island* N Sao Tome and Principe, Africa, E Atlantic Ocean
Tinhosa Pequena *54 E1 island* N Sao Tome and Principe, Africa, E Atlantic Ocean
Tinian *122 B1 island* S Northern Mariana Islands
Tínos *83 D6* Tínos, Kykládes, Greece, Aegean Sea
Tínos *83 D6 anc.* Tenos. *island* Kykládes, Greece, Aegean Sea
Tip *79 E6* Papua, E Indonesia
Tipitapa *30 D3* Managua, W Nicaragua
Tip Top Mountain *16 C4 mountain* Ontario, S Canada
Tirana *see* Tiranë
Tiranë *79 C6 var.* Tirana. *country capital* (Albania) Tiranë, C Albania
Tiraspol *86 D4 Rus.* Tiraspol'. E Moldova
Tiraspol' *see* Tiraspol
Tiree *66 B3 island* W Scotland, United Kingdom
Tîrgoviște *see* Târgoviște
Tîrgu Jiu *see* Targu Jiu
Tîrgu Mures *see* Târgu Mureș
Tîrgu-Neamț *see* Târgu-Neamț
Tîrgu Ocna *see* Târgu Ocna
Tîrgu Secuiesc *see* Târgu Secuiesc
Tirlemont *see* Tienen
Tírnavos *see* Týrnavos
Tirnovo *see* Veliko Tarnovo
Tirol *73 C7 off.* Land Tirol, *var.* Tyrol, *It.* Tirolo. *state* W Austria
Tirol, Land *see* Tirol
Tirolo *see* Tirol
Tirreno, Mare *see* Tyrrhenian Sea
Tiruchchirāppalli *110 C3 prev.* Trichinopoly. Tamil Nādu, SE India
Tiruppattūr *110 C2* Tamil Nādu, SE India
Tiruvantapuram *see* Thiruvanathapuram
Tisa *see* Tisza
Tisza *81 F1 Ger.* Theiss, *Rom./Slvn./Croatian* Tisa, *Rus.* Tissa, *Ukr.* Tysa. *river* SE Europe
Tiszakécske *77 D7* Bács-Kiskun, C Hungary
Titano, Monte *74 E1 mountain* C San Marino
Tīti *129 A8 var.* Muttonbird Islands. *island group* SW New Zealand
Titicaca, Lake *39 E4 lake* Bolivia/Peru
Titograd *see* Podgorica
Titose *see* Chitose
Titova Mitrovica *see* Mitrovicë
Titovo Užice *see* Užice
Titu *86 C5* Dâmbovița, S Romania
Titule *55 D5* Orientale, N Dem. Rep. Congo
Tiverton *67 C7* SW England, United Kingdom
Tivoli *74 C4 anc.* Tiber. Lazio, C Italy
Tizimín *29 H3* Yucatán, SE Mexico
Tizi Ouzou *49 E1 var.* Tizi-Ouzou. N Algeria
Tizi-Ouzou *see* Tizi Ouzou
Tiznit *48 B3* SW Morocco
Tjilatjap *see* Cilacap
Tjirebon *see* Cirebon
Tlaquepaque *28 D4* Jalisco, C Mexico
Tlascala *see* Tlaxcala
Tlaxcala *29 F4 var.* Tlascala, Tlaxcala de Xicohténcatl. Tlaxcala, C Mexico
Tlaxcala de Xicohténcatl *see* Tlaxcala
Tlemcen *48 D2 var.* Tilimsen, Tlemsen. NW Algeria
Tlemsen *see* Tlemcen
Toamasina *57 G3 var.* Tamatave. Toamasina, E Madagascar
Toba, Danau *116 B3 lake* Sumatera, W Indonesia
Tobago *33 H5 island* NE Trinidad and Tobago
Toba Kakar Range *112 B2 mountain range* NW Pakistan
Tobol *92 C4 Kaz.* Tobyl. *river* Kazakhstan/Russia
Tobol'sk *92 C3* Tyumenskaya Oblast', C Russia
Tobruch/Tobruk *see* Ṭubruq
Tobyl *see* Tobol
Tocantins *41 E3 off.* Estado do Tocantins. *state/region* C Brazil
Tocantins, Estado do *see* Tocantins
Tocantins, Rio *41 F2 river* N Brazil
Tocoa *30 D2* Colón, N Honduras
Tocopilla *42 B2* Antofagasta, N Chile
Todi *74 C4* Umbria, C Italy
Todos os Santos, Baía de *41 G3 bay* E Brazil
Toetoes Bay *129 B8 bay* South Island, New Zealand
Tofua *123 E4 island* Ha'apai Group, C Tonga
Togo *53 E4 off.* Togolese Republic; *prev.* French Togoland. *country* W Africa
Togolese Republic *see* Togo
Tojikiston, Jumhurii *see* Tajikistan
Tokanui *129 B7* Southland, South Island, New Zealand
Tokar *50 C3 var.* Ṭawkar. Red Sea, NE Sudan
Tokat *94 D3* Tokat, N Turkey (Türkiye)
Tokelau *123 E3 NZ overseas territory* W Polynesia
Tőketerebes *see* Trebišov
Tokio *see* Tōkyō
Tokmak *101 G2 Kir.* Tokmok. Chuyskaya Oblast', N Kyrgyzstan
Tokmak *87 G4 var.* Velykyy Tokmak. Zaporiz'ka Oblast', SE Ukraine
Tokmok *see* Tokmak
Tokoroa *128 D3* Waikato, North Island, New Zealand
Tokounou *52 C4* C Guinea
Tokushima *109 C6 var.* Tokusima. Tokushima, Shikoku, SW Japan
Tōkyō *108 A1 var.* Tokio. *country capital* (Japan) Tōkyō, Honshū, S Japan
Tōkyō-wan *108 A2 bay* S Japan
Tolbukhin *see* Dobrich
Toledo *70 D3 anc.* Toletum. Castilla-La Mancha, C Spain
Toledo *18 D3* Ohio, N USA
Toledo Bend Reservoir *27 G3 reservoir* Louisiana/Texas, SW USA
Toletum *see* Toledo
Toliara *57 F4 var.* Toliary; *prev.* Tuléar. Toliara, SW Madagascar
Toliary *see* Toliara
Tolmein *see* Tolmin
Tolmin *73 D7 Ger.* Tolmein, *It.* Tolmino. W Slovenia
Tolmino *see* Tolmin
Tolna *77 C7 Ger.* Tolnau. Tolna, S Hungary
Tolnau *see* Tolna
Tolochin *see* Talachyn
Tolosa *71 E1* País Vasco, N Spain
Tolosa *see* Toulouse
Toluca *29 E4 var.* Toluca de Lerdo. México, S Mexico
Toluca de Lerdo *see* Toluca
Tol'yatti *89 C6 prev.* Stavropol'. Samarskaya Oblast', W Russia
Tomah *18 B2* Wisconsin, N USA
Tomakomai *108 D2* Hokkaidō, NE Japan
Tomar *70 B3* Santarém, W Portugal
Tomaschow *see* Tomaszów Mazowiecki
Tomaschow *see* Tomaszów Lubelski
Tomaszów *see* Tomaszów Mazowiecki
Tomaszów Lubelski *76 E4 Ger.* Tomaschow. Lubelskie, E Poland
Tomaszów Mazowiecka *see* Tomaszów Mazowiecki
Tomaszów Mazowiecki *76 D4 var.* Tomaszów Mazowiecka; *prev.* Tomaszów, *Ger.* Tomaschow. Łódzkie, C Poland
Tombigbee River *20 C3 river* Alabama/Mississippi, S USA
Tombouctou *53 E3 Eng.* Timbuktu. *var.* Taoudenni, N Mali
Tombua *56 A2 Port.* Porto Alexandre. Namibe, SW Angola
Tomelloso *71 E3* Castilla-La Mancha, C Spain
Tomini, Gulf of *see* Teluk Tomini
Tomini, Teluk *117 E4 Eng.* Gulf of Tomini; *prev.* Teluk Gorontalo. *bay* Sulawesi, C Indonesia
Tomsk *92 D4* Tomskaya Oblast', C Russia
Tömür Feng *104 B3 pre. Pik Pobedy; Kyrg.* Jengish Chokusu. *mountain* China/Kyrgyzstan
Tonezh *see* Tonyezh
Tonga *123 E4 off.* Kingdom of Tonga, *var.* Friendly Islands. *country* SW Pacific Ocean
Tonga, Kingdom of *see* Tonga
Tongatapu *123 E5 island* Tongatapu Group, S Tonga
Tongatapu Group *123 E5 island group* S Tonga
Tonga Trench *121 E3 trench* S Pacific Ocean
Tongchuan *106 C4* Shaanxi, C China
Tongeren *65 D6 Fr.* Tongres. Limburg, NE Belgium
Tongking, Gulf of *see* Tonkin, Gulf of
Tongliao *105 G2* Nei Mongol Zizhiqu, N China
Tongres *see* Tongeren
Tongshan *see* Xuzhou, Jiangsu, China
Tongtian He *104 C4 river* C China
Tonj *51 B5* Warab, C South Sudan
Tonkin, Gulf of *106 B7 var.* Tongking, *Gulf of, Chin.* Beibu Wan, *Vtn.* Vinh Bắc Bộ. *gulf* China/Vietnam
Tônlé Sap *115 D5 Eng.* Great Lake. *lake* W Cambodia
Tonopah *25 C6* Nevada, W USA
Tonyezh *85 C7 Rus.* Tonezh. Homyel'skaya Voblasts', SE Belarus
Tooele *22 B4* Utah, W USA
Toowoomba *127 E5* Queensland, E Australia
Topeka *23 F4 state capital* Kansas, C USA
Toplicza *see* Toplița
Toplița *86 C3 Ger.* Töplitz, *Hung.* Maroshevíz; *prev.* Toplița Română, *Hung.* Oláh-Toplicza, Toplicza. Harghita, C Romania
Toplița Română/Töplitz *see* Toplița
Topoľčany *77 C6 Hung.* Nagytapolcsány. Nitriansky Kraj, W Slovakia
Topolovgrad *82 D3 prev.* Kavakli. Haskovo, S Bulgaria
Topolya *see* Bačka Topola
Top Springs Roadhouse *124 E3* Northern Territory, N Australia
Tor *132 C2 Norwegian research station* Antarctica
Torda *see* Turda
Torez *see* Chystiakove
Torgau *72 D4* Sachsen, E Germany
Torhout *65 A5* West-Vlaanderen, W Belgium
Torino *74 A2 Eng.* Turin. Piemonte, NW Italy
Tornacum *see* Tournai
Torneå *see* Tornio
Torneträsk *62 C3 lake* N Sweden
Tornio *62 D4 Swe.* Torneå. Lappi, NW Finland
Tornionjoki *62 D3 river* Finland/Sweden
Toro *70 D2* Castilla y León, N Spain
Toronto *16 D5 province capital* Ontario, S Canada
Toros Dağları *94 C4 Eng.* Taurus Mountains. *mountain range* S Turkey (Türkiye)
Torquay *67 C7* SW England, United Kingdom
Torrance *24 D2* California, W USA
Torre, Alto da *70 B3 mountain* C Portugal
Torre del Greco *75 D5* Campania, S Italy
Torrejón de Ardoz *71 E3* Madrid, C Spain
Torrelavega *70 D1* Cantabria, N Spain
Torrens, Lake *127 A6 salt lake* South Australia
Torrent *71 F3 Cas.* Torrente, *var.* Torrent de l'Horta. Comunitat Valenciana, E Spain
Torrent de l'Horta/Torrente *see* Torrent
Torreón *28 D3* Coahuila, NE Mexico
Torres Strait *126 C1 strait* Australia/Papua New Guinea
Torres Vedras *70 B3* Lisboa, C Portugal
Torrington *22 D3* Wyoming, C USA
Tórshavn *61 F5 Dan.* Thorshavn. *Dependent territory capital* Faroe Islands
To'rtkok'l *100 D2 var.* Türtkül, *Rus.* Turtkul'; *prev.* Petroaleksandrovsk. Qoraqalpog'iston Respublikasi, W Uzbekistan
Tortoise Islands *see* Galápagos Islands
Tortosa *71 F2 anc.* Dertosa. Cataluña, E Spain
Tortue, Montagne *37 H3 mountain range* C French Guiana
Tortuga, Isla *see* La Tortuga, Isla

Toruń 76 C3 *Ger.* Thorn. Toruń, Kujawsko-pomorskie, C Poland
Törva 84 D3 *Ger.* Törwa. Valgamaa, S Estonia
Törwa *see* Törva
Torzhok 88 B4 Tverskaya Oblast', W Russia
Tosa-wan 109 B7 *bay* SW Japan
Toscana 74 B3 *Eng.* Tuscany. *region* C Italy
Toscano, Archipelago 74 B4 *Eng.* Tuscan Archipelago. *island group* C Italy
Toshkent 101 E2 *Eng./Rus.* Tashkent. *country capital* (Uzbekistan) Toshkent Viloyati, E Uzbekistan
Totana 71 E4 Murcia, SE Spain
Tot'ma *see* Sukhona
Totness 37 G3 Coronie, N Suriname
Tottori 109 B6 Tottori, Honshū, SW Japan
Touâjîl 52 C2 Tiris Zemmour, N Mauritania
Touggourt 49 E2 NE Algeria
Toukoto 52 C3 Kayes, W Mali
Toul 68 D3 Meurthe-et-Moselle, NE France
Toulon 69 D6 *anc.* Telo Martius, Tilio Martius. Var, SE France
Toulouse 69 B6 *anc.* Tolosa. Haute-Garonne, S France
Toungoo *see* Taungoo
Touraine 68 B4 *cultural region* C France
Tourane *see* Đà Nẵng
Tourcoing 68 C2 Nord, N France
Tournai 65 A6 *var.* Tournay, *Dut.* Doornik; *anc.* Tornacum. Hainaut, SW Belgium
Tournay *see* Tournai
Tours 68 B4 *anc.* Caesarodunum, Turoni. Indre-et-Loire, C France
Tovarkovskiy 89 B5 Tul'skaya Oblast', W Russia
Tower Island *see* Genovesa, Isla
Townsville 126 D3 Queensland, NE Australia
Towoeti Meer *see* Towuti, Danau
Towraghoudī 100 D4 Herāt, NW Afghanistan
Towson 19 F4 Maryland, NE USA
Towuti, Danau 117 E4 *Dut.* Towoeti Meer. *lake* Sulawesi, C Indonesia
Toyama 109 C5 Toyama, Honshū, SW Japan
Toyama-wan 109 B5 *bay* W Japan
Toyohara *see* Yuzhno-Sakhalinsk
Toyota 109 C6 Aichi, Honshū, SW Japan
Tozeur 49 F2 *var.* Tawzar. W Tunisia
Trâblous *see* Tripoli
Trabzon 95 E2 *Eng.* Trebizond; *anc.* Trapezus. Trabzon, NE Turkey (Türkiye)
Traiectum ad Mosam/Traiectum Tungorum *see* Maastricht
Traiskirchen 73 E6 Niederösterreich, NE Austria
Trajani Portus *see* Civitavecchia
Trajectum ad Rhenum *see* Utrecht
Trakai 85 C5 *Ger.* Traken, *Pol.* Troki. Vilnius, SE Lithuania
Traken *see* Trakai
Tralee 67 A6 *Ir.* Trá Lí. SW Ireland
Trá Lí *see* Tralee
Tralles Aydin *see* Aydın
Tran 82 B2 *var.* Trŭn. Pernik, W Bulgaria
Trang 115 C7 Trang, S Thailand
Transantarctic Mountains 132 B3 *mountain range* Antarctica
Transilvania *see* Transylvania
Transilvaniei, Alpi *see* Carpaţii Meridionali
Transjordan *see* Jordan
Transnistria 86 D3 *disputed territory* NE Moldova
Transsylvanische Alpen/Transylvanian Alps *see* Carpaţii Meridionali
Transylvania 86 B4 *Eng.* Ardeal, Transilvania, *Ger.* Siebenbürgen, *Hung.* Erdély. *cultural region* NW Romania
Trapani 75 B7 *anc.* Drepanum. Sicilia, Italy, C Mediterranean Sea
Trâpeăng Vêng *see* Kampong Thom
Trapezus *see* Trabzon
Traralgon 127 C7 Victoria, SE Australia
Trasimenischersee *see* Trasimeno, Lago
Trasimeno, Lago 74 C4 *Eng.* Lake of Perugia, *Ger.* Trasimenischersee. *lake* C Italy
Traù *see* Trogir
Traverse City 18 C2 Michigan, N USA
Tra Vinh 115 D6 *var.* Phu Vinh. Tra Vinh, S Vietnam
Travis, Lake 27 F3 *reservoir* Texas, SW USA
Travnik 78 C4 Federacija Bosna I Hercegovina, C Bosnia and Herzegovina
Trbovlje 73 E7 *Ger.* Trifail. C Slovenia
Treasure State *see* Montana
Třebíč 77 B5 *Ger.* Trebitsch. Vysočina, C Czechia (Czech Republic)
Trebinje 79 C5 Republika Srpska, S Bosnia and Herzegovina
Trebišov 77 D6 *Hung.* Tőketerebes. Košický Kraj, E Slovakia
Trebitsch *see* Třebíč
Trebnitz *see* Trzebnica
Tree Planters State *see* Nebraska
Trélazé 68 B4 Maine-et-Loire, NW France
Trelew 43 C6 Chubut, SE Argentina
Tremelo 65 C5 Vlaams Brabant, C Belgium
Trenčín 77 C5 *Ger.* Trentschin, *Hung.* Trencsén. Trenčiansky Kraj, W Slovakia
Trencsén *see* Trenčín
Trengganu, Kuala *see* Kuala Terengganu
Trenque Lauquen 42 C4 Buenos Aires, E Argentina
Trent *see* Trento
Trento 74 C2 *Eng.* Trent, *Ger.* Trient; *anc.* Tridentum. Trentino-Alto Adige, N Italy
Trenton 19 F4 *state capital* New Jersey, NE USA
Trentschin *see* Trenčín
Tres Arroyos 43 D5 Buenos Aires, E Argentina
Treskavica 78 C4 *mountain range* SE Bosnia and Herzegovina
Tres Tabernae *see* Saverne
Treves/Trèves *see* Trier
Treviso 74 C2 *anc.* Tarvisium. Veneto, NE Italy
Trichinopoly *see* Tiruchchirāppalli
Trichūr *see* Thrissur
Tridentum/Trient *see* Trento
Trier 73 A5 *Eng.* Treves, *Fr.* Trèves; *anc.* Augusta Treverorum. Rheinland-Pfalz, SW Germany
Triesen 72 E2 SW Liechtenstein Europe
Triesenberg 72 E2 SW Liechtenstein
Trieste 74 D2 *Slvn.* Trst. Friuli-Venezia Giulia, NE Italy
Trifail *see* Trbovlje
Tríkala 82 B4 *prev.* Trikkala. Thessalía, C Greece
Trikkala *see* Tríkala
Trimontium *see* Plovdiv
Trinacria *see* Sicilia
Trincomalee 110 D3 *var.* Trinkomali. Eastern Province, NE Sri Lanka
Trindade, Ilha da 45 C5 *island* Brazil, W Atlantic Ocean
Trinidad 39 F3 Beni, N Bolivia
Trinidad 42 D4 Flores, S Uruguay
Trinidad 22 D5 Colorado, C USA
Trinidad 33 H5 *island* C Trinidad and Tobago
Trinidad and Tobago 33 H5 *off.* Republic of Trinidad and Tobago. *country* SE West Indies
Trinidad and Tobago, Republic of *see* Trinidad and Tobago
Trinité, Montagnes de la 37 H3 *mountain range* C French Guiana
Trinity River 27 G3 *river* Texas, SW USA
Trinkomali *see* Trincomalee
Trípoli 83 B6 *prev.* Trípolis. Pelopónnisos, S Greece
Tripoli 96 B4 *var.* Tarābulus, Ţarābulus ash Shām, Trâblous; *anc.* Tripolis. N Lebanon
Tripoli *see* Ţarābulus
Trípolis *see* Trípoli, Greece
Tripolis *see* Tripoli, Lebanon
Tripolitania 49 F3 *region* NW Libya
Tristan da Cunha *see* St Helena, Ascension and Tristan da Cunha
Triton Island 106 B7 *island* S Paracel Islands
Trivandrum *see* **Thiruvananthapuram**
Trnava 77 C6 *Ger.* Tyrnau, *Hung.* Nagyszombat. Trnavský Kraj, W Slovakia
Trnovo *see* Veliko Tarnovo
Trogir 78 B4 *It.* Traù. Split-Dalmacija, S Croatia
Troglav 78 B4 *mountain* Bosnia and Herzegovina/Croatia
Trois-Rivières 17 E4 Québec, SE Canada
Troki *see* Trakai
Troll 132 C2 *Norwegian research station* Antarctica
Trollhättan 63 B6 Västra Götaland, S Sweden
Tromsø 62 C2 *Fin.* Tromssa. Troms, N Norway
Tromssa *see* Tromsø
Trondheim 62 B4 *Ger.* Drontheim; *prev.* Nidaros, Trondhjem. Sør-Trøndelag, S Norway
Trondheimsfjorden 62 B4 *fjord* S Norway
Trondhjem *see* Trondheim
Troódos 80 C5 *var.* Troodos Mountains. *mountain range* C Cyprus
Troodos Mountains *see* Troódos
Troppau *see* Opava
Troy 20 D3 Alabama, S USA
Troy 19 F3 New York, NE USA
Troyan 82 C2 Lovech, N Bulgaria
Troyes 68 D3 *anc.* Augustobona Tricassium. Aube, N France
Trst *see* Trieste
Trstenik 78 E4 Serbia, C Serbia
Trucial States *see* United Arab Emirates
Trujillo 30 D2 Colón, NE Honduras
Trujillo 38 B3 La Libertad, NW Peru
Trujillo 70 C3 Extremadura, W Spain
Truk Islands *see* Chuuk Islands
Trŭn *see* Tran
Truro 17 F4 Nova Scotia, SE Canada
Truro 67 C7 SW England, United Kingdom
Trzcianka 76 B3 *Ger.* Schönlanke. Pila, Wielkopolskie, C Poland
Trzebnica 76 C4 *Ger.* Trebnitz. Dolnośląskie, SW Poland
Tsalka 95 F2 S Georgia Asia
Tsamkong *see* Zhanjiang
Tsangpo *see* Brahmaputra
Tsarevo 82 E2 *prev.* Michurin. Burgas, E Bulgaria
Tsarigrad *see* İstanbul
Tsaritsyn *see* Volgograd
Tschakathurn *see* Čakovec
Tschaslau *see* Čáslav
Tschenstochau *see* Częstochowa
Tsefat 97 B5 *var.* Safed, *Ar.* Safad; *prev.* Zefat. Northern, N Israel
Tselinograd *see* Nur-Sultan
Tsetsen Khan *see* Öndörhaan
Tsetserleg 104 D2 Arhangay, C Mongolia
Tshela 55 B6 Bas-Congo, W Dem. Rep. Congo
Tshikapa 55 C7 Kasai-Occidental, SW Dem. Rep. Congo
Tshuapa 55 D6 *river* C Dem. Rep. Congo
Tsinan *see* Jinan
Tsing Hai *see* Qinghai Hu, China
Tsinghai *see* Qinghai, China
Tsingtao/Tsingtau *see* Qingdao
Tsinkiang *see* Quanzhou
Tsintao *see* Qingdao
Tsitsihar *see* Qiqihar
Tsu 109 C6 *var.* Tu. Mie, Honshū, SW Japan
Tsugaru-kaikyo 108 C3 *strait* N Japan
Tsumeb 56 B3 Otjikoto, N Namibia
Tsuruga 109 C6 *var.* Turuga. Fukui, Honshū, SW Japan
Tsiurupynsk *see* Oleshky
Tsuruoka 108 D4 *var.* Turuoka. Yamagata, Honshū, C Japan
Tsushima 109 A7 *var.* Tsushima-tô, Tusima. *island group* SW Japan
Tsushima-tō *see* Tsushima
Tsyerakhowka 85 D8 *Rus.* Terekhovka. Homyel'skaya Voblasts', SE Belarus
Tu *see* Tsu
Tuamotu, Archipel des *see* Tuamotu, Îles
Tuamotu Fracture Zone 121 H3 *fracture zone* E Pacific Ocean
Tuamotu, Îles 123 H4 *var.* Archipel des Tuamotu, Dangerous Archipelago, Tuamotu Islands. *island group* N French Polynesia
Tuamotu Islands *see* Tuamotu, Îles
Tuapi 31 E2 Región Autónoma Atlántico Norte, NE Nicaragua
Tuapse 89 A7 Krasnodarskiy Kray, SW Russia
Tuba City 26 B1 Arizona, SW USA
Ţūbās 97 E6 N West Bank, Middle East
Tubbergen 64 E3 Overijssel, E Netherlands
Tubeke *see* Tubize
Tubize 65 B6 *Dut.* Tubeke. Walloon Brabant, C Belgium
Tubmanburg 52 C5 NW Liberia
Ţubruq 49 H2 *Eng.* Tobruk, *It.* Tobruch. NE Libya
Tubuai, Îles/Tubuai Islands *see* Australes, Îles
Tucker's Town 20 B5 E Bermuda
Tuckum *see* Tukums
Tucson 26 B3 Arizona, SW USA
Tucumán *see* San Miguel de Tucumán
Tucumcari 27 E2 New Mexico, SW USA
Tucupita 37 E2 Delta Amacuro, NE Venezuela
Tucuruí, Represa de 41 F2 *reservoir* NE Brazil
Tudela 71 E2 *Basq.* Tutera; *anc.* Tutela. Navarra, N Spain
Tuguegarao 117 E1 Luzon, N Philippines
Tuktoyaktuk 15 E3 Northwest Territories, NW Canada
Tukums 84 C3 *Ger.* Tuckum. Tukums, W Latvia
Tula 89 B5 Tul'skaya Oblast', W Russia
Tulancingo 29 E4 Hidalgo, C Mexico
Tulare Lake Bed 25 C7 *salt flat* California, W USA
Tulcán 38 B1 Carchi, N Ecuador
Tulcea 86 D5 Tulcea, E Romania
Tul'chin *see* Tulchyn
Tulchyn 86 D3 *Rus.* Tul'chin. Vinnyts'ka Oblast', C Ukraine
Tuléar *see* Toliara
Tulia 27 E2 Texas, SW USA
Ţūlkarm 97 D6 West Bank, Middle East
Tulle 69 C5 *anc.* Tutela. Corrèze, C France
Tulln 73 E6 *var.* Oberhollabrunn. Niederösterreich, NE Austria
Tully 126 D3 Queensland, NE Australia
Tulsa 27 G1 Oklahoma, C USA
Tuluá 36 B3 Valle del Cauca, W Colombia
Tulun 93 E4 Irkutskaya Oblast', S Russia
Tumaco 36 A4 Nariño, SW Colombia
Tumakūru 110 C2 *prev.* Tumkūr. Karnātaka, W India
Tumba, Lac *see* Ntomba, Lac
Tumbes 38 A2 Tumbes, NW Peru
Tumkūr *see* Tumakūru
Tumuc-Humac Mountains 41 E1 *var.* Serra Tumucumaque. *mountain range* N South America
Tumucumaque, Serra *see* Tumuc-Humac Mountains
Tunca Nehri *see* Tundzha
Tunduru 51 C8 Ruvuma, S Tanzania
Tundzha 82 D3 *Turk.* Tunca Nehri. *river* Bulgaria/Turkey (Türkiye)
Tungabhadra Reservoir 110 C2 *lake* S India
Tungaru 123 E2 *prev.* Gilbert Islands. *island group* W Kiribati
T'ung-shan *see* Xuzhou
Tungsten 14 D4 Northwest Territories, W Canada
Tung-t'ing Hu *see* Dongting Hu
Tunis 49 E1 *var.* Tūnis. *country capital* (Tunisia) NE Tunisia
Tunis, Golfe de 80 D3 *Ar.* Khalīj Tūnis. *gulf* NE Tunisia
Tunisia 49 F2 *off.* Republic of Tunisia, *Ar.* Al Jumhūrīyah at Tūnisīyah, *Fr.* République Tunisienne. *country* N Africa
Tunisia, Republic of *see* Tunisia
Tunisienne, République *see* Tunisia
Tūnisīyah, Al Jumhūrīyah at *see* Tunisia
Tūnis, Khalīj *see* Tunis, Golfe de
Tunja 36 B3 Boyacá, C Colombia
Tuong Buong *see* Tương Đương
Tương Đương 114 D4 *var.* Tuong Buong. Nghê An, N Vietnam
Tüp *see* Tyup
Tupelo 20 C2 Mississippi, S USA
Tupiza 39 G5 Potosí, S Bolivia
Turabah 99 B5 Makkah, W Saudi Arabia
Türanganui-a-Kiwa 128 E4 *var.* Poverty bay. *inlet* North Island, New Zealand
Turangi 128 D4 Waikato, North Island, New Zealand
Turan Lowland 100 C2 *var.* Turan Plain, *Kaz.* Turan Oypaty, *Rus.* Turanskaya Nizmennost', *Turk.* Turan Pesligi, *Uzb.* Turan Pasttekisligi. *plain* C Asia
Turan Oypaty/Turan Pesligi/Turan Plain/Turanskaya Nizmennost' *see* Turan Lowland
Turan Pasttekisligi *see* Turan Lowland
Ţurayf 98 A3 Al Ḩudūd ash Shamālīyah, NW Saudi Arabia
Turba *see* Teruel
Turbat 112 A3 Baluchistan, SW Pakistan
Turčianský Svätý Martin *see* Martin
Turda 86 B4 *Ger.* Thorenburg, *Hung.* Torda. Cluj, NW Romania
Turek 76 C3 Wielkopolskie, C Poland
Turfan *see* Turpan
Turin *see* Torino
Turkana, Lake 51 C6 *var.* Lake Rudolf. *lake* N Kenya
Turkestan *see* Türkistan/Turkistan
Turkey 94 B3 *off.* Republic of Türkiye, *var.* Türkiye, *Turk.* Türkiye Cumhuriyeti. *country* SW Asia
Turkish Republic of Northern Cyprus 80 D5 *disputed territory administered by Turkey* Cyprus
Türkistan/Turkistan 92 B5 *prev.* Turkestan. Türkistan/Turkestan, S Kazakhstan
Türkiye *see* Turkey
Türkiye, Republic of *see* Turkey
Türkiye Cumhuriyeti *see* Turkey
Türkmenabat 100 D3 *prev. Rus.* Chardzhev, Chardzhou, Chardzhui, Lenin-Turkmenski, *Turkm.* Chärjew. Lebap Welaýaty, E Turkmenistan
Türkmen Aylagy 100 B2 *Rus.* Turkmenskiy Zaliv. *lake gulf* W Turkmenistan
Turkmenbashi *see* Türkmenbaşy
Türkmenbaşy 100 B2 *Rus.* Turkmenbashi; *prev.* Krasnovodsk. Balkan Welaýaty, W Turkmenistan
Türkmenbasy Aylagy 100 A2 *prev. Rus.* Krasnovodskiy Zaliv, *Turkm.* Krasnowodsk Aylagy. *lake Gulf* W Turkmenistan
Turkmenistan 100 B2 ; *prev.* Turkmenskaya Soviet Socialist Republic. *country* C Asia
Turkmenskaya Soviet Socialist Republic *see* Turkmenistan
Turkmenskiy Zaliv *see* Türkmen Aylagy
Turks and Caicos Islands 33 E2 *UK Overseas Territory* N West Indies
Turku 63 D6 *Swe.* Åbo. Varsinais-Suomi, SW Finland
Turlock 25 B6 California, W USA
Turnagain, Cape 128 D4 *headland* North Island, New Zealand
Turnau *see* Turnov
Turnhout 65 C5 Antwerpen, N Belgium
Turnov 76 B4 *Ger.* Turnau. Liberecký Kraj, N Czechia (Czech Republic)
Tŭrnovo *see* Veliko Tarnovo
Turnu Măgurele 86 B5 *var.* Turnu-Măgurele. Teleorman, S Romania
Turnu Severin *see* Drobeta-Turnu Severin
Turócszentmárton *see* Martin
Turoni *see* Tours
Turpan 104 C3 *var.* Turfan. Xinjiang Uygur Zizhiqu, NW China
Turpan Depression *see* Turpan Pendi
Turpan Pendi 104 C3 *Eng.* Turpan Depression. *depression* NW China
Turpentine State *see* North Carolina
Türtkül/Turtkul' *see* To'rtko'l
Turuga *see* Tsuruga
Turuoka *see* Tsuruoka
Tuscaloosa 20 C2 Alabama, S USA
Tuscan Archipelago *see* Toscano, Archipelago
Tuscany *see* Toscana
Tusima *see* Tsushima
Tutela *see* Tulle, France
Tutela *see* Tudela, Spain
Tutera *see* Tudela
Tuticorin 110 C3 *var.* Thoothukudi. Tamil Nādu, SE India
Tutrakan 82 D1 Silistra, NE Bulgaria
Tutuila 123 F4 *island* W American Samoa
Tuvalu 123 E3 *prev.* Ellice Islands. *country* SW Pacific Ocean
Tuwayq, Jabal 99 C5 *mountain range* C Saudi Arabia
Tuxpan 28 D4 Jalisco, C Mexico
Tuxpán 29 F4 *var.* Tuxpán de Rodríguez Cano. Veracruz-Llave, E Mexico
Tuxpán de Rodríguez Cano *see* Tuxpán
Tuxtepec 29 F4 *var.* San Juan Bautista Tuxtepec. Oaxaca, S Mexico
Tuxtla 29 G5 *var.* Tuxtla Gutiérrez. Chiapas, SE Mexico
Tuxtla *see* San Andrés Tuxtla
Tuxtla Gutiérrez *see* Tuxtla
Tuy Hoa 115 E5 Phu Yên, S Vietnam
Tuz, Lake 94 C3 *lake* C Turkey (Türkiye)
Tver' 88 B4 *prev.* Kalinin. Tverskaya Oblast', W Russia
Tverya 97 B5 *var.* Tiberias; *prev.* Teverya. Northern, N Israel
Twin Falls 24 D4 Idaho, NW USA
Tyan'-Shan' *see* Tien Shan
Tychy 77 D5 *Ger.* Tichau. Śląskie, S Poland
Tyler 27 G3 Texas, SW USA
Tylos *see* Bahrain
Tympáki 83 C8 *var.* Timbaki; *prev.* Timbákion. Kríti, Greece, E Mediterranean Sea
Tynda 93 F4 Amurskaya Oblast', SE Russia
Tyne 66 D4 *river* N England, United Kingdom
Tyôsi *see* Chōshi
Tyras *see* Bilhorod-Dnistrovskyi
Tyras *see* Dniester
Tyre *see* Soûr
Tyrnau *see* Trnava
Týrnavos 82 B4 *var.* Tírnavos. Thessalía, C Greece
Tyrol *see* Tirol
Tyros *see* Bahrain
Tyrrhenian Sea 75 B6 *It.* Mare Tirreno. *sea* N Mediterranean Sea
Tyumen' 92 C3 Tyumenskaya Oblast', C Russia
Tyup 101 G2 *Kir.* Tüp. Issyk-Kul'skaya Oblast', NE Kyrgyzstan
Tywyn 67 C6 W Wales, United Kingdom
Tzekung *see* Zigong
Tziá *see* Kéa

U

Uaco Cungo 56 B1 C Angola
UAE *see* United Arab Emirates
Uanle Uen *see* Wanlaweyn
Uaupés, Rio *see* Vaupés, Río
Ubangi-Shari *see* Central African Republic
Ube 109 B7 Yamaguchi, Honshū, SW Japan
Ubeda 71 E4 Andalucía, S Spain
Uberaba 41 F4 Minas Gerais, SE Brazil
Uberlândia 41 F4 Minas Gerais, SE Brazil
Ubol Rajadhani/Ubol Ratchathani *see* Ubon Ratchathani
Ubon Ratchathani 115 D5 *var.* Muang Ubon, Ubol Rajadhani, Ubol Ratchathani, Udon Ratchathani. Ubon Ratchathani, E Thailand
Ubrique 70 D5 Andalucía, S Spain
Ubsu-Nur, Ozero *see* Uvs Nuur
Ucayali, Río 38 D3 *river* C Peru
Uchiura-wan 108 D3 *bay* NW Pacific Ocean
Uchkuduk *see* Uchquduq
Uchquduq 100 D2 *Rus.* Uchkuduk. Navoiy Viloyati, N Uzbekistan
Uchtagan Gumy/Uchtagan, Peski *see* Uçtagan Gumy
Uçtagan Gumy 100 C2 *var.* Uchtagan Gumy, *Rus.* Peski Uchtagan. *desert* NW Turkmenistan
Udaipur 112 C3 *prev.* Oodeypore. Rājasthān, N India
Uddevalla 63 B6 Västra Götaland, S Sweden
Udine 74 D2 *anc.* Utina. Friuli-Venezia Giulia, NE Italy
Udintsev Fracture Zone 132 A5 *tectonic feature* S Pacific Ocean
Udipi *see* Udupi
Udon Ratchathani *see* Ubon Ratchathani
Udon Thani 114 C4 *var.* Ban Mak Khaeng, Udorndhani. Udon Thani, N Thailand
Udorndhani *see* Udon Thani
Udupi 110 B2 *var.* Udipi. Karnātaka, SW India
Uele 55 D5 *var.* Welle. *river* NE Dem. Rep. Congo
Uelzen 72 C3 Niedersachsen, N Germany
Ufa 89 D6 Respublika Bashkortostan, W Russia
Ugāle 84 C2 Ventspils, NW Latvia
Uganda 51 B6 *off.* Republic of Uganda. *country* E Africa
Uganda, Republic of *see* Uganda
Uhorshchyna *see* Hungary
Uhuru Peak *see* Kilimanjaro
Uíge 56 B1 *Port.* Carmona, Vila Marechal Carmona. Uíge, NW Angola
Uinta Mountains 22 B4 *mountain range* Utah, W USA
Uitenhage *see* Kariega
Uithoorn 64 C3 Noord-Holland, C Netherlands
Ujda *see* Oujda
Ujelang Atoll 122 C1 *var.* Wujlān. *atoll* Ralik Chain, W Marshall Islands
Újgradiska *see* Nova Gradiška
Újmoldova *see* Moldova Nouă
Ujungpandang *see* Makassar
Ujung Salang *see* Phuket
Újvidék *see* Novi Sad
UK *see* United Kingdom
Ukhta 92 C3 Respublika Komi, NW Russia
Ukiah 25 B5 California, W USA
Ukmergė 84 C4 *Pol.* Wiłkomierz. Vilnius, C Lithuania
Ukraine 86 C2 *off.* Ukraine, *Rus.* Ukraina, *Ukr.* Ukrayina; *prev.* Ukrainian Soviet Socialist Republic, Ukrainskay S.S.R. *country* SE Europe
Ukraine *see* Ukraine
Ukrainian Soviet Socialist Republic *see* Ukraine
Ukrainskay S.S.R/Ukrayina *see* Ukraine
Ulaanbaatar 105 E2 *Eng.* Ulan Bator; *prev.* Urga. *country capital* (Mongolia) Töv, C Mongolia
Ulaangom 104 C2 Uvs, NW Mongolia
Ulan Bator *see* Ulaanbaatar
Ulanhad *see* Chifeng
Ulan-Ude 93 E4 *prev.* Verkhneudinsk. Respublika Buryatiya, S Russia
Uleälv *see* Oulujoki
Uleträsk *see* Oulujärvi
Ulft 64 E4 Gelderland, E Netherlands
Ulianovka *see* Blahovishchenske
Ullapool 66 C3 N Scotland, United Kingdom
Ulm 73 B6 Baden-Württemberg, S Germany
Ulsan 107 E4 *Jap.* Urusan. SE South Korea
Ulster 67 B5 *province* Northern Ireland, United Kingdom/Ireland
Ulungur Hu 104 B2 *lake* NW China
Uluṟu 125 D5 *var.* Ayers Rock. *monolith* Northern Territory, C Australia
Ulyanivka *see* Blahovishchenske
Ul'yanovsk 89 C5 *prev.* Simbirsk. Ul'yanovskaya Oblast', W Russia
Umán 29 H3 Yucatán, SE Mexico
Uman 87 E3 Cherkas'ka Oblast', C Ukraine
Umanak/Umanaq *see* Uummannaq
'Umān, Khalīj *see* Oman, Gulf of
'Umān, Salṭanat *see* Oman
Umbrian-Machigian Mountains *see* Umbro-Marchigiano, Appennino
Umbro-Marchigiano, Appennino 74 C3 *Eng.* Umbrian-Machigian Mountains. *mountain range* C Italy
Umeå 62 C4 Västerbotten, N Sweden
Umeälven 62 C4 *river* N Sweden
Umiat 14 D2 Alaska, USA
Umm Buru 50 A4 Western Darfur, W Sudan
Umm Durmān *see* Omdurman
Umm Ruwaba 50 C4 *var.* Umm Ruwābah, Um Ruwāba. Northern Kordofan, C Sudan
Umm Ruwābah *see* Umm Ruwaba
Umnak Island 14 A3 *island* Aleutian Islands, Alaska, USA
Um Ruwāba *see* Umm Ruwaba
Umtali *see* Mutare
Umtata *see* Mthatha
Una 78 B3 *river* Bosnia and Herzegovina/Croatia
Unac 78 B3 *river* W Bosnia and Herzegovina
Unalaska Island 14 A3 *island* Aleutian Islands, Alaska, USA
'Unayzah 98 B4 *var.* Anaiza. Al Qaşīm, C Saudi Arabia
Unci *see* Almería
Uncía 39 F4 Potosí, C Bolivia
Uncompahgre Peak 22 B5 *mountain* Colorado, C USA
Undur Khan *see* Öndörhaan
Ungaria *see* Hungary
Ungarisches Erzgebirge *see* Slovenské rudohorie
Ungarn *see* Hungary
Ungava Bay 17 E1 *bay* Québec, E Canada
Ungava Peninsula *see* Ungava, Péninsule d'
Ungava, Péninsule d' 16 D1 *Eng.* Ungava Peninsula. *peninsula* Québec, SE Canada
Ungeny *see* Ungheni
Ungheni 86 D3 *Rus.* Ungeny. W Moldova
Unguja *see* Zanzibar
Üngüz Angyrsyndaky Garagum 100 C2 *Rus.* Zaunguzskiye Garagumy. *desert* N Turkmenistan
Ungvár *see* Uzhhorod
Unimak Island 14 B3 *island* Aleutian Islands, Alaska, USA
Union 21 E1 South Carolina, SE USA
Union City 20 C1 Tennessee, S USA
Union of Myanmar *see* Myanmar
United Arab Emirates 99 C5 *Ar.* Al Imārāt al 'Arabīyah al Muttaḩidah, *abbrev.* UAE; *prev.* Trucial States. *country* SW Asia
United Arab Republic *see* Egypt
United Kingdom 67 B5 *off.* United Kingdom of Great Britain and Northern Ireland, *abbrev.* UK. *country* NW Europe
United Kingdom of Great Britain and Northern Ireland *see* United Kingdom
United Mexican States *see* Mexico
United Provinces *see* Uttar Pradesh
United States of America 13 B5 *off.* United States of America, *var.* America, The States, *abbrev.* U.S., USA. *country* North America
Unst 66 D1 *island* NE Scotland, United Kingdom
Ünye 94 D2 Ordu, W Turkey (Türkiye)
Upala 30 D4 Alajuela, NW Costa Rica
Upata 37 E2 Bolívar, E Venezuela
Upemba, Lac 55 D7 *lake* SE Dem. Rep. Congo
Upernavik 60 C2 *var.* Upernivik. Avannaata, C Greenland
Upernivik *see* Upernavik
Upington 56 C4 Northern Cape, W South Africa
'Upolu 123 F4 *island* SE Samoa
Upper Klamath Lake 24 A4 *lake* Oregon, NW USA
Upper Lough Erne 67 A5 *lake* SW Northern Ireland, United Kingdom
Upper Red Lake 23 F1 *lake* Minnesota, N USA
Upper Volta *see* Burkina
Uppsala 63 C6 Uppsala, C Sweden
Uqsuqtuuq *see* Gjoa Haven
Ural 90 B3 *Kaz.* Zhayyk. *river* Kazakhstan/Russia
Ural Mountains *see* Ural'skiye Gory
Ural'sk *see* Oral
Ural'skiye Gory 92 C3 *var.* Ural'skiy Khrebet, *Eng.* Ural Mountains. *mountain range* Kazakhstan/Russia
Ural'skiy Khrebet *see* Ural'skiye Gory
Uraricoera 40 D1 Roraima, N Brazil
Ura-Tyube *see* Istaravshan
Urbandale 23 F3 Iowa, C USA
Urdun *see* Jordan
Uren' 89 C5 Nizhegorodskaya Oblast', W Russia
Urga *see* Ulaanbaatar
Urganch 100 D2 *Rus.* Urgench; *prev.* Novo-Urgench. Xorazm Viloyati, W Uzbekistan
Urgench *see* Urganch

Urgut *101 E3* Samarqand Viloyati, C Uzbekistan
Urmia, Lake *see* Orūmīyeh, Daryācheh-ye
Uroševac *see* Ferizaj
Ūroteppa *see* Istaravshan
Uruapan *29 E4 var.* Uruapan del Progreso. Michoacán, SW Mexico
Uruapan del Progreso *see* Uruapan
Uruguai, Rio *see* Uruguay
Uruguay *42 D4 off.* Oriental Republic of Uruguay; *prev.* La Banda Oriental. *country* E South America
Uruguay *42 D3 var.* Rio Uruguai, Río Uruguay. *river* E South America
Uruguay, Oriental Republic of *see* Uruguay
Uruguay, Río *see* Uruguay
Urumchi *see* Ürümqi
Urumi Yeh *see* Orūmīyeh, Daryācheh-ye
Ürümqi *104 C3 var.* Tihwa, Urumchi, Urumqi, Urumtsi, Wu-lu-k'o-mu-shi, Wu-lu-mu-ch'i; *prev.* Ti-hua. Xinjiang Uygur Zizhiqu, NW China
Urumtsi *see* Ürümqi
Urundi *see* Burundi
Urup, Ostrov *93 H4 island* Kuril'skiye Ostrova, SE Russia
Urusan *see* Ulsan
Urziceni *86 C5* Ialomiţa, SE Romania
Usa *88 E3 river* NW Russia
Uşak *94 B3 prev.* Ushak. Uşak, W Turkey (Türkiye)
Ushak *see* Uşak
Ushant *see* Ouessant, Île d'
Ushuaia *43 B8* Tierra del Fuego, S Argentina
Usinsk *88 E3* Respublika Komi, NW Russia
Üsküb/Üsküp *see* Skopje
Usmas Ezers *84 B3 lake* NW Latvia
Usol'ye-Sibirskoye *93 E4* Irkutskaya Oblast', C Russia
Ussel *69 C5* Corrèze, C France
Ussuriysk *93 G5 prev.* Nikol'sk, Nikol'sk-Ussuriyskiy, Voroshilov. Primorskiy Kray, SE Russia
Ustica *75 B6 island* S Italy
Ust'-Ilimsk *93 E4* Irkutskaya Oblast', C Russia
Ústí nad Labem *76 A4 Ger.* Aussig. Ústecký Kraj, NW Czechia (Czech Republic)
Ustinov *see* Izhevsk
Ustka *76 C2 Ger.* Stolpmünde. Pomorskie, N Poland
Ust'-Kamchatsk *93 H2* Kamchatskaya Oblast', E Russia
Ust'-Kamenogorsk *see* Öskemen
Ust'-Kut *93 E4* Irkutskaya Oblast', C Russia
Ust'-Olenëk *93 E3* Respublika Sakha (Yakutiya), NE Russia
Ustrzyki Dolne *77 E5* Podkarpackie, SE Poland
Ust'-Sysol'sk *see* Syktyvkar
Ust Urt *see* Ustyurt Plateau
Ustyurt Plateau *100 B1 var.* Ust Urt, *Uzb.* Ustyurt Platosi. *plateau* Kazakhstan/Uzbekistan
Ustyurt Platosi *see* Ustyurt Plateau
Usulután *30 C3* Usulután, SE El Salvador
Usumacinta, Río *30 B1 river* Guatemala/Mexico
Usumbura *see* Bujumbura
U.S./USA *see* United States of America
Utah *22 B4 off.* State of Utah, *also known as* Beehive State, Mormon State. *state* W USA
Utah Lake *22 B4 lake* Utah, W USA
Utena *84 C4* Utena, E Lithuania
Utica *19 F3* New York, NE USA
Utina *see* Udine
Utrecht *64 C4 Lat.* Trajectum ad Rhenum. Utrecht, C Netherlands
Utsunomiya *109 D5 var.* Utunomiya. Tochigi, Honshū, S Japan
Uttarakhand *113 E2 cultural region* N India
Uttar Pradesh *113 E3 prev.* United Provinces, United Provinces of Agra and Oudh. *cultural region* N India
Utunomiya *see* Utsunomiya
Uulu *84 D2* Pärnumaa, SW Estonia
Uummannaq *60 C3 var.* Umanak, Umanaq. Avannaata, C Greenland
Uummannarsuaq *see* Nunap Isua
Uvalde *27 F4* Texas, SW USA
Uvaravichy *85 D7 Rus.* Uvarovichi. Homyel'skaya Voblasts', SE Belarus
Uvarovichi *see* Uvaravichy
Uvea, Île *123 E4 island* N Wallis and Futuna
Uviéu *see* Oviedo
Uvs Nuur *104 C1 var.* Ozero Ubsu-Nur. *lake* Mongolia/Russia
'Uwaynāt, Jabal al *66 A3 var.* Jebel Uweinat. *mountain* Libya/Sudan
Uweinat, Jebel *see* 'Uwaynāt, Jabal al
Uyo *53 G5* Akwa Ibom, S Nigeria
Uyuni *39 F5* Potosí, W Bolivia
Uzbekistan *100 D2 off.* Republic of Uzbekistan. *country* C Asia
Uzbekistan, Republic of *see* Uzbekistan
Uzhgorod *see* Uzhhorod
Uzhhorod *86 B2 Rus.* Uzhgorod; *prev.* Ungvár. Zakarpats'ka Oblast', W Ukraine
Užice *78 D4 prev.* Titovo Užice. Serbia, W Serbia

V

Vaal *56 D4 river* C South Africa
Vaals *65 D6* Limburg, SE Netherlands
Vaasa *63 D5 Swe.* Vasa; *prev.* Nikolainkaupunki. Österbotten, W Finland
Vaassen *64 D3* Gelderland, E Netherlands
Vác *77 C6 Ger.* Waitzen. Pest, N Hungary
Vadodara *112 C4 prev.* Baroda. Gujarāt, W India
Vaduz *72 E2 country capital* (Liechtenstein) W Liechtenstein
Vág *see* Váh
Vágbeszterce *see* Považská Bystrica
Váh *77 C5 Ger.* Waag, *Hung.* Vág. *river* W Slovakia
Váhtjer *see* Gällivare
Väinameri *84 C2 prev.* Muhu Väin, *Ger.* Moon-Sund. *sea* E Baltic Sea
Vajdahunyad *see* Hunedoara
Valachia *see* Wallachia
Valday *88 B4* Novgorodskaya Oblast', W Russia
Valdecañas, Embalse de *70 D3 reservoir* W Spain
Valdepeñas *71 E4* Castilla-La Mancha, C Spain
Valdez *14 C3* Alaska, USA
Valdia *see* Weldiya
Valdivia *43 B5* Los Lagos, C Chile
Val-d'Or *16 D4* Québec, SE Canada
Valdosta *21 E3* Georgia, SE USA
Valence *69 D5 anc.* Valentia, Valentia Julia, Ventia. Drôme, E France
València *71 F3 var.* Valencia. Comunitat Valenciana, E Spain
Valencia *see* València
Valencia *24 D1* California, USA
Valencia *36 D1* Carabobo, N Venezuela
Valencia, Gulf of *71 F3 var.* Gulf of Valencia. *gulf* E Spain
Valencia, Gulf of *see* Valencia, Golfo de
Valencia/València *see* Valenciana, Comunitat
Valenciana, Comunitat *71 F3 var.* Valencia, *Cat.* València; *anc.* Valentia. *autonomous community* NE Spain
Valenciennes *68 D2* Nord, N France
Valentia *see* Valence, France
Valentia *see* Valenciana, Comunitat
Valentia Julia *see* Valence
Valentine State *see* Oregon
Valera *36 C2* Trujillo, NW Venezuela
Valetta *see* Valletta
Valga *84 D3 Ger.* Walk, *Latv.* Valka. Valgamaa, S Estonia
Valira *69 A8 river* Andorra/Spain Europe
Valjevo *78 C4* Serbia, W Serbia
Valjok *see* Válljohka
Valka *84 D3 Ger.* Walk. Valka, N Latvia
Valka *see* Valga
Valkenswaard *65 D5* Noord-Brabant, S Netherlands
Valladolid *29 H3* Yucatán, SE Mexico
Valladolid *70 D2* Castilla y León, NW Spain
Vall D'Uxó *see* La Vall d'Uixó
Valle de La Pascua *36 D2* Guárico, N Venezuela
Valledupar *36 B1* Cesar, N Colombia
Vallejo *25 B6* California, W USA
Vallenar *42 B3* Atacama, N Chile
Valletta *75 C8 prev.* Valetta. *country capital* (Malta) E Malta
Valley City *23 E2* North Dakota, N USA
Válljohka *62 D2 var.* Valjok. Finnmark, N Norway
Valls *71 G2* Cataluña, NE Spain
Valmiera *84 D3 Est.* Volmari, *Ger.* Wolmar. Valmiera, N Latvia
Valona *see* Vlorë
Valozhyn *85 C5 Pol.* Wołożyn, *Rus.* Volozhin. Minskaya Voblasts', C Belarus
Valparaíso *42 B4* Valparaíso, C Chile
Valparaiso *18 C3* Indiana, N USA
Valverde del Camino *70 C4* Andalucía, S Spain
Van *95 F3* Van, E Turkey (Türkiye)
Vanadzor *95 F2 prev.* Kirovakan. N Armenia
Vancouver *14 D5* British Columbia, SW Canada
Vancouver *24 B3* Washington, NW USA
Vancouver Island *14 D5 island* British Columbia, SW Canada
Vanda *see* Vantaa
Van Diemen Gulf *124 D2 gulf* Northern Territory, N Australia
Van Diemen's Land *see* Tasmania
Vaner, Lake *see* Vänern
Vänern *63 B6 Eng.* Lake Vaner; *prev.* Lake Vener. *lake* S Sweden
Vangaindrano *57 G4* Fianarantsoa, SE Madagascar
Van Gölü *95 F3 Eng.* Lake Van; *anc.* Thospitis. *salt lake* E Turkey (Türkiye)
Van Horn *26 D3* Texas, SW USA
Van, Lake *see* Van Gölü
Vannes *68 A3 anc.* Dariorigum. Morbihan, NW France
Vantaa *63 D6 Swe.* Vanda. Uusimaa, S Finland
Vanua Levu *123 E4 island* N Fiji
Vanuatu *122 C4 off.* Republic of Vanuatu; *prev.* New Hebrides. *country* SW Pacific Ocean
Vanuatu, Republic of *see* Vanuatu
Van Wert *18 C4* Ohio, N USA
Vapincum *see* Gap
Varakļāni *84 D4* Madona, C Latvia
Vārānasi *113 E3 prev.* Banaras, Benares, *hist.* Kasi. Uttar Pradesh, N India
Varangerfjorden *62 E2 Lapp.* Várjjatvuotna. *fjord* N Norway
Varangerhalvøya *62 D2 Lapp.* Várnjárga. *peninsula* N Norway
Varannó *see* Vranov nad Topl'ou
Varasd *see* Varaždin
Varaždin *78 B2 Ger.* Warasdin, *Hung.* Varasd. Varaždin, N Croatia
Varberg *63 B7* Halland, S Sweden
Vardar *79 E6 Gk.* Áxios. *river* North Macedonia/Greece
Varde *63 A7* Ribe, W Denmark
Vareia *see* Logroño
Varėna *85 B5 Pol.* Orany. Alytus, S Lithuania
Varese *74 B2* Lombardia, N Italy
Vârful Moldoveanu *86 B4 var.* Moldoveanul; *prev.* Vîrful Moldoveanu. *mountain* C Romania
Várjjatvuotna *see* Varangerfjorden
Varkaus *63 E5* Pohjois-Savo, C Finland
Varna *82 E2 prev.* Stalin; *anc.* Odessus. Varna, E Bulgaria
Varnenski Zaliv *82 E2 prev.* Stalinski Zaliv. *bay* E Bulgaria
Várnjárga *see* Varangerhalvøya
Varshava *see* Warszawa
Vasa *see* Vaasa
Vasilikí *83 A5* Lefkáda, Iónia Nisiá, Greece, C Mediterranean Sea
Vasilishki *85 B5 Pol.* Wasiliszki. Hrodzyenskaya Voblasts', W Belarus
Vasil'kov *see* Vasylkiv
Vaslui *86 D4* Vaslui, C Romania
Västerås *63 C6* Västmanland, C Sweden
Vasylkiv *87 E2 var.* Vasil'kov. Kyivska Oblast, N Ukraine
Vatican City *75 A7 off.* Vatican City State. *country* S Europe
Vatican City State *see* Vatican City
Vatnajökull *61 E5 glacier* SE Iceland
Vatter, Lake *see* Vättern
Vättern *63 B6 Eng.* Lake Vatter; *prev.* Lake Vetter. *lake* S Sweden
Vaughn *26 D2* New Mexico, SW USA
Vaupés, Río *36 C4 var.* Rio Uaupés. *river* Brazil/Colombia
Vava'u Group *123 E4 island group* N Tonga
Vavuniya *110 D3* Northern Province, N Sri Lanka
Vawkavysk *85 B6 Pol.* Wołkowysk, *Rus.* Volkovysk. Hrodzyenskaya Voblasts', W Belarus
Växjö *63 C7 var.* Vexiö. Kronoberg, S Sweden
Vaygach, Ostrov *88 E2 island* NW Russia
Veendam *64 E2* Groningen, NE Netherlands
Veenendaal *64 D4* Utrecht, C Netherlands
Vega *62 B4 island* C Norway
Veglia *see* Krk
Veisiejai *85 B5* Alytus, S Lithuania
Vejer de la Frontera *70 C5* Andalucía, S Spain
Veldhoven *65 D5* Noord-Brabant, S Netherlands
Velebit *78 A3 mountain range* C Croatia
Velenje *73 E7 Ger.* Wöllan. N Slovenia
Veles *79 E6 Turk.* Köprülü. C North Macedonia
Velho *see* Porto Velho
Velika Kikinda *see* Kikinda
Velika Morava *78 D4 var.* Glavn'a Morava, Morava, *Ger.* Grosse Morava. *river* C Serbia
Velikaya *91 G2 river* NE Russia
Veliki Bečkerek *see* Zrenjanin
Velikiye Luki *88 A4* Pskovskaya Oblast', W Russia
Velikiy Novgorod *88 B4 prev.* Novgorod. Novgorodskaya Oblast', W Russian Federation
Veliko Tarnovo *82 D2 prev.* Tirnovo, Trnovo, Tŭrnovo. Veliko Tarnovo, N Bulgaria
Velingrad *82 C3* Pazardzhik, C Bulgaria
Veľký Krtíš *77 D6* Banskobystrický Kraj, C Slovakia
Vellore *110 C2* Tamil Nādu, SE India
Velobriga *see* Viana do Castelo
Velsen *see* Velsen-Noord
Velsen-Noord *64 C3 var.* Velsen. Noord-Holland, W Netherlands
Vel'sk *88 C4 var.* Velsk. Arkhangel'skaya Oblast', NW Russia
Velvendós *see* Velvéntos
Velvéntos *82 B4 var.* Velvendós. C Greece
Velykyy Tokmak *see* Tokmak
Vendôme *68 C4* Loir-et-Cher, C France
Venedig *see* Venezia
Vener, Lake *see* Vänern
Venetia *see* Venezia
Venezia *74 C2 Eng.* Venice, *Fr.* Venise, *Ger.* Venedig; *anc.* Venetia. Veneto, NE Italy
Venezia, Golfo di *see* Venice, Gulf of
Venezuela *36 D2 off.* Bolivarian Republic of Venezuela; *prev.* Republic of Venezuela, Estados Unidos de Venezuela, United States of Venezuela. *country* N South America
Venezuela, Estados Unidos de *see* Venezuela
Venezuela, Gulf of *36 C1 Eng.* Gulf of Maracaibo, Gulf of Venezuela. *gulf* NW Venezuela
Venezuela, Gulf of *see* Venezuela, Golfo de
Venezuelan Basin *34 B1 undersea basin* E Caribbean Sea
Venezuela, Bolivarian Republic of *see* Venezuela
Venezuela, Republic of *see* Venezuela
Venezuela, United States of *see* Venezuela
Venice *20 C4* Louisiana, S USA
Venice *see* Venezia
Venice, Gulf of *74 C2 It.* Golfo di Venezia, *Slvn.* Beneški Zaliv. *gulf* N Adriatic Sea
Venise *see* Venezia
Venlo *65 D5 prev.* Venloo. Limburg, SE Netherlands
Venloo *see* Venlo
Venta *84 B3 Ger.* Windau. *river* Latvia/Lithuania
Venta Belgarum *see* Winchester
Ventia *see* Valence
Ventimiglia *74 A3* Liguria, NW Italy
Ventspils *84 B2 Ger.* Windau. Ventspils, NW Latvia
Vera *42 D3* Santa Fe, C Argentina
Veracruz *29 F4 var.* Veracruz Llave. Veracruz-Llave, E Mexico
Veracruz Llave *see* Veracruz
Vercellae *see* Vercelli
Vercelli *74 A2 anc.* Vercellae. Piemonte, NW Italy
Verdal *see* Verdalsøra
Verdalsøra *62 B4 var.* Verdal. Nord-Trøndelag, C Norway
Verde, Cabo *see* Cape Verde
Verde, Costa *70 D1 coastal region* N Spain
Verden *72 B3* Niedersachsen, NW Germany
Veria *see* Véroia
Verkhnedvinsk *see* Vyerkhnyadzvinsk
Verkhneudinsk *see* Ulan-Ude
Verkhoyanskiy Khrebet *93 F3 mountain range* NE Russia
Vermillion *23 F3* South Dakota, N USA
Vermont *19 F2 off.* State of Vermont, *also known as* Green Mountain State. *state* NE USA
Vernadsky *132 A2 Ukrainian research station* Antarctica
Vernal *22 B4* Utah, W USA
Vernon *27 F2* Texas, SW USA
Verőcze *see* Virovitica
Véroia *82 B4 var.* Veria, Vérroia, *Turk.* Karaferiye. Kentrikí Makedonía, N Greece
Verona *74 C2* Veneto, NE Italy
Vérroia *see* Véroia
Versailles *68 D1* Yvelines, N France
Versecz *see* Vršac
Verulamium *see* St Albans
Verviers *65 D6* Liège, E Belgium
Vesdre *65 D6 river* E Belgium
Veselinovo *82 D2* Shumen, NE Bulgaria
Vesontio *see* Besançon
Vesoul *68 D4 anc.* Vesulium, Vesulum. Haute-Saône, E France
Vesterålen *62 B2 island group* N Norway
Vestfjorden *62 C3 fjord* C Norway
Vestmannaeyjar *61 E5* Sudhurland, S Iceland
Vesulium/Vesulum *see* Vesoul
Vesuna *see* Périgueux
Vesuvio *75 D5 Eng.* Vesuvius. *volcano* S Italy
Vesuvius *see* Vesuvio
Veszprém *77 C7 Ger.* Veszprim. Veszprém, W Hungary
Veszprim *see* Veszprém
Vetrino *see* Suvorovo
Vetrino *see* Vyetryna
Vetter, Lake *see* Vättern
Veurne *65 A5 var.* Furnes. West-Vlaanderen, W Belgium
Vexiö *see* Växjö
Viacha *39 F4* La Paz, W Bolivia
Viana de Castelo *see* Viana do Castelo
Viana do Castelo *70 B2 var.* Viana de Castelo; *anc.* Velobriga. Viana do Castelo, NW Portugal
Vianen *64 C4* Utrecht, C Netherlands
Viangchan *114 C4 Eng./Fr.* Vientiane. *country capital* (Laos) C Laos
Viangphoukha *114 C3 var.* Vieng Pou Kha. Louang Namtha, N Laos
Viareggio *74 B3* Toscana, C Italy
Viborg *63 A7* Viborg, NW Denmark
Vic *71 G2 var.* Vich; *anc.* Ausa, Vicus Ausonensis. Cataluña, NE Spain
Vicentia *see* Vicenza
Vicenza *74 C2 anc.* Vicentia. Veneto, NE Italy
Vich *see* Vic
Vichy *69 C5* Allier, C France
Vicksburg *20 B2* Mississippi, S USA
Victoria *57 H1 country capital* (Seychelles) Mahé, SW Seychelles
Victoria *14 D5 province capital* Vancouver Island, British Columbia, SW Canada
Victoria *80 A5 var.* Rabat. Gozo, NW Malta
Victoria *27 G4* Texas, SW USA
Victoria *127 C7 state* SE Australia
Victoria *see* Masvingo, Zimbabwe
Victoria Bank *see* Vitória Seamount
Victoria de Durango *see* Durango
Victoria de las Tunas *see* Las Tunas
Victoria Falls *56 C3* Matabeleland North, W Zimbabwe
Victoria Falls *56 C2 waterfall* Zambia/Zimbabwe
Victoria Falls *see* Iguaçu, Saltos do
Victoria Island *15 F3 island* Northwest Territories/Nunavut, NW Canada
Victoria, Lake *51 B6 var.* Victoria Nyanza. *lake* E Africa
Victoria Land *132 C4 physical region* Antarctica
Victoria Nyanza *see* Victoria, Lake
Victoria River *124 D3 river* Northern Territory, N Australia
Victorville *25 C7* California, W USA
Vicus Ausonensis *see* Vic
Vicus Elbii *see* Viterbo
Vidalia *21 E2* Georgia, SE USA
Videm-Krško *see* Krško
Vídeň *see* Wien
Vidin *82 B1 anc.* Bononia. Vidin, NW Bulgaria
Vidzy *85 C5* Vitsyebskaya Voblasts', NW Belarus
Viedma *43 C5* Río Negro, E Argentina
Vieja, Sierra *26 D3 mountain range* Texas, SW USA
Vieng Pou Kha *see* Viangphoukha
Vienna *see* Wien, Austria
Vienna *see* Vienne, France
Vienne *69 D5 anc.* Vienna. Isère, E France
Vienne *68 B4 river* W France
Vientiane *see* Viangchan
Vientos, Paso de los *see* Windward Passage
Vierzon *68 C4* Cher, C France
Viesīte *84 C4 Ger.* Eckengraf. Jēkabpils, S Latvia
Vietnam *114 D4 off.* Socialist Republic of Vietnam, *Vtn.* Công Hoa Xa Hôi Chu Nghia Viêt Nam. *country* SE Asia
Vietnam, Socialist Republic of *see* Vietnam
Vietri *see* Viêt Tri
Viêt Tri *114 D3 var.* Vietri. Vinh Phu, N Vietnam
Vieux Fort *33 F2* S Saint Lucia
Vigo *70 B2* Galicia, NW Spain
Viipuri *see* Vyborg
Vijayawāda *110 D1 prev.* Bezwada. Andhra Pradesh, SE India
Vila *see* Port-Vila
Vila Artur de Paiva *see* Cubango
Vila da Ponte *see* Cubango
Vila de João Belo *see* Xai-Xai
Vila de Mocímboa da Praia *see* Mocímboa da Praia
Vila do Conde *70 B2* Porto, NW Portugal
Vila do Zumbo *56 D2 prev.* Vila do Zumbu, Zumbo. Tete, NW Mozambique
Vila do Zumbu *see* Vila do Zumbo
Vilafranca del Penedès *71 G2 var.* Villafranca del Panadés. Cataluña, NE Spain
Vila General Machado *see* Camacupa
Vila Henrique de Carvalho *see* Saurimo
Viļaka *84 D4 Ger.* Marienhausen. Balvi, NE Latvia
Vilalba *70 C1* Galicia, NW Spain
Vila Marechal Carmona *see* Uíge
Vila Nova de Gaia *70 B2* Porto, NW Portugal
Vila Nova de Portimão *see* Portimão
Vila Pereira de Eça *see* N'Giva
Vila Real *70 C2 var.* Vila Rial. Vila Real, N Portugal
Vila Rial *see* Vila Real
Vila Robert Williams *see* Caála
Vila Salazar *see* N'Dalatando
Vila Serpa Pinto *see* Menongue
Vileyka *see* Vilyeyka
Vilhelmina *62 C4* Västerbotten, N Sweden
Vilhena *40 D3* Rondônia, W Brazil
Vília *83 C5* Attikí, C Greece
Viliya *85 C5 Lith.* Neris. *river* W Belarus
Viliya *see* Neris
Viljandi *84 D2 Ger.* Fellin. Viljandimaa, S Estonia
Vilkaviškis *84 B4 Pol.* Wyłkowyszki. Marijampolė, SW Lithuania
Villa Acuña *28 D2 var.* Ciudad Acuña. Coahuila, NE Mexico
Villa Bella *39 F2* Beni, N Bolivia
Villacarrillo *71 E4* Andalucía, S Spain
Villa Cecilia *see* Ciudad Madero
Villach *73 D7 Slvn.* Beljak. Kärnten, S Austria
Villacidro *75 A5* Sardegna, Italy, C Mediterranean Sea
Villa Concepción *see* Concepción
Villa del Pilar *see* Pilar
Villafranca de los Barros *70 C4* Extremadura, W Spain
Villafranca del Panadés *see* Vilafranca del Penedès
Villahermosa *29 G4 prev.* San Juan Bautista. Tabasco, SE Mexico
Villajoyosa *see* La Vila Joiosa
Villa María *42 C4* Córdoba, C Argentina
Villa Martín *39 F5* Potosí, SW Bolivia
Villa Mercedes *42 C4* San Juan, Argentina
Villanueva *28 D3* Zacatecas, C Mexico
Villanueva de la Serena *70 C3* Extremadura, W Spain
Villanueva de los Infantes *71 E4* Castilla-La Mancha, C Spain
Villarrica *42 D2* Guairá, SE Paraguay
Villavicencio *36 B3* Meta, C Colombia
Villaviciosa *70 D1* Asturias, N Spain
Villazón *39 G5* Potosí, S Bolivia
Villena *71 F4* Comunitat Valenciana, E Spain
Villeurbanne *69 D5* Rhône, E France
Villingen-Schwenningen *73 B6* Baden-Württemberg, S Germany
Villmanstrand *see* Lappeenranta
Vilna *see* Vilnius
Vilnius *85 C5 Pol.* Wilno, *Ger.* Wilna; *prev. Rus.* Vilna. *country capital* (Lithuania) Vilnius, SE Lithuania
Vilvoorde *65 C6 Fr.* Vilvorde. Vlaams Brabant, C Belgium
Vilvorde *see* Vilvoorde
Vilyeyka *85 C5 Pol.* Wilejka, *Rus.* Vileyka. Minskaya Voblasts', NW Belarus
Vilyuy *93 F3 river* NE Russia
Viña del Mar *42 B4* Valparaíso, C Chile
Vinaròs *71 F3* Comunitat Valenciana, E Spain
Vincennes *18 B4* Indiana, N USA
Vindhya Mountains *see* Vindhya Range
Vindhya Range *112 D4 var.* Vindhya Mountains. *mountain range* N India
Vindobona *see* Wien
Vineland *19 F4* New Jersey, NE USA
Vinh *114 D4* Nghê An, N Vietnam
Vinh Loi *see* Bac Liêu
Vinishte *82 C2* Montana, NW Bulgaria
Vinita *27 G1* Oklahoma, C USA
Vinkovci *78 C3 Ger.* Winkowitz, *Hung.* Vinkovcze. Vukovar-Srijem, E Croatia
Vinkovcze *see* Vinkovci
Vinnitsa *see* Vinnytsia
Vinnytsia *86 D2 prev.* Vinnytsya, *Rus.* Vinnitsa. Vinnyts'ka Oblast', C Ukraine
Vinnytsya *see* Vinnytsia
Vinogradov *see* Vynohradiv
Vinson Massif *132 A3 mountain* Antarctica
Viranşehir *95 E4* Şanlıurfa, SE Turkey (Türkiye)
Vîrful Moldoveanu *see* Vârful Moldoveanu
Virginia *23 G1* Minnesota, N USA
Virginia *19 E5 off.* Commonwealth of Virginia, *also known as* Mother of Presidents, Mother of States, Old Dominion. *state* NE USA
Virginia Beach *19 F5* Virginia, NE USA
Virgin Islands *see* British Virgin Islands
Virgin Islands (US) *33 F3 var.* Virgin Islands of the United States; *prev.* Danish West Indies. *US unincorporated territory* E West Indies
Virgin Islands of the United States *see* Virgin Islands (US)
Virôchey *115 E5* Ratanakiri, NE Cambodia
Virovitica *78 C2 Ger.* Virovititz, *Hung.* Verőcze; *prev. Ger.* Werowitz. Virovitica-Podravina, NE Croatia
Virovititz *see* Virovitica
Virton *65 D8* Luxembourg, SE Belgium
Virtsu *84 D2 Ger.* Werder. Läänemaa, W Estonia
Vis *78 B4 It.* Lissa; *anc.* Issa. *island* S Croatia
Vis *see* Fish
Visaginas *84 C4 prev.* Sniečkus. Utena, E Lithuania
Visākhapatnam *113 E5 var.* Vishakhapatnam. Andhra Pradesh, SE India
Visalia *25 C6* California, W USA
Visby *63 C7 Ger.* Wisby. Gotland, SE Sweden
Viscount Melville Sound *15 F2 prev.* Melville Sound. *sound* Northwest Territories, N Canada
Visé *65 D6* Liège, E Belgium
Viseu *70 C2 prev.* Vizeu. Viseu, N Portugal
Vishakhapatnam *see* Visākhapatnam
Vislinskiy Zaliv *see* Vistula Lagoon
Visoko *78 C4* Federacija Bosna I Hercegovina, C Bosnia and Herzegovina
Visttasjohka *62 D3 river* N Sweden
Vistula *76 C2 Eng.* Vistula, *Ger.* Weichsel. *river* C Poland
Vistula *see* Wisła
Vistula Lagoon *76 C2 Ger.* Frisches Haff, *Pol.* Zalew Wiślany, *Rus.* Vislinskiy Zaliv. *lagoon* Poland/Russia
Vitebsk *see* Vitsyebsk
Viterbo *74 C4 anc.* Vicus Elbii. Lazio, C Italy
Viti *see* Fiji
Viti Levu *123 E4 island* W Fiji
Vitim *93 F4 river* C Russia
Vitória *41 F4 state capital* Espírito Santo, SE Brazil
Vitoria *see* Vitoria-Gasteiz
Vitoria Bank *see* Vitória Seamount
Vitória da Conquista *41 F3* Bahia, E Brazil
Vitoria-Gasteiz *71 E1 var.* Vitoria, *Eng.* Vittoria. País Vasco, N Spain
Vitória Seamount *45 B5 var.* Victoria Bank, Vitoria Bank. *seamount* C Atlantic Ocean
Vitré *68 B3* Ille-et-Vilaine, NW France
Vitsyebsk *85 E5 Rus.* Vitebsk. Vitsyebskaya Voblasts', NE Belarus
Vittoria *75 C7* Sicilia, Italy, C Mediterranean Sea
Vittoria *see* Vitoria-Gasteiz
Vizcaya, Golfo de *see* Biscay, Bay of
Vizianagaram *113 E5 var.* Vizianagram. Andhra Pradesh, E India
Vizianagram *see* Vizianagaram
Vjosës, Lumi i *79 C7 var.* Vijosa, Vijosë, *Gk.* Aóos. *river* Albania/Greece
Vlaanderen *see* Flanders
Vlaardingen *64 B4* Zuid-Holland, SW Netherlands
Vladikavkaz *89 B8 prev.* Dzaudzhikau, Ordzhonikidze. Respublika Severnaya Osetiya, SW Russia
Vladimir *89 B5* Vladimirskaya Oblast', W Russia
Vladimirovka *see* Yuzhno-Sakhalinsk
Vladimir-Volynskiy *see* Volodymyr-Volynskyi
Vladivostok *93 G5* Primorskiy Kray, SE Russia
Vlagtwedde *64 E2* Groningen, NE Netherlands
Vlasotince *79 E5* Serbia, SE Serbia
Vlieland *64 C1 Fris.* Flylân. *island* Waddeneilanden, N Netherlands
Vlijmen *64 C4* Noord-Brabant, S Netherlands
Vlissingen *65 B5 Eng.* Flushing, *Fr.* Flessingue. Zeeland, SW Netherlands
Vlodava *see* Włodawa
Vlonë/Vlora *see* Vlorë
Vlorë *79 C7 prev.* Vlonë, *It.* Valona, Vlora. Vlorë, SW Albania
Vlotslavsk *see* Włocławek
Vöcklabruck *73 D6* Oberösterreich, NW Austria
Vogelkop *see* Doberai, Jazirah
Vohimena, Tanjona *57 F4 Fr.* Cap Sainte Marie. *headland* S Madagascar
Voiron *69 D5* Isère, E France
Vojvodina *78 D3 Ger.* Wojwodina. Vojvodina, N Serbia
Volga *89 B7 river* NW Russia
Volga Uplands *see* Privolzhskaya Vozvyshennost'
Volgodonsk *89 B7* Rostovskaya Oblast', SW Russia
Volgograd *89 B7 prev.* Stalingrad, Tsaritsyn. Volgogradskaya Oblast', SW Russia
Volkhov *88 B4* Leningradskaya Oblast', NW Russia
Volkoysk *see* Vawkavysk

Volmari see Valmiera
Volnovakha 87 G3 Donets'ka Oblast', SE Ukraine
Volodymyr-Volynskyi 86 C1 prev. Volodymyr-Volyns'kyy, Pol. Włodzimierz, Rus. Vladimir-Volynskiy. Volyns'ka Oblast', NW Ukraine
Volodymyr-Volyns'kyy see Volodymyr-Volynskyi
Vologda 88 B4 Vologodskaya Oblast', W Russia
Vólos 83 B5 Thessalía, C Greece
Volozhin see Valozhyn
Vol'sk 89 C6 Saratovskaya Oblast', W Russia
Volta 53 E5 river SE Ghana
Volta Blanche see White Volta
Volta, Lake 53 E5 reservoir SE Ghana
Volta Noire see Black Volta
Volturno 75 D5 river S Italy
Volunteer Island see Starbuck Island
Volzhskiy 89 B6 Volgogradskaya Oblast', SW Russia
Võnnu 84 E3 Ger. Wendau. Tartumaa, SE Estonia
Voorst 64 D3 Gelderland, E Netherlands
Voranava 85 C5 Pol. Werenów, Rus. Voronovo. Hrodzyenskaya Voblasts', W Belarus
Vorderrhein 73 B7 river SE Switzerland
Vóreies Sporádes 83 C5 var. Vóreioi Sporádes, Vórioi Sporádhes, Eng. Northern Sporades. island group E Greece
Vóreioi Sporádes see Vóreies Sporádes
Vórioi Sporádhes see Vóreies Sporádes
Vorkuta 92 C2 Respublika Komi, NW Russia
Vormsi 84 C2 var. Vormsi Saar, Ger. Worms, Swed. Ormsö. island W Estonia
Vormsi Saar see Vormsi
Voronezh 89 B6 Voronezhskaya Oblast', W Russia
Voronovo see Voranava
Voroshilov see Ussuriysk
Voroshilovgrad see Luhansk
Voroshilovsk see Stavropol', Russia
Võru 84 D3 Ger. Werro. Võrumaa, SE Estonia
Vosges 68 E4 mountain range NE France
Vostochno-Sibirskoye More 93 F1 Eng. East Siberian Sea. sea Arctic Ocean
Vostochnyy Sayan 93 E4 Mong. Dzüün Soyonï Nuruu, Eng. Eastern Sayans. mountain range Mongolia/Russia
Vostock Island see Vostok Island
Vostok 132 C3 Russian research station Antarctica
Vostok Island 123 G3 var. Vostock Island; prev. Stavers Island. island Line Islands, SE Kiribati
Voznesensk 87 E3 Mykolayivs'ka Oblast', S Ukraine
Vranje 79 E5 Serbia, SE Serbia
Vranov see Vranov nad Topl'ou
Vranov nad Topl'ou 77 D5 var. Vranov, Hung. Varannó. Prešovský Kraj, E Slovakia
Vratsa 82 C2 Vratsa, NW Bulgaria
Vrbas 78 C3 Vojvodina, NW Serbia
Vrbas 78 C3 river N Bosnia and Herzegovina
Vršac 78 E3 Ger. Werschetz, Hung. Versecz. Vojvodina, NE Serbia
Vsetín 77 C5 Ger. Wsetin. Zlínský Kraj, E Czechia (Czech Republic)
Vučitrn see Vushtrri
Vukovar 78 C3 Hung. Vukovár. Vukovar-Srijem, E Croatia
Vulcano, Isola 75 C7 island Isole Eolie, S Italy
Vung Tau 115 E6 prev. Fr. Cape Saint Jacques, Cap Saint-Jacques. Ba Ria-Vung Tau, S Vietnam
Vushtrri 79 D5 Serb. Vučitrn. N Kosovo
Vyatka 89 C5 river NW Russia
Vyatka see Kirov
Vyborg 88 B3 Fin. Viipuri. Leningradskaya Oblast', NW Russia
Vyerkhnyadzvinsk 85 D5 Rus. Verkhnedvinsk. Vitsyebskaya Voblasts', N Belarus
Vyetryna 85 D5 Rus. Vetrino. Vitsyebskaya Voblasts', N Belarus
Vynohradiv 86 B3 Cz. Sevluš, Hung. Nagyszöllös, Rus. Vinogradov; prev. Sevlyush. Zakarpats'ka Oblast', W Ukraine

W

Wa 53 E4 NW Ghana
Waag see Váh
Waagbistritz see Považská Bystrica
Waal 64 C4 river S Netherlands
Wabash 18 C4 Indiana, N USA
Wabash River 18 B5 river N USA
Waco 27 G3 Texas, SW USA
Wad Al-Hajarah see Guadalajara
Waddān 49 F3 NW Libya
Waddeneilanden 64 C1 Eng. West Frisian Islands. island group N Netherlands
Waddenzee 64 C1 var. Wadden Zee. sea SE North Sea
Wadden Zee see Waddenzee
Waddington, Mount 14 D5 mountain British Columbia, SW Canada
Wādī as Sīr 97 B6 var. Wadi es Sir. 'Ammān, NW Jordan
Wadi es Sir see Wādī as Sīr
Wadi Halfa 50 B3 var. Wādī Ḥalfā'. Northern, N Sudan
Wādī Mūsā 97 B7 var. Petra. Ma'ān, S Jordan
Wad Madanī see Wad Medani
Wad Medani 50 C4 var. Wad Madanī. Gezira, C Sudan
Waflia 117 F4 Pulau Buru, E Indonesia
Wagadugu see Ouagadougou
Wagga Wagga 127 C7 New South Wales, SE Australia
Wagin 125 B7 Western Australia
Wah 112 C1 Punjab, NE Pakistan
Wahai 117 F4 Pulau Seram, E Indonesia
Wahaybah, Ramlat Al see Wahībah, Ramlat Āl
Wahiawā 25 A8 var. Wahiawa. O'ahu, Hawaii, USA, C Pacific Ocean
Wahībah, Ramlat Ahl see Wahībah, Ramlat Āl
Wahībah Sands 99 E5 var. Ramlat Ahl Wahībah, Ramlat Al Wahaybah, Eng. Wahībah Sands. desert N Oman
Wahībah Sands see Wahībah, Ramlat Āl
Wahpeton 23 F2 North Dakota, N USA
Wahran see Oran
Waiau 129 A7 river South Island, New Zealand
Waigeo, Pulau 117 G4 island Maluku, E Indonesia
Waikaremoana, Lake 128 E4 lake North Island, New Zealand
Wailuku 25 B8 Maui, Hawaii, USA, C Pacific Ocean
Waimate 129 B6 Canterbury, South Island, New Zealand
Waiouru 128 D4 Manawatu-Wanganui, North Island, New Zealand
Waipara 129 C6 Canterbury, South Island, New Zealand
Waipawa 128 E4 Hawke's Bay, North Island, New Zealand
Waipukurau 128 D4 Hawke's Bay, North Island, New Zealand
Wairau 129 C5 river South Island, New Zealand
Wairoa 128 E4 Hawke's Bay, North Island, New Zealand
Wairoa 128 D2 river North Island, New Zealand
Waitaki 129 B6 river South Island, New Zealand
Waitara 128 D4 Taranaki, North Island, New Zealand
Waitzen see Vác
Waiuku 128 D3 Auckland, North Island, New Zealand
Wakasa-wan 109 C6 bay C Japan
Wakatipu, Lake 129 A7 lake South Island, New Zealand
Wakayama 109 C6 Wakayama, Honshū, SW Japan
Wake Island 130 C2 US unincorporated territory NW Pacific Ocean
Wake Island 120 D1 atoll NW Pacific Ocean
Wakkanai 108 C1 Hokkaidō, NE Japan
Walachei/Walachia see Wallachia
Wałbrzych 76 B4 Ger. Waldenburg, Waldenburg in Schlesien. Dolnośląskie, SW Poland
Walcourt 65 C7 Namur, S Belgium
Wałcz 76 B3 Ger. Deutsch Krone. Zachodnio-pomorskie, NW Poland
Waldenburg/Waldenburg in Schlesien see Wałbrzych
Waldia see Weldiya
Wales 14 C2 Alaska, USA
Wales 67 C6 Wel. Cymru. cultural region Wales, United Kingdom
Walgett 127 D5 New South Wales, SE Australia
Walk see Valga, Estonia
Walk see Valka, Latvia
Walker Lake 25 C5 lake Nevada, W USA
Wallachia 86 B5 var. Walachia, Ger. Walachei, Rom. Valachia. cultural region S Romania
Walla Walla 24 C2 Washington, NW USA
Wallenthal see Haţeg
Wallis and Futuna 123 E4 Fr. Territoire de Wallis et Futuna. French overseas collectivity C Pacific Ocean
Wallis et Futuna, Territoire de see Wallis and Futuna
Walnut Ridge 20 B1 Arkansas, C USA
Waltenberg see Zalău
Walthamstow 67 B7 Waltham Forest, SE England, United Kingdom
Walvisbaai see Walvis Bay
Walvis Bay 56 A4 Afr. Walvisbaai. Erongo, NW Namibia
Walvish Ridge see Walvis Ridge
Walvis Ridge 47 B7 var. Walvish Ridge. undersea ridge E Atlantic Ocean
Wan see Anhui
Wanaka 129 B6 Otago, South Island, New Zealand
Wanaka, Lake 129 A6 lake South Island, New Zealand
Wanchuan see Zhangjiakou
Wandel Sea 61 E1 sea Arctic Ocean
Wandsworth 67 A8 Wandsworth, SE England, United Kingdom
Wanganui 128 D4 Manawatu-Wanganui, North Island, New Zealand
Wangaratta 127 C7 Victoria, SE Australia
Wankie see Hwange
Wanki, Río see Coco, Río
Wanlaweyn 51 D6 var. Wanle Weyn, It. Uanle Uen. Shabeellaha Hoose, SW Somalia
Wanle Weyn see Wanlaweyn
Wanxian see Wanzhou
Wanzhou 106 B5 var. Wanxian. Chongqing, C China
Warangal 113 E5 Telangana, C India
Warasdin see Varaždin
Warburg 72 B4 Nordrhein-Westfalen, W Germany
Ware 15 E4 British Columbia, W Canada
Waremme 65 C6 Liège, E Belgium
Waren 72 C3 Mecklenburg-Vorpommern, NE Germany
Wargla see Ouargla
Warkworth 128 D2 Auckland, North Island, New Zealand
Warnemünde 72 C2 Mecklenburg-Vorpommern, NE Germany
Warner 27 G1 Oklahoma, C USA
Warnes 39 G4 Santa Cruz, C Bolivia
Warrego River 127 C5 seasonal river New South Wales/Queensland, E Australia
Warren 18 D3 Michigan, N USA
Warren 18 D3 Ohio, N USA
Warren 19 E3 Pennsylvania, NE USA
Warri 53 F5 Delta, S Nigeria
Warrnambool 127 B7 Victoria, SE Australia
Warsaw/Warschau see Warszawa
Warszawa 76 D3 Eng. Warsaw, Ger. Warschau, Rus. Varshava. country capital (Poland) Mazowieckie, C Poland
Warta 76 B3 Ger. Warthe. river W Poland
Wartberg see Senec
Warthe see Warta
Warwick 127 E5 Queensland, E Australia
Washington 22 A2 NE England, United Kingdom
Washington D.C. 19 E4 country capital (USA) District of Columbia, NE USA
Washington Island see Teraina
Washington, Mount 19 G2 mountain New Hampshire, NE USA
Wash, The 67 E6 inlet E England, United Kingdom
Wasiliszki see Vasilishki
Waspam 31 E2 var. Waspán. Región Autónoma Atlántico Norte, NE Nicaragua
Waspán see Waspam
Watampone 117 E4 var. Bone. Sulawesi, C Indonesia
Watenstedt-Salzgitter see Salzgitter
Waterbury 19 F3 Connecticut, NE USA
Waterford 67 B6 Ir. Port Láirge. Waterford, S Ireland
Waterloo 23 G3 Iowa, C USA
Watertown 19 F2 New York, NE USA
Watertown 23 F2 South Dakota, N USA
Waterville 19 G2 Maine, NE USA
Watford 67 A7 E England, United Kingdom
Watlings Island see San Salvador
Watsa 55 E5 Orientale, NE Dem. Rep. Congo
Watts Bar Lake 20 D1 reservoir Tennessee, S USA
Wau 51 B5 var. Wāw. Western Bahr el Ghazal, C South Sudan
Waukegan 18 B3 Illinois, N USA
Waukesha 18 B3 Wisconsin, N USA
Wausau 18 B2 Wisconsin, N USA
Waverly 23 G3 Iowa, C USA
Wavre 65 C6 Walloon Brabant, C Belgium
Wāw see Wau
Wawa 16 C4 Ontario, S Canada
Waycross 21 E3 Georgia, SE USA
Wearmouth see Sunderland
Webfoot State see Oregon
Webster City 23 F3 Iowa, C USA
Weddell Plain 132 A2 abyssal plain SW Atlantic Ocean
Weddell Sea 132 A2 sea SW Atlantic Ocean
Weener 72 A3 Niedersachsen, NW Germany
Weert 65 D5 Limburg, SE Netherlands
Weesp 64 C3 Noord-Holland, C Netherlands
Węgorzewo 76 D2 Ger. Angerburg. Warmińsko-Mazurskie, NE Poland
Weimar 72 C4 Thüringen, C Germany
Weissenburg see Alba Iulia, Romania
Weissenburg in Bayern 73 C6 Bayern, SE Germany
Weissenstein see Paide
Weisskirchen see Bela Crkva
Weiswampach 65 D7 Diekirch, N Luxembourg
Wejherowo 76 C2 Pomorskie, NW Poland
Welchman Hall 33 G1 C Barbados
Weldiya 50 C4 var. Waldia, It. Valdia. Āmara, N Ethiopia
Welkom 56 D4 Free State, C South Africa
Welle see Uele
Wellesley Islands 126 B2 island group Queensland, N Australia
Wellington 129 D5 country capital (New Zealand) Wellington, North Island, New Zealand
Wellington 23 F5 Kansas, C USA
Wellington see Wellington, Isla
Wellington, Isla 43 A7 var. Wellington. island S Chile
Wells 24 D4 Nevada, W USA
Wellsford 128 D2 Auckland, North Island, New Zealand
Wells, Lake 125 C5 lake Western Australia
Wels 73 D6 anc. Ovilava. Oberösterreich, N Austria
Wembley 67 A8 Alberta, W Canada
Wemmel 65 B6 Vlaams Brabant, C Belgium
Wenatchee 24 B2 Washington, NW USA
Wenchi 53 E4 W Ghana
Wen-chou/Wenchow see Wenzhou
Wendau see Võnnu
Wenden see Cēsis
Wenzhou 106 D5 var. Wen-chou, Wenchow. Zhejiang, SE China
Werda 56 C4 Kgalagadi, S Botswana
Werder see Virtsu
Werenów see Voranava
Werkendam 64 C4 Noord-Brabant, S Netherlands
Werowitz see Virovitica
Werro see Võru
Werschetz see Vršac
Wesenberg see Rakvere
Weser 72 B3 river NW Germany
Wessel Islands 126 B1 island group Northern Territory, N Australia
West Antarctica 132 A3 var. Lesser Antarctica. physical region Antarctica
West Australian Basin see Wharton Basin
West Bank 97 A6 disputed region SW Asia
West Bend 18 B3 Wisconsin, N USA
West Bengal 113 F4 cultural region NE India
West Cape 129 A7 headland South Island, New Zealand
West Des Moines 23 F3 Iowa, C USA
Westerland 72 B2 Schleswig-Holstein, N Germany
Western Australia 124 B4 state W Australia
Western Bug see Bug
Western Carpathians 77 E7 mountain range W Romania Europe
Western Desert see Ṣaḥrā' al Gharbīyah
Western Dvina 63 E7 Bel. Dzvina, Ger. Düna, Latv. Daugava, Rus. Zapadnaya Dvina. river W Europe
Western Ghats 112 C5 mountain range SW India
Western Isles see Outer Hebrides
Western Punjab see Punjab
Western Sahara 48 B3 disputed territory administered by Morocco N Africa
Western Samoa see Samoa
Western Samoa, Independent State of see Samoa
Western Sayans see Zapadnyy Sayan
Western Scheldt see Westerschelde
Western Sierra Madre see Madre Occidental, Sierra
Westerschelde 65 B5 Eng. Western Scheldt; prev. Honte. inlet S North Sea
West Falkland 43 C7 var. Gran Malvina, Isla Gran Malvina. island W Falkland Islands
West Fargo 23 F2 North Dakota, N USA
West Frisian Islands see Waddeneilanden
West Irian see Papua
Westliche Morava see Zapadna Morava
West Mariana Basin 120 B1 var. Perece Vela Basin. undersea feature W Pacific Ocean
West Memphis 20 B1 Arkansas, C USA
West New Guinea see Papua
Weston-super-Mare 67 D7 SW England, United Kingdom
West Palm Beach 21 F4 Florida, SE USA
West Papua see Papua
Westport 129 C5 West Coast, South Island, New Zealand
West Punjab see Punjab
West River see Xi Jiang
West Siberian Plain see Zapadno-Sibirskaya Ravnina
West Virginia 18 D4 off. State of West Virginia, also known as Mountain State. state NE USA
Wetar, Pulau 117 F5 island Kepulauan Damar, E Indonesia
Wetzlar 73 B5 Hessen, W Germany
Wevok 14 C2 var. Wewuk. Alaska, USA
Wewuk see Wevok
Wexford 67 B6 Ir. Loch Garman. SE Ireland
Weyburn 15 F5 Saskatchewan, S Canada
Weymouth 67 D7 S England, United Kingdom
Wezep 64 D3 Gelderland, E Netherlands
Whakatane 128 E3 Bay of Plenty, North Island, New Zealand
Whale Cove 15 G3 var. Tikiarjuaq. Nunavut, C Canada
Whangarei 128 D2 Northland, North Island, New Zealand
Wharton Basin 119 D5 var. West Australian Basin. undersea feature E Indian Ocean
Whataroa 129 B6 West Coast, South Island, New Zealand
Wheatland 22 D3 Wyoming, C USA
Wheeler Peak 26 D1 mountain New Mexico, SW USA
Wheeling 18 D4 West Virginia, NE USA
Whenua Hou see Codfish Island
Whitby 67 D5 N England, United Kingdom
Whitefish 22 B1 Montana, NW USA
Whitehaven 67 C5 NW England, United Kingdom
Whitehorse 14 D4 territory capital Yukon, W Canada
White Nile 50 B4 Ar. Al Baḥr al Abyaḍ, An Nīl al Abyaḍ, Bahr el Jebel. river C South Sudan
White River 22 D3 river South Dakota, N USA
White Sea see Beloye More
White Volta 53 E4 var. Nakambé, Fr. Volta Blanche. river Burkina/Ghana
Whitianga 128 D2 Waikato, North Island, New Zealand
Whitney, Mount 25 C6 mountain California, W USA
Whitsunday Group 126 D3 island group Queensland, E Australia
Whyalla 127 B6 South Australia
Wichita 23 F5 Kansas, C USA
Wichita Falls 27 F2 Texas, SW USA
Wichita River 27 F2 river Texas, SW USA
Wickenburg 26 B2 Arizona, SW USA
Wicklow 67 B6 Ir. Cill Mhantáin. county E Ireland
Wicklow Mountains 67 B6 Ir. Sléibhte Chill Mhantáin. mountain range E Ireland
Wieliczka 77 D5 Małopolskie, S Poland
Wieluń 76 C4 Sieradz, C Poland
Wien 73 E6 Eng. Vienna, Hung. Bécs, Slvk. Vídeň, Slvn. Dunaj; anc. Vindobona. country capital (Austria) Wien, NE Austria
Wiener Neustadt 73 E6 Niederösterreich, E Austria
Wierden 64 E3 Overijssel, E Netherlands
Wiesbaden 73 B5 Hessen, W Germany
Wieselburg and Ungarisch-Altenburg/ Wieselburg-Ungarisch-Altenburg see Mosonmagyaróvár
Wiesenhof see Ostrołęka
Wight, Isle of 67 D7 island United Kingdom
Wigorna Ceaster see Worcester
Wijchen 64 D4 Gelderland, SE Netherlands
Wijk bij Duurstede 64 D4 Utrecht, C Netherlands
Wilcannia 127 C6 New South Wales, SE Australia
Wilejka see Vilyeyka
Wilhelm, Mount 122 B3 mountain C Papua New Guinea
Wilhelm-Pieck-Stadt see Guben
Wilhelmshaven 72 B3 Niedersachsen, NW Germany
Wilia/Wilja see Neris
Wilkes Barre 19 F3 Pennsylvania, NE USA
Wilkes Land 132 C4 physical region Antarctica
Wiłkomierz see Ukmergė
Willard 26 D2 New Mexico, SW USA
Willcox 26 C3 Arizona, SW USA
Willebroek 65 B5 Antwerpen, C Belgium
Willemstad 33 E5 dependent territory capital (Curaçao) Lesser Antilles, S Caribbean Sea
Williston 22 D1 North Dakota, N USA
Wilmington 19 F4 Delaware, NE USA
Wilmington 21 F2 North Carolina, SE USA
Wilmington 18 C4 Ohio, N USA
Wilna/Wilno see Vilnius
Wilrijk 65 C5 Antwerpen, N Belgium
Winchester 67 D7 hist. Wintanceaster, Lat. Venta Belgarum. S England, United Kingdom
Winchester 19 E4 Virginia, NE USA
Windau see Ventspils, Latvia
Windau see Venta, Latvia/Lithuania
Windhoek 56 B3 Ger. Windhuk. country capital (Namibia) Khomas, C Namibia
Windhuk see Windhoek
Windorah 126 C4 Queensland, C Australia
Windsor 16 C5 Ontario, S Canada
Windsor 67 D7 S England, United Kingdom
Windsor 19 G3 Connecticut, NE USA
Windward Islands 33 H4 island group E West Indies
Windward Islands see Barlavento, Ilhas de, Cape Verde (Cabo Verde)
Windward Passage 32 D3 Sp. Paso de los Vientos. channel Cuba/Haiti
Winisk 16 C2 river Ontario, C Canada
Winkowitz see Vinkovci
Winnebago, Lake 18 B2 lake Wisconsin, N USA
Winnemucca 25 C5 Nevada, W USA
Winnipeg 15 G5 province capital Manitoba, S Canada
Winnipeg, Lake 15 G5 lake Manitoba, C Canada
Winnipegosis, Lake 16 A3 lake Manitoba, C Canada
Winona 23 G3 Minnesota, N USA
Winschoten 64 E2 Groningen, NE Netherlands
Winsen 72 B3 Niedersachsen, N Germany
Winston Salem 21 E1 North Carolina, SE USA
Winsum 64 D1 Groningen, NE Netherlands
Wintanceaster see Winchester
Winterswijk 64 E4 Gelderland, E Netherlands
Winterthur 73 B7 Zürich, NE Switzerland
Winton 126 C4 Queensland, E Australia
Winton 129 A7 Southland, South Island, New Zealand
Wisby see Visby
Wisconsin 18 A2 off. State of Wisconsin, also known as Badger State. state N USA
Wisconsin Rapids 18 B2 Wisconsin, N USA
Wisconsin River 18 B3 river Wisconsin, N USA
Wisła 76 C2 var. Vistula. river N Poland
Wiślany, Zalew see Vistula Lagoon
Wismar 72 C2 Mecklenburg-Vorpommern, N Germany
Wittenberge 72 C3 Brandenburg, N Germany
Wittlich 73 A5 Rheinland-Pfalz, SW Germany
Wittstock 72 C3 Brandenburg, NE Germany
W. J. van Blommesteinmeer 37 G3 reservoir E Suriname
Władysławowo 76 C2 Pomorskie, N Poland
Włocławek 76 C3 Ger./Rus. Vlotslavsk. Kujawsko-pomorskie, C Poland
Włodawa 76 E4 Rus. Vlodava. Lubelskie, SE Poland
Włodzimierz see Volodymyr-Volynskyi
Wlotzkasbaken 56 B3 Erongo, W Namibia
Wodonga 127 C7 Victoria, SE Australia
Wodzisław Śląski 77 C5 Ger. Loslau. Śląskie, S Poland
Wojerecy see Hoyerswerda
Wōjjā see Wotje Atoll
Wojwodina see Vojvodina
Woking 67 D7 SE England, United Kingdom
Wolf, Isla 38 A4 island Galápagos Islands, W Ecuador South America
Wolfsberg 73 D7 Kärnten, SE Austria
Wolfsburg 72 C3 Niedersachsen, N Germany
Wolgast 72 D2 Mecklenburg-Vorpommern, NE Germany
Wołkowysk see Vawkavysk
Wöllan see Velenje
Wollaston Lake 15 F4 Saskatchewan, C Canada
Wollongong 127 D6 New South Wales, SE Australia
Wolmar see Valmiera
Wołożyn see Valozhyn
Wolvega 64 D2 Fris. Wolvegea. Fryslân, N Netherlands
Wolvegea see Wolvega
Wolverhampton 67 D6 C England, United Kingdom
Wolverine State see Michigan
Wŏnsan 107 E3 SE North Korea
Woodburn 24 B3 Oregon, NW USA
Woodland 25 B5 California, W USA
Woodruff 18 B2 Wisconsin, N USA
Woods, Lake of the 16 A3 Fr. Lac des Bois. lake Canada/USA
Woodville 128 D4 Manawatu-Wanganui, North Island, New Zealand
Woodward 27 F1 Oklahoma, C USA
Worcester 56 C5 Western Cape, SW South Africa
Worcester 67 D6 hist. Wigorna Ceaster. W England, United Kingdom
Worcester 19 G3 Massachusetts, NE USA
Workington 67 C5 NW England, United Kingdom
Worland 22 C3 Wyoming, C USA
Wormatia see Worms
Worms 73 B5 anc. Augusta Vangionum, Borbetomagus, Wormatia. Rheinland-Pfalz, SW Germany
Worms see Vormsi
Worthington 23 F3 Minnesota, N USA
Wotje Atoll 122 D1 var. Wōjjā. atoll Ratak Chain, E Marshall Islands
Woudrichem 64 C4 Noord-Brabant, S Netherlands
Wrangel Island 93 F1 Eng. Wrangel Island. island NE Russia
Wrangel Island see Vrangelya, Ostrov
Wrangel Plain 133 B2 undersea feature Arctic Ocean
Wrocław 76 C4 Eng./Ger. Breslau. Dolnośląskie, SW Poland
Września 76 C3 Wielkopolskie, C Poland
Wsetin see Vsetín
Wuchang see Wuhan
Wuday'ah 99 C6 spring/well S Saudi Arabia
Wuhai 105 E3 var. Haibowan. Nei Mongol Zizhiqu, N China
Wuhan 106 C5 var. Han-kou, Han-k'ou, Hanyang, Wuchang, Wu-han; prev. Hankow. province capital Hubei, C China
Wu-han see Wuhan
Wuhsien see Suzhou
Wuhsi/Wu-his see Wuxi
Wuhu 106 D5 var. Wu-na-mu. Anhui, E China
Wujlān see Ujelang Atoll
Wukari 53 G4 Taraba, E Nigeria
Wuliang Shan 106 A6 mountain range SW China
Wu-lu-k'o-mu-shi/Wu-lu-mu-ch'i see Ürümqi
Wu-na-mu see Wuhu
Wuppertal 72 A4 prev. Barmen-Elberfeld. Nordrhein-Westfalen, W Germany
Würzburg 73 B5 Bayern, SW Germany
Wusih see Wuxi
Wuxi 106 D5 var. Wuhsi, Wu-hsi, Wusih. Jiangsu, E China
Wuyi Shan 103 E3 mountain range SE China
Wye 67 C6 Wel. Gwy. river England/Wales, United Kingdom
Wyłkowyszki see Vilkaviškis
Wyndham 124 D3 Western Australia
Wyoming 18 C3 Michigan, N USA
Wyoming 22 B3 off. State of Wyoming, also known as Equality State. state C USA
Wyszków 76 D3 Ger. Probstberg. Mazowieckie, NE Poland

X

Xaafuun, Raas 50 E4 var. Ras Hafun. cape NE Somalia
Xaçmaz 95 H2 Rus. Khachmas. N Azerbaijan
Xaignabouli 114 C4 prev. Muang Xaignabouri, Fr. Sayaboury. Xaignabouli, N Laos
Xai-Xai 57 E4 prev. João Belo, Vila de João Belo. Gaza, S Mozambique
Xalapa 29 F4 Veracruz-Llave, Mexico
Xam Nua 114 D3 var. Sam Neua. Houaphan, N Laos
Xankändi 95 G3 Rus. Khankendi; prev. Stepanakert. SW Azerbaijan
Xánthi 82 C3 Anatolikí Makedonía kai Thráki, NE Greece
Xàtiva 71 F3 Cas. Xátiva; anc. Setabis, var. Jativa. Comunitat Valenciana, E Spain
Xauen see Chefchaouen
Xäzär Dänizi see Caspian Sea
Xeres see Jeréz de la Frontera
Xiaguan see Dali
Xiamen 106 D6 var. Hsia-men; prev. Amoy. Fujian, SE China
Xi'an 106 C4 var. Changan, Sian, Signan, Siking, Singan, Xian. province capital Shaanxi, C China
Xiang see Hunan
Xiangkhoang see Phônsavan
Xiangtan 106 C5 var. Hsiang-t'an, Siangtan. Hunan, S China
Xiao Hinggan Ling 106 D2 Eng. Lesser Khingan Range. mountain range NE China

Xichang *106 B5* Sichuan, C China
Xieng Khouang *see* Phônsaven
Xieng Ngeun *see* Muong Xiang Ngeun
Xigazê *104 C5 var.* Jih-k'a-tse, Shigatse, Xigaze. Xizang Zizhiqu, W China
Xi Jiang *102 D3 var.* Hsi Chiang, *Eng.* West River. *river* S China
Xilinhot *105 F2 var.* Silinhot. Nei Mongol Zizhiqu, N China
Xilokastro *see* Xylókastro
Xin *see* Xinjiang Uygur Zizhiqu
Xingkai Hu *see* Khanka, Lake
Xingu, Rio *41 E2 river* C Brazil
Xingxingxia *104 D3* Xinjiang Uygur Zizhiqu, NW China
Xining *105 E4 var.* Hsining, Hsi-ning, Sining. *province capital* Qinghai, C China
Xinjiang *see* Xinjiang Uygur Zizhiqu
Xinjiang Uygur Zizhiqu *104 B3 var.* Sinkiang, Sinkiang Uighur Autonomous Region, Xin, Xinjiang. *autonomous region* NW China
Xinpu *see* Lianyungang
Xinxiang *106 C4* Henan, C China
Xinyang *106 C5 var.* Hsin-yang, Sinyang. Henan, C China
Xinzo de Limia *70 C2* Galicia, NW Spain
Xiqing Shan *102 D2 mountain range* C China
Xiva *100 D2 Rus.* Khiva, Khiwa. Xorazm Viloyati, W Uzbekistan
Xixón *see* Gijón
Xizang *see* Xizang Zizhiqu
Xizang Gaoyuan *see* Qingzang Gaoyuan
Xizang Zizhiqu *104 B4 var.* Thibet, Tibetan Autonomous Region, Xizang, *Eng.* Tibet. *autonomous region* W China
Xolotlán *see* Managua, Lago de
Xucheng *see* Xuwen
Xuddur *51 D5 var.* Hudur, *It.* Oddur. Bakool, SW Somalia
Xuwen *106 C7 var.* Xucheng. Guangdong, S China
Xuzhou *106 D4 var.* Hsu-chou, Suchow, Tongshan; *prev.* T'ung-shan. Jiangsu, E China
Xylókastro *83 B5 var.* Xilokastro. Pelopónnisos, S Greece

Y

Ya'an *106 B5 var.* Yaan. Sichuan, C China
Yabēlo *51 C5* Oromīya, C Ethiopia
Yablis *31 E2* Región Autónoma Atlántico Norte, NE Nicaragua
Yablonovyy Khrebet *93 F4 mountain range* S Russia
Yabrai Shan *105 E3 mountain range* NE China
Yafran *49 F2* NW Libya
Yaghan Basin *45 B7 undersea feature* SE Pacific Ocean
Yagotin *see* Yahotyn
Yahotyn *87 E2 Rus.* Yagotin. Kyivska Oblast, N Ukraine
Yahualica *28 D4* Jalisco, SW Mexico
Yakima *24 B2* Washington, NW USA
Yakima River *24 B2 river* Washington, NW USA
Yakoruda *82 C3* Blagoevgrad, SW Bulgaria
Yaku-shima *109 B8 island* Nansei-shotō, SW Japan
Yakutat *14 D4* Alaska, USA
Yakutsk *93 F3* Respublika Sakha (Yakutiya), NE Russia
Yakymivka *87 F4* Zaporizka Oblast, S Ukraine
Yala *115 C7* Yala, SW Thailand
Yalizava *85 D6 Rus.* Yelizovo. Mahilyowskaya Voblasts', E Belarus
Yalong Jiang *106 A5 river* C China
Yalova *94 B3* Yalova, NW Turkey (Türkiye)
Yalpug, Ozero *see* Yalpuh, Ozero
Yalpuh, Ozero *86 D4 Rus.* Ozero Yalpug. *lake* SW Ukraine
Yalta *87 F5* Respublika Krym, S Ukraine
Yalu *103 E2 Chin.* Yalu Jiang, *Jap.* Oryokko, *Kor.* Amnok-kang. *river* China/North Korea
Yalu Jiang *see* Yalu
Yamaguchi *109 B7 var.* Yamaguti. Yamaguchi, Honshū, SW Japan
Yamal, Poluostrov *92 D2 peninsula* N Russia
Yamanīyah, Al Jumhūrīyah al *see* Yemen
Yambio *51 B5 var.* Yambiyo. Western Equatoria, S South Sudan
Yambiyo *see* Yambio
Yambol *82 D2 Turk.* Yanboli. Yambol, E Bulgaria
Yamdena, Pulau *117 G5 prev.* Jamdena. *island* Kepulauan Tanimbar, E Indonesia
Yamoussoukro *52 D5 country capital* (Ivory Coast) C Ivory Coast
Yamuna *112 D3 prev.* Jumna. *river* N India
Yana *93 F2 river* NE Russia
Yanboli *see* Yambol
Yanbu 'al Baḩr *99 A5* Al Madīnah, W Saudi Arabia
Yangambi *55 D5* Orientale, N Dem. Rep. Congo
Yangchow *see* Yangzhou
Yangiyo'l *101 E2 Rus.* Yangiyul'. Toshkent Viloyati, E Uzbekistan
Yangiyul' *see* Yangiyo'l
Yangku *see* Taiyuan
Yangon *114 B4 Eng.* Rangoon. Yangon, S Myanmar (Burma)
Yangtze *106 B5 var.* Yangtze Kiang, *Eng.* Yangtze. *river* C China
Yangtze *see* Chang Jiang
Yangtze Kiang *see* Chang Jiang
Yangzhou *106 D5 var.* Yangchow. Jiangsu, E China
Yankton *23 E3* South Dakota, N USA
Yany Kapu *87 F4 Rus.* Krasnoperekopsk. Respublika Krym, S Ukraine
Yannina *see* Ioánnina
Yanskiy Zaliv *91 F2 bay* N Russia
Yantai *106 D4 var.* Yan-t'ai; *prev.* Chefoo, Chih-fu. Shandong, E China
Yaoundé *55 B5 var.* Yaunde. *country capital* (Cameroon) Centre, S Cameroon
Yap *122 A1 island* Caroline Islands, W Micronesia
Yapanskoye More East Sea/Japan, Sea of
Yapen, Pulau *117 G4 prev.* Japen. *island* E Indonesia
Yap Trench *120 B2 var.* Yap Trough. *undersea feature* SE Philippine Sea
Yap Trough *see* Yap Trench
Yapurá *see* Caquetá, Río, Brazil/Colombia
Yapurá *see* Japurá, Rio, Brazil/Colombia
Yaqui, Río *28 C2 river* NW Mexico
Yaransk *89 C5* Kirovskaya Oblast', NW Russia
Yarega *88 D4* Respublika Komi, NW Russia
Yaren *122 D2 de facto country capital* (Nauru) Nauru, SW Pacific
Yarkant *see* Shache
Yarlung Zangbo Jiang *see* Brahmaputra
Yarmouth *17 F5* Nova Scotia, SE Canada
Yarmouth *see* Great Yarmouth
Yaroslav *see* Jarosław
Yaroslavl' *88 B4* Yaroslavskaya Oblast', W Russia
Yarumal *36 B2* Antioquia, NW Colombia
Yasyel'da *85 B7 river* Brestskaya Voblasts', SW Belarus Europe
Yatsushiro *109 A7 var.* Yatusiro. Kumamoto, Kyūshū, SW Japan
Yatusiro *see* Yatsushiro
Yaunde *see* Yaoundé
Yavari *see* Javari, Rio
Yavari, Rio *40 C2 var.* Yavarí. *river* Brazil/Peru
Yaviza *31 H5* Darién, SE Panama
Yavoriv *86 B2 Pol.* Jaworów, *Rus.* Yavorov. L'vivs'ka Oblast', NW Ukraine
Yavorov *see* Yavoriv
Yazd *98 D3 var.* Yezd. Yazd, C Iran
Yazoo City *20 B2* Mississippi, S USA
Yding Skovhøj *63 A7 hill* C Denmark
Ýdra *83 C6 var.* Ídhra. *island* Ýdra, S Greece
Ye *115 B5* Mon State, S Myanmar (Burma)
Yecheng *104 A3 var.* Kargilik. Xinjiang Uygur Zizhiqu, NW China
Yedy Kuiu *87 G5 prev.* Lenine, *Rus.* Lenino. Respublika Krym, S Ukraine
Yefremov *89 B5* Tul'skaya Oblast', W Russia
Yekaterinburg *92 C3 prev.* Sverdlovsk. Sverdlovskaya Oblast', C Russia
Yekaterinodar *see* Krasnodar
Yekaterinoslav *see* Dnipro
Yelets *89 B5* Lipetskaya Oblast', W Russia
Yelisavetpol *see* Gäncä
Yelizavetgrad *see* Kropyvnytskyi
Yelizovo *see* Yalizava
Yell *66 D1 island* NE Scotland, United Kingdom
Yellowhammer State *see* Alabama
Yellowknife *15 E4 territory capital* Northwest Territories, W Canada
Yellow River *see* Huang He
Yellow Sea *106 D4 Chin.* Huang Hai, *Kor.* Hwang-Hae. *sea* E Asia
Yellowstone River *22 C2 river* Montana/Wyoming, NW USA
Yel'sk *85 C7* Homyel'skaya Voblasts', SE Belarus
Yelwa *53 F4* Kebbi, W Nigeria
Yemen *99 C7 off.* Republic of Yemen, *Ar.* Al Jumhūrīyah al Yamanīyah, Al Yaman. *country* SW Asia
Yemen, Republic of *see* Yemen
Yemva *88 D4 prev.* Zheleznodorozhnyy. Respublika Komi, NW Russia
Yenakiieve *87 G3 prev.* Yenakiyeve, Ordzhonikidze, Rykovo; *Rus.* Yenakiyevo. Donets'ka Oblast', E Ukraine
Yenakiyeve *see* Yenakiieve
Yenakiyevo *see* Yenakiieve
Yenangyaung *114 A3* Magway, W Myanmar (Burma)
Yendi *53 E4* NE Ghana
Yengisar *104 A3* Xinjiang Uygur Zizhiqu, NW China
Yenierenköy *80 D4 var.* Yialousa, *Gk.* Agialoúsa. NE Cyprus
Yenipazar *see* Novi Pazar
Yenisey *92 D3 river* Mongolia/Russia
Yenping *see* Nanping
Yeovil *67 D7* SW England, United Kingdom
Yeppoon *126 D4* Queensland, E Australia
Yerevan *95 F3 Eng.* Erivan. *country capital* (Armenia) C Armenia
Yeriho *see* Jericho
Yerushalayim *see* Jerusalem
Yeso *see* Hokkaidō
Yeu, Île d' *68 A4 island* NW France
Yevlakh *see* Yevlax
Yevlax *95 G2 Rus.* Yevlakh. C Azerbaijan
Yevpatoriia *87 F5 prev.* Yevpatoriya. Respublika Krym, S Ukraine
Yevpatoriya *see* Yevpatoriia
Yeya *87 H4 river* SW Russia
Yezerishche *see* Yezyaryshcha
Yezo *see* Hokkaidō
Yezyaryshcha *85 E5 Rus.* Yezerishche. Vitsyebskaya Voblasts', NE Belarus
Yialousa *see* Yenierenköy
Yiannitsá *see* Giannitsá
Yichang *106 C5* Hubei, C China
Yıldızeli *94 D3* Sivas, N Turkey (Türkiye)
Yinchuan *106 B4 var.* Yinch'uan, Yin-ch'uan, Yinchwan. *province capital* Ningxia, N China
Yinchwan *see* Yinchuan
Yindu He *see* Indus
Yin-hsien *see* Ningbo
Yining *104 B2 var.* I-ning, *Uigh.* Gulja, Kuldja. Xinjiang Uygur Zizhiqu, NW China
Yin-tu Ho *see* Indus
Yisrael/Yisra'el *see* Israel
Yíthion *see* Gýtheio
Yogyakarta *116 C5 prev.* Djokjakarta, Jogjakarta, Jokyakarta. Jawa, C Indonesia
Yokohama *109 D5* Aomori, Honshū, C Japan
Yokohama *108 A2* Kanagawa, Honshū, S Japan
Yokote *108 D4* Akita, Honshū, C Japan
Yola *53 H4* Adamawa, E Nigeria
Yonago *109 B6* Tottori, Honshū, SW Japan
Yong'an *106 D6 var.* Yongan. Fujian, SE China
Yongzhou *107 C6 var.* Lengshuitan. Hunan, S China
Yonkers *19 F3* New York, NE USA
Yonne *68 C4 river* C France
Yopal *36 C3 var.* El Yopal. Casanare, C Colombia
York *67 D5 anc.* Eboracum, Eburacum. N England, United Kingdom
York *23 E4* Nebraska, C USA
York, Cape *126 C1 headland* Queensland, NE Australia
York, Kap *see* Innaanganeq
Yorkton *15 F5* Saskatchewan, S Canada
Yoro *30 C2* Yoro, C Honduras
Yoshkar-Ola *89 C5* Respublika Mariy El, W Russia
Yösönbulag *see* Altay
Youngstown *18 D4* Ohio, N USA
Youth, Isle of *see* Juventud, Isla de la
Ypres *see* Ieper
Yreka *24 B4* California, W USA
Yrendagüé *see* General Eugenio A. Garay
Yssel *see* IJssel
Ysyk-Köl *see* Issyk-Kul', Ozero
Ysyk-Köl *see* Balykchy
Yu *see* Henan
Yuan *see* Red River
Yuan Jiang *see* Red River
Yuba City *25 B5* California, W USA
Yucatán, Canal de *see* Yucatan Channel
Yucatan Channel *29 H3 Sp.* Canal de Yucatán. *channel* Cuba/Mexico
Yucatan Peninsula *see* Yucatán, Península de
Yucatán, Península de *13 C7 Eng.* Yucatan Peninsula. *peninsula* Guatemala/Mexico
Yuci *see* Jinzhong
Yue *see* Guangdong
Yue Shan, Tai *see* Lantau Island
Yueyang *106 C5* Hunan, S China
Yugoslavia *see* Serbia
Yukhavichy *85 D5 Rus.* Yukhovichi. Vitsyebskaya Voblasts', N Belarus
Yukhovichi *see* Yukhavichy
Yukon *14 D3 prev.* Yukon Territory, *Fr.* Territoire du Yukon. *territory* NW Canada
Yukon River *14 C2 river* Canada/USA
Yukon, Territoire du *see* Yukon
Yukon Territory *see* Yukon
Yulin *106 C6* Guangxi Zhuangzu Zizhiqu, S China
Yuma *26 A2* Arizona, SW USA
Yun *see* Yunnan
Yungki *see* Jilin
Yung-ning *see* Nanning
Yunjinghong *see* Jinghong
Yunki *see* Jilin
Yunnan *106 A6 var.* Yun, Yunnan Sheng, Yünnan, Yun-nan. *province* SW China
Yunnan *see* Kunming
Yunnan Sheng *see* Yunnan
Yünnan/Yun-nan *see* Yunnan
Yurev *see* Tartu
Yurihonjō *see* Honjō
Yuruá, Río *see* Juruá, Rio
Yury'ev *see* Tartu
Yushu *104 D4 var.* Gyêgu. Qinghai, C China
Yuty *42 D3* Caazapá, S Paraguay
Yuzhno-Sakhalinsk *93 H4 Jap.* Toyohara; *prev.* Vladimirovka. Ostrov Sakhalin, Sakhalinskaya Oblast', SE Russia
Yuzhnyy Bug *see* Pivdennyi Buh
Yuzhou *see* Chongqing
Ýylanly *see* Gurbansoltan Eje

Z

Zaandam *see* Zaanstad
Zaanstad *64 C3 prev.* Zaandam. Noord-Holland, C Netherlands
Zabaykal'sk *93 F5* Chitinskaya Oblast', S Russia
Zabern *see* Saverne
Zabīd *99 B7* W Yemen
Żabinka *see* Zhabinka
Ząbkowice *see* Ząbkowice Śląskie
Ząbkowice Śląskie *76 B4 var.* Ząbkowice, *Ger.* Frankenstein, Frankenstein in Schlesien. Dolnośląskie, SW Poland
Zábřeh *77 C5 Ger.* Hohenstadt. Olomoucký Kraj, E Czechia (Czech Republic)
Zacapa *30 B2* Zacapa, E Guatemala
Zacatecas *28 D3* Zacatecas, C Mexico
Zacatepec *29 E4* Morelos, S Mexico
Zacháro *83 B6 var.* Zaharo, Zakháro. Dytikí Ellás, S Greece
Zadar *78 A3 It.* Zara; *anc.* Iader. Zadar, SW Croatia
Zadetkyi Kyun *115 B6 var.* St.Matthew's Island. *island* Myeik Archipelago, S Myanmar (Burma)
Zafra *70 C4* Extremadura, W Spain
Żagań *76 B4 var.* Zagań, Żegań, *Ger.* Sagan. Lubuskie, W Poland
Zagazig *see* Az Zaqāzīq
Zágráb *see* Zagreb
Zagreb *78 B2 Ger.* Agram, *Hung.* Zágráb. *country capital* (Croatia) Zagreb, N Croatia
Zagros Mountains *98 C3 Eng.* Zagros Mountains. *mountain range* W Iran
Zagros Mountains *see* Zāgros, Kūhhā-ye
Zaharo *see* Zacháro
Zāhedān *98 E4 var.* Zahidan; *prev.* Duzdab. Sīstān va Balūchestān, SE Iran
Zahidan *see* Zāhedān
Zaḩlah *see* Zahlé
Zahlé *96 B4 var.* Zaḩlah. C Lebanon
Záhony *77 E6* Szabolcs-Szatmár-Bereg, NE Hungary
Zaire *see* Congo (river)
Zaire *see* Congo (Democratic Republic of)
Zaječar *78 E4* Serbia, E Serbia
Zakataly *see* Zaqatala
Zakháro *see* Zacháro
Zakhidnyi Buh/Zakhodni Buh *see* Bug
Zākhō *98 B2 var.* Zākhū, Zaxo. Dahūk/Dihok, N Iraq
Zākhū *see* Zākhō
Zakopane *77 D5* Małopolskie, S Poland
Zákynthos *83 A6 var.* Zákinthos, *It.* Zante. *island* Iónia Nísoi, Greece, C Mediterranean Sea
Zalaegerszeg *77 B7* Zala, W Hungary
Zalău *86 B3 Ger.* Waltenberg, *Hung.* Zilah; *prev. Ger.* Zillenmarkt. Sălaj, NW Romania
Ẓalim *99 B5* Makkah, W Saudi Arabia
Zambesi/Zambeze *see* Zambezi
Zambezi *56 C2* North Western, W Zambia
Zambezi *56 D2 var.* Zambesi, *Port.* Zambeze. *river* S Africa
Zambia *56 C2 off.* Republic of Zambia; *prev.* Northern Rhodesia. *country* S Africa
Zambia, Republic of *see* Zambia
Zamboanga *117 E3 off.* Zamboanga City. Mindanao, S Philippines
Zamboanga City *see* Zamboanga
Zambrów *76 E3* Łomża, E Poland
Zamora *70 D2* Castilla y León, NW Spain
Zamora de Hidalgo *28 D4* Michoacán, SW Mexico
Zamość *76 E4 Rus.* Zamoste. Lubelskie, E Poland
Zamoste *see* Zamość
Zancle *see* Messina
Zanda *104 A4* Xizang Zizhiqu, W China
Zanesville *18 D4* Ohio, N USA
Zanjān *98 C2 var.* Zenjan, Zinjan. Zanjān, NW Iran
Zante *see* Zákynthos
Zanthus *125 C6* Western Australia, S Australia Oceania
Zanzibar *51 D7* Zanzibar, E Tanzania
Zanzibar *51 C7 Swa.* Unguja. *island* E Tanzania
Zaozhuang *106 D4* Shandong, E China
Zapadna Morava *78 D4 Ger.* Westliche Morava. *river* C Serbia
Zapadnaya Dvina *88 A4* Tverskaya Oblast', W Russia
Zapadnaya Dvina *see* Western Dvina
Zapadno-Sibirskaya Ravnina *92 C3 Eng.* West Siberian Plain. *plain* C Russia
Zapadnyy Bug *see* Bug
Zapadnyy Sayan *92 D4 Eng.* Western Sayans. *mountain range* S Russia
Zapala *43 B5* Neuquén, W Argentina
Zapiola Ridge *45 B6 undersea feature* SW Atlantic Ocean
Zapolyarnyy *88 C2* Murmanskaya Oblast', NW Russia
Zaporizhzhia *87 F3 prev.* Aleksandrovsk, Zaporizhzhya; *Rus.* Zaporozh'ye. Zaporiz'ka Oblast', SE Ukraine
Zaporizhzhya *see* Zaporizhzhia
Zaporozh'ye *see* Zaporizhzhia
Zapotiltic *28 D4* Jalisco, SW Mexico
Zaqatala *95 G2 Rus.* Zakataly. NW Azerbaijan
Zara *94 D3* Sivas, C Turkey (Türkiye)
Zara *see* Zadar
Zarafshan *see* Zarafshon
Zarafshon *100 D2 Rus.* Zarafshan. Navoiy Viloyati, N Uzbekistan
Zarafshon *see* Zeravshan
Zaragoza *71 F2 Eng.* Saragossa; *anc.* Caesaraugusta, Salduba. Aragón, NE Spain
Zarand *98 D3* Kermān, C Iran
Zaranj *100 D5* Nīmrūz, SW Afghanistan
Zarasai *84 C4* Utena, E Lithuania
Zárate *42 D4 prev.* General José F.Uriburu. Buenos Aires, E Argentina
Zarautz *71 E1 var.* Zara
Zaxo *see* Zākhō
Zelenogradsk *84 A4 Ger.* Cranz, Kranz. Kaliningradskaya Oblast', W Russia
Zelle *see* Celle
Zel'va *85 B6 Pol.* Zelwa. Hrodzyenskaya Voblasts', W Belarus
Zelwa *see* Zel'va
Zelzate *65 B5 var.* Selzaete. Oost-Vlaanderen, NW Belgium
Žemaičių Aukštumas *84 B4 physical region* W Lithuania
Zemst *65 C5* Vlaams Brabant, C Belgium
Zemun *78 D3* Serbia, N Serbia
Zengg *see* Senj
Zenica *78 C4* Federacija Bosna I Hercegovina, C Bosnia and Herzegovina
Zenta *see* Senta
Zeravshan *101 E3 Taj./Uzb.* Zarafshon. *river* Tajikistan/Uzbekistan
Zevenaar *64 D4* Gelderland, SE Netherlands
Zevenbergen *64 C4* Noord-Brabant, S Netherlands
Zeya *91 E3 river* SE Russia
Zgerzh *see* Zgierz
Zgierz *76 C4 Ger.* Neuhof, *Rus.* Zgerzh. Łódź, C Poland
Zgorzelec *76 B4 Ger.* Görlitz. Dolnośląskie, SW Poland
Zhabinka *85 A6 Pol.* Żabinka. Brestskaya Voblasts', SW Belarus
Zhambyl *see* Taraz
Zhanaozen *see* Zhangaözen
Zhangaözen *92 A4 var.* Zhanaozen; *prev.* Novyy Uzen'. Mangistau, W Kazakhstan
Zhangaqazaly *see* Ayteke Bi
Zhang-chia-k'ou *see* Zhangjiakou
Zhangdian *see* Zibo
Zhangjiakou *106 C3 var.* Changkiakow, Zhang-chia-k'ou, *Eng.* Kalgan; *prev.* Wanchuan. Hebei, E China
Zhangzhou *106 D6* Fujian, SE China
Zhanjiang *106 C7 var.* Chanchiang, Chan-chiang, *Cant.* Tsamkong, *Fr.* Fort-Bayard. Guangdong, S China
Zhaoqing *106 C6* Guangdong, S China
Zhayyk *see* Ural
Zhdanov *see* Mariupol
Zhe *see* Zhejiang
Zhejiang *106 D5 var.* Che-chiang, Chekiang, Zhe, Zhejiang Sheng. *province* SE China
Zhejiang Sheng *see* Zhejiang
Zheleznodorozhnyy *84 A4* Kaliningradskaya Oblast', W Russia
Zheleznodorozhnyy *see* Yemva
Zheleznogorsk *89 A5* Kurskaya Oblast', W Russia
Zhëltyye Vody *see* Zhovti Vody
Zhengzhou *106 C4 var.* Ch'eng-chou, Chengchow; *prev.* Chenghsien. *province capital* Henan, C China
Zhezkazgan *see* Zhezqazghan
Zhezqazghan *92 C4 var.* Zhezkazgan; *prev.* Dzhezkazgan. Karagandy, C Kazakhstan
Zhidachov *see* Zhydachiv
Zhitkovichi *see* Zhytkavichy
Zhitomir *see* Zhytomyr
Zhlobin *85 D7* Homyel'skaya Voblasts', SE Belarus
Zhmerinka *see* Zhmerynka
Zhmerynka *86 D2 Rus.* Zhmerinka. Vinnyts'ka Oblast', C Ukraine
Zhodino *see* Zhodzina
Zhodzina *85 D6 Rus.* Zhodino. Minskaya Voblasts', C Belarus
Zholkev/Zholkva *see* Zhovkva
Zhonghua Renmin Gongheguo *see* China
Zhongshan *132 D3 Chinese research station* Antarctica
Zhosaly *92 B4 prev.* Dzhusaly. Kzylorda, SW Kazakhstan
Zhovkva *86 B2 Pol.* Żółkiew, *Rus.* Zholkev, Zholkva; *prev.* Nesterov. L'vivs'ka Oblast', NW Ukraine
Zhovti Vody *87 F3 Rus.* Zhëltyye Vody. Dnipropetrovska Oblast, E Ukraine
Zhovtnevoye *see* Zhovtneve
Zhydachiv *86 B2 Pol.* Żydaczów, *Rus.* Zhidachov. L'vivs'ka Oblast', NW Ukraine
Zhytkavichy *85 C7 Rus.* Zhitkovichi. Homyel'skaya Voblasts', SE Belarus
Zhytomyr *86 D2 Rus.* Zhitomir. Zhytomyrs'ka Oblast', NW Ukraine
Zibo *106 D4 var.* Zhangdian. Shandong, E China
Zichenau *see* Ciechanów
Zielona Góra *76 B4 Ger.* Grünberg, Grünberg in Schlesien, Grüneberg. Lubuskie, W Poland
Zierikzee *64 B4* Zeeland, SW Netherlands
Zigong *106 B5 var.* Tzekung. Sichuan, C China
Ziguinchor *52 B3* SW Senegal
Zilah *see* Zalău
Žilina *77 C5 Ger.* Sillein, *Hung.* Zsolna. Zlínský Kraj, N Slovakia
Zillenmarkt *see* Zalău
Zimbabwe *56 D3 off.* Republic of Zimbabwe; *prev.* Rhodesia. *country* S Africa
Zimbabwe, Republic of *see* Zimbabwe
Zimnicea *86 C5* Teleorman, S Romania
Zimovniki *89 B7* Rostovskaya Oblast', SW Russia
Zinder *53 G3* Zinder, S Niger
Zinov'yevsk *see* Kropyvnytskyi
Zintenhof *see* Sindi
Zipaquirá *36 B3* Cundinamarca, C Colombia
Zittau *72 D4* Sachsen, E Germany
Zlatni Pyasatsi *82 E2 var.* Zlatni Pyasŭtsi. Dobrich, NE Bulgaria
Zlatni Pyasŭtsi *see* Zlatni Pyasatsi
Zlín *77 C5 prev.* Gottwaldov. Zlínský Kraj, E Czechia (Czech Republic)
Złoczów *see* Zolochev
Złotów *76 C3* Wielkopolskie, C Poland
Znamenka *see* Znamianka
Znamianka *87 F3 prev.* Znam"yanka, *Rus.* Znamenka. Kirovohrads'ka Oblast', C Ukraine
Znam"yanka *see* Znamianka
Żnin *76 C3* Kujawsko-pomorskie, C Poland
Zoetermeer *64 C4* Zuid-Holland, W Netherlands
Żółkiew *see* Zhovkva
Zolochev *86 C2 Pol.* Złoczów, *Rus.* Zolochiv. L'vivs'ka Oblast', W Ukraine
Zolochiv *87 G2 Rus.* Zolochev. Kharkivs'ka Oblast', E Ukraine
Zolochiv *see* Zolochev
Zolote *87 H3 Rus.* Zolotoye. Luhans'ka Oblast', E Ukraine
Zolotonosha *87 E2* Cherkas'ka Oblast', C Ukraine
Zolotoye *see* Zolote
Zólyom *see* Zvolen
Zomba *57 E2* Southern, S Malawi
Zombor *see* Sombor
Zongo *55 C5* Equateur, N Dem. Rep. Congo
Zonguldak *94 C2* Zonguldak, NW Turkey (Türkiye)
Zonhoven *65 D6* Limburg, NE Belgium
Zoppot *see* Sopot
Żory *77 C5 var.* Zory, *Ger.* Sohrau. Śląskie, S Poland
Zouar *54 C2* Borkou-Ennedi-Tibesti, N Chad
Zouérat *52 C2 var.* Zouérate, Zouîrât. Tiris Zemmour, N Mauritania
Zouérate *see* Zouérat
Zouîrât *see* Zouérat
Zrenjanin *78 D3 prev.* Petrovgrad, Veliki Bečkerek, *Ger.* Grossbetschkerek, *Hung.* Nagybecskerek. Vojvodina, N Serbia
Zsil/Zsily *see* Jiu
Zsolna *see* Žilina
Zsombolya *see* Jimbolia
Zsupanya *see* Županja
Zubov Seamount *45 D5 undersea feature* E Atlantic Ocean
Zueila *see* Zawilah
Zug *73 B7 Fr.* Zoug. Zug, C Switzerland
Zugspitze *73 C7 mountain* S Germany
Zuid-Beveland *65 B5 var.* South Beveland. *island* SW Netherlands
Zuider Zee *see* IJsselmeer
Zuidhorn *64 E1* Groningen, NE Netherlands
Zuidlaren *64 E2* Drenthe, NE Netherlands
Zula *50 C4* E Eritrea
Züllichau *see* Sulechów
Zumbo *see* Vila do Zumbo
Zundert *65 C5* Noord-Brabant, S Netherlands
Zunyi *106 B5* Guizhou, S China
Županja *78 C3 Hung.* Zsupanya. Vukovar-Srijem, E Croatia
Zürich *73 B7 Eng./Fr.* Zurich, *It.* Zurigo. Zürich, N Switzerland
Zurich, Lake *see* Zürichsee
Zürichsee *73 B7 Eng.* Lake Zurich. *lake* NE Switzerland
Zurigo *see* Zürich
Zutphen *64 D3* Gelderland, E Netherlands
Zuwārah *49 F2* NW Libya
Zuwaylah *see* Zawilah
Zuyevka *89 D5* Kirovskaya Oblast', NW Russia
Zvenigorodka *see* Zvenyhorodka
Zvenyhorodka *87 E2 Rus.* Zvenigorodka. Cherkas'ka Oblast', C Ukraine
Zvishavane *56 D3 prev.* Shabani. Matabeleland South, S Zimbabwe
Zvolen *77 C6 Ger.* Altsohl, *Hung.* Zólyom. Banskobystrický Kraj, C Slovakia
Zvornik *78 C4* E Bosnia and Herzegovina
Zwedru *52 D5 var.* Tchien. E Liberia
Zwettl *73 E6* Wien, NE Austria
Zwevegem *65 A6* West-Vlaanderen, W Belgium
Zwickau *73 C5* Sachsen, E Germany
Zwolle *64 D3* Overijssel, E Netherlands
Żydaczów *see* Zhydachiv
Zyōetu *see* Jōetsu
Żyrardów *76 D3* Mazowieckie, C Poland
Zyryanovsk *92 D5* Vostochnyy Kazakhstan, E Kazakhstan